COLLEGE TRIGONOMETRY

FIFTH EDITION

COLLEGE TRIGONOMETRY

Richard N. Aufmann

Vernon C. Barker

Richard D. Nation

Palomar College

Houghton Mifflin Company

Boston   New York

Publisher: Jack Shira
Senior Sponsoring Editor: Lynn Cox
Senior Development Editor: Dawn Nuttall
Associate Editor: Jennifer King
Assistant Editor: Melissa Parkin
Senior Project Editor: Tamela Ambush
Editorial Assistant: Lisa Sullivan
Senior Production/Design Coordinator: Carol Merrigan
Manufacturing Manager: Florence Cadran
Marketing Manager: Danielle Potvin
Marketing Associate: Nicole Mollica

Cover photograph: © Jon Arnold / Getty Images

PHOTO CREDITS

Chapter 1: n/a. **Chapter 2:** *p. 126* Courtesy of *NASA* and *JPL*; *p. 140* (Eiffle Tower) Tony Craddock / Getty Images, (Petronas Towers) Art; *p. 159* Reuters / NewMedia Inc. / CORBIS. **Chapter 3:** *p. 243* Courtesy of Richard Nation; *p. 263* Art. **Chapter 4:** *p. 279* Massimo Listr / Corbis; *p. 295* MacDuff Everton / CORBIS. **Chapter 5:** *p. 321* Stephen Johnson / Getty Images. **Chapter 6:** *p. 349* John Gay, US Navy / Getty Images; *p. 358* 1998 International Conference on Quality Control by Artificial Vision-QCAV'98. Kagawa Convention Center, Takamatsu, Kagawa, Japan, November 10–12, 1998, pp. 521–528; *p. 359* Reprinted with permission of Iam Morison, Jodrell Bank Observatory; *p. 375* Hugh Rooney / Eye Ubiquitous / CORBIS; *p. 383* The Granger Collection. **Chapter 7:** *p. 433* Chris McLaughlin / CORBIS; *p. 440* Bettmann / CORBIS; *p. 446* Charles O'Rear / CORBIS; *p. 447* David James / Getty Images; *p. 450* Bettmann / CORBIS; *p. 496* Tom Brakefield / CORBIS; *p. 503* Bettmann / CORBIS; *p. 508* AP / Wide World Photos.

Printed in the U.S.A.

Library of Congress Control Number: 2003110131

ISBN:
Student's Edition: 0-618-38804-4

1 2 3 4 5 6 7 8 9-DOW-08 07 06 05 04

CONTENTS

PREFACE

With each successive edition of *College Trigonometry* we strive to enhance and refine our instructional materials. In this edition we have continued our emphasis on *doing* mathematics rather than duplicating mathematics through extensive drill. Students are urged to investigate concepts, apply those concepts, and then present their findings. We are ever cognizant of the motivating influence that contemporary, relevant applications have on students. As a result, we have added many new application exercises and deleted those that are of little student interest.

Technology is introduced naturally to illustrate or enhance the current topic. We integrate technology into a discussion when it can be used to foster and promote a better understanding of a concept. The optional *Integrating Technology* boxes and graphing calculator exercises are designed to instill in students an appreciation of both the power and the limitations of technology.

In this edition, we have retained our basic philosophy, which is to deliver a comprehensive and mathematically sound treatment of the topics considered essential for a college trigonometry course. To help students master these concepts, we have tried to maintain a balance among theory, application, modeling, and drill. Carefully developed mathematics is complemented by abundant, creative applications that are both contemporary and representative of a wide range of disciplines. Many application exercises are accompanied by a diagram that helps the student visualize the mathematics of the application.

Changes for the Fifth Edition

Many new student support features have been added to this edition:

A *Focus on Problem Solving* feature has been added at the beginning of each chapter. This feature reviews and demonstrates the strategies used by successful problem-solvers.

Review Notes, identified by ⇄ , direct students to the page where a previously discussed concept can be reviewed. Students needing to review this concept can do so before proceeding with new material.

Prepare for Next Section Exercises, found in each section's exercise set (except the last section of a chapter), allow students to practice prerequisite skills and concepts needed to be successful in the next section. Most of these exercises reference the section of the text that contains the concepts related to the question for students to easily review. All answers are provided in the student answer section.

Cumulative Review Exercises have been added to every chapter, after Chapter 1. These 20 Cumulative Review Exercises allow students to review material from previous chapters. The answers for all of these exercises are included in the student answer section. Next to each answer is a section reference that directs the student to the section of the text from which the exercise was taken.

Answers to All Chapter Review Exercises are now included in the student answer section. Next to each answer is a section reference that directs the student to the section of the text from which the exercise was taken.

Some specific chapter changes are indicated below.

Chapter 1, *Functions and Graphs,* includes a completely new discussion on composition of functions that is motivated through an application problem. Concise reviews of vertical and horizontal translations have been added. Many new application exercises were added.

In Chapter 2, *Trigonometric Functions,* new applications from astronomy, aviation, and physics have been included.

Chapter 4, *Applications of Trigonometry,* now includes a discussion of how one of Mollweide's formulas can be used to check whether a triangle has been solved correctly.

In Chapter 6, *Topics in Analytic Geometry,* we have included several new applications of a more contemporary nature. We have continued the use of graphing utilities to illustrate important concepts.

Chapter 7, *Exponential and Logarithmic Functions,* now includes additional contemporary application exercises. For instance, some exercises illustrate an exponential relationship between various water temperatures and the time required for a scuba diver to reach hypothermia.

CHAPTER OPENER FEATURES

CHAPTER OPENER

Each chapter begins with a **Chapter Opener** that illustrates a specific application of a concept from the chapter. There is a reference to a particular exercise within the chapter that asks the student to solve a problem related to the chapter opener topic.

The icons at the bottom of the page let students know of additional resources available on CD, video/DVD, in the *Student Study Guide*, and online at math.college.hmco.com/students.

Page 434

Page 433

CHAPTER 7

EXPONENTIAL AND LOGARITHMIC FUNCTIONS

7.1 EXPONENTIAL FUNCTIONS AND THEIR APPLICATIONS
7.2 LOGARITHMIC FUNCTIONS AND THEIR APPLICATIONS
7.3 LOGARITHMS AND LOGARITHMIC SCALES
7.4 EXPONENTIAL AND LOGARITHMIC EQUATIONS
7.5 EXPONENTIAL GROWTH AND DECAY
7.6 MODELING DATA WITH EXPONENTIAL AND LOGARITHMIC FUNCTIONS

Modeling Data with an Exponential Function

The following table shows the time, in hours, before the body of a scuba diver, wearing a 5-millimeter-thick wet suit, reaches hypothermia (95°F) for various water temperatures.

Water Temperature, °F	Time, hours
36	1.5
41	1.8
46	2.6
50	3.1
55	4.9

Source: Data extracted from the *American Journal of Physics*, vol. 71, no. 4 (April 2003), Fig. 3, p. 336.

The following function, which is an example of an exponential function, closely models the data in the table:

$$T(F) = 0.1509(1.0639)^F$$

In this function F represents the Fahrenheit temperature of the water, and T represents the time in hours. A diver can use the function to determine the time it takes to reach hypothermia for water temperatures that are not included in the table. See Exercise 21, page 513. The function $T(F)$ was determined by using exponential regression, which is one of the topics in Section 7.6.

---

21. HYPOTHERMIA The following table shows the time T, in hours, before a scuba diver wearing a 3-millimeter-thick wet suit reaches hypothermia (95°F) for various water temperatures F, in degrees Fahrenheit.

Water temperature, °F	Time T, hours
41	1.1
46	1.4
50	1.8
59	3.7

a. Find an exponential regression model for the data. Round the constants a and b to the nearest hundred thousandth.

b. Use the model from part a. to estimate the time it takes for the diver to reach hypothermia in water that has a temperature of 65°F. Round to the nearest tenth of an hour.

Page 513

FOCUS ON PROBLEM SOLVING

Use Two Methods to Solve and Compare Results

Sometimes it is possible to solve a problem in two or more ways. In such situations it is recommended that you use at least two methods to solve the problem, and compare your results. Here is an example of an application that can be solved in more than one way.

Example

In a league of eight basketball teams, each team plays every other team in the league exactly once. How many league games will take place?

Solution

Method 1: *Use an analytic approach.* Each of the eight teams must play the other seven teams. Using this information, you might be tempted to conclude that there will be $8 \cdot 7 = 56$ games, but this result is too large because it counts each game between two individual teams as two different games. Thus the number of league games will be

$$\frac{8 \cdot 7}{2} = \frac{56}{2} = 28$$

Method 2: *Make an organized list.* Use the letters A, B, C, D, E, F, G, and H to represent the eight teams. Use the notation AB to represent the game between team A and team B. Do not include BA in your list because it represents the same game between team A and team B.

AB AC AD AE AF AG AH
    BC BD BE BF BG BH
        CD CE CF CG CH
            DE DF DG DH
                EF EG EH
                    FG FH
                        GH

The list shows that there will be 28 league games.

The procedure of using two different solution methods and comparing results is employed often in this chapter. For instance, see Example 2, page 490. In this example, a solution is found by applying algebraic procedures and also by graphing. Notice that both methods produce the same result.

NEW! *FOCUS ON PROBLEM SOLVING*

A **Focus on Problem Solving** follows the Chapter Opener. This feature highlights and demonstrates a problem-solving strategy that may be used to successfully solve some of the problems presented in the chapter.

AUFMANN INTERACTIVE METHOD (AIM)

INTERACTIVE PRESENTATION

College Trigonometry is written in a style that encourages the student to interact with the textbook.

EXAMPLES

Each section contains a variety of worked examples. Examples are **titled** so that the student can see at a glance the type of problem being illustrated, often accompanied by **annotations** that assist the student in moving from step to step; and offers the **final answer in color** so that it is readily identifiable.

TRY EXERCISES

Following every example is a suggested **Try Exercise** from that section's exercise set for the student to work. The exercises are color coded by number in the exercise set and the *complete solution* of that exercise can be found in an appendix to the text.

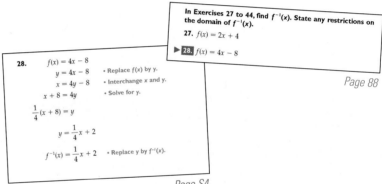

Page 83

EXAMPLE 3   Find the Inverse of a Function

Find the inverse of $f(x) = 3x + 8$.

Solution

$$f(x) = 3x + 8$$
$$y = 3x + 8 \qquad \text{• Replace } f(x) \text{ by } y.$$
$$x = 3y + 8 \qquad \text{• Interchange } x \text{ and } y.$$
$$x - 8 = 3y \qquad \text{• Solve for } y.$$
$$\frac{x - 8}{3} = y$$
$$\frac{1}{3}x - \frac{8}{3} = f^{-1}(x) \qquad \text{• Replace } y \text{ by } f^{-1}(x).$$

The inverse function is given by $f^{-1}(x) = \frac{1}{3}x - \frac{8}{3}$.

▶ **TRY EXERCISE 28, PAGE 88**

In Exercises 27 to 44, find $f^{-1}(x)$. State any restrictions on the domain of $f^{-1}(x)$.

27. $f(x) = 2x + 4$

▶ **28.** $f(x) = 4x - 8$

Page 88

28. $f(x) = 4x - 8$
$$y = 4x - 8 \qquad \text{• Replace } f(x) \text{ by } y.$$
$$x = 4y - 8 \qquad \text{• Interchange } x \text{ and } y.$$
$$x + 8 = 4y \qquad \text{• Solve for } y.$$
$$\frac{1}{4}(x + 8) = y$$
$$y = \frac{1}{4}x + 2$$
$$f^{-1}(x) = \frac{1}{4}x + 2 \qquad \text{• Replace } y \text{ by } f^{-1}(x).$$

Page S4

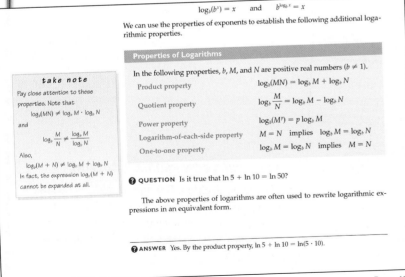

$$\log_b(b^x) = x \qquad \text{and} \qquad b^{\log_b x} = x$$

We can use the properties of exponents to establish the following additional logarithmic properties.

Properties of Logarithms

In the following properties, b, M, and N are positive real numbers ($b \neq 1$).

Product property $\log_b(MN) = \log_b M + \log_b N$

Quotient property $\log_b \dfrac{M}{N} = \log_b M - \log_b N$

Power property $\log_b(M^p) = p \log_b M$

Logarithm-of-each-side property $M = N$ implies $\log_b M = \log_b N$

One-to-one property $\log_b M = \log_b N$ implies $M = N$

take note

Pay close attention to these properties. Note that
$$\log_b(MN) \neq \log_b M \cdot \log_b N$$
and
$$\log_b \frac{M}{N} \neq \frac{\log_b M}{\log_b N}$$
Also,
$$\log_b(M + N) \neq \log_b M + \log_b N$$
In fact, the expression $\log_b(M + N)$ cannot be expanded at all.

❓ **QUESTION** Is it true that $\ln 5 + \ln 10 = \ln 50$?

The above properties of logarithms are often used to rewrite logarithmic expressions in an equivalent form.

❓ **ANSWER** Yes. By the product property, $\ln 5 + \ln 10 = \ln(5 \cdot 10)$.

Page 463

QUESTION/ANSWER

In every section, we pose at least one **Question** to the student about the material being read. This question encourages the reader to pause and think about the current discussion and to answer the question. To make sure the student does not miss important information, the **Answer** to the question is provided as a footnote on the same page.

REAL DATA AND APPLICATIONS

APPLICATIONS

One way to motivate an interest in mathematics is through applications. Applications require the student to use problem-solving strategies, along with the skills covered in a section, to solve practical problems. This careful integration of applications generates student awareness of the value of trigonometry as a real-life tool.

Applications are taken from many disciplines including agriculture, business, chemistry, construction, Earth science, education, economics, manufacturing, nutrition, real estate, and sociology.

Page 516

---

496   Chapter 7   Exponential and Logarithmic Functions

In the following example we determine a logistic growth model for a coyote population.

EXAMPLE 8   Find and Use a Logistic Model

At the beginning of 2002, the coyote population in a wilderness area was estimated at 200. By the beginning of 2004, the coyote population had increased to 250. A park ranger estimates that the carrying capacity of the wilderness area is 500 coyotes.

a. Use the given data to determine the growth rate constant for the logistic model of this coyote population.

b. Use the logistic model determined in part **a.** to predict the year in which the coyote population will first reach 400.

Solution

a. If we represent the beginning of the year 2002 by $t = 0$, then the beginning of the year 2004 will be represented by $t = 2$. In the logistic model, make the following substitutions: $P(2) = 250$, $c = 500$, and

$$a = \frac{c - P_0}{P_0} = \frac{500 - 200}{200} = 1.5.$$

$$P(t) = \frac{c}{1 + ae^{-bt}}$$

$$P(2) = \frac{500}{1 + 1.5e^{-b \cdot 2}} \qquad \bullet \text{ Substitute the given values.}$$

$$250 = \frac{500}{1 + 1.5e^{-b \cdot 2}}$$

$$250(1 + 1.5e^{-b \cdot 2}) = 500 \qquad \bullet \text{ Solve for the growth rate constant } b.$$

$$1 + 1.5e^{-b \cdot 2} = \frac{500}{250}$$

$$1.5e^{-b \cdot 2} = 2 - 1$$

$$e^{-b \cdot 2} = \frac{1}{1.5}$$

$$-2b = \ln\left(\frac{1}{1.5}\right)$$

$$b = -\frac{1}{2}\ln\left(\frac{1}{1.5}\right)$$

$$b \approx 0.20273255$$

Page 496

---

516   Chapter 7   Exponential and Logarithmic Functions

Time t (minutes)	0	5	10	15	20	25
Coffee temp. T (°F)	165°	140°	121°	107°	97°	89°
$T - 70°$	95°	70°	51°	37°	27°	19°

a. Use a graphing utility to find an exponential model for the difference $T - 70°$ as a function of t.

b. Use the model to predict how long it will take (to the nearest minute) for the coffee to cool to 80°F.

35. OLYMPIC DISTANCES The following table shows the winning Olympic distances for the men's shot put for the years 1948 to 2000.

Men's Olympic Shot Put, 1948 to 2000

Year	Distance	Year	Distance
1948	56 ft 2 in.	1976	69 ft $\frac{3}{4}$ in.
1952	57 ft 1$\frac{1}{2}$ in.	1980	70 ft $\frac{1}{2}$ in.
1956	60 ft 11 in.	1984	69 ft 9 in.
1960	64 ft 6$\frac{3}{4}$ in.	1988	73 ft 8$\frac{3}{4}$ in.
1964	66 ft 8$\frac{1}{4}$ in.	1992	71 ft 2$\frac{1}{2}$ in.
1968	67 ft 4$\frac{3}{4}$ in.	1996	70 ft 11$\frac{1}{4}$ in.
1972	69 ft 6 in.	2000	69 ft 10$\frac{1}{4}$ in.

Source: Time Almanac 2002

Represent the year 1948 by $t = 48$.

a. Use the regression features of a graphing utility to determine a logistic growth model and a logarithmic model for the data.

b. Use graphs of the models in part **a.** to determine which model provides the better fit for the data.

c. Use the model you selected in part **b.** to predict the men's shot put distance for the year 2008. Round to the nearest hundredth of a foot.

36. WORLD POPULATION The following table lists the years in which the world's population first reached 3, 4, 5, and 6 billion.

World Population Milestones

Year	Population
1960	3 billion
1974	4 billion
1987	5 billion
1999	6 billion

Source: Time Almanac 2002, p. 708.

a. Find a logistic growth model, $P(t)$, for the data in the table. Let t represent the number of years after 1960 ($t = 0$ represents the year 1960).

b. According to the logistic growth model, what will the world's population approach as $t \to \infty$? Round to the nearest billion.

37. DESALINATION The following table shows the amount of fresh water w (in cubic yards) produced from saltwater after t hours of a desalination process.

t	1	2.5	3.5	4.0	5.1	6.5
w	18.2	46.6	57.4	61.5	68.7	76.2

a. Use a graphing utility to find a linear model and a logarithmic model for the data.

b. Examine the correlation coefficients of the two regression models to determine which model provides the better fit for the data. State the correlation coefficient r for each model.

c. Use the model you selected in part **b.** to predict the amount of fresh water that will be produced after 10 hours of the desalination process. Round to the nearest tenth of a cubic yard.

38. A CORRELATION COEFFICIENT OF 1 A scientist uses a graphing utility to model the data set $\{(2, 5), (4, 6)\}$ with a logarithmic function. The following display shows the results.

What is the significance of the fact that the correlation coefficient for the regression equation is $r = 1$?

```
LnReg
y=a+blnx
a=4
b=1.442695041
r²=1
r=1
```

REAL DATA

Real data examples and exercises, identified by ⚫, ask students to analyze and create mathematical models from actual situations. Students are often required to work with tables, graphs, and charts drawn from a variety of disciplines.

TECHNOLOGY

INTEGRATING TECHNOLOGY

The **Integrating Technology** feature contains optional discussions that can be used to further explore a concept using technology. Some introduce technology as an alternative way to solve certain problems and others provide suggestions for using a calculator to solve certain problems and applications. Additionally, optional graphing calculator examples and exercises (identified by ) are presented throughout the text.

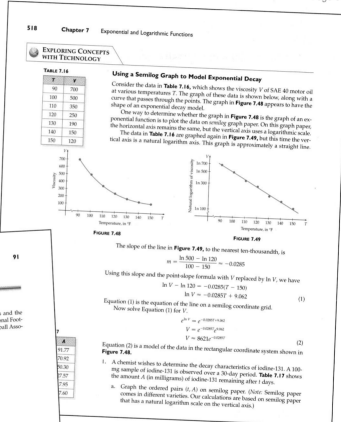

INTEGRATING TECHNOLOGY

The graph of $y^2 = -8x$ is shown in **Figure 6.5.** Note that the graph is not the graph of a function. To graph $y^2 = -8x$ with a graphing utility, we first solve for y to produce $y = \pm\sqrt{-8x}$. From this equation we can see that for any $x < 0$, there are two values of y. For example, when $x = -2$,

$$y = \pm\sqrt{(-8)(-2)} = \pm\sqrt{16} = \pm 4$$

The graph of $y^2 = -8x$ in **Figure 6.5** was drawn by graphing both $y_1 = \sqrt{-8x}$ and $y_2 = -\sqrt{-8x}$ in the same window.

FIGURE 6.5

Page 353

EXPLORING CONCEPTS WITH TECHNOLOGY

A special end-of-chapter feature, **Exploring Concepts with Technology**, extends ideas introduced in the text by using technology (graphing calculator, CAS, etc.) to investigate extended applications or mathematical topics. These explorations can serve as group projects, class discussions, or extra-credit assignments.

518   Chapter 7   Exponential and Logarithmic Functions

EXPLORING CONCEPTS WITH TECHNOLOGY

Using a Semilog Graph to Model Exponential Decay

Consider the data in **Table 7.16,** which shows the viscosity V of SAE 40 motor oil at various temperatures T. The graph of these data is shown below, along with a curve that passes through the points. The graph in **Figure 7.48** appears to have the shape of an exponential decay model.

One way to determine whether the graph in **Figure 7.48** is the graph of an exponential function is to plot the data on *semilog* graph paper. On this graph paper, the horizontal axis remains the same, but the vertical axis uses a logarithmic scale.

The data in **Table 7.16** are graphed again in **Figure 7.49,** but this time the vertical axis is a natural logarithm axis. This graph is approximately a straight line.

TABLE 7.16

T	V
90	700
100	500
110	350
120	250
130	190
140	150
150	120

FIGURE 7.48

FIGURE 7.49

The slope of the line in **Figure 7.49,** to the nearest ten-thousandth, is

$$m = \frac{\ln 500 - \ln 120}{100 - 150} \approx -0.0285$$

Using this slope and the point-slope formula with V replaced by $\ln V$, we have

$$\ln V - \ln 120 = -0.0285(T - 150)$$
$$\ln V \approx -0.0285T + 9.062 \qquad (1)$$

Equation (1) is the equation of the line on a semilog coordinate grid. Now solve Equation (1) for V.

$$e^{\ln V} = e^{-0.0285T + 9.062}$$
$$V = e^{-0.0285T}e^{9.062}$$
$$V = 8621e^{-0.0285T} \qquad (2)$$

Equation (2) is a model of the data in the rectangular coordinate system shown in **Figure 7.48.**

1. A chemist wishes to determine the decay characteristics of iodine-131. A 100-mg sample of iodine-131 is observed over a 30-day period. **Table 7.17** shows the amount A (in milligrams) of iodine-131 remaining after t days.

   a. Graph the ordered pairs (t, A) on semilog paper. (*Note:* Semilog paper comes in different varieties. Our calculations are based on semilog paper that has a natural logarithm scale on the vertical axis.)

	A
	91.77
	70.92
	50.30
	27.57
	7.95
	7.60

Page 518

Page 91

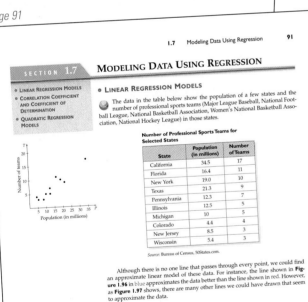

SECTION 1.7   MODELING DATA USING REGRESSION

- LINEAR REGRESSION MODELS
- CORRELATION COEFFICIENT AND COEFFICIENT OF DETERMINATION
- QUADRATIC REGRESSION MODELS

LINEAR REGRESSION MODELS

The data in the table below show the population of a few states and the number of professional sports teams (Major League Baseball, National Football League, National Basketball Association, Women's National Basketball Association, National Hockey League) in those states.

Number of Professional Sports Teams for Selected States

State	Population (in millions)	Number of Teams
California	34.5	17
Florida	16.4	11
New York	19.0	10
Texas	21.3	9
Pennsylvania	12.3	7
Illinois	12.5	5
Michigan	10	5
Colorado	4.4	4
New Jersey	8.5	3
Wisconsin	5.4	3

Source: Bureau of Census, 50States.com.

Although there is no one line that passes through every point, we could find an approximate linear model of these data. For instance, the line shown in **Figure 1.96** in blue approximates the data better than the line shown in red. However, as **Figure 1.97** shows, there are many other lines we could have drawn that seem to approximate the data.

FIGURE 1.96   **FIGURE 1.97**

MODELING

Special modeling sections, which rely heavily on the use of a graphing calculator, are incorporated throughout the text. These optional sections introduce the idea of a mathematical model using various real-world data sets that further motivate students and help them see the relevance of mathematics.

STUDENT PEDAGOGY

TOPIC LIST

At the beginning of each section is a list of the major topics covered in the section.

KEY TERMS AND CONCEPTS

Key terms, in bold, emphasize important terms. Key concepts are presented in blue boxes in order to highlight these important concepts and to provide for easy reference.

MATH MATTERS

These margin notes contain interesting sidelights about mathematics, its history, or its application.

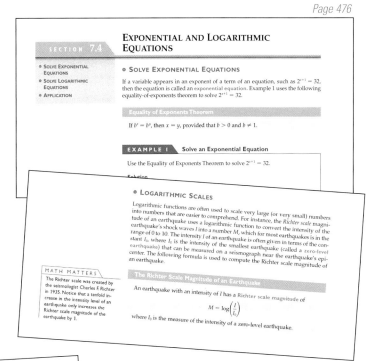

Page 476

SECTION 7.4

EXPONENTIAL AND LOGARITHMIC EQUATIONS

- Solve Exponential Equations
- Solve Logarithmic Equations
- Application

Solve Exponential Equations

If a variable appears in an exponent of a term of an equation, such as $2^{x+1} = 32$, then the equation is called an exponential equation. Example 1 uses the following equality-of-exponents theorem to solve $2^{x+1} = 32$.

Equality of Exponents Theorem

If $b^x = b^y$, then $x = y$, provided that $b > 0$ and $b \neq 1$.

EXAMPLE 1   Solve an Exponential Equation

Use the Equality of Exponents Theorem to solve $2^{x+1} = 32$.

Logarithmic Scales

Logarithmic functions are often used to scale very large (or very small) numbers into numbers that are easier to comprehend. For instance, the *Richter scale* magnitude of an earthquake uses a logarithmic function to convert the intensity of the earthquake's shock waves I into a number M, which for most earthquakes is in the range of 0 to 10. The intensity I of an earthquake is often given in terms of the constant I_0, where I_0 is the intensity of the smallest earthquake (called a *zero-level earthquake*) that can be measured on a seismograph near the earthquake's epicenter. The following formula is used to compute the Richter scale magnitude of an earthquake.

MATH MATTERS

The Richter scale was created by the seismologist Charles F. Richter in 1935. Notice that a tenfold increase in the intensity level of an earthquake only increases the Richter scale magnitude of the earthquake by 1.

The Richter Scale Magnitude of an Earthquake

An earthquake with an intensity of I has a Richter scale magnitude of

$$M = \log\left(\frac{I}{I_0}\right)$$

where I_0 is the measure of the intensity of a zero-level earthquake.

Page 466

Page 352

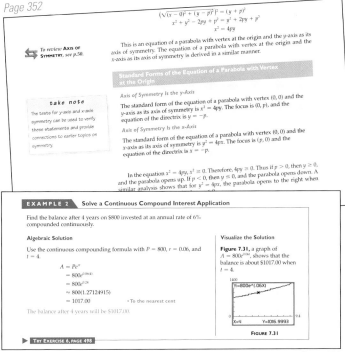

$$(\sqrt{(x-0)^2 + (y-p)^2})^2 = (y+p)^2$$
$$x^2 + y^2 - 2py + p^2 = y^2 + 2py + p^2$$
$$x^2 = 4py$$

To review **AXIS OF SYMMETRY**, see p.50.

This is an equation of a parabola with vertex at the origin and the y-axis as its axis of symmetry. The equation of a parabola with vertex at the origin and the x-axis as its axis of symmetry is derived in a similar manner.

Standard Forms of the Equation of a Parabola with Vertex at the Origin

Axis of Symmetry Is the y-Axis

The standard form of the equation of a parabola with vertex $(0, 0)$ and the y-axis as its axis of symmetry is $x^2 = 4py$. The focus is $(0, p)$, and the equation of the directrix is $y = -p$.

Axis of Symmetry Is the x-Axis

The standard form of the equation of a parabola with vertex $(0, 0)$ and the x-axis as its axis of symmetry is $y^2 = 4px$. The focus is $(p, 0)$ and the equation of the directrix is $x = -p$.

take note

The tests for y-axis and x-axis symmetry can be used to verify these statements and provide connections to earlier topics on symmetry.

In the equation $x^2 = 4py$, $x^2 \geq 0$. Therefore, $4py \geq 0$. Thus if $p > 0$, then $y \geq 0$, and the parabola opens up. If $p < 0$, then $y \leq 0$, and the parabola opens down. A similar analysis shows that for $y^2 = 4px$, the parabola opens to the right when

EXAMPLE 2   Solve a Continuous Compound Interest Application

Find the balance after 4 years on $800 invested at an annual rate of 6% compounded continuously.

Algebraic Solution

Use the continuous compounding formula with $P = 800$, $r = 0.06$, and $t = 4$.

$$A = Pe^{rt}$$
$$= 800e^{0.06(4)}$$
$$= 800e^{0.24}$$
$$\approx 800(1.27124915)$$
$$\approx 1017.00 \qquad \text{• To the nearest cent}$$

The balance after 4 years will be $1017.00.

Visualize the Solution

Figure 7.31, a graph of $A = 800e^{0.06x}$, shows that the balance is about $1017.00 when $t = 4$.

Y=800e^(.06X)

X=4   Y=1016.9993

FIGURE 7.31

▶ TRY EXERCISE 6, PAGE 498

Page 490

NEW!   *REVIEW NOTES*

A ⬌ directs the student to the place in the text where the student can review a concept that was previously discussed.

TAKE NOTE

These margin notes alert students to a point requiring special attention or are used to amplify the concept under discussion.

VISUALIZE THE SOLUTION

For appropriate examples within the text, we have provided both an algebraic solution and a graphical representation of the solution. This approach creates a link between the algebraic and visual components of a solution.

EXERCISES

TOPICS FOR DISCUSSION

These special exercises provide questions related to key concepts in the section. Instructors can use these to initiate class discussions or to ask students to write about concepts presented.

EXERCISES

The exercise sets in *College Trigonometry* were carefully developed to provide a wide variety of exercises. The exercises range from drill and practice to interesting challenges. They were chosen to illustrate the many facets of topics discussed in the text. Each exercise set emphasizes skill building, skill maintenance, and, as appropriate, applications. **Icons** identify appropriate writing , group, data analysis, web, and graphing calculator exercises.

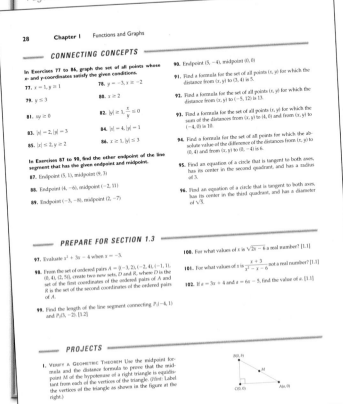

Page 28

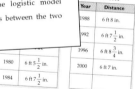

Included in each exercise set are **Connecting Concepts** exercises. These exercises extend some of the concepts discussed in the section and require students to connect ideas studied earlier with new concepts.

NEW! *EXERCISES TO PREPARE FOR THE NEXT SECTION*

Every section's exercise set (except for the last section of a chapter) contains exercises that allow students to practice the previously-learned skills they will need to be successful in the next section. Next to each question, in brackets, is a reference to the section of the text that contains the concepts related to the question for students to easily review. All answers are provided in the Answer Appendix.

PROJECTS

Projects are provided at the end of each exercise set. They are designed to encourage students to do research and write about what they have learned. These Projects generally emphasize critical thinking skills and can be used as collaborative learning exercises or as extra-credit assignments.

Page 515

END OF CHAPTER

Page 519

CHAPTER SUMMARY

At the end of each chapter there is a Chapter Summary that provides a concise section-by-section review of the chapter topics.

CHAPTER 7 SUMMARY

7.1 Exponential Functions and Their Applications

- For all positive real numbers b, $b \neq 1$, the exponential function defined by $f(x) = b^x$ has the following properties:
  1. f has the set of real numbers as its domain.
  2. f has the set of positive real numbers as its range.
  3. f has a graph with a y-intercept of $(0, 1)$.
  4. f has a graph asymptotic to the x-axis.

  5. f is a one-to-one function.
  6. f is an increasing function if $b > 1$.
  7. f is a decreasing function if $0 < b < 1$.
- As n increases without bound, $(1 + 1/n)^n$ approaches an irrational number denoted by e. The value of e accurate to eight decimal places is 2.71828183.
- The function defined by $f(x) = e^x$ is called the natural ex-

TRUE/FALSE EXERCISES

Following each chapter summary are true/false exercises. These exercises are intended to help students understand concepts and can be used to initiate class discussions.

CHAPTER 7 TRUE/FALSE EXERCISES

In Exercises 1 to 14, answer true or false. If the statement is false, give an example or state a reason to demonstrate that the statement is false.

1. If $7^x = 40$, then $\log_7 40 = x$.
2. If $\log_4 x = 3.1$, then $4^{3.1} = x$.
3. If $f(x) = \log x$ and $g(x) = 10^x$, then $f[g(x)] = x$ for all real numbers x.
4. If $f(x) = \log x$ and $g(x) = 10^x$, then $g[f(x)] = x$ for all real numbers x.
5. The exponential function $h(x) = b^x$ is an increasing function.
6. The logarithmic function $j(x) = \log_b x$ is an increasing function.
7. The exponential function $h(x) = b^x$ is a one-to-one function.
8. The logarithmic function $j(x) = \log_b x$ is a one-to-one function.
9. The graph of $f(x) = \dfrac{2^x + 2^{-x}}{2}$ is symmetric with respect to the y-axis.
10. The graph of $f(x) = \dfrac{2^x - 2^{-x}}{2}$ is symmetric with respect to the origin.
11. If $x > 0$ and $y > 0$, then $\log(x + y) = \log x + \log y$.
12. If $x > 0$, then $\log x^2 = 2 \log x$.
13. If M and N are positive real numbers, then
$$\ln \frac{M}{N} = \ln M - \ln N$$
14. For all $p > 0$, $e^{\ln p} = p$.

Page 521

Page 521

CHAPTER 7

In Exercises 1 to 12, solve each equation. Do not use a calculator.

1. $\log_3 25 = x$
2. $\log_5 81 = x$
3. $\ln e^3 = x$
4. $\ln e^x = x$
5. $3^{2x+7} = 27$
6. $5^{x-4} = 625$
7. $2^x = \dfrac{1}{8}$
8. $27(3^x) = 3^{-1}$
9. $\log x^2 = 6$
10. $\dfrac{1}{2} \log |x| = 5$
11. $10^{\log 2x} = 14$
12. $e^{\ln x^2} = 64$

In Exercises 13 to 22, sketch the graph of each function.

13. $f(x) = (2.5)^x$
14. $f(x) = \left(\dfrac{1}{4}\right)^x$
15. $f(x) = 3^{x}$
16. $f(x) = 4^{-x}$
17. $f(x) = 2^x - 3$
18. $f(x) = 2^{x-3}$
19. $f(x) = \dfrac{1}{3} \log x$
20. $f(x) = 3 \log x^{1/3}$
21. $f(x) = -\dfrac{1}{2} \ln x$
22. $f(x) = -\ln |x|$

In Exercises 23 and 24, use a graphing utility to graph each function.

23. $f(x) = \dfrac{4^x + 4^{-x}}{2}$
24. $f(x) = \dfrac{3^x - 3^{-x}}{2}$

In Exercises 25 to 28, change each logarithmic equation to its exponential form.

25. $\log_4 64 = 3$
26. $\log_{1/2} 8 = -3$
27. $\log_{1/2} 4 = 4$
28. $\ln 1 = 0$

CHAPTER REVIEW EXERCISES

Review exercises are found at the end of each chapter. These exercises are selected to help the student integrate all of the topics presented in the chapter.

CHAPTER TEST

The Chapter Test exercises are designed to simulate a possible test of the material in the chapter.

NEW! CUMULATIVE REVIEW EXERCISES

Cumulative Review Exercises, which appear at the end of each chapter (except Chapter 1), help students maintain skills learned in previous chapters.

NEW! The answers to all **Chapter Review Exercises,** all **Chapter Test Exercises,** and all **Cumulative Review Exercises** are given in the Answer Section. Along with the answer, there is a reference to the section that pertains to each exercise.

CHAPTER 7 TEST

1. **a.** Write $\log_4(5x - 3) = c$ in exponential form.
   b. Write $3^{1/2} = y$ in logarithmic form.
2. Write $\log_b \dfrac{z^2}{y^3 \sqrt{x}}$ in terms of logarithms of x, y, and z.
3. Write $\log(2x + 3) - 3 \log(x - 2)$ as a single logarithm with a coefficient of 1.
4. Use the change-of-base formula and a calculator to approximate $\log_4 12$. Round your result to the nearest ten thousandth.
5. Graph: $f(x) = 3^{-1/2}$
6. Graph: $f(x) = -\ln(x + 1)$

the ratio of the larger intensity to the smaller intensity. Round to the nearest whole number.

14. **a.** Find the exponential growth function for a city whose population was 34,600 in 1996 and 39,800 in 1999. Use $t = 0$ to represent the year 1996.
    b. Use the growth function to predict the population of the city in 2006. Round to the nearest thousand.
15. Determine, to the nearest ten years, the age of a bone if it now contains 92% of its original amount of carbon-14. The half-life of carbon-14 is 5730 years.
16. **a.** Use a graphing utility to find the exponential regression function for the following data.
    $\{(2.5, 16), (3.7, 48), (5.0, 155), (6.5, 571), (6.9, 896)\}$

Page 524

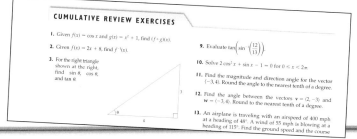

CUMULATIVE REVIEW EXERCISES

1. Given $f(x) = \cos x$ and $g(x) = x^2 + 1$, find $(f \circ g)(x)$.
2. Given $f(x) = 2x + 8$, find $f^{-1}(x)$.
3. For the right triangle shown at the right, find $\sin \theta$, $\cos \theta$, and $\tan \theta$.
9. Evaluate $\tan\left(\sin^{-1}\left(\dfrac{12}{13}\right)\right)$.
10. Solve $2 \cos^2 x + \sin x - 1 = 0$ for $0 \le x < 2\pi$.
11. Find the magnitude and direction angle for the vector $\langle -3, 4\rangle$. Round the angle to the nearest tenth of a degree.
12. Find the angle between the vectors $\mathbf{v} = \langle 2, -3\rangle$ and $\mathbf{w} = \langle -3, 4\rangle$. Round to the nearest tenth of a degree.
13. An airplane is traveling with an airspeed of 400 mph at a heading of 48°. A wind of 55 mph is blowing at a heading of 115°. Find the ground speed and the course

Page 525

END OF CHAPTER

Instructor Resources

College Trigonometry has a complete set of support materials for the instructor.

Instructor's Solutions Manual The *Instructor's Solutions Manual* contains worked-out solutions for all exercises in the text.

Instructor's Resource Manual with Testing This resource includes six ready-to-use printed *Chapter Tests* per chapter, and a *Printed Test Bank* providing a printout of one example of each of the algorithmic items on the *HM Testing* CD-ROM program.

HM ClassPrep w/ HM Testing CD-ROM *HM ClassPrep* contains a multitude of text-specific resources for instructors to use to enhance the classroom experience. These resources can be easily accessed by chapter or resource type and also can link you to the text's website. *HM Testing* is our computerized test generator and contains a database of algorithmic test items, as well as providing **online testing** and **gradebook** functions.

Instructor Text-specific Website The resources available on the *ClassPrep CD* are also available on the instructor website at math.college.hmco.com/instructors. Appropriate items are password protected. Instructors also have access to the student part of the text's website.

Student Resources

Student Study Guide The *Student Study Guide* contains complete solutions to all odd-numbered exercises in the text, as well as study tips and a practice test for each chapter.

Math Study Skills Workbook *by Paul D. Nolting* This workbook is designed to reinforce skills and minimize frustration for students in any math class, lab, or study skills course. It offers a wealth of study tips and sound advice on note taking, time management, and reducing math anxiety. In addition, numerous opportunities for self assessment enable students to track their own progress.

HM Eduspace® Online Learning Environment *Eduspace* is a text-specific, web-based learning environment that combines an algorithmic tutorial program with homework capabilities. Specific content is available 24 hours a day to help you further understand your textbook.

HM mathSpace® Tutorial CD-ROM This tutorial CD-ROM allows students to practice skills and review concepts as many times as necessary by providing algorithmically-generated exercises and step-by-step solutions for practice.

SMARTHINKING™ Live, Online Tutoring Houghton Mifflin has partnered with SMARTHINKING to provide an easy-to-use and effective online tutorial service. **Whiteboard Simulations** and **Practice Area** promote real-time visual interaction. Three levels of service are offered.

- **Text-specific Tutoring** provides real-time, one-on-one instruction with a specially qualified 'e-structor.'
- **Questions Any Time** allows students to submit questions to the tutor outside the scheduled hours and receive a reply within 24 hours.

- **Independent Study Resources** connect students with around-the-clock access to additional educational services, including interactive websites, diagnostic tests, and Frequently Asked Questions posed to SMARTHINKING e-structors.

Houghton Mifflin Instructional Videos and DVDs   Text-specific videos and DVDs, hosted by Dana Mosely, cover all sections of the text and provide a valuable resource for further instruction and review.

Student Text-specific Website   Online student resources can be found at this text's website at math.college.hmco.com/students.

Acknowledgments

The authors would like to thank the people who have reviewed this manuscript and provided many valuable suggestions.

Ioannis K. Argyros, *Cameron University, OK*
Peter Arvanites, *Rockland Community College, NY*
Linda Berg, *University of Great Falls, MT*
Paul Bialek, *Trinity College, IL*
Zhixiong Cai, *Barton College, NC*
Cheryl F. Cavaliero, *Butler County Community College, PA*
Jennie Cox, *Central Georgia Technical College, GA*
Marilyn Danchanko, *Cambria County Area Community College, PA*
Laura Davis, *Garland County Community College, AR*
Sylvia Dorminey, *Savannah Technical College, GA*
Gay Grubbs, *Griffin Technical College, GA*
Kathryn Hodge, *Midland College, TX*
Clement S. Lam, *Mission College, CA*
Dr. Helen Medley
Carla Monticelli, *Camden County College, NJ*
J. Reid Mowrer, *University of New Mexico–Valencia Campus, NM*
Sue Neal, *Wichita State University, KS*
Georgie O'Leary, *Warner Southern College, FL*
Suzanne Pauly, *Delaware Technical and Community College, DE*
Lauri Semarne
Mike Shirazi, *Germanna Community College, VA*
Anthony Tongen, *Trinity International University, IL*
Hanson Umoh, *Delaware State University, DE*
Rebecca Wells, *Henderson Community College, KY*

Special thanks to Christi Verity for her diligent preparation of the solutions manuals and to Sandy Doerfel for her contribution to this project.

FUNCTIONS AND GRAPHS

Tax Brackets Versus a Flat Tax

A perennial issue in Congress is the equity of the income tax laws. Currently, there are six tax brackets. As a person's income increases, the percent of adjusted gross income (income after allowable deductions) that the person pays in income tax increases. In 2003, Congress passed the Tax Relief Reconciliation Act of 2003, which changed the laws relating to income taxes. An alternative proposal before Congress that year was to replace these brackets with one bracket so that all taxpayers paid 20% of their income as tax. This type of tax structure is called a *flat tax* and is used in some states and countries. The flat-tax proposal was not passed by Congress.

The graph below shows the amount of tax a single person would pay using tax brackets and a flat tax. (Only the first three tax brackets are shown.)

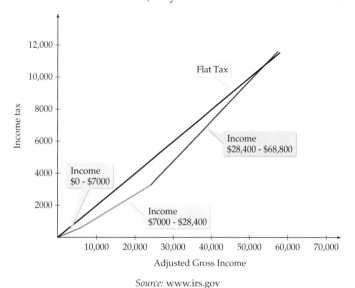

Source: www.irs.gov

The amount of income tax a taxpayer pays can be found by evaluating a piecewise function, a topic of this chapter. See **Exercise 48 on page 44.**

Difference Tables

When devising a plan to solve a problem, it may be helpful to organize information in a table. One particular type of table that can be used to discern some patterns is called a *difference table.* For instance, suppose that we want to determine the number of square tiles in the tenth figure of a pattern whose first four figures are

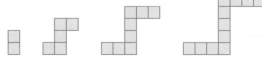

We begin by creating a difference table by listing the number of tiles in each figure and the differences between the numbers of tiles on the next line.

Tiles		2		5		8		11
Differences			3		3		3	

From the difference table, note that each succeeding figure has three more tiles. Therefore, we can find the number of tiles in the tenth figure by extending the difference table.

Tiles	2		5		8		11		14		17		20		23		26		29
Differences		3		3		3		3		3		3		3		3		3	

There are 29 tiles in the tenth figure.

Sometimes the first differences are not constant, as was the case for the preceding example. In this case we find the second differences. For instance, consider the pattern at the right.

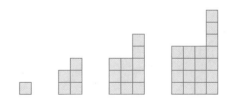

The difference table is shown below.

Tiles		1		5		11		19
First differences			4		6		8	
Second differences				2		2		

In this case the first differences are not constant, but the second differences are constant. With this information we can determine the number of tiles in succeeding figures.

Tiles	1		5		11		19		29		41		55		71		89		109
First differences		4		6		8		10		12		14		16		18		20	
Second differences			2		2		2		2		2		2		2		2		

In this case there are 109 tiles in the tenth figure.

If second differences are not constant, try the third differences.[1] If third differences are not constant, try fourth differences, and so on.

---

[1] Not all lists of numbers will end with a difference row of constants. For instance, consider 1, 1, 2, 3, 5, 8, 13, 21, 34,....

SECTION 1.1 EQUATIONS AND INEQUALITIES

- THE REAL NUMBERS
- ABSOLUTE VALUE AND DISTANCE
- LINEAR AND QUADRATIC EQUATIONS
- INEQUALITIES
- SOLVING INEQUALITIES BY THE CRITICAL VALUE METHOD
- ABSOLUTE VALUE INEQUALITIES

MATH MATTERS

Archimedes (c. 287–212 B.C.) was the first to calculate π with any degree of precision. He was able to show that

$$3\frac{10}{71} < \pi < 3\frac{1}{7}$$

from which we get the approximation $3\frac{1}{7} \approx \pi$. The use of the symbol π for this quantity was introduced by Leonhard Euler (1707–1783) in 1739, approximately 2000 years after Archimedes.

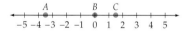

FIGURE 1.1

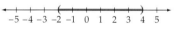

The open interval (−2, 4)

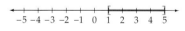

The closed interval [1, 5]

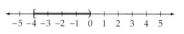

The half-open interval [−4, 0)

The half-open interval (−5, −2]

FIGURE 1.2

THE REAL NUMBERS

The real numbers are used extensively in mathematics. The set of real numbers is quite comprehensive and contains several unique sets of numbers.

The **integers** are the set of numbers

$$\{\ldots, -4, -3, -2, -1, 0, 1, 2, 3, 4, \ldots\}$$

Recall that the brace symbols, { }, are used to identify a set. The positive integers are called **natural numbers.**

The **rational numbers** are the set of numbers of the form $\frac{a}{b}$, where a and b are integers and $b \neq 0$. Thus the rational numbers include $-\frac{3}{4}$ and $\frac{5}{2}$. Because each integer can be expressed in the form $\frac{a}{b}$ with denominator $b = 1$, the integers are included in the set of rational numbers. Every rational number can be written as either a terminating or a repeating decimal.

A number written in decimal form that does not repeat or terminate is called an **irrational number.** Some examples of irrational numbers are 0.141141114..., $\sqrt{2}$, and π. These numbers cannot be expressed as quotients of integers. The set of **real numbers** is the union of the sets of rational and irrational numbers.

A real number can be represented geometrically on a **coordinate axis** called a **real number line.** Each point on this line is associated with a real number called the **coordinate** of the point. Conversely, each real number can be associated with a point on a real number line. In **Figure 1.1,** the coordinate of A is $-\frac{7}{2}$, the coordinate of B is 0, and the coordinate of C is $\sqrt{2}$.

Given any two real numbers a and b, we say that a is **less than** b, denoted by $a < b$, if $a - b$ is a negative number. Similarly, we say that a is **greater than** b, denoted by $a > b$, if $a - b$ is a positive number. When a **equals** b, $a - b$ is zero. The symbols $<$ and $>$ are called **inequality symbols.** Two other inequality symbols, $\leq$ (less than or equal to) and $\geq$ (greater than or equal to), are also used.

The inequality symbols can be used to designate sets of real numbers. If $a < b$, the **interval notation** (a, b) is used to indicate the set of real numbers between a and b. This set of numbers also can be described using **set-builder notation:**

$$(a, b) = \{x \mid a < x < b\}$$

When reading a set written in set-builder notation, we read $\{x \mid$ as "the set of x such that." The expression that follows the vertical bar designates the elements in the set.

The set (a, b) is called an **open interval.** The graph of the open interval consists of all the points on the real number line between a and b, not including a and b. A **closed interval,** denoted by $[a, b]$, consists of all points between a and b, including a and b. We can also discuss **half-open intervals.** An example of each type of interval is shown in **Figure 1.2.**

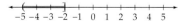

take note

The interval notation $[1, \infty)$ represents all real numbers greater than or equal to 1. The interval notation $(-\infty, 4)$ represents all real numbers less than 4.

$(-2, 4) = \{x \mid -2 < x < 4\}$	An open interval
$[1, 5] = \{x \mid 1 \le x \le 5\}$	A closed interval
$[-4, 0) = \{x \mid -4 \le x < 0\}$	A half-open interval
$(-5, -2] = \{x \mid -5 < x \le -2\}$	A half-open interval

● ABSOLUTE VALUE AND DISTANCE

The *absolute value* of a real number is a measure of the distance from zero to the point associated with the number on a real number line. Therefore, the absolute value of a real number is always positive or zero. We now give a more formal definition of absolute value.

Absolute Value

For a real number a, the **absolute value** of a, denoted by $|a|$, is

$$|a| = \begin{cases} a & \text{if } a \ge 0 \\ -a & \text{if } a < 0 \end{cases}$$

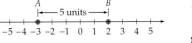

FIGURE 1.3

The distance d between the points A and B with coordinates -3 and 2, respectively, on a real number line is the absolute value of the difference between the coordinates. See **Figure 1.3.**

$$d = |2 - (-3)| = 5$$

Because the absolute value is used, we could also write

$$d = |(-3) - 2| = 5$$

In general, we define the *distance* between any two points A and B on a real number line as the absolute value of the difference between the coordinates of the points.

Distance Between Two Points on a Coordinate Line

Let a and b be the coordinates of the points A and B, respectively, on a real number line. Then the **distance** between A and B, denoted $d(A, B)$, is

$$d(A, B) = |a - b|$$

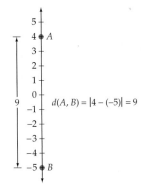

FIGURE 1.4

This formula applies to any real number line. It can be used to find the distance between two points on a vertical real number line, as shown in **Figure 1.4.**

• LINEAR AND QUADRATIC EQUATIONS

An **equation** is a statement about the equality of two expressions. Examples of equations follow.

$$7 = 2 + 5 \qquad x^2 = 4x + 5 \qquad 3x - 2 = 2(x + 1) + 3$$

The values of the variable that make an equation a true statement are the **roots** or **solutions** of the equation. To **solve** an equation means to find the solutions of the equation. The number 2 is said to satisfy the equation $2x + 1 = 5$, because substituting 2 for x produces $2(2) + 1 = 5$, which is a true statement.

Definition of a Linear Equation

A **linear equation** in the single variable x is an equation of the form $ax + b = 0$, where $a \neq 0$.

To solve a linear equation in one variable, isolate the variable on one side of the equals sign.

EXAMPLE 1   Solve a Linear Equation

Solve: $3x - 5 = 2$

Solution

$$3x - 5 = 2$$
$$3x - 5 + 5 = 2 + 5 \qquad \text{• Add \textbf{5} to each side of the equation.}$$
$$3x = 7$$
$$\frac{3x}{3} = \frac{7}{3} \qquad \text{• Divide each side of the equation by \textbf{3}.}$$
$$x = \frac{7}{3}$$

The solution is $\frac{7}{3}$.

▶ **TRY EXERCISE 6, PAGE 12**

MATH MATTERS

The term *quadratic* is derived from the Latin word *quadrāre*, which means "to make square." Because the area of a square that measures x units on each side is x^2, we refer to equations that can be written in the form $ax^2 + bx + c = 0$ as equations that are quadratic in x.

Definition of a Quadratic Equation

An equation of the form $ax^2 + bx + c = 0$, $a \neq 0$, is a **quadratic equation** in x.

MATH MATTERS

There is a general procedure to solve "by radicals" the general cubic

$$ax^3 + bx^2 + cx + d = 0$$

and the general quartic

$$ax^4 + bx^3 + cx^2 + dx + e = 0$$

However, it has been proved that there are no general procedures that can solve "by radicals" general equations of degree 5 or larger.

A quadratic equation can be solved using the **quadratic formula.**

The Quadratic Formula

The solution of the quadratic equation $ax^2 + bx + c = 0$, $a \neq 0$, is given by

$$x = \frac{-b \pm \sqrt{b^2 - 4ac}}{2a}$$

❓ **QUESTION** For $2x^2 - 3x - 1 = 0$, what are the values of a, b, and c?

EXAMPLE 2 Solve a Quadratic Equation

Solve by using the quadratic formula: $2x^2 - 4x + 1 = 0$

Solution

We have $a = 2$, $b = -4$, and $c = 1$.

$$x = \frac{-(-4) \pm \sqrt{(-4)^2 - 4(2)(1)}}{2(2)} = \frac{4 \pm \sqrt{16 - 8}}{4}$$

$$= \frac{4 \pm \sqrt{8}}{4} = \frac{4 \pm 2\sqrt{2}}{4} = \frac{2 \pm \sqrt{2}}{2}$$

The solutions are $\dfrac{2 + \sqrt{2}}{2}$ and $\dfrac{2 - \sqrt{2}}{2}$.

▶ **TRY EXERCISE 22, PAGE 12**

Although every quadratic equation can be solved using the quadratic formula, it is sometimes easier to factor and use the **zero product property.**

Zero Product Property

If a and b are algebraic expressions, then $ab = 0$ if and only if $a = 0$ or $b = 0$.

For example, to solve $2x^2 + x - 6 = 0$, first factor the polynomial.

$$2x^2 + x - 6 = 0$$
$$(2x - 3)(x + 2) = 0$$
$$2x - 3 = 0 \quad \text{or} \quad x + 2 = 0 \qquad \text{• Zero product property}$$
$$x = \frac{3}{2} \quad \text{or} \qquad x = -2$$

The solutions are $\dfrac{3}{2}$ and -2.

❓ **ANSWER** $a = 2$, $b = -3$, $c = -1$

EXAMPLE 3	Solve by Using the Zero Product Property

Solve: $(2x - 1)(x - 3) = x^2 + x - 4$

Solution

$$(2x - 1)(x - 3) = x^2 + x - 4$$
$$2x^2 - 7x + 3 = x^2 + x - 4 \qquad \text{• Expand the binomial product.}$$
$$x^2 - 8x + 7 = 0 \qquad \text{• Write as } ax^2 + bx + c = 0.$$
$$(x - 7)(x - 1) = 0 \qquad \text{• Factor.}$$
$$x - 7 = 0 \quad \text{or} \quad x - 1 = 0 \qquad \text{• Apply the zero product property.}$$
$$x = 7 \qquad\qquad x = 1$$

The solutions are 1 and 7.

▶ TRY EXERCISE 36, PAGE 12

● **INEQUALITIES**

A statement that contains the symbol $<, >, \leq,$ or $\geq$ is called an **inequality.** An inequality expresses the relative order of two mathematical expressions. The **solution set of an inequality** is the set of real numbers each of which, when substituted for the variable, results in a true inequality. The inequality $x > 4$ is true for any value of x greater than 4. For instance, 5, $\sqrt{21}$, and $\dfrac{17}{3}$ are all solutions of $x > 4$.

The solution set of the inequality can be written in set-builder notation as $\{x \,|\, x > 4\}$ or in interval notation as $(4, \infty)$.

  Equivalent inequalities have the same solution set. We solve an inequality by producing *simpler* but equivalent inequalities until the solutions are found. To produce these simpler but equivalent inequalities, we apply the following properties.

Properties of Inequalities

Let a, b, and c be real numbers.

1. *Addition Property*   Adding the same real number to each side of an inequality preserves the direction of the inequality symbol.

    $a < b$ and $a + c < b + c$ are equivalent inequalities.

2. *Multiplication Properties*
   a.   Multiplying each side of an inequality by the same *positive* real number *preserves* the direction of the inequality symbol.

       If $c > 0$, then $a < b$ and $ac < bc$ are equivalent inequalities.

   b.   Multiplying each side of an inequality by the same *negative* real number *changes* the direction of the inequality symbol.

       If $c < 0$, then $a < b$ and $ac > bc$ are equivalent inequalities.

  Note the difference between Property 2a and Property 2b. Property 2a states that an equivalent inequality is produced when each side of a given inequality is multiplied by the same *positive* real number and that the direction of the inequality symbol is *not* changed. By contrast, Property 2b states that when each side of a given inequality is multiplied by a *negative* real number, we must *reverse* the direction of the inequality symbol to produce an equivalent inequality.

For instance, $-2b < 6$ and $b > -3$ are equivalent inequalities. (We multiplied each side of the first inequality by $-\dfrac{1}{2}$, and we changed the "less than" symbol to a "greater than" symbol.)

Because subtraction is defined in terms of addition, subtracting the same real number from each side of an inequality does not change the direction of the inequality symbol.

Because division is defined in terms of multiplication, dividing each side of an inequality by the same *positive* real number does *not* change the direction of the inequality symbol, and dividing each side of an inequality by a *negative* real number *changes* the direction of the inequality symbol.

| EXAMPLE 4 | Solve an Inequality |

Solve: $2(x + 3) < 4x + 10$. Write the solution set in set-builder notation.

Solution

$2(x + 3) < 4x + 10$

$\quad 2x + 6 < 4x + 10$ • **Use the distributive property.**

$\qquad -2x < 4$ • **Subtract 4x and 6 from each side of the inequality.**

$\qquad\quad x > -2$ • **Divide each side by −2 and reverse the inequality symbol.**

Thus the original inequality is true for all real numbers greater than -2. The solution set is $\{x \mid x > -2\}$.

▶ **TRY EXERCISE 50, PAGE 12**

> *take note*
>
> Solutions of inequalities are often stated using set-builder notation or interval notation. For instance, the real numbers that are solutions of the inequality in Example 4 can be written in set notation as $\{x \mid x > -2\}$ or in interval notation as $(-2, \infty)$.

● SOLVING INEQUALITIES BY THE CRITICAL VALUE METHOD

Any value of x that causes a polynomial in x to equal zero is called a **zero of the polynomial**. For example, -4 and 1 are both zeros of the polynomial $x^2 + 3x - 4$, because $(-4)^2 + 3(-4) - 4 = 0$ and $1^2 + 3 \cdot 1 - 4 = 0$.

A Sign Property of Polynomials

Nonzero polynomials in x have the property that for any value of x between two consecutive real zeros, either all values of the polynomial are positive or all values of the polynomial are negative.

In our work with inequalities that involve polynomials, the real zeros of the polynomial are also referred to as **critical values of the inequality**, because on a number line they separate the real numbers that make an inequality involving a polynomial true from those that make it false. In Example 5 we use critical values and the sign property of polynomials to solve an inequality.

EXAMPLE 5   Solve a Polynomial Inequality

Solve: $x^2 + 3x - 4 < 0$

Solution

Factoring the polynomial $x^2 + 3x - 4$ produces the equivalent inequality

$$(x + 4)(x - 1) < 0$$

FIGURE 1.5

Thus the zeros of the polynomial $x^2 + 3x - 4$ are -4 and 1. They are the critical values of the inequality $x^2 + 3x - 4 < 0$. They separate the real number line into the three intervals shown in **Figure 1.5.**

To determine the intervals on which $x^2 + 3x - 4 < 0$, pick a number called a **test value** from each of the three intervals and then determine whether $x^2 + 3x - 4 < 0$ for each of these test values. For example, in the interval $(-\infty, -4)$, pick a test value of, say, -5. Then

$$x^2 + 3x - 4 = (-5)^2 + 3(-5) - 4 = 6$$

Because 6 is not less than 0, by the sign property of polynomials, no number in the interval $(-\infty, -4)$ makes $x^2 + 3x - 4 < 0$.

Now pick a test value from the interval $(-4, 1)$, say, 0. When $x = 0$,

$$x^2 + 3x - 4 = 0^2 + 3(0) - 4 = -4$$

Because -4 is less than 0, by the sign property of polynomials, all numbers in the interval $(-4, 1)$ make $x^2 + 3x - 4 < 0$.

If we pick a test value of 2 from the interval $(1, \infty)$, then

$$x^2 + 3x - 4 = (2)^2 + 3(2) - 4 = 6$$

Because 6 is not less than 0, by the sign property of polynomials, no number in the interval $(1, \infty)$ makes $x^2 + 3x - 4 < 0$.

The following table is a summary of our work.

Interval	Test Value x	$x^2 + 3x - 4 \overset{?}{<} 0$
$(-\infty, -4)$	-5	$(-5)^2 + 3(-5) - 4 < 0$ $6 < 0$ False
$(-4, 1)$	0	$(0)^2 + 3(0) - 4 < 0$ $-4 < 0$ True
$(1, \infty)$	2	$(2)^2 + 3(2) - 4 < 0$ $6 < 0$ False

FIGURE 1.6

In interval notation the solution set of $x^2 + 3x - 4 < 0$ is $(-4, 1)$. The solution set is graphed in **Figure 1.6.** Note that in this case the critical values -4 and 1 are not included in the solution set because they do not make $x^2 + 3x - 4$ less than 0.

▶ **TRY EXERCISE 58, PAGE 12**

To avoid the arithmetic in Example 5, we often use a *sign diagram*. For example, note that the factor $(x + 4)$ is negative for all $x < -4$ and positive for all $x > -4$. The factor $(x - 1)$ is negative for all $x < 1$ and positive for all $x > 1$. These results are shown in **Figure 1.7.**

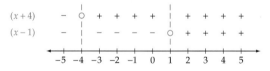

FIGURE 1.7

To determine on which intervals the product $(x + 4)(x - 1)$ is negative, we examine the sign diagram to see where the factors have opposite signs. This occurs only on the interval $(-4, 1)$, where $(x + 4)$ is positive and $(x - 1)$ is negative, so the original inequality is true only on the interval $(-4, 1)$.

Following is a summary of the steps used to solve polynomial inequalities by the critical value method.

Solving a Polynomial Inequality by the Critical Value Method

1. Write the inequality so that one side of the inequality is a nonzero polynomial and the other side is 0.

2. Find the real zeros of the polynomial. They are the critical values of the original inequality.

3. Use test values to determine which of the intervals formed by the critical values are to be included in the solution set.

4. Any critical value that satisfies the original inequality is an element of the solution set.

● ABSOLUTE VALUE INEQUALITIES

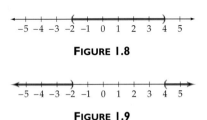

FIGURE 1.8

The solution set of the absolute value inequality $|x - 1| < 3$ is the set of all real numbers whose distance from 1 is *less than* 3. Therefore, the solution set consists of all numbers between -2 and 4. See **Figure 1.8.** In interval notation, the solution set is $(-2, 4)$.

The solution set of the absolute value inequality $|x - 1| > 3$ is the set of all real numbers whose distance from 1 is *greater than* 3. Therefore, the solution set consists of all real numbers less than -2 *or* greater than 4. See **Figure 1.9.** In interval notation, the solution set is $(-\infty, -2) \cup (4, \infty)$.

The following properties are used to solve absolute value inequalities.

Properties of Absolute Value Inequalities

For any variable expression E and any nonnegative real number k,

$$|E| \leq k \quad \text{if and only if} \quad -k \leq E \leq k$$

$$|E| \geq k \quad \text{if and only if} \quad E \leq -k \quad \text{or} \quad E \geq k$$

EXAMPLE 6   **Solve an Absolute Value Inequality**

Solve: $|2 - 3x| < 7$

Solution

$|2 - 3x| < 7$ implies $-7 < 2 - 3x < 7$. Solve this compound inequality.

$$-7 < 2 - 3x < 7$$

$$-9 < \quad -3x \quad < 5$$
     • Subtract 2 from each of the three parts of the inequality.

$$3 > \quad x \quad > -\frac{5}{3}$$
    • Multiply each part of the inequality by $-\frac{1}{3}$ and reverse the inequality symbols.

FIGURE 1.10

In interval notation, the solution set is given by $\left(-\dfrac{5}{3}, 3\right)$. See **Figure 1.10.**

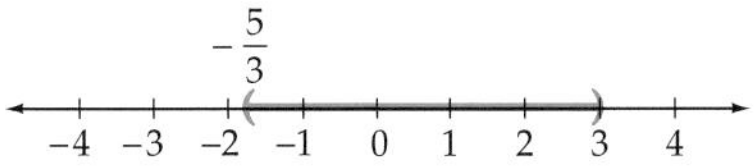

▶ TRY EXERCISE 70, PAGE 12

---

take note

Some inequalities have a solution set that consists of all real numbers. For example, $|x + 9| \geq 0$ is true for all values of x. Because an absolute value is always nonnegative, the equation is always true.

EXAMPLE 7   **Solve an Absolute Value Inequality**

Solve: $|4x - 3| \geq 5$

Solution

$|4x - 3| \geq 5$ implies $4x - 3 \leq -5$ or $4x - 3 \geq 5$. Solving each of these inequalities produces

$$4x - 3 \leq -5 \qquad \text{or} \qquad 4x - 3 \geq 5$$
$$4x \leq -2 \qquad\qquad\qquad 4x \geq 8$$
$$x \leq -\frac{1}{2} \qquad\qquad\qquad x \geq 2$$

FIGURE 1.11

Therefore, the solution set is $\left(-\infty, -\dfrac{1}{2}\right] \cup [2, \infty)$. See **Figure 1.11.**

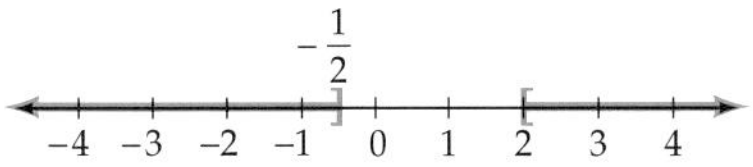

▶ TRY EXERCISE 72, PAGE 12

TOPICS FOR DISCUSSION

1. Discuss the similarities and differences among natural numbers, integers, rational numbers, and real numbers.

2. Discuss the differences among an equation, an inequality, and an expression.

3. Is it possible for an equation to have no solution? If not, explain why. If so, give an example of an equation with no solution.

4. Is the statement $|x| = -x$ ever true? Explain why or why not.

5. How do quadratic equations in one variable differ from linear equations in one variable? Explain how the method used to solve an equation depends on whether it is a linear or a quadratic equation.

EXERCISE SET 1.1

In Exercises 1 to 18, solve and check each equation.

1. $2x + 10 = 40$

2. $-3y + 20 = 2$

3. $5x + 2 = 2x - 10$

4. $4x - 11 = 7x + 20$

5. $2(x - 3) - 5 = 4(x - 5)$

▶ **6.** $6(5s - 11) - 12(2s + 5) = 0$

7. $\dfrac{3}{4}x + \dfrac{1}{2} = \dfrac{2}{3}$

8. $\dfrac{x}{4} - 5 = \dfrac{1}{2}$

9. $\dfrac{2}{3}x - 5 = \dfrac{1}{2}x - 3$

10. $\dfrac{1}{2}x + 7 - \dfrac{1}{4}x = \dfrac{19}{2}$

11. $0.2x + 0.4 = 3.6$

12. $0.04x - 0.2 = 0.07$

13. $\dfrac{3}{5}(n + 5) - \dfrac{3}{4}(n - 11) = 0$

14. $-\dfrac{5}{7}(p + 11) + \dfrac{2}{5}(2p - 5) = 0$

15. $3(x + 5)(x - 1) = (3x + 4)(x - 2)$

16. $5(x + 4)(x - 4) = (x - 3)(5x + 4)$

17. $0.08x + 0.12(4000 - x) = 432$

18. $0.075y + 0.06(10,000 - y) = 727.50$

In Exercises 19 to 32, solve by using the quadratic formula.

19. $x^2 - 2x - 15 = 0$

20. $x^2 - 5x - 24 = 0$

21. $x^2 + x - 1 = 0$

▶ **22.** $x^2 + x - 2 = 0$

23. $2x^2 + 4x + 1 = 0$

24. $2x^2 + 4x - 1 = 0$

25. $3x^2 - 5x - 3 = 0$

26. $3x^2 - 5x - 4 = 0$

27. $\dfrac{1}{2}x^2 + \dfrac{3}{4}x - 1 = 0$

28. $\dfrac{2}{3}x^2 - 5x + \dfrac{1}{2} = 0$

29. $\sqrt{2}x^2 + 3x + \sqrt{2} = 0$

30. $2x^2 + \sqrt{5}x - 3 = 0$

31. $x^2 = 3x + 5$

32. $-x^2 = 7x - 1$

In Exercises 33 to 40, solve each quadratic equation by factoring and applying the zero product property.

33. $x^2 - 2x - 15 = 0$

34. $y^2 + 3y - 10 = 0$

35. $8y^2 + 189y - 72 = 0$

▶ **36.** $12w^2 - 41w + 24 = 0$

37. $3x^2 - 7x = 0$

38. $5x^2 = -8x$

39. $(x - 5)^2 - 9 = 0$

40. $(3x + 4)^2 - 16 = 0$

In Exercises 41 to 50, use the properties of inequalities to solve each inequality. Write answers using interval notation.

41. $2x + 3 < 11$

42. $3x - 5 > 16$

43. $x + 4 > 3x + 16$

44. $5x + 6 < 2x + 1$

45. $-6x + 1 \geq 19$

46. $-5x + 2 \leq 37$

47. $-3(x + 2) \leq 5x + 7$

48. $-4(x - 5) \geq 2x + 15$

49. $-4(3x - 5) > 2(x - 4)$

▶ **50.** $3(x + 7) \leq 5(2x - 8)$

In Exercises 51 to 58, use the critical value method to solve each inequality. Use interval notation to write each solution set.

51. $x^2 + 7x > 0$

52. $x^2 - 5x \leq 0$

53. $x^2 + 7x + 10 < 0$

54. $x^2 + 5x + 6 < 0$

55. $x^2 - 3x \geq 28$

56. $x^2 < -x + 30$

57. $6x^2 - 4 \leq 5x$

▶ **58.** $12x^2 + 8x \geq 15$

In Exercises 59 to 76, use interval notation to express the solution set of each inequality.

59. $|x| < 4$

60. $|x| > 2$

61. $|x - 1| < 9$

62. $|x - 3| < 10$

63. $|x + 3| > 30$

64. $|x + 4| < 2$

65. $|2x - 1| > 4$

66. $|2x - 9| < 7$

67. $|x + 3| \geq 5$

68. $|x - 10| \geq 2$

69. $|3x - 10| \leq 14$

▶ **70.** $|2x - 5| \geq 1$

71. $|4 - 5x| \geq 24$

▶ **72.** $|3 - 2x| \leq 5$

73. $|x - 5| \geq 0$

74. $|x - 7| \geq 0$

75. $|x - 4| \leq 0$

76. $|2x + 7| \leq 0$

77. **GEOMETRY** The perimeter of a rectangle is 27 centimeters, and its area is 35 square centimeters. Find the length and the width of the rectangle.

78. GEOMETRY The perimeter of a rectangle is 34 feet and its area is 60 square feet. Find the length and the width of the rectangle.

79. RECTANGULAR ENCLOSURE A gardener wishes to use 600 feet of fencing to enclose a rectangular region and subdivide the region into two smaller rectangles. The total enclosed area is 15,000 square feet. Find the dimensions of the enclosed region.

80. RECTANGULAR ENCLOSURE A farmer wishes to use 400 yards of fencing to enclose a rectangular region and subdivide the region into three smaller rectangles. If the total enclosed area is 400 square yards, find the dimensions of the enclosed region.

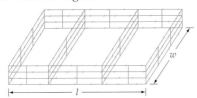

81. PERSONAL FINANCE A bank offers two checking account plans. The monthly fee and charge per check for each plan are shown below. Under what conditions is it less expensive to use the LowCharge plan?

Account Plan	Monthly Fee	Charge per Check
LowCharge	$5.00	$.01
FeeSaver	$1.00	$.08

82. PERSONAL FINANCE You can rent a car for the day from company A for $29.00 plus $0.12 a mile. Company B charges $22.00 plus $0.21 a mile. Find the number of miles m (to the nearest mile) per day for which it is cheaper to rent from company A.

83. PERSONAL FINANCE A sales clerk has a choice between two payment plans. Plan A pays $100.00 a week plus $8.00 a sale. Plan B pays $250.00 a week plus $3.50 a sale. How many sales per week must be made for plan A to yield the greater paycheck?

84. PERSONAL FINANCE A video store offers two rental plans. The yearly membership fee and the daily charge per video are shown below. How many videos can be rented per year if the No-fee plan is to be the less expensive of the plans?

THE VIDEO STORE		
Rental Plan	Yearly Fee	Daily Charge per Video
Low-rate	$15.00	$1.49
No-fee	None	$1.99

85. AVERAGE TEMPERATURES The average daily minimum-to-maximum temperatures for the city of Palm Springs during the month of September are 68°F to 104°F. What is the corresponding temperature range measured on the Celsius temperature scale?

CONNECTING CONCEPTS

86. A GOLDEN RECTANGLE The ancient Greeks defined a rectangle as a "golden rectangle" if its length l and its width w satisfied the equation

$$\frac{l}{w} = \frac{w}{l - w}$$

a. Solve this formula for w.

b. If the length of a golden rectangle is 101 feet, determine its width. Round to the nearest hundredth.

87. SUM OF NATURAL NUMBERS The sum S of the first n natural numbers $1, 2, 3, \ldots, n$ is given by the formula

$$S = \frac{n}{2}(n + 1)$$

How many consecutive natural numbers starting with 1 produce a sum of 253?

88. NUMBER OF DIAGONALS The number of diagonals D of a polygon with n sides is given by the formula

$$D = \frac{n}{2}(n - 3)$$

a. Determine the number of sides of a polygon with 464 diagonals.

b. Can a polygon have 12 diagonals? Explain.

89. REVENUE The monthly revenue R for a product is given by $R = 420x - 2x^2$, where x is the price in dollars of each unit produced. Find the interval in terms of x for which the monthly revenue is greater than zero.

90. Write an absolute value inequality to represent all real numbers within

a. 8 units of 3

b. k units of j (assume $k > 0$)

91. HEIGHT OF A PROJECTILE The equation

$$s = -16t^2 + v_0 t + s_0$$

gives the height s, in feet above ground level, at the time t seconds after an object is thrown directly upward from a height s_0 feet above the ground with an initial velocity of v_0 feet per second. A ball is thrown directly upward from ground level with an initial velocity of 64 feet per second. Find the time interval during which the ball has a height of more than 48 feet.

92. HEIGHT OF A PROJECTILE A ball is thrown directly upward from a height of 32 feet above the ground with an initial velocity of 80 feet per second. Find the time interval during which the ball will be more than 96 feet above the ground. (*Hint:* See Exercise 91.)

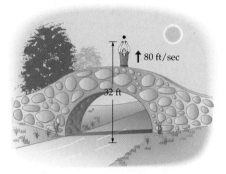

93. GEOMETRY The length of the side of a square has been measured accurately to within 0.01 foot. This measured length is 4.25 feet.

a. Write an absolute value inequality that describes the relationship between the actual length of each side of the square s and its measured length.

b. Solve the absolute value inequality you found in part **a.** for s.

PREPARE FOR SECTION 1.2

94. Evaluate $\dfrac{x_1 + x_2}{2}$ when $x_1 = 4$ and $x_2 = -7$.

95. Simplify $\sqrt{50}$.

96. Is $y = 3x - 2$ a true equation when $y = 5$ and $x = -1$? [1.1]

97. If $y = x^2 - 3x - 2$, find y when $x = -3$. [1.1]

98. Evaluate $|-x - y|$ when $x = 3$ and $y = -1$. [1.1]

99. Evaluate $\sqrt{b^2 - 4ac}$ when $a = -2$, $b = -3$, and $c = 2$.

PROJECTS

1. TEACHING MATHEMATICS Prepare a lesson that you could use to explain to someone how to solve linear and quadratic equations. Be sure to include an explanation of the differences between these two types of equations and the different methods that are used to solve them.

2. CUBIC EQUATIONS Write an essay on the development of the solution of the cubic equation. An excellent source of information is the chapter "Cardano and the Solution of the Cubic" in *Journey Through Genius* by William Dunham (New York: Wiley, 1990). Another excellent source is *A History of Mathematics: An Introduction* by Victor J. Katz (New York: Harper Collins, 1993).

A TWO-DIMENSIONAL COORDINATE SYSTEM AND GRAPHS

take note

Abscissa comes from the same root word as scissors. An open pair of scissors looks like an x.

MATH MATTERS

The concepts of *analytic geometry* developed over an extended period of time, culminating in 1637 with the publication of two works: *Discourse on the Method for Rightly Directing One's Reason and Searching for Truth in the Sciences* by René Descartes (1596–1650) and *Introduction to Plane and Solid Loci* by Pierre de Fermat. Each of these works was an attempt to integrate the study of geometry with the study of algebra. Of the two mathematicians, Descartes is usually given most of the credit for developing analytic geometry. In fact, Descartes became so famous in La Haye, the city in which he was born, that it was renamed La Haye-Descartes.

• CARTESIAN COORDINATE SYSTEMS

Each point on a coordinate axis is associated with a number called its **coordinate**. Each point on a flat, two-dimensional surface, called a **coordinate plane** or *xy*-plane, is associated with an **ordered pair** of numbers called **coordinates** of the point. Ordered pairs are denoted by (a, b), where the real number a is the **x-coordinate** or **abscissa** and the real number b is the **y-coordinate** or **ordinate**.

The coordinates of a point are determined by the point's position relative to a horizontal coordinate axis called the **x-axis** and a vertical coordinate axis called the **y-axis**. The axes intersect at the point $(0, 0)$, called the **origin**. In **Figure 1.12,** the axes are labeled such that positive numbers appear to the right of the origin on the *x*-axis and above the origin on the *y*-axis. The four regions formed by the axes are called **quadrants** and are numbered counterclockwise. This two-dimensional coordinate system is referred to as a **Cartesian coordinate system** in honor of René Descartes.

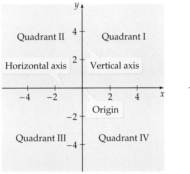

FIGURE 1.12

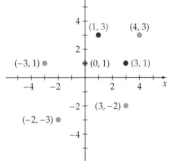

FIGURE 1.13

To **plot a point** $P(a, b)$ means to draw a dot at its location in the coordinate plane. In **Figure 1.13** we have plotted the points $(4, 3)$, $(-3, 1)$, $(-2, -3)$, $(3, -2)$, $(0, 1)$, $(1, 3)$, and $(3, 1)$. The order in which the coordinates of an ordered pair are listed is important. **Figure 1.13** shows that $(1, 3)$ and $(3, 1)$ do not denote the same point.

Data often are displayed in visual form as a set of points called a *scatter diagram* or *scatter plot*. For instance, the scatter diagram in **Figure 1.14** shows the number of Internet virus incidents from 1993 to 2003. The point whose coordinates are approximately $(7, 21{,}000)$ means that in the year 2000 there were approximately 21,000 Internet virus incidents. The line segments that connect the points in **Figure 1.14** help illustrate trends.

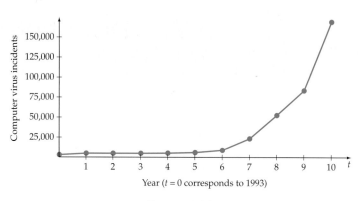

FIGURE 1.14

Source: www.cert.org

? **QUESTION** If the trend in **Figure 1.14** continues, will the number of virus incidents in 2004 be more or less than 200,000?

In some instances, it is important to know when two ordered pairs are equal.

Equality of Ordered Pairs

The ordered pairs (a, b) and (c, d) are equal if and only if $a = c$ and $b = d$.

For instance, if $(3, y) = (x, -2)$, then $x = 3$ and $y = -2$.

• THE DISTANCE AND MIDPOINT FORMULAS

The Cartesian coordinate system makes it possible to combine the concepts of algebra and geometry into a branch of mathematics called *analytic geometry*.

The distance between two points on a horizontal line is the absolute value of the difference between the x-coordinates of the two points. The distance between two points on a vertical line is the absolute value of the difference between the y-coordinates of the two points. For example, as shown in **Figure 1.15,** the distance d between the points with coordinates $(1, 2)$ and $(1, -3)$ is $d = |2 - (-3)| = 5$.

If two points are not on a horizontal or vertical line, then a *distance formula* for the distance between the two points can be developed as follows.

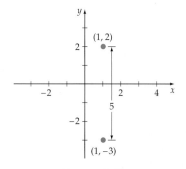

FIGURE 1.15

? **ANSWER** More. The increase between 2002 and 2003 was more than 70,000. If this trend continues, the increase between 2003 and 2004 will be at least 70,000 more than 150,000. That is, the number of virus incidents in 2004 will be at least 220,000.

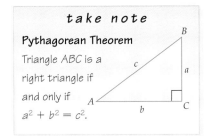

Pythagorean Theorem

Triangle ABC is a
right triangle if
and only if
$a^2 + b^2 = c^2$.

The distance between the points $P_1(x_1, y_1)$ and $P_2(x_2, y_2)$ in **Figure 1.16** is the length of the hypotenuse of a right triangle whose sides are horizontal and vertical line segments that measure $|x_2 - x_1|$ and $|y_2 - y_1|$, respectively. Applying the Pythagorean Theorem to this triangle produces

$$d^2 = |x_2 - x_1|^2 + |y_2 - y_1|^2$$
$$d = \sqrt{|x_2 - x_1|^2 + |y_2 - y_1|^2}$$

• The square root theorem. Because d is nonnegative, the negative root is not listed.

$$= \sqrt{(x_2 - x_1)^2 + (y_2 - y_1)^2}$$

• Because $|x_2 - x_1|^2 = (x_2 - x_1)^2$ and $|y_2 - y_1|^2 = (y_2 - y_1)^2$

Thus we have established the following theorem.

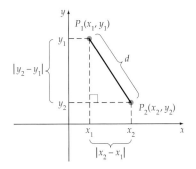

FIGURE 1.16

The Distance Formula

The distance d between the points $P_1(x_1, y_1)$ and $P_2(x_2, y_2)$ is

$$d = \sqrt{(x_2 - x_1)^2 + (y_2 - y_1)^2}$$

The distance d between the points whose coordinates are $P_1(x_1, y_1)$ and $P_2(x_2, y_2)$ is denoted by $d(P_1, P_2)$. To find the distance $d(P_1, P_2)$ between the points $P_1(-3, 4)$ and $P_2(7, 2)$, we apply the distance formula with $x_1 = -3$, $y_1 = 4$, $x_2 = 7$, and $y_2 = 2$.

$$d(P_1, P_2) = \sqrt{(x_2 - x_1)^2 + (y_2 - y_1)^2}$$
$$= \sqrt{[7 - (-3)]^2 + (2 - 4)^2}$$
$$= \sqrt{104} = 2\sqrt{26} \approx 10.2$$

The **midpoint** M of a line segment is the point on the line segment that is equidistant from the endpoints $P_1(x_1, y_1)$ and $P_2(x_2, y_2)$ of the segment. See **Figure 1.17**.

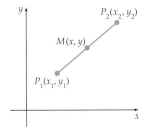

FIGURE 1.17

The Midpoint Formula

The midpoint M of the line segment from $P_1(x_1, y_1)$ to $P_2(x_2, y_2)$ is given by

$$\left(\frac{x_1 + x_2}{2}, \frac{y_1 + y_2}{2} \right)$$

The midpoint formula states that the x-coordinate of the midpoint of a line segment is the *average* of the x-coordinates of the endpoints of the line segment and that the y-coordinate of the midpoint of a line segment is the *average* of the y-coordinates of the endpoints of the line segment.

The midpoint M of the line segment connecting $P_1(-2, 6)$ and $P_2(3, 4)$ is

$$M = \left(\frac{x_1 + x_2}{2}, \frac{y_1 + y_2}{2} \right) = \left(\frac{(-2) + 3}{2}, \frac{6 + 4}{2} \right) = \left(\frac{1}{2}, 5 \right)$$

EXAMPLE 1   **Find the Midpoint and Length of a Line Segment**

Find the midpoint and the length of the line segment connecting the points whose coordinates are $P_1(-4, 3)$ and $P_2(4, -2)$.

Solution

$$\text{Midpoint} = \left(\frac{x_1 + x_2}{2}, \frac{y_1 + y_2}{2} \right)$$

$$= \left(\frac{-4 + 4}{2}, \frac{3 + (-2)}{2} \right)$$

$$= \left(0, \frac{1}{2} \right)$$

$$d(P_1, P_2) = \sqrt{(x_2 - x_1)^2 + (y_2 - y_1)^2}$$

$$= \sqrt{(4 - (-4))^2 + (-2 - 3)^2} = \sqrt{(8)^2 + (-5)^2}$$

$$= \sqrt{64 + 25} = \sqrt{89}$$

▶ TRY EXERCISE 6, PAGE 26

● GRAPH OF AN EQUATION

The equations below are equations in two variables.

$$y = 3x^3 - 4x + 2 \qquad x^2 + y^2 = 25 \qquad y = \frac{x}{x + 1}$$

The solution of an equation in two variables is an ordered pair (x, y) whose coordinates satisfy the equation. For instance, the ordered pairs $(3, 4)$, $(4, -3)$, and $(0, 5)$ are some of the solutions of $x^2 + y^2 = 25$. Generally, there are an infinite number of solutions of an equation in two variables. These solutions can be displayed in a *graph*.

Graph of an Equation

The **graph of an equation** in the two variables x and y is the set of all points whose coordinates satisfy the equation.

Consider $y = 2x - 1$. Substituting various values of x into the equation and solving for y produces some of the ordered pairs of the equation. It is convenient to record the results in a table similar to the one shown on the following page. The graph of the ordered pairs is shown in **Figure 1.18.**

x	y = 2x − 1	y	(x, y)
−2	2(−2) − 1	−5	(−2, −5)
−1	2(−1) − 1	−3	(−1, −3)
0	2(0) − 1	−1	(0, −1)
1	2(1) − 1	1	(1, 1)
2	2(2) − 1	3	(2, 3)

Choosing some noninteger values of x produces more ordered pairs to graph, such as $\left(-\dfrac{3}{2}, -4\right)$ and $\left(\dfrac{5}{2}, 4\right)$, as shown in **Figure 1.19.** Using still other values of x would result in more and more ordered pairs being graphed. The result would be so many dots that the graph would appear as the straight line shown in **Figure 1.20,** which is the graph of $y = 2x - 1$.

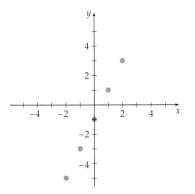

FIGURE 1.18

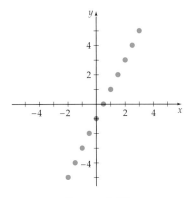

FIGURE 1.19

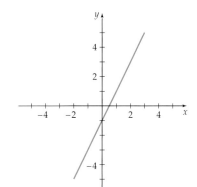

FIGURE 1.20

EXAMPLE 2   Draw a Graph by Plotting Points

Graph: $-x^2 + y = 1$

Solution

Solve the equation for y.
$$y = x^2 + 1$$

Select values of x and use the equation to calculate y. Choose enough values of x so that an accurate graph can be drawn. Plot the points and draw a curve through them. See **Figure 1.21.**

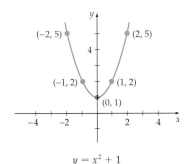

$y = x^2 + 1$

FIGURE 1.21

x	y = x² + 1	y	(x, y)
−2	(−2)² + 1	5	(−2, 5)
−1	(−1)² + 1	2	(−1, 2)
0	(0)² + 1	1	(0, 1)
1	(1)² + 1	2	(1, 2)
2	(2)² + 1	5	(2, 5)

▶ TRY **EXERCISE 26, PAGE 27**

MATH MATTERS

Maria Agnesi (1718–1799) wrote *Foundations of Analysis for the Use of Italian Youth*, one of the most successful textbooks of the 18th century. The French Academy authorized a translation into French in 1749, noting that "there is no other book, in any language, which would enable a reader to penetrate as deeply, or as rapidly, into the fundamental concepts of analysis." A curve that she discusses in her text is given by the equation $y = \dfrac{a^3}{x^2 + a^2}$.

Unfortunately, due to a translation error from Italian to English, the curve became known as the "witch of Agnesi."

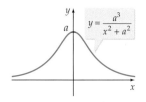

INTEGRATING TECHNOLOGY

Some graphing calculators, such as the *TI-83*, have a TABLE feature that allows you to create a table similar to the one shown in Example 2. Enter the equation to be graphed, the first value for x, and the increment (the difference between successive values of x). For instance, entering $y_1 = x^2 + 1$, an initial value of x as -2, and an increment of 1 yields a display similar to the one in **Figure 1.22**. Changing the initial value to -6 and the increment to 2 gives the table in **Figure 1.23**.

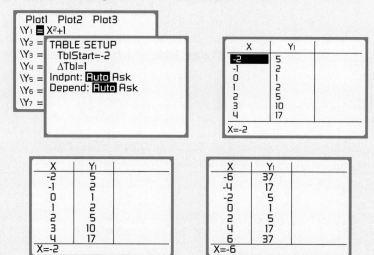

FIGURE 1.22   **FIGURE 1.23**

With some calculators, you may scroll through the table by using the up- or down-arrow keys. In this way, you can determine many more ordered pairs of the graph.

EXAMPLE 3   **Graph by Plotting Points**

Graph: $y = |x - 2|$

Solution

This equation is already solved for y, so start by choosing an x value and using the equation to determine the corresponding y value. For example, if $x = -3$, then $y = |(-3) - 2| = |-5| = 5$. Continuing in this manner produces the following table:

When x is	-3	-2	-1	0	1	2	3	4	5
y is	5	4	3	2	1	0	1	2	3

Now plot the points listed in the table. Connecting the points forms a V shape, as shown in **Figure 1.24**.

▶ **TRY EXERCISE 30, PAGE 27**

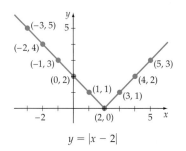

$y = |x - 2|$

FIGURE 1.24

EXAMPLE 4   **Graph by Plotting Points**

Graph: $y^2 = x$

Solution

Solve the equation for y.

$$y^2 = x$$
$$y = \pm\sqrt{x}$$

Choose several x values, and use the equation to determine the corresponding y values.

When x is	0	1	4	9	16
y is	0	±1	±2	±3	±4

Plot the points as shown in **Figure 1.25.** The graph is a *parabola*.

 TRY EXERCISE 32, PAGE 27

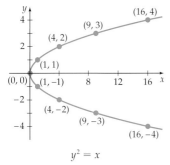

$y^2 = x$

FIGURE 1.25

🖩 **INTEGRATING TECHNOLOGY**

A graphing calculator or computer graphing software can be used to draw the graphs in Examples 3 and 4. These graphing utilities graph a curve in much the same way as you would, by selecting values of x and calculating the corresponding values of y. A curve is then drawn through the points.

If you use a graphing utility to graph $y = |x - 2|$, you will need to use the *absolute value* function that is built into the utility. The equation you enter will look similar to Y₁=abs(X–2).

To graph the equation in Example 4, you will enter two equations. The equations you enter will be similar to

$$Y_1 = \sqrt{(X)}$$
$$Y_2 = -\sqrt{(X)}$$

The graph of the first equation will be the top half of the parabola; the graph of the second equation will graph the bottom half.

● **INTERCEPTS**

Any point that has an x- or a y-coordinate of zero is called an **intercept** of the graph of an equation because it is at these points that the graph intersects the x- or the y-axis.

Definition of x-Intercepts and y-Intercepts

If $(x_1, 0)$ satisfies an equation, then the point $(x_1, 0)$ is called an **x-intercept** of the graph of the equation.

If $(0, y_1)$ satisfies an equation, then the point $(0, y_1)$ is called a **y-intercept** of the graph of the equation.

To find the x-intercepts of the graph of an equation, let $y = 0$, and solve the equation for x. To find the y-intercepts of the graph of an equation, let $x = 0$, and solve the equation for y.

EXAMPLE 5   **Find x- and y-Intercepts**

Find the x- and y-intercepts of the graph of $y = x^2 - 2x - 3$.

Algebraic Solution

To find the y-intercept, let $x = 0$ and solve for y.

$$y = 0^2 - 2(0) - 3 = -3$$

To find the x-intercepts, let $y = 0$ and solve for x.

$$0 = x^2 - 2x - 3$$
$$0 = (x - 3)(x + 1)$$
$$(x - 3) = 0 \quad \text{or} \quad (x + 1) = 0$$
$$x = 3 \quad \text{or} \quad x = -1$$

Because $y = -3$ when $x = 0$, $(0, -3)$ is a y-intercept. Because $x = 3$ or -1 when $y = 0$, $(3, 0)$ and $(-1, 0)$ are x-intercepts. **Figure 1.26** confirms that these three points are intercepts.

Visualize the Solution

The graph of $y = x^2 - 2x - 3$ is shown below. Observe that the graph intersects the x-axis at $(-1, 0)$ and $(3, 0)$, the x-intercepts. The graph also intersects the y-axis at $(0, -3)$, the y-intercept.

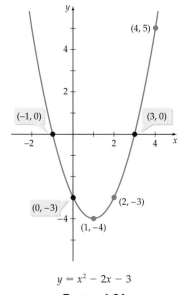

$$y = x^2 - 2x - 3$$

FIGURE 1.26

▶ TRY EXERCISE 40, PAGE 27

2. SOLVE A QUADRATIC EQUATION GEOMETRICALLY In the 17th century, Descartes (and others) solved equations by using both algebra and geometry. This project outlines the method Descartes used to solve certain quadratic equations.

a. Consider the equation $x^2 = 2ax + b^2$. Construct a right triangle ABC with $d(A, C) = a$ and $d(C, B) = b$. Now draw a circle with center at A and radius a. Let P be the point at which the circle intersects the hypotenuse of the right triangle and Q the point at which an extension of the hypotenuse intersects the circle. Your drawing should be similar to the one at the right.

b. Show that a solution of the equation $x^2 = 2ax + b^2$ is $d(Q, B)$.

c. Show that $d(P, B)$ is a solution of the equation $x^2 = -2ax + b^2$.

d. Construct a line parallel to AC and passing through B. Let S and T be the points at which the line intersects the circle. Show that $d(S, B)$ and $d(T, B)$ are solutions of the equation $x^2 = 2ax - b^2$.

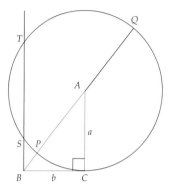

INTRODUCTION TO FUNCTIONS

● RELATIONS

In many situations in science, business, and mathematics, a correspondence exists between two sets. The correspondence is often defined by a *table*, an *equation*, or a *graph*, each of which can be viewed from a mathematical perspective as a set of ordered pairs. In mathematics, any set of ordered pairs is called a **relation.**

Table 1.1 defines a correspondence between a set of percent scores and a set of letter grades. For each score from 0 to 100, there corresponds only one letter grade. The score 94% corresponds to the letter grade of A. Using ordered-pair notation, we record this correspondence as (94, A).

The *equation* $d = 16t^2$ indicates that the distance d that a rock falls (neglecting air resistance) corresponds to the time t that it has been falling. For each nonnegative value t, the equation assigns only one value for the distance d. According to this equation, in 3 seconds a rock will fall 144 feet, which we record as (3, 144). Some of the other ordered pairs determined by $d = 16t^2$ are (0, 0), (1, 16), (2, 64), and (2.5, 100).

TABLE 1.1

Score	Grade
[90, 100]	A
[80, 90)	B
[70, 80)	C
[60, 70)	D
[0, 60)	F

$$\text{Equation:} \qquad d = 16t^2$$
$$\text{If } t = 3, \text{ then } \quad d = 16(3)^2 = 144$$

The *graph* in **Figure 1.30** defines a correspondence between the length of a pendulum and the time it takes the pendulum to complete one oscillation. For each nonnegative pendulum length, the graph yields only one time. According to the graph, a pendulum length of 2 feet yields an oscillation time of 1.6 seconds, and a length of 4 feet yields an oscillation time of 2.2 seconds, where the time is measured to the nearest tenth of a second. These results can be recorded as the ordered pairs (2, 1.6) and (4, 2.2).

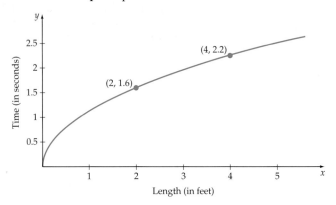

Graph: A pendulum's oscillation time

FIGURE 1.30

● FUNCTIONS

The preceding table, equation, and graph each determines a special type of relation called a *function*.

Definition of a Function
A **function** is a set of ordered pairs in which no two ordered pairs have the same first coordinate and different second coordinates.

Although every function is a relation, not every relation is a function. For instance, consider (94, A) from the grading correspondence. The first coordinate, 94, is paired with a second coordinate of A. It would not make sense to have 94 paired with A, (94, A), and 94 paired with B, (94, B). The same first coordinate would be paired with two different second coordinates. This would mean that two students with the same score received different grades, one student an A and the other a B!

Functions may have ordered pairs with the same second coordinate. For instance, (94, A) and (95, A) are both ordered pairs that belong to the function defined by **Table 1.1.** A function may have different first coordinates and the same second coordinate.

The equation $d = 16t^2$ represents a function because for each value of t there is only one value of d. Not every equation, however, represents a function. For instance, $y^2 = 25 - x^2$ does not represent a function. The ordered pairs $(-3, 4)$ and $(-3, -4)$ are both solutions of the equation. But these ordered pairs do not satisfy the definition of a function; there are two ordered pairs with the same first coordinate but *different* second coordinates.

❓ **QUESTION**   Does the set $\{(0, 0), (1, 0), (2, 0), (3, 0), (4, 0)\}$ define a function?

The **domain** of a function is the set of all the first coordinates of the ordered pairs. The **range** of a function is the set of all the second coordinates. In the func-

---

❓ **ANSWER**   Yes. There are no two ordered pairs with the same first coordinate that have different second coordinates.

tion determined by the grading correspondence in **Table 1.1,** the domain is the interval [0, 100]. The range is {A, B, C, D, F}. In a function, each domain element is paired with one and only one range element.

If a function is defined by an equation, the variable that represents elements of the domain is the independent variable. The variable that represents elements of the range is the dependent variable. In the free-fall experiment, we used the equation $d = 16t^2$. The elements of the domain represented the time the rock fell, and the elements of the range represented the distance the rock fell. Thus, in $d = 16t^2$, the independent variable is t and the dependent variable is d.

The specific letters used for the independent and the dependent variable are not important. For example, $y = 16x^2$ represents the same function as $d = 16t^2$. Traditionally, x is used for the independent variable and y for the dependent variable. Anytime we use the phrase "y is a function of x" or a similar phrase with different letters, the variable that follows "function of" is the independent variable.

● FUNCTIONAL NOTATION

Functions can be named by using a letter or a combination of letters, such as f, g, A, log, or tan. If x is an element of the domain of f, then $f(x)$, which is read "f of x" or "the value of f at x," is the element in the range of f that corresponds to the domain element x. The notation "f" and the notation "$f(x)$" mean different things. "f" is the name of the function, whereas "$f(x)$" is the value of the function at x. Finding the value of $f(x)$ is referred to as *evaluating f at x*. To evaluate $f(x)$ at $x = a$, substitute a for x, and simplify.

EXAMPLE 1   Evaluate Functions

Let $f(x) = x^2 - 1$, and evaluate.

a. $f(-5)$ b. $f(3b)$ c. $3f(b)$ d. $f(a + 3)$ e. $f(a) + f(3)$

Solution

a. $f(-5) = (-5)^2 - 1 = 25 - 1 = 24$ • Substitute **−5** for x, and simplify.

b. $f(3b) = (3b)^2 - 1 = 9b^2 - 1$ • Substitute **3b** for x, and simplify.

c. $3f(b) = 3(b^2 - 1) = 3b^2 - 3$ • Substitute **b** for x, and simplify.

d. $f(a + 3) = (a + 3)^2 - 1$ • Substitute **a + 3** for x.
$\qquad\quad = a^2 + 6a + 8$ • Simplify.

e. $f(a) + f(3) = (a^2 - 1) + (3^2 - 1)$ • Substitute **a** for x; substitute **3** for x.
$\qquad\qquad\quad = a^2 + 7$ • Simplify.

▶ **TRY EXERCISE 2, PAGE 42**

take note

In Example 1, observe that
$$f(3b) \neq 3f(b)$$
and that
$$f(a + 3) \neq f(a) + f(3)$$

Piecewise-defined functions are functions represented by more than one expression. The function shown below is an example of a piecewise-defined function.

$$f(x) = \begin{cases} 2x, & x < -2 \\ x^2, & -2 \le x < 1 \\ 4 - x, & x \ge 1 \end{cases}$$

• This function is made up of different *pieces*, $2x$, x^2, and $4 - x$, depending on the value of x.

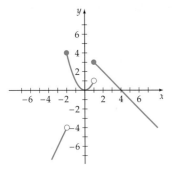

FIGURE 1.31

The expression that is used to evaluate this function depends on the value of x. For instance, to find $f(-3)$, we note that $-3 < -2$ and therefore use the expression $2x$ to evaluate the function.

$$f(-3) = 2(-3) = -6 \qquad \bullet \textbf{When } x < -2, \text{ use the expression } 2x.$$

Here are some additional instances of evaluating this function:

$$f(-1) = (-1)^2 = 1 \qquad \bullet \textbf{When } x \text{ satisfies } -2 \le x < 1,$$
$$\text{use the expression } x^2.$$

$$f(4) = 4 - 4 = 0 \qquad \bullet \textbf{When } x \ge 1, \text{ use the expression } 4 - x.$$

The graph of this function is shown in **Figure 1.31.** Note the use of the open and closed circles at the endpoints of the intervals. These circles are used to show the evaluation of the function at the endpoints of each interval. For instance, because -2 is in the interval $-2 \le x < 1$, the value of the function at -2 is 4 $[f(-2) = (-2)^2 = 4]$. Therefore a closed dot is placed at $(-2, 4)$. Similarly, when $x = 1$, because 1 is in the interval $x \ge 1$, the value of the function at 1 is 3 $(f(1) = 4 - 1 = 3)$.

❓ QUESTION   Evaluate the function f defined at the bottom of page 31 when $x = 0.5$.

The absolute value function is another example of a piecewise-defined function. Below is the definition of this function, which is sometimes abbreviated $\text{abs}(x)$. Its graph **(Figure 1.32)** is shown at the left.

$$\text{abs}(x) = \begin{cases} -x, & x < 0 \\ x, & x \ge 0 \end{cases}$$

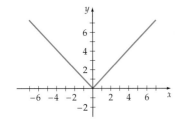

FIGURE 1.32

| **EXAMPLE 2** | **Evaluate a Piecewise-Defined Function** |

The number of monthly spam email attacks is shown in **Figure 1.33.**

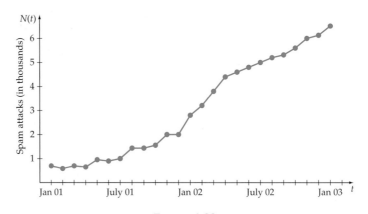

FIGURE 1.33

Source: www.brightmail.com

❓ ANSWER   0.5 is in the interval $-2 \le x < 1$. Therefore, $f(0.5) = 0.5^2 = 0.25$.

The data in the graph can be approximated by

$$N(t) = \begin{cases} 24.68t^2 - 170.47t + 957.73, & 0 \le t < 17 \\ 196.9t + 1164.6, & 17 \le t \le 26 \end{cases}$$

where $N(t)$ is the number of spam attacks in thousands for month t, where $t = 0$ corresponds to January 2001. Use this function to estimate, to the nearest hundred thousand, the number of monthly spam attacks for the following months.

a. October 2001   b. December 2002

Solution

a. The month October 2001 corresponds to $t = 9$. Because $t = 9$ is in the interval $0 \le t < 17$, evaluate $24.68t^2 - 170.47t + 957.73$ at $t = 9$.

$$24.68t^2 - 170.47t + 957.73$$
$$24.68(9)^2 - 170.47(9) + 957.73 = 1422.58$$

There were approximately 1,423,000 spam attacks in October 2001.

b. The month December 2002 corresponds to $t = 23$. Because $t = 23$ is in the interval $17 \le t \le 26$, evaluate $196.9t + 1164.6$ at 23.

$$196.9t + 1164.6$$
$$196.9(23) + 1164.6 = 5693.3$$

There were approximately 5,693,000 spam attacks in December 2002.

▶ **TRY EXERCISE 10, PAGE 43**

● IDENTIFYING FUNCTIONS

Recall that although every function is a relation, not every relation is a function. In the next example we examine four relations to determine which are functions.

EXAMPLE 3   **Identify Functions**

Which relations define y as a function of x?

a. $\{(2, 3), (4, 1), (4, 5)\}$ b. $3x + y = 1$ c. $-4x^2 + y^2 = 9$

d. The correspondence between the x values and the y values in **Figure 1.34.**

Solution

a. There are two ordered pairs, $(4, 1)$ and $(4, 5)$, with the same first coordinate and different second coordinates. This set does not define y as a function of x.

b. Solving $3x + y = 1$ for y yields $y = -3x + 1$. Because $-3x + 1$ is a unique real number for each x, this equation defines y as a function of x.

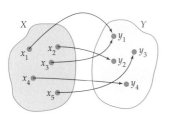

FIGURE 1.34

Continued ▶

c. Solving $-4x^2 + y^2 = 9$ for y yields $y = \pm\sqrt{4x^2 + 9}$. The right side $\pm\sqrt{4x^2 + 9}$ produces two values of y for each value of x. For example, when $x = 0$, $y = 3$ or $y = -3$. Thus $-4x^2 + y^2 = 9$ does not define y as a function of x.

d. Each x is paired with one and only one y. The correspondence in **Figure 1.34** defines y as a function of x.

TRY EXERCISE 14, PAGE 43

take note

You may indicate the domain of a function using set notation or interval notation. For instance, the domain of $f(x) = \sqrt{x - 3}$ may be given in each of the following ways:

Set notation: $\{x \mid x \geq 3\}$

Interval notation: $[3, \infty)$

Sometimes the domain of a function is stated explicitly. For example, each of f, g, and h below is given by an equation, followed by a statement that indicates the domain of the function.

$$f(x) = x^2, x > 0 \qquad g(t) = \frac{1}{t^2 + 4}, 0 \leq t \leq 5 \qquad h(x) = x^2, x = 1, 2, 3$$

Although f and h have the same equation, they are different functions because they have different domains. If the domain of a function is not explicitly stated, then its domain is determined by the following convention.

Domain of a Function

Unless otherwise stated, the domain of a function is the set of all real numbers for which the function makes sense and yields real numbers.

EXAMPLE 4 Determine the Domain of a Function

Determine the domain of each function.

a. $G(t) = \dfrac{1}{t - 4}$ b. $f(x) = \sqrt{x + 1}$

c. $A(s) = s^2$, where $A(s)$ is the area of a square whose sides are s units.

Solution

a. The number 4 is not an element of the domain because G is undefined when the denominator $t - 4$ equals 0. The domain of G is all real numbers except 4. In interval notation the domain is $(-\infty, 4) \cup (4, \infty)$.

b. The radical $\sqrt{x + 1}$ is a real number only when $x + 1 \geq 0$ or when $x \geq -1$. Thus, in set notation, the domain of f is $\{x \mid x \geq -1\}$.

c. Because s represents the length of the side of a square, s must be positive. In interval notation the domain of A is $(0, \infty)$.

TRY EXERCISE 28, PAGE 43

• GRAPHS OF FUNCTIONS

If a is an element of the domain of a function, then $(a, f(a))$ is an ordered pair that belongs to the function.

Graph of a Function

The **graph of a function** is the graph of all the ordered pairs that belong to the function.

EXAMPLE 5   **Graph a Function by Plotting Points**

Graph each function. State the domain and the range of each function.

a. $f(x) = |x - 1|$ b. $n(x) = \begin{cases} 2, & \text{if } x \leq 1 \\ x, & \text{if } x > 1 \end{cases}$

Solution

a.   The domain of f is the set of all real numbers. Evaluate the function for several domain values. We have used $x = -3, -2, -1, 0, 1, 2, 3,$ and 4.

x	−3	−2	−1	0	1	2	3	4
$y = f(x) = |x - 1|$	4	3	2	1	0	1	2	3

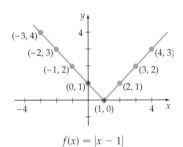

$f(x) = |x - 1|$

FIGURE 1.35

Plot the points determined by the ordered pairs. Connect the points to form the graph in **Figure 1.35**.

Because $|x - 1| \geq 0$, we can conclude that the graph of f extends from a height of 0 upward, so the range is $\{y \mid y \geq 0\}$.

b.   The domain is the union of the inequalities $x \leq 1$ and $x > 1$. Thus the domain of n is the set of all real numbers. For $x \leq 1$, graph $n(x) = 2$. This results in the horizontal ray in **Figure 1.36**. The solid circle indicates that the point $(1, 2)$ *is* part of the graph. For $x > 1$, graph $n(x) = x$. This produces the second ray in **Figure 1.36**. The open circle indicates that the point $(1, 1)$ is *not* part of the graph.

Examination of the graph shows that it includes only points whose y values are greater than 1. Thus the range of n is $\{y \mid y > 1\}$.

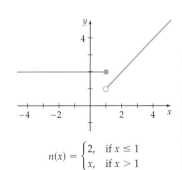

$n(x) = \begin{cases} 2, & \text{if } x \leq 1 \\ x, & \text{if } x > 1 \end{cases}$

FIGURE 1.36

▶ **TRY EXERCISE 40, PAGE 43**

INTEGRATING TECHNOLOGY

A graphing utility also can be used to draw the graph of a function. For instance, to graph $f(x) = x^2 - 1$, you will enter an equation similar to Y₁=x²–1. The graph is shown in **Figure 1.37.**

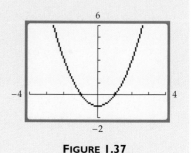

FIGURE 1.37

The definition that a function is a set of ordered pairs in which no two ordered pairs that have the same first coordinate have different second coordinates implies that any vertical line intersects the graph of a function at no more than one point. This is known as the *vertical line test*.

The Vertical Line Test for Functions

A graph is the graph of a function if and only if no vertical line intersects the graph at more than one point.

EXAMPLE 6   Apply the Vertical Line Test

Which of the following graphs are graphs of functions?

a.
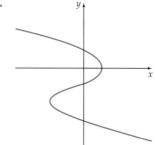

b.
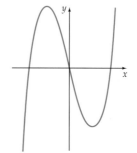

Solution

a. This graph *is not* the graph of a function because some vertical lines intersect the graph in more than one point.

b. This graph *is* the graph of a function because every vertical line intersects the graph in at most one point.

▶ **TRY EXERCISE 50, PAGE 44**

Consider the graph in **Figure 1.38.** As a point on the graph moves from left to right, this graph falls for values of $x \le -2$, it remains the same height from

FIGURE 1.38

$x = -2$ to $x = 2$, and it rises for $x \geq 2$. The function represented by the graph is said to be *decreasing* on the interval $(-\infty, -2]$, *constant* on the interval $[-2, 2]$, and *increasing* on the interval $[2, \infty)$.

Definition of Increasing, Decreasing, and Constant Functions

If a and b are elements of an interval I that is a subset of the domain of a function f, then

- f is **increasing** on I if $f(a) < f(b)$ whenever $a < b$.
- f is **decreasing** on I if $f(a) > f(b)$ whenever $a < b$.
- f is **constant** on I if $f(a) = f(b)$ for all a and b.

Recall that a function is a relation in which no two ordered pairs that have the same first coordinate have different second coordinates. This means that given any x, there is only one y that can be paired with that x. A **one-to-one function** satisfies the additional condition that given any y, there is only one x that can be paired with that given y. In a manner similar to applying the vertical line test, we can apply a *horizontal line test* to identify one-to-one functions.

Horizontal Line Test for a One-To-One Function

If every horizontal line intersects the graph of a function at most once, then the graph is the graph of a one-to-one function.

For example, some horizontal lines intersect the graph in **Figure 1.39** at more than one point. It is *not* the graph of a one-to-one function. Every horizontal line intersects the graph in **Figure 1.40** at most once. This is the graph of a one-to-one function.

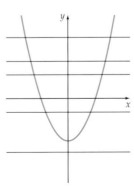

FIGURE 1.39
Some horizontal lines intersect this graph at more than one point. It is *not* the graph of a one-to-one function.

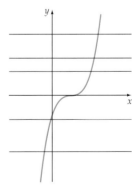

FIGURE 1.40
Every horizontal line intersects this graph at most once. It is the graph of a one-to-one function.

● THE GREATEST INTEGER FUNCTION (FLOOR FUNCTION)

To this point, the graphs of the functions have not had any breaks or gaps. These functions whose graphs can be drawn without lifting the pencil off the paper are called *continuous functions*. The graphs of some functions do have breaks or *discontinuities*. One such function is the **greatest integer function** or **floor function**. This function is denoted by various symbols such as $[\![x]\!]$, $\lfloor x \rfloor$, and int(x).

The value of the greatest integer function at x is the greatest integer that is less than or equal to x. For instance,

$$\lfloor -1.1 \rfloor = -2 \qquad [\![-3]\!] = -3 \qquad \text{int}\left(\frac{5}{2}\right) = 2 \qquad \lfloor 5 \rfloor = 5 \qquad [\![\pi]\!] = 3$$

❓ QUESTION Evaluate. **a.** $\text{int}\left(-\frac{3}{2}\right)$ **b.** $\lfloor 2 \rfloor$

INTEGRATING TECHNOLOGY

Many graphing calculators use the notation int(x) for the greatest integer function. Here is a screen from a TI-83 Plus.

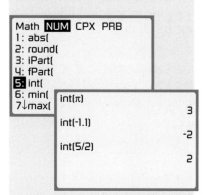

To graph the floor function, first observe that the value of the floor function is constant between any two consecutive integers. For instance, between the integers 1 and 2, we have

$$\text{int}(1.1) = 1 \qquad \text{int}(1.35) = 1 \qquad \text{int}(1.872) = 1 \qquad \text{int}(1.999) = 1$$

Between -3 and -2, we have

$$\text{int}(-2.98) = -3 \qquad \text{int}(-2.4) = -3 \qquad \text{int}(-2.35) = -3 \qquad \text{int}(-2.01) = -3$$

Using this property of the floor function, we can create a table of values and then graph the floor function.

x	$y = \text{int}(x)$
$-5 \le x < 4$	-5
$-4 \le x < -3$	-4
$-3 \le x < -2$	-3
$-2 \le x < -1$	-2
$-1 \le x < 0$	-1
$0 \le x < 1$	0
$1 \le x < 2$	1
$2 \le x < 3$	2
$3 \le x < 4$	3
$4 \le x < 5$	4

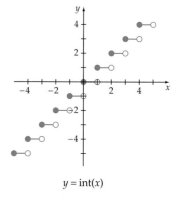

$y = \text{int}(x)$

FIGURE 1.41

The graph of the floor function has discontinuities (breaks) whenever x is an integer. The domain of the floor function is the set of real numbers; the range is the set of integers. Because the graph appears to be a series of steps, sometimes the floor function is referred to as a **step function**.

❓ ANSWER **a.** Because -2 is the greatest integer that is less than $-\frac{3}{2}$, $\text{int}\left(-\frac{3}{2}\right) = -2$.

  b. Because 2 is the greatest integer less than or *equal* to 2, $\lfloor 2 \rfloor = 2$.

INTEGRATING TECHNOLOGY

Many graphing calculators use the notation int(x) for the floor function. The screens below are from a TI-83 Plus graphing calculator.

A graphing calculator also can be used to graph the floor function. The graph in **Figure 1.42** was drawn in "connected" mode. This graph does not show the discontinuities that occur whenever x is an integer.

The graph in **Figure 1.43** was constructed by graphing the floor function in "dot" mode. In this case the discontinuities at the integers are apparent.

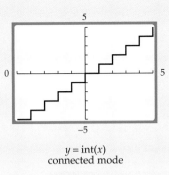

$y = \text{int}(x)$
connected mode

FIGURE 1.42

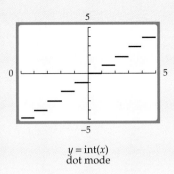

$y = \text{int}(x)$
dot mode

FIGURE 1.43

EXAMPLE 7   **Use the Greatest Integer Function to Model Expenses**

The cost of parking in a garage is $3 for the first hour or any part of the hour and $2 for each additional hour or any part of the hour thereafter. If x is the time in hours that you park your car, then the cost is given by

$$C(x) = 3 - 2\,\text{int}(1 - x), \quad x > 0$$

a. Evaluate $C(2)$ and $C(2.5)$.   b. Graph $y = C(x)$ for $0 < x \le 5$.

Solution

a. $C(2) = 3 - 2\,\text{int}(1 - 2)$ $C(2.5) = 3 - 2\,\text{int}(1 - 2.5)$
$\quad\quad = 3 - 2\,\text{int}(-1)$ $\quad\quad\quad = 3 - 2\,\text{int}(-1.5)$
$\quad\quad = 3 - 2(-1)$ $\quad\quad\quad = 3 - 2(-2)$
$\quad\quad = \$5$ $\quad\quad\quad = \$7$

b. To graph $C(x)$ for $0 < x \le 5$, consider the value of $\text{int}(1 - x)$ for each of the intervals $0 < x \le 1$, $1 < x \le 2$, $2 < x \le 3$, $3 < x \le 4$, and $4 < x \le 5$. For instance, when $0 < x \le 1$, $0 \le 1 - x < 1$. Thus $\text{int}(1 - x) = 0$ when $0 < x \le 1$. Now consider $1 < x \le 2$. When $1 < x \le 2$, $1 \le 1 - x < 2$. Thus $\text{int}(1 - x) = 1$ when $1 < x \le 2$. Applying the same reasoning to

Continued ▶

each of the other intervals gives the following table of values and the corresponding graph of C.

x	C(x) = 3 − 2 int(1 − x)
0 < x ≤ 1	3
1 < x ≤ 2	5
2 < x ≤ 3	7
3 < x ≤ 4	9
4 < x ≤ 5	11

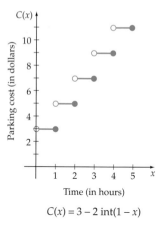

$C(x) = 3 - 2 \text{ int}(1 - x)$

FIGURE 1.44

Because $C(1) = 3$, $C(2) = 5$, $C(3) = 7$, $C(4) = 9$, and $C(5) = 11$, we can use a solid circle at the right endpoint of each "step" and an open circle at each left endpoint.

▶  **TRY EXERCISE 48, PAGE 44**

INTEGRATING TECHNOLOGY

Example 7 illustrates that a graphing calculator may not produce a graph that is a good representation of a function. You may be required to *make adjustments* in the MODE, SET UP, or WINDOW of the graphing calculator so that it will produce a better representation of the function. Some graphs may also require some *fine tuning*, such as open or solid circles at particular points, to accurately represent the function.

● APPLICATIONS

EXAMPLE 8   **Solve an Application**

A car was purchased for $16,500. Assuming the car depreciates at a constant rate of $2200 per year (*straight-line depreciation*) for the first 7 years, write the value v of the car as a function of time, and calculate the value of the car 3 years after purchase.

Solution

Let t represent the number of years that have passed since the car was purchased. Then $2200t$ is the amount that the car has depreciated after t years. The value of the car at time t is given by

$$v(t) = 16{,}500 - 2200t, \quad 0 \le t \le 7$$

When $t = 3$, the value of the car is

$$v(3) = 16{,}500 - 2200(3) = 16{,}500 - 6600 = \$9900$$

▶ **Try Exercise 66, page 45**

Often in applied mathematics, formulas are used to determine the functional relationship that exists between two variables.

EXAMPLE 9 Solve an Application

A lighthouse is 2 miles south of a port. A ship leaves port and sails east at a rate of 7 mph. Express the distance d between the ship and the lighthouse as a function of time, given that the ship has been sailing for t hours.

Solution

Draw a diagram and label it as shown in **Figure 1.45.** Note that because distance = (rate)(time) and the rate is 7, in t hours the ship has sailed a distance of $7t$.

$$[d(t)]^2 = (7t)^2 + 2^2 \qquad \bullet \textbf{ The Pythagorean Theorem}$$
$$[d(t)]^2 = 49t^2 + 4$$
$$d(t) = \sqrt{49t^2 + 4} \qquad \bullet \textbf{ The } \pm \textbf{ sign is not used because}$$
$$\textit{d} \textbf{ must be nonnegative.}$$

▶ **Try Exercise 72, page 46**

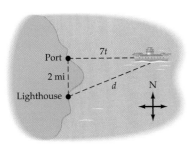

FIGURE 1.45

EXAMPLE 10 Solve an Application

An open box is to be made from a square piece of cardboard that measures 40 inches on each side. To construct the box, squares that measure x inches on each side are cut from each corner of the cardboard as shown in **Figure 1.46.**

a. Express the volume V of the box as a function of x.

b. Determine the domain of V.

Solution

a. The length l of the box is $40 - 2x$. The width w is also $40 - 2x$. The height of the box is x. The volume V of a box is the product of its length, its width, and its height. Thus

$$V = (40 - 2x)^2 x$$

b. The squares that are cut from each corner require x to be larger than 0 inches but less than 20 inches. Thus the domain is $\{x \mid 0 < x < 20\}$.

▶ **Try Exercise 68, page 45**

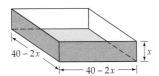

FIGURE 1.46

TOPICS FOR DISCUSSION

1. Discuss the definition of *function*. Give some examples of relationships that are functions and some that are not functions.

2. What is the difference between the domain and range of a function?

3. How many *y*-intercepts can a function have? How many *x*-intercepts can a function have?

4. Discuss how the vertical line test is used to determine whether or not a graph is the graph of a function. Explain why the vertical line test works.

5. What is the domain of $f(x) = \dfrac{\sqrt{1-x}}{x^2 - 9}$? Explain.

6. Is 2 in the range of $g(x) = \dfrac{6x-5}{3x+1}$? Explain the process you used to make your decision.

7. Suppose that f is a function and that $f(a) = f(b)$. Does this imply that $a = b$? Explain your answer.

EXERCISE SET 1.3

In Exercises 1 to 8, evaluate each function.

1. Given $f(x) = 3x - 1$, find

  a. $f(2)$ **b.** $f(-1)$ **c.** $f(0)$

  d. $f\left(\dfrac{2}{3}\right)$ **e.** $f(k)$ **f.** $f(k + 2)$

▶**2.** Given $g(x) = 2x^2 + 3$, find

  a. $g(3)$ **b.** $g(-1)$ **c.** $g(0)$

  d. $g\left(\dfrac{1}{2}\right)$ **e.** $g(c)$ **f.** $g(c + 5)$

3. Given $A(w) = \sqrt{w^2 + 5}$, find

  a. $A(0)$ **b.** $A(2)$ **c.** $A(-2)$

  d. $A(4)$ **e.** $A(r + 1)$ **f.** $A(-c)$

4. Given $J(t) = 3t^2 - t$, find

  a. $J(-4)$ **b.** $J(0)$ **c.** $J\left(\dfrac{1}{3}\right)$

  d. $J(-c)$ **e.** $J(x + 1)$ **f.** $J(x + h)$

5. Given $f(x) = \dfrac{1}{|x|}$, find

  a. $f(2)$ **b.** $f(-2)$ **c.** $f\left(\dfrac{-3}{5}\right)$

  d. $f(2) + f(-2)$ **e.** $f(c^2 + 4)$ **f.** $f(2 + h)$

6. Given $T(x) = 5$, find

  a. $T(-3)$ **b.** $T(0)$ **c.** $T\left(\dfrac{2}{7}\right)$

  d. $T(3) + T(1)$ **e.** $T(x + h)$ **f.** $T(3k + 5)$

7. Given $s(x) = \dfrac{x}{|x|}$, find

  a. $s(4)$ **b.** $s(5)$ **c.** $s(-2)$

  d. $s(-3)$ **e.** $s(t), t > 0$ **f.** $s(t), t < 0$

8. Given $r(x) = \dfrac{x}{x + 4}$, find

  a. $r(0)$ **b.** $r(-1)$ **c.** $r(-3)$

d. $r\left(\dfrac{1}{2}\right)$ **e.** $r(0.1)$ **f.** $r(10{,}000)$

In Exercises 9 and 10, evaluate each piecewise-defined function for the indicated values.

9. $P(x) = \begin{cases} 3x + 1, & \text{if } x < 2 \\ -x^2 + 11, & \text{if } x \geq 2 \end{cases}$

  a. $P(-4)$ **b.** $P\left(\sqrt{5}\right)$

  c. $P(c), \quad c < 2$ **d.** $P(k + 1), \quad k \geq 1$

▶ **10.** $Q(t) = \begin{cases} 4, & \text{if } 0 \leq t \leq 5 \\ -t + 9, & \text{if } 5 < t \leq 8 \\ \sqrt{t - 7}, & \text{if } 8 < t \leq 11 \end{cases}$

  a. $Q(0)$ **b.** $Q(e), \quad 6 < e < 7$

  c. $Q(n), \quad 1 < n < 2$ **d.** $Q(m^2 + 7), \quad 1 < m \leq 2$

In Exercises 11 to 20, identify the equations that define y as a function of x.

11. $2x + 3y = 7$ **12.** $5x + y = 8$

13. $-x + y^2 = 2$ ▶ **14.** $x^2 - 2y = 2$

15. $y = 4 \pm \sqrt{x}$ **16.** $x^2 + y^2 = 9$

17. $y = \sqrt[3]{x}$ **18.** $y = |x| + 5$

19. $y^2 = x^2$ **20.** $y^3 = x^3$

In Exercises 21 to 26, identify the sets of ordered pairs (x, y) that define y as a function of x.

21. $\{(2, 3), (5, 1), (-4, 3), (7, 11)\}$

22. $\{(5, 10), (3, -2), (4, 7), (5, 8)\}$

23. $\{(4, 4), (6, 1), (5, -3)\}$

24. $\{(2, 2), (3, 3), (7, 7)\}$

25. $\{(1, 0), (2, 0), (3, 0)\}$

26. $\left\{\left(-\dfrac{1}{3}, \dfrac{1}{4}\right), \left(-\dfrac{1}{4}, \dfrac{1}{3}\right), \left(\dfrac{1}{4}, \dfrac{2}{3}\right)\right\}$

In Exercises 27 to 38, determine the domain of the function represented by the given equation.

27. $f(x) = 3x - 4$ ▶ **28.** $f(x) = -2x + 1$

29. $f(x) = x^2 + 2$ **30.** $f(x) = 3x^2 + 1$

31. $f(x) = \dfrac{4}{x + 2}$ **32.** $f(x) = \dfrac{6}{x - 5}$

33. $f(x) = \sqrt{7 + x}$ **34.** $f(x) = \sqrt{4 - x}$

35. $f(x) = \sqrt{4 - x^2}$ **36.** $f(x) = \sqrt{12 - x^2}$

37. $f(x) = \dfrac{1}{\sqrt{x + 4}}$ **38.** $f(x) = \dfrac{1}{\sqrt{5 - x}}$

In Exercises 39 to 46, graph each function. Insert solid circles or hollow circles where necessary to indicate the true nature of the function.

39. $f(x) = \begin{cases} |x|, & \text{if } x \leq 1 \\ 2, & \text{if } x > 1 \end{cases}$

▶ **40.** $g(x) = \begin{cases} -4, & \text{if } x \leq 0 \\ x^2 - 4, & \text{if } 0 < x \leq 1 \\ -x, & \text{if } x > 1 \end{cases}$

41. $J(x) = \begin{cases} 4, & \text{if } x \leq -1 \\ x^2, & \text{if } -1 < x < 1 \\ -x + 5, & \text{if } x \geq 1 \end{cases}$

42. $K(x) = \begin{cases} 1, & \text{if } x \leq -2 \\ x^2 - 4, & \text{if } -2 < x < 2 \\ \dfrac{1}{2}x, & \text{if } x \geq 2 \end{cases}$

43. $L(x) = \left[\!\left[\dfrac{1}{3}x\right]\!\right] \quad \text{for } -6 \leq x \leq 6$

44. $M(x) = [\![x]\!] + 2 \quad \text{for } 0 \leq x \leq 4$

45. $N(x) = \text{int}(-x) \quad \text{for } -3 \leq x \leq 3$

46. $P(x) = \text{int}(x) + x \quad \text{for } 0 \leq x \leq 4$

47. FIRST-CLASS MAIL   In 2003, the cost to mail a first-class letter is given by

$$C(w) = 0.37 - 0.34\,\text{int}(1 - w), \quad w > 0$$

where C is in dollars and w is the weight of the letter in ounces.

  a. What is the cost to mail a letter that weighs 2.8 ounces?

  b. Graph C for $0 < w \leq 5$.

▶ **48.** **INCOME TAX** The amount of federal income tax $T(x)$ a person owed in 2003 is given by

$$T(x) = \begin{cases} 0.10x, & 0 \le x < 6000 \\ 0.15(x - 6000) + 600, & 6000 \le x < 27{,}950 \\ 0.27(x - 27{,}950) + 3892.50, & 27{,}950 \le x < 67{,}700 \\ 0.30(x - 67{,}700) + 14{,}625, & 67{,}700 \le x < 141{,}250 \\ 0.35(x - 141{,}250) + 36{,}690, & 141{,}250 \le x < 307{,}050 \\ 0.386(x - 307{,}050) + 94{,}720, & x \ge 307{,}050 \end{cases}$$

where x is the adjusted gross income tax of the taxpayer.

a. What is the domain of this function?

b. Find the income tax owed by a taxpayer whose adjusted gross income was $31,250.

c. Find the income tax owed by a taxpayer whose adjusted gross income was $72,000.

49. Use the vertical line test to determine which of the following graphs are graphs of functions.

a.
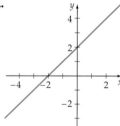

b.
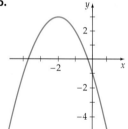

c.
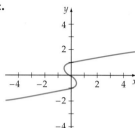

d.
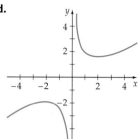

▶ **50.** Use the vertical line test to determine which of the following graphs are graphs of functions.

a.
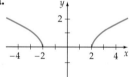

b.
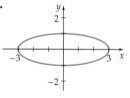

c.
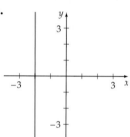

d.
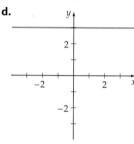

In Exercises 51 to 60, use the indicated graph to identify the intervals over which the function is increasing, constant, or decreasing.

51.
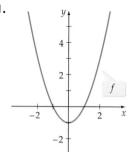

52.
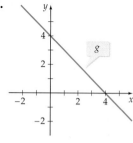

53.
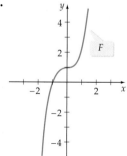

54.
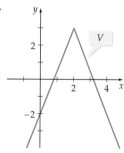

55.
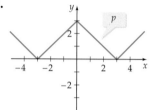

56.
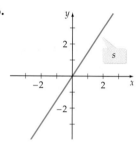

57.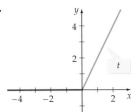

58.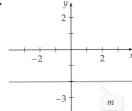

59.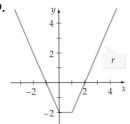

60.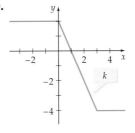

61. Use the horizontal line test to determine which of the following functions are one-to-one.

f as shown in Exercise 51
g as shown in Exercise 52
F as shown in Exercise 53
V as shown in Exercise 54
p as shown in Exercise 55

62. Use the horizontal line test to determine which of the following functions are one-to-one.

s as shown in Exercise 56
t as shown in Exercise 57
m as shown in Exercise 58
r as shown in Exercise 59
k as shown in Exercise 60

63. A rectangle has a length of l feet and a perimeter of 50 feet.

a. Write the width w of the rectangle as a function of its length.

b. Write the area A of the rectangle as a function of its length.

64. The sum of two numbers is 20. Let x represent one of the numbers.

a. Write the second number y as a function of x.

b. Write the product P of the two numbers as a function of x.

65. DEPRECIATION A bus was purchased for $80,000. Assuming the bus depreciates at a rate of $6500 per year (*straight-line depreciation*) for the first 10 years, write the value v of the bus as a function of the time t (measured in years) for $0 \le t \le 10$.

▶ **66.** DEPRECIATION A boat was purchased for $44,000. Assuming the boat depreciates at a rate of $4200 per year (*straight-line depreciation*) for the first 8 years, write the value v of the boat as a function of the time t (measured in years) for $0 \le t \le 8$.

67. COST, REVENUE, AND PROFIT A manufacturer produces a product at a cost of $22.80 per unit. The manufacturer has a fixed cost of $400.00 per day. Each unit retails for $37.00. Let x represent the number of units produced in a 5-day period.

a. Write the total cost C as a function of x.

b. Write the revenue R as a function of x.

c. Write the profit P as a function of x. (*Hint:* The profit function is given by $P(x) = R(x) - C(x)$.)

▶ **68.** VOLUME OF A BOX An open box is to be made from a square piece of cardboard having dimensions 30 inches by 30 inches by cutting out squares of area x^2 from each corner, as shown in the figure.

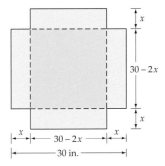

a. Express the volume V of the box as a function of x.

b. State the domain of V.

69. HEIGHT OF AN INSCRIBED CYLINDER A cone has an altitude of 15 centimeters and a radius of 3 centimeters. A right circular cylinder of radius r and height h is inscribed in the cone as shown in the figure. Use similar triangles to write h as a function of r.

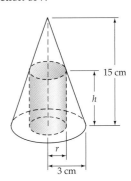

70. **VOLUME OF WATER** Water is flowing into a conical drinking cup that has an altitude of 4 inches and a radius of 2 inches, as shown in the figure.

a. Write the radius r of the water as a function of its depth h.

b. Write the volume V of the water as a function of its depth h.

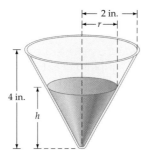

71. **DISTANCE FROM A BALLOON** For the first minute of flight, a hot air balloon rises vertically at a rate of 3 meters per second. If t is the time in seconds that the balloon has been airborne, write the distance d between the balloon and a point on the ground 50 meters from the point of lift-off as a function of t.

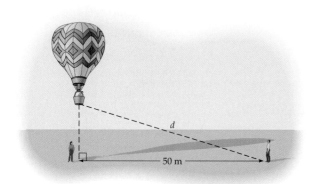

72. **TIME FOR A SWIMMER** An athlete swims from point A to point B at the rate of 2 mph and runs from point B to point C at a rate of 8 mph. Use the dimensions in the figure to write the time t required to reach point C as a function of x.

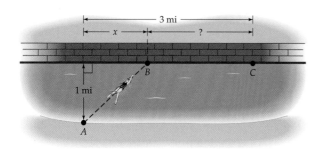

73. **DISTANCE BETWEEN SHIPS** At 12:00 noon Ship A is 45 miles due south of ship B and is sailing north at a rate of 8 mph. Ship B is sailing east at a rate of 6 mph. Write the distance d between the ships as a function of the time t, where $t = 0$ represents 12:00 noon.

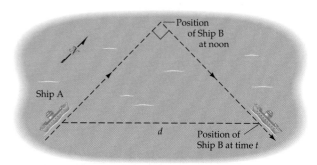

74. **AREA** A rectangle is bounded by the x- and y-axes and the graph of $y = -\dfrac{1}{2}x + 4$.

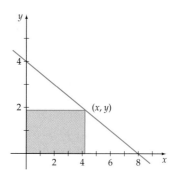

a. Find the area of the rectangle as a function of x.

b. Complete the table below.

x	Area
1	
2	
4	
6	
7	

c. What is the domain of this function?

75. **AREA** A piece of wire 20 centimeters long is cut at a point x centimeters from the left end. The left-hand piece is formed into the shape of a circle and the right-hand piece is formed into a square, as shown in the following figure.

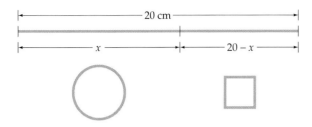

a. Find the area enclosed by the two figures as a function of x.

b. Complete the table below. Round the area to the nearest hundredth.

x	Total Area Enclosed
0	
4	
8	
12	
16	
20	

c. What is the domain of this function?

76. AREA A triangle is bounded by the x- and y-axes and its hypotenuse must pass through $P(2, 2)$, as shown below.

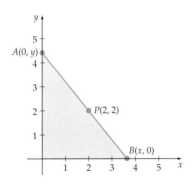

a. Find the area of the triangle as a function of x. (*Suggestion:* The slope of the line between points A and P equals the slope of the line between P and B.)

b. What is the domain of the function you found in part **a.**?

77. LENGTH Two guy wires are attached to utility poles that are 40 feet apart, as shown in the following diagram.

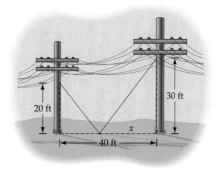

a. Find the total length of the two guy wires as a function of x.

b. Complete the table below. Round the length to the nearest hundredth.

x	Total Length of Wires
0	
10	
20	
30	
40	

c. What is the domain of this function?

78. SALES VS. PRICE A business finds that the number of feet f of pipe it can sell per week is a function of the price p in cents per foot as given by

$$f(p) = \frac{320{,}000}{p + 25}, \quad 40 \le p \le 90$$

Complete the following table by evaluating f (to the nearest hundred feet) for the indicated values of p.

p	40	50	60	75	90
$f(p)$					

79. MODEL YIELD The yield Y of apples per tree is related to the amount x of a particular type of fertilizer applied (in pounds per year) by the function

$$Y(x) = 400[1 - 5(x - 1)^{-2}], \quad 5 \le x \le 20$$

Complete the following table by evaluating Y (to the nearest apple) for the indicated applications.

x	5	10	12.5	15	20
$Y(x)$					

80. MODEL COST A manufacturer finds that the cost C in dollars of producing x items of a product is given by

$$C(x) = \left(225 + 1.4\sqrt{x}\right)^2, \quad 100 \le x \le 1000$$

Complete the following table by evaluating C (to the nearest dollar) for the indicated numbers of items.

x	100	200	500	750	1000
$C(x)$					

81. If $f(x) = x^2 - x - 5$ and $f(c) = 1$, find c.

82. If $g(x) = -2x^2 + 4x - 1$ and $g(c) = -4$, find c.

83. Determine whether 1 is in the range of $f(x) = \dfrac{x-1}{x+1}$.

84. Determine whether 0 is in the range of $g(x) = \dfrac{1}{x-3}$.

In Exercises 85 to 90, use a graphing utility.

85. Graph $f(x) = \dfrac{[\![x]\!]}{|x|}$ for $-4.7 \le x \le 4.7$ and $x \ne 0$.

86. Graph $f(x) = \dfrac{[\![2x]\!]}{|x|}$ for $-4 \le x \le 4$ and $x \ne 0$.

87. Graph: $f(x) = x^2 - 2|x| - 3$

88. Graph: $f(x) = x^2 - |2x - 3|$

89. Graph: $f(x) = |x^2 - 1| - |x - 2|$

90. Graph: $f(x) = |x^2 - 2x| - 3$

—— **CONNECTING CONCEPTS** ——

The notation $f(x)\big|_a^b$ is used to denote the difference $f(b) - f(a)$. That is,

$$f(x)\big|_a^b = f(b) - f(a)$$

In Exercises 91 to 94, evaluate $f(x)\big|_a^b$ for the given function f and the indicated values of a and b.

91. $f(x) = x^2 - x; f(x)\big|_2^3$

92. $f(x) = -3x + 2; f(x)\big|_4^7$

93. $f(x) = 2x^3 - 3x^2 - x; f(x)\big|_0^2$

94. $f(x) = \sqrt{8 - x}; f(x)\big|_0^8$

In Exercises 95 to 98, each function has two or more independent variables.

95. Given $f(x, y) = 3x + 5y - 2$, find

  a. $f(1, 7)$ **b.** $f(0, 3)$ **c.** $f(-2, 4)$

  d. $f(4, 4)$ **e.** $f(k, 2k)$ **f.** $f(k + 2, k - 3)$

96. Given $g(x, y) = 2x^2 - |y| + 3$, find

  a. $g(3, -4)$ **b.** $g(-1, 2)$

  c. $g(0, -5)$ **d.** $g\left(\dfrac{1}{2}, -\dfrac{1}{4}\right)$

  e. $g(c, 3c), c > 0$ **f.** $g(c + 5, c - 2), c < 0$

97. AREA OF A TRIANGLE The area of a triangle with sides a, b, and c is given by the function

$$A(a, b, c) = \sqrt{s(s - a)(s - b)(s - c)}$$

where s is the semiperimeter

$$s = \frac{a + b + c}{2}$$

Find $A(5, 8, 11)$.

98. COST OF A PAINTER The cost in dollars to hire a house painter is given by the function

$$C(h, g) = 15h + 14g$$

where h is the number of hours it takes to paint the house and g is the number of gallons of paint required to paint the house. Find $C(18, 11)$.

A *fixed point* of a function is a number a such that $f(a) = a$. In Exercises 99 and 100, find all fixed points for the given function.

99. $f(x) = x^2 + 3x - 3$

100. $g(x) = \dfrac{x}{x + 5}$

In Exercises 101 and 102, sketch the graph of the piecewise-defined function.

101. $s(x) = \begin{cases} 1 & \text{if } x \text{ is an integer} \\ 2 & \text{if } x \text{ is not an integer} \end{cases}$

102. $v(x) = \begin{cases} 2x - 2 & \text{if } x \neq 3 \\ 1 & \text{if } x = 3 \end{cases}$

PREPARE FOR SECTION 1.4

103. For $f(x) = \dfrac{3x^4}{x^2 + 1}$, show that $f(-3) = f(3)$. [1.3]

104. For $f(x) = 2x^3 - 5x$, show that $f(-2) = -f(2)$. [1.3]

105. Let $f(x) = x^2$ and $g(x) = x + 3$. Find $f(a) - g(a)$ for $a = -2, -1, 0, 1, 2$. [1.3]

106. What is the midpoint of the line segment between $P(-a, b)$ and $Q(a, b)$? [1.2]

107. What is the midpoint of the line segment between $P(-a, -b)$ and $Q(a, b)$? [1.2]

PROJECTS

1. **DAY OF THE WEEK** A formula known as Zeller's Congruence makes use of the greatest integer function $[\![x]\!]$ to determine the day of the week on which a given day fell or will fall. To use Zeller's Congruence, we first compute the integer z given by

$$z = \left[\!\!\left[\frac{13m - 1}{5}\right]\!\!\right] + \left[\!\!\left[\frac{y}{4}\right]\!\!\right] + \left[\!\!\left[\frac{c}{4}\right]\!\!\right] + d + y - 2c$$

The variables c, y, d, and m are defined as follows:

$c = $ the century
$y = $ the year of the century
$d = $ the day of the month
$m = $ the month, using 1 for March, 2 for April,..., 10 for December. January and February are assigned the values 11 and 12 of the previous year.

For example, for the date September 12, 2001, we use $c = 20$, $y = 1$, $d = 12$, and $m = 7$. The remainder of z divided by 7 gives the day of the week. A remainder of 0 represents a Sunday, a remainder of 1 a Monday,..., and a remainder of 6 a Saturday.

a. Verify that December 7, 1941 was a Sunday.

b. Verify that January 1, 2010 will fall on a Friday.

c. Determine on what day of the week Independence Day (July 4, 1776) fell.

d. Determine on what day of the week you were born.

PROPERTIES OF GRAPHS

• SYMMETRY

The graph in **Figure 1.47** is symmetric with respect to the line *l*. Note that the graph has the property that if the paper is folded along the dotted line *l*, the point *A'* will coincide with the point *A*, the point *B'* will coincide with the point *B*, and the point *C'* will coincide with the point *C*. One part of the graph is a *mirror image* of the rest of the graph across the line *l*.

A graph is **symmetric with respect to the y-axis** if, whenever the point given by (x, y) is on the graph, then $(-x, y)$ is also on the graph. The graph in **Figure 1.48** is symmetric with respect to the y-axis. A graph is **symmetric with respect to the x-axis** if, whenever the point given by (x, y) is on the graph, then $(x, -y)$ is also on the graph. The graph in **Figure 1.49** is symmetric with respect to the x-axis.

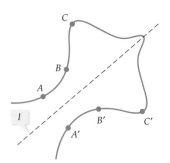

FIGURE 1.47

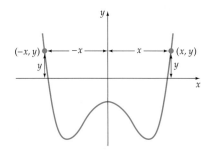

FIGURE 1.48
Symmetry with respect to the *y*-axis

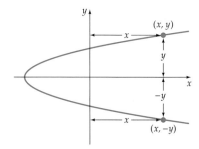

FIGURE 1.49
Symmetry with respect to the *x*-axis

Tests for Symmetry with Respect to a Coordinate Axis

The graph of an equation is symmetric with respect to

- the *y*-axis if the replacement of *x* with $-x$ leaves the equation unaltered.
- the *x*-axis if the replacement of *y* with $-y$ leaves the equation unaltered.

❓ **QUESTION** Which of the graphs below, I, II, or III, is **a.** symmetric with respect to the *x*-axis? **b.** symmetric with respect to the *y*-axis?

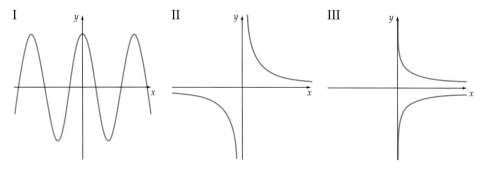

❓ **ANSWER** **a.** I is symmetric with respect to the *y*-axis.
  b. III is symmetric with respect to the *x*-axis.

EXAMPLE 1   **Determine Symmetries of a Graph**

Determine whether the graph of the given equation has symmetry with respect to either the x- or the y-axis.

a. $y = x^2 + 2$ **b.** $x = |y| - 2$

Solution

a. The equation $y = x^2 + 2$ *is unaltered* by the replacement of x with $-x$. That is, the simplification of $y = (-x)^2 + 2$ yields the original equation $y = x^2 + 2$. Thus the graph of $y = x^2 + 2$ is symmetric with respect to the y-axis. However, the equation $y = x^2 + 2$ *is altered* by the replacement of y with $-y$. That is, the simplification of $-y = x^2 + 2$, which is $y = -x^2 - 2$, *does not* yield the original equation $y = x^2 + 2$. The graph of $y = x^2 + 2$ is not symmetric with respect to the x-axis. See **Figure 1.50.**

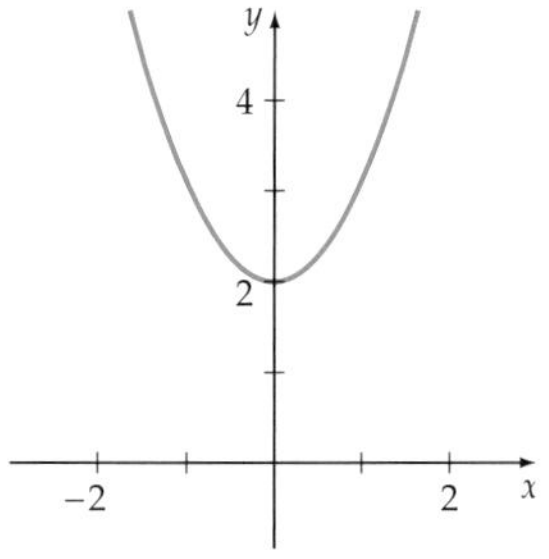

b. The equation $x = |y| - 2$ *is altered* by the replacement of x with $-x$. That is, the simplification of $-x = |y| - 2$, which is $x = -|y| + 2$, *does not* yield the original equation $x = |y| - 2$. This implies that the graph of $x = |y| - 2$ is not symmetric with respect to the y-axis. However, the equation $x = |y| - 2$ *is unaltered* by the replacement of y with $-y$. That is, the simplification of $x = |-y| - 2$ yields the original equation $x = |y| - 2$. The graph of $x = |y| - 2$ is symmetric with respect to the x-axis. See **Figure 1.51.**

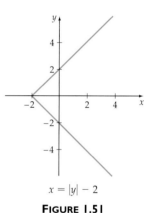

$x = |y| - 2$

FIGURE 1.51

▶ TRY EXERCISE 14, PAGE 60

$y = x^2 + 2$

FIGURE 1.50

Symmetry with Respect to a Point

A graph is **symmetric with respect to a point** Q if for each point P on the graph there is a point P' on the graph such that Q is the midpoint of the line segment PP'.

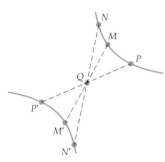

FIGURE 1.52

The graph in **Figure 1.52** is symmetric with respect to the point Q. For any point P on the graph, there exists a point P' on the graph such that Q is the midpoint of $P'P$.

When we discuss symmetry with respect to a point, we frequently use the origin. A graph is symmetric with respect to the origin if, whenever the point given by (x, y) is on the graph, then $(-x, -y)$ is also on the graph. The graph in **Figure 1.53** is symmetric with respect to the origin.

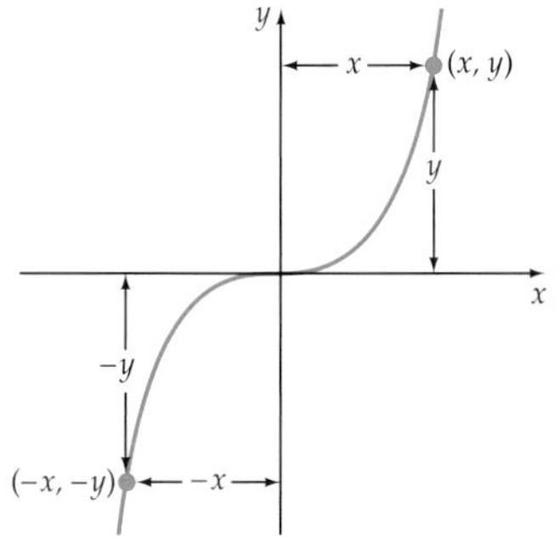

Symmetry with respect to the origin

FIGURE 1.53

Test for Symmetry with Respect to the Origin

The graph of an equation is symmetric with respect to the origin if the replacement of x with $-x$ and of y with $-y$ leaves the equation unaltered.

EXAMPLE 2   **Determine Symmetry with Respect to the Origin**

Determine whether the graph of each equation has symmetry with respect to the origin.

a. $xy = 4$ b. $y = x^3 + 1$

Solution

a.   The equation $xy = 4$ is unaltered by the replacement of x with $-x$ and of y with $-y$. That is, the simplification of $(-x)(-y) = 4$ yields the original equation $xy = 4$. Thus the graph of $xy = 4$ is symmetric with respect to the origin. See **Figure 1.54.**

b.   The equation $y = x^3 + 1$ *is altered* by the replacement of x with $-x$ and of y with $-y$. That is, the simplification of $-y = (-x)^3 + 1$, which is $y = x^3 - 1$, *does not* yield the original equation $y = x^3 + 1$. Thus the graph of $y = x^3 + 1$ is not symmetric with respect to the origin. See **Figure 1.55.**

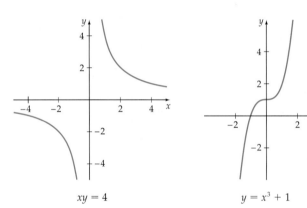

$xy = 4$

FIGURE 1.54

$y = x^3 + 1$

FIGURE 1.55

▶ **TRY EXERCISE 24, PAGE 60**

Some graphs have more than one symmetry. For example, the graph of $|x| + |y| = 2$ has symmetry with respect to the x-axis, the y-axis, and the origin. **Figure 1.56** is the graph of $|x| + |y| = 2$.

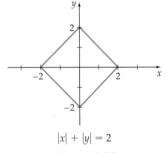

$|x| + |y| = 2$

FIGURE 1.56

● **EVEN AND ODD FUNCTIONS**

Some functions are classified as either *even* or *odd*.

Definition of Even and Odd Functions

The function f is an **even function** if

$$f(-x) = f(x) \quad \text{for all } x \text{ in the domain of } f$$

The function f is an **odd function** if

$$f(-x) = -f(x) \quad \text{for all } x \text{ in the domain of } f$$

EXAMPLE 3   **Identify Even or Odd Functions**

Determine whether each function is even, odd, or neither.

a. $f(x) = x^3$ b. $F(x) = |x|$ c. $h(x) = x^4 + 2x$

Solution

Replace x with $-x$ and simplify.

a. $f(-x) = (-x)^3 = -x^3 = -(x^3) = -f(x)$
Because $f(-x) = -f(x)$, this function is an odd function.

b. $F(-x) = |-x| = |x| = F(x)$
Because $F(-x) = F(x)$, this function is an even function.

c. $h(-x) = (-x)^4 + 2(-x) = x^4 - 2x$
This function is neither an even nor an odd function **because**

$$h(-x) = x^4 - 2x,$$

which is not equal to either $h(x)$ or $-h(x)$.

▶ **TRY EXERCISE 44, PAGE 60**

The following properties are a result of the tests for symmetry:

- The graph of an even function is symmetric with respect to the y-axis.

- The graph of an odd function is symmetric with respect to the origin.

The graph of f in **Figure 1.57** is symmetric with respect to the y-axis. It is the graph of an even function. The graph of g in **Figure 1.58** is symmetric with respect to the origin. It is the graph of an odd function. The graph of h in **Figure 1.59** is not symmetric with respect to the y-axis and is not symmetric with respect to the origin. It is neither an even nor an odd function.

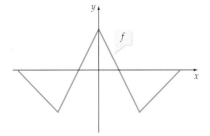

FIGURE 1.57
The graph of an even function is symmetric with respect to the y-axis.

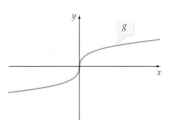

FIGURE 1.58
The graph of an odd function is symmetric with respect to the origin.

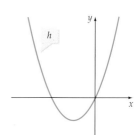

FIGURE 1.59
If the graph of a function is not symmetric to the y-axis or to the origin, then the function is neither even nor odd.

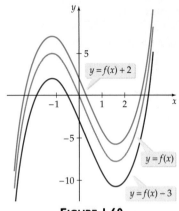

FIGURE 1.60

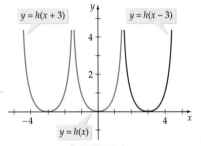

FIGURE 1.61

● TRANSLATIONS OF GRAPHS

The shape of a graph may be exactly the same as the shape of another graph; only their positions in the xy-plane may differ. For example, the graph of $y = f(x) + 2$ is the graph of $y = f(x)$ with each point moved up vertically 2 units. The graph of $y = f(x) - 3$ is the graph of $y = f(x)$ with each point moved down vertically 3 units. See **Figure 1.60.**

The graphs of $y = f(x) + 2$ and $y = f(x) - 3$ in **Figure 1.60** are called *vertical translations* of the graph of $y = f(x)$.

Vertical Translations

If f is a function and c is a positive constant, then the graph of

● $y = f(x) + c$ is the graph of $y = f(x)$ shifted up *vertically* c units.

● $y = f(x) - c$ is the graph of $y = f(x)$ shifted down *vertically* c units.

In **Figure 1.61,** the graph of $y = h(x + 3)$ is the graph of $y = h(x)$ with each point shifted to the left horizontally 3 units. Similarly, the graph of $y = h(x - 3)$ is the graph of $y = h(x)$ with each point shifted to the right horizontally 3 units.

The graphs of $y = h(x + 3)$ and $y = h(x - 3)$ in **Figure 1.61** are called *horizontal translations* of the graph of $y = h(x)$.

Horizontal Translations

If f is a function and c is a positive constant, then the graph of

● $y = f(x + c)$ is the graph of $y = f(x)$ shifted left *horizontally* c units.

● $y = f(x - c)$ is the graph of $y = f(x)$ shifted right *horizontally* c units.

 INTEGRATING TECHNOLOGY

A graphing calculator can be used to draw the graphs of a *family* of functions. For instance, $f(x) = x^2 + c$ constitutes a family of functions with **parameter** c. The only feature of the graph that changes is the value of c.

A graphing calculator can be used to produce the graphs of a family of curves for specific values of the parameter. The LIST feature of the calculator can be used. For instance, to graph $f(x) = x^2 + c$ for $c = -2, 0$, and 1, we will create a list and use that list to produce the family of curves. The keystrokes for a TI-83 calculator are given below.

2nd { -2 , 0 , 1 2nd } STO 2nd L1

Now use the Y= key to enter

Y= X x² + 2nd L1 ZOOM 6

Sample screens for the keystrokes and graphs are shown here. You can use similar keystrokes for Exercises 75–82 of this section.

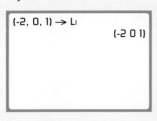

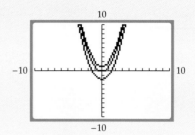

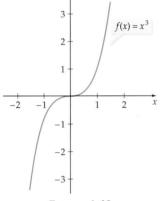

FIGURE 1.62

EXAMPLE 4   **Graph by Using Translations**

Use vertical and horizontal translations of the graph of $f(x) = x^3$, shown in **Figure 1.62,** to graph

a. $g(x) = x^3 - 2$ b. $h(x) = (x + 1)^3$

Solution

a.   The graph of $g(x) = x^3 - 2$ is the graph of $f(x) = x^3$ shifted down vertically 2 units. See **Figure 1.63.**

b.   The graph of $h(x) = (x + 1)^3$ is the graph of $f(x) = x^3$ shifted to the left horizontally 1 unit. See **Figure 1.64.**

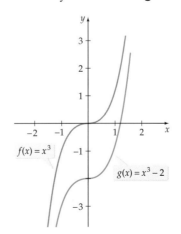

FIGURE 1.63

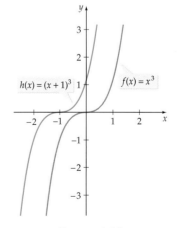

FIGURE 1.64

▶ TRY EXERCISE 58, PAGE 61

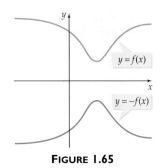

FIGURE 1.65

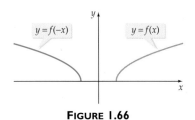

FIGURE 1.66

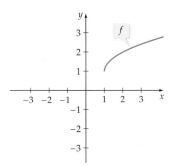

FIGURE 1.67

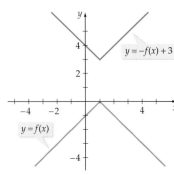

FIGURE 1.70

● **REFLECTIONS OF GRAPHS**

The graph of $y = -f(x)$ cannot be obtained from the graph of $y = f(x)$ by a combination of vertical and/or horizontal shifts. **Figure 1.65** illustrates that the graph of $y = -f(x)$ is the reflection of the graph of $y = f(x)$ across the x-axis.

The graph of $y = f(-x)$ is the reflection of the graph of $y = f(x)$ across the y-axis as, shown in **Figure 1.66.**

Reflections

The graph of

● $y = -f(x)$ is the graph of $y = f(x)$ reflected across the x-axis.

● $y = f(-x)$ is the graph of $y = f(x)$ reflected across the y-axis.

EXAMPLE 5   **Graph by Using Reflections**

Use reflections of the graph of $f(x) = \sqrt{x-1} + 1$, shown in **Figure 1.67,** to graph

a. $g(x) = -\left(\sqrt{x-1} + 1\right)$ **b.** $h(x) = \sqrt{-x-1} + 1$

Solution

a. Because $g(x) = -f(x)$, the graph of g is the graph of f reflected across the x-axis. See **Figure 1.68.**

b. Because $h(x) = f(-x)$, the graph of h is the graph of f reflected across the y-axis. See **Figure 1.69.**

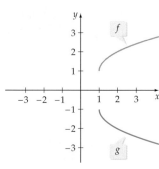

FIGURE 1.68

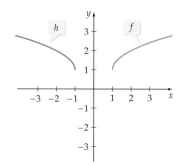

FIGURE 1.69

▶ **TRY EXERCISE 68, PAGE 62**

Some graphs of functions can be constructed by using a combination of translations and reflections. For instance, the graph of $y = -f(x) + 3$ in **Figure 1.70** was obtained by reflecting the graph of $y = f(x)$ in **Figure 1.70** across the x-axis and then shifting that graph up vertically 3 units.

● COMPRESSING AND STRETCHING OF GRAPHS

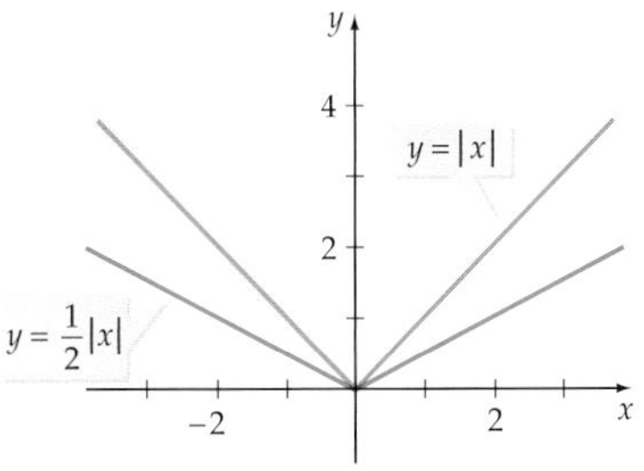

The graph of the equation $y = c \cdot f(x)$ for $c \neq 1$ vertically compresses or stretches the graph of $y = f(x)$. To determine the points on the graph of $y = c \cdot f(x)$, multiply each y-coordinate of the points on the graph of $y = f(x)$ by c. For example, **Figure 1.71** shows that the graph of $y = \frac{1}{2}|x|$ can be obtained by plotting points that have a y-coordinate that is one-half of the y-coordinate of those found on the graph of $y = |x|$.

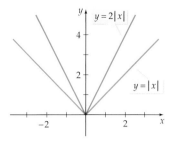

If $0 < c < 1$, then the graph of $y = c \cdot f(x)$ is obtained by *compressing* the graph of $y = f(x)$. **Figure 1.71** illustrates the vertical compressing of the graph of $y = |x|$ toward the x-axis to form the graph of $y = \frac{1}{2}|x|$.

If $c > 1$, then the graph of $y = c \cdot f(x)$ is obtained by *stretching* the graph of $y = f(x)$. For example, if $f(x) = |x|$, then we obtain the graph of

$$y = 2f(x) = 2|x|$$

by stretching the graph of f away from the x-axis. See **Figure 1.72**.

FIGURE 1.71

FIGURE 1.72

Vertical Stretching and Compressing of Graphs

If f is a function and c is a positive constant, then

- if $c > 1$, the graph of $y = c \cdot f(x)$ is the graph of $y = f(x)$ *stretched* vertically by a factor of c away from the x-axis.

- if $0 < c < 1$, the graph of $y = c \cdot f(x)$ is the graph of $y = f(x)$ *compressed* vertically by a factor of c toward the x-axis.

EXAMPLE 6   **Graph by Using Vertical Compressing and Shifting**

Graph: $H(x) = \frac{1}{4}|x| - 3$

Solution

The graph of $y = |x|$ has a V shape that has its lowest point at $(0, 0)$ and passes through $(4, 4)$ and $(-4, 4)$. The graph of $y = \frac{1}{4}|x|$ is a compressing of the graph of $y = |x|$. The y-coordinates of the ordered pairs $(0, 0)$, $(4, 1)$, and $(-4, 1)$ are obtained by multiplying the y-coordinates of the ordered pairs $(0, 0)$, $(4, 4)$, and $(-4, 4)$ by $\frac{1}{4}$. To find the points on the graph of H, we still need to subtract 3 from each y-coordinate. Thus the graph of H is a V shape that has its lowest point at $(0, -3)$ and passes through $(4, -2)$ and $(-4, -2)$. See **Figure 1.73**.

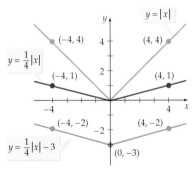

FIGURE 1.73

▶ TRY EXERCISE 70, PAGE 62

Some functions can be graphed by using a horizontal compressing or stretching of a given graph. The procedure makes use of the following concept.

Horizontal Compressing and Stretching of Graphs

If f is a function and c is a positive constant, then

- if $c > 1$, the graph of $y = f(c \cdot x)$ is the graph of $y = f(x)$ *compressed* horizontally by a factor of $\dfrac{1}{c}$ toward the y-axis.

- if $0 < c < 1$, the graph of $y = f(c \cdot x)$ is the graph of $y = f(x)$ *stretched* horizontally by a factor of $\dfrac{1}{c}$ away from the y-axis.

If the point (x, y) is on the graph of $y = f(x)$, then the graph of $y = f(cx)$ will contain the point $\left(\dfrac{1}{c} x, y \right)$.

EXAMPLE 7 **Graph by Using Horizontal Compressing and Stretching**

Use the graph of $y = f(x)$ shown in **Figure 1.74** to graph

a. $y = f(2x)$ **b.** $y = f\left(\dfrac{1}{3} x \right)$

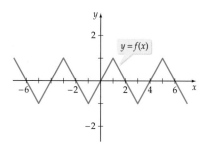

FIGURE 1.74

Solution

a. Because $2 > 1$, the graph of $y = f(2x)$ is a horizontal compression of the graph of $y = f(x)$ by a factor of $\dfrac{1}{2}$. For example, the point $(2, 0)$ on the graph of $y = f(x)$ becomes the point $(1, 0)$ on the graph of $y = f(2x)$. See **Figure 1.75.**

b. Since $0 < \dfrac{1}{3} < 1$, the graph of $y = f\left(\dfrac{1}{3} x \right)$ is a horizontal stretching of the graph of $y = f(x)$ by a factor of 3. For example, the point $(1, 1)$ on the

graph of $y = f(x)$ becomes the point $(3, 1)$ on the graph of $y = f\left(\dfrac{1}{3}x\right)$.
See **Figure 1.76.**

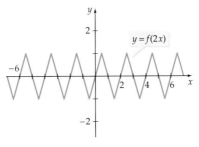

| FIGURE 1.75 | FIGURE 1.76 |

 TRY EXERCISE 72, PAGE 62

TOPICS FOR DISCUSSION

1. Discuss the meaning of symmetry of a graph with respect to a line. How do you determine whether a graph has symmetry with respect to the x-axis? with respect to the y-axis?

2. Discuss the meaning of symmetry of a graph with respect to a point. How do you determine whether a graph has symmetry with respect to the origin?

3. What does it mean to reflect a graph across the x-axis or across the y-axis?

4. Explain how the graphs of $y_1 = 2x^3 - x^2$ and $y_2 = 2(-x)^3 - (-x)^2$ are related.

5. Given the graph of $y_3 = f(x)$, explain how to obtain the graph of $y_4 = f(x - 3) + 1$.

6. The graph of the *step function* $y_5 = [\![x]\!]$ has steps that are 1 unit wide. Determine how wide the steps are in the graph of $y_6 = \left[\!\left[\dfrac{1}{3}x\right]\!\right]$.

EXERCISE SET 1.4

In Exercises I to 6, plot the image of the given point with respect to
a. **the y-axis. Label this point A.**
b. **the x-axis. Label this point B.**
c. **the origin. Label this point C.**

1. $P(5, -3)$ **2.** $Q(-4, 1)$ **3.** $R(-2, 3)$

4. $S(-5, 3)$ **5.** $T(-4, -5)$ **6.** $U(5, 1)$

In Exercises 7 and 8, sketch a graph that is symmetric to the given graph with respect to the x-axis.

7.
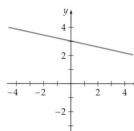

8.
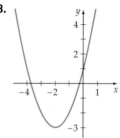

In Exercises 9 and 10, sketch a graph that is symmetric to the given graph with respect to the y-axis.

9.
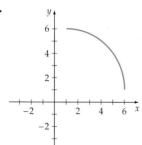

10.
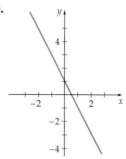

In Exercises 11 and 12, sketch a graph that is symmetric to the given graph with respect to the origin.

11.
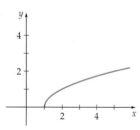

12.
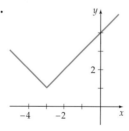

In Exercises 13 to 21, determine whether the graph of each equation is symmetric with respect to the a. **x-axis,** b. **y-axis.**

13. $y = 2x^2 - 5$ ▶ **14.** $x = 3y^2 - 7$ **15.** $y = x^3 + 2$

16. $y = x^5 - 3x$ **17.** $x^2 + y^2 = 9$ **18.** $x^2 - y^2 = 10$

19. $x^2 = y^4$ **20.** $xy = 8$ **21.** $|x| - |y| = 6$

In Exercises 22 to 30, determine whether the graph of each equation is symmetric with respect to the origin.

22. $y = x + 1$ **23.** $y = 3x - 2$ ▶ **24.** $y = x^3 - x$

25. $y = -x^3$ **26.** $y = \dfrac{9}{x}$ **27.** $x^2 + y^2 = 10$

28. $x^2 - y^2 = 4$ **29.** $y = \dfrac{x}{|x|}$ **30.** $|y| = |x|$

In Exercises 31 to 42, graph the given equations. Label each intercept. Use the concept of symmetry to confirm that the graph is correct.

31. $y = x^2 - 1$ **32.** $x = y^2 - 1$

33. $y = x^3 - x$ **34.** $y = -x^3$

35. $xy = 4$ **36.** $xy = -8$

37. $y = 2|x - 4|$ **38.** $y = |x - 2| - 1$

39. $y = (x - 2)^2 - 4$ **40.** $y = (x - 1)^2 - 4$

41. $y = x - |x|$ **42.** $|y| = |x|$

In Exercises 43 to 56, identify whether the given function is an even function, an odd function, or neither.

43. $g(x) = x^2 - 7$ ▶ **44.** $h(x) = x^2 + 1$

45. $F(x) = x^5 + x^3$ **46.** $G(x) = 2x^5 - 10$

47. $H(x) = 3|x|$ **48.** $T(x) = |x| + 2$

49. $f(x) = 1$ **50.** $k(x) = 2 + x + x^2$

51. $r(x) = \sqrt{x^2 + 4}$ **52.** $u(x) = \sqrt{3 - x^2}$

53. $s(x) = 16x^2$ **54.** $v(x) = 16x^2 + x$

55. $w(x) = 4 + \sqrt[3]{x}$

56. $z(x) = \dfrac{x^3}{x^2 + 1}$

57. Use the graph of f to sketch the graph of

    a. $y = f(x) + 3$

    b. $y = f(x - 3)$

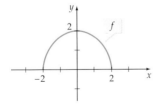

▶ **58.** Use the graph of g to sketch the graph of

    a. $y = g(x) - 2$

    b. $y = g(x - 3)$

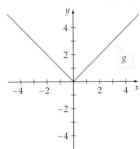

59. Use the graph of f to sketch the graph of

    a. $y = f(x + 2)$

    b. $y = f(x) + 2$

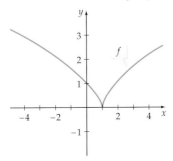

60. Use the graph of g to sketch the graph of

    a. $y = g(x - 1)$

    b. $y = g(x) - 1$

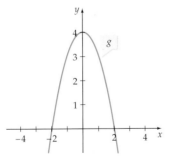

61. Let f be a function such that $f(-2) = 5$, $f(0) = -2$, and $f(1) = 0$. Give the coordinates of three points on the graph of

    a. $y = f(x + 3)$

    b. $y = f(x) + 1$

62. Let g be a function such that $g(-3) = -1$, $g(1) = -3$, and $g(4) = 2$. Give the coordinates of three points on the graph of

    a. $y = g(x - 2)$

    b. $y = g(x) - 2$

63. Use the graph of f to sketch the graph of

    a. $y = f(-x)$

    b. $y = -f(x)$

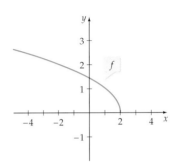

64. Use the graph of g to sketch the graph of

    a. $y = -g(x)$

    b. $y = g(-x)$

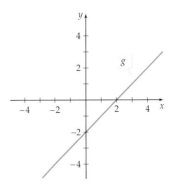

65. Let f be a function such that $f(-1) = 3$ and $f(2) = -4$. Give the coordinates of two points on the graph of

    a. $y = f(-x)$

    b. $y = -f(x)$

66. Let g be a function such that $g(4) = -5$ and $g(-3) = 2$. Give the coordinates of two points on the graph of

    a. $y = -g(x)$

    b. $y = g(-x)$.

67. Use the graph of F to sketch the graph of

a. $y = -F(x)$ **b.** $y = F(-x)$

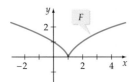

▶ **68.** Use the graph of E to sketch the graph of

a. $y = -E(x)$ **b.** $y = E(-x)$

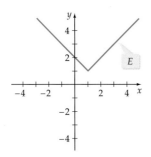

69. Use the graph of $m(x) = x^2 - 2x - 3$ to sketch the graph of $y = -\dfrac{1}{2}m(x) + 3$.

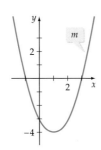

▶ **70.** Use the graph of $n(x) = -x^2 - 2x + 8$ to sketch the graph of $y = \dfrac{1}{2}n(x) + 1$.

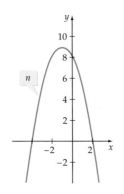

71. Use the graph of $y = f(x)$ to sketch the graph of

a. $y = f(2x)$ **b.** $y = f\left(\dfrac{1}{3}x\right)$

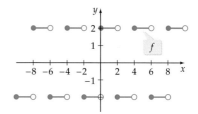

▶ **72.** Use the graph of $y = g(x)$ to sketch the graph of

a. $y = g(2x)$ **b.** $y = g\left(\dfrac{1}{2}x\right)$

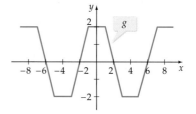

73. Use the graph of $y = h(x)$ to sketch the graph of

a. $y = h(2x)$ **b.** $y = h\left(\dfrac{1}{2}x\right)$

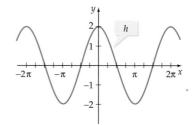

74. Use the graph of $y = j(x)$ to sketch the graph of

a. $y = j(2x)$ **b.** $y = j\left(\dfrac{1}{3}x\right)$

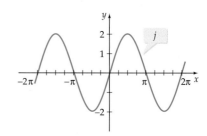

In Exercises 75 to 82, use a graphing utility.

75. On the same coordinate axes, graph
$$G(x) = \sqrt[3]{x} + c$$
for $c = 0, -1$, and 3.

76. On the same coordinate axes, graph
$$H(x) = \sqrt[3]{x + c}$$
for $c = 0, -1$, and 3.

77. On the same coordinate axes, graph
$$J(x) = |2(x + c) - 3| - |x + c|$$
for $c = 0, -1$, and 2.

78. On the same coordinate axes, graph
$$K(x) = |x - 1| - |x| + c$$
for $c = 0, -1$, and 2.

79. On the same coordinate axes, graph
$$L(x) = cx^2$$
for $c = 1, \frac{1}{2}$, and 2.

80. On the same coordinate axes, graph
$$M(x) = c\sqrt{x^2 - 4}$$
for $c = 1, \frac{1}{3}$, and 3.

81. On the same coordinate axes, graph
$$S(x) = c(|x - 1| - |x|)$$
for $c = 1, \frac{1}{4}$, and 4.

82. On the same coordinate axes, graph
$$T(x) = c\left(\frac{x}{|x|}\right)$$
for $c = 1, \frac{2}{3}$, and $\frac{3}{2}$.

83. Graph $V(x) = [\![cx]\!], 0 \le x \le 6$, for each value of c.

   a. $c = 1$ **b.** $c = \frac{1}{2}$ **c.** $c = 2$

84. Graph $W(x) = [\![cx]\!] - cx, 0 \le x \le 6$, for each value of c.

   a. $c = 1$ **b.** $c = \frac{1}{3}$ **c.** $c = 3$

CONNECTING CONCEPTS

85. Use the graph of $f(x) = 2/(x^2 + 1)$ to determine an equation for the graphs shown in **a.** and **b.**

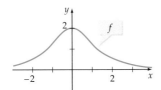

a.
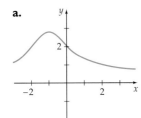

b.
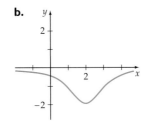

86. Use the graph of $f(x) = x\sqrt{2 + x}$ to determine an equation for the graphs shown in **a.** and **b.**

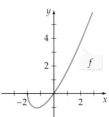

a.
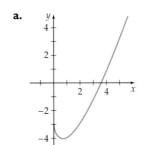

b.
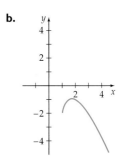

—— *PREPARE FOR SECTION 1.5* ——

87. Subtract: $(2x^2 + 3x - 4) - (x^2 + 3x - 5)$

88. Multiply: $(3x^2 - x + 2)(2x - 3)$

In Exercises 89 and 90, find each of the following for $f(x) = 2x^2 - 5x + 2$.

89. $f(3a)$ [1.3]

90. $f(2 + h)$ [1.3]

In Exercises 91 and 92, find the domain of each function.

91. $F(x) = \dfrac{x}{x - 1}$ [1.3]

92. $r(x) = \sqrt{2x - 8}$ [1.3]

—— *PROJECTS* ——

1. **DIRICHLET FUNCTION** We owe our present-day definition of a function to the German mathematician Peter Gustav Dirichlet (1805–1859). He created the following unusual function, which is now known as the *Dirichlet function*.

$$f(x) = \begin{cases} 0, & \text{if } x \text{ is a rational number} \\ 1, & \text{if } x \text{ is an irrational number} \end{cases}$$

Answer the following questions about the Dirichlet function.

a. What is its domain?   **b.** What is its range?

c. What are its x-intercepts?

d. What is its y-intercept?

e. Is it an even or an odd function?

f. Explain why a graphing calculator cannot be used to produce an accurate graph of the function.

g. Write a sentence or two that describes its graph.

2.   **ISOLATED POINT** Consider the function given by

$$y = \sqrt{(x - 1)^2(x - 2)} + 1$$

Verify that the point $(1, 1)$ is a solution of the equation. Now use a graphing utility to graph the function. Does your graph include the isolated point at $(1, 1)$, as shown at the right? If the graphing utility you used failed to include the point $(1, 1)$, explain at least one reason for the omission of this isolated point.

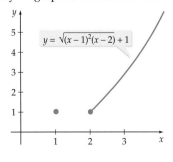

$y = \sqrt{(x-1)^2(x-2)} + 1$

3. 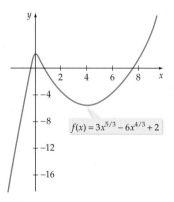 **A LINE WITH A HOLE** The function

$$f(x) = \dfrac{(x - 2)(x + 1)}{(x - 2)}$$

graphs as a line with a y-intercept of 1, a slope of 1, and a hole at $(2, 3)$. Use a graphing utility to graph f. Explain why a graphing utility might not show the hole at $(2, 3)$.

4. **FINDING A COMPLETE GRAPH** Use a graphing utility to graph the function $f(x) = 3x^{5/3} - 6x^{4/3} + 2$ for $-2 \le x \le 10$. Compare your graph with the graph below. Does your graph include the part to the left of the y-axis? If not, how might you enter the function in such a way that the graphing utility you used would include this part?

$f(x) = 3x^{5/3} - 6x^{4/3} + 2$

SECTION 1.5

THE ALGEBRA OF FUNCTIONS

- OPERATIONS ON FUNCTIONS
- THE DIFFERENCE QUOTIENT
- COMPOSITION OF FUNCTIONS

● OPERATIONS ON FUNCTIONS

Functions can be defined in terms of other functions. For example, the function defined by $h(x) = x^2 + 8x$ is the sum of

$$f(x) = x^2 \quad \text{and} \quad g(x) = 8x$$

Thus, if we are given any two functions f and g, we can define the four new functions $f + g$, $f - g$, fg, and $\dfrac{f}{g}$ as follows.

Operations on Functions

For all values of x for which both $f(x)$ and $g(x)$ are defined, we define the following functions.

Sum	$(f + g)(x) = f(x) + g(x)$
Difference	$(f - g)(x) = f(x) - g(x)$
Product	$(fg)(x) = f(x) \cdot g(x)$
Quotient	$\left(\dfrac{f}{g}\right)(x) = \dfrac{f(x)}{g(x)}, \quad g(x) \neq 0$

Domain of $f + g$, $f - g$, fg, f/g

For the given functions f and g, the domains of $f + g$, $f - g$, and $f \cdot g$ consist of all real numbers formed by the intersection of the domains of f and g. The domain of $\dfrac{f}{g}$ is the set of all real numbers formed by the intersection of the domains of f and g, except for those real numbers x such that $g(x) = 0$.

EXAMPLE I   Determine the Domain of a Function

If $f(x) = \sqrt{x - 1}$ and $g(x) = x^2 - 4$, find the domain of $f + g$, of $f - g$, of fg, and of $\dfrac{f}{g}$.

Solution

Note that f has the domain $\{x \mid x \geq 1\}$ and g has the domain of all real numbers. Therefore, the domain of $f + g$, $f - g$, and fg is $\{x \mid x \geq 1\}$. Because $g(x) = 0$ when $x = -2$ or $x = 2$, neither -2 nor 2 is in the domain of $\dfrac{f}{g}$. The domain of $\dfrac{f}{g}$ is $\{x \mid x \geq 1 \text{ and } x \neq 2\}$.

▶ **TRY EXERCISE 10, PAGE 73**

EXAMPLE 2   **Evaluate Functions**

Let $f(x) = x^2 - 9$ and $g(x) = 2x + 6$. Find

a. $(f + g)(5)$ b. $(fg)(-1)$ c. $\left(\dfrac{f}{g}\right)(4)$

Solution

a. $(f + g)(x) = f(x) + g(x) = (x^2 - 9) + (2x + 6) = x^2 + 2x - 3$
    Therefore, $(f + g)(5) = (5)^2 + 2(5) - 3 = 25 + 10 - 3 = 32.$

b. $(fg)(x) = f(x) \cdot g(x) = (x^2 - 9)(2x + 6) = 2x^3 + 6x^2 - 18x - 54$
    Therefore, $(fg)(-1) = 2(-1)^3 + 6(-1)^2 - 18(-1) - 54$
    $$= -2 + 6 + 18 - 54 = -32.$$

c. $\left(\dfrac{f}{g}\right)(x) = \dfrac{f(x)}{g(x)} = \dfrac{x^2 - 9}{2x + 6} = \dfrac{(x + 3)(x - 3)}{2(x + 3)} = \dfrac{x - 3}{2}, \quad x \neq -3$
    Therefore, $\left(\dfrac{f}{g}\right)(4) = \dfrac{4 - 3}{2} = \dfrac{1}{2}.$

▶ **TRY EXERCISE 14, PAGE 73**

● **THE DIFFERENCE QUOTIENT**

take note

The difference quotient is an important concept that plays a fundamental role in calculus.

The expression

$$\frac{f(x + h) - f(x)}{h}, \quad h \neq 0$$

is called the **difference quotient** of f. It enables us to study the manner in which a function changes in value as the independent variable changes.

EXAMPLE 3   **Determine a Difference Quotient**

Determine the difference quotient of $f(x) = x^2 + 7$.

Solution

$\dfrac{f(x + h) - f(x)}{h} = \dfrac{[(x + h)^2 + 7] - [x^2 + 7]}{h}$ • Apply the difference quotient.

$$= \frac{[x^2 + 2xh + h^2 + 7] - [x^2 + 7]}{h}$$

$$= \frac{x^2 + 2xh + h^2 + 7 - x^2 - 7}{h}$$

$$= \frac{2xh + h^2}{h} = \frac{h(2x + h)}{h} = 2x + h$$

▶ **TRY EXERCISE 30, PAGE 73**

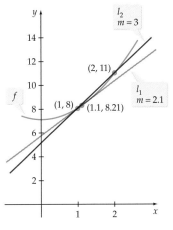

FIGURE 1.77

The difference quotient $2x + h$ of $f(x) = x^2 + 7$ from Example 3 is the slope of the secant line through the points

$$(x, f(x)) \quad \text{and} \quad (x + h, f(x + h))$$

For instance, let $x = 1$ and $h = 1$. Then the difference quotient is

$$2x + h = 2(1) + 1 = 3$$

This is the slope of the secant line l_2 through $(1, 8)$ and $(2, 11)$, as shown in **Figure 1.77.** If we let $x = 1$ and $h = 0.1$, then the difference quotient is

$$2x + h = 2(1) + 0.1 = 2.1$$

This is the slope of the secant line l_1 through $(1, 8)$ and $(1.1, 8.21)$.
The difference quotient

$$\frac{f(x + h) - f(x)}{h}$$

can be used to compute *average velocities*. In such cases it is traditional to replace f with s (for distance), the variable x with the variable a (for the time at the start of an observed interval of time), and the variable h with Δt (read as "delta t"), where Δt is the difference between the time at the end of an interval and the time at the start of the interval. For example, if an experiment is observed over the time interval from $t = 3$ seconds to $t = 5$ seconds, then the time interval is denoted as $[3, 5]$ with $a = 3$ and $\Delta t = 5 - 3 = 2$. Thus if the distance traveled by a ball that rolls down a ramp is given by $s(t)$, where t is the time in seconds after the ball is released (see **Figure 1.78**), then the average velocity of the ball over the interval $t = a$ to $t = a + \Delta t$ is the difference quotient

$$\frac{s(a + \Delta t) - s(a)}{\Delta t}$$

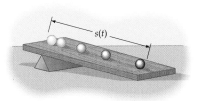

FIGURE 1.78

EXAMPLE 4   **Evaluate Average Velocities**

The distance traveled by a ball rolling down a ramp is given by $s(t) = 4t^2$, where t is the time in seconds after the ball is released, and $s(t)$ is measured in feet. Evaluate the average velocity of the ball for each time interval.

a. $[3, 5]$ b. $[3, 4]$ c. $[3, 3.5]$ d. $[3, 3.01]$

Solution

a. In this case, $a = 3$ and $\Delta t = 2$. Thus the average velocity over this interval is

$$\frac{s(a + \Delta t) - s(a)}{\Delta t} = \frac{s(3 + 2) - s(3)}{2} = \frac{s(5) - s(3)}{2} = \frac{100 - 36}{2}$$

$$= 32 \text{ feet per second}$$

b. Let $a = 3$ and $\Delta t = 4 - 3 = 1$.

$$\frac{s(a + \Delta t) - s(a)}{\Delta t} = \frac{s(3 + 1) - s(3)}{1} = \frac{s(4) - s(3)}{1} = \frac{64 - 36}{1}$$

$$= 28 \text{ feet per second}$$

Continued ▶

   c.   Let $a = 3$ and $\Delta t = 3.5 - 3 = 0.5$.

$$\frac{s(a + \Delta t) - s(a)}{\Delta t} = \frac{s(3 + 0.5) - s(3)}{0.5} = \frac{49 - 36}{0.5} = 26 \text{ feet per second}$$

   d.   Let $a = 3$ and $\Delta t = 3.01 - 3 = 0.01$.

$$\frac{s(a + \Delta t) - s(a)}{\Delta t} = \frac{s(3 + 0.01) - s(3)}{0.01} = \frac{36.2404 - 36}{0.01}$$

$$= 24.04 \text{ feet per second}$$

▶ **TRY EXERCISE 72, PAGE 75**

● COMPOSITION OF FUNCTIONS

Composition of functions is another way in which functions can be combined. This method of combining functions uses the output of one function as the input for a second function.

   Suppose that the spread of oil from a leak in a tanker can be approximated by a circle with the tanker at its center. The radius r (in feet) of the spill t hours after the leak begins is given by $r(t) = 150\sqrt{t}$. The area of the spill is the area of a circle and is given by the formula $A(r) = \pi r^2$. To find the area of the spill 4 hours after the leak begins, we first find the radius of the spill and then use that number to find the area of the spill.

$$r(t) = 150\sqrt{t} \qquad\qquad\qquad A(r) = \pi r^2$$
$$r(4) = 150\sqrt{4} \quad \bullet\, t = 4 \text{ hours} \qquad A(300) = \pi(300^2) \quad \bullet\, r = 300 \text{ feet}$$
$$= 150(2) \qquad\qquad\qquad\qquad = 90{,}000\pi$$
$$= 300 \qquad\qquad\qquad\qquad\quad \approx 283{,}000$$

The area of the spill after 4 hours is approximately 283,000 square feet.

   There is an alternative way to solve this problem. Because the area of the spill depends on the radius and the radius depends on the time, there is a relationship between area and time. We can determine this relationship by evaluating the formula for the area of a circle using $r(t) = 150\sqrt{t}$. This will give the area of the spill as a function of time.

$$A(r) = \pi r^2$$
$$A[r(t)] = \pi[r(t)]^2 \qquad \bullet \textbf{ Replace } r \textbf{ by } r(t).$$
$$= \pi\left[150\sqrt{t}\,\right]^2 \qquad \bullet\, r(t) = 150\sqrt{t}$$
$$A(t) = 22{,}500\pi t \qquad \bullet \textbf{ Simplify.}$$

The area of the spill as a function of time is $A(t) = 22{,}500\pi t$. To find the area of the oil spill after 4 hours, evaluate this function at $t = 4$.

$$A(t) = 22{,}500\pi t$$
$$A(4) = 22{,}500\pi(4) \qquad \bullet\, t = 4 \text{ hours}$$
$$= 90{,}000\pi$$
$$\approx 283{,}000$$

This is the same result we calculated earlier.

The function $A(t) = 22{,}500\pi t$ is referred to as the *composition* of A with r. The notation $A \circ r$ is used to denote this composition of functions. That is,

$$(A \circ r)(t) = 22{,}500\pi t$$

Definition of the Composition of Two Functions

Let f and g be two functions such that $g(x)$ is in the domain of f for all x in the domain of g. Then the composition of the two functions, denoted by $f \circ g$, is the function whose value at x is given by $(f \circ g)(x) = f[g(x)]$.

The function defined by $(f \circ g)(x)$ is also called the *composite* of f and g. We read $(f \circ g)(x)$ as "f circle g of x" and $f[g(x)]$ as "f of g of x."

Consider the functions $f(x) = 2x - 1$ and $g(x) = x^2 - 3$. The expression $(f \circ g)(-1)$ (or, equivalently, $f[g(-1)]$) means to evaluate the function f at $g(-1)$.

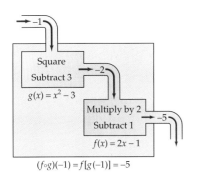

$(f \circ g)(-1) = f[g(-1)] = -5$

FIGURE 1.79

$$g(x) = x^2 - 3$$
$$g(-1) = (-1)^2 - 3 \qquad \text{• Evaluate } g \text{ at } -1.$$
$$= -2$$

$$f(x) = 2x - 1$$
$$f(-2) = 2(-2) - 1 = -5 \qquad \text{• Evaluate } f \text{ at } g(-1) = -2.$$

A graphical depiction of the composition $(f \circ g)(-1)$ would look something like **Figure 1.79.**

The requirement in the definition of the composition of two functions that $g(x)$ be in the domain of f for all x in the domain of g is important. For instance, let

$$f(x) = \frac{1}{x - 1} \qquad \text{and} \qquad g(x) = 3x - 5$$

When $x = 2$,

$$g(2) = 3(2) - 5 = 1$$
$$f[g(2)] = f(1) = \frac{1}{1 - 1} = \frac{1}{0} \qquad \text{• Undefined}$$

In this case, $g(2)$ is not in the domain of f. Thus the composition $(f \circ g)(x)$ is not defined at 2.

We can find a general expression for $f[g(x)]$ by evaluating f at $g(x)$. For instance, using $f(x) = 2x - 1$ and $g(x) = x^2 - 3$ as in **Figure 1.79,** we have

$$f(x) = 2x - 1$$
$$f[g(x)] = 2[g(x)] - 1 \qquad \text{• Replace } x \text{ by } g(x).$$
$$= 2[x^2 - 3] - 1 \qquad \text{• Replace } g(x) \text{ by } x^2 - 3.$$
$$= 2x^2 - 7 \qquad \text{• Simplify.}$$

In general, the composition of functions is not a commutative operation. That is, $(f \circ g)(x) \neq (g \circ f)(x)$. To verify this, we will compute the composition

$(g \circ f)(x) = g[f(x)]$, again using the functions $f(x) = 2x - 1$ and $g(x) = x^2 - 3$.

$$g(x) = x^2 - 3$$
$$g[f(x)] = [f(x)]^2 - 3 \qquad \text{• Replace } x \text{ by } f(x).$$
$$= [2x - 1]^2 - 3 \qquad \text{• Replace } f(x) \text{ by } 2x - 1.$$
$$= 4x^2 - 4x - 2 \qquad \text{• Simplify.}$$

Thus $f[g(x)] = 2x^2 - 7$, which is not equal to $g[f(x)] = 4x^2 - 4x - 2$. Therefore, $(f \circ g)(x) \neq (g \circ f)(x)$ and composition is not a commutative operation.

❓ QUESTION Let $f(x) = x - 1$ and $g(x) = x + 1$. Then $f[g(x)] = g[f(x)]$. (You should verify this statement.) Does this contradict the statement we made that composition is not a commutative operation?

EXAMPLE 5   **Form Composite Functions**

If $f(x) = x^2 - 3x$ and $g(x) = 2x + 1$, find

a. $(g \circ f)$ b. $(f \circ g)$

Solution

a.
$$(g \circ f) = g[f(x)] = 2(f(x)) + 1 \qquad \text{• Substitute } f(x) \text{ for } x \text{ in } g.$$
$$= 2(x^2 - 3x) + 1 \qquad \text{• } f(x) = x^2 - 3x$$
$$= 2x^2 - 6x + 1$$

b.
$$(f \circ g) = f[g(x)] = (g(x))^2 - 3(g(x)) \qquad \text{• Substitute } g(x) \text{ for } x \text{ in } f.$$
$$= (2x + 1)^2 - 3(2x + 1) \qquad \text{• } g(x) = 2x + 1$$
$$= 4x^2 - 2x - 2$$

▶ **TRY EXERCISE 38, PAGE 74**

Note that in this example $(f \circ g) \neq (g \circ f)$. In general, the composition of functions is not a commutative operation.

Caution   Some care must be used when forming the composition of functions. For instance, if $f(x) = x + 1$ and $g(x) = \sqrt{x - 4}$, then

$$(g \circ f)(2) = g[f(2)] = g(3) = \sqrt{3 - 4} = \sqrt{-1}$$

which is not a real number. We can avoid this problem by imposing suitable restrictions on the domain of f so that the range of f is part of the domain of g. If the

---

❓ ANSWER   No. When we say that composition is not a commutative operation, we mean that generally, given any two functions, $(f \circ g)(x) \neq (g \circ f)(x)$. However, there may be particular instances in which $(f \circ g)(x) = (g \circ f)(x)$. It turns out that these particular instances are quite important, as we shall see later.

domain of f is restricted to $[3, \infty)$, then the range of f is $[4, \infty)$. But this is precisely the domain of g. Note that $2 \notin [3, \infty)$, and thus we avoid the problem of $(g \circ f)(2)$ not being a real number.

To evaluate $(f \circ g)(c)$ for some constant c, you can use either of the following methods.

Method 1   First evaluate $g(c)$. Then substitute this result for x in $f(x)$.

Method 2   First determine $f[g(x)]$ and then substitute c for x.

EXAMPLE 6   Evaluate a Composite Function

Evaluate $(f \circ g)(3)$, where $f(x) = 2x - 7$ and $g(x) = x^2 + 4$.

Solution

Method 1 $\begin{aligned}(f \circ g)(3) &= f[g(3)] \\ &= f[(3)^2 + 4] \qquad \text{• Evaluate } g(3). \\ &= f(13) \\ &= 2(13) - 7 = 19 \qquad \text{• Substitute 13 for } x \text{ in } f.\end{aligned}$

Method 2 $\begin{aligned}(f \circ g)(x) &= 2[g(x)] - 7 \qquad \text{• Form } f[g(x)]. \\ &= 2[x^2 + 4] - 7 \\ &= 2x^2 + 1 \\ (f \circ g)(3) &= 2(3)^2 + 1 = 19 \qquad \text{• Substitute 3 for } x.\end{aligned}$

▶ **TRY EXERCISE 50, PAGE 74**

---

take note

In Example 6, both Method 1 and Method 2 produce the same result. Although Method 2 is longer, it is the better method if you must evaluate $(f \circ g)(x)$ for several values of x.

Figures 1.80 and **1.81** graphically illustrate the difference between Method 1 and Method 2.

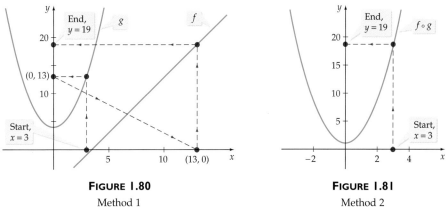

FIGURE 1.80
Method 1

FIGURE 1.81
Method 2

EXAMPLE 7 **Use a Composite Function to Solve an Application**

A graphic artist has drawn a 3-inch by 2-inch rectangle on a computer screen. The artist has been scaling the size of the rectangle for t seconds in such a way that the upper right corner of the original rectangle is moving to the right at the rate of 0.5 inch per second and downward at the rate of 0.2 inch per second. See **Figure 1.82.**

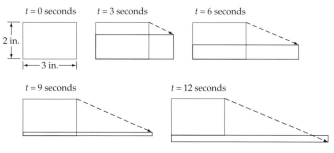

FIGURE 1.82

a. Write the length l and the width w of the scaled rectangles as functions of t.

b. Write the area A of the scaled rectangle as a function of t.

c. Find the intervals on which A is an increasing function for $0 \leq t \leq 14$. Also find the intervals on which A is a decreasing function.

d. Find the value of t (where $0 \leq t \leq 14$) that maximizes $A(t)$.

Solution

a. Because *distance* = *rate* · *time*, we see that the change in l is given by $0.5t$. Therefore, the length at any time t is $l = 3 + 0.5t$. For $0 \leq t \leq 10$, the width is given by $w = 2 - 0.2t$. For $10 < t \leq 14$, the width is $w = -2 + 0.2t$. In either case the width can be determined by finding $w = |2 - 0.2t|$. (The absolute value symbol is needed to keep the width positive for $10 < t \leq 14$.)

b. $A = lw = (3 + 0.5t)|2 - 0.2t|$

c. Use a graphing utility to determine that A is increasing on $[0, 2]$ and on $[10, 14]$ and that A is decreasing on $[2, 10]$. See **Figure 1.83.**

d. The highest point on the graph of A occurs when $t = 14$ seconds. See **Figure 1.83.**

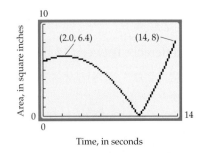

$A = (3 + 0.5t)|2 - 0.2t|$

FIGURE 1.83

▶ **TRY EXERCISE 66, PAGE 74**

You may be inclined to think that if the area of a rectangle is decreasing, then its perimeter is also decreasing, but this is not always the case. For example, the area of the scaled rectangle in Example 7 was shown to decrease on $[2, 10]$ even though its perimeter is always increasing. See Exercise 68 on page 75.

TOPICS FOR DISCUSSION

1. The domain of $f + g$ consists of all real numbers formed by the *union* of the domain of f and the domain of g. Do you agree?

2. Given $f(x) = 3x - 2$ and $g(x) = \dfrac{1}{3}x + \dfrac{2}{3}$, determine $f \circ g$ and $g \circ f$. Does this show that composition of functions is a commutative operation?

3. A tutor states that the difference quotient of $f(x) = x^2$ and the difference quotient of $g(x) = x^2 + 4$ are the same. Do you agree?

4. A classmate states that the difference quotient of any linear function $f(x) = mx + b$ is always m. Do you agree?

5. When we use a difference quotient to determine an average velocity, we generally replace the variable h with the variable Δt. What does Δt represent?

EXERCISE SET 1.5

In Exercises 1 to 12, use the given functions f and g to find $f + g, f - g, fg,$ and $\dfrac{f}{g}$. State the domain of each.

1. $f(x) = x^2 - 2x - 15$, $g(x) = x + 3$

2. $f(x) = x^2 - 25$, $g(x) = x - 5$

3. $f(x) = 2x + 8$, $g(x) = x + 4$

4. $f(x) = 5x - 15$, $g(x) = x - 3$

5. $f(x) = x^3 - 2x^2 + 7x$, $g(x) = x$

6. $f(x) = x^2 - 5x - 8$, $g(x) = -x$

7. $f(x) = 2x^2 + 4x - 7$, $g(x) = 2x^2 + 3x - 5$

8. $f(x) = 6x^2 + 10$, $g(x) = 3x^2 + x - 10$

9. $f(x) = \sqrt{x - 3}$, $g(x) = x$

▶ **10.** $f(x) = \sqrt{x - 4}$, $g(x) = -x$

11. $f(x) = \sqrt{4 - x^2}$, $g(x) = 2 + x$

12. $f(x) = \sqrt{x^2 - 9}$, $g(x) = x - 3$

In Exercises 13 to 28, evaluate the indicated function, where $f(x) = x^2 - 3x + 2$ and $g(x) = 2x - 4$.

13. $(f + g)(5)$

▶ **14.** $(f + g)(-7)$

15. $(f + g)\left(\dfrac{1}{2}\right)$

16. $(f + g)\left(\dfrac{2}{3}\right)$

17. $(f - g)(-3)$

18. $(f - g)(24)$

19. $(f - g)(-1)$

20. $(f - g)(0)$

21. $(fg)(7)$

22. $(fg)(-3)$

23. $(fg)\left(\dfrac{2}{5}\right)$

24. $(fg)(-100)$

25. $\left(\dfrac{f}{g}\right)(-4)$

26. $\left(\dfrac{f}{g}\right)(11)$

27. $\left(\dfrac{f}{g}\right)\left(\dfrac{1}{2}\right)$

28. $\left(\dfrac{f}{g}\right)\left(\dfrac{1}{4}\right)$

In Exercises 29 to 36, find the difference quotient of the given function.

29. $f(x) = 2x + 4$

▶ **30.** $f(x) = 4x - 5$

31. $f(x) = x^2 - 6$

32. $f(x) = x^2 + 11$

33. $f(x) = 2x^2 + 4x - 3$

34. $f(x) = 2x^2 - 5x + 7$

35. $f(x) = -4x^2 + 6$

36. $f(x) = -5x^2 - 4x$

In Exercises 37 to 48, find $g \circ f$ and $f \circ g$ for the given functions f and g.

37. $f(x) = 3x + 5, \quad g(x) = 2x - 7$

▶ **38.** $f(x) = 2x - 7, \quad g(x) = 3x + 2$

39. $f(x) = x^2 + 4x - 1, \quad g(x) = x + 2$

40. $f(x) = x^2 - 11x, \quad g(x) = 2x + 3$

41. $f(x) = x^3 + 2x, \quad g(x) = -5x$

42. $f(x) = -x^3 - 7, \quad g(x) = x + 1$

43. $f(x) = \dfrac{2}{x + 1}, \quad g(x) = 3x - 5$

44. $f(x) = \sqrt{x + 4}, \quad g(x) = \dfrac{1}{x}$

45. $f(x) = \dfrac{1}{x^2}, \quad g(x) = \sqrt{x - 1}$

46. $f(x) = \dfrac{6}{x - 2}, \quad g(x) = \dfrac{3}{5x}$

47. $f(x) = \dfrac{3}{|5 - x|}, \quad g(x) = -\dfrac{2}{x}$

48. $f(x) = |2x + 1|, \quad g(x) = 3x^2 - 1$

In Exercises 49 to 64, evaluate each composite function, where $f(x) = 2x + 3$, $g(x) = x^2 - 5x$, and $h(x) = 4 - 3x^2$.

49. $(g \circ f)(4)$

▶ **50.** $(f \circ g)(4)$

51. $(f \circ g)(-3)$

52. $(g \circ f)(-1)$

53. $(g \circ h)(0)$

54. $(h \circ g)(0)$

55. $(f \circ f)(8)$

56. $(f \circ f)(-8)$

57. $(h \circ g)\left(\dfrac{2}{5}\right)$

58. $(g \circ h)\left(-\dfrac{1}{3}\right)$

59. $(g \circ f)\left(\sqrt{3}\right)$

60. $(f \circ g)\left(\sqrt{2}\right)$

61. $(g \circ f)(2c)$

62. $(f \circ g)(3k)$

63. $(g \circ h)(k + 1)$

64. $(h \circ g)(k - 1)$

65. WATER TANK A water tank has the shape of a right circular cone, with height 16 feet and radius 8 feet. Water is running into the tank so that the radius r (in feet) of the surface of the water is given by $r = 1.5t$, where t is the time (in minutes) that the water has been running.

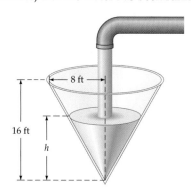

a. The area A of the surface of the water is $A = \pi r^2$. Find $A(t)$ and use it to determine the area of the surface of the water when $t = 2$ minutes.

b. The volume V of the water is given by $V = \dfrac{1}{3}\pi r^2 h$.

Find $V(t)$ and use it to determine the volume of the water when $t = 3$ minutes. (*Hint:* The height of the water in the cone is always twice the radius of the water.)

▶ **66.** ▦ **SCALING A RECTANGLE** Work Example 7 of this section with the scaling as follows. The upper right corner of the original rectangle is pulled to the *left* at 0.5 inch per second and downward at 0.2 inch per second.

67. TOWING A BOAT A boat is towed by a rope that runs through a pulley that is 4 feet above the point where the rope is tied to the boat. The length (in feet) of the rope from the boat to the pulley is given by $s = 48 - t$, where t is the time in seconds that the boat has been in tow. The horizontal distance from the pulley to the boat is d.

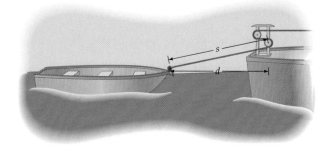

a. Find $d(t)$.   **b.** Evaluate $s(35)$ and $d(35)$.

68. 🖩 **PERIMETER OF A SCALED RECTANGLE** Show by a graph that the perimeter

$$P = 2(3 + 0.5t) + 2|2 - 0.2t|$$

of the scaled rectangle in Example 7 of this section is an increasing function over $0 \le t \le 14$.

69. **CONVERSION FUNCTIONS** The function $F(x) = \dfrac{x}{12}$ converts x inches to feet. The function $Y(x) = \dfrac{x}{3}$ converts x feet to yards. Explain the meaning of $(Y \circ F)(x)$.

70. **CONVERSION FUNCTIONS** The function $F(x) = 3x$ converts x yards to feet. The function $I(x) = 12x$ converts x feet to inches. Explain the meaning of $(I \circ F)(x)$.

71. 🥧 **CONCENTRATION OF A MEDICATION** The concentration $C(t)$ (in milligrams per liter) of a medication in a patient's blood is given by the data in the following table.

Concentration of Medication in Patient's Blood

t hours	C(t) mg/l
0	0
0.25	47.3
0.50	78.1
0.75	94.9
1.00	99.8
1.25	95.7
1.50	84.4
1.75	68.4
2.00	50.1
2.25	31.6
2.50	15.6
2.75	4.3

The **average rate of change** of the concentration over the time interval from $t = a$ to $t = a + \Delta t$ is

$$\frac{C(a + \Delta t) - C(a)}{\Delta t}$$

Use the data in the table to evaluate the average rate of change for each of the following time intervals.

a. $[0, 1]$ (*Hint:* In this case, $a = 0$ and $\Delta t = 1$.) Compare this result to the slope of the line through $(0, C(0))$ and $(1, C(1))$.

b. $[0, 0.5]$ **c.** $[1, 2]$ **d.** $[1, 1.5]$ **e.** $[1, 1.25]$

f. The data in the table can be modeled by the function $Con(t) = 25t^3 - 150t^2 + 225t$. Use $Con(t)$ to verify that the average rate of change over $[1, 1 + \Delta t]$ is $-75(\Delta t) + 25(\Delta t)^2$. What does the average rate of change over $[1, 1 + \Delta t]$ seem to approach as Δt approaches 0?

▶ **72.** **BALL ROLLING ON A RAMP** The distance traveled by a ball rolling down a ramp is given by $s(t) = 6t^2$, where t is the time in seconds after the ball is released, and $s(t)$ is measured in feet. The ball travels 6 feet in 1 second and it travels 24 feet in 2 seconds. Use the difference quotient for average velocity given on page 67 to evaluate the average velocity for each of the following time intervals.

a. $[2, 3]$ (*Hint:* In this case, $a = 2$ and $\Delta t = 1$.) Compare this result to the slope of the line through $(2, s(2))$ and $(3, s(3))$.

b. $[2, 2.5]$ **c.** $[2, 2.1]$ **d.** $[2, 2.01]$ **e.** $[2, 2.001]$

f. Verify that the average velocity over $[2, 2 + \Delta t]$ is $24 + 6(\Delta t)$. What does the average velocity seem to approach as Δt approaches 0?

───── *CONNECTING CONCEPTS* ─────

In Exercises 73 to 76, show that $(f \circ g)(x) = (g \circ f)(x)$.

73. $f(x) = 2x + 3;$ $g(x) = 5x + 12$

74. $f(x) = 4x - 2;$ $g(x) = 7x - 4$

75. $f(x) = \dfrac{6x}{x - 1};$ $g(x) = \dfrac{5x}{x - 2}$

76. $f(x) = \dfrac{5x}{x + 3};$ $g(x) = -\dfrac{2x}{x - 4}$

In Exercises 77 to 82, show that

$$(g \circ f)(x) = x \quad \text{and} \quad (f \circ g)(x) = x$$

77. $f(x) = 2x + 3, \quad g(x) = \dfrac{x - 3}{2}$

78. $f(x) = 4x - 5, \quad g(x) = \dfrac{x + 5}{4}$

79. $f(x) = \dfrac{4}{x + 1}, \quad g(x) = \dfrac{4 - x}{x}$

80. $f(x) = \dfrac{2}{1 - x}, \quad g(x) = \dfrac{x - 2}{x}$

81. $f(x) = x^3 - 1, \quad g(x) = \sqrt[3]{x + 1}$

82. $f(x) = -x^3 + 2, \quad g(x) = \sqrt[3]{2 - x}$

PREPARE FOR SECTION 1.6

83. Solve $2x + 5y = 15$ for y.

84. Solve $x = \dfrac{y + 1}{y}$ for y.

85. Given $f(x) = \dfrac{2x^2}{x - 1}$, find $f(-1)$. [1.3]

86. Suppose $a < b$. Replace the question mark in $5 - 4a \, ? \, 5 - 4b$ to make a true statement. [1.1]

87. Suppose $a < b$. What other condition on b must be true so that $a^2 > b^2$? [1.1]

88. What is the domain of $f(x) = \sqrt{x + 2}$? [1.3]

PROJECTS

1.   **A GRAPHING UTILITY PROJECT** For any two different real numbers x and y, the larger of the two numbers is given by

$$\text{Maximum}(x, y) = \frac{x + y}{2} + \frac{|x - y|}{2} \qquad (1)$$

a. Verify Equation (1) for $x = 5$ and $y = 9$.

b. Verify Equation (1) for $x = 201$ and $y = 80$.

For any two different functional values $f(x)$ and $g(x)$, the larger of the two is given by

$$\text{Maximum}(f(x), g(x)) = \frac{f(x) + g(x)}{2} + \frac{|f(x) - g(x)|}{2} \qquad (2)$$

To illustrate how we might make use of Equation (2), consider the functions $y_1 = x^2$ and $y_2 = \sqrt{x}$ on the interval from Xmin $= -1$ to Xmax $= 6$. The graphs of y_1 and y_2 are shown at the right.

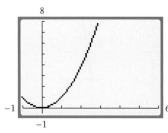

$y_1 = x^2$

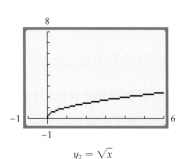

$y_2 = \sqrt{x}$

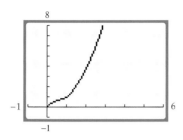

$$y_3 = (y_1 + y_2)/2 + (\text{abs}\,(y_1 - y_2))/2$$

Now consider the function

$$y_3 = (y_1 + y_2)/2 + (\text{abs}(y_1 - y_2))/2$$

where "abs" represents the absolute value function. The graph of y_3 is shown above.

c. ✎ Write a sentence or two that explains why the graph of y_3 is as shown.

d. ✎ What is the domain of y_1? of y_2? of y_3? Write a sentence that explains how to determine the domain of y_3, given the domain of y_1 and the domain of y_2.

e. Determine a formula for the function Minimum($f(x), g(x)$).

2. 🖩 👥 THE NEVER-NEGATIVE FUNCTION The author J. D. Murray describes a function f_+ that is defined in the following manner.[2]

$$f_+ = \begin{cases} f & \text{if } f \ge 0 \\ 0 & \text{if } f < 0 \end{cases}$$

We will refer to this function as a **never-negative** function. Never-negative functions can be graphed by using Equation (2) in Project 1. For example, if we let $g(x) = 0$, then Equation (2) simplifies to

$$\text{Maximum}(f(x), 0) = \frac{f(x)}{2} + \frac{|f(x)|}{2} \qquad (3)$$

The graph of $y = \text{Maximum}(f(x), 0)$ is the graph of $y = f(x)$ provided that $f(x) \ge 0$, and it is the graph of $y = 0$ provided that $f(x) < 0$.

An Application The mosquito population per acre of a large resort is controlled by spraying on a monthly basis. A biologist has determined that the mosquito population can be approximated by the never-negative function M_+ with

$$M(t) = -35{,}400(t - \text{int}(t))^2 + 35{,}400(t - \text{int}(t)) - 4000$$

Here t represents the month, and $t = 0$ corresponds to June 1, 2004.

a. Use a graphing utility to graph M for $0 \le t \le 3$.

b. Use a graphing utility to graph M_+ for $0 \le t \le 3$.

c. ✎ Write a sentence or two that explains how the graph of M_+ differs from the graph of M.

d. What is the maximum mosquito population per acre for $0 \le t \le 3$? When does this maximum mosquito population occur?

e. ✎ Explain when would be the best time to visit the resort, provided that you wished to minimize your exposure to mosquitos.

[2]*Mathematical Biology* (New York: Springer-Verlag, 1989), p. 101.

SECTION 1.6

INVERSE FUNCTIONS

● INTRODUCTION TO INVERSE FUNCTIONS

Consider the "doubling" function $f(x) = 2x$ that doubles every input. Some of the ordered pairs of this function are

$$\left\{ (-4, -8), (-1.5, -3), (1, 2), \left(\frac{5}{3}, \frac{10}{3} \right), (7, 14) \right\}$$

Now consider the "halving" function $g(x) = \frac{1}{2}x$ that takes one-half of every input. Some of the ordered pairs of this function are

$$\left\{ (-8, -4), (-3, -1.5), (2, 1), \left(\frac{10}{3}, \frac{5}{3} \right), (14, 7) \right\}$$

Observe that the coordinates of the ordered pairs of g are the reverse of the coordinates of the ordered pairs of f. This is always the case for f and g. Here are two more examples.

$$f(5) = 2(5) = 10 \qquad\qquad g(10) = \frac{1}{2}(10) = 5$$

Ordered pair: (5, 10)   **Ordered pair: (10, 5)**

$$f(a) = 2(a) = 2a \qquad\qquad g(2a) = \frac{1}{2}(2a) = a$$

Ordered pair: (a, 2a)   **Ordered pair: (2a, a)**

For these functions, f and g are called *inverse functions* of one another.

Inverse Function

If the coordinates of the ordered pairs of a function g are the reverse of the coordinates of the ordered pairs of a function f, then g is said to be the **inverse function** of f.

---

take note

It is important to remember the information in the paragraph at the right. If f is a function and g is the inverse of f, then

Domain of g = range of f

and

Range of g = domain of f

Because the coordinates of the ordered pairs of the inverse function g are the reverse of the coordinates of the ordered pairs of the function f, the domain of g is the range of f, and the range of g is the domain of f.

Not all functions have an inverse that is a function. Consider, for instance, the "square" function $S(x) = x^2$. Some of the ordered pairs of S are

$$\{ (-3, 9), (-1, 1), (0, 0), (1, 1), (3, 9), (5, 25) \}$$

If we reverse the coordinates of the ordered pairs, we have

$$\{ (9, -3), (1, -1), (0, 0), (1, 1), (9, 3), (25, 5) \}$$

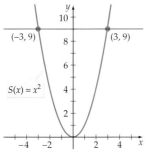

FIGURE 1.84

This set of ordered pairs is not a function because there are ordered pairs, for instance $(9, -3)$ and $(9, 3)$, with the same first coordinate and different second coordinates. In this case, S has an inverse *relation* but not an inverse *function*.

A graph of S is shown in **Figure 1.84.** Note that $x = -3$ and $x = 3$ produce the same value of y. Thus the graph of S fails the horizontal line test, and therefore S is not a one-to-one function. This observation is used in the following theorem.

Condition for an Inverse Function

A function f has an inverse function if and only if f is a one-to-one function.

Recall that increasing functions or decreasing functions are one-to-one functions. Thus we can state the following theorem.

Alternative Condition for an Inverse Function

If f is an increasing function or a decreasing function, then f has an inverse function.

? QUESTION   Which of the functions graphed below has an inverse function?

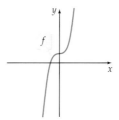
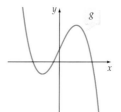
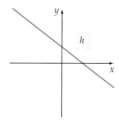

take note

$f^{-1}(x)$ *does not mean* $\dfrac{1}{f(x)}$. For

$f(x) = 2x$, $f^{-1}(x) = \dfrac{1}{2}x$ but

$\dfrac{1}{f(x)} = \dfrac{1}{2x}$.

If a function g is the inverse of a function f, we usually denote the inverse function by f^{-1} rather than g. For the doubling and halving functions f and g discussed on page 78, we write

$$f(x) = 2x \qquad f^{-1}(x) = \frac{1}{2}x$$

• GRAPHS OF INVERSE FUNCTIONS

Because the coordinates of the ordered pairs of the inverse of a function f are the reverse of the coordinates of f, we can use them to create a graph of f^{-1}.

? ANSWER   The graph of f is the graph of an increasing function. Therefore, f is a one-to-one function and has an inverse function. The graph of h is the graph of a decreasing function. Therefore, h is a one-to-one function and has an inverse function. The graph of g is not the graph of a one-to-one function. g does not have an inverse function.

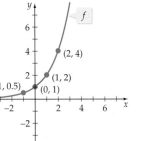

FIGURE 1.85

EXAMPLE 1 **Sketch the Graph of the Inverse of a Function**

Sketch the graph of f^{-1} given that f is the function shown in **Figure 1.85.**

Solution

Because the graph of f passes through $(-1, 0.5)$, $(0, 1)$, $(1, 2)$, and $(2, 4)$, the graph of f^{-1} must pass through $(0.5, -1)$, $(1, 0)$, $(2, 1)$, and $(4, 2)$. Plot the points and then draw a smooth graph through the points, as shown in **Figure 1.86.**

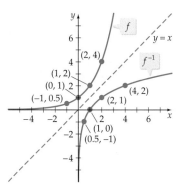

FIGURE 1.86

▶ TRY EXERCISE 10, PAGE 87

The graph from the solution to Example 1 is shown again in **Figure 1.87.** Note that the graph of f^{-1} is symmetric to the graph of f with respect to the graph of $y = x$. If the graph were folded along the dashed line, the graph of f would lie on top of the graph of f^{-1}. This is a characteristic of all graphs of functions and their inverses. In **Figure 1.88,** although S does not have an inverse that is a function, the graph of the inverse relation S^{-1} is symmetric to S with respect to the graph of $y = x$.

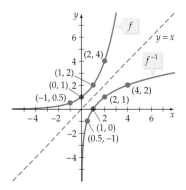

FIGURE 1.87

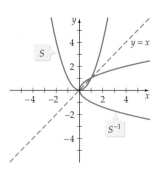

FIGURE 1.88

● COMPOSITION OF A FUNCTION AND ITS INVERSE

Observe the effect, as shown below, of taking the composition of functions that are inverses of one another.

$$f(x) = 2x \qquad\qquad g(x) = \frac{1}{2}x$$

$$f[g(x)] = 2\left[\frac{1}{2}x\right] \quad \text{• Replace } x \qquad g[f(x)] = \frac{1}{2}[2x] \quad \text{• Replace } x$$
$$\hphantom{f[g(x)] = 2\left[\frac{1}{2}x\right]} \quad \text{by } g(x). \qquad\qquad\qquad \text{by } f(x).$$
$$f[g(x)] = x \qquad\qquad\qquad g[f(x)] = x$$

This property of the composition of inverse functions always holds true. When taking the composition of inverse functions, the inverse function reverses the effect of the original function. For the two functions above, f doubles a number, and g halves a number. If you double a number and then take one-half of the result, you are back to the original number.

---

take note

If we think of a function as a machine, then the Composition of Inverse Functions Property can be represented as shown below. Take any input x for f. Use the output of f as the input for f^{-1}. The result is the original input, x.

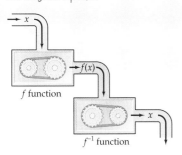

f function

f^{-1} function

Composition of Inverse Functions Property

If f is a one-to-one function, then f^{-1} is the inverse function of f if and only if

$$(f \circ f^{-1})(x) = f[f^{-1}(x)] = x \qquad \text{for all } x \text{ in the domain of } f^{-1}$$

and

$$(f^{-1} \circ f)(x) = f^{-1}[f(x)] = x \qquad \text{for all } x \text{ in the domain of } f.$$

EXAMPLE 2   **Use the Composition of Inverse Functions Property**

Use composition of functions to show that $f^{-1}(x) = 3x - 6$ is the inverse function of $f(x) = \frac{1}{3}x + 2$.

Solution

We must show that $f[f^{-1}(x)] = x$ and $f^{-1}[f(x)] = x$.

$$f(x) = \frac{1}{3}x + 2 \qquad\qquad f^{-1}(x) = 3x - 6$$

$$f[f^{-1}(x)] = \frac{1}{3}[3x - 6] + 2 \qquad f^{-1}[f(x)] = 3\left[\frac{1}{3}x + 2\right] - 6$$

$$f[f^{-1}(x)] = x \qquad\qquad\qquad f^{-1}[f(x)] = x$$

▶ **TRY EXERCISE 20, PAGE 88**

INTEGRATING TECHNOLOGY

In the standard viewing window of a calculator, the distance between two tic marks on the x-axis is not equal to the distance between two tic marks on the y-axis. As a result, the graph of $y = x$ does not appear to bisect the first and third quadrants. See **Figure 1.89.** This anomaly is important if a graphing calculator is being used to check whether two functions are inverses of one another. Because the graph of $y = x$ does not appear to bisect the first and third quadrants, the graphs of f and f^{-1} will not appear to be symmetric about the graph of $y = x$. The graphs of $f(x) = \dfrac{1}{3}x + 2$ and $f^{-1}(x) = 3x - 6$ from Example 2 are shown in **Figure 1.90.** Notice that the graphs do not appear to be quite symmetric about the graph of $y = x$.

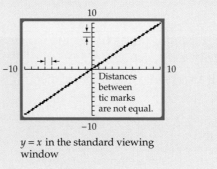

$y = x$ in the standard viewing window

FIGURE 1.89

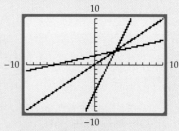

f, f^{-1}, and $y = x$ in the standard viewing window

FIGURE 1.90

To get a better view of a function and its inverse, it is necessary to use the SQUARE viewing window, as in **Figure 1.91.** In this window, the distance between two tic marks on the x-axis is equal to the distance between two tic marks on the y-axis.

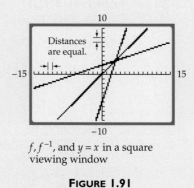

f, f^{-1}, and $y = x$ in a square viewing window

FIGURE 1.91

● FIND AN INVERSE FUNCTION

If a one-to-one function f is defined by an equation, then we can use the following method to find the equation for f^{-1}.

take note

If the ordered pairs of f are given by (x, y), then the ordered pairs of f^{-1} are given by (y, x). That is, x and y are interchanged. This is the reason for Step 2 at the right.

Steps for Finding the Inverse of a Function

To find the equation of the inverse f^{-1} of the one-to-one function f:

1. Substitute y for $f(x)$.

2. Interchange x and y.

3. Solve, if possible, for y in terms of x.

4. Substitute $f^{-1}(x)$ for y.

EXAMPLE 3 **Find the Inverse of a Function**

Find the inverse of $f(x) = 3x + 8$.

Solution

$$f(x) = 3x + 8$$
$$y = 3x + 8 \qquad \text{• Replace } f(x) \text{ by } y.$$
$$x = 3y + 8 \qquad \text{• Interchange } x \text{ and } y.$$
$$x - 8 = 3y \qquad \text{• Solve for } y.$$
$$\frac{x - 8}{3} = y$$
$$\frac{1}{3}x - \frac{8}{3} = f^{-1}(x) \qquad \text{• Replace } y \text{ by } f^{-1}(x).$$

The inverse function is given by $f^{-1}(x) = \dfrac{1}{3}x - \dfrac{8}{3}$.

▶ **TRY EXERCISE 28, PAGE 88**

EXAMPLE 4 **Find the Inverse of a Function**

Find the inverse of $f(x) = \dfrac{2x + 1}{x}, x \neq 0$.

Continued ▶

Solution

$$f(x) = \frac{2x + 1}{x}$$

$$y = \frac{2x + 1}{x}$$ • **Replace $f(x)$ by y.**

$$x = \frac{2y + 1}{y}$$ • **Interchange x and y.**

$$xy = 2y + 1$$ • **Solve for y.**

$$xy - 2y = 1$$

$$y(x - 2) = 1$$ • **Factor the left side.**

$$y = \frac{1}{x - 2}$$

$$f^{-1}(x) = \frac{1}{x - 2}, x \neq 2$$ • **Replace y by $f^{-1}(x)$.**

▶ **TRY EXERCISE 34, PAGE 88**

❓ **QUESTION** If f is a one-to-one function and $f(4) = 5$, what is $f^{-1}(5)$?

The graph of $f(x) = x^2 + 4x + 3$ is shown in **Figure 1.92a.** The function f is not a one-to-one function and therefore does not have an inverse function. However, the function given by $G(x) = x^2 + 4x + 3$, shown in **Figure 1.92b,** for which the domain is restricted to $\{x \mid x \geq -2\}$, is a one-to-one function and has an inverse function G^{-1}. This is shown in Example 5.

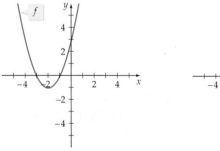

FIGURE 1.92a   **FIGURE 1.92b**

❓ **ANSWER** Because f^{-1} is the inverse function of f, the coordinates of the ordered pairs of f^{-1} are the reverse of the coordinates of the ordered pairs of f. Therefore, $f^{-1}(5) = 4$.

| EXAMPLE 5 | Find the Inverse of a Function with a Restricted Domain |

Find the inverse of $G(x) = x^2 + 4x + 3$, where the domain of G is $\{x \mid x \geq -2\}$.

Solution

$$G(x) = x^2 + 4x + 3$$

$y = x^2 + 4x + 3$	• Replace $G(x)$ by y.
$x = y^2 + 4y + 3$	• Interchange x and y.
$x = (y^2 + 4y + 4) - 4 + 3$	• Solve for y by completing the square of $y^2 + 4y$.
$x = (y + 2)^2 - 1$	• Factor.
$x + 1 = (y + 2)^2$	• Add 1 to each side of the equation.
$\sqrt{x + 1} = \sqrt{(y + 2)^2}$	• Take the square root of each side of the equation.
$\pm\sqrt{x + 1} = y + 2$	• Recall that if $a^2 = b$, then $a = \pm\sqrt{b}$.
$\pm\sqrt{x + 1} - 2 = y$	

Because the domain of G is $\{x \mid x \geq -2\}$, the range of G^{-1} is $\{y \mid y \geq -2\}$. This means that we must choose the positive value of $\pm\sqrt{x + 1}$. Thus $G^{-1}(x) = \sqrt{x + 1} - 2$. See **Figure 1.93.**

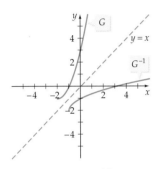

FIGURE 1.93

take note

Recall that the range of a function f is the domain of f^{-1}, and the domain of f is the range of f^{-1}.

▶ **TRY EXERCISE 40, PAGE 88**

● APPLICATION

There are practical applications of finding the inverse of a function. Here is one in which a shirt size in the United States is converted to a shirt size in Italy. Finding the inverse function gives the function that converts a shirt size in Italy to a shirt size in the United States.

| EXAMPLE 6 | Solve an Application |

 The function $IT(x) = 2x + 8$ converts a men's shirt size x in the United States to the equivalent shirt size in Italy.

a. Use IT to determine the equivalent Italian shirt size for a size 16.5 U.S. shirt.

b. Find IT^{-1} and use IT^{-1} to determine the U.S. men's shirt size that is equivalent to an Italian shirt size of 36.

Solution

a. $IT(16.5) = 2(16.5) + 8 = 33 + 8 = 41$

A size 16.5 U.S. shirt is equivalent to a size 41 Italian shirt.

Continued ▶

b. To find the inverse function, begin by substituting y for $IT(x)$.

$$IT(x) = 2x + 8$$
$$y = 2x + 8$$
$$x = 2y + 8 \qquad \text{• Interchange } x \text{ and } y.$$
$$x - 8 = 2y \qquad \text{• Solve for } y.$$
$$\frac{x - 8}{2} = y$$

In inverse notation, the above equation can be written as

$$IT^{-1}(x) = \frac{x - 8}{2} \qquad \text{or} \qquad IT^{-1}(x) = \frac{1}{2}x - 4$$

Substitute 36 for x to find the equivalent U.S. shirt size.

$$IT^{-1}(36) = \frac{1}{2}(36) - 4 = 18 - 4 = 14$$

A size 36 Italian shirt is equivalent to a size 14 U.S. shirt.

▶ **TRY EXERCISE 50, PAGE 89**

INTEGRATING TECHNOLOGY

Some graphing utilities can be used to draw the graph of the inverse of a function without the user having to find the inverse function. For instance, **Figure 1.94** shows the graph of $f(x) = 0.1x^3 - 4$. The graphs of f and f^{-1} are both shown in **Figure 1.95**, along with the graph of $y = x$. Note that the graph of f^{-1} is the reflection of the graph of f with respect to the graph of $y = x$. The display shown in **Figure 1.95** was produced on a TI-83 graphing calculator by using the DrawInv command, which is in the DRAW menu.

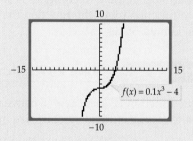

FIGURE 1.94

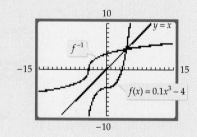

FIGURE 1.95

 TOPICS FOR DISCUSSION

1. If $f(x) = 3x + 1$, what are the values of $f^{-1}(2)$ and $[f(2)]^{-1}$?

2. How are the domain and range of a one-to-one function f related to the domain and range of the inverse function of f?

3. How is the graph of the inverse of a function f related to the graph of f?

4. The function $f(x) = -x$ is its own inverse. Find at least two other functions that are their own inverses.

5. What are the steps in finding the inverse of a one-to-one function?

EXERCISE SET 1.6

In Exercises 1 to 4, assume that the given function has an inverse function.

1. Given $f(3) = 7$, find $f^{-1}(7)$.

2. Given $g(-3) = 5$, find $g^{-1}(5)$.

3. Given $h^{-1}(-3) = -4$, find $h(-4)$.

4. Given $f^{-1}(7) = 0$, find $f(0)$.

5. If 3 is in the domain of f^{-1}, find $f[f^{-1}(3)]$.

6. If f is a one-to-one function and $f(0) = 5$, $f(1) = 2$, and $f(2) = 7$, find:

   a. $f^{-1}(5)$ **b.** $f^{-1}(2)$

7. The domain of the inverse function f^{-1} is the _____ of f.

8. The range of the inverse function f^{-1} is the _____ of f.

In Exercises 9 to 16, draw the graph of the inverse relation. Is the inverse relation a function?

9.

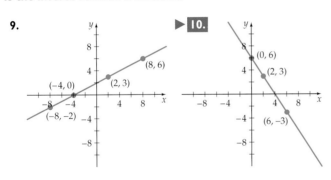

▶ 10.

11.

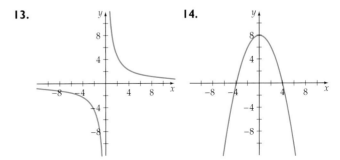

12.

13.

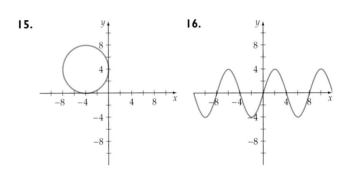

14.

15.

16.

In Exercises 17 to 22, use composition of functions to determine whether f and g are inverses of one another.

17. $f(x) = 4x$; $g(x) = \dfrac{x}{4}$

18. $f(x) = 3x$; $g(x) = \dfrac{1}{3x}$

19. $f(x) = 4x - 1$; $g(x) = \dfrac{1}{4}x + \dfrac{1}{4}$

▶ **20.** $f(x) = \dfrac{1}{2}x - \dfrac{3}{2}$; $g(x) = 2x + 3$

21. $f(x) = -\dfrac{1}{2}x - \dfrac{1}{2}$; $g(x) = -2x + 1$

22. $f(x) = 3x + 2$; $g(x) = \dfrac{1}{3}x - \dfrac{2}{3}$

In Exercises 23 to 26, find the inverse of the function. If the function does not have an inverse function, write "no inverse function."

23. $\{(-3, 1), (-2, 2), (1, 5), (4, -7)\}$

24. $\{(-5, 4), (-2, 3), (0, 1), (3, 2), (7, 11)\}$

25. $\{(0, 1), (1, 2), (2, 4), (3, 8), (4, 16)\}$

26. $\{(1, 0), (10, 1), (100, 2), (1000, 3), (10{,}000, 4)\}$

In Exercises 27 to 44, find $f^{-1}(x)$. State any restrictions on the domain of $f^{-1}(x)$.

27. $f(x) = 2x + 4$

▶ **28.** $f(x) = 4x - 8$

29. $f(x) = 3x - 7$

30. $f(x) = -3x - 8$

31. $f(x) = -2x + 5$

32. $f(x) = -x + 3$

33. $f(x) = \dfrac{2x}{x - 1}$, $x \neq 1$

▶ **34.** $f(x) = \dfrac{x}{x - 2}$, $x \neq 2$

35. $f(x) = \dfrac{x - 1}{x + 1}$, $x \neq -1$

36. $f(x) = \dfrac{2x - 1}{x + 3}$, $x \neq -3$

37. $f(x) = x^2 + 1$, $x \geq 0$

38. $f(x) = x^2 - 4$, $x \geq 0$

39. $f(x) = \sqrt{x - 2}$, $x \geq 2$

▶ **40.** $f(x) = \sqrt{4 - x}$, $x \leq 4$

41. $f(x) = x^2 + 4x$, $x \geq -2$

42. $f(x) = x^2 - 6x$, $x \leq 3$

43. $f(x) = x^2 + 4x - 1$, $x \leq -2$

44. $f(x) = x^2 - 6x + 1$, $x \geq 3$

45. **GEOMETRY** The volume of a cube is given by $V(x) = x^3$, where x is the measure of the length of a side of the cube. Find $V^{-1}(x)$ and explain what it represents.

46. **UNIT CONVERSIONS** The function $f(x) = 12x$ converts feet, x, into inches, $f(x)$. Find $f^{-1}(x)$ and explain what it determines.

47. **UNIT CONVERSIONS** A conversion function such as the one in Exercise 46 converts a measurement in one unit into another unit. Is a conversion function always a one-to-one function? Does a conversion function always have an inverse function? Explain your answer.

48. **GRADING SCALE** Does the grading scale function given below have an inverse function? Explain your answer.

Score	Grade
90–100	A
80–89	B
70–79	C
60–69	D
0–59	F

49.  **FASHION** The function $s(x) = 2x + 24$ can be used to convert a U.S. women's shoe size into an Italian women's shoe size. Determine the function $s^{-1}(x)$ that can be used to convert an Italian women's shoe size to its equivalent U.S. shoe size.

▶ **50.** **FASHION** The function $K(x) = 1.3x - 4.7$ converts a men's shoe size in the United States to the equivalent shoe size in the United Kingdom. Determine the function $K^{-1}(x)$ that can be used to convert a United Kingdom men's shoe size to its equivalent U.S. shoe size.

51. **COMPENSATION** The monthly earnings $E(s)$, in dollars, of a software sales executive is given by $E(s) = 0.05s + 2500$, where s is the value, in dollars, of the software sold by the executive during the month. Find $E^{-1}(s)$, and explain how the executive could use this function.

52. **POSTAGE** Does the first-class postage rate function given below have an inverse function? Explain your answer.

Weight (in ounces)	Cost
$0 < w \le 1$	\$.37
$1 < w \le 2$	\$.60
$2 < w \le 3$	\$.83
$3 < w \le 4$	\$1.06

53. **INTERNET COMMERCE** Functions and their inverses can be used to create secret codes that are used to secure business transactions made over the Internet. Let A = 10, B = 11, ..., and Z = 35. Let $f(x) = 2x - 1$ define a coding function. Code the word MATH (M—22, A—10, T—29, H—17), which is 22102917, by finding $f(22102917)$. Now find the inverse of f and show that applying f^{-1} to the output of f returns the original word.

54. **CRYPTOGRAPHY** A friend is using the letter-number correspondence in Exercise 53 and the coding function $f(x) = 2x + 3$. Suppose this friend sends you the coded message 5658602671. Decode this message.

In Exercises 55 to 60, answer the question without finding the equation of the linear function.

55. Suppose that f is a linear function, $f(2) = 7$, and $f(5) = 12$. If $f(4) = c$, then is c less than 7, between 7 and 12, or greater than 12? Explain your answer.

56. Suppose that f is a linear function, $f(1) = 13$, and $f(4) = 9$. If $f(3) = c$, then is c less than 9, between 9 and 13, or greater than 13? Explain your answer.

57. Suppose that f is a linear function, $f(2) = 3$, and $f(5) = 9$. Between which two numbers is $f^{-1}(6)$?

58. Suppose that f is a linear function, $f(5) = -1$, and $f(9) = -3$. Between which two numbers is $f^{-1}(-2)$?

59. Suppose that g is a linear function, $g^{-1}(3) = 4$, and $g^{-1}(7) = 8$. Between which two numbers is $g(5)$?

60. Suppose that g is a linear function, $g^{-1}(-2) = 5$, and $g^{-1}(0) = -3$. Between which two numbers is $g(0)$?

CONNECTING CONCEPTS

In Exercises 61 and 62, find the inverse of the given function.

61. $f(x) = ax + b, \quad a \ne 0$

62. $f(x) = ax^2 + bx + c, \quad a \ne 0, \quad x \ge -\dfrac{b}{2a}$

63. Use a graph of $f(x) = -x + 3$ to explain why f is its own inverse.

64. Use a graph of $f(x) = \sqrt{16 - x^2}$, with $0 \le x \le 4$, to explain why f is its own inverse.

Only one-to-one functions have inverses that are functions. In Exercises 65 to 68, determine if the given function is a one-to-one function.

65. $p(t) = \sqrt{9 - t}$

66. $v(t) = \sqrt{16 + t}$

67. $F(x) = |x| + x$

68. $T(x) = |x^2 - 6|, \quad x \geq 0$

PREPARE FOR SECTION 1.7

In Exercises 69 and 70, find the slope and y-intercept of the graph of the equation.

69. $y = -\dfrac{x}{3} + 4$

70. $3x - 4y = 12$

71. Find the equation of the line that has a slope of -0.45 and a y-intercept of $(0, 2.3)$.

72. Find the equation of the line that passes through $P(3, -4)$ and has a slope of $-\dfrac{2}{3}$.

73. If $f(x) = 3x^2 + 4x - 1$, find $f(2)$. [1.3]

74. You are given $P_1(2, -1)$ and $P_2(4, 14)$. If $f(x) = x^2 - 3$, find $|f(x_1) - y_1| + |f(x_2) - y_2|$. [1.3]

PROJECTS

1. **INTERSECTION POINTS FOR THE GRAPHS OF f AND f^{-1}** For each of the following, graph f and its inverse.

i. $f(x) = 2x - 4$

ii. $f(x) = -x + 2$

iii. $f(x) = x^3 + 1$

iv. $f(x) = x - 3$

v. $f(x) = -3x + 2$

vi. $f(x) = \dfrac{1}{x}$

a. Do the graphs of a function and its inverse always intersect?

b. If the graphs of a function and its inverse intersect at one point, what is true about the coordinates of the point of intersection?

c. Can the graphs of a function and its inverse intersect at more than one point?

MODELING DATA USING REGRESSION

LINEAR REGRESSION MODELS

The data in the table below show the population of a few states and the number of professional sports teams (Major League Baseball, National Football League, National Basketball Association, Women's National Basketball Association, National Hockey League) in those states.

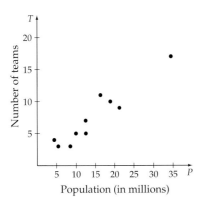

Number of Professional Sports Teams for Selected States

State	Population (in millions)	Number of Teams
California	34.5	17
Florida	16.4	11
New York	19.0	10
Texas	21.3	9
Pennsylvania	12.3	7
Illinois	12.5	5
Michigan	10	5
Colorado	4.4	4
New Jersey	8.5	3
Wisconsin	5.4	3

Source: Bureau of Census, 50States.com.

Although there is no one line that passes through every point, we could find an approximate linear model of these data. For instance, the line shown in **Figure 1.96** in blue approximates the data better than the line shown in red. However, as **Figure 1.97** shows, there are many other lines we could have drawn that seem to approximate the data.

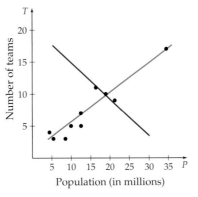

FIGURE 1.96

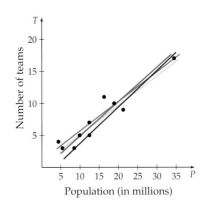

FIGURE 1.97

To find the line that "best" approximates the data, **regression analysis** is used. This analysis produces the linear function whose graph is called the **line of best fit** or the **least-squares regression line.**[3]

Definition of the Least-Squares Regression Line

The **least-squares regression line** is the line that minimizes the sum of the squares of the vertical deviations of all data points from the line.

To help understand this definition, consider the data set $S = \{(1, 2), (2, 3), (3, 3), (4, 4), (5, 7)\}$ as shown in **Figure 1.98.** As we will show later, the least-squares line for this data set is $y = 1.1x + 0.5$. If we evaluate this function at the x-coordinates of the data set S, we obtain the set of ordered pairs $T = \{(1, 1.6), (2, 2.7), (3, 3.8), (4, 4.9), (5, 6)\}$. The vertical deviations are the differences between the y-coordinates in S and the y-coordinates in T. From the definition, we must calculate the sum of the squares of these deviations.

$$(2 - 1.6)^2 + (3 - 2.7)^2 + (3 - 3.8)^2 + (4 - 4.9)^2 + (7 - 6)^2 = 2.7$$

Because $y = 1.1x + 0.5$ is the least squares regression line, for no other line is the sum of the squares of the deviations less than 2.7. For instance, if we consider the equation $y = 1.25x + 0.75$, which is the equation of the line through the two points $P_1(1, 2)$ and $P_2(5, 7)$ of the data set as shown in **Figure 1.99,** the sum of the squared deviations is larger than 2.7.

$$(2 - 2)^2 + (3 - 3.25)^2 + (3 - 4.5)^2 + (4 - 5.75)^2 + (7 - 7)^2 = 5.375$$

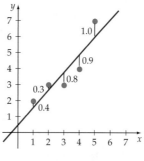

FIGURE 1.98

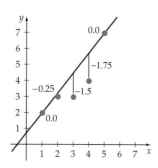

FIGURE 1.99

---

[3]The least-squares regression line is also called the *least-squares line* and the *regression line.*

INTEGRATING TECHNOLOGY

The equations used to calculate a regression line are somewhat cumbersome. Fortunately, these equations are preprogrammed into most graphing calculators. We will now illustrate the technique for a TI-83 calculator using data set S given on page 92.

Press STAT . Select EDIT.
Press ENTER .

```
EDIT CALC TESTS
1:Edit...
2:SortA(
3:SortD(
4:ClrList
5:SetUpEditor
```

If necessary, check any data in L1 and L2. Enter the given data.

```
L1     L2     L3    2
1      2     ------
2      3
3      3
4      4
5      7
-----  ▮▮▮▮▮
L2(6) =
```

Press STAT . Select CALC.
Select 4, LinReg(ax+b).
Press ENTER .

```
EDIT CALC TESTS
1:1-Var Stats
2:1-Var Stats
3:Med-Med
4:LinReg(ax+b)
5:QuadReg
6:CubicReg
7↓QuartReg
```

Press VARS .

```
LinReg(ax+b)
```

Press Y-VARS.
Press ENTER .

```
VARS Y-VARS
1:Function...
2:Parametric...
3:Polar...
4:On/Off...
```

Select 1. Press ENTER .

```
FUNCTION
1:Y1
2:Y2
3:Y3
4:Y4
5:Y5
6:Y6
7↓Y7
```

Press ENTER .

```
LinReg(ax+b) Y1
```

View the results.

```
LinReg
y=ax+b
a=1.1
b=.5
r²=.8175675676
r=.9041944302
```

From the last screen, the equation of the regression line is $y = 1.1x + 0.5$. Your last screen may not look exactly like ours. The information provided on our screen requires that DiagnosticsOn be enabled. This is accomplished using the following keystrokes:

2ND CATALOG (Scroll to DiagnosticsOn) ENTER

With DiagnosticsOn enabled, besides the values for the regression equation, two other values, r^2 and r, are given. We will discuss these values later in this section.

   If you used the keystrokes we have shown above, the regression line will be stored in **Y1**. This is helpful if you wish to graph the regression line. However, if it is not necessary to graph the regression line, then instead of pressing VARS at step 4, just press ENTER . The result will be the last screen showing the results of the regression calculations.

EXAMPLE 1   **Find a Regression Equation**

Find the regression equation for the data on the population of a state and the number of professional sports teams in that state. How many sports teams are predicted for North Carolina, whose population is approximately 8.2 million? Round to the nearest whole number.

Continued ▶

INTEGRATING TECHNOLOGY

If you followed the steps we gave on page 93 and stored the regression equation in Y1, then you can evaluate the regression equation using the following keystrokes:

Solution

Using your calculator, enter the data from the table. Then have the calculator produce the values for the regression equation. Your results should be similar to those shown at the right. The equation of the regression line is

$$y = 0.4704143697x + 0.6119206457$$

```
LinReg
y=ax+b
a=.4704143697
b=.6119206457
r²=.9031742558
r=.9503548052
```

To find the number of sports teams the regression equation predicts for North Carolina, evaluate the regression equation for $x = 8.2$.

$$y = 0.4704143697x + 0.6119206457$$
$$= 0.4704143697(8.2) + 0.6119206457$$
$$\approx 4.46$$

The equation predicts that North Carolina should have 4 sports teams.

▶ **TRY EXERCISE 18, PAGE 99**

● **CORRELATION COEFFICIENT AND COEFFICIENT OF DETERMINATION**

The scatter plot of a state's population and the corresponding number of professional sports teams is shown in **Figure 1.100,** along with the graph of the regression line. Note that the slope of the regression line is positive. This indicates that as a state's population increases, the number of teams increases. Note also that for these data the value of r on the regression calculation screen was positive, $r \approx 0.9504$.

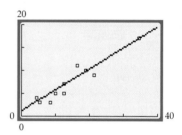

FIGURE 1.100

Now consider the data in the table below, which shows the trade-in value of a 2001 Corvette for various odometer readings.

The scatter diagram below is based on the trade-in value table. The graph of the regression line is also shown.

take note

The data for the Corvette were created assuming that the condition of the car was excellent. The only variable that changed was the odometer reading.

```
LinReg
y=ax+b
a=-223.0405405
b=32959.45946
r²=.9912312349
r=-.9956059637
```

Trade-in Value of 2001 Corvette Coupe, April 2003

Odometer Reading, in thousands	Trade-in Value
25	27,200
30	26,325
35	25,475
40	23,900
50	21,750

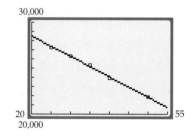

Source: Kelley Blue Book Web site, April 2003.

In this case the slope of the regression line is negative. This means that as the odometer reading increases, the trade-in value of the car decreases. Note also that the value of r is negative, $r \approx -0.996$.

Linear Correlation Coefficient

The **linear correlation coefficient** r is a measure of how close the points of a data set can be modeled by a straight line. If $r = -1$, then the points of the data set can be modeled *exactly* by a straight line with negative slope. If $r = 1$, then the data set can be modeled *exactly* by a straight line with positive slope. For all data sets, $-1 \leq r \leq 1$.

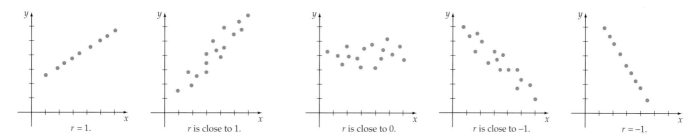

$r = 1$. r is close to 1. r is close to 0. r is close to -1. $r = -1$.

If $r \neq 1$ or $r \neq -1$, then the data set *cannot* be modeled exactly by a straight line. The further the value of r is from 1 or -1 (or in other words, the closer the value of r to zero), the more the ordered pairs of the data set deviate from a straight line.

The graphs below show the points of the data sets and the graphs of the regression lines for the state population/sports teams data and the odometer reading/trade-in data. Note the values of r and the closeness of the data points to the regression lines.

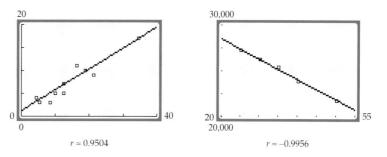

$r \approx 0.9504$ $r \approx -0.9956$

A researcher calculates a regression line to determine a relationship between two variables. The researcher wants to know whether a change in one variable produces a predictable change in the second variable. The value of r^2 tells the researcher the extent of that relationship.

Coefficient of Determination

The **coefficient of determination** is r^2. It measures the proportion of the variation in the dependent variable that is explained by the regression line.

For the population/sports team data, $r^2 \approx 0.90$. This means that approximately 90% of the total variation in the dependent variable (number of teams) can be attributed to the state population. This also means that population alone does not predict with certainty the number of sports teams. Other factors, such as climate, are also involved in the number of sports teams.

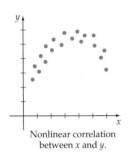

Nonlinear correlation
between x and y.

❷ QUESTION What is the coefficient of determination for the odometer
reading/trade-in value data (see page 94), and what is its
significance?

● QUADRATIC REGRESSION MODELS

To this point our focus has been *linear* regression equations. However, there may
be a nonlinear relationship between two quantities. The accompanying scatter
diagram to the left suggests that a quadratic function might be a better model of
the data than a linear model. As we proceed through this text, various functional
models will be discussed.

EXAMPLE 2   Find a Quadratic Regression Model

The data in the table below were collected on five successive Saturdays.
They show the average number of cars entering a shopping center
parking lot. The value of t is the number of minutes after 9:00 A.M. The value
of N is the number of cars that entered the parking lot in the 10 minutes
prior to the value of t. Find a regression model for this data.

**Average Number of Cars Entering a
Shopping Center Parking Lot**

t	N	t	N
20	70	140	301
40	135	160	298
60	178	180	284
80	210	200	286
100	260	220	260
120	280	240	195

Solution

1. **Construct a scatter diagram for these
   data.** Enter the data into your calculator
   as explained on page 93.

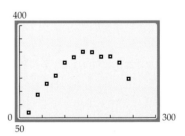

From the scatter diagram, it appears that there is a nonlinear relationship
between the variables.

Continued ▶

---

❷ ANSWER $r^2 \approx 0.991$. This means that 99.1% of the total variation in trade-in value can
be attributed to the odometer reading.

2. **Find the regression equation.** Try a quadratic regression model. For a TI-83 calculator, press $\boxed{\text{STAT}}\ \blacktriangleright\ \boxed{\text{2ND}}\ \text{CALC 5}\ \boxed{\text{ENTER}}$.

```
QuadReg
y=ax²+bx+c
a=-.0124881369
b=3.904433067
c=-7.25
R²=.9840995401
```

take note

In the case of nonlinear regression calculations, the value of *r* is not shown on a TI-83 graphing calculator. In these cases, the coefficient of determination is used to determine how well the data fit the model.

3. **Examine the coefficient of determination.** The coefficient of determination is approximately 0.984. Because this number is fairly close to 1, the regression equation $y = -0.0124881369x^2 + 3.904433067x - 7.25$ provides a good model of the data.

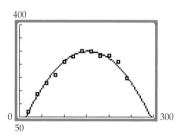

▶ **TRY EXERCISE 32, PAGE 102**

```
LinReg
y=ax+b
a=.6575174825
b=144.2727273
r²=.4193509866
r=.6475731515
```

For Example 2, we could have calculated the *linear* regression line for the data. The results are shown at the left. Note that the coefficient of determination for this calculation is approximately 0.419. Because this number is less than the coefficient of determination for the quadratic model, we choose a quadratic model of the data rather than a linear model.

Now for a final note: The regression line equation does not *prove* that the changes in the dependent variable are *caused* by the independent variable. For instance, suppose various cities throughout the United States were randomly selected and the numbers of gas stations (independent variable) and restaurants (dependent variable) were recorded in a table. If we calculated the regression equation for these data, we would find that *r* would be close to 1. However, this does not mean that gas stations *cause* restaurants to be built. The primary cause is that there are fewer gas stations and restaurants in cities with small populations and greater numbers of gas stations and restaurants in cities with large populations.

TOPICS FOR DISCUSSION

1. What is the purpose of calculating the equation of a regression line?

2. Discuss the implications of the following correlation coefficients: $r = -1$, $r = 0$, and $r = 1$.

3. Discuss the coefficient of determination and what its value says about a data set.

4. What are the implications of $r^2 = 1$ for a nonlinear regression equation?

EXERCISE SET 1.7

 Use a graphing calculator for this Exercise Set.

In Exercises 1 to 4, determine whether the scatter diagram suggests a linear relationship between x and y, a nonlinear relationship between x and y, or no relationship between x and y.

1.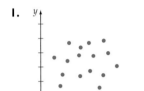

2.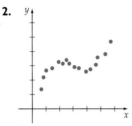

3.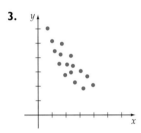

4.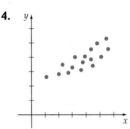

In Exercises 5 and 6, determine for which scatter diagram, A or B, the coefficient of determination is closer to 1.

5.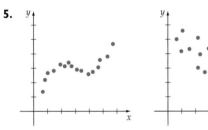

FIGURE A   FIGURE B

6.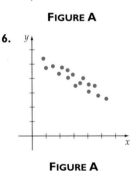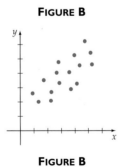

FIGURE A   FIGURE B

In Exercises 7 to 12, find the linear regression equation for the given set.

7. $\{(2, 6), (3, 6), (4, 8), (6, 11), (8, 18)\}$

8. $\{(2, -3), (3, -4), (4, -9), (5, -10), (7, -12)\}$

9. $\{(-3, 11.8), (-1, 9.5), (0, 8.6), (2, 8.7), (5, 5.4)\}$

10. $\{(-7, -11.7), (-5, -9.8), (-3, -8.1), (1, -5.9), (2, -5.7)\}$

11. $\{(1.3, -4.1), (2.6, -0.9), (5.4, 1.2), (6.2, 7.6), (7.5, 10.5)\}$

12. $\{(-1.5, 8.1), (-0.5, 6.2), (3.0, -2.3), (5.4, -7.1), (6.1, -9.6)\}$

In Exercises 13 to 16, find a quadratic model of the given data.

13. $\{(1, -1), (2, 1), (4, 8), (5, 14), (6, 25)\}$

14. $\{(-2, -5), (-1, 0), (0, 1), (1, 4), (2, 4)\}$

15. $\{(1.5, -2.2), (2.2, -4.8), (3.4, -11.2), (5.1, -20.6),$
$(6.3, -28.7)\}$

16. $\{(-2, -1), (-1, -3.1), (0, -2.9), (1, 0.8), (2, 6.8), (3, 15.9)\}$

In Exercises 17 to 32, determine a regression model of the data.

17. ⬤ **ARCHEOLOGY** The data below show the length, in centimeters, of the humerus and the total wingspan, in centimeters, of several pterosaurs, which are extinct flying reptiles of the order Pterosauria. (*Source: Southwest Educational Development Laboratory.*)

Pterosaur Data

Humerus	Wingspan	Humerus	Wingspan
24	600	20	500
32	750	27	570
22	430	15	300
17	370	15	310
13	270	9	240
4.4	68	4.4	55
3.2	53	2.9	50
1.5	24		

a. Compute the linear regression equation for these data.

b. On the basis of this model, what is the projected wingspan of the pterosaur *Quetzalcoatlus northropi*, which is thought to have been the largest of the prehistoric birds, if its humerus is 54 centimeters?

▶ **18.** ● CONSUMER SCIENCE The table below shows the trade-in value for a 2-door, 1996 Ford Explorer in excellent condition for various odometer readings in thousands of miles. (*Source:* Kelley Blue Book web site, May–June 2000 Edition)

Trade-in Value of 1996 Ford Explorer, May 2000

Odometer	Trade-in	Odometer	Trade-in
45	11,635	70	9,710
50	11,435	75	9,460
60	10,735	80	8,985
68	10,060	95	8,260

a. Compute the linear regression equation for these data.

b. On the basis of this model, what is the expected trade-in value of a similar Ford Explorer with 55,000 miles on the odometer?

19. ● BOTANY The data in the table below are based on a study by R. A. Fisher of various flowers of the iris family. The width, in centimeters, and length, in centimeters, of the petal for selected flowers are shown in the table.

Iris Petal Data

Width	Length	Width	Length
2	14	24	56
23	51	10	36
20	52	19	51
13	45	16	47
17	45	14	47
16	31	17	45
14	47	16	31

a. Compute the linear regression equation for these data.

b. On the basis of this model, what is the estimated length of an iris petal if the iris has a petal width of 18 centimeters?

20. ● BOTANY The study by R. A. Fisher (see Exercise 19) also included the width, in centimeters, and length, in centimeters, of the sepal for these flowers. Some of the data are shown below.

Iris Sepal Data

Width	Length	Width	Length
33	50	31	67
31	69	36	46
30	65	27	58
28	57	33	63
25	49	32	70
31	48	25	63
32	70	25	63

a. Compute the linear correlation coefficient for these data.

b. On the basis of the value of the linear correlation coefficient, is a linear model of the data reasonable?

21. ● HEALTH The body mass index (BMI) of a person is a measure of the person's ideal body weight. The table below shows the BMI for different weights for a person 5 feet 6 inches tall. (*Source:* San Diego *Union-Tribune*, May 31, 2000.)

BMI Data for Person 5′ 6″ Tall

Weight (lb)	BMI	Weight (lb)	BMI
110	17	160	25
120	19	170	27
125	20	180	29
135	21	190	30
140	22	200	32
145	23	205	33
150	24	215	34

a. Compute the linear regression equation for these data.

b. On the basis of the model, what is the estimated BMI for a person 5 feet 6 inches tall whose weight is 158 pounds?

22. **HEALTH** The BMI (see Exercise 21) of a person depends on height as well as weight. The table below shows the changes in BMI for a 150-pound person as height (in inches) changes. (*Source: San Diego Union-Tribune*, May 31, 2000.)

BMI Data for 150-Pound Person

Height (in.)	BMI	Height (in.)	BMI
60	29	71	21
62	27	72	20
64	25	73	19
66	24	74	19
67	23	75	18
68	23	76	18
70	21		

a. Compute the linear regression equation for these data.

b. On the basis of the model, what is the estimated BMI for a 150-pound person who is 5 feet 8 inches tall?

23. INDUSTRIAL ENGINEERING Permanent-magnet direct-current motors are used in a variety of industrial applications. For these motors to be effective, there must be a strong linear relationship between the current (in amps, A) supplied to the motor and the resulting torque (in newton-centimeters, N-cm) produced by the motor. A randomly selected motor is chosen from a production line and tested, with the following results.

Direct-Current Motor Data at 12 Volts

Current, in A	Torque, in N-cm	Current, in A	Torque, in N-cm
7.3	9.4	8.5	8.6
11.9	2.8	7.9	4.3
5.6	5.6	14.5	9.5
14.2	4.9	12.7	8.3
7.9	7.0	10.6	4.7

Based on the data in this table, is the chosen motor effective? Explain.

24. **HEALTH SCIENCES** The average remaining lifetime for men in the United States is given in the table below. (*Source: National Institutes of Health.*)

Average Remaining Lifetime for Men

Age	Years	Age	Years
0	73.6	65	15.9
15	59.4	75	9.9
35	40.8		

Based on the data in this table, is there a strong correlation between a man's age and the average remaining lifetime for that man? Explain.

25. **HEALTH SCIENCES** The average remaining lifetime for women in the United States is given in the table below. (*Source: National Institutes of Health.*)

Average Remaining Lifetime for Women

Age	Years	Age	Years
0	79.4	65	19.2
15	65.1	75	12.1
35	45.7		

a. Based on the data in this table, is there a strong correlation between a woman's age and the average remaining lifetime for that woman?

b. Compute the linear regression equation for these data.

c. On the basis of the model, what is the estimated remaining lifetime of a woman of age 25?

26. **BIOLOGY** The table below gives the body lengths, in centimeters, and the highest observed flying speeds, in meters per second, of various animals.

Species	Length	Flying speed
Horsefly	1.3	6.6
Hummingbird	8.1	11.2
Dragonfly	8.5	10.0
Willow warbler	11	12.0
Common pintail	56	22.8

Based on these data, what is the flying speed of a Whimbrel whose length is 41 centimeters? Round to the nearest whole number. (*Source: Based on data from Leiva, Algebra 2: Explorations and Applications, p. 76, McDougall Littell, Boston; copyright 1997.*)

27. AUTOMOTIVE TECHNOLOGY The data in the table below show the horsepower and EPA fuel economy rating for *Car and Driver* magazine's 10 best cars for 2003.

Car	Horsepower	EPA
Acura RSX	160	27
BMW 3-series/M3	184	21
Cheverolet Corvette	350	19
Ford Focus	110	28
Honda Accord	160	26
Infiniti G35	260	20
Mazda 6 S	220	20
Nissan 350Z	287	20
Porsche Boxster	228	19
Subaru Impreza WRX	227	20

Find the linear coefficient of determination for these data, and write a sentence that explains the meaning of the coefficient of determination in the context of this problem. (*Source: www.caranddriver.com.*)

28. HEALTH The table below shows the number of calories burned in 1 hour when running at various speeds.

Running Speed (mph)	Calories Burned
10	1126
10.9	1267
5	563
5.2	633
6	704
6.7	774
7	809
8	950
8.6	985
9	1056
7.5	880

a. Are the data positively or negatively correlated?

b. How many calories does this model predict a person will burn who runs at 9.5 mph for 1 hour? Round to the nearest whole number.

29. BIOLOGY The survival of certain larvae after hatching depends on the temperature (in degrees Celsius) of the surrounding environment. The table below shows the number of larvae that survive at various temperatures. Find a quadratic model of these data.

Larvae Surviving for Various Temperatures

Temp.	Number Surviving	Temp.	Number Surviving
20	40	26	68
21	47	27	67
22	52	28	64
23	61	29	62
24	64	30	61
25	64		

30. METEOROLOGY The temperature at various times on a summer day at a resort in southern California is given in the following table. The variable t is the number of minutes after 6:00 A.M., and the variable T is the temperature in degrees Fahrenheit.

Temperatures at a Resort

Time, t	Temp., T	Time, t	Temp., T
20	59	240	86
40	65	280	88
80	71	320	86
120	78	360	85
160	81	400	80
200	83		

a. Find a quadratic model for these data.

b. Use the model to predict the temperature at 1:00 P.M.

31. AUTOMOTIVE ENGINEERING The fuel efficiency, in miles per gallon, for a certain midsize car at various speeds, in miles per hour, is given in the table below.

Fuel Efficiency of a Midsize Car

mph	mpg	mph	mpg
25	29	55	31
30	32	60	28
35	33	65	24
40	35	70	19
45	34	75	17
50	33		

a. Find a quadratic model for these data.

b. Use the model to predict the fuel efficiency of this car when it is traveling at a speed of 50 mph.

▶ **32.** BIOLOGY The data in the table at the right show the oxygen consumption, in milliliters per minute, of a bird flying level at various speeds in kilometers per hour.

a. Find a quadratic model for these data.

b. Use the model to determine the speed at which the bird has minimum oxygen consumption.

Oxygen Consumption

Speed	Consumption
20	32
25	27
28	22
35	21
42	26
50	34

CONNECTING CONCEPTS

33. PHYSICS Galileo (1564–1642) studied the acceleration due to gravity by allowing balls of various weights to roll down an incline. This allowed him to time the descent of a ball more accurately than by just dropping the ball. The data in the table show some possible results of such an experiment using balls of different masses. Time, t, is measured in seconds; distance, s, is measured in centimeters.

Distance Traveled for Balls of Various Weights

5-Pound Ball		10-Pound Ball		15-Pound Ball	
t	s	t	s	t	s
2	2	3	5	3	5
4	10	6	22	5	15
6	22	9	49	7	30
8	39	12	87	9	49
10	61	15	137	11	75
12	86	18	197	13	103
14	120			15	137
16	156				

a. Find a quadratic model for each of the balls.

b. On the basis of a similar experiment, Galileo concluded that if air resistance is excluded, all falling objects fall with the same acceleration. Explain how one could make such a conclusion from the regression equations.

34. ASTRONOMY In 1929, Edwin Hubble published a paper that revolutionized astronomy ("A Relationship Between Distance and Radial Velocity Among Extra-Galactic Nebulae," *Proceedings of the National Academy of Science*, 168). His paper dealt with the distance an extra-galactic nebula was from the Milky Way galaxy and the nebula's velocity with respect to the Milky Way. The data are given in the table below. Distance is measured in megaparsecs (1 megaparsec equals 1.918×10^{19} miles), and velocity (called the *recession velocity*) is measured in kilometers per second. A negative velocity means the nebula is moving toward the Milky Way; a positive velocity means the nebula is moving away from the Milky Way.

Recession Velocities

Distance	Velocity	Distance	Velocity
0.032	170	0.9	650
0.034	290	0.9	150
0.214	−130	0.9	500
0.263	−70	1.0	920
0.275	−185	1.1	450
0.275	−220	1.1	500
0.45	200	1.4	500
0.5	290	1.7	960
0.5	270	2.0	500
0.63	200	2.0	850
0.8	300	2.0	800
0.9	−30	2.0	1090

a. Find the linear regression model for these data.

b. On the basis of this model, what is the recession velocity of a nebula that is 1.5 megaparsecs from the Milky Way?

35. The data in the table at the right were collected on five successive Saturdays. They show the average number of cars entering a shopping center parking lot. The value of t is the number of minutes after 9:00 A.M. The value of N is the number of cars that entered the parking lot in the 10 minutes prior to the value of t. Does a linear model or a quadratic regression model better fit these data? Explain.

Average Number of Cars Entering a Parking Lot

t	N	t	N
20	70	140	301
40	135	160	298
60	178	180	284
80	210	200	286
100	260	220	260
120	280	240	195

PROJECTS

MEDIAN–MEDIAN LINE Another linear model of data is called the **median–median line**. This line employs *summary points* calculated using the medians of subsets of the independent and dependent variables. The **median** of a data set is the middle number or the average of the two middle numbers for a data set arranged in numerical order. For instance, to find the median of $\{8, 12, 6, 7, 9\}$, first arrange the data in numerical order.

$$6, 7, 8, 9, 12$$

The median is 8, the number in the middle. To find the median of $\{15, 12, 20, 9, 13, 10\}$, arrange the numbers in numerical order.

$$9, 10, 12, 13, 15, 20$$

The median is 12.5, the average of the two middle numbers.

$$\text{Median} = \frac{12 + 13}{2} = 12.5$$

The median–median line is determined by dividing a data set into three equal groups. (If the set cannot be divided into three equal groups, the first and third groups should be equal. For instance, if there are 11 data points, divide the set into groups of 4, 3, and 4.) The slope of the median–median line is the slope of the line through the x-medians and y-medians of the first and third sets of points. The median–median line passes through the average of the x- and y-medians of all three sets.

A graphing calculator can be used to find the median–median line. This line, along with the linear regression line, is shown in the next column for the data in the accompanying table.

1. Find the median–median line for the data in Exercise 17 on page 98.

2. Find the median–median line for the data in Exercise 18 on page 99.

x	y
2	3
3	5
4	4
5	7
6	8
7	9
8	12
9	12
10	14
11	15
12	14

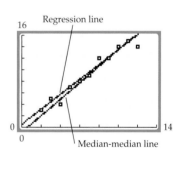

3. Consider the data set $\{(1, 3), (2, 5), (3, 7), (4, 9), (5, 11), (6, 13), (7, 15), (8, 17)\}$.

a. Find the linear regression line for these data.

b. Find the median–median line for these data.

c. What conclusion might you draw from the answers to parts **a.** and **b.**?

4. For this exercise, use the data in the table in Project 1.

a. Calculate the median–median line and the linear regression line.

b. Change the entry (12, 14) to (12, 1) and then recalculate the median–median line and the linear regression line.

c. Explain why there is more change in the linear regression line than in the median–median line.

**EXPLORING CONCEPTS
WITH TECHNOLOGY**

Graphing Piecewise Functions with a Graphing Calculator

A graphing calculator can be used to graph piecewise functions by including as part of the function the interval on which each piece of the function is defined. The method is based on the fact that a graphing calculator "evaluates" inequalities. For purposes of this Exploration, we will use keystrokes for a TI-83 calculator.

For instance, store 3 in **X** by pressing 3 $\boxed{\text{STO►}}$ $\boxed{\text{X,T,Θ,}n}$ $\boxed{\text{ENTER}}$. Now enter the inequality $x > 4$ by pressing $\boxed{\text{X,T,Θ,}n}$ $\boxed{\text{2ND}}$ TEST 3 4 $\boxed{\text{ENTER}}$. Your screen should look like the one at the left. Note that the value of the inequality is 0. This occurs because the calculator replaced **X** by 3 and then determined whether the inequality $3 > 4$ was true or false. The calculator expresses the fact that the inequality is false by placing a zero on the screen. If we repeat the sequence of steps above, except that we store 5 in **X** instead of 3, the calculator will determine that the inequality is true and place a 1 on the screen.

This property of calculators is used to graph piecewise functions. Graphs of these functions work best when Dot mode rather than Connected mode is used. To switch to Dot mode, select $\boxed{\text{MODE}}$, use the arrow keys to highlight $\boxed{\text{DOT}}$, and then press $\boxed{\text{ENTER}}$.

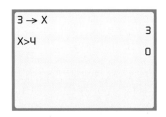

Now we will graph the piecewise function defined by $f(x) = \begin{cases} x, & x \leq -2 \\ x^2, & x > -2 \end{cases}$.

Enter the function[4] as Y₁=X*(X≤-2)+X²*(X>-2) and graph this in the standard viewing window. Note that you are multiplying each piece of the function by its domain. The graph will appear as shown at the left.

To understand how the graph is drawn, we will consider two values of x, -8 and 2, and evaluate Y₁ for each of these values.

Y₁=X*(X≤-2)+X²*(X>-2)

$$= -8(-8 \leq -2) + (-8)^2(-8 > -2)$$

$$= -8(1) + 64(0) = -8 \qquad \bullet \textbf{When } x = -8\textbf{, the value assigned to } -8 \leq -2 \textbf{ is 1; the value assigned to } -8 > -2 \textbf{ is 0.}$$

Y₁=X*(X≤-2)+X²*(X>-2)

$$= 2(2 \leq -2) + 2^2(2 > -2)$$

$$= 2(0) + 4(1) = 4 \qquad \bullet \textbf{When } x = 2\textbf{, the value assigned to } 2 \leq -2 \textbf{ is 0; the value assigned to } 2 > -2 \textbf{ is 1.}$$

In a similar manner, for any value of x for which $x \leq -2$, the value assigned to (X≤-2) is 1 and the value assigned to (X>-2) is 0. Thus Y₁=X*1+X²*0=X on that interval. This means that only the $f(x) = x$ piece of the function is graphed. When $x > -2$, the value assigned to (X≤-2) is 0 and the value assigned to (X>-2) is 1. Thus Y₁=X*0+X²*1=X² on that interval. This means that only the $f(x) = x^2$ piece of the function is graphed on that interval.

---

[4]Note that pressing $\boxed{\text{2ND}}$ TEST will display the inequality menu.

1. Graph: $f(x) = \begin{cases} x^2, & x < 2 \\ -x, & x \geq 2 \end{cases}$ **2.** Graph: $f(x) = \begin{cases} x^2 - x, & x < 2 \\ -x + 4, & x \geq 2 \end{cases}$

3. Graph: $f(x) = \begin{cases} -x^2 + 1, & x < 0 \\ x^2 - 1, & x \geq 0 \end{cases}$ **4.** Graph: $f(x) = \begin{cases} x^3 - 4x, & x < 1 \\ x^2 - x + 2, & x \geq 1 \end{cases}$

CHAPTER 1 SUMMARY

1.1 Equations and Inequalities

- The *integers* are the set $\{\ldots, -3, -2, -1, 0, 1, 2, 3, \ldots\}$. The positive integers are called *natural numbers*. The *rational numbers* are $\left\{ \dfrac{a}{b} \,\middle|\, a, b \text{ integers and } b \neq 0 \right\}$. *Irrational numbers* are nonterminating and nonrepeating decimals. The *real numbers* are the set of rational and irrational numbers.

- The interval of real numbers $(a, b) = \{x \mid a < x < b\}$.

- A *linear equation* is one that can be written in the form $ax + b = 0$, $a \neq 0$. A *quadratic equation* is one that can be written in the form $ax^2 + bx + c = 0$, $a \neq 0$.

- An *inequality* is a statement that contains the symbol $<$, $\leq$, $>$, or $\geq$.

1.2 A Two-Dimensional Coordinate System and Graphs

- *The Distance Formula*   The distance d between the points represented by (x_1, y_1) and (x_2, y_2) is

$$d = \sqrt{(x_2 - x_1)^2 + (y_2 - y_1)^2}$$

- The midpoint of the line segment from $P_1(x_1, y_1)$ to $P_2(x_2, y_2)$ is

$$\left(\frac{x_1 + x_2}{2}, \frac{y_1 + y_2}{2} \right)$$

- The standard form of the equation of a circle with center at (h, k) and radius r is $(x - h)^2 + (y - k)^2 = r^2$.

1.3 Introduction to Functions

- *Definition of a Function*   A function is a set of ordered pairs in which no two ordered pairs that have the same first coordinate have different second coordinates.

- A graph is the graph of a function if and only if no vertical line intersects the graph at more than one point. If every horizontal line intersects the graph of a function at most once, then the graph is the graph of a one-to-one function.

1.4 Properties of Graphs

- The graph of an equation is symmetric with respect to

  the y-axis if the replacement of x with $-x$ leaves the equation unaltered.

  the x-axis if the replacement of y with $-y$ leaves the equation unaltered.

  the origin if the replacement of x with $-x$ and y with $-y$ leaves the equation unaltered.

- If f is a function and c is a positive constant, then

  $y = f(x) + c$ is the graph of $y = f(x)$ shifted up *vertically* c units

  $y = f(x) - c$ is the graph of $y = f(x)$ shifted down *vertically* c units

  $y = f(x + c)$ is the graph of $y = f(x)$ shifted left *horizontally* c units

  $y = f(x - c)$ is the graph of $y = f(x)$ shifted right *horizontally* c units

- The graph of

  $y = -f(x)$ is the graph of $y = f(x)$ reflected across the x-axis.

  $y = f(-x)$ is the graph of $y = f(x)$ reflected across the y-axis.

1.5 The Algebra of Functions

- For all values of x for which both $f(x)$ and $g(x)$ are defined, we define the following functions.

  Sum $(f + g)(x) = f(x) + g(x)$

  Difference $(f - g)(x) = f(x) - g(x)$

  Product $(fg)(x) = f(x) \cdot g(x)$

  Quotient $\left(\dfrac{f}{g} \right)(x) = \dfrac{f(x)}{g(x)}, \quad g(x) \neq 0$

- The expression

$$\frac{f(x + h) - f(x)}{h}, \quad h \neq 0$$

is called the difference quotient of f. The difference quotient is an important function because it can be used to compute the *average rate of change* of f over the time interval $[x, x + h]$.

- For the functions f and g, the composite function, or composition, of f by g is given by $(g \circ f)(x) = g[f(x)]$ for all x in the domain of f such that $f(x)$ is in the domain of g.

1.6 Inverse Functions

- If f is a one-to-one function with domain X and range Y, and g is a function with domain Y and range X, then g is the inverse function of f if and only if $(f \circ g)(x) = x$ for all x in the domain of g and $(g \circ f)(x) = x$ for all x in the domain of f.

- A function f has an inverse function if and only if it is a one-to-one function. The graph of a function f and the graph of its inverse function f^{-1} are symmetric with respect to the line given by $y = x$.

1.7 Modeling Data Using Regression

- Regression analysis is used to find a mathematical model of collected data.

- The least-squares regression line is the line that minimizes the sum of the squares of the vertical deviations of all data points from the line.

- The linear correlation coefficient r is a measure of how closely the points of a data set can be modeled by a straight line. If $r = -1$, then the points of the data set can be modeled *exactly* by a straight line with negative slope. If $r = 1$, then the data set can be modeled *exactly* by a straight line with positive slope. For all data sets, $-1 \leq r \leq 1$.

- The coefficient of determination is r^2. It measures the percent of the total variation in the dependent variable that is explained by the regression line.

- It is possible to find both linear and nonlinear mathematical models of data.

CHAPTER 1 TRUE/FALSE EXERCISES

In Exercises 1 to 13, answer true or false. If the statement is false, state a reason or give an example to show that the statement is false.

1. Let f be any function. Then $f(a) = f(b)$ implies that $a = b$.

2. Every function has an inverse function.

3. If $(f \circ g)(a) = a$ and $(g \circ f)(a) = a$ for some constant a, then f and g are inverse functions.

4. Let f be a function such that $f(x) = f(x + 4)$ for all real numbers x. If $f(2) = 3$, then $f(18) = 3$.

5. For all functions f, $[f(x)]^2 = f[f(x)]$.

6. Let f be any function. Then for all a and b in the domain of f such that $f(b) \neq 0$ and $b \neq 0$,

$$\frac{f(a)}{f(b)} = \frac{a}{b}$$

7. The identity function $f(x) = x$ is its own inverse.

8. If f is defined by $f(x) = |x|$, then $f(a + b) = f(a) + f(b)$ for all real numbers a and b.

9. If f is defined by $f(x) = |x|$, then $f(ab) = f(a)f(b)$ for all real numbers a and b.

10. If f is a one-to-one function and a and b are real numbers in the domain of f with $a \neq b$, then $f(a) \neq f(b)$.

11. The coordinates of a point on the graph of $y = f(x)$ are (a, b). If k is a positive constant, then (a, kb) are the coordinates of a point on the graph of $y = kf(x)$.

12. For every function f, the real number c is a solution of $f(x) = 0$ if and only if $(c, 0)$ is an x-intercept of the graph of $y = f(x)$.

13. If the correlation coefficient r equals -1, then all of the points in the data set lie on a straight line.

CHAPTER 1 REVIEW EXERCISES

In Exercises 1 to 14, solve each equation or inequality.

1. $3 - 4z = 12$

2. $4y - 3 = 6y + 5$

3. $2x - 3(2 - 3x) = 14x$

4. $5 - 2(3m + 2) = 3(1 - m)$

5. $y^2 - 3y - 18 = 0$

6. $2z^2 - 9z + 4 = 0$

7. $3v^2 + v = 1$

8. $3s = 4 - 2s^2$

9. $3c - 5 \le 5c + 7$

10. $7a > 5 - 2(3a - 4)$

11. $x^2 - x - 12 \ge 0$

12. $2x^2 - x < 1$

13. $|2x - 5| > 3$

14. $|1 - 3x| \le 4$

In Exercises 15 to 18, use the Pythagorean Theorem, $c^2 = a^2 + b^2$, to find the unknown. The letters a and b represent the lengths of the legs of a right triangle, and c represents the length of the hypotenuse.

15. $a = 6$, $b = 8$. Find c.

16. $a = 11$, $c = 20$. Find b.

17. $b = 12$, $c = 13$. Find a.

18. $a = 7$, $b = 14$. Find c.

In Exercises 19 and 20, find the distance between the points whose coordinates are given.

19. $(-3, 2)$ $(7, 11)$

20. $(5, -4)$ $(-3, -8)$

In Exercises 21 and 22, find the midpoint of the line segment with the given endpoints.

21. $(2, 8)$ $(-3, 12)$

22. $(-4, 7)$ $(8, -11)$

In Exercises 23 and 24, complete the graph so that it is symmetric with respect to $a.$ the x-axis, $b.$ the y-axis, and $c.$ the origin.

23.

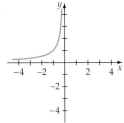

24.

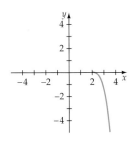

In Exercises 25 to 32, determine whether the graph of each equation is symmetric with respect to the $a.$ x-axis, $b.$ y-axis, $c.$ origin.

25. $y = x^2 - 7$

26. $x = y^2 + 3$

27. $y = x^3 - 4x$

28. $y^2 = x^2 + 4$

29. $\dfrac{x^2}{3^2} + \dfrac{y^2}{4^2} = 1$

30. $xy = 8$

31. $|y| = |x|$

32. $|x + y| = 4$

In Exercises 33 and 34, determine the center and radius of the circle with the given equation.

33. $(x - 3)^2 + (y + 4)^2 = 81$

34. $x^2 + y^2 + 10x + 4y + 20 = 0$

In Exercises 35 and 36, find the equation in standard form of a circle that satisfies the given conditions.

35. Center $C = (2, -3)$, radius $r = 5$

36. Center $C = (-5, 1)$, passing through $(3, 1)$

37. If $f(x) = 3x^2 + 4x - 5$, find

   a. $f(1)$ **b.** $f(-3)$ **c.** $f(t)$

   d. $f(x + h)$ **e.** $3f(t)$ **f.** $f(3t)$

38. If $g(x) = \sqrt{64 - x^2}$, find

   a. $g(3)$ **b.** $g(-5)$ **c.** $g(8)$

   d. $g(-x)$ **e.** $2g(t)$ **f.** $g(2t)$

39. If $f(x) = x^2 + 4x$ and $g(x) = x - 8$, find

   a. $(f \circ g)(3)$ **b.** $(g \circ f)(-3)$

   c. $(f \circ g)(x)$ **d.** $(g \circ f)(x)$

40. If $f(x) = 2x^2 + 7$ and $g(x) = |x - 1|$, find

   a. $(f \circ g)(-5)$ **b.** $(g \circ f)(-5)$

   c. $(f \circ g)(x)$ **d.** $(g \circ f)(x)$

41. If $f(x) = 4x^2 - 3x - 1$, find the difference quotient

$$\frac{f(x + h) - f(x)}{h}$$

42. If $g(x) = x^3 - x$, find the difference quotient

$$\frac{g(x + h) - g(x)}{h}$$

In Exercises 43 to 46, determine the domain of the function represented by the given equation.

43. $f(x) = -2x^2 + 3$

44. $f(x) = \sqrt{6 - x}$

45. $f(x) = \sqrt{25 - x^2}$

46. $f(x) = \dfrac{3}{x^2 - 2x - 15}$

In Exercises 47 to 52, sketch the graph of f. Find the interval(s) in which f is a. increasing, b. constant, c. decreasing.

47. $f(x) = |x - 3| - 2$

48. $f(x) = x^2 - 5$

49. $f(x) = |x + 2| - |x - 2|$

50. $f(x) = [\![x + 3]\!]$

51. $f(x) = \dfrac{1}{2}x - 3$

52. $f(x) = \sqrt[3]{x}$

In Exercises 53 to 58, sketch the graph of g. a. Find the domain and the range of g. b. State whether g is even, odd, or neither even nor odd.

53. $g(x) = -x^2 + 4$

54. $g(x) = -2x - 4$

55. $g(x) = |x - 2| + |x + 2|$

56. $g(x) = \sqrt{16 - x^2}$

57. $g(x) = x^3 - x$

58. $g(x) = 2[\![x]\!]$

In Exercises 59 and 60, use the given functions f and g to find $f + g$, $f - g$, fg, and $\dfrac{f}{g}$. State the domain of each.

59. $f(x) = x^2 - 9$, $g(x) = x + 3$

60. $f(x) = x^3 + 8$, $g(x) = x^2 - 2x + 4$

In Exercises 61 to 64, determine whether the given functions are inverses.

61. $F(x) = 2x - 5$ $G(x) = \dfrac{x + 5}{2}$

62. $h(x) = \sqrt{x}$ $k(x) = x^2$ $x \geq 0$

63. $l(x) = \dfrac{x + 3}{x}$ $m(x) = \dfrac{3}{x - 1}$

64. $p(x) = \dfrac{x - 5}{2x}$ $q(x) = \dfrac{2x}{x - 5}$

In Exercises 65 to 68, find the inverse of the function. Sketch the graph of the function and its inverse on the same set of coordinate axes.

65. $f(x) = 3x - 4$

66. $g(x) = -2x + 3$

67. $h(x) = -\dfrac{1}{2}x - 2$

68. $k(x) = \dfrac{1}{x}$

69. Find two numbers whose sum is 50 and whose product is a maximum.

70. Find two numbers whose difference is 10 and the sum of whose squares is a minimum.

71. The roadway of the Golden Gate Bridge is 220 feet above the water. The height h of a rock dropped from the bridge is a function of the time t it has fallen. If h is measured in feet and t is measured in seconds, then the function is given by $h(t) = -16t^2 + 220$. Use the function to find the time it will take the rock to hit the water.

72. The suspension cables of the Golden Gate Bridge approximate the shape of a parabola. If h is the height of the cables above the roadway in feet, and $|x|$ is the distance in feet from the center of the bridge, then the parabolic shape of the cables is represented by

$$h(x) = \frac{1}{8820}x^2 + 25, \quad -2100 \leq x \leq 2100$$

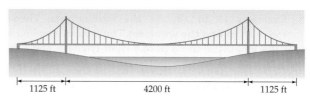

| 1125 ft | 4200 ft | 1125 ft |

a. Find the height of the cables 1050 feet from the center of the bridge.

b. The towers that support the cables are 2100 feet from the center of the bridge. Find the height (above the roadway) of the towers that support the cables.

73. **COMPUTER SCIENCE** A test of an Internet service provider showed the following download times (in seconds) for files of various sizes (in kilobytes).

Download times

Size	Time	Size	Time
10.5	0.20	110	2.01
12.9	0.24	156	2.68
15	0.27	163	2.87
20	0.36	175	3.10
60	1.09	200	3.64
75	1.42	250	4.61

a. Find a linear regression model for this data.

b. From the value of r, is a linear model of this data a reasonable model? Explain.

c. On the basis of the model, what is the expected download time of a file 100 kilobytes in size?

74. **Physics** The rate at which water will escape from the bottom of a can depends on a number of factors, which include the height of the water, the size of the hole, and the diameter of the can. The table below shows the height (in millimeters) of water in a can after t seconds.

Water Escaping a Ruptured Can

Height	Time	Height	Time
0	180	60	93
10	163	70	81
20	147	80	70
30	133	90	60
40	118	100	50
50	105	110	48

a. Find the quadratic regression model for these data.

b. On the basis of this model, will the can ever empty?

c. Explain why there seems to be a contradiction between the model and reality in that we know the can will eventually run out of water.

CHAPTER 1 TEST

1. Solve: $4x - 2(2 - x) = 5 - 3(2x + 1)$

2. Solve: $6 - 3x \geq 3 - 4(2 - 2x)$

3. Solve: $2x^2 - 3x = 2$ **4.** Solve: $3x^2 - x = 2$

5. Solve: $|4 - 5x| > 6$

6. Find the distance between the points $(-2, 5)$ and $(4, -2)$.

7. Find the midpoint and the length of the line segment with endpoints $(-2, 3)$ and $(4, -1)$.

8. Determine the x- and y-intercepts, and then graph the equation $x = 2y^2 - 4$.

9. Graph the equation $y = |x + 2| + 1$.

10. Find the center and radius of the circle that has the general form $x^2 - 4x + y^2 + 2y - 4 = 0$.

11. Given $f(x) = -\sqrt{25 - x^2}$, evaluate $f(-3)$.

12. Determine the domain of the function defined by
$$f(x) = -\sqrt{x^2 - 16}.$$

13. Graph $f(x) = -2|x - 2| + 1$. Identify the intervals over which the function is

a. increasing   **b.** constant   **c.** decreasing

14. Use the graph of $f(x) = |x|$ to graph $y = -f(x + 2) - 1$.

15. Which of the following define odd functions?

a. $f(x) = x^4 - x^2$ **b.** $f(x) = x^3 - x$

c. $f(x) = x - 1$

16. Let $f(x) = x^2 - 1$ and $g(x) = x - 2$. Find $(f + g)$ and $\left(\dfrac{f}{g}\right)$.

17. Find the difference quotient of the function
$$f(x) = x^2 + 1$$

18. Evaluate $(f \circ g)$, where
$$f(x) = x^2 - 2x + 1 \quad \text{and} \quad g(x) = \sqrt{x - 2}$$

19. Find the inverse of the function given by the equation
$$f(x) = \frac{x}{x + 1}$$

20. The table below shows the percent of water and the number of calories in various canned soups to which 100 grams of water are added.

Percent Water in Soups

% Water	Calories
93.2	28
92.3	26
91.9	39
89.5	56

% Water	Calories
89.6	56
90.5	36
91.9	32
91.7	32

a. Find the linear regression line for these data.

b. Using the linear model from part **a.,** find the expected number of calories in a soup that is 89% water.

CHAPTER 2

TRIGONOMETRIC FUNCTIONS

Applications of Trigonometric Functions

In this chapter we introduce trigonometric functions. The following illustrations are from three of the applications of trigonometric functions that are considered in this chapter. The path of a satellite is modeled by a trigonometric function in **Exercise 77, page 161.**

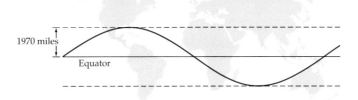

In **Exercise 77, page 140,** trigonometric functions are used to find the height and the length of the skybridge between the Petronas Towers. See the figure at the left.

In **Exercise 70, page 139,** a trigonometric function is used to calculate the distance from Venus to the sun.

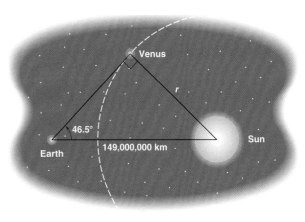

AB = 412 feet
$\angle CAB = 53.6°$
$\overline{AB}$ is at ground level
$\angle CAD = 15.5°$

Generalizing Problem-Solving Strategies

Many times in mathematics a strategy for solving one type of problem can be applied in other circumstances. For instance, earlier in this text we discussed graphing techniques that involved compressing, stretching, and reflecting the graphs of various types of functions. For instance, suppose that f is defined by the graph at the right.

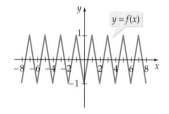

Then, using the techniques described earlier, the graphs of $y = f(2x)$, $y = f\left(\dfrac{1}{2}x\right)$, and $y = -f(x)$ are

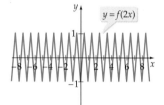 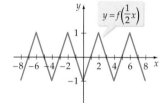 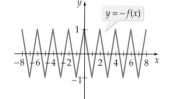

In this chapter we begin the study of trigonometric functions. One such trigonometric function is the sine function. As you will learn in this chapter, the graph of the sine function is as shown at the right.

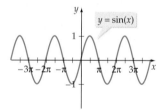

Applying the graphing techniques presented earlier, we have

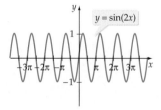 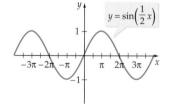 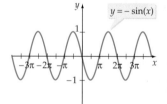

For exercises similar to these, see Exercises 33 to 38 on page 187.

ANGLES AND ARCS

A point P on a line separates the line into two parts, each of which is called a **half-line.** The union of point P and the half-line formed by P that includes point A is called a **ray,** and it is represented as $\overrightarrow{PA}$. The point P is the **endpoint** of ray $\overrightarrow{PA}$. **Figure 2.1** shows the ray $\overrightarrow{PA}$ and a second ray $\overrightarrow{QR}$.

In geometry, an *angle* is defined simply as the union of two rays that have a common endpoint. In trigonometry and many advanced mathematics courses, it is beneficial to define an angle in terms of a rotation.

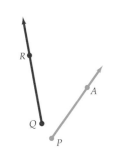

FIGURE 2.1

Definition of an Angle

An **angle** is formed by rotating a given ray about its endpoint to some terminal position. The original ray is the **initial side** of the angle, and the second ray is the **terminal side** of the angle. The common endpoint is the **vertex** of the angle.

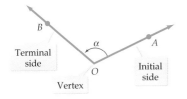

FIGURE 2.2

There are several methods used to name an angle. One way is to employ Greek letters. For example, the angle shown in **Figure 2.2** can be designated as α or as $\angle\alpha$. It also can be named $\angle O$, $\angle AOB$, or $\angle BOA$. If you name an angle by using three points, such as $\angle AOB$, it is traditional to list the vertex point between the other two points.

Angles formed by a counterclockwise rotation are considered **positive angles,** and angles formed by a clockwise rotation are considered **negative angles.** See **Figure 2.3.**

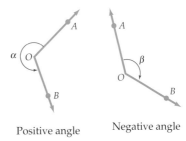

Positive angle   Negative angle

FIGURE 2.3

● DEGREE MEASURE

The **measure** of an angle is determined by the amount of rotation of the initial ray. The concept of measuring angles in *degrees* grew out of the belief of the early Sumerians and Babylonians that the seasons repeated every 360 days.

Definition of Degree

One **degree** is the measure of an angle formed by rotating a ray $\frac{1}{360}$ of a complete revolution. The symbol for degree is °.

The angle shown in **Figure 2.4** has a measure of 1°. The angle β shown in **Figure 2.5** has a measure of 30°. We will use the notation $\beta = 30°$ to denote that the measure of angle β is 30°. The protractor shown in **Figure 2.6** can be used to measure an angle in degrees or to draw an angle with a given degree measure.

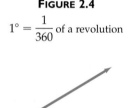

FIGURE 2.4

$1° = \frac{1}{360}$ of a revolution

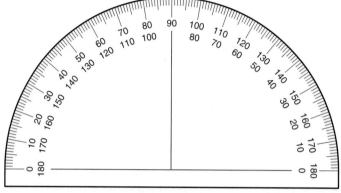

FIGURE 2.5

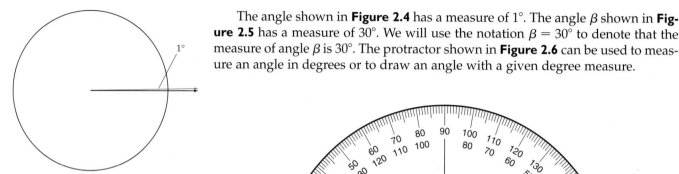

FIGURE 2.6

Protractor for measuring angles in degrees

● CLASSIFICATION OF ANGLES

Angles are often classified according to their measure.

● 180° angles are **straight angles**. See **Figure 2.7a.**

● 90° angles are **right angles**. See **Figure 2.7b.**

● Angles that have a measure greater than 0° but less than 90° are **acute angles**. See **Figure 2.7c.**

● Angles that have a measure greater than 90° but less than 180° are **obtuse angles**. See **Figure 2.7d.**

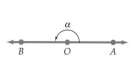

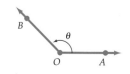

a. Straight angle ($\alpha = 180°$)   **b.** Right angle ($\beta = 90°$)   **c.** Acute angle ($0° < \theta < 90°$)   **d.** Obtuse angle ($90° < \theta < 180°$)

FIGURE 2.7

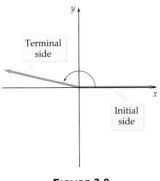

FIGURE 2.8
An angle in standard position

An angle superimposed in a Cartesian coordinate system is in **standard position** if its vertex is at the origin and its initial side is on the positive x-axis. See **Figure 2.8.**

Two positive angles are **complementary angles** **(Figure 2.9a)** if the sum of the measures of the angles is 90°. Each angle is the *complement* of the other angle. Two positive angles are **supplementary angles** **(Figure 2.9b)** if the sum of the measures of the angles is 180°. Each angle is the *supplement* of the other angle.

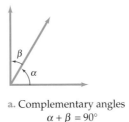

a. Complementary angles
$\alpha + \beta = 90°$

b. Supplementary angles
$\alpha + \beta = 180°$

FIGURE 2.9

EXAMPLE 1   **Find the Measure of the Complement and the Supplement of an Angle**

For each angle, find the measure (if possible) of its complement and of its supplement.

a. $\theta = 40°$ b. $\theta = 125°$

Solution

a. **Figure 2.10** shows $\angle\theta = 40°$ in standard position. The measure of its complement is $90° - 40° = 50°$. The measure of its supplement is $180° - 40° = 140°$.

b. **Figure 2.11** shows $\angle\theta = 125°$ in standard position. Angle θ does not have a complement because there is no positive number x such that

$$x° + 125° = 90°$$

The measure of its supplement is $180° - 125° = 55°$.

▶ **TRY EXERCISE 2, PAGE 124**

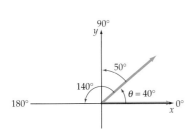

FIGURE 2.10

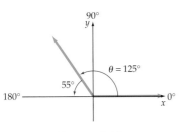

FIGURE 2.11

❓ **QUESTION** Are the two acute angles of any right triangle complementary angles? Explain.

Some angles have a measure greater than 360°. See **Figure 2.12a** and **Figure 2.12b.** The angle shown in **Figure 2.12c** has a measure less than −360°,

❓ **ANSWER** Yes. The sum of the measures of the angles of any triangle is 180°. The right angle has a measure of 90°. Thus the measure of the sum of the two acute angles must be $180° - 90° = 90°$.

because it is formed by a clockwise rotation of more than one revolution of the initial ray.

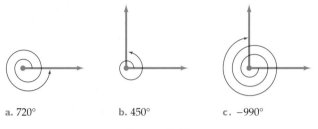

a. 720°   **b.** 450°   **c.** −990°

FIGURE 2.12

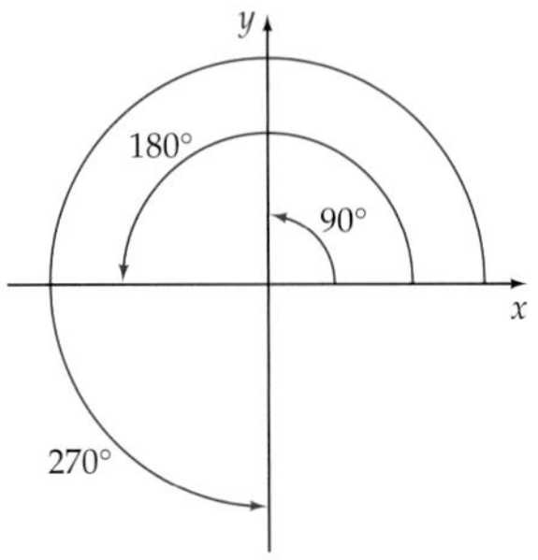

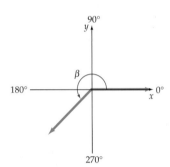

FIGURE 2.13

If the terminal side of an angle in standard position lies on a coordinate axis, then the angle is classified as a **quadrantal angle.** For example, the 90° angle, the 180° angle, and the 270° angle shown in **Figure 2.13** are all quadrantal angles.

If the terminal side of an angle in standard position does not lie on a coordinate axis, then the angle is classified according to the quadrant that contains the terminal side. For example, $\angle\beta$ in **Figure 2.14** is a Quadrant III angle.

Angles in standard position that have the same terminal sides are **coterminal angles.** Every angle has an unlimited number of coterminal angles. **Figure 2.15** shows $\angle\theta$ and two of its coterminal angles, labeled $\angle 1$ and $\angle 2$.

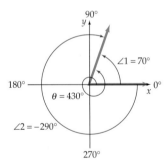

FIGURE 2.14

Measures of Coterminal Angles

Given $\angle\theta$ in standard position with measure $x°$, then the measures of the angles that are coterminal with $\angle\theta$ are given by

$$x° + k \cdot 360°$$

where k is an integer.

This theorem states that the measures of any two coterminal angles differ by an integer multiple of 360°. For instance, in **Figure 2.15**, $\theta = 430°$,

$$\angle 1 = 430° + (-1) \cdot 360° = 70°, \quad \text{and}$$
$$\angle 2 = 430° + (-2) \cdot 360° = -290°$$

If we add positive multiples of 360° to 430°, we find that the angles with measures 790°, 1150°, 1510°, ... are also coterminal with $\angle\theta$.

FIGURE 2.15

EXAMPLE 2   Classify by Quadrant and Find a Coterminal Angle

Assume the following angles are in standard position. Classify each angle by quadrant, and then determine the measure of the positive angle with measure less than 360° that is coterminal with the given angle.

a. $\alpha = 550°$ **b.** $\beta = -225°$ **c.** $\gamma = 1105°$

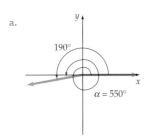

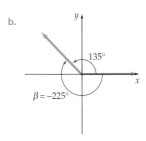

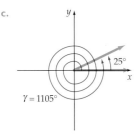

FIGURE 2.16

Solution

a. Because $550° = 190° + 360°$, $\angle\alpha$ is coterminal with an angle that has measure of $190°$. $\angle\alpha$ is a Quadrant III angle. See **Figure 2.16a.**

b. Because $-225° = 135° + (-1)\cdot 360°$, $\angle\beta$ is coterminal with an angle that has measure of $135°$. $\angle\beta$ is a Quadrant II angle. See **Figure 2.16b.**

c. $1105° \div 360° = 3\dfrac{5}{72}$. Thus $\angle\gamma$ is an angle formed by three complete counterclockwise rotations, plus $\dfrac{5}{72}$ of a rotation. To convert $\dfrac{5}{72}$ of a rotation to degrees, multiply $\dfrac{5}{72}$ times $360°$.

$$\frac{5}{72}\cdot 360° = 25°$$

Thus $1105° = 25° + 3\cdot 360°$. Hence $\angle\gamma$ is coterminal with an angle that has a measure of $25°$. $\angle\gamma$ is a Quadrant I angle. See **Figure 2.16c.**

▶ **TRY EXERCISE 14, PAGE 124**

● CONVERSION BETWEEN UNITS

There are two popular methods for representing a fractional part of a degree. One is the decimal degree method. For example, the measure $29.76°$ is a decimal degree. It means

$$29° \text{ plus } 76 \text{ hundredths of } 1°$$

A second method of measurement is known as the DMS (**D**egree, **M**inute, **S**econd) method. In the DMS method, a degree is subdivided into 60 equal parts, each of which is called a *minute,* denoted by ′. Thus $1° = 60'$. Furthermore, a minute is subdivided into 60 equal parts, each of which is called a *second,* denoted by ″. Thus $1' = 60''$ and $1° = 3600''$. The fractions

$$\frac{1°}{60'} = 1, \qquad \frac{1'}{60''} = 1, \quad \text{and} \quad \frac{1°}{3600''} = 1$$

are another way of expressing the relationships among degrees, minutes, and seconds. Each of the fractions is known as a **unit fraction** or a **conversion factor.** Because all conversion factors are equal to 1, you can multiply a magnitude by a conversion factor and not change the magnitude, even though you change the units used to express the magnitude. The following illustrates the process of multiplying by conversion factors to write $126°12'27''$ as a decimal degree.

$$126°12'27'' = 126° + 12' + 27''$$

$$= 126° + 12'\left(\frac{1°}{60'}\right) + 27''\left(\frac{1°}{3600''}\right)$$

$$= 126° + 0.2° + 0.0075° = 126.2075°$$

INTEGRATING TECHNOLOGY

Many graphing calculators can be used to convert a decimal degree measure to its equivalent DMS measure, and vice versa. For instance, **Figure 2.17** shows that 31.57° is equivalent to 31°34′12″. On a TI-83 graphing calculator, the degree symbol, °, and the DMS function are in the ANGLE menu.

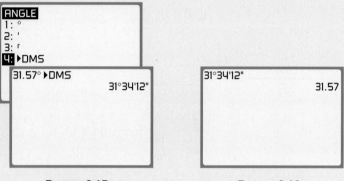

FIGURE 2.17   **FIGURE 2.18**

To convert a DMS measure to its equivalent decimal degree measure, enter the DMS measure and press $\boxed{\text{ENTER}}$. The calculator screen in **Figure 2.18** shows that 31°34′12″ is equivalent to 31.57°. A TI-83 needs to be in degree mode to produce the results displayed in **Figures 2.17** and **2.18**. On a TI-83, the degree symbol, °, and the minute symbol, ′, are both in the ANGLE menu; however, the second symbol, ″, is entered by pressing $\boxed{\text{ALPHA}}$ +.

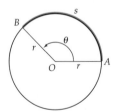

FIGURE 2.19

● RADIAN MEASURE

Another commonly used angle measurement is the *radian*. To define a radian, first consider a circle of radius r and two radii $\overline{OA}$ and $\overline{OB}$. The angle θ formed by the two radii is a **central angle.** The portion of the circle between A and B is an **arc** of the circle and is written $\overset{\frown}{AB}$. We say that $\overset{\frown}{AB}$ *subtends* the angle θ. The length of $\overset{\frown}{AB}$ is s (see **Figure 2.19**).

Definition of Radian

One **radian** is the measure of the central angle subtended by an arc of length r on a circle of radius r. See **Figure 2.20.**

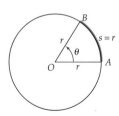

FIGURE 2.20

Central angle θ has a measure of 1 radian.

Figure 2.21 shows a protractor that can be used to measure angles in radians or to construct angles given in radian measure.

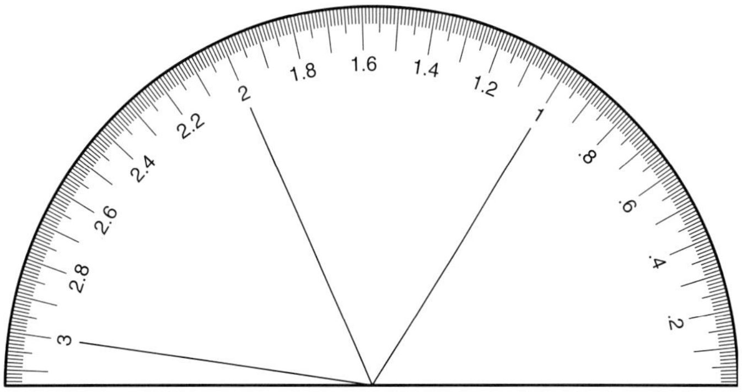

FIGURE 2.21

Protractor for measuring angles in radians

Radian Measure

Given an arc of length s on a circle of radius r, the measure of the central angle subtended by the arc is $\theta = \dfrac{s}{r}$ radians.

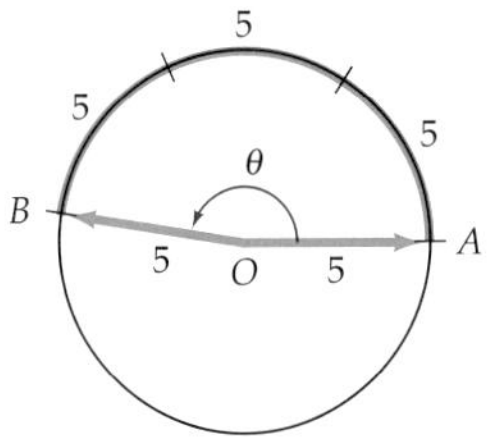

FIGURE 2.22

Central angle θ has a
measure of 3 radians.

As an example, consider that an arc of length 15 centimeters on a circle with a radius of 5 centimeters subtends an angle of 3 radians, as shown in **Figure 2.22.** The same result can be found by dividing 15 centimeters by 5 centimeters.

To find the measure in radians of any central angle θ, divide the length s of the arc that subtends θ by the length of the radius of the circle. Using the formula for radian measure, we find that an arc of length 12 centimeters on a circle of radius 8 centimeters subtends a central angle θ whose measure is

$$\theta = \frac{s}{r} \text{ radians} = \frac{12 \text{ centimeters}}{8 \text{ centimeters}} \text{ radians} = \frac{3}{2} \text{ radians}$$

Note that the centimeter units are *not* part of the final result. The radian measure of a central angle formed by an arc of length 12 miles on a circle of radius 8 miles would be the same, $\dfrac{3}{2}$ radians. We say that radian is a *dimensionless* quantity because there are no units of measurement associated with a radian.

Recall that the circumference of a circle is given by the equation $C = 2\pi r$. The radian measure of the central angle θ subtended by the circumference is $\theta = \dfrac{2\pi r}{r} = 2\pi$. In degree measure, the central angle θ subtended by the circumference is 360°. Thus we have the relationship 360° = 2π radians. Dividing each side of the equation by 2 gives 180° = π radians. From this last equation, we can establish the following conversion factors.

INTEGRATING TECHNOLOGY

A calculator shows that

1 radian $\approx 57.29577951°$

and

$1° \approx 0.017453293$ radian

Radian-Degree Conversion

• To convert from radians to degrees, multiply by $\left(\dfrac{180°}{\pi \text{ radians}}\right)$.

• To convert from degrees to radians, multiply by $\left(\dfrac{\pi \text{ radians}}{180°}\right)$.

EXAMPLE 3   **Convert Degree Measure to Radian Measure**

Convert 300° to radians.

Solution

$$300° = 300°\left(\frac{\pi \text{ radians}}{180°}\right)$$

$$= \frac{5}{3}\pi \text{ radians} \qquad \bullet \textbf{ Exact answer}$$

$$\approx 5.23598776 \text{ radians} \qquad \bullet \textbf{Approximate answer}$$

▶ **TRY EXERCISE 32, PAGE 124**

In Example 3, note that $\dfrac{5}{3}\pi$ radians is an exact result. Many times it will be convenient to leave the measure of an angle in terms of π and not change it to a decimal approximation.

EXAMPLE 4   **Convert Radian Measure to Degree Measure**

Convert $-\dfrac{3}{4}\pi$ radians to degrees.

Solution

$$-\frac{3}{4}\pi \text{ radians} = -\frac{3}{4}\pi \text{ radians}\left(\frac{180°}{\pi \text{ radians}}\right) = -135°$$

▶ **TRY EXERCISE 40, PAGE 124**

Table 2.1 lists the degree and radian measures of selected angles. **Figure 2.23** illustrates each angle as measured from the positive x-axis.

TABLE 2.1

Degrees	Radians
0	0
30	$\pi/6$
45	$\pi/4$
60	$\pi/3$
90	$\pi/2$
120	$2\pi/3$
135	$3\pi/4$
150	$5\pi/6$
180	π
210	$7\pi/6$
225	$5\pi/4$
240	$4\pi/3$
270	$3\pi/2$
300	$5\pi/3$
315	$7\pi/4$
330	$11\pi/6$
360	2π

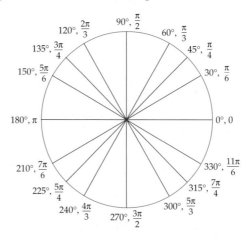

FIGURE 2.23

Degree and radian measures of selected angles

FIGURE 2.24

FIGURE 2.25

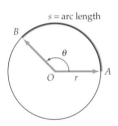

FIGURE 2.26
$s = r\theta$

INTEGRATING TECHNOLOGY

A graphing calculator can convert degree measure to radian measure, and vice versa. For example, the calculator display in **Figure 2.24** shows that 100° is approximately 1.74533 radians. The calculator must be in radian mode to convert from degrees to radians. The display in **Figure 2.25** shows that 2.2 radians is approximately 126.051°. The calculator must be in degree mode to convert from radians to degrees.

On a TI-83, the symbol for radian measure is r, and it is in the ANGLE menu.

• ARCS AND ARC LENGTH

Consider a circle of radius r. By solving the formula $\theta = \dfrac{s}{r}$ for s, we have an equation for arc length.

Arc Length Formula

Let r be the length of the radius of a circle and θ the nonnegative radian measure of a central angle of the circle. Then the length of the arc s that subtends the central angle is $s = r\theta$. See **Figure 2.26.**

EXAMPLE 5   Find the Length of an Arc

Find the length of an arc that subtends a central angle of 120° in a circle of radius 10 centimeters.

Solution

The formula $s = r\theta$ requires that θ be expressed in radians. We first convert 120° to radian measure and then use the formula $s = r\theta$.

$$\theta = 120° = 120°\left(\frac{\pi \text{ radians}}{180°}\right) = \frac{2\pi}{3}\text{ radians} = \frac{2\pi}{3}$$

$$s = r\theta = (10 \text{ centimeters})\left(\frac{2\pi}{3}\right) = \frac{20\pi}{3}\text{ centimeters}$$

▶ **TRY EXERCISE 62, PAGE 125**

EXAMPLE 6   Solve an Application

A pulley with a radius of 10 inches uses a belt to drive a pulley with a radius of 6 inches. Find the angle through which the smaller pulley turns as the 10-inch pulley makes one revolution. State your answer in radians and also in degrees.

Continued ▶

take note

The formula $s = r\theta$ is valid only when θ is expressed in radians.

FIGURE 2.27

Solution

Use the formula $s = r\theta$. As the 10-inch pulley turns through an angle θ_1, a point on that pulley moves s_1 inches, where $s_1 = 10\theta_1$. See **Figure 2.27.** At the same time, the 6-inch pulley turns through an angle of θ_2 and a point on that pulley moves s_2 inches, where $s_2 = 6\theta_2$. Assuming that the belt does not slip on the pulleys, we have $s_1 = s_2$. Thus

$$10\theta_1 = 6\theta_2$$

$$10(2\pi) = 6\theta_2 \qquad \bullet \text{ Solve for } \theta_2, \text{ when } \theta_1 = 2\pi \text{ radians.}$$

$$\frac{10}{3}\pi = \theta_2$$

The 6-inch pulley turns through an angle of $\dfrac{10}{3}\pi$ radians, or $600°$.

▶ **TRY EXERCISE 66, PAGE 125**

● **LINEAR AND ANGULAR SPEED**

A car traveling at a speed of 55 miles per hour covers a distance of 55 miles in 1 hour. **Linear speed** v is *distance* traveled per unit time. In equation form,

$$v = \frac{s}{t}$$

where v is the linear speed, s is the distance traveled, and t is the time.

The floppy disk in a computer disk drive revolving at 300 revolutions per minute makes 300 complete revolutions in 1 minute. **Angular speed** ω is the *angle* through which a point on a circle moves per unit time. In equation form,

$$\omega = \frac{\theta}{t}$$

where ω is the angular speed, θ is the measure (in radians) of the angle through which a point has moved, and t is the time. Some common units of angular speed are revolutions per second, revolutions per minute, radians per second, and radians per minute.

EXAMPLE 7 **Convert an Angular Speed**

A hard disk in a computer rotates at 3600 revolutions per minute. Find the angular speed of the disk in radians per second.

Solution

As a point on the disk rotates 1 revolution (rev), the angle through which the point moves is 2π radians. Thus $\dfrac{2\pi \text{ radians}}{1 \text{ rev}}$ will be the unit fraction we will use to convert from revolutions to radians. To convert from minutes to seconds, use the unit fraction $\dfrac{1 \text{ minute}}{60 \text{ seconds}}$.

$$3600 \text{ rev/minute} = \frac{3600 \text{ rev}}{1 \text{ minute}}\left(\frac{2\pi \text{ radians}}{1 \text{ rev}}\right)\left(\frac{1 \text{ minute}}{60 \text{ seconds}}\right)$$

$$= 120\pi \text{ radians/second} \qquad \bullet \textbf{ Exact answer}$$

$$\approx 377 \text{ radians/second} \qquad \bullet \textbf{ Approximate answer}$$

▶ **TRY EXERCISE 68, PAGE 125**

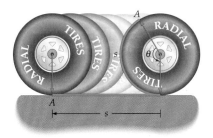

FIGURE 2.28

The tire on a car traveling along a road has both linear speed and angular speed. The relationship between linear and angular speed can be expressed by an equation.

Assume that the wheel in **Figure 2.28** is rolling without slipping. As the wheel moves a distance s, point A moves through an angle θ. The arc length subtending angle θ is also s, the distance traveled by the wheel. From the equations for linear and angular speed, we have

$$v = \frac{s}{t} = \frac{r\theta}{t} = r\frac{\theta}{t} \qquad \bullet\, s = r\theta$$

$$v = r\omega \qquad\qquad \bullet\, \omega = \frac{\theta}{t}$$

The equation $v = r\omega$ gives the linear speed of a point on a rotating body in terms of distance r from the axis of rotation and the angular speed ω, provided that ω is in radians per unit of time.

EXAMPLE 8   **Find Linear Speed**

A wind machine is used to generate electricity. The wind machine has propeller blades that are 12 feet in length (see **Figure 2.29**). If the propeller is rotating at 3 revolutions per second, what is the linear speed in feet per second of the tips of the blades?

Solution

Convert the angular speed $\omega = 3$ revolutions per second into radians per second, and then use the formula $v = r\omega$.

$$\omega = \frac{3 \text{ revolutions}}{1 \text{ second}} = \left(\frac{3 \text{ revolutions}}{1 \text{ second}}\right)\left(\frac{2\pi \text{ radians}}{1 \text{ revolution}}\right) = \frac{6\pi \text{ radians}}{1 \text{ second}}$$

Thus

$$v = r\omega = (12 \text{ feet})\left(\frac{6\pi \text{ radians}}{1 \text{ second}}\right)$$

$$= 72\pi \text{ feet per second} \approx 226 \text{ feet per second}$$

▶ **TRY EXERCISE 74, PAGE 125**

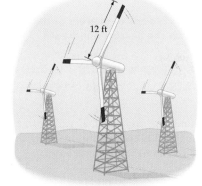

FIGURE 2.29

TOPICS FOR DISCUSSION

1. The measure of a radian differs depending on the length of the radius of the circle used. Do you agree? Explain.

2. The measure of 1 radian is over 100 times larger than the measure of 1 degree. Do you agree? Explain.

3. What are the necessary conditions for an angle to be in standard position?

4. Is the supplement of an obtuse angle an acute angle?

5. Do all acute angles have a positive measure?

EXERCISE SET 2.1

In Exercises 1 to 12, find the measure (if possible) of the complement and the supplement of each angle.

1. $15°$ ▶ **2.** $87°$ 3. $70°15'$

4. $22°43'$ 5. $56°33'15''$ 6. $19°42'05''$

7. 1 8. 0.5 9. $\dfrac{\pi}{4}$

10. $\dfrac{\pi}{3}$ 11. $\dfrac{2\pi}{5}$ 12. $\dfrac{\pi}{6}$

In Exercises 13 to 18, classify each angle by quadrant, and state the measure of the positive angle with measure less than 360° that is coterminal with the given angle.

13. $\alpha = 610°$ ▶ **14.** $\alpha = 765°$ 15. $\alpha = -975°$

16. $\alpha = -872°$ 17. $\alpha = 2456°$ 18. $\alpha = -3789°$

In Exercises 19 to 24, use a calculator to convert each decimal degree measure to its equivalent DMS measure.

19. $24.56°$ 20. $110.24°$ 21. $64.158°$

22. $18.96°$ 23. $3.402°$ 24. $224.282°$

In Exercises 25 to 30, use a calculator to convert each DMS measure to its equivalent decimal degree measure.

25. $25°25'12''$ 26. $63°29'42''$ 27. $183°33'36''$

28. $141°6'9''$ 29. $211°46'48''$ 30. $19°12'18''$

In Exercises 31 to 39, convert the degree measure to exact radian measure.

31. $30°$ ▶ **32.** $-45°$ 33. $90°$

34. $15°$ 35. $165°$ 36. $315°$

37. $420°$ 38. $630°$ 39. $585°$

In Exercises 40 to 48, convert the radian measure to exact degree measure.

▶ **40.** $\dfrac{\pi}{4}$ 41. $\dfrac{\pi}{5}$ 42. $-\dfrac{2\pi}{3}$

43. $\dfrac{\pi}{6}$ 44. $\dfrac{\pi}{9}$ 45. $\dfrac{3\pi}{8}$

46. $\dfrac{11\pi}{18}$ 47. $\dfrac{11\pi}{3}$ 48. $\dfrac{6\pi}{5}$

In Exercises 49 to 54, convert radians to degrees or degrees to radians. Round answers to the nearest hundredth.

49. 1.5 50. -2.3 51. $133°$

52. $427°$ 53. 8.25 54. $-90°$

In Exercises 55 to 58, find the measure in radians and degrees of the central angle of a circle subtended by the given arc. Round answers to the nearest hundredth.

55. $r = 2$ inches, $s = 8$ inches

56. $r = 7$ feet, $s = 4$ feet

57. $r = 5.2$ centimeters, $s = 12.4$ centimeters

58. $r = 35.8$ meters, $s = 84.3$ meters

In Exercises 59 to 62, find the measure of the intercepted arc of a circle with the given radius and central angle. Round answers to the nearest hundredth.

59. $r = 8$ inches, $\theta = \dfrac{\pi}{4}$ **60.** $r = 3$ feet, $\theta = \dfrac{7\pi}{2}$

61. $r = 25$ centimeters, $\theta = 42°$

▶ **62.** $r = 5$ meters, $\theta = 144°$

63. Find the number of radians in $1\dfrac{1}{2}$ revolutions.

64. Find the number of radians in $\dfrac{3}{8}$ revolution.

65. ANGULAR ROTATION OF TWO PULLEYS A pulley with a radius of 14 inches uses a belt to drive a pulley with a radius of 28 inches. The 14-inch pulley turns through an angle of 150°. Find the angle through which the 28-inch pulley turns.

▶ **66.** ANGULAR ROTATION OF TWO PULLEYS A pulley with a diameter of 1.2 meters uses a belt to drive a pulley with a diameter of 0.8 meter. The 1.2-meter pulley turns through an angle of 240°. Find the angle through which the 0.8-meter pulley turns.

67. ANGULAR SPEED Find the angular speed, in radians per second, of the second hand on a clock.

▶ **68.** ANGULAR SPEED Find the angular speed, in radians per second, of a point on the equator of the earth.

69. ANGULAR SPEED A wheel is rotating at 50 revolutions per minute. Find the angular speed in radians per second.

70. ANGULAR SPEED A wheel is rotating at 200 revolutions per minute. Find the angular speed in radians per second.

71. ANGULAR SPEED The turntable of a record player turns at $33\frac{1}{3}$ revolutions per minute. Find the angular speed in radians per second.

72. ANGULAR SPEED A car with a wheel of radius 14 inches is moving with a speed of 55 mph. Find the angular speed of the wheel in radians per second.

73. LINEAR SPEED OF A CAR Each tire on a car has a radius of 15 inches. The tires are rotating at 450 revolutions per

minute. Find the speed of the automobile to the nearest mile per hour.

▶ **74.** LINEAR SPEED OF A TRUCK Each tire on a truck has a radius of 18 inches. The tires are rotating at 500 revolutions per minute. Find the speed of the truck to the nearest mile per hour.

75. BICYCLE GEARS The chain wheel of Emma's bicycle has a radius of 3.5 inches. The rear gear has a radius of 1.75 inches, and the back tire has a radius of 12 inches. If Emma pedals for 150 revolutions of the chain wheel, how far will she travel? Round to the nearest foot.

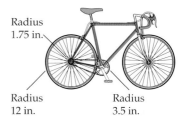

Radius 1.75 in.

Radius 12 in.

Radius 3.5 in.

76. ROTATION VERSUS LIFT DISTANCE A winch with a 6-inch radius is used to lift a container. The winch is designed so that as it is rotated, the cable stays in contact with the surface of the winch. That is, the cable does not wrap on top of itself.

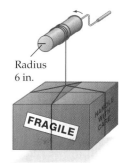

Radius 6 in.

a. Find the distance the container is lifted as the winch is rotated through an angle of $\dfrac{5\pi}{6}$ radians.

b. Determine the angle, in radians, through which the winch must be rotated to lift the container a distance of 2 feet.

77. SPEED OF THE CONCORDE During the time the Concorde was flying, it could fly from London to New York City, a distance of 3460 miles, in 2 hours and 59 minutes.

a. What was the average linear speed of the Concorde in miles per hour during one of these flights? Round to the nearest mile per hour.

b. What was the average angular speed of the Concorde in radians per hour during one of these flights? Assume that the Concorde maintains an altitude of 10 miles and that the radius of Earth is 3960 miles. Round to the nearest hundredth of a radian per hour.

c. If the Concorde left London at 1 P.M., what time would it be expected to arrive in New York City? (*Hint:* New York City is five time zones to the west of London.)

78. RACING THE SUN A pilot is flying a supersonic jet plane from east to west along a path over the equator. How fast, in miles per hour, does the pilot need to fly to keep the sun in the same relative position to the airplane? Assume that the plane is flying at an altitude of 2 miles above Earth and that Earth has a radius of 3960 miles. Round to the nearest mile per hour.

79. ASTRONOMY At a time when the earth was 93,000,000 miles from the sun, you observed through a tinted glass that the diameter of the sun occupied an arc of 31'. Determine, to the nearest ten thousand miles, the diameter of the sun.

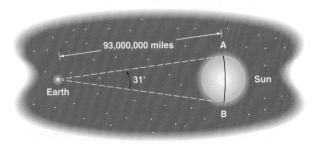

(*Hint:* Because the radius of arc *AB* is large and its central angle is small, the length of the diameter of the sun is approximately the length of the arc *AB*.)

80. ANGLE OF ROTATION AND DISTANCE The minute hand on the clock atop city hall measures 6 feet 3 inches from its tip to its axle.

a. Through what angle (in radians) does the minute hand pass between 9:12 A.M. and 9:48 A.M.?

b. What distance, to the nearest tenth of a foot, does the tip of the minute hand travel during this period?

81. VELOCITY OF THE HUBBLE SPACE TELE-SCOPE On April 25, 1990, the Hubble Space Telescope (HST) was deployed into a circular orbit 625 kilometers above the surface of the earth. The HST completes an earth orbit every 1.61 hours.

a. Find the angular velocity, with respect to the center of the earth, of the HST. Round your answer to the nearest 0.1 radian per hour.

b. Find the linear velocity of the HST. (*Hint:* The radius of the earth is about 6370 kilometers.) Round your answer to the nearest 100 kilometers per hour.

82. ESTIMATING THE RADIUS OF THE EARTH Eratosthenes, the fifth librarian of Alexandria (230 B.C.), was able to estimate the radius of the earth from the following data: The distance between the Egyptian cities of Alexandria and Syrene was 5000 stadia (520 miles). Syrene was located directly south of Alexandria. One summer, at noon, the sun was directly overhead at Syrene, whereas at the same time in Alexandria, the sun was at a 7.5° angle from the zenith.

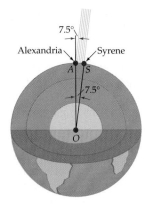

Eratosthenes reasoned that because the sun is far away, the rays of sunlight that reach the earth must be nearly parallel. From this assumption he concluded that the measure of ∠*AOS* in the accompanying figure must be 7.5°. Use this information to estimate the radius (to the nearest 10 miles) of the earth.

83.  **VELOCITY COMPARISONS** Assume that the bicycle in the figure is moving forward at a constant rate. Point *A* is on the edge of the 30-inch rear tire, and point *B* is on the edge of the 20-inch front tire.

a. Which point (*A* or *B*) has the greater angular velocity? Explain.

b. Which point (*A* or *B*) has the greater linear velocity? Explain.

84. Given that s, r, θ, t, v, and ω are as defined in Section 2.1, determine which of the following formulas are valid.

$$s = r\theta \qquad r = \frac{s}{\theta} \qquad v = \frac{r\theta}{t}$$

$$v = r\omega \qquad v = \frac{s}{t} \qquad \omega = \frac{\theta}{t}$$

85. NAUTICAL MILES AND STATUTE MILES A nautical mile is the length of an arc, on the earth's equator, that subtends a 1' central angle. The equatorial radius of the earth is about 3960 statute miles.

a. Convert 1 nautical mile to statute miles. Round to the nearest hundredth of a statute mile.

b. Determine what percent (to the nearest 1 percent) of the earth's circumference is covered by a trip from Los Angeles, California to Honolulu, Hawaii (a distance of 2217 nautical miles).

86. PHOTOGRAPHY The field of view for a camera with a 200-millimeter lens is 12°. A photographer takes a photograph of a large building that is 485 feet in front of the camera. What is the approximate width, to the nearest foot, of the building that will appear in the photograph? (*Hint:* If the radius of an arc AB is large and its central angle is small, then the length of the chord AB is approximately the length of the arc AB.)

CONNECTING CONCEPTS

A *sector* of a circle is the region bounded by radii **OA** and **OB** and the intercepted arc **AB**. See the following figure. The area of the sector is given by

$$A = \frac{1}{2}r^2\theta$$

where *r* is the radius of the circle and *θ* is the measure of the central angle in radians.

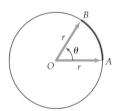

In Exercises 87 to 90, find the area, to the nearest square unit, of the sector of a circle with the given radius and central angle.

87. $r = 5$ inches, $\theta = \dfrac{\pi}{3}$ radians

88. $r = 2.8$ feet, $\theta = \dfrac{5\pi}{2}$ radians

89. $r = 120$ centimeters, $\theta = 0.65$ radian

90. $r = 30$ feet, $\theta = 62°$

91. AREA OF A CIRCULAR SEGMENT Find the area of the shaded portion of the circle. The radius of the circle is 9 inches.

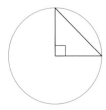

Latitude describes the position of a point on the earth's surface in relation to the equator. A point on the equator has a latitude of 0°. The north pole has a latitude of 90°. The radius of the earth is approximately 3960 miles.

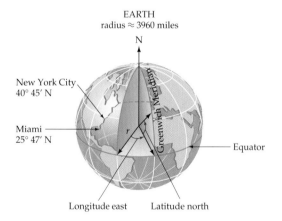

92. GEOGRAPHY The city of New York has a latitude of 40°45′N. How far north, to the nearest 10 miles, is it from the equator? Use 3960 miles as the radius of the earth.

93. GEOGRAPHY The city of Miami has a latitude of 25°47′N. How far north, to the nearest 10 miles, is it from the equator? Use 3960 miles as the radius of the earth.

94. GEOGRAPHY Assuming that the earth is a perfect sphere, and expressing your answer to three significant digits, find the distance along the earth's surface (in miles) that subtends a central angle of

a. 1°   **b.** 1′   **c.** 1″

PREPARE FOR SECTION 2.2
---

95. Rationalize the denominator of $\dfrac{1}{\sqrt{3}}$.

96. Rationalize the denominator of $\dfrac{2}{\sqrt{2}}$.

97. Simplify: $a \div \left(\dfrac{a}{2}\right)$

98. Simplify: $\left(\dfrac{a}{2}\right) \div \left(\dfrac{\sqrt{3}}{2}a\right)$

99. Solve $\dfrac{\sqrt{2}}{2} = \dfrac{x}{5}$ for x. Round your answer to the nearest hundredth. [1.1]

100. Solve $\dfrac{\sqrt{3}}{3} = \dfrac{x}{18}$ for x. Round your answer to the nearest hundredth. [1.1]

PROJECTS
---

1. **CONVERSION OF UNITS** You are traveling in a foreign country. You discover that the currency used consists of lollars, mollars, nollars, and tollars.

$$5 \text{ lollars} = 1 \text{ mollar}$$
$$3 \text{ mollars} = 5 \text{ nollars}$$
$$4 \text{ nollars} = 7 \text{ tollars}$$

The fare for a taxi is 14 tollars.

a. How much is the fare in mollars? (*Hint:* Make use of unit fractions to convert from tollars to mollars.)

b. If you have only mollars and lollars, how many of each could you use to pay the fare?

c. Explain to a classmate the concept of unit fractions. Also explain how you knew which of the unit fractions

$$\left(\dfrac{4 \text{ nollars}}{7 \text{ tollars}}\right) \qquad \left(\dfrac{7 \text{ tollars}}{4 \text{ nollars}}\right)$$

should be used in the conversion in part **a.**

d. If you wish to convert x degrees to radians, you need to multiply x by which of the following unit fractions?

$$\left(\dfrac{\pi}{180°}\right) \qquad \left(\dfrac{180°}{\pi}\right)$$

2. **SPACE SHUTTLE** The rotational period of the earth is 23.933 hours. A space shuttle revolves around the earth's equator every 2.231 hours. Both are rotating in the same direction. At the present time, the space shuttle is directly above the Galápagos Islands. How long will it take for the space shuttle to circle the earth and return to a position directly above the Galápagos Islands?

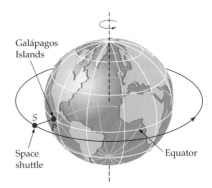

TRIGONOMETRIC FUNCTIONS OF ACUTE ANGLES

• TRIGONOMETRIC FUNCTIONS OF ACUTE ANGLES

The study of trigonometry, which means "triangle measurement," began more than 2000 years ago, partially as a means of solving surveying problems. Early trigonometry used the length of a chord of a circle as the value of a *trigonometric function*. In the sixteenth century, right triangles were used to define a trigonometric function. We will use a modification of this approach.

When working with right triangles, it is convenient to refer to the side *opposite* an angle or the side *adjacent* to (next to) an angle. **Figure 2.30a** shows the sides opposite and adjacent to the angle α. For angle β, the opposite and adjacent sides are as shown in **Figure 2.30b.** In both cases, the hypotenuse remains the same.

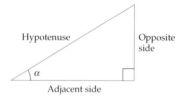

a. Adjacent and opposite sides of $\angle\alpha$ **b.** Adjacent and opposite sides of $\angle\beta$

FIGURE 2.30

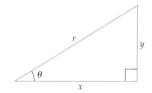

FIGURE 2.31

Consider an angle θ in the right triangle shown in **Figure 2.31.** Let x and y represent the lengths, respectively, of the adjacent and opposite sides of the triangle, and let r be the length of the hypotenuse. Six possible ratios can be formed:

$$\frac{y}{r} \quad \frac{x}{r} \quad \frac{y}{x} \quad \frac{r}{y} \quad \frac{r}{x} \quad \frac{x}{y}$$

Each ratio defines a value of a trigonometric function of the acute angle θ. The functions are **sine** (sin), **cosine** (cos), **tangent** (tan), **cosecant** (csc), **secant** (sec), and **cotangent** (cot).

Trigonometric Functions of an Acute Angle

Let θ be an acute angle of a right triangle. The values of the six trigonometric functions of θ are

$$\sin \theta = \frac{\text{length of opposite side}}{\text{length of hypotenuse}} = \frac{y}{r} \qquad \cos \theta = \frac{\text{length of adjacent side}}{\text{length of hypotenuse}} = \frac{x}{r}$$

$$\tan \theta = \frac{\text{length of opposite side}}{\text{length of adjacent side}} = \frac{y}{x} \qquad \cot \theta = \frac{\text{length of adjacent side}}{\text{length of opposite side}} = \frac{x}{y}$$

$$\sec \theta = \frac{\text{length of hypotenuse}}{\text{length of adjacent side}} = \frac{r}{x} \qquad \csc \theta = \frac{\text{length of hypotenuse}}{\text{length of opposite side}} = \frac{r}{y}$$

We will write opp, adj, and hyp as abbreviations for *the length of the* opposite side, adjacent side, and hypotenuse, respectively.

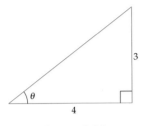

FIGURE 2.32

EXAMPLE I   Evaluate Trigonometric Functions

Find the values of the six trigonometric functions of θ for the triangle given in **Figure 2.32.**

Solution

Use the Pythagorean Theorem to find the length of the hypotenuse.

$$r = \sqrt{3^2 + 4^2} = \sqrt{25} = 5$$

From the definitions of the trigonometric functions,

$$\sin \theta = \frac{\text{opp}}{\text{hyp}} = \frac{3}{5} \qquad \cos \theta = \frac{\text{adj}}{\text{hyp}} = \frac{4}{5} \qquad \tan \theta = \frac{\text{opp}}{\text{adj}} = \frac{3}{4}$$

$$\cot \theta = \frac{\text{adj}}{\text{opp}} = \frac{4}{3} \qquad \sec \theta = \frac{\text{hyp}}{\text{adj}} = \frac{5}{4} \qquad \csc \theta = \frac{\text{hyp}}{\text{opp}} = \frac{5}{3}$$

▶ **TRY EXERCISE 6, PAGE 136**

Given the value of one trigonometric function of the acute angle θ, it is possible to find the value of any of the remaining trigonometric functions of θ.

EXAMPLE 2   Find the Value of a Trigonometric Function

Given that θ is an acute angle and $\cos \theta = \dfrac{5}{8}$, find $\tan \theta$.

Solution

$$\cos \theta = \frac{5}{8} = \frac{\text{adj}}{\text{hyp}}$$

Sketch a right triangle with one leg of length 5 units and a hypotenuse of length 8 units. Label as θ the acute angle that has the leg of length 5 units as its adjacent side (see **Figure 2.33**). Use the Pythagorean Theorem to find the length of the opposite side.

$$(\text{opp})^2 + 5^2 = 8^2$$
$$(\text{opp})^2 + 25 = 64$$
$$(\text{opp})^2 = 39$$
$$\text{opp} = \sqrt{39}$$

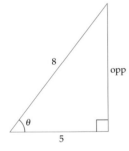

FIGURE 2.33

Therefore, $\tan \theta = \dfrac{\text{opp}}{\text{adj}} = \dfrac{\sqrt{39}}{5}$.

▶ **TRY EXERCISE 20, PAGE 136**

● TRIGONOMETRIC FUNCTIONS OF SPECIAL ANGLES

In Example 1, the lengths of the legs of the triangle were given, and you were asked to find the values of the six trigonometric functions of the angle θ. Often we will want to find the value of a trigonometric function when we are given *the mea-*

sure of an angle rather than the measure of the sides of a triangle. For most angles, advanced mathematical methods are required to evaluate a trigonometric function. For some *special angles,* however, the value of a trigonometric function can be found by geometric methods. These special acute angles are 30°, 45°, and 60°.

First, we will find the values of the six trigonometric functions of 45°. (This discussion is based on angles measured in degrees. Radian measure could have been used without changing the results.) **Figure 2.34** shows a right triangle with angles 45°, 45°, and 90°. Because $\angle A = \angle B$, the lengths of the sides opposite these angles are equal. Let the length of each equal side be denoted by a. From the Pythagorean Theorem,

$$r^2 = a^2 + a^2 = 2a^2$$
$$r = \sqrt{2a^2} = a\sqrt{2}$$

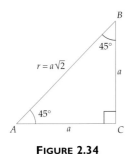

FIGURE 2.34

The values of the six trigonometric functions of 45° are

$$\sin 45° = \frac{a}{a\sqrt{2}} = \frac{1}{\sqrt{2}} = \frac{\sqrt{2}}{2} \qquad \cos 45° = \frac{a}{a\sqrt{2}} = \frac{1}{\sqrt{2}} = \frac{\sqrt{2}}{2}$$

$$\tan 45° = \frac{a}{a} = 1 \qquad\qquad \cot 45° = \frac{a}{a} = 1$$

$$\sec 45° = \frac{a\sqrt{2}}{a} = \sqrt{2} \qquad\qquad \csc 45° = \frac{a\sqrt{2}}{a} = \sqrt{2}$$

The values of the trigonometric functions of the special angles 30° and 60° can be found by drawing an equilateral triangle and bisecting one of the angles, as **Figure 2.35** shows. The angle bisector also bisects one of the sides. Thus the length of the side opposite the 30° angle is one-half the length of the hypotenuse of triangle OAB.

Let a denote the length of the hypotenuse. Then the length of the side opposite the 30° angle is $\frac{a}{2}$. The length of the side adjacent to the 30° angle, h, is found by using the Pythagorean Theorem.

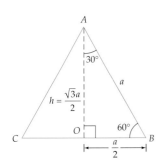

FIGURE 2.35

$$a^2 = \left(\frac{a}{2}\right)^2 + h^2$$

$$a^2 = \frac{a^2}{4} + h^2$$

$$\frac{3a^2}{4} = h^2 \qquad\qquad \bullet \text{ Subtract } \frac{a^2}{4} \text{ from each side.}$$

$$h = \frac{\sqrt{3}a}{2} \qquad\qquad \bullet \text{ Solve for } h.$$

The values of the six trigonometric functions of 30° are

$$\sin 30° = \frac{a/2}{a} = \frac{1}{2} \qquad\qquad\qquad \cos 30° = \frac{\sqrt{3}a/2}{a} = \frac{\sqrt{3}}{2}$$

$$\tan 30° = \frac{a/2}{\sqrt{3}a/2} = \frac{1}{\sqrt{3}} = \frac{\sqrt{3}}{3} \qquad \cot 30° = \frac{\sqrt{3}a/2}{a/2} = \sqrt{3}$$

$$\sec 30° = \frac{a}{\sqrt{3}a/2} = \frac{2}{\sqrt{3}} = \frac{2\sqrt{3}}{3} \qquad \csc 30° = \frac{a}{a/2} = 2$$

The values of the trigonometric functions of 60° can be found by again using **Figure 2.35.** The length of the side opposite the 60° angle is $\dfrac{\sqrt{3}a}{2}$, and the length of the side adjacent to the 60° angle is $\dfrac{a}{2}$. The values of the trigonometric functions of 60° are

$$\sin 60° = \frac{\sqrt{3}a/2}{a} = \frac{\sqrt{3}}{2} \qquad \cos 60° = \frac{a/2}{a} = \frac{1}{2}$$

$$\tan 60° = \frac{\sqrt{3}a/2}{a/2} = \sqrt{3} \qquad \cot 60° = \frac{a/2}{\sqrt{3}a/2} = \frac{1}{\sqrt{3}} = \frac{\sqrt{3}}{3}$$

$$\sec 60° = \frac{a}{a/2} = 2 \qquad \csc 60° = \frac{a}{\sqrt{3}a/2} = \frac{2}{\sqrt{3}} = \frac{2\sqrt{3}}{3}$$

Table 2.2 summarizes the values of the trigonometric functions of the special angles 30° ($\pi/6$), 45° ($\pi/4$), and 60° ($\pi/3$).

take note

Memorizing the values given in Table 2.2 will prove to be extremely useful in the remaining trigonometry sections.

TABLE 2.2   Trigonometric Functions of Special Angles

θ	$\sin\theta$	$\cos\theta$	$\tan\theta$	$\csc\theta$	$\sec\theta$	$\cot\theta$
$30°; \dfrac{\pi}{6}$	$\dfrac{1}{2}$	$\dfrac{\sqrt{3}}{2}$	$\dfrac{\sqrt{3}}{3}$	2	$\dfrac{2\sqrt{3}}{3}$	$\sqrt{3}$
$45°; \dfrac{\pi}{4}$	$\dfrac{\sqrt{2}}{2}$	$\dfrac{\sqrt{2}}{2}$	1	$\sqrt{2}$	$\sqrt{2}$	1
$60°; \dfrac{\pi}{3}$	$\dfrac{\sqrt{3}}{2}$	$\dfrac{1}{2}$	$\sqrt{3}$	$\dfrac{2\sqrt{3}}{3}$	2	$\dfrac{\sqrt{3}}{3}$

❓ QUESTION What is the measure, in degrees, of the acute angle θ for which $\sin\theta = \cos\theta$, $\tan\theta = \cot\theta$, and $\sec\theta = \csc\theta$?

EXAMPLE 3   Evaluate a Trigonometric Expression

Find the *exact* value of $\sin^2 45° + \cos^2 60°$.

Solution

Substitute the values of $\sin 45°$ and $\cos 60°$ and simplify.

Note: $\sin^2\theta = (\sin\theta)(\sin\theta) = (\sin\theta)^2$ and $\cos^2\theta = (\cos\theta)(\cos\theta) = (\cos\theta)^2$.

$$\sin^2 45° + \cos^2 60° = \left(\frac{\sqrt{2}}{2}\right)^2 + \left(\frac{1}{2}\right)^2 = \frac{2}{4} + \frac{1}{4} = \frac{3}{4}$$

▶ **TRY EXERCISE 38, PAGE 137**

take note

The patterns in the following chart can be used to memorize the sine and cosine of 30°, 45°, and 60°.

$\sin 30° = \dfrac{\sqrt{1}}{2} \qquad \cos 30° = \dfrac{\sqrt{3}}{2}$

$\sin 45° = \dfrac{\sqrt{2}}{2} \qquad \cos 45° = \dfrac{\sqrt{2}}{2}$

$\sin 60° = \dfrac{\sqrt{3}}{2} \qquad \cos 60° = \dfrac{\sqrt{1}}{2}$

❓ ANSWER 45°

From the definition of the sine and cosecant functions,

$$(\sin \theta)(\csc \theta) = \frac{y}{r} \cdot \frac{r}{y} = 1 \quad \text{or} \quad (\sin \theta)(\csc \theta) = 1$$

By rewriting the last equation, we find

$$\sin \theta = \frac{1}{\csc \theta} \quad \text{and} \quad \csc \theta = \frac{1}{\sin \theta}, \text{ provided } \sin \theta \neq 0$$

The sine and cosecant functions are called **reciprocal functions.** The cosine and secant are also reciprocal functions, as are the tangent and cotangent functions. **Table 2.3** shows each trigonometric function and its reciprocal. These relationships hold for all values of θ for which both of the functions are defined.

TABLE 2.3   Trigonometric Functions and Their Reciprocals

$\sin \theta = \dfrac{1}{\csc \theta}$	$\cos \theta = \dfrac{1}{\sec \theta}$	$\tan \theta = \dfrac{1}{\cot \theta}$
$\csc \theta = \dfrac{1}{\sin \theta}$	$\sec \theta = \dfrac{1}{\cos \theta}$	$\cot \theta = \dfrac{1}{\tan \theta}$

INTEGRATING TECHNOLOGY

Some graphing calculators will allow you to display the degree symbol. For example, the TI-83 display

was produced by entering

sin 30 2nd ANGLE

ENTER)

Displaying the degree symbol will cause the calculator to evaluate a trigonometric function using degree mode even if the calculator is in radian mode.

INTEGRATING TECHNOLOGY

When evaluating a trigonometric function by using a graphing calculator, be sure the calculator is in the correct mode. If the measure of an angle is written with the degree symbol, then make sure the calculator is in degree mode. If the measure of an angle is given in radians (no degree symbol is used), then make sure the calculator is in radian mode. *Many errors are made because the correct mode is not selected.*

Some graphing calculators can be used to construct a table of functional values. For instance, the TI-83 keystrokes shown below generate the table in **Figure 2.36,** in which the first column lists the domain values

$$0°, 1°, 2°, 3°, \ldots$$

and the second column lists the range values

$$\sin 0°, \sin 1°, \sin 2°, \sin 3°, \ldots$$

TI-83 Keystrokes

2nd TBLSET 0

ENTER 1 Y= CLEAR sin

X,T,θ) 2nd TABLE

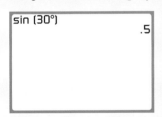

FIGURE 2.36

The graphing calculator must be in degree mode to produce this table.

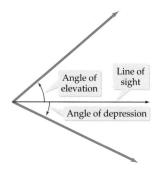

FIGURE 2.37

● APPLICATIONS INVOLVING RIGHT TRIANGLES

One of the major reasons for the development of trigonometry was to solve application problems. In this section we will consider some applications involving right triangles. In some application problems a horizontal line of sight is used as a reference line. An angle measured above the line of sight is called an **angle of elevation,** and an angle measured below the line of sight is called an **angle of depression.** See **Figure 2.37.**

EXAMPLE 4	Solve an Angle-of-Elevation Problem

From a point 115 feet from the base of a redwood tree, the angle of elevation to the top of the tree is 64.3°. Find the height of the tree to the nearest foot.

Solution

From **Figure 2.38,** the length of the adjacent side of the angle is known (115 feet). Because we need to determine the height of the tree (length of the opposite side), we use the tangent function. Let h represent the length of the opposite side.

$$\tan 64.3° = \frac{\text{opp}}{\text{adj}} = \frac{h}{115}$$
$$h = 115 \tan 64.3° \approx 238.952 \qquad \bullet \text{ Use a calculator to evaluate } \tan 64.3°.$$

The height of the tree is approximately 239 feet.

▶ **TRY EXERCISE 64, PAGE 138**

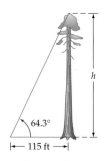

FIGURE 2.38

Because the cotangent function involves the sides adjacent to and opposite an angle, we could have solved Example 4 by using the cotangent function. The solution would have been

$$\cot 64.3° = \frac{\text{adj}}{\text{opp}} = \frac{115}{h}$$
$$h = \frac{115}{\cot 64.3°} \approx 238.952 \text{ feet}$$

The accuracy of a calculator is sometimes beyond the limits of measurement. In the last example the distance from the base of the tree was given as 115 feet (three significant digits), whereas the height of the tree was shown to be 238.952 feet (six significant digits). When using approximate numbers, we will use the conventions given below for calculating with trigonometric functions.

take note

The significant digits of an approximate number are

● every nonzero digit.
● the digit 0, provided it is between two nonzero digits or it is to the right of a nonzero digit in a number which includes a decimal point.

For example, the approximate number:

502 has 3 significant digits.
3700 has 2 significant digits.
47.0 has 3 significant digits.
0.0023 has 2 significant digits.
0.00840 has 3 significant digits.

A Rounding Convention: Significant Digits for Trigonometric Calculations

Angle Measure to the Nearest	*Significant Digits of the Lengths*
Degree	Two
Tenth of a degree	Three
Hundredth of a degree	Four

EXAMPLE 5   Solve an Angle-of-Depression Problem

DME (Distance Measuring Equipment) is standard avionic equipment on a commercial airplane. This equipment measures the distance from a plane to a radar station. If the distance from a plane to a radar station is 160 miles and the angle of depression is 33°, find the number of ground miles from a point directly below the plane to the radar station.

Solution

From **Figure 2.39**, the length of the hypotenuse is known (160 miles). The length of the side opposite the angle of 57° is unknown. The sine function involves the hypotenuse and the opposite side, x, of the 57° angle.

$$\sin 57° = \frac{x}{160}$$

$$x = 160 \sin 57° \approx 134.1873$$

Rounded to two significant digits, the plane is 130 ground miles from the radar station.

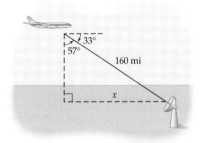

FIGURE 2.39

▶ **TRY EXERCISE 66, PAGE 138**

EXAMPLE 6   Solve an Angle-of-Elevation Problem

An observer notes that the angle of elevation from point A to the top of a space shuttle is 27.2°. From a point 17.5 meters further from the space shuttle, the angle of elevation is 23.9°. Find the height of the space shuttle.

Solution

From **Figure 2.40**, let x denote the distance from point A to the base of the space shuttle, and let y denote the height of the space shuttle. Then

$$(1) \quad \tan 27.2° = \frac{y}{x} \quad \text{and} \quad (2) \quad \tan 23.9° = \frac{y}{x + 17.5}$$

Solving Equation (1) for x, $x = \dfrac{y}{\tan 27.2°} = y \cot 27.2°$, and substituting into Equation (2), we have

$$\tan 23.9° = \frac{y}{y \cot 27.2° + 17.5}$$

$$y = (\tan 23.9°)(y \cot 27.2° + 17.5) \qquad \text{• Solve for } y.$$

$$y - y \tan 23.9° \cot 27.2° = (\tan 23.9°)(17.5)$$

$$y = \frac{(\tan 23.9°)(17.5)}{1 - \tan 23.9° \cot 27.2°} \approx 56.2993$$

To three significant digits, the height of the space shuttle is 56.3 meters.

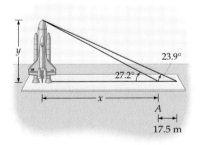

FIGURE 2.40

take note

The intermediate calculations in Example 6 were not rounded off. This ensures better accuracy for the final result. Using the rounding convention stated on page 134, we round off only the last result.

▶ **TRY EXERCISE 74, PAGE 139**

TOPICS FOR DISCUSSION

1. If θ is an acute angle of a right triangle for which $\cos \theta = \frac{3}{8}$, then it must be the case that $\sin \theta = \frac{5}{8}$. Do you agree? Explain.

2. A tutor claims that $\tan 30° = \cot 60°$. Do you agree?

3. Does $\sin 2\theta = 2 \sin \theta$? Explain.

4. How many significant digits are in each of the following measurements?
   a. 0.0042 inches   b. 5.03 inches   c. 62.00 inches

5. A student claims that $\sin^2 30° = (\sin 30°)^2$. Do you agree? Explain.

EXERCISE SET 2.2

In Exercises 1 to 14, find the values of the six trigonometric functions of θ for the right triangle with the given sides.

1.
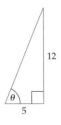

2.

3.

4.
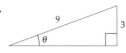

5.

▶ **6.**
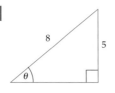

7.

8.

9.
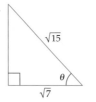

10.
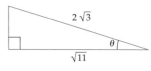

11.

12.

13.

14.

In Exercises 15 to 17, let θ be an acute angle of a right triangle for which $\sin \theta = \frac{3}{5}$. Find

15. $\tan \theta$ 16. $\sec \theta$ 17. $\cos \theta$

In Exercises 18 to 20, let θ be an acute angle of a right triangle for which $\tan \theta = \frac{4}{3}$. Find

18. $\sin \theta$ 19. $\cot \theta$ ▶ **20.** $\sec \theta$

In Exercises 21 to 23, let β be an acute angle of a right triangle for which $\sec \beta = \dfrac{13}{12}$. Find

21. $\cos \beta$ **22.** $\cot \beta$ **23.** $\csc \beta$

In Exercises 24 to 26, let θ be an acute angle of a right triangle for which $\cos \theta = \dfrac{2}{3}$. Find

24. $\sin \theta$ **25.** $\sec \theta$ **26.** $\tan \theta$

In Exercises 27 to 42, find the *exact* value of each expression.

27. $\sin 45° + \cos 45°$

28. $\csc 45° - \sec 45°$

29. $\sin 30° \cos 60° - \tan 45°$

30. $\csc 60° \sec 30° + \cot 45°$

31. $\sin 30° \cos 60° + \tan 45°$

32. $\sec 30° \cos 30° - \tan 60° \cot 60°$

33. $2 \sin 60° - \sec 45° \tan 60°$

34. $\sec 45° \cot 30° + 3 \tan 60°$

35. $\sin \dfrac{\pi}{3} + \cos \dfrac{\pi}{6}$ **36.** $\csc \dfrac{\pi}{6} - \sec \dfrac{\pi}{3}$

37. $\sin \dfrac{\pi}{4} + \tan \dfrac{\pi}{6}$ ▶ **38.** $\sin \dfrac{\pi}{3} \cos \dfrac{\pi}{4} - \tan \dfrac{\pi}{4}$

39. $\sec \dfrac{\pi}{3} \cos \dfrac{\pi}{3} - \tan \dfrac{\pi}{6}$

40. $\cos \dfrac{\pi}{4} \tan \dfrac{\pi}{6} + 2 \tan \dfrac{\pi}{3}$

41. $2 \csc \dfrac{\pi}{4} - \sec \dfrac{\pi}{3} \cos \dfrac{\pi}{6}$

42. $3 \tan \dfrac{\pi}{4} + \sec \dfrac{\pi}{6} \sin \dfrac{\pi}{3}$

In Exercises 43 to 56, use a calculator to find the value of the trigonometric function to four decimal places.

43. $\tan 32°$ **44.** $\sec 88°$ **45.** $\cos 63°20'$

46. $\cot 55°50'$ **47.** $\cos 34.7°$ **48.** $\tan 81.3°$

49. $\sec 5.9°$ **50.** $\sin \dfrac{\pi}{5}$ **51.** $\tan \dfrac{\pi}{7}$

52. $\sec \dfrac{3\pi}{8}$ **53.** $\csc 1.2$ **54.** $\sin 0.45$

55. $\cos 1.25$ **56.** $\tan \dfrac{3}{4}$

57. **VERTICAL HEIGHT FROM SLANT HEIGHT** A 12-foot ladder is resting against a wall and makes an angle of 52° with the ground. Find the height to which the ladder will reach on the wall.

58. **DISTANCE ACROSS A MARSH** Find the distance AB across the marsh shown in the accompanying figure.

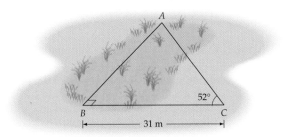

59. **SLOPE OF A LINE** Show that the slope of a line that makes an angle θ with the positive x-axis equals $\tan \theta$.

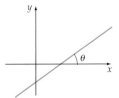

60. **WIDTH OF A SCREEN** Television screens are measured by the length of the diagonal of the screen. Find the width of a 19-inch television screen if the diagonal makes an angle of 38° with the base of the screen.

61. **CLOSEST APPROACH** A boat is 40 kilometers due east of a lighthouse and traveling in a direction that is 30° south of an east-west line as shown in the following figure. How close will the boat come to the lighthouse? Assume the 40 kilometer distance is measured to the nearest kilometer and the 30° angle is measured to the nearest degree.

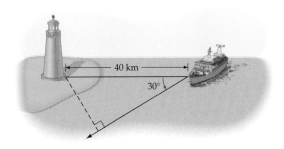

62. TIME OF CLOSEST APPROACH At 3:00 P.M., a boat is 12.5 miles due west of a radar station and traveling at 11 mph in a direction that is 57.3° south of an east-west line. At what time will the boat be closest to the radar station?

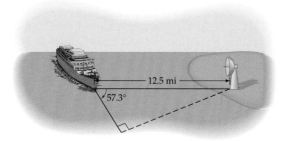

63. PLACEMENT OF A LIGHT For best illumination of a piece of art, a lighting specialist for an art gallery recommends that a ceiling-mounted light be 6 feet from the piece of art and that the angle of depression of the light be 38°. How far from a wall should the light be placed so that the recommendations of the specialist are met? Notice that the art extends outward 4 inches from the wall.

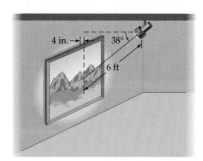

▶ **64.** HEIGHT OF THE EIFFEL TOWER The angle of elevation from a point 116 meters from the base of the Eiffel Tower to the top of the tower is 68.9°. Find the approximate height of the tower.

65. DISTANCE OF A DESCENT An airplane traveling at 240 mph is descending at an angle of depression of 6°. How many miles will the plane descend in 4 minutes?

▶ **66.** TIME OF A DESCENT A submarine traveling at 9.0 mph is descending at an angle of depression of 5°. How many minutes, to the nearest tenth, does it take the submarine to reach a depth of 80 feet?

67. HEIGHT OF AN AQUEDUCT From a point 300 feet from the base of a Roman aqueduct in southern France, the angle of elevation to the top of the aqueduct is 78°. Find the height of the aqueduct.

68. WIDTH OF A LAKE The angle of depression to one side of a lake, measured from a balloon 2500 feet above the lake as shown in the accompanying figure, is 43°. The angle of depression to the opposite side of the lake is 27°. Find the width of the lake.

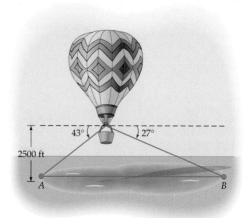

69. ASTRONOMY The moon Europa rotates in a nearly circular orbit around Jupiter. The orbital radius of Europa is approximately 670,900 kilometers. During a revolution of Europa around Jupiter, an astronomer found that the maximum value of the angle θ formed by Europa, Earth, and Jupiter was 0.056°. See the following figure on page 139. Find the distance d between Earth and Jupiter at the time the astronomer found the maximum value of θ. Round to the nearest million kilometers.

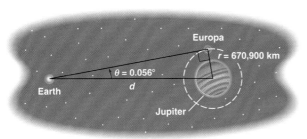

Not drawn to scale.

70. ASTRONOMY Venus rotates in a nearly circular orbit around the sun. The largest angle formed by Venus, Earth, and the sun is 46.5°. The distance from Earth to the sun is approximately 149,000,000 kilometers. What is the orbital radius r of Venus? Round to the nearest million kilometers.

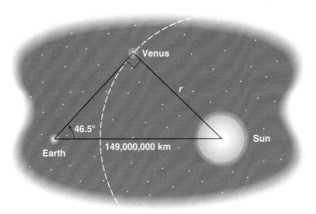

71. Consider the following *isosceles* triangle. The length of each of the two equal sides of the triangle is a, and each of the base angles has a measure of θ. Verify that the area of the triangle is $A = a^2 \sin \theta \cos \theta$.

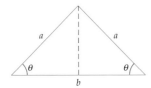

72. Find the area of the following hexagon. (*Hint:* The area consists of six isosceles triangles. Use the formula from Exercise 71 to compute the area of one of the triangles and multiply by 6.)

73. HEIGHT OF A PYRAMID The angle of elevation to the top of the Egyptian pyramid Cheops is 36.4°, measured from a point 350 feet from the base of the pyramid. The angle of elevation from the base of a face of the pyramid is 51.9°. Find the height of Cheops.

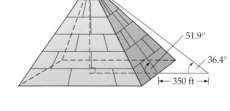

▶ **74.** HEIGHT OF A BUILDING Two buildings are 240 feet apart. The angle of elevation from the top of the shorter building to the top of the other building is 22°. If the shorter building is 80 feet high, how high is the taller building?

75. HEIGHT OF THE WASHINGTON MONUMENT From a point A on a line from the base of the Washington Monument, the angle of elevation to the top of the monument is 42.0°. From a point 100 feet away and on the same line, the angle to the top is 37.8°. Find the approximate height of the Washington Monument.

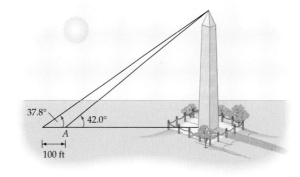

76. HEIGHT OF A BUILDING The angle of elevation to the top of a radio antenna on the top of a building is 53.4°. After moving 200 feet closer to the building, the angle of elevation is 64.3°. Find the height of the building if the height of the antenna is 180 feet.

77. THE PETRONAS TOWERS The Petronas Towers in Kuala Lumpur, Malaysia, are the world's tallest twin towers. Each tower is 1483 feet in height. The towers are connected by a skybridge at the forty-first floor. Note the information given in the accompanying figure.

a. Determine the height of the skybridge.

b. Determine the length of the skybridge.

$AB = 412$ feet
$\angle CAB = 53.6°$
$\overline{AB}$ is at ground level
$\angle CAD = 15.5°$

78. AN EIFFEL TOWER REPLICA Use the information in the accompanying figure to estimate the height of the Eiffel Tower replica that stands in front of the Paris Las Vegas Hotel in Las Vegas, Nevada.

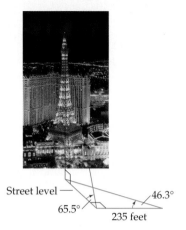

Street level ⎯   46.3°
65.5°
235 feet

CONNECTING CONCEPTS

79. A circle is inscribed in a regular hexagon with each side 6.0 meters long. Find the radius of the circle.

80. Show that the area A of the triangle given in the figure is $A = \dfrac{1}{2}ab\sin\theta$.

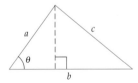

81. DETERMINE A RANGE OF HEIGHTS If an angle of 27° has been measured to the nearest degree, then the actual measure of the angle θ is such that $26.5° \le \theta < 27.5$. From a distance of exactly 100 meters from the base of a tree, the angle of elevation is measured as 27°, to the nearest degree. Find the range in which the height h of the tree must fall.

82. HEIGHT OF A TOWER Let B denote the base of a clock tower. The angle of elevation from a point A, on the ground, to the top of the clock tower is 56.3°. On a line on the

ground that is perpendicular to AB and 25 feet from A, the angle of elevation is 53.3°. Find the height of the clock tower.

83. FIND A MAXIMUM LENGTH Find the length of the longest piece of wood that can be slid around the corner of the hallway in the figure.

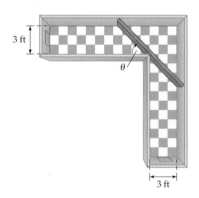

3 ft

θ

3 ft

84. FIND A MAXIMUM LENGTH In Exercise 83, suppose that the hall is 8 feet high. Find the length of the longest piece of wood that can be taken around the corner. Round to the nearest 0.1 foot.

───── *PREPARE FOR SECTION 2.3* ─────

85. Find the reciprocal of $-\dfrac{3}{4}$.

86. Find the reciprocal of $\dfrac{2\sqrt{5}}{5}$.

87. Find: $|120 - 180|$ [1.1]

88. Find: $2\pi - \dfrac{9\pi}{5}$

89. Find: $\dfrac{3}{2}\pi - \dfrac{\pi}{2}$

90. Find: $\sqrt{(-3)^2 + (-5)^2}$ [1.2]

───── *PROJECTS* ─────

1. a. PERIMETER OF A REGULAR n-GON Show that the perimeter P of a regular n-sided polygon (n-gon) inscribed in a circle of radius 1 is $P = 2n \sin\dfrac{180°}{n}$.

b. Let P_n denote the perimeter of a regular n-gon inscribed in a circle of radius 1. Use the result from part **a.** to complete the following table.

n	10	50	100	1000	10,000
P_n					

Write a few sentences explaining why P_n approaches 2π as n increases without bound.

2. a. AREA OF A REGULAR n-GON Show that the area A of a regular n-gon inscribed in a circle of radius 1 is $A = \dfrac{n}{2} \sin\dfrac{360°}{n}$.

b. Let A_n denote the area of a regular n-gon inscribed in a circle of radius 1. Use the result from part **a.** to complete the following table.

n	10	50	100	1000	10,000
A_n					

Write a few sentences explaining why A_n approaches π as n increases without bound.

TRIGONOMETRIC FUNCTIONS OF ANY ANGLE

- TRIGONOMETRIC FUNCTIONS OF QUADRANTAL ANGLES
- SIGNS OF TRIGONOMETRIC FUNCTIONS
- THE REFERENCE ANGLE

The applications of trigonometry would be quite limited if all angles had to be acute angles. Fortunately, this is not the case. In this section we extend the definition of a trigonometric function to include any angle.

Consider angle θ in **Figure 2.41** in standard position and a point $P(x, y)$ on the terminal side of the angle. We define the trigonometric functions of any angle according to the following definitions.

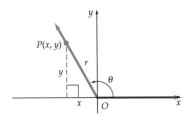

FIGURE 2.41

The Trigonometric Functions of Any Angle

Let $P(x, y)$ be any point, except the origin, on the terminal side of an angle θ in standard position. Let $r = d(O, P)$, the distance from the origin to P. The six trigonometric functions of θ are

$$\sin \theta = \frac{y}{r} \qquad\qquad \cos \theta = \frac{x}{r} \qquad\qquad \tan \theta = \frac{y}{x}, \quad x \neq 0$$

$$\csc \theta = \frac{r}{y}, \quad y \neq 0 \qquad \sec \theta = \frac{r}{x}, \quad x \neq 0 \qquad \cot \theta = \frac{x}{y}, \quad y \neq 0$$

where $r = \sqrt{x^2 + y^2}$.

---

take note

The measure of angle θ can be positive or negative. Note from the following figure that $\sin 120° = \sin(-240°)$ because $P(x, y)$ is on the terminal side of each angle. In a similar way, the value of any trigonometric function of 120° is equal to the value of that function of −240°.

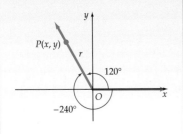

The value of a trigonometric function is independent of the point chosen on the terminal side of the angle. Consider any two points on the terminal side of an angle θ in standard position, as shown in **Figure 2.42.** The right triangles formed are similar triangles, so the ratios of the corresponding sides are equal. Thus, for example, $\dfrac{b}{a} = \dfrac{b'}{a'}$. Because $\tan \theta = \dfrac{b}{a} = \dfrac{b'}{a'}$, we have $\tan \theta = \dfrac{b'}{a'}$. Therefore, the value of the tangent function is independent of the point chosen on the terminal side of the angle. By a similar argument, we can show that the value of any trigonometric function is independent of the point chosen on the terminal side of the angle.

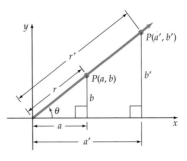

FIGURE 2.42

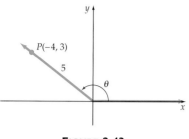

FIGURE 2.43

Any point in a rectangular coordinate system (except the origin) can determine an angle in standard position. For example, $P(-4, 3)$ in **Figure 2.43** is a point in the second quadrant and determines an angle θ in standard position with $r = \sqrt{(-4)^2 + 3^2} = 5$. The values of the trigonometric functions of θ are

$$\sin \theta = \frac{3}{5} \qquad \cos \theta = \frac{-4}{5} = -\frac{4}{5} \qquad \tan \theta = \frac{3}{-4} = -\frac{3}{4}$$

$$\csc \theta = \frac{5}{3} \qquad \sec \theta = \frac{5}{-4} = -\frac{5}{4} \qquad \cot \theta = \frac{-4}{3} = -\frac{4}{3}$$

EXAMPLE 1   Evaluate Trigonometric Functions

Find the value of each of the six trigonometric functions of an angle θ in standard position whose terminal side contains the point $P(-3, -2)$.

Solution

The angle is sketched in **Figure 2.44**. Find r by using the equation $r = \sqrt{x^2 + y^2}$, where $x = -3$ and $y = -2$.

$$r = \sqrt{(-3)^2 + (-2)^2} = \sqrt{9 + 4} = \sqrt{13}$$

Now use the definitions of the trigonometric functions.

$$\sin \theta = \frac{-2}{\sqrt{13}} = -\frac{2\sqrt{13}}{13} \qquad \cos \theta = \frac{-3}{\sqrt{13}} = -\frac{3\sqrt{13}}{13} \qquad \tan \theta = \frac{-2}{-3} = \frac{2}{3}$$

$$\csc \theta = \frac{\sqrt{13}}{-2} = -\frac{\sqrt{13}}{2} \qquad \sec \theta = \frac{\sqrt{13}}{-3} = -\frac{\sqrt{13}}{3} \qquad \cot \theta = \frac{-3}{-2} = \frac{3}{2}$$

▶ **TRY EXERCISE 6, PAGE 149**

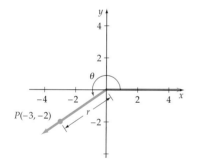

FIGURE 2.44

● TRIGONOMETRIC FUNCTIONS OF QUADRANTAL ANGLES

Recall that a quadrantal angle is an angle whose terminal side coincides with the x- or y-axis. The value of a trigonometric function of a quadrantal angle can be found by choosing any point on the terminal side of the angle and then applying the definition of that trigonometric function.

The terminal side of $0°$ coincides with the positive x-axis. Let $P(x, 0)$, $x > 0$, be any point on the x-axis, as shown in **Figure 2.45**. Then $y = 0$ and $r = x$. The values of the six trigonometric functions of $0°$ are

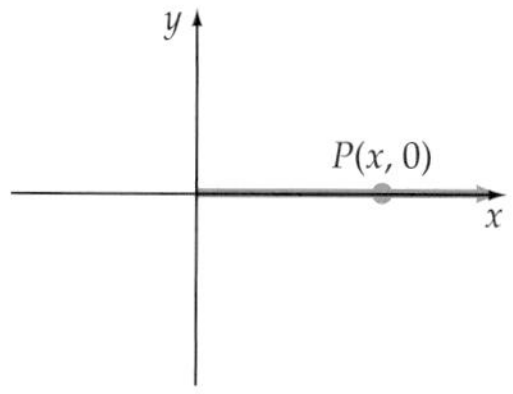

$$\sin 0° = \frac{0}{r} = 0 \qquad \cos 0° = \frac{x}{r} = \frac{x}{x} = 1 \qquad \tan 0° = \frac{0}{x} = 0$$

$$\csc 0° \text{ is undefined.} \qquad \sec 0° = \frac{r}{x} = \frac{x}{x} = 1 \qquad \cot 0° \text{ is undefined.}$$

FIGURE 2.45

❷ QUESTION Why are csc 0° and cot 0° undefined?

In like manner, the values of the trigonometric functions of the other quadrantal angles can be found. The results are shown in **Table 2.4.**

TABLE 2.4 Values of Trigonometric Functions for Quadrantal Angles

θ	$\sin \theta$	$\cos \theta$	$\tan \theta$	$\csc \theta$	$\sec \theta$	$\cot \theta$
0°	0	1	0	undefined	1	undefined
90°	1	0	undefined	1	undefined	0
180°	0	−1	0	undefined	−1	undefined
270°	−1	0	undefined	−1	undefined	0

● SIGNS OF TRIGONOMETRIC FUNCTIONS

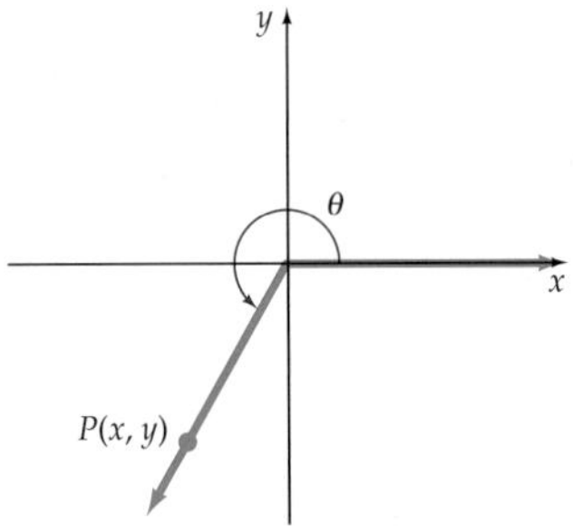

FIGURE 2.46

The sign of a trigonometric function depends on the quadrant in which the terminal side of the angle lies. For example, if θ is an angle whose terminal side lies in Quadrant III and $P(x, y)$ is on the terminal side of θ, then both x and y are negative, and therefore, $\dfrac{y}{x}$ and $\dfrac{x}{y}$ are positive. See **Figure 2.46.** Because $\tan \theta = \dfrac{y}{x}$ and $\cot \theta = \dfrac{x}{y}$, the values of the tangent and cotangent functions are positive for any Quadrant III angle. The values of the other four trigonometric functions of any Quadrant III angle are all negative.

Table 2.5 lists the signs of the six trigonometric functions in each quadrant. **Figure 2.47** is a graphical display of the contents of **Table 2.5.**

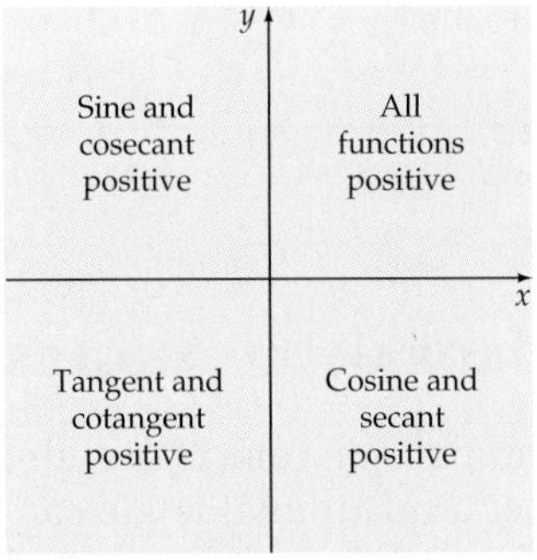

FIGURE 2.47

TABLE 2.5 Signs of the Trigonometric Functions

Sign of	Terminal Side of θ in Quadrant			
	I	II	III	IV
$\sin \theta$ and $\csc \theta$	positive	positive	negative	negative
$\cos \theta$ and $\sec \theta$	positive	negative	negative	positive
$\tan \theta$ and $\cot \theta$	positive	negative	positive	negative

In the next example we are asked to evaluate two trigonometric functions of the angle θ. A key step is to use our knowledge about trigonometric functions and their signs to determine that θ is a Quadrant IV angle.

EXAMPLE 2   Evaluate Trigonometric Functions

Given $\tan \theta = -\dfrac{7}{5}$ and $\sin \theta < 0$, find $\cos \theta$ and $\csc \theta$.

❷ ANSWER $P(x, 0)$ is a point on the terminal side of 0°. Thus $\csc 0° = \dfrac{r}{0}$, which is undefined. Similarly, $\cot 0° = \dfrac{x}{0}$, which is undefined.

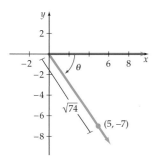

FIGURE 2.48

Solution

The terminal side of angle θ must lie in Quadrant IV; that is the only quadrant for which $\sin \theta$ and $\tan \theta$ are both negative. Because

$$\tan \theta = -\frac{7}{5} = \frac{y}{x} \tag{1}$$

and the terminal side of θ is in Quadrant IV, we know that y must be negative and x must be positive. Thus Equation (1) is true for $y = -7$ and $x = 5$. Now $r = \sqrt{5^2 + (-7)^2} = \sqrt{74}$. See **Figure 2.48.** Hence

$$\cos \theta = \frac{x}{r} = \frac{5}{\sqrt{74}} = \frac{5\sqrt{74}}{74} \quad \text{and} \quad \csc \theta = \frac{r}{y} = \frac{\sqrt{74}}{-7} = -\frac{\sqrt{74}}{7}$$

▶ **TRY EXERCISE 18, PAGE 149**

> **take note**
>
> In this section r is a distance, and hence it is nonnegative.

● **THE REFERENCE ANGLE**

We will often find it convenient to evaluate trigonometric functions by making use of the concept of a *reference angle*.

> **take note**
>
> The reference angle is a very important concept that will be used time and time again in the remaining trigonometry sections.

Reference Angle

Given $\angle \theta$ in standard position, its **reference angle** θ' is the smallest positive angle formed by the terminal side of $\angle \theta$ and the x-axis.

Figure 2.49 shows $\angle \theta$ and its reference angle θ' for four cases. In every case the reference angle θ' is formed by the terminal side of $\angle \theta$ and the x-axis (never the y-axis). The process of determining the measure of $\angle \theta'$ varies according to what quadrant contains the terminal side of $\angle \theta$.

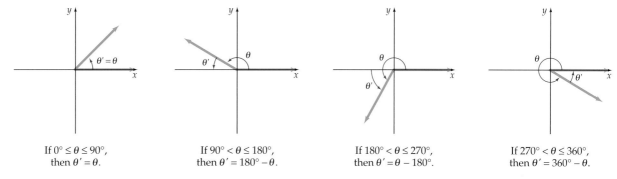

If $0° \leq \theta \leq 90°$, then $\theta' = \theta$.   If $90° < \theta \leq 180°$, then $\theta' = 180° - \theta$.   If $180° < \theta \leq 270°$, then $\theta' = \theta - 180°$.   If $270° < \theta \leq 360°$, then $\theta' = 360° - \theta$.

FIGURE 2.49

EXAMPLE 3   **Find the Measure of the Reference Angle**

For each of the following, sketch the given angle θ (in standard position) and its reference angle θ'. Then determine the measure of θ'.

a. $\theta = 120°$ **b.** $\theta = 345°$ **c.** $\theta = 924°$

d. $\theta = \dfrac{9}{5}\pi$ **e.** $\theta = -4$ **f.** $\theta = 17$

Solution

a.

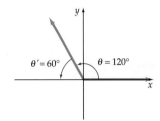

$\theta' = 180° - 120° = 60°$

b.

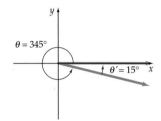

$\theta' = 360° - 345° = 15°$

c.

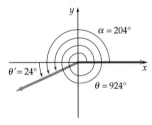

Because $\theta = 924° > 360°$, we first determine the coterminal angle, $\alpha = 204°$.

$\theta' = 204° - 180° = 24°$

d.

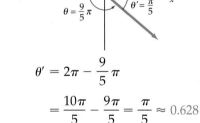

$\theta' = 2\pi - \dfrac{9}{5}\pi$

$= \dfrac{10\pi}{5} - \dfrac{9\pi}{5} = \dfrac{\pi}{5} \approx 0.628$

e.

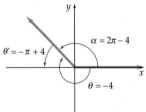

-4 radians is coterminal with $\alpha = 2\pi - 4$.

$\theta' = \pi - (2\pi - 4)$
$\quad = -\pi + 4 \approx 0.8584$

f.

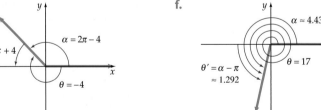

Because $17 \div (2\pi) \approx 2.70563$, a coterminal angle is $\alpha = 0.70563(2\pi) \approx 4.434$.

Because α is a Quadrant III angle, $\theta' = \alpha - \pi \approx 1.292$

▶ **TRY EXERCISE 26, PAGE 149**

take note

The Reference Angle Theorem is also valid if the sine function is replaced by any other trigonometric function.

The following theorem states an important relationship that exists between sin θ and sin θ', where θ' is the reference angle for angle θ.

Reference Angle Theorem

To evaluate sin θ, determine sin θ'. Then use either sin θ' or its opposite as the answer, depending on which has the correct sign.

In the following example, we illustrate how to evaluate a trigonometric function of θ by first evaluating the trigonometric function of θ'.

EXAMPLE 4   **Use the Reference Angle Theorem to Evaluate Trigonometric Functions**

Evaluate each function.

a. sin 210°   b. cos 405°   c. tan 300°

Solution

a. We know that sin 210° is negative (the sign chart is given in **Table 2.5**). The reference angle for $\theta = 210°$ is $\theta' = 30°$. By the Reference Angle Theorem, we know that sin 210° equals either

$$\sin 30° = \frac{1}{2} \quad \text{or} \quad -\sin 30° = -\frac{1}{2}$$

Thus sin 210° $= -\frac{1}{2}$.

b. Because $\theta = 405°$ is a Quadrant I angle, we know that cos 405° > 0. The reference angle for $\theta = 405°$ is $\theta' = 45°$. By the Reference Angle Theorem, cos 405° equals either

$$\cos 45° = \frac{\sqrt{2}}{2} \quad \text{or} \quad -\cos 45° = -\frac{\sqrt{2}}{2}$$

Thus cos 405° $= \frac{\sqrt{2}}{2}$.

c. Because $\theta = 300°$ is a Quadrant IV angle, tan 300° < 0. The reference angle for $\theta = 300°$ is $\theta' = 60°$. Hence tan 300° equals either

$$\tan 60° = \sqrt{3} \quad \text{or} \quad -\tan 60° = -\sqrt{3}$$

Thus tan 300° $= -\sqrt{3}$.

▶ **TRY EXERCISE 38, PAGE 149**

INTEGRATING TECHNOLOGY

In many applications a calculator is used to evaluate a trigonometric function of the angle. Use *degree mode* to evaluate a trigonometric function of an angle given in degrees. Use *radian mode* to evaluate a trigonometric function of an angle given in radians. **Figure 2.50** shows the sine of 137.4 degrees, and **Figure 2.51** shows the cotangent of 3.4 radians.

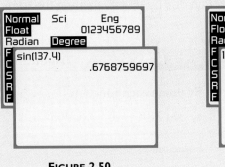

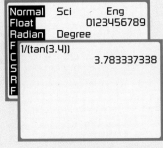

FIGURE 2.50   **FIGURE 2.51**

 ## TOPICS FOR DISCUSSION

1. Is every reference angle an acute angle? Explain.

2. If θ' is the reference angle for the angle θ, then $\sin \theta = \sin \theta'$. Do you agree? Explain.

3. If $\sin \theta < 0$ and $\cos \theta > 0$, then the terminal side of the angle θ lies in which quadrant?

4. A student claims that if $\theta = 160$, then $\theta' = 20$. Explain why the student is not correct.

5. Explain how to find the measure of the reference angle θ' for the angle $\theta = \dfrac{19}{5}\pi$.

EXERCISE SET 2.3

In Exercises 1 to 8, find the value of each of the six trigonometric functions for the angle whose terminal side passes through the given point.

1. $P(2, 3)$

2. $P(3, 7)$

3. $P(-2, 3)$

4. $P(-3, 5)$

5. $P(-8, -5)$

▶ **6.** $P(-6, -9)$

7. $P(-5, 0)$

8. $P(0, 2)$

In Exercises 9 to 14, let θ be an angle in standard position. State the quadrant in which the terminal side of θ lies.

9. $\sin \theta > 0, \quad \cos \theta > 0$

10. $\tan \theta < 0, \quad \sin \theta < 0$

11. $\cos \theta > 0, \quad \tan \theta < 0$

12. $\sin \theta < 0, \quad \cos \theta > 0$

13. $\sin \theta < 0, \quad \cos \theta < 0$

14. $\tan \theta < 0, \quad \cos \theta < 0$

In Exercises 15 to 24, find the value of each expression.

15. $\sin \theta = -\dfrac{1}{2}, 180° < \theta < 270°$; find $\tan \theta$.

16. $\cot \theta = -1, 90° < \theta < 180°$; find $\cos \theta$.

17. $\csc \theta = \sqrt{2}, \dfrac{\pi}{2} < \theta < \pi$; find $\cot \theta$.

▶ **18.** $\sec \theta = \dfrac{2\sqrt{3}}{3}, \dfrac{3\pi}{2} < \theta < 2\pi$; find $\sin \theta$.

19. $\sin \theta = -\dfrac{1}{2}$ and $\cos \theta > 0$; find $\tan \theta$.

20. $\tan \theta = 1$ and $\sin \theta < 0$; find $\cos \theta$.

21. $\cos \theta = \dfrac{1}{2}$ and $\tan \theta = \sqrt{3}$; find $\csc \theta$.

22. $\tan \theta = 1$ and $\sin \theta = \dfrac{\sqrt{2}}{2}$; find $\sec \theta$.

23. $\cos \theta = -\dfrac{1}{2}$ and $\sin \theta = \dfrac{\sqrt{3}}{2}$; find $\cot \theta$.

24. $\sec \theta = \dfrac{2\sqrt{3}}{3}$ and $\sin \theta = -\dfrac{1}{2}$; find $\cot \theta$.

In Exercises 25 to 36, find the measure of the reference angle θ' for the given angle θ.

25. $\theta = 160°$

▶ **26.** $\theta = 255°$

27. $\theta = 351°$

28. $\theta = 48°$

29. $\theta = \dfrac{11}{5}\pi$

30. $\theta = -6$

31. $\theta = \dfrac{8}{3}$

32. $\theta = \dfrac{18}{7}\pi$

33. $\theta = 1406°$

34. $\theta = 840°$

35. $\theta = -475°$

36. $\theta = -650°$

In Exercises 37 to 48, use the Reference Angle Theorem to find the exact value of each trigonometric function.

37. $\sin 225°$

▶ **38.** $\cos 300°$

39. $\tan 405°$

40. $\sec 150°$

41. $\csc \dfrac{4}{3}\pi$

42. $\cot \dfrac{7}{6}\pi$

43. $\cos \dfrac{17\pi}{4}$

44. $\tan \left(-\dfrac{\pi}{3}\right)$

45. $\sec 765°$

46. $\csc (-510°)$

47. $\cot 540°$

48. $\cos 570°$

In Exercises 49 to 60, use a calculator to approximate the given trigonometric functions to six significant digits.

49. $\sin 127°$

50. $\sin (-257°)$

51. $\cos (-116°)$

52. $\cot 398°$

53. $\sec 578°$

54. $\sec 740°$

55. $\sin \left(-\dfrac{\pi}{5}\right)$

56. $\cos \dfrac{3\pi}{7}$

57. $\csc \dfrac{9\pi}{5}$

58. $\tan (-4.12)$

59. $\sec (-4.45)$

60. $\csc 0.34$

In Exercises 61 to 68, find (without using a calculator) the exact value of each expression.

61. $\sin 210° - \cos 330° \tan 330°$

62. $\tan 225° + \sin 240° \cos 60°$

63. $\sin^2 30° + \cos^2 30°$

64. $\cos \pi \sin \dfrac{7\pi}{4} - \tan \dfrac{11\pi}{6}$

65. $\sin\left(\dfrac{3\pi}{2}\right)\tan\left(\dfrac{\pi}{4}\right)-\cos\left(\dfrac{\pi}{3}\right)$

66. $\cos\left(\dfrac{7\pi}{4}\right)\tan\left(\dfrac{4\pi}{3}\right)+\cos\left(\dfrac{7\pi}{6}\right)$

67. $\sin^2\left(\dfrac{5\pi}{4}\right)+\cos^2\left(\dfrac{5\pi}{4}\right)$

68. $\tan^2\left(\dfrac{7\pi}{4}\right)-\sec^2\left(\dfrac{7\pi}{4}\right)$

CONNECTING CONCEPTS

In Exercises 69 to 74, find two values of θ, $0° \le \theta < 360°$, that satisfy the given trigonometric equation.

69. $\sin\theta=\dfrac{1}{2}$

70. $\tan\theta=-\sqrt{3}$

71. $\cos\theta=-\dfrac{\sqrt{3}}{2}$

72. $\tan\theta=1$

73. $\csc\theta=-\sqrt{2}$

74. $\cot\theta=-1$

In Exercises 75 to 80, find two values of θ, $0 \le \theta < 2\pi$, that satisfy the given trigonometric equation.

75. $\tan\theta=-1$

76. $\cos\theta=\dfrac{1}{2}$

77. $\tan\theta=-\dfrac{\sqrt{3}}{3}$

78. $\sec\theta=-\dfrac{2\sqrt{3}}{3}$

79. $\sin\theta=\dfrac{\sqrt{3}}{2}$

80. $\cos\theta=-\dfrac{1}{2}$

If $P(x, y)$ is a point on the terminal side of an acute angle θ in standard position and $r = \sqrt{x^2 + y^2}$, then $\sin\theta = \dfrac{y}{r}$ and $\cos\theta = \dfrac{x}{r}$. Using these definitions, we find that

$$\cos^2\theta+\sin^2\theta=\left(\dfrac{x}{r}\right)^2+\left(\dfrac{y}{r}\right)^2=\dfrac{x^2}{r^2}+\dfrac{y^2}{r^2}=\dfrac{x^2+y^2}{r^2}$$

$$=\dfrac{r^2}{r^2}=1$$

Hence $\cos^2\theta+\sin^2\theta=1$ for all acute angles θ. This important identity is actually true for all angles θ. We will show this later. In the meantime, use the definitions of the trigonometric functions to prove the identities in Exercises 81 to 90 for the acute angle θ.

81. $1+\tan^2\theta=\sec^2\theta$

82. $\cot^2\theta+1=\csc^2\theta$

83. $\tan\theta=\dfrac{\sin\theta}{\cos\theta}$

84. $\cot\theta=\dfrac{\cos\theta}{\sin\theta}$

85. $\cos(90°-\theta)=\sin\theta$

86. $\sin(90°-\theta)=\cos\theta$

87. $\tan(90°-\theta)=\cot\theta$

88. $\cot(90°-\theta)=\tan\theta$

89. $\sin(\theta+\pi)=-\sin\theta$

90. $\cos(\theta+\pi)=-\cos\theta$

PREPARE FOR SECTION 2.4

91. Determine whether the point $(0, 1)$ is a point on the circle defined by $x^2 + y^2 = 1$. [1.1/1.2]

92. Determine whether the point $\left(\dfrac{1}{2},\dfrac{\sqrt{3}}{2}\right)$ is a point on the circle defined by $x^2 + y^2 = 1$. [1.1/1.2]

93. Determine whether the point $\left(\dfrac{\sqrt{2}}{2},\dfrac{\sqrt{3}}{2}\right)$ is a point on the circle defined by $x^2 + y^2 = 1$. [1.1/1.2]

94. Determine the circumference of a circle with a radius of 1.

95. Determine whether $f(x) = x^2 - 3$ is an even function, an odd function, or a function that is neither even nor odd. [1.4]

96. Determine whether $f(x) = x^3 - x^2$ is an even function, an odd function, or a function that is neither even nor odd. [1.4]

PROJECTS

1. **FIND SUMS OR PRODUCTS** Determine the following sums or products. Do not use a calculator. (*Hint:* The Reference Angle Theorem may be helpful.) Explain to a classmate how you know you are correct.

    a. $\cos 0° + \cos 1° + \cos 2° + \cdots + \cos 178° + \cos 179° + \cos 180°$

    b. $\sin 0° + \sin 1° + \sin 2° + \cdots + \sin 358° + \sin 359° + \sin 360°$

    c. $\cot 1° + \cot 2° + \cot 3° + \cdots + \cot 177° + \cot 178° + \cot 179°$

    d. $(\cos 1°)(\cos 2°)(\cos 3°) \cdots (\cos 177°)(\cos 178°)(\cos 179°)$

    e. $\cos^2 1° + \cos^2 2° + \cos^2 3° + \cdots + \cos^2 357° + \cos^2 358° + \cos^2 359°$

SECTION 2.4

TRIGONOMETRIC FUNCTIONS OF REAL NUMBERS

- THE WRAPPING FUNCTION
- PROPERTIES OF TRIGONOMETRIC FUNCTIONS OF REAL NUMBERS
- TRIGONOMETRIC IDENTITIES
- AN APPLICATION INVOLVING A TRIGONOMETRIC FUNCTION OF A REAL NUMBER

• THE WRAPPING FUNCTION

In the seventeenth century, applications of trigonometry were extended to problems in physics and engineering. These kinds of problems required trigonometric functions whose domains were sets of real numbers rather than sets of angles. During this time, the definitions of trigonometric functions were extended to real numbers by using a correspondence between an angle and a real number.

Consider a circle given by the equation $x^2 + y^2 = 1$, called a **unit circle,** and a vertical coordinate line l tangent to the unit circle at $(1, 0)$. We define a function W that pairs a real number t on the coordinate line with a point $P(x, y)$ on the unit circle. This function is called the *wrapping function* because it is analogous to wrapping a line around a circle.

As shown in **Figure 2.52,** the positive part of the coordinate line is wrapped around the unit circle in a counterclockwise direction. The negative part of the coordinate line is wrapped around the circle in a clockwise direction. The wrapping function is defined by the equation $W(t) = P(x, y)$, where t is a real number and $P(x, y)$ is the point on the unit circle that corresponds to t.

Through the wrapping function, each real number t defines an arc $\overset{\frown}{AP}$ that subtends a central angle with a measure of θ radians. The length of the arc $\overset{\frown}{AP}$ is t (see **Figure 2.53**). From the equation $s = r\theta$ for the arc length of a circle, we have (with $t = s$) $t = r\theta$. For a unit circle, $r = 1$, and the equation becomes $t = \theta$. Thus, on a unit circle, *the measure of a central angle and the length of its arc can be represented by the same real number t.*

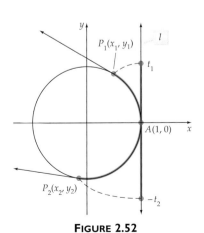

FIGURE 2.52

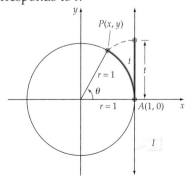

FIGURE 2.53

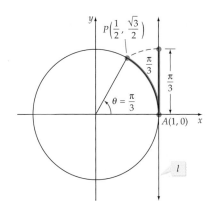

FIGURE 2.54

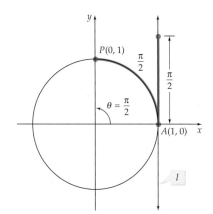

FIGURE 2.55

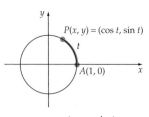

$x = \cos t, y = \sin t$
FIGURE 2.56

| EXAMPLE I | **Evaluate the Wrapping Function** |

Find the values of x and y such that $W\left(\dfrac{\pi}{3}\right) = P(x, y)$.

Solution

The point $\dfrac{\pi}{3}$ on line l is shown in **Figure 2.54**. From the wrapping function, $W\left(\dfrac{\pi}{3}\right)$ is the point P on the unit circle for which arc $\overset{\frown}{AP}$ subtends an angle θ, the measure of which is $\dfrac{\pi}{3}$ radians. The coordinates of P can be determined from the definitions of $\cos\theta$ and $\sin\theta$ given in Section 2.3 and from **Table 2.2**, page 132.

$$\cos\theta = \frac{x}{r} \qquad\qquad \sin\theta = \frac{y}{r}$$

$$\cos\frac{\pi}{3} = \frac{x}{1} = x \qquad \sin\frac{\pi}{3} = \frac{y}{1} = y \qquad \bullet\, \theta = \frac{\pi}{3};\, r = 1$$

$$\frac{1}{2} = x \qquad\qquad \frac{\sqrt{3}}{2} = y \qquad \bullet\, \cos\frac{\pi}{3} = \frac{1}{2};\, \sin\frac{\pi}{3} = \frac{\sqrt{3}}{2}$$

From these equations, $x = \dfrac{1}{2}$ and $y = \dfrac{\sqrt{3}}{2}$. Therefore, $W\left(\dfrac{\pi}{3}\right) = P\left(\dfrac{1}{2}, \dfrac{\sqrt{3}}{2}\right)$.

▶ **TRY EXERCISE 10, PAGE 160**

To determine $W\left(\dfrac{\pi}{2}\right)$, recall that the circumference of a unit circle is 2π. One-fourth the circumference is $\dfrac{1}{4}(2\pi) = \dfrac{\pi}{2}$ (see **Figure 2.55**). Thus $W\left(\dfrac{\pi}{2}\right) = P(0, 1)$.

Note from the last two examples that for the given real number t, $\cos t = x$ and $\sin t = y$. That is, for a real number t and $W(t) = P(x, y)$, the value of the cosine of t is the x-coordinate of P, and the value of the sine of t is the y-coordinate of P. See **Figure 2.56**.

❓ **QUESTION** What is the point defined by $W\left(\dfrac{\pi}{4}\right)$?

The following definition makes use of the wrapping function $W(t)$ to define trigonometric functions of real numbers.

Definition of the Trigonometric Functions of Real Numbers

Let W be the wrapping function, t be a real number, and $W(t) = P(x, y)$. Then

$$\sin t = y \qquad\qquad \cos t = x \qquad\qquad \tan t = \frac{y}{x},\, x \neq 0$$

$$\csc t = \frac{1}{y},\ y \neq 0 \qquad \sec t = \frac{1}{x},\ x \neq 0 \qquad \cot t = \frac{x}{y},\ y \neq 0$$

❓ **ANSWER** $\left(\dfrac{\sqrt{2}}{2}, \dfrac{\sqrt{2}}{2}\right)$

Trigonometric functions of real numbers are frequently called *circular functions* to distinguish them from trigonometric functions of angles.

The *trigonometric functions of real numbers* (or circular functions) look remarkably like the trigonometric functions defined in the last section. The difference between the two is that of domain: In one case, the domains are sets of *real numbers*; in the other case, the domains are sets of *angles.* However, there are similarities between the two functions.

Consider an angle θ (in radians) in standard position, as shown in **Figure 2.57.** Let $P(x, y)$ and $P'(x', y')$ be two points on the terminal side of θ, where $x^2 + y^2 = 1$ and $(x')^2 + (y')^2 = r^2$. Let t be the length of the arc from $A(1, 0)$ to $P(x, y)$. Then

$$\sin\theta = \frac{y'}{r} = \frac{y}{1} = \sin t$$

Thus the value of the sine function of θ, measured in radians, is equal to the value of the sine of the real number t. Similar arguments can be given to show corresponding results for the other five trigonometric functions. With this in mind, we can assert that *the value of a trigonometric function at the real number t is its value at an angle of t radians.*

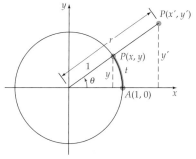

FIGURE 2.57

EXAMPLE 2   **Evaluate Trigonometric Functions of Real Numbers**

Find the exact value of each function.

a. $\cos\dfrac{\pi}{4}$ b. $\sin\left(-\dfrac{7\pi}{6}\right)$ c. $\tan\left(-\dfrac{5\pi}{4}\right)$ d. $\sec\dfrac{5\pi}{3}$

Solution

The value of a trigonometric function at the real number t is its value at an angle of t radians. Using **Table 2.2** on page 132, we have

a. $\cos\dfrac{\pi}{4} = \dfrac{\sqrt{2}}{2}$

b. $\sin\left(-\dfrac{7\pi}{6}\right) = \sin\dfrac{\pi}{6} = \dfrac{1}{2}$ • Reference angle for $-\dfrac{7\pi}{6}$ is $\dfrac{\pi}{6}$ and $\sin t > 0$ in Quadrant II.

c. $\tan\left(-\dfrac{5\pi}{4}\right) = -\tan\dfrac{\pi}{4} = -1$ • Reference angle for $-\dfrac{5\pi}{4}$ is $\dfrac{\pi}{4}$ and $\tan t < 0$ in Quadrant II.

d. $\sec\dfrac{5\pi}{3} = \sec\dfrac{\pi}{3} = 2$ • Reference angle for $\dfrac{5\pi}{3}$ is $\dfrac{\pi}{3}$ and $\sec t > 0$ in Quadrant IV.

▶ TRY EXERCISE 16, PAGE 160

● **PROPERTIES OF TRIGONOMETRIC FUNCTIONS OF REAL NUMBERS**

To review DOMAIN AND RANGE, *see p. 30.*

The domain and range of the trigonometric functions can be found from the definitions of these functions. If t is any real number and $P(x, y)$ is the point corresponding to $W(t)$, then by definition $\cos t = x$ and $\sin t = y$. Thus the domain of the sine and cosine functions is the set of real numbers.

Because the radius of the unit circle is 1, we have

$$-1 \le x \le 1 \quad \text{and} \quad -1 \le y \le 1$$

Therefore, with $x = \cos t$ and $y = \sin t$, we have

$$-1 \le \cos t \le 1 \quad \text{and} \quad -1 \le \sin t \le 1$$

The range of the cosine and sine functions is $[-1, 1]$.

Using the definitions of tangent and secant,

$$\tan t = \frac{y}{x} \quad \text{and} \quad \sec t = \frac{1}{x}$$

The domain of the tangent function is all real numbers t except those for which the x-coordinate of $W(t)$ is zero. The x-coordinate is zero when $t = \pm\frac{\pi}{2}$, $t = \pm\frac{3\pi}{2}$, and $t = \pm\frac{5\pi}{2}$ and in general when $t = \frac{(2n + 1)\pi}{2}$, where n is an integer. Thus the domain of the tangent function is the set of all real numbers t except $t = \frac{(2n + 1)\pi}{2}$, where n is an integer. The range of the tangent function is all real numbers.

Similar methods can be used to find the domain and range of the cotangent, secant, and cosecant functions. The results are summarized in **Table 2.6.**

TABLE 2.6   Domain and Range of the Trigonometric Functions
(n is an integer)

Function	Domain	Range
$y = \sin t$	$\{t \mid -\infty < t < \infty\}$	$\{y \mid -1 \le y \le 1\}$
$y = \cos t$	$\{t \mid -\infty < t < \infty\}$	$\{y \mid -1 \le y \le 1\}$
$y = \tan t$	$\left\{t \mid -\infty < t < \infty, t \ne \frac{(2n + 1)\pi}{2}\right\}$	$\{y \mid -\infty < y < \infty\}$
$y = \csc t$	$\{t \mid -\infty < t < \infty, t \ne n\pi\}$	$\{y \mid y \ge 1, y \le -1\}$
$y = \sec t$	$\left\{t \mid -\infty < t < \infty, t \ne \frac{(2n + 1)\pi}{2}\right\}$	$\{y \mid y \ge 1, y \le -1\}$
$y = \cot t$	$\{t \mid -\infty < t < \infty, t \ne n\pi\}$	$\{y \mid -\infty < y < \infty\}$

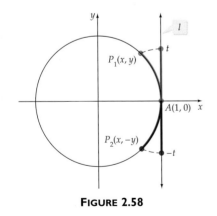

FIGURE 2.58

Consider the points t and $-t$ on the coordinate line l tangent to the unit circle at the point $(1, 0)$. The points $W(t)$ and $W(-t)$ are symmetric with respect to the x-axis. Therefore, if $P_1(x, y)$ are the coordinates of $W(t)$, then $P_2(x, -y)$ are the coordinates of $W(-t)$. See **Figure 2.58.**

From the definitions of the trigonometric functions, we have

$$\sin t = y \quad \text{and} \quad \sin(-t) = -y \quad \text{and} \quad \cos t = x \quad \text{and} \quad \cos(-t) = x$$

Substituting $\sin t$ for y and $\cos t$ for x yields

$$\sin(-t) = -\sin t \quad \text{and} \quad \cos(-t) = \cos t$$

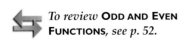

To review **ODD AND EVEN FUNCTIONS,** *see p. 52.*

Thus the sine is an odd function, and the cosine is an even function. Because $\csc t = \dfrac{1}{\sin t}$ and $\sec t = \dfrac{1}{\cos t}$, it follows that

$$\csc(-t) = -\csc t \quad \text{and} \quad \sec(-t) = \sec t$$

These equations state that the cosecant is an odd function and the secant is an even function.

From the definition of the tangent function, we have $\tan t = \dfrac{y}{x}$ and $\tan(-t) = -\dfrac{y}{x}$. Substituting $\tan t$ for $\dfrac{y}{x}$ yields $\tan(-t) = -\tan t$. Because $\cot t = \dfrac{1}{\tan t}$, it follows that $\cot(-t) = -\cot t$. Thus the tangent and cotangent are odd functions.

Even and Odd Trigonometric Functions

The odd trigonometric functions are $y = \sin t$, $y = \csc t$, $y = \tan t$, and $y = \cot t$. The even trigonometric functions are $y = \cos t$ and $y = \sec t$.

Thus for all t in their domain,

$$\sin(-t) = -\sin t \quad \cos(-t) = \cos t \quad \tan(-t) = -\tan t$$
$$\csc(-t) = -\csc t \quad \sec(-t) = \sec t \quad \cot(-t) = -\cot t$$

EXAMPLE 3 **Determine Whether a Function Is Even, Odd, or Neither**

Is the function defined by $f(x) = x - \tan x$ an even function, an odd function, or neither?

Solution

Find $f(-x)$ and compare it to $f(x)$.

$$f(-x) = (-x) - \tan(-x) = -x + \tan x \qquad \bullet \ \tan(-x) = -\tan x$$
$$= -(x - \tan x)$$
$$= -f(x)$$

The function $f(x) = x - \tan x$ is an odd function.

▶ **TRY EXERCISE 44, PAGE 160**

Let W be the wrapping function, t be a point on the coordinate line tangent to the unit circle at $(1, 0)$, and $W(t) = P(x, y)$. Because the circumference of the unit circle is 2π, $W(t + 2\pi) = W(t) = P(x, y)$. Thus the value of the wrapping function repeats itself in 2π units. *The wrapping function is periodic, and the period is 2π.*

Recall the definitions of $\cos t$ and $\sin t$:

$$\cos t = x \quad \text{and} \quad \sin t = y$$

where $W(t) = P(x, y)$. Because $W(t + 2\pi) = W(t) = P(x, y)$ for all t,

$$\cos(t + 2\pi) = x \qquad \text{and} \qquad \sin(t + 2\pi) = y$$

Thus $\cos t$ and $\sin t$ have period 2π. Because

$$\sec t = \frac{1}{\cos t} = \frac{1}{\cos(t + 2\pi)} = \sec(t + 2\pi) \qquad \text{and}$$

$$\csc t = \frac{1}{\sin t} = \frac{1}{\sin(t + 2\pi)} = \csc(t + 2\pi)$$

$\sec t$ and $\csc t$ have a period of 2π.

Period of cos t, sin t, sec t, and csc t

The period of $\cos t$, $\sin t$, $\sec t$, and $\csc t$ is 2π.

Although it is true that $\tan t = \tan(t + 2\pi)$, the period of $\tan t$ is not 2π. Recall that the period of a function is the *smallest* value of p for which $f(t) = f(t + p)$.

If W is the wrapping function (see **Figure 2.59**) and $W(t) = P(x, y)$, then $W(t + \pi) = P(-x, -y)$. Because

$$\tan t = \frac{y}{x} \qquad \text{and} \qquad \tan(t + \pi) = \frac{-y}{-x} = \frac{y}{x} = \tan t$$

we have $\tan(t + \pi) = \tan t$ for all t. A similar argument applies to $\cot t$.

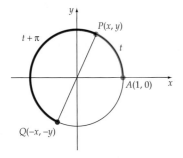

FIGURE 2.59

Period of tan t and cot t

The period of $\tan t$ and $\cot t$ is π.

● **TRIGONOMETRIC IDENTITIES**

Recall that any equation that is true for every number in the domain of the equation is an identity. The statement

$$\csc t = \frac{1}{\sin t}, \quad \sin t \neq 0$$

is an identity because the two expressions produce the same result for all values of t for which both functions are defined.

The **ratio identities** are obtained by writing the tangent and cotangent functions in terms of the sine and cosine functions.

$$\tan t = \frac{y}{x} = \frac{\sin t}{\cos t} \qquad \text{and} \qquad \cot t = \frac{x}{y} = \frac{\cos t}{\sin t} \qquad \bullet\ x = \cos t \text{ and } y = \sin t$$

The **Pythagorean identities** are based on the equation of a unit circle, $x^2 + y^2 = 1$, and on the definitions of the sine and cosine functions.

$$x^2 + y^2 = 1$$
$$\cos^2 t + \sin^2 t = 1 \qquad \bullet\ \textbf{Replace } x \textbf{ by cos } t \textbf{ and } y \textbf{ by sin } t.$$

Dividing each term of $\cos^2 t + \sin^2 t = 1$ by $\cos^2 t$, we have

$$\frac{\cos^2 t}{\cos^2 t} + \frac{\sin^2 t}{\cos^2 t} = \frac{1}{\cos^2 t} \qquad \bullet\ \cos t \neq 0$$

$$1 + \tan^2 t = \sec^2 t \qquad \bullet\ \frac{\sin t}{\cos t} = \tan t$$

Dividing each term of $\cos^2 t + \sin^2 t = 1$ by $\sin^2 t$, we have

$$\frac{\cos^2 t}{\sin^2 t} + \frac{\sin^2 t}{\sin^2 t} = \frac{1}{\sin^2 t} \qquad \bullet\ \sin t \neq 0$$

$$\cot^2 t + 1 = \csc^2 t \qquad \bullet\ \frac{\cos t}{\sin t} = \cot t$$

Here is a summary of the Fundamental Trigonometric Identities:

Fundamental Trigonometric Identities

The reciprocal identities are

$$\sin t = \frac{1}{\csc t} \qquad \cos t = \frac{1}{\sec t} \qquad \tan t = \frac{1}{\cot t}$$

The ratio identities are

$$\tan t = \frac{\sin t}{\cos t} \qquad \cot t = \frac{\cos t}{\sin t}$$

The Pythagorean identities are

$$\cos^2 t + \sin^2 t = 1 \qquad 1 + \tan^2 t = \sec^2 t \qquad 1 + \cot^2 t = \csc^2 t$$

EXAMPLE 4   **Use the Unit Circle to Verify an Identity**

By using the unit circle and the definitions of the trigonometric functions, show that $\sin (t + \pi) = -\sin t$.

Solution

Sketch a unit circle, and let P be the point on the unit circle such that $W(t) = P(x, y)$, as shown in **Figure 2.60.** Draw a diameter from P, and label the endpoint Q. For any line through the origin, if $P(x, y)$ is a point on the line, then $Q(-x, -y)$ is also a point on the line. Because PQ is a diameter, the length of arc PQ is π. Thus the length of arc AQ is $t + \pi$. Therefore, $W(t + \pi) = Q(-x, -y)$. From the definition of $\sin t$, we have

$$\sin t = y \qquad \text{and} \qquad \sin (t + \pi) = -y$$

Thus $\sin (t + \pi) = -\sin t$.

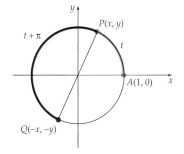

FIGURE 2.60

▶ **TRY EXERCISE 50, PAGE 161**

Using identities and basic algebra concepts, we can rewrite trigonometric expressions in different forms.

EXAMPLE 5   **Simplify a Trigonometric Expression**

Write the expression $\dfrac{1}{\sin^2 t} + \dfrac{1}{\cos^2 t}$ as a single term.

Solution

Express each fraction in terms of a common denominator. The common denominator is $\sin^2 t \cos^2 t$.

$$\frac{1}{\sin^2 t} + \frac{1}{\cos^2 t} = \frac{1}{\sin^2 t}\frac{\cos^2 t}{\cos^2 t} + \frac{1}{\cos^2 t}\frac{\sin^2 t}{\sin^2 t}$$

$$= \frac{\cos^2 t + \sin^2 t}{\sin^2 t \cos^2 t} = \frac{1}{\sin^2 t \cos^2 t} \qquad \bullet \; \cos^2 t + \sin^2 t = 1$$

▶ **TRY EXERCISE 68, PAGE 161**

take note

Because

$$\frac{1}{\sin^2 t \cos^2 t} = \frac{1}{\sin^2 t} \cdot \frac{1}{\cos^2 t}$$

$$= (\csc^2 t)(\sec^2 t)$$

we could have written the answer to Example 5 in terms of the cosecant and secant functions.

EXAMPLE 6   **Write a Trigonometric Expression in Terms of a Given Function**

For $\dfrac{\pi}{2} < t < \pi$, write $\tan t$ in terms of $\sin t$.

Solution

Write $\tan t = \dfrac{\sin t}{\cos t}$. Now solve $\cos^2 t + \sin^2 t = 1$ for $\cos t$.

$$\cos^2 t + \sin^2 t = 1$$

$$\cos^2 t = 1 - \sin^2 t$$

$$\cos t = \pm\sqrt{1 - \sin^2 t}$$

Because $\dfrac{\pi}{2} < t < \pi$, $\cos t$ is negative. Therefore, $\cos t = -\sqrt{1 - \sin^2 t}$.

Thus

$$\tan t = -\frac{\sin t}{\sqrt{1 - \sin^2 t}} \qquad \bullet \; \frac{\pi}{2} < t < \pi$$

▶ **TRY EXERCISE 74, PAGE 161**

● **AN APPLICATION INVOLVING A TRIGONOMETRIC FUNCTION OF A REAL NUMBER**

EXAMPLE 7   **Determine a Height as a Function of Time**

The Millennium Wheel, in London, is the world's largest Ferris wheel. It has a diameter of 450 feet. When the Millennium Wheel is in uniform motion, it completes one revolution every 30 minutes. The height h, in feet above the Thames River, of a person riding on the Millennium Wheel can be estimated by

$$h(t) = 255 - 225 \cos\left(\frac{\pi}{15}t\right)$$

where t is the time in minutes since the person started the ride.

a.   How high is the person at the start of the ride ($t = 0$)?

b.   How high is the person after 18.0 minutes?

Solution

a. $h(0) = 255 - 225 \cos\left(\frac{\pi}{15} \cdot 0\right)$ **b.** $h(18.0) = 255 - 225 \cos\left(\frac{\pi}{15} \cdot 18.0\right)$

$= 255 - 225$ $\approx 255 - (-182)$

$= 30$ $= 437$

At the start of the ride, the   After 18.0 minutes, the person is
person is 30 feet above   about 437 feet above the Thames.
the Thames.

The Millennium Wheel, on the
banks of the Thames River,
London.

 TRY EXERCISE 78, PAGE 161

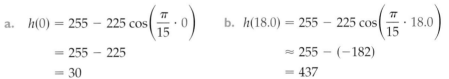

▨ ✎ **TOPICS FOR DISCUSSION**

1.   Is $W(t)$ a number? Explain.

2.   Explain how to find the exact value of $\cos\left(\frac{13\pi}{6}\right)$.

3.   Explain why the equation $\cos^2 t + \sin^2 t = 1$ is called a Pythagorean identity.

4.   Is $f(x) = \cos^3 x$ an even function or an odd function? Explain how you made your decision.

5.   Explain how to make use of a unit circle to show that $\sin(-t) = -\sin t$.

EXERCISE SET 2.4

In Exercises 1 to 12, find $W(t)$ for each given t.

1. $t = \dfrac{\pi}{6}$ **2.** $t = \dfrac{\pi}{4}$ **3.** $t = \dfrac{7\pi}{6}$

4. $t = \dfrac{4\pi}{3}$ **5.** $t = \dfrac{5\pi}{3}$ **6.** $t = -\dfrac{\pi}{6}$

7. $t = \dfrac{11\pi}{6}$ **8.** $t = 0$ **9.** $t = \pi$

▶ **10.** $t = -\dfrac{7\pi}{4}$ **11.** $t = -\dfrac{2\pi}{3}$ **12.** $t = -\pi$

In Exercises 13 to 22, find the exact value of each function.

13. $\tan\left(\dfrac{11\pi}{6}\right)$ **14.** $\cot\left(\dfrac{2\pi}{3}\right)$

15. $\cos\left(-\dfrac{2\pi}{3}\right)$ ▶ **16.** $\sec\left(-\dfrac{5\pi}{6}\right)$

17. $\csc\left(-\dfrac{\pi}{3}\right)$ **18.** $\tan(12\pi)$

19. $\sin\left(\dfrac{3\pi}{2}\right)$ **20.** $\cos\left(\dfrac{7\pi}{3}\right)$

21. $\sec\left(-\dfrac{7\pi}{6}\right)$ **22.** $\sin\left(-\dfrac{5\pi}{3}\right)$

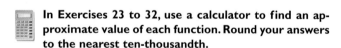 **In Exercises 23 to 32, use a calculator to find an approximate value of each function. Round your answers to the nearest ten-thousandth.**

23. $\sin 1.22$ **24.** $\cos 4.22$

25. $\csc(-1.05)$ **26.** $\sin(-0.55)$

27. $\tan\left(\dfrac{11\pi}{12}\right)$ **28.** $\cos\left(\dfrac{2\pi}{5}\right)$

29. $\cos\left(-\dfrac{\pi}{5}\right)$ **30.** $\csc 8.2$

31. $\sec 1.55$ **32.** $\cot 2.11$

In Exercises 33 to 40, use the unit circle in the next column to estimate the following values to the nearest tenth.

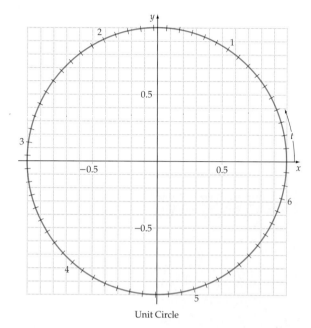

Unit Circle

33. a. $\sin 2$ **b.** $\cos 2$

34. a. $\sin 3$ **b.** $\cos 3$

35. a. $\sin 5.4$ **b.** $\cos 5.4$

36. a. $\sin 4.1$ **b.** $\cos 4.1$

37. All real numbers t between 0 and 2π for which $\sin t = 0.4$.

38. All real numbers t between 0 and 2π for which $\cos t = 0.8$.

39. All real numbers t between 0 and 2π for which $\sin t = -0.3$.

40. All real numbers t between 0 and 2π for which $\cos t = -0.7$.

In Exercises 41 to 48, determine whether the function defined by each equation is even, odd, or neither.

41. $f(x) = -4\sin x$ **42.** $f(x) = -2\cos x$

43. $G(x) = \sin x + \cos x$ ▶ **44.** $F(x) = \tan x + \sin x$

45. $S(x) = \dfrac{\sin x}{x}, x \neq 0$ **46.** $C(x) = \dfrac{\cos x}{x}, x \neq 0$

47. $v(x) = 2\sin x \cos x$ **48.** $w(x) = x \tan x$

In Exercises 49 to 56, use the unit circle to verify each identity.

49. $\cos(-t) = \cos t$

▶ **50.** $\tan(t - \pi) = \tan t$

51. $\cos(t + \pi) = -\cos t$

52. $\sin(-t) = -\sin t$

53. $\sin(t - \pi) = -\sin t$

54. $\sec(-t) = \sec t$

55. $\csc(-t) = -\csc t$

56. $\tan(-t) = -\tan t$

In Exercises 57 to 72, use the trigonometric identities to write each expression in terms of a single trigonometric function or a constant. Answers may vary.

57. $\tan t \cos t$

58. $\cot t \sin t$

59. $\dfrac{\csc t}{\cot t}$

60. $\dfrac{\sec t}{\tan t}$

61. $1 - \sec^2 t$

62. $1 - \csc^2 t$

63. $\tan t - \dfrac{\sec^2 t}{\tan t}$

64. $\dfrac{\csc^2 t}{\cot t} - \cot t$

65. $\dfrac{1 - \cos^2 t}{\tan^2 t}$

66. $\dfrac{1 - \sin^2 t}{\cot^2 t}$

67. $\dfrac{1}{1 - \cos t} + \dfrac{1}{1 + \cos t}$

▶ **68.** $\dfrac{1}{1 - \sin t} + \dfrac{1}{1 + \sin t}$

69. $\dfrac{\tan t + \cot t}{\tan t}$

70. $\dfrac{\csc t - \sin t}{\csc t}$

71. $\sin^2 t(1 + \cot^2 t)$

72. $\cos^2 t(1 + \tan^2 t)$

73. Write $\sin t$ in terms of $\cos t$, $0 < t < \dfrac{\pi}{2}$.

▶ **74.** Write $\tan t$ in terms of $\sec t$, $\dfrac{3\pi}{2} < t < 2\pi$.

75. Write $\csc t$ in terms of $\cot t$, $\dfrac{\pi}{2} < t < \pi$.

76. Write $\sec t$ in terms of $\tan t$, $\pi < t < \dfrac{3\pi}{2}$.

77. PATH OF A SATELLITE A satellite is launched into space from Cape Canaveral. The directed distance, in miles, that the satellite is above or below the equator is

$$d(t) = 1970 \cos\left(\dfrac{\pi}{64} t\right)$$

where t is the number of minutes since liftoff. A negative d value indicates that the satellite is south of the equator.

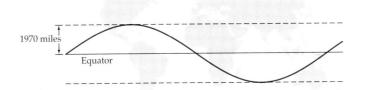

What distance, to the nearest 10 miles, is the satellite north of the equator 24 minutes after liftoff?

▶ **78.** AVERAGE HIGH TEMPERATURE The average high temperature T, in degrees Fahrenheit, for Fairbanks, Alaska, is given by

$$T(t) = -41 \cos\left(\dfrac{\pi}{6} t\right) + 36$$

where t is the number of months after January 5. Use the formula to estimate (to the nearest 0.1 degree Fahrenheit) the average high temperature in Fairbanks for March 5 and July 20.

In Exercises 79 to 90, perform the indicated operation and simplify.

79. $\cos t - \dfrac{1}{\cos t}$

80. $\tan t + \dfrac{1}{\tan t}$

81. $\cot t + \dfrac{1}{\cot t}$

82. $\sin t - \dfrac{1}{\sin t}$

83. $(1 - \sin t)^2$

84. $(1 - \cos t)^2$

85. $(\sin t - \cos t)^2$

86. $(\sin t + \cos t)^2$

87. $(1 - \sin t)(1 + \sin t)$

88. $(1 - \cos t)(1 + \cos t)$

89. $\dfrac{\sin t}{1 + \cos t} + \dfrac{1 + \cos t}{\sin t}$

90. $\dfrac{1 - \sin t}{\cos t} - \dfrac{1}{\tan t + \sec t}$

In Exercises 91 to 98, factor the expression.

91. $\cos^2 t - \sin^2 t$

92. $\sec^2 t - \csc^2 t$

93. $\tan^2 t - \tan t - 6$

94. $\cos^2 t + 3 \cos t - 4$

95. $2 \sin^2 t - \sin t - 1$

96. $4 \cos^2 t + 4 \cos t + 1$

97. $\cos^4 t - \sin^4 t$

98. $\sec^4 t - \csc^4 t$

CONNECTING CONCEPTS

In Exercises 99 to 102, use the trigonometric identities to find the value of the function.

99. Given $\csc t = \sqrt{2}$, $0 < t < \dfrac{\pi}{2}$, find $\cos t$.

100. Given $\cos t = \dfrac{1}{2}$, $\dfrac{3\pi}{2} < t < 2\pi$, find $\sin t$.

101. Given $\sin t = \dfrac{1}{2}$, $\dfrac{\pi}{2} < t < \pi$, find $\tan t$.

102. Given $\cot t = \dfrac{\sqrt{3}}{3}$, $\pi < t < \dfrac{3\pi}{2}$, find $\cos t$.

In Exercises 103 to 106, simplify the first expression to the second expression.

103. $\dfrac{\sin^2 t + \cos^2 t}{\sin^2 t}$; $\csc^2 t$

104. $\dfrac{\sin^2 t + \cos^2 t}{\cos^2 t}$; $\sec^2 t$

105. $(\cos t - 1)(\cos t + 1)$; $-\sin^2 t$

106. $(\sec t - 1)(\sec t + 1)$; $\tan^2 t$

PREPARE FOR SECTION 2.5
---

107. Estimate, to the nearest tenth, $\sin \dfrac{3\pi}{4}$. [2.3]

108. Estimate, to the nearest tenth, $\cos \dfrac{5\pi}{4}$. [2.3]

109. Explain how to use the graph of $y = f(x)$ to produce the graph of $y = -f(x)$. [1.4]

110. Explain how to use the graph of $y = f(x)$ to produce the graph of $y = f(2x)$. [1.4]

111. Simplify: $\dfrac{2\pi}{1/3}$

112. Simplify: $\dfrac{2\pi}{2/5}$

PROJECTS
---

1. VISUAL INSIGHT

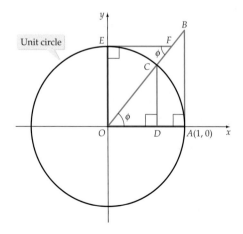

$OD = \cos \phi \quad DC = \sin \phi$

 Make use of the circle and similar triangles at the left to explain why the length of line segment

a. AB is equal to $\tan \phi$ **b.** EF is equal to $\cot \phi$

c. OB is equal to $\sec \phi$ **d.** OF is equal to $\csc \phi$

2. PERIODIC FUNCTIONS

a. A function f is periodic with a period of 3. If $f(2) = -1$, determine $f(14)$.

b. If g is a periodic function with period 2 and h is a periodic function with period 3, determine the period of $f + g$.

c. Find two periodic functions f and g such that the function $f + g$ is not a periodic function.

3. Consider a square as shown. Start at the point $(1, 0)$ and travel counterclockwise around the square for a distance t ($t \geq 0$).

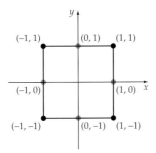

Let $WSQ(t) = P(x, y)$ be the point on the square determined by traveling counterclockwise a distance of t units from $(1, 0)$. For instance,

$$WSQ(0.5) = (1, 0.5)$$
$$WSQ(1.75) = (0.25, 1)$$

Find $WSQ(4.2)$ and $WSQ(6.4)$. We define the square sine of t, denoted by ssin t, to be the y-value of point P. The square cosine of t, denoted by scos t, is defined to be the x-value of point P. For example,

$$\text{ssin } 0.4 = 0.4 \qquad \text{scos } 0.4 = 1$$
$$\text{scos } 1.2 = 0.8 \qquad \text{scos } 5.3 = -0.7$$

The square tangent of t, denoted by stan t, is defined as

$$\text{stan } t = \frac{\text{ssin } t}{\text{scos } t} \qquad \text{scos } t \neq 0$$

Find each of the following.

a. ssin 3.2   **b.** scos 4.4   **c.** stan 5.5

d. ssin 11.2   **e.** scos -5.2 **f.** stan -6.5

SECTION 2.5

GRAPHS OF THE SINE AND COSINE FUNCTIONS

- THE GRAPH OF THE SINE FUNCTION
- THE GRAPH OF THE COSINE FUNCTION

• THE GRAPH OF THE SINE FUNCTION

The trigonometric functions can be graphed on a rectangular coordinate system by plotting the points whose coordinates belong to the function. We begin with the graph of the sine function.

Table 2.7 on the following page lists some ordered pairs (x, y), where $y = \sin x, 0 \leq x \leq 2\pi$. In **Figure 2.61,** the points are plotted and a smooth curve is drawn through the points.

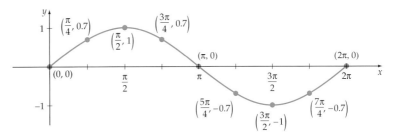

$y = \sin x, 0 \leq x \leq 2\pi$

FIGURE 2.61

TABLE 2.7

x	y = sin x
0	0
$\dfrac{\pi}{4}$	≈ 0.7
$\dfrac{\pi}{2}$	1
$\dfrac{3\pi}{4}$	≈ 0.7
π	0
$\dfrac{5\pi}{4}$	≈ -0.7
$\dfrac{3\pi}{2}$	-1
$\dfrac{7\pi}{4}$	≈ -0.7
2π	0

Because the domain of the sine function is the real numbers and the period is 2π, the graph of $y = \sin x$ is drawn by repeating the portion shown in **Figure 2.61.** Any part of the graph that corresponds to one period (2π) is one cycle of the graph of $y = \sin x$ (see **Figure 2.62**).

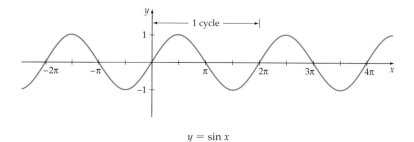

$$y = \sin x$$

FIGURE 2.62

The maximum value M reached by $\sin x$ is 1, and the minimum value m is -1. The amplitude of the graph of $y = \sin x$ is given by

$$\text{Amplitude} = \frac{1}{2}(M - m)$$

❷ QUESTION What is the amplitude of $y = \sin x$?

Recall that the graph of $y = a \cdot f(x)$ is obtained by *stretching* ($|a| > 1$) or *shrinking* ($0 < |a| < 1$) the graph of $y = f(x)$. **Figure 2.63** shows the graph of $y = 3 \sin x$ that was drawn by stretching the graph of $y = \sin x$. The amplitude of $y = 3 \sin x$ is 3 because

$$\text{Amplitude} = \frac{1}{2}(M - m) = \frac{1}{2}[3 - (-3)] = 3$$

Note that for $y = \sin x$ and $y = 3 \sin x$, the amplitude of the graph is the coefficient of $\sin x$. This suggests the following theorem.

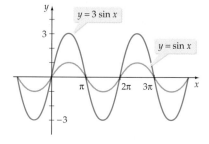

FIGURE 2.63

Amplitude of y = a sin x

The amplitude of $y = a \sin x$ is $|a|$.

❷ ANSWER $\text{Amplitude} = \dfrac{1}{2}(M - m) = \dfrac{1}{2}[1 - (-1)] = \dfrac{1}{2}(2) = 1$

EXAMPLE I Graph $y = a \sin x$

Graph: $y = -2 \sin x$

Solution

The amplitude of $y = -2 \sin x$ is 2. The graph of $y = -f(x)$ is a *reflection* across the x-axis of $y = f(x)$. Thus the graph of $y = -2 \sin x$ is a reflection across the x-axis of $y = 2 \sin x$. See **Figure 2.64.**

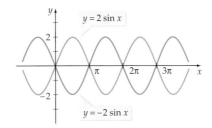

FIGURE 2.64

▶ **TRY EXERCISE 20, PAGE 170**

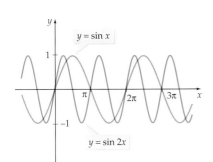

FIGURE 2.65

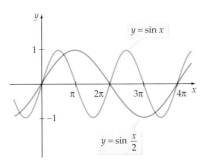

FIGURE 2.66

The graphs of $y = \sin x$ and $y = \sin 2x$ are shown in **Figure 2.65.** Because one cycle of the graph of $y = \sin 2x$ is completed in an interval of length π, the period of $y = \sin 2x$ is π.

The graphs of $y = \sin x$ and $y = \sin\left(\dfrac{x}{2}\right)$ are shown in **Figure 2.66.** Because one cycle of the graph of $y = \sin\left(\dfrac{x}{2}\right)$ is completed in an interval of length 4π, the period of $y = \sin\left(\dfrac{x}{2}\right)$ is 4π.

Generalizing the last two examples, one cycle of $y = \sin bx$, $b > 0$, is completed as bx varies from 0 to 2π. Algebraically, one cycle of $y = \sin bx$ is completed as bx varies from 0 to 2π. Therefore,

$$0 \le bx \le 2\pi$$

$$0 \le x \le \frac{2\pi}{b}$$

The length of the interval, $\dfrac{2\pi}{b}$, is the period of $y = \sin bx$. Now we consider the case when the coefficient of x is negative. If $b > 0$, then using the fact that the sine function is an odd function, we have $y = \sin(-bx) = -\sin bx$, and thus the period is still $\dfrac{2\pi}{b}$. This gives the following theorem.

Period of $y = \sin bx$

The period of $y = \sin bx$ is $\dfrac{2\pi}{|b|}$.

Table **2.8** gives the amplitude and period of several sine functions.

TABLE 2.8

Function	$y = a \sin bx$	$y = 3 \sin(-2x)$	$y = -\sin \dfrac{x}{3}$	$y = -2 \sin \dfrac{3x}{4}$
Amplitude	$|a|$	$|3| = 3$	$|-1| = 1$	$|-2| = 2$
Period	$\dfrac{2\pi}{|b|}$	$\dfrac{2\pi}{2} = \pi$	$\dfrac{2\pi}{1/3} = 6\pi$	$\dfrac{2\pi}{3/4} = \dfrac{8\pi}{3}$

EXAMPLE 2 Graph y = sin bx

Graph: $y = \sin \pi x$

Solution

$$\text{Amplitude} = 1 \qquad \text{Period} = \frac{2\pi}{b} = \frac{2\pi}{\pi} = 2 \qquad \bullet\, b = \pi$$

The graph is sketched in **Figure 2.67.**

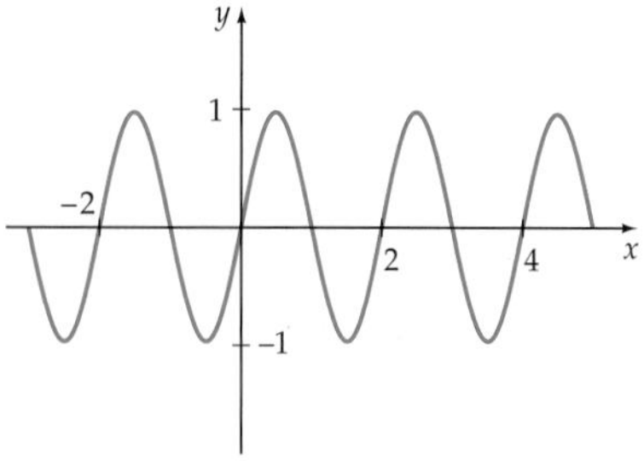

▶ **TRY EXERCISE 30, PAGE 170**

$y = \sin \pi x$

FIGURE 2.67

Figure 2.68 shows the graph of $y = a \sin bx$ for both a and b positive. Note from the graph that

- The amplitude is a.

- The period is $\dfrac{2\pi}{b}$.

- For $0 \le x \le \dfrac{2\pi}{b}$, the zeros are 0, $\dfrac{\pi}{b}$, and $\dfrac{2\pi}{b}$.

- The maximum value is a when $x = \dfrac{\pi}{2b}$, and the minimum value is $-a$ when $x = \dfrac{3\pi}{2b}$.

- If $a < 0$, the graph is reflected across the x-axis.

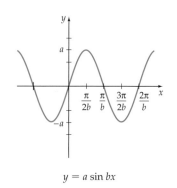

$y = a \sin bx$

FIGURE 2.68

EXAMPLE 3 Graph y = a sin bx

Graph: $y = -\dfrac{1}{2} \sin \dfrac{x}{3}$

Solution

$$\text{Amplitude} = \left|-\frac{1}{2}\right| = \frac{1}{2} \qquad \text{Period} = \frac{2\pi}{1/3} = 6\pi \qquad \bullet\, b = \frac{1}{3}$$

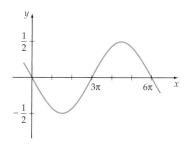

$$y = -\frac{1}{2} \sin \frac{x}{3}$$

FIGURE 2.69

The zeros in the interval $0 \le x \le 6\pi$ are $0, \dfrac{\pi}{1/3} = 3\pi$, and $\dfrac{2\pi}{1/3} = 6\pi$.

Because $-\dfrac{1}{2} < 0$, the graph is the graph of $y = \dfrac{1}{2} \sin \dfrac{x}{3}$ reflected across the x-axis as shown in **Figure 2.69**.

▶ **Try Exercise 38, page 170**

● **THE GRAPH OF THE COSINE FUNCTION**

Table 2.9 lists some of the ordered pairs of $y = \cos x$, $0 \le x \le 2\pi$. In **Figure 2.70,** the points are plotted and a smooth curve is drawn through the points.

TABLE 2.9

x	y = cos x
0	1
$\dfrac{\pi}{4}$	≈ 0.7
$\dfrac{\pi}{2}$	0
$\dfrac{3\pi}{4}$	≈ -0.7
π	-1
$\dfrac{5\pi}{4}$	≈ -0.7
$\dfrac{3\pi}{2}$	0
$\dfrac{7\pi}{4}$	≈ 0.7
2π	1

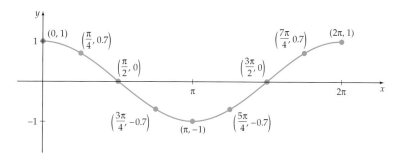

$y = \cos x$, $0 \le x \le 2\pi$

FIGURE 2.70

Because the domain of $y = \cos x$ is the real numbers and the period is 2π, the graph of $y = \cos x$ is drawn by repeating the portion shown in **Figure 2.70**. Any part of the graph corresponding to one period (2π) is one cycle of $y = \cos x$ (see **Figure 2.71**).

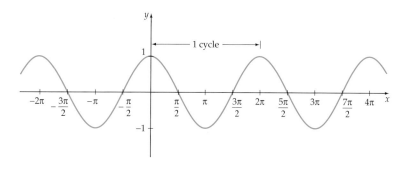

$y = \cos x$

FIGURE 2.71

The two theorems on the following page concerning cosine functions can be developed using methods that are analogous to those we used to determine the amplitude and period of a sine function.

Amplitude of y = a cos x

The amplitude of $y = a \cos x$ is $|a|$.

Period of y = cos bx

The period of $y = \cos bx$ is $\dfrac{2\pi}{|b|}$.

Table 2.10 gives the amplitude and period of some cosine functions.

TABLE 2.10

Function	$y = a \cos bx$	$y = 2 \cos 3x$	$y = -3 \cos \dfrac{2x}{3}$
Amplitude	$|a|$	$|2| = 2$	$|-3| = 3$
Period	$\dfrac{2\pi}{|b|}$	$\dfrac{2\pi}{3}$	$\dfrac{2\pi}{2/3} = 3\pi$

EXAMPLE 4 **Graph y = cos bx**

Graph: $y = \cos \dfrac{2\pi}{3} x$

Solution

$$\text{Amplitude} = 1 \qquad \text{Period} = \frac{2\pi}{b} = \frac{2\pi}{2\pi/3} = 3 \qquad \bullet\, b = \frac{2\pi}{3}$$

The graph is shown in **Figure 2.72.**

▶ **TRY EXERCISE 32, PAGE 170**

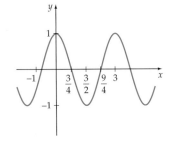

$y = \cos \dfrac{2\pi}{3} x$

FIGURE 2.72

Figure 2.73 shows the graph of $y = a \cos bx$ for both a and b positive. Note from the graph that

● The amplitude is a.

● The period is $\dfrac{2\pi}{b}$.

● For $0 \le x \le \dfrac{2\pi}{b}$, the zeros are $\dfrac{\pi}{2b}$ and $\dfrac{3\pi}{2b}$.

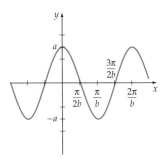

$y = a \cos bx$

FIGURE 2.73

• The maximum value is a when $x = 0$, and the minimum value is $-a$ when

$$x = \frac{\pi}{b}.$$

• If $a < 0$, then the graph is reflected across the x-axis.

EXAMPLE 5   **Graph a Cosine Function**

Graph: $y = -2 \cos \dfrac{\pi x}{4}$

Solution

$$\text{Amplitude} = |-2| = 2 \qquad \text{Period} = \frac{2\pi}{\pi/4} = 8 \qquad • \, b = \frac{\pi}{4}$$

The zeros in the interval $0 \le x \le 8$ are $\dfrac{\pi}{2\pi/4} = 2$ and $\dfrac{3\pi}{2\pi/4} = 6$. Because

$-2 < 0$, the graph is the graph of $y = 2 \cos \dfrac{\pi x}{4}$ reflected across the x-axis

as shown in **Figure 2.74.**

▶ TRY EXERCISE 46, PAGE 170

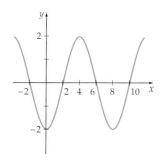

$$y = -2 \cos \frac{\pi x}{4}$$

FIGURE 2.74

EXAMPLE 6   **Graph the Absolute Value of the Cosine Function**

Graph $y = |\cos x|$, where $0 \le x \le 2\pi$.

Solution

Because $|\cos x| \ge 0$, the graph of $y = |\cos x|$ is drawn by reflecting the negative portion of the graph of $y = \cos x$ across the x-axis. The graph is the one shown in dark blue and light blue in **Figure 2.75.**

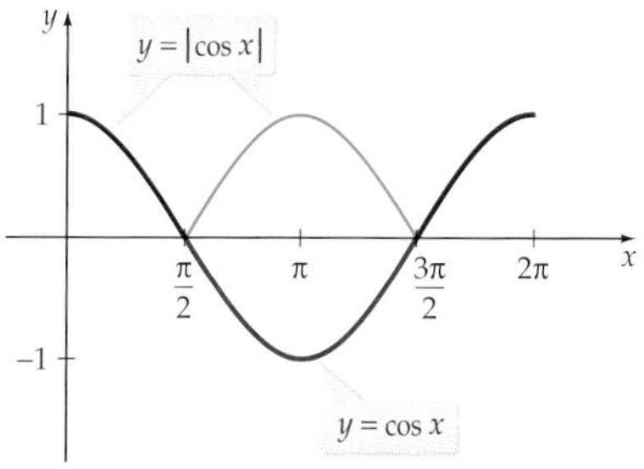

▶ TRY EXERCISE 52, PAGE 170

FIGURE 2.75

TOPICS FOR DISCUSSION

1. Is the graph of $f(x) = |\sin x|$ the same as the graph of $y = \sin|x|$? Explain.

2. Explain how the graph of $y = \cos 2x$ differs from the graph of $y = \cos x$.

3. Does the graph of $y = \sin(-2x)$ have the same period as the graph of $y = \sin 2x$? Explain.

4. The function $h(x) = a \sin bt$ has an amplitude of 3 and a period of 4. What are the possible values of a? What are the possible values of b?

EXERCISE SET 2.5

In Exercises 1 to 16, state the amplitude and period of the function defined by each equation.

1. $y = 2 \sin x$

2. $y = -\dfrac{1}{2} \sin x$

3. $y = \sin 2x$

4. $y = \sin \dfrac{2x}{3}$

5. $y = \dfrac{1}{2} \sin 2\pi x$

6. $y = 2 \sin \dfrac{\pi x}{3}$

7. $y = -2 \sin \dfrac{x}{2}$

8. $y = -\dfrac{1}{2} \sin \dfrac{x}{2}$

9. $y = \dfrac{1}{2} \cos x$

10. $y = -3 \cos x$

11. $y = \cos \dfrac{x}{4}$

12. $y = \cos 3x$

13. $y = 2 \cos \dfrac{\pi x}{3}$

14. $y = \dfrac{1}{2} \cos 2\pi x$

15. $y = -3 \cos \dfrac{2x}{3}$

16. $y = \dfrac{3}{4} \cos 4x$

In Exercises 17 to 54, graph at least one full period of the function defined by each equation.

17. $y = \dfrac{1}{2} \sin x$

18. $y = \dfrac{3}{2} \cos x$

19. $y = 3 \cos x$

▶ 20. $y = -\dfrac{3}{2} \sin x$

21. $y = -\dfrac{7}{2} \cos x$

22. $y = 3 \sin x$

23. $y = -4 \sin x$

24. $y = -5 \cos x$

25. $y = \cos 3x$

26. $y = \sin 4x$

27. $y = \sin \dfrac{3x}{2}$

28. $y = \cos \pi x$

29. $y = \cos \dfrac{\pi}{2} x$

▶ 30. $y = \sin \dfrac{3\pi}{4} x$

31. $y = \sin 2\pi x$

▶ 32. $y = \cos 3\pi x$

33. $y = 4 \cos \dfrac{x}{2}$

34. $y = 2 \cos \dfrac{3x}{4}$

35. $y = -2 \cos \dfrac{x}{3}$

36. $y = -\dfrac{4}{3} \cos 3x$

37. $y = 2 \sin \pi x$

▶ 38. $y = \dfrac{1}{2} \sin \dfrac{\pi x}{3}$

39. $y = \dfrac{3}{2} \cos \dfrac{\pi x}{2}$

40. $y = \cos \dfrac{\pi x}{3}$

41. $y = 4 \sin \dfrac{2\pi x}{3}$

42. $y = 3 \cos \dfrac{3\pi x}{2}$

43. $y = 2 \cos 2x$

44. $y = \dfrac{1}{2} \sin 2.5x$

45. $y = -2 \sin 1.5x$

▶ 46. $y = -\dfrac{3}{4} \cos 5x$

47. $y = \left| 2 \sin \dfrac{x}{2} \right|$

48. $y = \left| \dfrac{1}{2} \sin 3x \right|$

49. $y = \left| -2 \cos 3x \right|$

50. $y = \left| -\dfrac{1}{2} \cos \dfrac{x}{2} \right|$

51. $y = -\left| 2 \sin \dfrac{x}{3} \right|$

▶ 52. $y = -\left| 3 \sin \dfrac{2x}{3} \right|$

53. $y = -\left| 3 \cos \pi x \right|$

54. $y = -\left| 2 \cos \dfrac{\pi x}{2} \right|$

In Exercises 55 to 60, find an equation of each graph.

55.

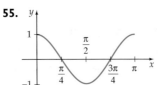

56.

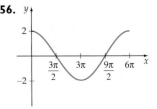

57.

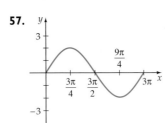

58.

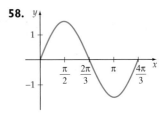

59.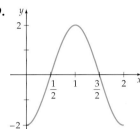

60.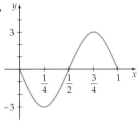

61. Sketch the graph of $y = 2 \sin \dfrac{2x}{3}$, $-3\pi \le x \le 6\pi$.

62. Sketch the graph of $y = -3 \cos \dfrac{3x}{4}$, $-2\pi \le x \le 4\pi$.

63. Sketch the graphs of

$$y_1 = 2 \cos \frac{x}{2} \quad \text{and} \quad y_2 = 2 \cos x$$

on the same set of axes for $-2\pi \le x \le 4\pi$.

64. Sketch the graphs of

$$y_1 = \sin 3\pi x \quad \text{and} \quad y_2 = \sin \frac{\pi x}{3}$$

on the same set of axes for $-2 \le x \le 4$.

In Exercises 65 to 72, use a graphing utility to graph each function.

65. $y = \cos^2 x$

66. $y = 3^{\cos^2 x} \cdot 3^{\sin^2 x}$

67. $y = \cos |x|$

68. $y = \sin |x|$

69. $y = \dfrac{1}{2} x \sin x$

70. $y = \dfrac{1}{2} x + \sin x$

71. $y = -x \cos x$

72. $y = -x + \cos x$

73. Graph $y = e^{\sin x}$. What is the maximum value of $e^{\sin x}$? What is the minimum value of $e^{\sin x}$? Is the function defined by $y = e^{\sin x}$ a periodic function? If so, what is the period?

74. Graph $y = e^{\cos x}$. What is the maximum value of $e^{\cos x}$? What is the minimum value of $e^{\cos x}$? Is the function defined by $y = e^{\cos x}$ a periodic function? If so, what is the period?

75. EQUATION OF A WAVE A tidal wave that is caused by an earthquake under the ocean is called a **tsunami**. A model of a tsunami is given by $f(t) = A \cos Bt$. Find the equation of a tsunami that has an amplitude of 60 feet and a period of 20 seconds.

76. EQUATION OF HOUSEHOLD CURRENT The electricity supplied to your home, called *alternating current*, can be modeled by $I = A \sin \omega t$, where I is the number of amperes of current at time t seconds. Write the equation of household current whose graph is given in the figure below. Calculate I when $t = 0.5$ second.

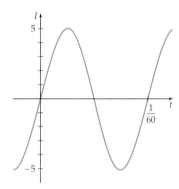

CONNECTING CONCEPTS

In Exercises 77 to 80, write an equation for a sine function using the given information.

77. Amplitude $= 2$; period $= 3\pi$

78. Amplitude $= 5$; period $= \dfrac{2\pi}{3}$

79. Amplitude $= 4$; period $= 2$

80. Amplitude $= 2.5$; period $= 3.2$

In Exercises 81 to 84, write an equation for a cosine function using the given information.

81. Amplitude $= 3$; period $= \dfrac{\pi}{2}$

82. Amplitude $= 0.8$; period $= 4\pi$

83. Amplitude $= 3$; period $= 2.5$

84. Amplitude $= 4.2$; period $= 1$

85. Estimate, to the nearest tenth, $\tan \dfrac{\pi}{3}$. [2.2]

86. Estimate, to the nearest tenth, $\cot \dfrac{\pi}{3}$. [2.2]

87. Explain how to use the graph of $y = f(x)$ to produce the graph of $y = 2f(x)$. [1.4]

88. Explain how to use the graph of $y = f(x)$ to produce the graph of $y = f(x - 2) + 3$. [1.4]

89. Simplify: $\dfrac{\pi}{1/2}$

90. Simplify: $\dfrac{\pi}{\left| -\dfrac{3}{4} \right|}$

─── *PROJECTS* ───

1.   **A TRIGONOMETRIC POWER FUNCTION**

a. Determine the domain and the range of $y = (\sin x)^{\cos x}$. Explain.

b. What is the amplitude of the function? Explain.

GRAPHS OF THE OTHER TRIGONOMETRIC FUNCTIONS

SECTION 2.6

- THE GRAPH OF THE TANGENT FUNCTION
- THE GRAPH OF THE COTANGENT FUNCTION
- THE GRAPH OF THE COSECANT FUNCTION
- THE GRAPH OF THE SECANT FUNCTION

● THE GRAPH OF THE TANGENT FUNCTION

Figure 2.76 shows the graph of $y = \tan x$ for $-\dfrac{\pi}{2} < x < \dfrac{\pi}{2}$. The lines $x = \dfrac{\pi}{2}$ and $x = -\dfrac{\pi}{2}$ are vertical asymptotes for the graph of $y = \tan x$. From Section 2.4, the period of $y = \tan x$ is π. Therefore, the portion of the graph shown in **Figure 2.76** is repeated along the x-axis, as shown in **Figure 2.77.**

Because the tangent function is unbounded, there is no amplitude for the tangent function. The graph of $y = a \tan x$ is drawn by stretching ($|a| > 1$) or shrinking ($|a| < 1$) the graph of $y = \tan x$. If $a < 0$, then the graph is reflected across the x-axis. **Figure 2.78** shows the graphs of three tangent functions. Because $\tan \dfrac{\pi}{4} = 1$, the point $\left(\dfrac{\pi}{4}, a \right)$ is convenient to plot as a guide for the graph of $y = a \tan x$.

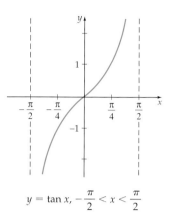

$$y = \tan x, \ -\frac{\pi}{2} < x < \frac{\pi}{2}$$

FIGURE 2.76

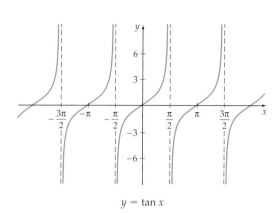

$$y = \tan x$$

FIGURE 2.77

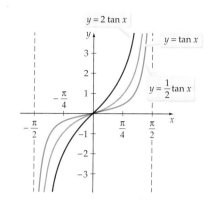

FIGURE 2.78

EXAMPLE I   **Graph $y = a \tan x$**

Graph: $y = -3 \tan x$

Solution

The graph of $y = -3 \tan x$ is the reflection across the x-axis of the graph of $y = 3 \tan x$, as shown in **Figure 2.79**.

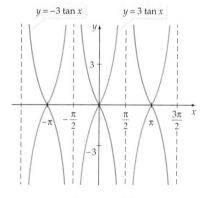

FIGURE 2.79

▶ **TRY EXERCISE 22, PAGE 178**

The period of $y = \tan x$ is π and the graph completes one cycle on the interval $-\frac{\pi}{2} < x < \frac{\pi}{2}$. The period of $y = \tan bx \ (b > 0)$ is $\frac{\pi}{b}$. The graph of $y = \tan bx$ completes one cycle on the interval $\left(-\frac{\pi}{2b}, \frac{\pi}{2b} \right)$.

Period of $y = \tan bx$

The period of $y = \tan bx$ is $\dfrac{\pi}{|b|}$.

? **QUESTION** What is the period of the graph of $y = \tan \pi x$?

EXAMPLE 2　　**Graph $y = a \tan bx$**

Graph: $y = 2 \tan \dfrac{x}{2}$

Solution

Period $= \dfrac{\pi}{b} = \dfrac{\pi}{1/2} = 2\pi$. Graph one cycle for values of x such that $-\pi < x < \pi$. This curve is repeated along the x-axis as shown in **Figure 2.80.**

▶ **TRY EXERCISE 30, PAGE 178**

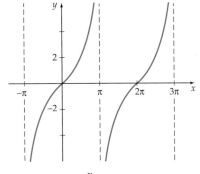

$y = 2 \tan \dfrac{x}{2}, \ -\pi < x < 3\pi$

FIGURE 2.80

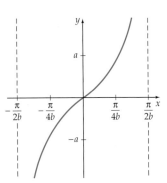

$y = a \tan bx, \ -\dfrac{\pi}{2b} < x < \dfrac{\pi}{2b}$

FIGURE 2.81

Figure 2.81 shows one cycle of the graph of $y = a \tan bx$ for both a and b positive. Note from the graph that

- The period is $\dfrac{\pi}{b}$.

- $x = 0$ is a zero.

- The graph passes through $\left(-\dfrac{\pi}{4b}, -a \right)$ and $\left(\dfrac{\pi}{4b}, a \right)$.

- If $a < 0$, the graph is reflected across the x-axis.

● **THE GRAPH OF THE COTANGENT FUNCTION**

Figure 2.82 shows the graph of $y = \cot x$ for $0 < x < \pi$. The lines $x = 0$ and $x = \pi$ are vertical asymptotes for the graph of $y = \cot x$. From Section 2.4, the period of $y = \cot x$ is π. Therefore, the graph cycle shown in **Figure 2.82** is repeated along the x-axis, as shown in **Figure 2.83**. As with the graph of $y = \tan x$, the graph of $y = \cot x$ is unbounded and there is no amplitude.

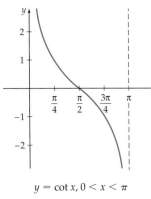

$y = \cot x, 0 < x < \pi$

FIGURE 2.82

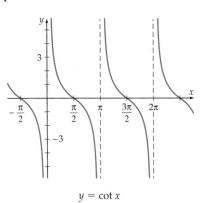

$y = \cot x$

FIGURE 2.83

? **ANSWER** 1

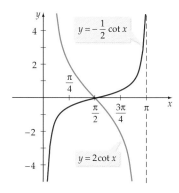

FIGURE 2.84

The graph of $y = a \cot x$ is drawn by stretching ($|a| > 1$) or shrinking ($|a| < 1$) the graph of $y = \cot x$. The graph is reflected across the x-axis when $a < 0$. **Figure 2.84** shows the graphs of two cotangent functions.

The period of $y = \cot x$ is π, and the period of $y = \cot bx$ is $\dfrac{\pi}{|b|}$. One cycle of the graph of $y = \cot bx$ is completed on the interval $\left(0, \dfrac{\pi}{b}\right)$.

Period of $y = \cot bx$

The period of $y = \cot bx$ is $\dfrac{\pi}{|b|}$.

EXAMPLE 3 Graph $y = a \cot bx$

Graph: $y = 3 \cot \dfrac{x}{2}$

Solution

Period $= \dfrac{\pi}{b} = \dfrac{\pi}{1/2} = 2\pi$. Sketch the graph for values of x for which $0 < x < 2\pi$. This curve is repeated along the x-axis, as shown in **Figure 2.85.**

▶ **TRY EXERCISE 32, PAGE 179**

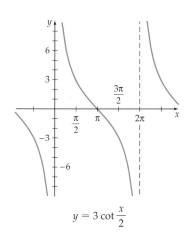

$y = 3 \cot \dfrac{x}{2}$

FIGURE 2.85

Figure 2.86 shows one cycle of the graph of $y = a \cot bx$ for both a and b positive. Note from the graph that

- The period is $\dfrac{\pi}{b}$.

- $x = \dfrac{\pi}{2b}$ is a zero.

- The graph passes through $\left(\dfrac{\pi}{4b}, a\right)$ and $\left(\dfrac{3\pi}{4b}, -a\right)$.

- If $a < 0$, the graph is reflected across the x-axis.

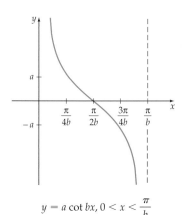

$y = a \cot bx, \; 0 < x < \dfrac{\pi}{b}$

FIGURE 2.86

THE GRAPH OF THE COSECANT FUNCTION

Because $\csc x = \dfrac{1}{\sin x}$, the value of $\csc x$ is the reciprocal of the value of $\sin x$. Therefore, $\csc x$ is undefined when $\sin x = 0$ or when $x = n\pi$, where n is an integer. The graph of $y = \csc x$ has vertical asymptotes at $n\pi$. Because $y = \csc x$ has period 2π, the graph will be repeated along the x-axis every 2π units. A graph of $y = \csc x$ is shown in **Figure 2.87.**

The graph of $y = \sin x$ is also shown in **Figure 2.87.** Note the relationships among the zeros of $y = \sin x$ and the asymptotes of $y = \csc x$. Also note that because $|\sin x| \le 1$, $\dfrac{1}{|\sin x|} \ge 1$. Thus the range of $y = \csc x$ is $\{y \mid y \ge 1, y \le -1\}$.

The general procedure for graphing $y = a \csc bx$ is first to graph $y = a \sin bx$. Then sketch the graph of the cosecant function by using y values equal to the product of a and the reciprocal values of $\sin bx$.

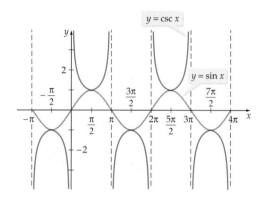

FIGURE 2.87

EXAMPLE 4   Graph $y = a \csc bx$

Graph: $y = 2 \csc \dfrac{\pi x}{2}$

Solution

First sketch the graph of $y = 2 \sin \dfrac{\pi x}{2}$ and draw vertical asymptotes

through the zeros. Now sketch the graph of $y = 2 \csc \dfrac{\pi x}{2}$, using the

asymptotes as guides for the graph, as shown in **Figure 2.88.**

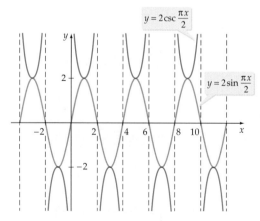

FIGURE 2.88

▶ **TRY EXERCISE 38, PAGE 179**

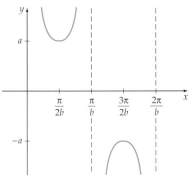

$$y = a \csc bx, 0 < x < \frac{2\pi}{b}$$

FIGURE 2.89

Figure 2.89 shows one cycle of the graph of $y = a \csc bx$ for both a and b positive. Note from the graph that

• The period is $\dfrac{2\pi}{b}$.

• The vertical asymptotes of $y = a \csc bx$ are located at the zeros of $y = a \sin bx$.

• The graph passes through $\left(\dfrac{\pi}{2b}, a\right)$ and $\left(\dfrac{3\pi}{2b}, -a\right)$.

• If $a < 0$, then the graph is reflected across the x-axis.

• THE GRAPH OF THE SECANT FUNCTION

Because $\sec x = \dfrac{1}{\cos x}$, the value of $\sec x$ is the reciprocal of the value of $\cos x$.

Therefore, $\sec x$ is undefined when $\cos x = 0$ or when $x = \dfrac{\pi}{2} + n\pi$, n an integer.

The graph of $y = \sec x$ has vertical asymptotes at $\dfrac{\pi}{2} + n\pi$. Because $y = \sec x$ has period 2π, the graph will be replicated along the x-axis every 2π units. A graph of $y = \sec x$ is shown in **Figure 2.90.**

The graph of $y = \cos x$ is also shown in **Figure 2.90.** Note the relationships among the zeros of $y = \cos x$ and the asymptotes of $y = \sec x$. Also note that because $|\cos x| \leq 1$, $\dfrac{1}{|\cos x|} \geq 1$. Thus the range of $y = \sec x$ is $\{y \,|\, y \geq 1, y \leq -1\}$. The general procedure for graphing $y = a \sec bx$ is first to graph $y = a \cos bx$. Then sketch the graph of the secant function by using y values equal to the product of a and the reciprocal values of $\cos bx$.

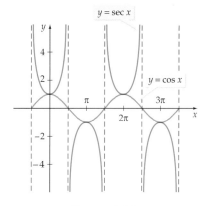

FIGURE 2.90

EXAMPLE 5   **Graph $y = a \sec bx$**

Graph: $y = -3 \sec \dfrac{x}{2}$

Solution

First sketch the graph of $y = -3 \cos \dfrac{x}{2}$ and draw vertical asymptotes through the zeros. Now sketch the graph of $y = -3 \sec \dfrac{x}{2}$, using the asymptotes as guides for the graph, as shown in **Figure 2.91.**

▶ **TRY EXERCISE 42, PAGE 179**

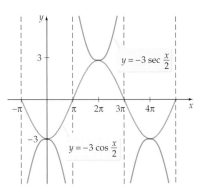

FIGURE 2.91

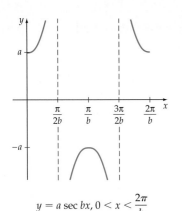

$$y = a \sec bx, \, 0 < x < \frac{2\pi}{b}$$

FIGURE 2.92

Figure **2.92** shows one cycle of the graph of $y = a \sec bx$ for both a and b positive. Note from the graph that

- The period is $\dfrac{2\pi}{b}$.

- The vertical asymptotes of $y = a \sec bx$ are located at the zeros of $y = a \cos bx$.

- The graph passes through $(0, a)$, $\left(\dfrac{\pi}{b}, -a\right)$, and $\left(\dfrac{2\pi}{b}, a\right)$.

- If $a < 0$, then the graph is reflected across the x-axis.

 TOPICS FOR DISCUSSION

1. What are the zeros of $y = \tan x$? Explain.

2. What are the zeros of $y = \sec x$? Explain.

3. The functions $f(x) = \tan x$ and $g(x) = \tan(-x)$ both have a period of π. Do you agree?

4. What is the amplitude of the function $k(x) = 4 \cot(\pi x)$? Explain.

EXERCISE SET 2.6

1. For what values of x is $y = \tan x$ undefined?

2. For what values of x is $y = \cot x$ undefined?

3. For what values of x is $y = \sec x$ undefined?

4. For what values of x is $y = \csc x$ undefined?

In Exercises 5 to 20, state the period of each function.

5. $y = \sec x$ **6.** $y = \cot x$ **7.** $y = \tan x$

8. $y = \csc x$ **9.** $y = 2 \tan \dfrac{x}{2}$ **10.** $y = \dfrac{1}{2} \cot 2x$

11. $y = \csc 3x$ **12.** $y = \csc \dfrac{x}{2}$

13. $y = -\tan 3x$ **14.** $y = -3 \cot \dfrac{2x}{3}$

15. $y = -3 \sec \dfrac{x}{4}$ **16.** $y = -\dfrac{1}{2} \csc 2x$

17. $y = \cot \pi x$ **18.** $y = \cot \dfrac{\pi x}{3}$

19. $y = 2 \csc \dfrac{\pi x}{2}$ **20.** $y = -3 \cot \pi x$

In Exercises 21 to 40, sketch one full period of the graph of each function.

21. $y = 3 \tan x$ ▶ **22.** $y = \dfrac{1}{3} \tan x$

23. $y = \dfrac{3}{2} \cot x$ **24.** $y = 4 \cot x$

25. $y = 2 \sec x$ **26.** $y = \dfrac{3}{4} \sec x$

27. $y = \dfrac{1}{2} \csc x$ **28.** $y = 2 \csc x$

29. $y = 2 \tan \dfrac{x}{2}$ ▶ **30.** $y = -3 \tan 3x$

31. $y = -3 \cot \dfrac{x}{2}$

▶ 32. $y = \dfrac{1}{2} \cot 2x$

33. $y = -2 \csc \dfrac{x}{3}$

34. $y = \dfrac{3}{2} \csc 3x$

35. $y = \dfrac{1}{2} \sec 2x$

36. $y = -3 \sec \dfrac{2x}{3}$

37. $y = -2 \sec \pi x$

▶ 38. $y = 3 \csc \dfrac{\pi x}{2}$

39. $y = 3 \tan 2\pi x$

40. $y = -\dfrac{1}{2} \cot \dfrac{\pi x}{2}$

41. Graph $y = 2 \csc 3x$ from -2π to 2π.

▶ 42. Graph $y = \sec \dfrac{x}{2}$ from -4π to 4π.

43. Graph $y = 3 \sec \pi x$ from -2 to 4.

44. Graph $y = \csc \dfrac{\pi x}{2}$ from -4 to 4.

45. Graph $y = 2 \cot 2x$ from $-\pi$ to π.

46. Graph $y = \dfrac{1}{2} \tan \dfrac{x}{2}$ from -4π to 4π.

47. Graph $y = 3 \tan \pi x$ from -2 to 2.

48. Graph $y = \cot \dfrac{\pi x}{2}$ from -4 to 4.

In Exercises 49 to 54, find an equation of each blue graph.

49.

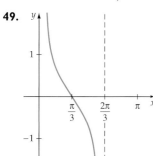

50.

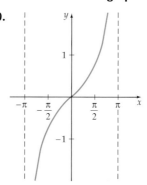

51.

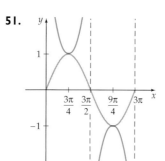

52.

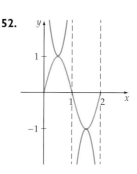

53.

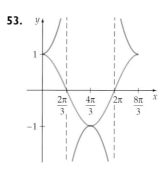

54.

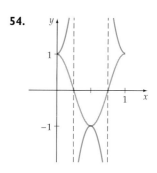

In Exercises 55 to 60, use a graphing utility to graph each equation. If needed, use open circles so that your graph is accurate.

55. $y = \tan |x|$

56. $y = \sec |x|$

57. $y = |\csc x|$

58. $y = |\cot x|$

59. $y = \tan x \cos x$

60. $y = \cot x \sin x$

61. Graph $y = \tan x$ and $x = \tan y$ on the same coordinate axes.

62. Graph $y = \sin x$ and $x = \sin y$ on the same coordinate axes.

CONNECTING CONCEPTS

In Exercises 63 to 70, write an equation that is of the form $y = \tan bx, y = \cot bx, y = \sec bx,$ or $y = \csc bx$ and satisfies the given conditions.

63. Tangent, period: $\dfrac{\pi}{3}$

64. Cotangent, period: $\dfrac{\pi}{2}$

65. Secant, period: $\dfrac{3\pi}{4}$

66. Cosecant, period: $\dfrac{5\pi}{2}$

67. Cotangent, period: 2

68. Tangent, period: 0.5

69. Cosecant, period: 1.5

70. Secant, period: 3

PREPARE FOR SECTION 2.7

71. Find the amplitude and period of the graph of $y = 2 \sin 2x$. [2.5]

72. Find the amplitude and period of the graph of $y = \dfrac{2}{3} \cos \dfrac{x}{3}$. [2.5]

73. Find the amplitude and the period of the graph of $y = -4 \sin 2\pi x$. [2.5]

74. What is the maximum value of $f(x) = 2 \sin x$? [2.5]

75. What is the minimum value of $f(x) = 3 \cos 2x$? [2.5]

76. Is the graph of $f(x) = \cos x$ symmetric with respect to the origin or with respect to the y-axis? [1.4/2.5]

PROJECTS

1. **A TECHNOLOGY QUESTION** A student's calculator shows the display at the right. Note that the domain values 4.7123 and 4.7124 are close together, but the range values are over 100,000 units apart. Explain how this is possible.

```
tan 4.7123
                11238.43194
tan 4.7124
              -90747.26955
```

2. **SOLUTIONS OF A TRIGONOMETRIC EQUATION** How many solutions does $\tan\left(\dfrac{1}{x}\right) = 0$ have on the interval $-1 \le x \le 1$? Explain.

GRAPHING TECHNIQUES

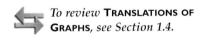

To review **TRANSLATIONS OF GRAPHS**, *see Section 1.4.*

TRANSLATION OF TRIGONOMETRIC FUNCTIONS

Recall that the graph of $y = f(x) \pm c$ is a *vertical translation* of the graph of $y = f(x)$. For $c > 0$, the graph of $y = f(x) - c$ is shifted c units down; the graph of $y = f(x) + c$ is shifted c units up. The graph in **Figure 2.93** is a graph of the equation $y = 2 \sin \pi x - 3$, which is a vertical translation of $y = 2 \sin \pi x$ down 3 units. Note that subtracting 3 from $y = 2 \sin \pi x$ changes neither its amplitude nor its period.

Also, the graph of $y = f(x \pm c)$ is a *horizontal translation* of the graph of $y = f(x)$. For $c > 0$, the graph of $y = f(x - c)$ is shifted c units to the right; the graph of $y = f(x + c)$ is shifted c units to the left. The graph in **Figure 2.94** is a graph of the equation $y = 2 \sin\left(x - \dfrac{\pi}{4}\right)$, which is the graph of $y = 2 \sin x$ translated $\dfrac{\pi}{4}$ units to the right. Note that neither the period nor the amplitude was affected. The horizontal shift of the graph of a trigonometric function is called its **phase shift**.

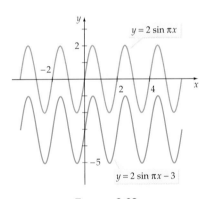

FIGURE 2.93

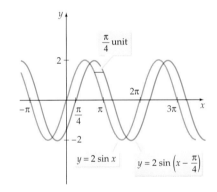

FIGURE 2.94

Because one cycle of $y = a \sin x$ is completed for $0 \le x \le 2\pi$, one cycle of the graph of $y = a \sin(bx + c)$, where $b > 0$, is completed for $0 \le bx + c \le 2\pi$. Solving this inequality for x, we have

$$0 \le bx + c \le 2\pi$$

$$-c \le bx \le -c + 2\pi$$

$$-\frac{c}{b} \le x \le -\frac{c}{b} + \frac{2\pi}{b}$$

The number $-\dfrac{c}{b}$ is the phase shift for $y = a \sin(bx + c)$. The graph of the equation $y = a \sin(bx + c)$ is the graph of $y = a \sin bx$ shifted $-\dfrac{c}{b}$ units horizontally. Similar arguments apply to the remaining trigonometric functions.

The Graphs of $y = a \sin(bx + c)$ and $y = a \cos(bx + c)$

The graphs of $y = a \sin(bx + c)$ and $y = a \cos(bx + c)$, with $b > 0$, have

$$\text{Amplitude: } |a| \quad \text{Period: } \frac{2\pi}{b} \quad \text{Phase shift: } -\frac{c}{b}$$

One cycle of each graph is completed on the interval

$$-\frac{c}{b} \leq x \leq -\frac{c}{b} + \frac{2\pi}{b}$$

❓ QUESTION What is the phase shift of the graph of $y = 3 \sin\left(\frac{1}{2}x - \frac{\pi}{6}\right)$?

EXAMPLE 1 **Graph $y = a \cos(bx + c)$**

Graph: $y = 3 \cos\left(2x + \frac{\pi}{3}\right)$

Solution

The phase shift is $-\dfrac{c}{b} = -\dfrac{\pi/3}{2} = -\dfrac{\pi}{6}$. The graph of the equation

$y = 3 \cos\left(2x + \dfrac{\pi}{3}\right)$ is the graph of $y = 3 \cos 2x$ shifted $\dfrac{\pi}{6}$ units to the left as shown in **Figure 2.95.**

▶ **TRY EXERCISE 20, PAGE 187**

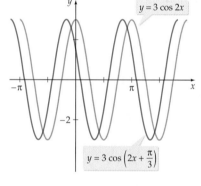

$y = 3 \cos 2x$

$y = 3 \cos\left(2x + \dfrac{\pi}{3}\right)$

FIGURE 2.95

EXAMPLE 2 **Graph $y = a \cot(bx + c)$**

Graph: $y = 2 \cot(3x - 2)$

Solution

The phase shift is

$$-\frac{c}{b} = -\frac{-2}{3} = \frac{2}{3} \qquad \bullet \; 3x - 2 = 3x + (-2)$$

The graph of $y = 2 \cot(3x - 2)$ is the graph of $y = 2 \cot(3x)$ shifted $\dfrac{2}{3}$ unit to the right as shown in **Figure 2.96.**

▶ **TRY EXERCISE 22, PAGE 187**

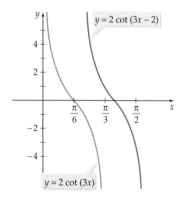

$y = 2 \cot(3x - 2)$

$y = 2 \cot(3x)$

FIGURE 2.96

❓ ANSWER $\dfrac{\pi}{3}$

The graph of a trigonometric function may be the combination of a vertical translation and a phase shift.

EXAMPLE 3   Graph $y = a \sin(bx + c) + d$

Graph: $y = \dfrac{1}{2} \sin\left(x - \dfrac{\pi}{4} \right) - 2$

Solution

The phase shift is $-\dfrac{c}{b} = -\dfrac{-\pi/4}{1} = \dfrac{\pi}{4}$. The vertical shift is 2 units down.

The graph of $y = \dfrac{1}{2} \sin\left(x - \dfrac{\pi}{4} \right) - 2$ is the graph of $y = \dfrac{1}{2} \sin x$ shifted $\dfrac{\pi}{4}$ units to the right and 2 units down as shown in **Figure 2.97.**

▶ **TRY EXERCISE 40, PAGE 187**

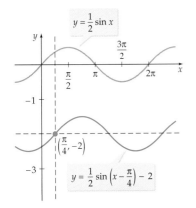

FIGURE 2.97

EXAMPLE 4   Graph $y = a \cos(bx + c) + d$

Graph: $y = -2 \cos\left(\pi x + \dfrac{\pi}{2} \right) + 1$

Solution

The phase shift is $-\dfrac{c}{b} = -\dfrac{\pi/2}{\pi} = -\dfrac{1}{2}$. The vertical shift is 1 unit up. The

graph of $y = -2 \cos\left(\pi x + \dfrac{\pi}{2} \right) + 1$ is the graph of $y = -2 \cos \pi x$ shifted

$\dfrac{1}{2}$ unit to the left and 1 unit up as shown in **Figure 2.98.**

▶ **TRY EXERCISE 42, PAGE 187**

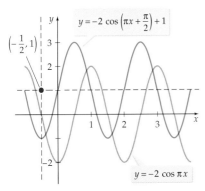

FIGURE 2.98

The following example involves a function of the form $y = \cos(bx + c) + d$.

EXAMPLE 5   A Mathematical Model of a Patient's Blood Pressure

The function $bp(t) = 32 \cos\left(\dfrac{10\pi}{3} t - \dfrac{\pi}{3} \right) + 112, \ 0 \leq t \leq 20$, gives the blood pressure, in millimeters of mercury (mm Hg), of a patient during a 20-second interval.

a. Find the phase shift and the period of bp.

b. Graph one period of bp.

c. What are the patient's maximum (*systolic*) and minimum (*diastolic*) blood pressure readings during the given time interval?

d. What is the patient's pulse rate, in beats per minute?

Continued ▶

Solution

a. Phase shift $= -\dfrac{c}{b} = -\dfrac{\left(-\dfrac{\pi}{3}\right)}{\left(\dfrac{10\pi}{3}\right)} = 0.1$

Period $= \dfrac{2\pi}{b} = \dfrac{2\pi}{\left(\dfrac{10\pi}{3}\right)} = 0.6$ second

b. The graph of bp is the graph of $y_1 = 32 \cos\left(\dfrac{10\pi}{3}t\right)$ shifted 0.1 unit to the right, shown by $y_2 = 32 \cos\left(\dfrac{10\pi}{3}t - \dfrac{\pi}{3}\right)$ in **Figure 2.99,** and upward 112 units.

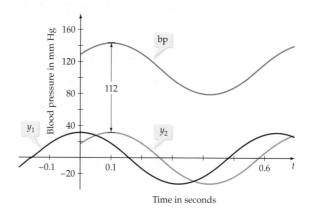

FIGURE 2.99

c. The function $y_2 = 32 \cos\left(\dfrac{10\pi}{3}t - \dfrac{\pi}{3}\right)$ has a maximum of 32 and a minimum of -32. Thus the patient's maximum blood pressure is $32 + 112 = 144$ mm Hg, and the patient's minimum blood pressure is $-32 + 112 = 80$ mm Hg.

d. From part **a.** we know that the patient has 1 heartbeat every 0.6 second. Therefore, during the given time interval, the patient has a pulse rate of

$$\left(\frac{1 \text{ heartbeat}}{0.6 \text{ second}}\right)\left(\frac{60 \text{ seconds}}{1 \text{ minute}}\right) = 100 \text{ heartbeats per minute}$$

▶ **TRY EXERCISE 52, PAGE 188**

Translation techniques also can be used to graph secant and cosecant functions.

EXAMPLE 6 **Graph a Cosecant Function**

Graph: $y = 2\csc(2x - \pi)$

Solution

The phase shift is $-\dfrac{c}{b} = -\dfrac{-\pi}{2} = \dfrac{\pi}{2}$. The graph of the equation

$y = 2\csc(2x - \pi)$ is the graph of $y = 2\csc 2x$ shifted $\dfrac{\pi}{2}$ units to the

right. Sketch the graph of the equation $y = 2\sin 2x$ shifted $\dfrac{\pi}{2}$ units to the

right. Use this graph to draw the graph of $y = 2\csc(2x - \pi)$ as shown in
Figure 2.100.

▶ TRY EXERCISE 48, PAGE 188

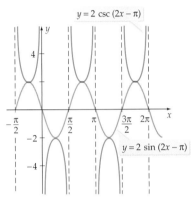

FIGURE 2.100

● **ADDITION OF ORDINATES**

Given two functions g and h, the sum of the functions is the function f defined by
$f(x) = g(x) + h(x)$. The graph of the sum f can be obtained by graphing g and h
separately and then geometrically adding the y-coordinates of each function for a
given value of x. It is convenient, when we are drawing the graph of the sum of
two functions, to pick zeros of the function.

EXAMPLE 7 **Graph the Sum of Two Functions**

Graph: $y = x + \cos x$

Solution

Graph $g(x) = x$ and $h(x) = \cos x$ on the same coordinate grid. Then add the
y-coordinates geometrically point by point. **Figure 2.101** shows the results of
adding, by using a ruler, the
y-coordinates of the two functions
for selected values of x.

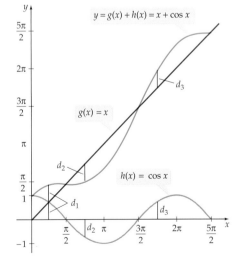

FIGURE 2.101

▶ TRY EXERCISE 54, PAGE 188

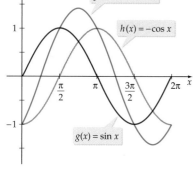

FIGURE 2.102

▶ **TRY EXERCISE 58, PAGE 188**

| EXAMPLE 8 | Graph the Difference of Two Functions |

Graph $y = \sin x - \cos x$ for $0 \leq x \leq 2\pi$.

Solution

Graph $g(x) = \sin x$ and $h(x) = -\cos x$ on the same coordinate grid. For selected values of x, add $g(x)$ and $h(x)$ geometrically. Now draw a smooth curve through the points. See **Figure 2.102.**

● DAMPING FACTOR

The factor $\frac{1}{4}x$ in $f(x) = \frac{1}{4}x \cos x$ is referred to as the *damping factor.* In the next example we analyze the role of the damping factor.

| EXAMPLE 9 | Graph the Product of Two Functions |

Use a graphing utility to graph $f(x) = \frac{1}{4}x \cos x$, $x \geq 0$, and analyze the role of the damping factor.

Solution

Figure 2.103 shows that the graph of f intersects

FIGURE 2.103

- the graph of $y = \frac{1}{4}x$ for $x = 0, 2\pi, 4\pi, \ldots$　　• **Because cos x = 1 for**
　　　　　　　　　　　　　　　　　　　　　　　　　x = 2nπ

- the graph of $y = -\frac{1}{4}x$ for $x = \pi, 3\pi, 5\pi, \ldots$　• **Because cos x = −1 for**
　　　　　　　　　　　　　　　　　　　　　　　　　x = (2n − 1)π

- the x-axis for $x = \frac{1}{2}\pi, \frac{3}{2}\pi, \frac{5}{2}\pi, \ldots$　　• **Because cos x = 0 for**
　　　　　　　　　　　　　　　　　　　　　　　　　$x = \frac{2n - 1}{2}\pi$

Figure 2.103 also shows that the graph of f lies on or between the lines

$$y = \frac{1}{4}x \quad \text{and} \quad y = -\frac{1}{4}x \quad$$ • **Because |cos x| ≤ 1 for all x**

▶ **TRY EXERCISE 78, PAGE 190**

---

take note

Replacing the damping factor can make a dramatic change in the graph of a function. For instance, the graph of $f(x) = 2^{-0.1x} \cos x$ approaches 0 as x approaches ∞.

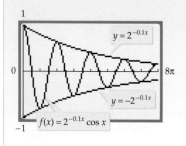

TOPICS FOR DISCUSSION

1.　The maximum value of $f(x) = (3 \sin x) + 4$ is 7. Thus f has an amplitude of 7. Do you agree? Explain.

2.　The graph of $y = \sec x$ has a period of 2π. What is the period of the graph of $y = |\sec x|$?

3. The zeros of $y = \sin x$ are the same as the zeros of $y = x \sin x$. Do you agree? Explain.

4. What is the phase shift of the graph of $y = \tan\left(3x - \dfrac{\pi}{6}\right)$?

EXERCISE SET 2.7

In Exercises 1 to 8, find the amplitude, phase shift, and period for the graph of each function.

1. $y = 2\sin\left(x - \dfrac{\pi}{2}\right)$

2. $y = -3\sin(x + \pi)$

3. $y = \cos\left(2x - \dfrac{\pi}{4}\right)$

4. $y = \dfrac{3}{4}\cos\left(\dfrac{x}{2} + \dfrac{\pi}{3}\right)$

5. $y = -4\sin\left(\dfrac{2x}{3} + \dfrac{\pi}{6}\right)$

6. $y = \dfrac{3}{2}\sin\left(\dfrac{x}{4} - \dfrac{3\pi}{4}\right)$

7. $y = \dfrac{5}{4}\cos(3x - 2\pi)$

8. $y = 6\cos\left(\dfrac{x}{3} - \dfrac{\pi}{6}\right)$

In Exercises 9 to 16, find the phase shift and the period for the graph of each function.

9. $y = 2\tan\left(2x - \dfrac{\pi}{4}\right)$

10. $y = \dfrac{1}{2}\tan\left(\dfrac{x}{2} - \pi\right)$

11. $y = -3\csc\left(\dfrac{x}{3} + \pi\right)$

12. $y = -4\csc\left(3x - \dfrac{\pi}{6}\right)$

13. $y = 2\sec\left(2x - \dfrac{\pi}{8}\right)$

14. $y = 3\sec\left(\dfrac{x}{4} - \dfrac{\pi}{2}\right)$

15. $y = -3\cot\left(\dfrac{x}{4} + 3\pi\right)$

16. $y = \dfrac{3}{2}\cot\left(2x - \dfrac{\pi}{4}\right)$

In Exercises 17 to 32, graph one full period of each function.

17. $y = \sin\left(x - \dfrac{\pi}{2}\right)$

18. $y = \sin\left(x + \dfrac{\pi}{6}\right)$

19. $y = \cos\left(\dfrac{x}{2} + \dfrac{\pi}{3}\right)$

▶ 20. $y = \cos\left(2x - \dfrac{\pi}{3}\right)$

21. $y = \tan\left(x + \dfrac{\pi}{4}\right)$

▶ 22. $y = \tan(x - \pi)$

23. $y = 2\cot\left(\dfrac{x}{2} - \dfrac{\pi}{8}\right)$

24. $y = \dfrac{3}{2}\cot\left(3x + \dfrac{\pi}{4}\right)$

25. $y = \sec\left(x + \dfrac{\pi}{4}\right)$

26. $y = \csc(2x + \pi)$

27. $y = \csc\left(\dfrac{x}{3} - \dfrac{\pi}{2}\right)$

28. $y = \sec\left(2x + \dfrac{\pi}{6}\right)$

29. $y = -2\sin\left(\dfrac{x}{3} - \dfrac{2\pi}{3}\right)$

30. $y = -\dfrac{3}{2}\sin\left(2x + \dfrac{\pi}{4}\right)$

31. $y = -3\cos\left(3x + \dfrac{\pi}{4}\right)$

32. $y = -4\cos\left(\dfrac{3x}{2} + 2\pi\right)$

In Exercises 33 to 50, graph each function using translations.

33. $y = \sin x + 1$

34. $y = -\sin x + 1$

35. $y = -\cos x - 2$

36. $y = 2\sin x + 3$

37. $y = \sin 2x - 2$

38. $y = -\cos\dfrac{x}{2} + 2$

39. $y = 4\cos(\pi x - 2) + 1$

▶ 40. $y = 2\sin\left(\dfrac{\pi x}{2} + 1\right) - 2$

41. $y = -\sin(\pi x + 1) - 2$

▶ 42. $y = -3\cos(2\pi x - 3) + 1$

43. $y = \sin\left(x - \dfrac{\pi}{2}\right) - \dfrac{1}{2}$

44. $y = -2\cos\left(x + \dfrac{\pi}{3}\right) + 3$

45. $y = \tan \dfrac{x}{2} - 4$

46. $y = \cot 2x + 3$

47. $y = \sec 2x - 2$

▶ **48.** $y = \csc \dfrac{x}{3} + 4$

49. $y = \csc \dfrac{x}{2} - 1$

50. $y = \sec\left(x - \dfrac{\pi}{2}\right) + 1$

51. RETAIL SALES The manager of a major department store finds that the number of men's suits S, in hundreds, that the store sells is given by

$$S = 4.1 \cos\left(\dfrac{\pi}{6}t - 1.25\pi\right) + 7$$

where t is time measured in months, with $t = 0$ representing January 1.

a. Find the phase shift and the period of S.

b. Graph one period of S.

c. Use the graph to determine in which month the store sells the most suits.

▶ **52. RETAIL SALES** The owner of a shoe store finds that the number of pairs of shoes S, in hundreds, that the store sells can be modeled by the function

$$S = 2.7 \cos\left(\dfrac{\pi}{6}t - \dfrac{7}{12}\pi\right) + 4$$

where t is time measured in months, with $t = 0$ representing January 1.

a. Find the phase shift and the period of S.

b. Graph one period of S.

c. Use the graph to determine in which month the store sells the most shoes.

In Exercises 53 to 58, graph the given function by using the addition-of-ordinates method.

53. $y = x - \sin x$

▶ **54.** $y = \dfrac{x}{2} + \cos x$

55. $y = x + \sin 2x$

56. $y = \dfrac{2x}{3} - \sin x$

57. $y = \sin x + \cos x$

▶ **58.** $y = -\sin x + \cos x$

In Exercises 59 to 64, find an equation of each blue graph.

59.

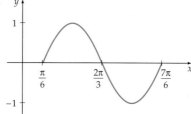

60.

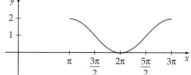

61.

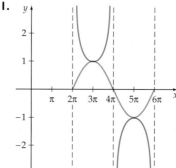

62.

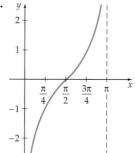

63.

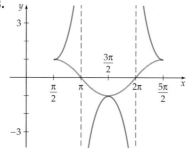

64.

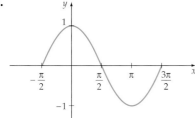

65. CARBON DIOXIDE LEVELS Because of seasonal changes in vegetation, carbon dioxide (CO_2) levels, as a product of photosynthesis, rise and fall during the year. Besides the naturally occurring CO_2 from plants, additional CO_2 is given off as a pollutant. A reasonable model of CO_2 levels in a city for the years 1982–2002 is given by $y = 2.3 \sin 2\pi t + 1.25t + 315$, where t is the number of years since 1982 and y is the concentration of CO_2 in parts per million (ppm). Find the difference in CO_2 levels between the beginning of 1982 and the beginning of 2002.

66. ENVIRONMENTAL SCIENCE Some environmentalists contend that the rate of growth of atmospheric CO_2 in parts per million is given by the equation $y = 2.54e^{0.112t} + \sin 2\pi t + 315$. See Exercise 65. Use this model to find the difference between CO_2 levels from the beginning of 1982 to the beginning of 2002.

67. HEIGHT OF A PADDLE The paddle wheel on a river boat is shown in the accompanying figure. Write an equation for the height of a paddle relative to the water at time t. The radius of the paddle wheel is 7 feet, and the distance from the center of the paddle wheel to the water is 5 feet. Assume that the paddle wheel rotates at 5 revolutions per minute and that the paddle is at its highest point at $t = 0$. Graph the equation for $0 \le t \le 0.20$ minute.

68. VOLTAGE AND AMPERAGE The graphs of the voltage and the amperage of an alternating household circuit are shown in the following figures, where t is measured in seconds. Note that there is a phase shift between the graph of the voltage and the graph of the current. The cur-

rent is said to *lag* the voltage by 0.005 second. Write an equation for the voltage and an equation for the current.

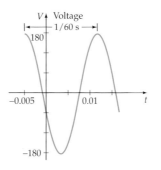
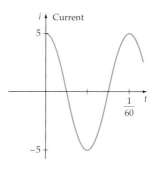

69. A LIGHTHOUSE BEACON The beacon of a lighthouse 400 meters from a straight sea wall rotates at 6 revolutions per minute. Using the accompanying figures, write an equation expressing the distance s, measured in meters, in terms of time t. Assume that when $t = 0$, the beam is perpendicular to the sea wall. Sketch a graph of the equation for $0 \le t \le 10$ seconds.

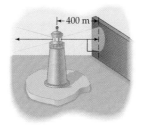
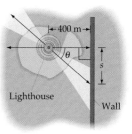

Side view   Top view

70. HOURS OF DAYLIGHT The duration of daylight for a region is dependent not only on the time of year but also on the latitude of the region. The graph gives the daylight hours for a one-year period at various latitudes. Assuming that a sine function can model these curves, write an equation for each curve.

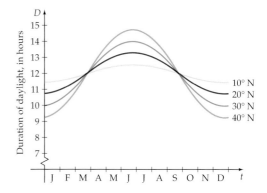

71. **TIDES** During a 24-hour day, the tides raise and lower the depth of water at a pier as shown in the figure below. Write an equation in the form $f(t) = A \cos Bt + d$, and find the depth of the water at 6 P.M.

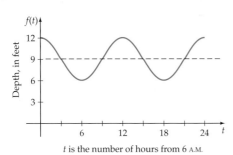

t is the number of hours from 6 A.M.

72. **TEMPERATURE** During a summer day, the ground temperature at a desert location was recorded and graphed as a function of time as shown in the following figure. The graph can be approximated by $f(t) = A \cos(bt + c) + d$. Find the equation, and approximate the temperature (to the nearest degree) at 1:00 P.M.

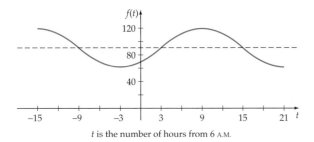

t is the number of hours from 6 A.M.

In Exercises 73 to 82, use a graphing utility to graph each function.

73. $y = \sin x - \cos \dfrac{x}{2}$ **74.** $y = 2 \sin 2x - \cos x$

75. $y = 2 \cos x + \sin \dfrac{x}{2}$ **76.** $y = -\dfrac{1}{2} \cos 2x + \sin \dfrac{x}{2}$

77. $y = \dfrac{x}{2} \sin x$ ▶ **78.** $y = x \cos x$

79. $y = x \sin \dfrac{x}{2}$ **80.** $y = \dfrac{x}{2} \cos \dfrac{x}{2}$

81. $y = x \sin\left(x + \dfrac{\pi}{2}\right)$ **82.** $y = x \cos\left(x - \dfrac{\pi}{2}\right)$

BEATS **When two sound waves have approximately the same frequency, the sound waves interfere with one another and produce phenomena called** *beats,* **which are heard as variations in the loudness of the sound. A piano tuner can use these phenomena to tune a piano. By striking a tuning fork and then tapping the corresponding key on a piano, the piano tuner listens for beats and adjusts the tension in the string until the beats disappear. Use a graphing utility to graph the functions in Exercises 83 to 86, which are based on beats.**

83. $y = \sin(5\pi x) \cdot \sin\left(-\dfrac{\pi}{2} x\right)$

84. $y = \sin(9\pi x) \cdot \sin\left(-\dfrac{\pi}{2} x\right)$

85. $y = \sin(13\pi x) \cdot \sin\left(-\dfrac{\pi}{2} x\right)$

86. $y = \sin(17\pi x) \cdot \sin\left(-\dfrac{\pi}{2} x\right)$

CONNECTING CONCEPTS

87. Find an equation of the sine function with amplitude 2, period π, and phase shift $\dfrac{\pi}{3}$.

88. Find an equation of the cosine function with amplitude 3, period 3π, and phase shift $-\dfrac{\pi}{4}$.

89. Find an equation of the tangent function with period 2π and phase shift $\dfrac{\pi}{2}$.

90. Find an equation of the cotangent function with period $\dfrac{\pi}{2}$ and phase shift $-\dfrac{\pi}{4}$.

91. Find an equation of the secant function with period 4π and phase shift $\dfrac{3\pi}{4}$.

92. Find an equation of the cosecant function with period $\dfrac{3\pi}{2}$ and phase shift $\dfrac{\pi}{4}$.

93. If $g(x) = \sin^2 x$ and $h(x) = \cos^2 x$, find $g(x) + h(x)$.

94. If $g(x) = 2\sin x - 3$ and $h(x) = 4\cos x + 2$, find the sum $g(x) + h(x)$.

95. If $g(x) = x^2 + 2$ and $h(x) = \cos x$, find $g[h(x)]$.

96. If $g(x) = \sin x$ and $h(x) = x^2 + 2x + 1$, find $h[g(x)]$.

 In Exercises 97 to 100, use a graphing utility to graph each function.

97. $y = \dfrac{\sin x}{x}$

98. $y = 2 + \sec \dfrac{x}{2}$

99. $y = |x| \sin x$

100. $y = |x| \cos x$

PREPARE FOR SECTION 2.8

101. Find the reciprocal of $\dfrac{2\pi}{3}$.

102. Find the reciprocal of $\dfrac{2}{5}$.

103. Evaluate $a \cos 2\pi t$ for $a = 4$ and $t = 0$. [2.4]

104. Evaluate $\sqrt{\dfrac{k}{m}}$ for $k = 18$ and $m = 2$.

105. Evaluate $4\cos\left(\sqrt{\dfrac{k}{m}}\,t\right)$ for $k = 16$, $m = 4$, and $t = 2\pi$. [2.4]

106. Write an equation of the form $y = a\cos bx$ whose graph has an amplitude of 4 and a period of 2. [2.5]

PROJECTS

1. **PREDATOR-PREY RELATIONSHIP** Predator-prey interactions can produce cyclic population growth for both the predator population and the prey population. Consider an animal reserve where the rabbit population r is given by

$$r(t) = 850 + 210\sin\left(\frac{\pi}{6}t\right)$$

and the wolf population w is given by

$$w(t) = 120 + 30\sin\left(\frac{\pi}{6}t - 2\right)$$

where t is the number of months after March 1, 2000. Graph $r(t)$ and $w(t)$ on the same coordinate system for $0 \le t \le 24$. Write a few sentences that explains a possible relationship between the two populations.

Write an equation that could be used to model the rabbit population shown by the graph at the right, where t is measured in months.

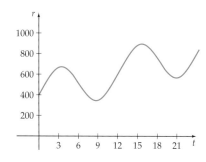

HARMONIC MOTION—AN APPLICATION OF THE SINE AND COSINE FUNCTIONS

- SIMPLE HARMONIC MOTION
- DAMPED HARMONIC MOTION

Many phenomena that occur in nature can be modeled by periodic functions, including vibrations of a swing or in a spring. These phenomena can be described by the *sinusoidal* functions, which are the sine and cosine functions or a sum of these two functions.

● SIMPLE HARMONIC MOTION

We will consider a mass on a spring to illustrate vibratory motion. Assume that we have placed a mass on a spring and allowed the spring to come to rest, as shown in **Figure 2.104.** The system is said to be in equilibrium when the mass is at rest. The point of rest is called the *origin* of the system. We consider the distance above the equilibrium point as positive and the distance below the equilibrium point as negative.

If the mass is now lifted a distance a and released, the mass will oscillate up and down in periodic motion. If there is no friction, the motion repeats itself in a certain period of time. The distance a is called the **displacement** from the origin. The number of times the mass oscillates in 1 unit of time is called the *frequency f* of the motion, and the time one oscillation takes is the *period p* of the motion. The motion is referred to as *simple harmonic motion.* **Figure 2.105** shows the displacement y of the mass for one oscillation for $t = 0$, $\dfrac{p}{4}$, $\dfrac{p}{2}$, $\dfrac{3p}{4}$, and p.

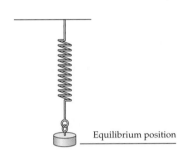

FIGURE 2.104

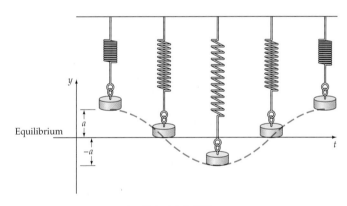

FIGURE 2.105

The frequency and the period are related by the formulas

$$f = \frac{1}{p} \quad \text{and} \quad p = \frac{1}{f}$$

The maximum displacement from the equilibrium position is called the *amplitude of the motion.* Vibratory motion can be quite complicated. However, the simple harmonic motion of the mass on the spring can be described by one of the following equations.

Simple Harmonic Motion

Simple harmonic motion can be modeled by one of the following functions:

$$y = a \cos 2\pi ft \quad (1) \qquad \text{or} \qquad y = a \sin 2\pi ft \quad (2)$$

where $|a|$ is the amplitude (maximum displacement), f is the frequency, $\dfrac{1}{f}$ is the period, y is the displacement, and t is the time.

❓ QUESTION   A simple harmonic motion has a frequency of 2 cycles per second. What is the period of the simple harmonic motion?

EXAMPLE 1   **Find the Equation of Motion of a Mass on a Spring**

A mass on a spring has been displaced 4 centimeters above the equilibrium point and released. The mass is vibrating with a frequency of $\dfrac{1}{2}$ cycle per second. Write the equation of simple harmonic motion, and graph three cycles of the displacement as a function of time.

Solution

Because the maximum displacement is 4 centimeters when $t = 0$, use $y = a \cos 2\pi ft$. See **Figure 2.106.**

$$y = a \cos 2\pi ft \qquad \text{• Equation for simple harmonic motion}$$
$$= 4 \cos 2\pi\left(\frac{1}{2}\right)t \qquad \text{• } a = 4, f = \frac{1}{2}$$
$$= 4 \cos \pi t$$

▶ **TRY EXERCISE 20, PAGE 196**

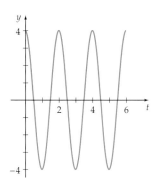

$y = 4 \cos \pi t$

FIGURE 2.106

From physical laws determined by experiment, the frequency of oscillation of a mass on a spring is given by

$$f = \frac{1}{2\pi} \sqrt{\frac{k}{m}}$$

where k is a spring constant determined by experiment and m is the mass. The simple harmonic motion of the mass on the spring (with maximum displacement

❓ ANSWER $\dfrac{1}{2}$ second per cycle

at $t = 0$) can then be described by

$$y = a \cos 2\,\pi f t = a \cos 2\pi\left(\frac{1}{2\pi}\,\sqrt{\frac{k}{m}}\right)t$$

$$= a \cos\,\sqrt{\frac{k}{m}}\,t \qquad\qquad\qquad (3)$$

The equation of the simple harmonic motion for zero displacement at $t = 0$ is

$$y = a \sin\,\sqrt{\frac{k}{m}}\,t \qquad\qquad\qquad (4)$$

EXAMPLE 2   **Find the Equation of Motion of a Mass on a Spring**

A mass of 2 units is in equilibrium suspended from a spring with a spring constant of $k = 18$. The mass is pulled down 0.5 unit and released. Find the period, frequency, and amplitude of the resulting simple harmonic motion. Write the equation of the motion, and graph two cycles of the displacement as a function of time.

Solution

At the start of the motion, the displacement is at a maximum but in the negative direction. The resulting motion is described by Equation (3), using $a = -0.5$, $k = 18$, and $m = 2$.

$$y = a \cos\,\sqrt{\frac{k}{m}}\,t = -0.5 \cos\,\sqrt{\frac{18}{2}}\,t \qquad \text{• Substitute for } a, k, \text{ and } m.$$

$$= -0.5 \cos 3t \qquad\qquad\qquad\qquad \text{• Equation of motion}$$

Period: $\dfrac{2\pi}{|b|} = \dfrac{2\pi}{3}$

Frequency: $\dfrac{1}{\text{period}} = \dfrac{3}{2\pi}$

Amplitude: $|a| = |-0.5| = 0.5$

See **Figure 2.107**.

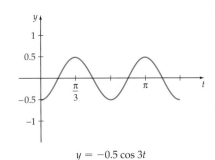

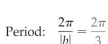

$y = -0.5 \cos 3t$

FIGURE 2.107

▶ TRY EXERCISE 28, PAGE 196

● **DAMPED HARMONIC MOTION**

The previous examples have assumed that there is no friction within the spring and no air resistance. If we consider friction and air resistance, then the motion of the mass tends to decrease as t increases. The motion is called **damped harmonic motion**.

> **EXAMPLE 3**   **Model Damped Harmonic Motion**

 A mass on a spring has been displaced 14 inches below the equilibrium point and released. The damped harmonic motion of the mass is given by

$$f(t) = -14e^{-0.4t} \cos 2t, \quad t \geq 0$$

where $f(t)$ is measured in inches and t is the time in seconds.

a. Find the values of t for which $f(t) = 0$.

b. Use a graphing utility to determine how long it will be until the absolute value of the displacement of the mass is always less than 0.01 inch.

Solution

a. $f(t) = 0$ if and only if $\cos 2t = 0$, and $\cos 2t = 0$ when $2t = \dfrac{\pi}{2} + n\pi$.

Therefore, $f(t) = 0$ when $t = \dfrac{\pi}{4} + n\left(\dfrac{\pi}{2}\right) \approx 0.79 + n(1.57)$. The displacement $f(t) = 0$ first occurs when $t \approx 0.79$ second. The graph of f in **Figure 2.108** confirms these results. **Figure 2.108** also shows that the graph of f lies on or between the graphs of $y = -14e^{-0.4t}$ and $y = 14e^{-0.4t}$. Recall from Section 2.7 that $-14e^{-0.4t}$ is the damping factor.

b. We need to find the smallest value of t for which the absolute value of the displacement of the mass is always less than 0.01. **Figure 2.109** shows a graph of f using a viewing window of $0 \leq t \leq 20$ and $-0.01 \leq f(t) \leq 0.01$. Use the TRACE or the INTERSECT feature to determine that $t \approx 17.59$ seconds.

take note

The motion in Example 3 is not periodic in a strict sense, because the motion does not repeat exactly but tends to diminish as t increases. However, because the motion cycles every π seconds, we call π the *pseudoperiod* of the motion. In general, the damped harmonic motion modeled by $f(t) = ae^{-kt} \cos \omega t$ has

pseudoperiod $p = \dfrac{2\pi}{\omega}$ and

frequency $f = \dfrac{\omega}{2\pi}$.

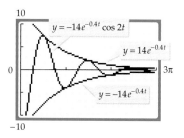

FIGURE 2.108

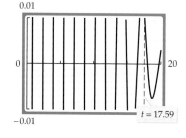

$$f(t) = -14e^{-0.4t} \cos 2t$$

FIGURE 2.109

▶ **TRY EXERCISE 30, PAGE 197**

 TOPICS FOR DISCUSSION

1. The period of a simple harmonic motion is the same as the frequency of the motion. Do you agree? Explain.

2. In the simple harmonic motion modeled by $y = 3 \cos 2\pi t$, does the displacement y approach 0 as t increases without bound? Explain.

3. Explain how you know whether to use

$$y = a \cos 2\pi ft \qquad \text{or} \qquad y = a \sin 2\pi ft$$

to model a particular harmonic motion.

4. If the mass on a spring is increased from m to $4m$, what effect will this have on the frequency of the simple harmonic motion of the mass?

EXERCISE SET 2.8

In Exercises 1 to 8, find the amplitude, period, and frequency of the simple harmonic motion.

1. $y = 2 \sin 2t$

2. $y = \dfrac{2}{3} \cos \dfrac{t}{3}$

3. $y = 3 \cos \dfrac{2t}{3}$

4. $y = 4 \sin 3t$

5. $y = 4 \cos \pi t$

6. $y = 2 \sin \dfrac{\pi t}{3}$

7. $y = \dfrac{3}{4} \sin \dfrac{\pi t}{2}$

8. $y = 5 \cos 2\pi t$

In Exercises 9 to 12, write an equation for the simple harmonic motion that satisfies the given conditions. Assume that the maximum displacement occurs at t = 0. Sketch a graph of the equation.

9. Frequency = 1.5 cycles per second, $a = 4$ inches

10. Frequency = 0.8 cycle per second, $a = 4$ centimeters

11. Period = 1.5 seconds, $a = \dfrac{3}{2}$ feet

12. Period = 0.6 second, $a = 1$ meter

In Exercises 13 to 18, write an equation for the simple harmonic motion with the given conditions. Assume zero displacement at t = 0. Sketch a graph of the equation.

13. Amplitude 2 centimeters, period π seconds

14. Amplitude 4 inches, period $\dfrac{\pi}{2}$ seconds

15. Amplitude 1 inch, period 2 seconds

16. Amplitude 3 centimeters, period 1 second

17. Amplitude 2 centimeters, frequency 1 second

18. Amplitude 4 inches, frequency 4 seconds

In Exercises 19 to 26, write an equation for simple harmonic motion. Assume that the maximum displacement occurs when t = 0.

19. Amplitude $\dfrac{1}{2}$ centimeter, frequency $\dfrac{2}{\pi}$ cycle per second

▶ **20.** Amplitude 3 inches, frequency $\dfrac{1}{\pi}$ cycle per second

21. Amplitude 2.5 inches, frequency 0.5 cycle per second

22. Amplitude 5 inches, frequency $\dfrac{1}{8}$ cycle per second

23. Amplitude $\dfrac{1}{2}$ inch, period 3 seconds

24. Amplitude 5 centimeters, period 5 seconds

25. Amplitude 4 inches, period $\dfrac{\pi}{2}$ seconds

26. Amplitude 2 centimeters, period π seconds

27. SIMPLE HARMONIC MOTION A mass of 32 units is in equilibrium suspended from a spring. The mass is pulled down 2 feet and released. Find the period, frequency, and amplitude of the resulting simple harmonic motion. Write an equation of the motion. Assume a spring constant of $k = 8$.

▶ **28.** SIMPLE HARMONIC MOTION A mass of 27 units is in equilibrium suspended from a spring. The mass is pulled down 1.5 feet and released. Find the period, frequency, and amplitude of the resulting simple harmonic motion. Write an equation of the motion. Assume a spring constant of $k = 3$.

In Exercises 29 to 36, each of the equations models a damped harmonic motion.

a. Find the number of complete oscillations that occur during the time interval $0 \le t \le 10$ seconds.

b. Use a graph to determine how long it will be (to the nearest 0.1 second) until the absolute value of the displacement of the mass is always less than 0.01.

29. $f(t) = 4e^{-0.1t} \cos 2t$

30. $f(t) = 12e^{-0.6t} \cos t$

31. $f(t) = -6e^{-0.09t} \cos 2\pi t$

32. $f(t) = -11e^{-0.4t} \cos \pi t$

33. $f(t) = e^{-0.5t} \cos 2\pi t$

34. $f(t) = e^{-0.2t} \cos 3\pi t$

35. $f(t) = e^{-0.75t} \cos 2\pi t$

36. $f(t) = e^{-t} \cos 2\pi t$

CONNECTING CONCEPTS

37. Assume that a mass of m pounds on the end of a spring is oscillating in simple harmonic motion. What effect will there be on the period of the motion if the mass is increased to $9m$?

38. A mass on a spring is displaced 9 inches below its equilibrium position and then released. The weight oscillates in simple harmonic motion with a frequency of 2 cycles per second. Find the period and the equation of the motion.

In Exercises 39 to 42, use a graphing utility to determine whether both of the given functions model the same damped harmonic motion.

39. $f(t) = \sqrt{2}\,e^{-0.2t} \sin\left(t + \dfrac{\pi}{4}\right)$

$g(t) = e^{-0.2t}(\cos t + \sin t)$

40. $f(t) = 5e^{-0.3t} \cos(t - 0.927295)$

$g(t) = e^{-0.2t}(3 \cos t + 4 \sin t)$

41. $f(t) = 13e^{-0.4t} \cos(t - 1.176005)$

$g(t) = e^{-0.4t}(5 \cos t + 12 \sin t)$

42. $f(t) = \sqrt{2}\,e^{-0.2t} \cos(t + 0.785398)$

$g(t) = e^{-0.2t}(\cos t - \sin t)$

PROJECTS

1. **THREE TYPES OF DAMPED HARMONIC MOTION** In some cases the damped harmonic motion of a mass on the end of a spring does not cycle about the equilibrium point. The following three functions illustrate different types of damped harmonic motion. Use a graphing utility to graph each function, and then write a few sentences that explain the major differences among the motions.

$$f(t) = (0.5t + 1)e^{-0.5t} \qquad g(t) = -2e^{-0.4t} + 5e^{-t}$$
$$h(t) = 4e^{-0.2t} \cos 2\pi t$$

2. **LOGARITHMIC DECREMENT** If a damped harmonic motion is modeled by

$$f(t) = ae^{-\alpha t} \cos \omega t$$

then the ratio of any two consecutive relative maxima of the motion is a constant γ.

a. Use a graphing utility to determine γ for the damped harmonic motion modeled by

$$f(t) = -14e^{-0.4t} \cos 2t, \quad t \ge 0$$

b. The constant $\Delta = \dfrac{2\pi\alpha}{\omega}$ is called the **logarithmic decrement** of the motion. Compute Δ for the damped harmonic motion in the equation in part **a.** How does $\ln \gamma$ compare with Δ?

EXPLORING CONCEPTS WITH TECHNOLOGY

Sinusoidal Families

Some graphing calculators have a feature that allows you to graph a family of functions easily. For instance, entering Y₁={2,4,6}sin(X) in the Y= menu and pressing the **GRAPH** key on a TI-83 calculator produces a graph of the three functions $y = 2 \sin x$, $y = 4 \sin x$, and $y = 6 \sin x$, all displayed in the same window.

1. Use a graphing calculator to graph Y₁={2,4,6}sin(X). Write a sentence that indicates the similarities and the differences among the three graphs.

2. Use a graphing calculator to graph Y₁=sin({π,2π,4π}X). Write a sentence that indicates the similarities and the differences among the three graphs.

3. Use a graphing calculator to graph Y₁=sin(X+{π/4, π/6, π/12}). Write a sentence that indicates the similarities and the differences among the three graphs.

4. A student has used a graphing calculator to graph Y₁=sin(X+{π,3π,5π}) and expects to see three graphs. However, the student sees only one graph displayed on the graph window. Has the calculator displayed all three graphs? Explain.

CHAPTER 2 SUMMARY

2.1 Angles and Arcs

• An angle is in standard position when its initial side is along the positive x-axis and its vertex is at the origin of the coordinate axes.

• Angle α is an acute angle when $0° < \alpha < 90°$; it is an obtuse angle when $90° < \alpha < 180°$.

• α and β are complementary angles when $\alpha + \beta = 90°$; they are supplementary angles when $\alpha + \beta = 180°$.

• The length of the arc s that subtends the central angle θ (in radians) on a circle of radius r is given by $s = r\theta$.

• Angular speed is given by $\omega = \dfrac{\theta}{t}$.

2.2 Trigonometric Functions of Acute Angles

• Let θ be an acute angle of a right triangle. The six trigonometric functions of θ are given by

$$\sin \theta = \frac{\text{opp}}{\text{hyp}} \qquad \csc \theta = \frac{\text{hyp}}{\text{opp}}$$

$$\cos \theta = \frac{\text{adj}}{\text{hyp}} \qquad \sec \theta = \frac{\text{hyp}}{\text{adj}}$$

$$\tan \theta = \frac{\text{opp}}{\text{adj}} \qquad \cot \theta = \frac{\text{adj}}{\text{opp}}$$

2.3 Trigonometric Functions of Any Angle

• Let $P(x, y)$ be a point, except the origin, on the terminal side of an angle θ in standard position. The six trigonometric functions of θ are

$$\sin \theta = \frac{y}{r} \qquad \csc \theta = \frac{r}{y}, \quad y \neq 0$$

$$\cos \theta = \frac{x}{r} \qquad \sec \theta = \frac{r}{x}, \quad x \neq 0$$

$$\tan \theta = \frac{y}{x}, \quad x \neq 0 \qquad \cot \theta = \frac{x}{y}, \quad y \neq 0$$

2.4 Trigonometric Functions of Real Numbers

• The wrapping function pairs a real number with a point on the unit circle.

- Let W be the wrapping function, t be a real number, and $W(t) = P(x, y)$. Then the trigonometric functions of the real number t are defined as follows:

$$\sin t = y \qquad\qquad \csc t = \frac{1}{y}, \quad y \neq 0$$

$$\cos t = x \qquad\qquad \sec t = \frac{1}{x}, \quad x \neq 0$$

$$\tan t = \frac{y}{x}, \quad x \neq 0 \qquad \cot t = \frac{x}{y}, \quad y \neq 0$$

- $\sin t$, $\csc t$, $\tan t$, and $\cot t$ are odd functions.

- $\cos t$ and $\sec t$ are even functions.

- $\sin t$, $\cos t$, $\sec t$, and $\csc t$ have period 2π.

- $\tan t$ and $\cot t$ have period π.

Domain and Range of Each Trigonometric Function (n is an integer)

Function	Domain	Range
$\sin t$	$\{t \mid -\infty < t < \infty\}$	$\{y \mid -1 \leq y \leq 1\}$
$\cos t$	$\{t \mid -\infty < t < \infty\}$	$\{y \mid -1 \leq y \leq 1\}$
$\tan t$	$\left\{t \mid -\infty < t < \infty, \; t \neq \dfrac{(2n+1)\pi}{2}\right\}$	$\{y \mid -\infty < y < \infty\}$
$\csc t$	$\{t \mid -\infty < t < \infty, \; t \neq n\pi\}$	$\{y \mid y \geq 1, y \leq -1\}$
$\sec t$	$\left\{t \mid -\infty < t < \infty, \; t \neq \dfrac{(2n+1)\pi}{2}\right\}$	$\{y \mid y \geq 1, y \leq -1\}$
$\cot t$	$\{t \mid -\infty < t < \infty, \; t \neq n\pi\}$	$\{y \mid -\infty < y < \infty\}$

2.5 Graphs of the Sine and Cosine Functions

- The graphs of $y = a \sin bx$ and $y = a \cos bx$ both have an amplitude of $|a|$ and a period of $\dfrac{2\pi}{|b|}$. The graph of each for $a > 0$ and $b > 0$ is given below.

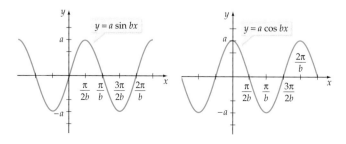

2.6 Graphs of the Other Trigonometric Functions

- The period of $y = a \tan bx$ and $y = a \cot bx$ is $\dfrac{\pi}{|b|}$.

- The period of $y = a \sec bx$ and $y = a \csc bx$ is $\dfrac{2\pi}{|b|}$.

2.7 Graphing Techniques

- Phase shift is a horizontal translation of the graph of a trigonometric function. If $y = f(bx + c)$, where f is a trigonometric function, then the phase shift is $-\dfrac{c}{b}$.

- The graphs of $y = a \sin(bx + c)$ and $y = a \cos(bx + c)$, $b > 0$, have amplitude $|a|$, period $\dfrac{2\pi}{b}$, and phase shift $-\dfrac{c}{b}$. One cycle of each graph is completed on the interval $-\dfrac{c}{b} \leq x \leq -\dfrac{c}{b} + \dfrac{2\pi}{b}$.

- Addition of ordinates is a method of graphing the sum of two functions by graphically adding the values of their y-coordinates.

- The factor $g(x)$ in $f(x) = g(x) \cos x$ is called a damping factor. The graph of f lies on or between the graphs of the equations $y = g(x)$ and $y = -g(x)$.

2.8 Harmonic Motion—An Application of the Sine and Cosine Functions

- The equations of simple harmonic motion are

$$y = a \cos 2\pi ft \qquad \text{and} \qquad y = a \sin 2\pi ft$$

where a is the amplitude and f is the frequency.

- Functions of the form $f(t) = ae^{-\alpha t} \cos \omega t$ are used to model some forms of damped harmonic motion.

CHAPTER 2 TRUE/FALSE EXERCISES

In Exercises 1 to 16, answer true or false. If the statement is false, give a reason or state an example to show that the statement is false.

1. An angle is in standard position when the vertex is at the origin of a coordinate system.

2. The angle θ in radians is in standard position with the terminal side in the second quadrant. The reference angle of θ is $\pi - \theta$.

3. In the formula $s = r\theta$, the angle θ must be measured in radians.

4. If $\tan \theta < 0$ and $\cos \theta > 0$, then the terminal side of θ is in Quadrant III.

5. $\sec^2 \theta + \tan^2 \theta = 1$ is an identity.

6. The amplitude of the graph of $y = 2 \tan x$ is 2.

7. The period of $y = \cos x$ is π.

8. The graph of $y = \sin x$ is symmetric to the origin.

9. For any acute angle θ, $\sin \theta + \cos(90° - \theta) = 1$.

10. $\sin (x + y) = \sin x + \sin y$

11. $\sin^2 x = \sin x^2$

12. The phase shift of $f(x) = 2 \sin \left(2x - \dfrac{\pi}{3} \right)$ is $\dfrac{\pi}{3}$.

13. The measure of one radian is more than 50 times the measure of one degree.

14. The measure of one radian differs depending on the radius of the circle used.

15. The graph of $y = 2^{-x} \cos x$ lies on or between the graphs of $y = 2^{-x}$ and $y = 2^x$.

16. The function $f(t) = e^{-0.1t} \cos t$, $t > 0$, models damped harmonic motion in which $|f(t)| \to 0$ as $t \to 0$.

CHAPTER 2 REVIEW EXERCISES

1. Find the complement and supplement of the angle θ whose measure is 65°.

2. Find the measure of the reference angle θ' for the angle θ whose measure is 980°.

3. Convert 2 radians to the nearest hundredth of a degree.

4. Convert 315° to radian measure.

5. Find the length (to the nearest 0.01 meter) of the arc on a circle of radius 3 meters that subtends an angle of 75°.

6. Find the radian measure of the angle subtended by an arc of length 12 centimeters on a circle whose radius is 40 centimeters.

7. A car with a 16-inch-radius wheel is moving with a speed of 50 mph. Find the angular speed (to the nearest radian per second) of the wheel in radians per second.

In Exercises 8 to 11, let θ be an acute angle of a right triangle and csc $\theta = \dfrac{3}{2}$. Evaluate each function.

8. $\cos \theta$ 9. $\cot \theta$ 10. $\sin \theta$ 11. $\sec \theta$

12. Find the values of the six trigonometric functions of an angle in standard position with the point $P(1, -3)$ on the terminal side of the angle.

13. Find the exact value of

   a. $\sec 150°$ b. $\tan\left(-\dfrac{3\pi}{4} \right)$

   c. $\cot(-225°)$ d. $\cos\left(\dfrac{2\pi}{3} \right)$

14. ▦ Find the value of each of the following to the nearest ten-thousandth.

   a. $\cos 123°$ b. $\cot 4.22$

   c. $\sec 612°$ d. $\tan \dfrac{2\pi}{5}$

15. Given $\cos \phi = -\dfrac{\sqrt{3}}{2}$, $180° < \phi < 270°$, find the exact value of

    a. $\sin \phi$ **b.** $\tan \phi$

16. Given $\tan \phi = -\dfrac{\sqrt{3}}{3}$, $90° < \phi < 180°$, find the exact value of

    a. $\sec \phi$ **b.** $\csc \phi$

17. Given $\sin \phi = -\dfrac{\sqrt{2}}{2}$, $270° < \phi < 360°$, find the exact value of

    a. $\cos \phi$ **b.** $\cot \phi$

18. Let W be the wrapping function. Evaluate

    a. $W(\pi)$ **b.** $W\left(-\dfrac{\pi}{3}\right)$ **c.** $W\left(\dfrac{5\pi}{4}\right)$ **d.** $W(28\pi)$

19. Is the function defined by $f(x) = \sin(x)\tan(x)$ even, odd, or neither?

In Exercises 20 and 21, use the unit circle to show that each equation is an identity.

20. $\cos(\pi + t) = -\cos t$ **21.** $\tan(-t) = -\tan t$

In Exercises 22 to 27, use trigonometric identities to write each expression in terms of a single trigonometric function or as a constant.

22. $1 + \dfrac{\sin^2 \phi}{\cos^2 \phi}$ **23.** $\dfrac{\tan \phi + 1}{\cot \phi + 1}$

24. $\dfrac{\cos^2 \phi + \sin^2 \phi}{\csc \phi}$ **25.** $\sin^2 \phi(\tan^2 \phi + 1)$

26. $1 + \dfrac{1}{\tan^2 \phi}$ **27.** $\dfrac{\cos^2 \phi}{1 - \sin^2 \phi} - 1$

In Exercises 28 to 33, state the amplitude (if there is one), period, and phase shift of the graph of each function.

28. $y = 3\cos(2x - \pi)$ **29.** $y = 2\tan 3x$

30. $y = -2\sin\left(3x + \dfrac{\pi}{3}\right)$ **31.** $y = \cos\left(2x - \dfrac{2\pi}{3}\right) + 2$

32. $y = -4\sec\left(4x - \dfrac{3\pi}{2}\right)$ **33.** $y = 2\csc\left(x - \dfrac{\pi}{4}\right) - 3$

In Exercises 34 to 51, graph each function.

34. $y = 2\cos \pi x$ **35.** $y = -\sin \dfrac{2x}{3}$

36. $y = 2\sin \dfrac{3x}{2}$ **37.** $y = \cos\left(x - \dfrac{\pi}{2}\right)$

38. $y = \dfrac{1}{2}\sin\left(2x + \dfrac{\pi}{4}\right)$ **39.** $y = 3\cos 3(x - \pi)$

40. $y = -\tan \dfrac{x}{2}$ **41.** $y = 2\cot 2x$

42. $y = \tan\left(x - \dfrac{\pi}{2}\right)$ **43.** $y = -\cot\left(2x + \dfrac{\pi}{4}\right)$

44. $y = -2\csc\left(2x - \dfrac{\pi}{3}\right)$ **45.** $y = 3\sec\left(x + \dfrac{\pi}{4}\right)$

46. $y = 3\sin 2x - 3$ **47.** $y = 2\cos 3x + 3$

48. $y = -\cos\left(3x + \dfrac{\pi}{2}\right) + 2$

49. $y = 3\sin\left(4x - \dfrac{2\pi}{3}\right) - 3$

50. $y = 2 - \sin 2x$

51. $y = \sin x - \sqrt{3}\cos x$

52. **INCREASE IN ALTITUDE** A car climbs a hill that has a constant angle of 4.5° for a distance of 1.14 miles. What is the car's increase in altitude?

53. **HEIGHT OF A TREE** A tree casts a shadow of 8.55 feet when the angle of elevation of the sun is 55.3°. Find the height of the tree.

54. Find the sine of the angle α formed by the intersection of a diagonal of a face of a cube and the diagonal of the cube originating from the same vertex.

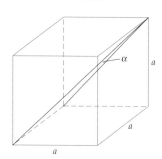

55. HEIGHT OF A BUILDING Find the height of a building if the angle of elevation to the top of the building changes from 18° to 37° as an observer moves a distance of 80 feet toward the building.

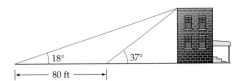

56. Find the amplitude, period, and frequency of the simple harmonic motion given by $y = 2.5 \sin 50t$.

57. SIMPLE HARMONIC MOTION A mass of 5 kilograms is in equilibrium suspended from a spring. The mass is pulled down 0.5 foot and released. Find the period, frequency, and amplitude of the motion, assuming the mass oscillates in simple harmonic motion. Write an equation of motion. Assume $k = 20$.

58. DAMPED HARMONIC MOTION Use a graphing utility to graph the damped harmonic motion that is modeled by

$$f(t) = 3e^{-0.75t} \cos \pi t$$

where t is in seconds. Use the graph to determine how long (to the nearest 0.1 second) it will be until the absolute value of the displacement of the mass is always less than 0.01.

CHAPTER 2 TEST

1. Convert 150° to exact radian measure.

2. Find the supplement of the angle whose radian measure is $\dfrac{11}{12}\pi$. Express your answer in terms of π.

3. Find the length (to the nearest 0.1 centimeter) of an arc that subtends a central angle of 75° in a circle of radius 10 centimeters.

4. A wheel is rotating at 6 revolutions per second. Find the angular speed in radians per second.

5. A wheel with a diameter of 16 centimeters is rotating at 10 radians per second. Find the linear speed (in centimeters per second) of a point on the edge of the wheel.

6. If θ is an acute angle of a right triangle and $\tan \theta = \dfrac{3}{7}$, find $\sec \theta$.

7. Use a calculator to find the value of $\csc 67°$ to the nearest ten-thousandth.

8. Find the exact value of $\tan \dfrac{\pi}{6} \cos \dfrac{\pi}{3} - \sin \dfrac{\pi}{2}$.

9. Find the exact coordinates of $W\left(\dfrac{11\pi}{6}\right)$.

10. Express $\dfrac{\sec^2 t - 1}{\sec^2 t}$ in terms of a single trigonometric function.

11. State the period of $y = -4 \tan 3x$.

12. State the amplitude, period, and phase shift for the function $y = -3 \cos\left(2x + \dfrac{\pi}{2}\right)$.

13. State the period and phase shift for the function $y = 2 \cot\left(\dfrac{\pi}{3}x + \dfrac{\pi}{6}\right)$.

14. Graph one full period of $y = 3 \cos \dfrac{1}{2}x$.

15. Graph one full period of $y = -2 \sec \dfrac{1}{2}x$.

16. Write a sentence that explains how to obtain the graph of $y = 2 \sin\left(2x - \dfrac{\pi}{2}\right) - 1$ from the graph of $y = 2 \sin 2x$.

17. Graph one full period of $y = 2 - \sin \dfrac{x}{2}$.

18. Graph one full period of $y = \sin x - \cos 2x$.

19. The angle of elevation from point A to the top of a tree is 42.2°. At point B, 5.24 meters from A and on a line through the base of the tree and A, the angle of elevation is 37.4°. Find the height of the tree.

20. Write the equation for simple harmonic motion, given that the amplitude is 13 feet, the period is 5 seconds, and the displacement is zero when $t = 0$.

CUMULATIVE REVIEW EXERCISES

1. Find the distance between the points $P(-3, 2)$ and $Q(4, 1)$.

2. The hypotenuse of a right triangle has a length of 1 and one of its legs has a length of $\frac{1}{2}$. Find the length of the other leg.

3. Find the x- and the y-intercept(s) of the graph of $f(x) = x^2 - 9$.

4. Determine whether $f(x) = \dfrac{x}{x^2 + 1}$ is an even function or an odd function.

5. Find the inverse of $f(x) = \dfrac{x}{2x - 3}$.

6. Use interval notation to state the domain of $f(x) = \dfrac{2}{x - 4}$.

7. Solve: $x^2 + x - 6 = 0$

8. Explain how to use the graph of $y = f(x)$ to produce the graph of $y = f(x - 3)$.

9. Explain how to use the graph of $y = f(x)$ to produce the graph of $y = f(-x)$.

10. Convert $300°$ to radians.

11. Convert $\dfrac{5\pi}{4}$ to degrees.

12. Evaluate $f(x) = \sin\left(x + \dfrac{\pi}{6}\right)$ for $x = \dfrac{\pi}{3}$.

13. Evaluate $f(x) = \sin x + \sin \dfrac{\pi}{6}$ for $x = \dfrac{\pi}{3}$.

14. Find the exact value of $\cos^2 45° + \sin^2 60°$.

15. Determine the sign of $\tan \theta$ given that $\dfrac{\pi}{2} < \theta < \pi$.

16. What is the measure of the reference angle for the angle $\theta = 210°$?

17. What is the measure of the reference angle for the angle $\theta = \dfrac{2\pi}{3}$?

18. Use interval notation to state the domain of $f(x) = \sin x$, where x is a real number.

19. Use interval notation to state the range of $f(x) = \cos x$, where x is a real number.

20. If θ is an acute angle of a right triangle and $\tan \theta = \dfrac{3}{4}$, find $\sin \theta$.

TRIGONOMETRIC IDENTITIES AND EQUATIONS

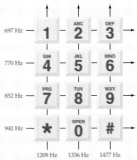

Source: Data in chart from
http://www.howstuffworks.
com/telephone2.htm and also found at
http://hyperarchive.lcs.mit.
edu/telecom-archives/
tribute/touch_tone_info.html

Touch-Tone Phones and Trigonometry

The dial tone emitted by a telephone is produced by adding a 350-hertz (cycles per second) sound to a 440-hertz sound. An equation that models the dial tone is

$$p_1(t) = \cos(2\pi \cdot 440t) + \cos(2\pi \cdot 350t)$$

where p is the pressure on the eardrum and t is the time in seconds. Concepts from this chapter can be used to show that the dial tone can also be modeled by

$$p_2(t) = 2\cos(790\pi t)\cos(90\pi t)$$

The equation $p_1(t) = p_2(t)$ is called an *identity* because the left side of the equation equals the right side for all domain values t. Use a graphing utility to graph p_1 and p_2 on the interval $[0, 0.1]$ to see that they appear to represent the same function.

Every tone made on a touch-tone phone is produced by adding a pair of sounds. The chart to the left shows the frequencies used for each key. For instance, the sound emitted by pressing 3 on the keypad is produced by adding a 1477-hertz sound to a 697-hertz sound. An equation that models this tone is

$$p(t) = \cos(2\pi \cdot 1477t) + \cos(2\pi \cdot 697t)$$

In **Exercises 77 and 78 on page 240,** you will determine trigonometric equations that can be used to model some of the other tones that can be produced on a touch-tone phone.

FOCUS ON PROBLEM SOLVING

The Importance of Experimentation

In this chapter you will need to verify several trigonometric identities. The equation $\sin 2\alpha = 2 \sin \alpha \cos \alpha$ is an example of a trigonometric identity. The equation is an identity because $\sin 2\alpha$ is equal to $2 \sin \alpha \cos \alpha$ for all values of α.

Trigonometry identities are very useful. In many applications a solution can be found by making use of an identity that allows you to replace a given function with an equivalent function.

To verify an identity, you need to show that one side of the identity can be rewritten in an equivalent form that is identical to the other side. The guidelines on pages 207–208 list several procedures that may be used in a verification of a trigonometric identity. The verification process is similar to the process of solving a crossword puzzle. You examine the clues and make a guess at a possible solution. If your first guess does not work, then you make additional guesses until a proper solution is found. In verifying an identity, try one or more of the procedures listed in the guidelines. If this does not produce a verification, then experiment, by trying another approach, until a verification is found.

VERIFICATION OF TRIGONOMETRIC IDENTITIES

- FUNDAMENTAL TRIGONOMETRIC IDENTITIES
- VERIFICATION OF TRIGONOMETRIC IDENTITIES

● FUNDAMENTAL TRIGONOMETRIC IDENTITIES

The domain of an equation consists of all values of the variable for which every term is defined. For example, the domain of

$$\frac{\sin x \cos x}{\sin x} = \cos x \tag{1}$$

includes all real numbers x except $x = n\pi$, where n is an integer, because $\sin x = 0$ for $x = n\pi$, and division by 0 is undefined. An **identity** is an equation that is true for all of its domain values. **Table 3.1** lists identities that were introduced earlier.

TABLE 3.1   Fundamental Trigonometric Identities

Reciprocal identities	$\sin x = \dfrac{1}{\csc x}$	$\cos x = \dfrac{1}{\sec x}$	$\tan x = \dfrac{1}{\cot x}$
Ratio identities	$\tan x = \dfrac{\sin x}{\cos x}$	$\cot x = \dfrac{\cos x}{\sin x}$	
Pythagorean identities	$\sin^2 x + \cos^2 x = 1$	$\tan^2 x + 1 = \sec^2 x$	$1 + \cot^2 x = \csc^2 x$
Odd-even identities	$\sin(-x) = -\sin x$ $\cos(-x) = \cos x$	$\tan(-x) = -\tan x$ $\cot(-x) = -\cot x$	$\sec(-x) = \sec x$ $\csc(-x) = -\csc x$

● VERIFICATION OF TRIGONOMETRIC IDENTITIES

To verify an identity, we show that one side of the identity can be rewritten in a form that is identical to the other side. There is no one method that can be used to verify every identity; however, the following guidelines should prove useful.

Guidelines for Verifying Trigonometric Identities

- If one side of the identity is more complex than the other, then it is generally best to try first to simplify the more complex side until it becomes identical to the other side.

- Perform indicated operations such as adding fractions or squaring a binomial. Also be aware of any factorization that may help you to achieve your goal of producing the expression on the other side.

- Make use of previously established identities that enable you to rewrite one side of the identity in an equivalent form.

- Rewrite one side of the identity so that it involves only sines and/or cosines.

Continued ▶

● Rewrite one side of the identity in terms of a single trigonometric function.

● Multiplying both the numerator and the denominator of a fraction by the same factor (such as the conjugate of the denominator or the conjugate of the numerator) may get you closer to your goal.

● Keep your goal in mind. Does it involve products, quotients, sums, radicals, or powers? Knowing exactly what your goal is may provide the insight you need to verify the identity.

EXAMPLE 1   **Determine Whether an Equation is an Identity**

Determine whether each equation is an identity. If the equation is an identity, then verify the identity. If the equation is not an identity, then find a value of the domain for which the left side of the equation is not equal to the right side.

a. $\sin\left(x + \dfrac{\pi}{6}\right) = \sin x + \sin\dfrac{\pi}{6}$

b. $(\sin x + \cos x)^2 = 2\sin x \cos x + 1$

Solution

a. The graphs of $f(x) = \sin\left(x + \dfrac{\pi}{6}\right)$ and $g(x) = \sin x + \sin\dfrac{\pi}{6}$ are shown in **Figure 3.1.** Because the graphs are not identical, we know that the equation is not an identity. This can be further confirmed by letting $x = \dfrac{\pi}{3}$ and observing that

$$f\left(\frac{\pi}{3}\right) = \sin\left(\frac{\pi}{3} + \frac{\pi}{6}\right) = \sin\frac{\pi}{2} = 1$$

whereas

$$g\left(\frac{\pi}{3}\right) = \sin\frac{\pi}{3} + \sin\frac{\pi}{6} = \frac{\sqrt{3}}{2} + \frac{1}{2} \approx 1.366$$

Thus $\sin\left(x + \dfrac{\pi}{6}\right) \neq \sin x + \sin\dfrac{\pi}{6}$ for $x = \dfrac{\pi}{3}$. Therefore, the equation is not an identity.

b. The graphs of $f(x) = (\sin x + \cos x)^2$ and $g(x) = 2\sin x \cos x + 1$ are shown in **Figure 3.2.** The graphs appear to be identical. To verify that $(\sin x + \cos x)^2 = 2\sin x \cos x + 1$ is an identity, we expand $(\sin x + \cos x)^2$ as shown below.

$$(\sin x + \cos x)^2 = \sin^2 x + 2\sin x \cos x + \cos^2 x$$
$$= 2\sin x \cos x + (\sin^2 x + \cos^2 x)$$
$$= 2\sin x \cos x + 1 \qquad \bullet\ \sin^2 x + \cos^2 x = 1$$

We have rewritten the left side in an equivalent form that is identical to the right side. Thus $(\sin x + \cos x)^2 = 2\sin x \cos x + 1$ is an identity.

▶ **TRY EXERCISE 2, PAGE 211**

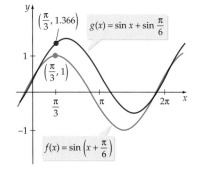

FIGURE 3.1

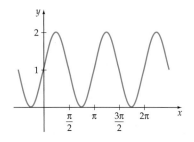

$f(x) = (\sin x + \cos x)^2$
$g(x) = 2\sin x \cos x + 1$

FIGURE 3.2

❷ QUESTION Is $\cos(-x) = \cos x$ an identity?

EXAMPLE 2   Verify an Identity

Verify the identity $1 - 2\sin^2 x = 2\cos^2 x - 1$.

Solution

Rewrite the right side of the equation.

$$2\cos^2 x - 1 = 2(1 - \sin^2 x) - 1 \qquad \bullet\ \cos^2 x = 1 - \sin^2 x$$
$$= 2 - 2\sin^2 x - 1$$
$$= 1 - 2\sin^2 x$$

▶ **TRY EXERCISE 22, PAGE 212**

---

take note

Each of the Pythagorean identities can be written in several different forms. For instance,

$$\sin^2 x + \cos^2 x = 1$$

also can be written as

$$\sin^2 x = 1 - \cos^2 x$$

and as

$$\cos^2 x = 1 - \sin^2 x$$

Figure 3.3 shows the graph of $f(x) = 1 - 2\sin^2 x$ and the graph of $g(x) = 2\cos^2 x - 1$ on the same coordinate axes. The fact that the graphs appear to be identical on the interval $[-2\pi, 2\pi]$ supports the verification in Example 2.

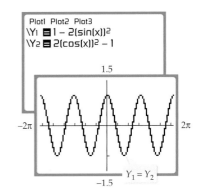

FIGURE 3.3

EXAMPLE 3   Factor to Verify an Identity

Verify the identity $\csc^2 x - \cos^2 x \csc^2 x = 1$.

Solution

Simplify the left side of the equation.

$$\csc^2 x - \cos^2 x \csc^2 x = \csc^2 x(1 - \cos^2 x) \qquad \bullet\ \text{Factor out } \csc^2 x.$$
$$= \csc^2 x \sin^2 x$$
$$= \frac{1}{\sin^2 x} \cdot \sin^2 x = 1 \qquad \bullet\ \csc^2 x = \frac{1}{\sin^2 x}$$

▶ **TRY EXERCISE 34, PAGE 212**

In the next example we make use of the guideline that indicates that it may be useful to multiply both the numerator and the denominator of a fraction by the same factor.

---

❷ ANSWER Yes, $\cos(-x) = \cos x$ is one of the odd-even identities shown in **Table 3.1.**

EXAMPLE 4 **Multiply by a Conjugate to Verify an Identity**

Verify the identity $\dfrac{\sin x}{1 + \cos x} = \dfrac{1 - \cos x}{\sin x}$.

Solution

Multiply the numerator and denominator of the left side of the identity by the conjugate of $1 + \cos x$, which is $1 - \cos x$.

$$\frac{\sin x}{1 + \cos x} = \frac{\sin x}{1 + \cos x} \cdot \frac{1 - \cos x}{1 - \cos x} = \frac{\sin x(1 - \cos x)}{1 - \cos^2 x}$$

$$= \frac{\sin x(1 - \cos x)}{\sin^2 x} = \frac{1 - \cos x}{\sin x}$$

> **take note**
>
> The sum $a + b$ and the difference $a - b$ are called conjugates of each other.

▶ **TRY EXERCISE 44, PAGE 212**

EXAMPLE 5 **Change to Sines and Cosines to Verify an Identity**

Verify the identity $\dfrac{\sin x + \tan x}{1 + \cos x} = \tan x$.

Solution

Rewrite the left side of the identity in terms of sines and cosines.

$$\frac{\sin x + \tan x}{1 + \cos x} = \frac{\sin x + \dfrac{\sin x}{\cos x}}{1 + \cos x} \qquad \bullet \tan x = \frac{\sin x}{\cos x}$$

$$= \frac{\dfrac{\sin x \cos x + \sin x}{\cos x}}{1 + \cos x} \qquad \bullet \textbf{Write the terms in the numerator with a common denominator.}$$

$$= \frac{\sin x \cos x + \sin x}{\cos x(1 + \cos x)} \qquad \bullet \textbf{Simplify.}$$

$$= \frac{\sin x(\cancel{1 + \cos x})}{\cos x(\cancel{1 + \cos x})}$$

$$= \tan x$$

▶ **TRY EXERCISE 54, PAGE 212**

 TOPICS FOR DISCUSSION

1. Explain why $\tan = \dfrac{\sin}{\cos}$ is not an identity.

2. Is $\cos |x| = |\cos x|$ an identity? Explain. What about $\cos |x| = \cos x$? Explain.

3. The identity $\sin^2 x + \cos^2 x = 1$ is one of the Pythagorean identities. What are the other two Pythagorean identities, and how are they derived?

4. The graph of $y = \sin\dfrac{x}{2}$ for $0 \le x \le 2\pi$ is shown on the left below. The graph

of $y = \sqrt{\dfrac{1 - \cos x}{2}}$ for $0 \le x \le 2\pi$ is shown on the right below.

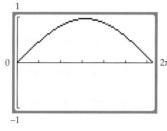

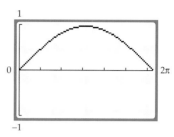

$$y = \sin\frac{x}{2}$$

$$y = \sqrt{\frac{1 - \cos x}{2}}$$

The graphs appear identical, but the equation

$$\sin\frac{x}{2} = \sqrt{\frac{1 - \cos x}{2}}$$

is not an identity. Explain.

EXERCISE SET 3.1

In Exercises 1 to 10, verify that the equation is *not* an identity by finding an *x*-value for which the left side of the equation is not equal to the right side.

1. $(\sin x + \cos x)^2 = \sin^2 x + \cos^2 x$

▶ **2.** $\tan 2x = 2 \tan x$

3. $\cos(x + 30°) = \cos x + \cos 30°$

4. $\sqrt{1 - \sin^2 x} = \cos x$

5. $\tan^4 x - \sec^4 x = \tan^2 x + \sec^2 x$

6. $\sqrt{1 + \tan^2 x} = \sec x$

7. $\tan^4 x - 1 = \sec^2 x$

8. $\sin^3 x + \cos^3 x = (\sin x + \cos x)(1 + \sin x \cos x)$

9. $\cot x \csc x \sec x = 1$

10. $\sin^4 x + \cos^2 x = 1 + \sin^2 x$

In Exercises 11 to 66, verify each identity.

11. $\tan x \csc x \cos x = 1$

12. $\sin x \cot x \sec x = 1$

13. $\dfrac{4 \sin^2 x - 1}{2 \sin x + 1} = 2 \sin x - 1$

14. $\dfrac{\sin^2 x - 2 \sin x + 1}{\sin x - 1} = \sin x - 1$

15. $(\sin x - \cos x)(\sin x + \cos x) = 1 - 2 \cos^2 x$

16. $(\tan x)(1 - \cot x) = \tan x - 1$

17. $\dfrac{1}{\sin x} - \dfrac{1}{\cos x} = \dfrac{\cos x - \sin x}{\sin x \cos x}$

18. $\dfrac{1}{\sin x} + \dfrac{3}{\cos x} = \dfrac{\cos x + 3 \sin x}{\sin x \cos x}$

19. $\dfrac{\cos x}{1 - \sin x} = \sec x + \tan x$

20. $\dfrac{\sin x}{1 - \cos x} = \csc x + \cot x$

21. $\dfrac{1 - \tan^4 x}{\sec^2 x} = 1 - \tan^2 x$

▶ **22.** $\sin^4 x - \cos^4 x = \sin^2 x - \cos^2 x$

23. $\dfrac{1 + \tan^3 x}{1 + \tan x} = 1 - \tan x + \tan^2 x$

24. $\dfrac{\cos x \tan x - \sin x}{\cot x} = 0$

25. $\dfrac{\sin x - 2 + \dfrac{1}{\sin x}}{\sin x - \dfrac{1}{\sin x}} = \dfrac{\sin x - 1}{\sin x + 1}$

26. $\dfrac{\sin x}{1 - \cos x} - \dfrac{\sin x}{1 + \cos x} = 2 \cot x$

27. $(\sin x + \cos x)^2 = 1 + 2 \sin x \cos x$

28. $(\tan x + 1)^2 = \sec^2 x + 2 \tan x$

29. $\dfrac{\cos x}{1 + \sin x} = \sec x - \tan x$

30. $\dfrac{\sin x}{1 + \cos x} = \csc x - \cot x$

31. $\csc x = \dfrac{\cot x + \tan x}{\sec x}$

32. $\sec x = \dfrac{\cot x + \tan x}{\csc x}$

33. $\dfrac{\cos x \tan x + 2 \cos x - \tan x - 2}{\tan x + 2} = \cos x - 1$

▶ **34.** $\dfrac{2 \sin x \cot x + \sin x - 4 \cot x - 2}{2 \cot x + 1} = \sin x - 2$

35. $\sec x - \tan x = \dfrac{1 - \sin x}{\cos x}$

36. $\cot x - \csc x = \dfrac{\cos x - 1}{\sin x}$

37. $\sin^2 x - \cos^2 x = 2 \sin^2 x - 1$

38. $\sin^2 x - \cos^2 x = 1 - 2 \cos^2 x$

39. $\dfrac{1}{\sin^2 x} + \dfrac{1}{\cos^2 x} = \csc^2 x \sec^2 x$

40. $\dfrac{1}{\tan^2 x} - \dfrac{1}{\cot^2 x} = \csc^2 x - \sec^2 x$

41. $\sec x - \cos x = \sin x \tan x$

42. $\tan x + \cot x = \sec x \csc x$

43. $\dfrac{\dfrac{1}{\sin x} + 1}{\dfrac{1}{\sin x} - 1} = \tan^2 x + 2 \tan x \sec x + \sec^2 x$

▶ **44.** $\dfrac{\dfrac{1}{\sin x} + \dfrac{1}{\cos x}}{\dfrac{1}{\sin x} - \dfrac{1}{\cos x}} = \dfrac{\cos^2 x - \sin^2 x}{1 - 2 \cos x \sin x}$

45. $\sin^4 x - \cos^4 x = 2 \sin^2 x - 1$

46. $\sin^6 x + \cos^6 x = \sin^4 x - \sin^2 x \cos^2 x + \cos^4 x$

47. $\dfrac{1}{1 - \cos x} = \dfrac{1 + \cos x}{\sin^2 x}$

48. $1 + \sin x = \dfrac{\cos^2 x}{1 - \sin x}$

49. $\dfrac{\sin x}{1 - \sin x} - \dfrac{\cos x}{1 - \sin x} = \dfrac{1 - \cot x}{\csc x - 1}$

50. $\dfrac{\tan x}{1 + \tan x} - \dfrac{\cot x}{1 + \tan x} = 1 - \cot x$

51. $\dfrac{1}{1 + \cos x} - \dfrac{1}{1 - \cos x} = -2 \cot x \csc x$

52. $\dfrac{1}{1 - \sin x} - \dfrac{1}{1 + \sin x} = 2 \tan x \sec x$

53. $\dfrac{\dfrac{1}{\sin x} + \csc x}{\dfrac{1}{\sin x} - \sin x} = \dfrac{2}{\cos^2 x}$

▶ **54.** $\dfrac{\dfrac{1}{\tan x} + \cot x}{\dfrac{1}{\tan x} + \tan x} = \dfrac{2}{\sec^2 x}$

55. $\dfrac{\cot x}{1 + \csc x} + \dfrac{1 + \csc x}{\cot x} = 2 \sec x$

56. $\sec^2 x - \csc^2 x = \dfrac{\tan x - \cot x}{\sin x \cos x}$

57. $\sqrt{\dfrac{1 + \sin x}{1 - \sin x}} = \dfrac{1 + \sin x}{\cos x}, \quad \cos x > 0$

58. $\dfrac{\cos x + \cot x \sin x}{\cot x} = 2 \sin x$

59. $\dfrac{\sin^3 x + \cos^3 x}{\sin x + \cos x} = 1 - \sin x \cos x$

60. $\dfrac{1 - \sin x}{1 + \sin x} - \dfrac{1 + \sin x}{1 - \sin x} = -4 \sec x \tan x$

61. $\dfrac{\sec x - 1}{\sec x + 1} - \dfrac{\sec x + 1}{\sec x - 1} = -4 \csc x \cot x$

62. $\dfrac{1}{1 - \cos x} - \dfrac{\cos x}{1 + \cos x} = 2 \csc^2 x - 1$

63. $\dfrac{1 + \sin x}{\cos x} - \dfrac{\cos x}{1 - \sin x} = 0$

64. $(\sin x + \cos x + 1)^2 = 2(\sin x + 1)(\cos x + 1)$

65. $\dfrac{\sec x + \tan x}{\sec x - \tan x} = \dfrac{(\sin x + 1)^2}{\cos^2 x}$

66. $\dfrac{\sin^3 x - \cos^3 x}{\sin x + \cos x} = \dfrac{\csc^2 x - \cot x - 2 \cos^2 x}{1 - \cot^2 x}$

In Exercises 67 to 74, compare the graphs of each side of the equation to predict whether the equation is an identity.

67. $\sin 2x = 2 \sin x \cos x$ **68.** $\sin^2 x + \cos^2 x = 1$

69. $\sin x + \cos x = \sqrt{2} \sin\left(x + \dfrac{\pi}{4}\right)$

70. $\cos 2x = 2 \cos^2 x - 1$

71. $\cos\left(x + \dfrac{\pi}{3}\right) = \cos x \cos \dfrac{\pi}{3} - \sin x \sin \dfrac{\pi}{3}$

72. $\cos\left(x - \dfrac{\pi}{4}\right) = \cos x \cos \dfrac{\pi}{4} + \sin x \sin \dfrac{\pi}{4}$

73. $\sin\left(x + \dfrac{\pi}{6}\right) = \sin x \sin \dfrac{\pi}{6} + \cos x \cos \dfrac{\pi}{6}$

74. $\sin\left(x - \dfrac{\pi}{3}\right) = \sin x \cos \dfrac{\pi}{3} + \cos x \sin \dfrac{\pi}{3}$

CONNECTING CONCEPTS

In Exercises 75 to 80, verify the identity.

75. $\dfrac{1 - \sin x + \cos x}{1 + \sin x + \cos x} = \dfrac{\cos x}{\sin x + 1}$

76. $\dfrac{1 - \tan x + \sec x}{1 + \tan x - \sec x} = \dfrac{1 + \sec x}{\tan x}$

77. $\dfrac{2 \sin^4 x + 2 \sin^2 x \cos^2 x - 3 \sin^2 x - 3 \cos^2 x}{2 \sin^2 x}$

$= 1 - \dfrac{3}{2} \csc^2 x$

78. $\dfrac{4 \tan x \sec^2 x - 4 \tan x - \sec^2 x + 1}{4 \tan^3 x - \tan^2 x} = 1$

79. $\dfrac{\sin x(\tan x + 1) - 2 \tan x \cos x}{\sin x - \cos x} = \tan x$

80. $\dfrac{\sin^2 x \cos x + \cos^3 x - \sin^3 x \cos x - \sin x \cos^3 x}{1 - \sin^2 x}$

$= \dfrac{\cos x}{1 + \sin x}$

81. Verify the identity $\sin^4 x + \cos^4 x = 1 - 2 \sin^2 x \cos^2 x$ by completing the square of the left side of the identity.

82. Verify the identity $\tan^4 x + \sec^4 x = 1 + 2 \tan^2 x \sec^2 x$ by completing the square of the left side of the identity.

─────── *PREPARE FOR SECTION 3.2* ───────────────────────

83. Compare $\cos(\alpha - \beta)$ and $\cos\alpha\cos\beta + \sin\alpha\sin\beta$ for $\alpha = \dfrac{\pi}{2}$ and $\beta = \dfrac{\pi}{6}$. [2.2]

84. Compare $\sin(\alpha + \beta)$ and $\sin\alpha\cos\beta + \cos\alpha\sin\beta$ for $\alpha = \dfrac{\pi}{2}$ and $\beta = \dfrac{\pi}{3}$. [2.2]

85. Compare $\sin(90° - \theta)$ and $\cos\theta$ for $\theta = 30°$, $\theta = 45°$, and $\theta = 120°$. [2.2]

86. Compare $\tan\left(\dfrac{\pi}{2} - \theta\right)$ and $\cot\theta$ for $\theta = \dfrac{\pi}{6}$, $\theta = \dfrac{\pi}{4}$, and $\theta = \dfrac{4\pi}{3}$. [2.2]

87. Compare $\tan(\alpha - \beta)$ and $\dfrac{\tan\alpha - \tan\beta}{1 + \tan\alpha\tan\beta}$ for $\alpha = \dfrac{\pi}{3}$ and $\beta = \dfrac{\pi}{6}$. [2.2]

88. Find the value of $\sin[(2k + 1)\pi]$, where k is any arbitrary integer. [2.2]

─────── *PROJECTS* ───────────────────────────────────

1. **GRADING A QUIZ** Suppose that you are a teacher's assistant. You are to assist the teacher of a trigonometry class by grading a four-question quiz. Each question asks the student to find a trigonometric expression that models a given application. The teacher has prepared an answer key. These answers are shown in the next column. A student gives as answers the expressions shown in the far right column. Determine for which problems the student has given a correct response.

Answer Key	**Student's Response**
1. $\csc x \sec x$	1. $\cot x + \tan x$
2. $\cos^2 x$	2. $(1 + \sin x)(1 - \sin x)$
3. $\cos x \cot x$	3. $\csc x - \sec x$
4. $\csc x \cot x$	4. $\sin x(\cot x + \cot^3 x)$

SUM, DIFFERENCE, AND COFUNCTION IDENTITIES

IDENTITIES THAT INVOLVE ($\alpha \pm \beta$)

Each identity in Section 3.1 involved only one variable. We now consider identities that involve a trigonometric function of the sum or difference of two variables.

Sum and Difference Identities

$$\cos(\alpha - \beta) = \cos \alpha \cos \beta + \sin \alpha \sin \beta$$

$$\cos(\alpha + \beta) = \cos \alpha \cos \beta - \sin \alpha \sin \beta$$

$$\sin(\alpha - \beta) = \sin \alpha \cos \beta - \cos \alpha \sin \beta$$

$$\sin(\alpha + \beta) = \sin \alpha \cos \beta + \cos \alpha \sin \beta$$

$$\tan(\alpha + \beta) = \frac{\tan \alpha + \tan \beta}{1 - \tan \alpha \tan \beta}$$

$$\tan(\alpha - \beta) = \frac{\tan \alpha - \tan \beta}{1 + \tan \alpha \tan \beta}$$

To establish the identity for $\cos(\alpha - \beta)$, we make use of the unit circle shown in **Figure 3.4.** The angles α and β are drawn in standard position, with OA and OB as the terminal sides of α and β, respectively. The coordinates of A are ($\cos \alpha, \sin \alpha$), and the coordinates of B are ($\cos \beta, \sin \beta$). The angle ($\alpha - \beta$) is formed by the terminal sides of the angles α and β (angle AOB).

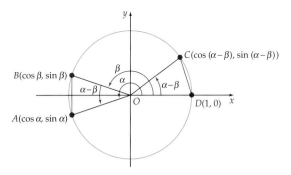

FIGURE 3.4

An angle equal in measure to angle ($\alpha - \beta$) is placed in standard position in the same figure (angle COD). From geometry, if two central angles of a circle have the same measure, then their chords are also equal in measure. Thus the chords AB and CD are equal in length. Using the distance formula, we can calculate the lengths of the chords AB and CD.

$$d(A, B) = \sqrt{(\cos \alpha - \cos \beta)^2 + (\sin \alpha - \sin \beta)^2}$$

$$d(C, D) = \sqrt{[\cos(\alpha - \beta) - 1]^2 + [\sin(\alpha - \beta) - 0]^2}$$

Because $d(A, B) = d(C, D)$, we have

$$\sqrt{(\cos \alpha - \cos \beta)^2 + (\sin \alpha - \sin \beta)^2} = \sqrt{[\cos(\alpha - \beta) - 1]^2 + [\sin(\alpha - \beta)]^2}$$

Squaring each side of the equation and simplifying, we obtain

$$(\cos \alpha - \cos \beta)^2 + (\sin \alpha - \sin \beta)^2 = [\cos(\alpha - \beta) - 1]^2 + [\sin(\alpha - \beta)]^2$$

$$\cos^2 \alpha - 2 \cos \alpha \cos \beta + \cos^2 \beta + \sin^2 \alpha - 2 \sin \alpha \sin \beta + \sin^2 \beta$$

$$= \cos^2(\alpha - \beta) - 2 \cos(\alpha - \beta) + 1 + \sin^2(\alpha - \beta)$$

$$\cos^2 \alpha + \sin^2 \alpha + \cos^2 \beta + \sin^2 \beta - 2 \cos \alpha \cos \beta - 2 \sin \alpha \sin \beta$$

$$= \cos^2(\alpha - \beta) + \sin^2(\alpha - \beta) + 1 - 2 \cos(\alpha - \beta)$$

Simplifying by using $\sin^2 \theta + \cos^2 \theta = 1$, we have

$$2 - 2 \sin \alpha \sin \beta - 2 \cos \alpha \cos \beta = 2 - 2 \cos(\alpha - \beta)$$

Solving for $\cos(\alpha - \beta)$ gives us

$$\cos(\alpha - \beta) = \cos \alpha \cos \beta + \sin \alpha \sin \beta$$

To derive an identity for $\cos(\alpha + \beta)$, write $\cos(\alpha + \beta)$ as $\cos[\alpha - (-\beta)]$.

$$\cos(\alpha + \beta) = \cos[\alpha - (-\beta)] = \cos \alpha \cos(-\beta) + \sin \alpha \sin(-\beta)$$

Recall that $\cos(-\beta) = \cos \beta$ and $\sin(-\beta) = -\sin \beta$. Substituting into the previous equation, we obtain the identity

$$\cos(\alpha + \beta) = \cos \alpha \cos \beta - \sin \alpha \sin \beta$$

EXAMPLE 1 **Evaluate a Trigonometric Expression**

Use an identity to find the *exact* value of $\cos(60° - 45°)$.

Solution

Use the identity $\cos(\alpha - \beta) = \cos \alpha \cos \beta + \sin \alpha \sin \beta$ with $\alpha = 60°$ and $\beta = 45°$.

$\cos(60° - 45°) = \cos 60° \cos 45° + \sin 60° \sin 45°$ • Substitute.

$$= \left(\frac{1}{2}\right)\left(\frac{\sqrt{2}}{2}\right) + \left(\frac{\sqrt{3}}{2}\right)\left(\frac{\sqrt{2}}{2}\right)$$ • Evaluate each factor.

$$= \frac{\sqrt{2}}{4} + \frac{\sqrt{6}}{4}$$ • Simplify.

$$= \frac{\sqrt{2} + \sqrt{6}}{4}$$

▶ **TRY EXERCISE 4, PAGE 222**

● **COFUNCTIONS**

Any pair of trigonometric functions f and g for which

$$f(x) = g(90° - x) \quad \text{and} \quad g(x) = f(90° - x)$$

are said to be **cofunctions**.

To visualize the cofunction identities, consider the right triangle shown in the following figure.

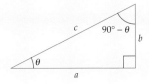

If θ is the degree measure of one of the acute angles, then the degree measure of the other acute angle is $(90° - \theta)$. Using the definitions of the trigonometric functions gives us

$$\sin \theta = \frac{b}{c} = \cos(90° - \theta)$$

$$\tan \theta = \frac{b}{a} = \cot(90° - \theta)$$

$$\sec \theta = \frac{c}{a} = \csc(90° - \theta)$$

These identities state that the value of a trigonometric function of θ is equal to the cofunction of the complement of θ.

Cofunction Identities

$$\sin(90° - \theta) = \cos \theta \qquad \cos(90° - \theta) = \sin \theta$$

$$\tan(90° - \theta) = \cot \theta \qquad \cot(90° - \theta) = \tan \theta$$

$$\sec(90° - \theta) = \csc \theta \qquad \csc(90° - \theta) = \sec \theta$$

If θ is in radian measure, replace 90° with $\dfrac{\pi}{2}$.

To verify that the sine function and the cosine function are cofunctions, we make use of the identity for $\cos(\alpha - \beta)$.

$$\cos(90° - \beta) = \cos 90° \cos \beta + \sin 90° \sin \beta$$
$$= 0 \cdot \cos \beta + 1 \cdot \sin \beta$$

which gives

$$\cos(90° - \beta) = \sin \beta$$

Thus the sine of an angle is equal to the cosine of its complement. Using $\cos(90° - \beta) = \sin \beta$ with $\beta = 90° - \alpha$, we have

$$\cos \alpha = \cos[90° - (90° - \alpha)] = \sin(90° - \alpha)$$

Therefore,

$$\cos \alpha = \sin(90° - \alpha)$$

We can use the ratio identities to show that the tangent and cotangent functions are cofunctions.

$$\tan(90° - \theta) = \frac{\sin(90° - \theta)}{\cos(90° - \theta)} = \frac{\cos \theta}{\sin \theta} = \cot \theta$$

$$\cot(90° - \theta) = \frac{\cos(90° - \theta)}{\sin(90° - \theta)} = \frac{\sin \theta}{\cos \theta} = \tan \theta$$

The secant and cosecant functions are also cofunctions.

EXAMPLE 2   Write an Equivalent Expression

Use a cofunction identity to write an equivalent expression for sin 20°.

Solution

The value of a given trigonometric function of θ, measured in degrees, is equal to its cofunction of $90° - \theta$. Thus

$$\sin 20° = \cos(90° - 20°)$$
$$= \cos 70°$$

▶ **TRY EXERCISE 20, PAGE 222**

● ADDITIONAL SUM AND DIFFERENCE IDENTITIES

We can use the cofunction identities to verify the remaining sum and difference identities. To derive an identity for $\sin(\alpha + \beta)$, substitute $\alpha + \beta$ for θ in the cofunction identity $\sin \theta = \cos(90° - \theta)$.

$$\sin \theta = \cos(90° - \theta)$$
$$\sin(\alpha + \beta) = \cos[90° - (\alpha + \beta)] \qquad \text{• Replace } \theta \text{ with } \alpha + \beta.$$
$$= \cos[(90° - \alpha) - \beta] \qquad \text{• Rewrite as the difference of two angles.}$$
$$= \cos(90° - \alpha) \cos \beta + \sin(90° - \alpha) \sin \beta$$
$$= \sin \alpha \cos \beta + \cos \alpha \sin \beta$$

Therefore,

$$\sin(\alpha + \beta) = \sin \alpha \cos \beta + \cos \alpha \sin \beta$$

We also can derive an identity for $\sin(\alpha - \beta)$ by rewriting $(\alpha - \beta)$ as $[\alpha + (-\beta)]$.

$$\sin(\alpha - \beta) = \sin[\alpha + (-\beta)]$$
$$= \sin \alpha \cos(-\beta) + \cos \alpha \sin(-\beta)$$
$$= \sin \alpha \cos \beta - \cos \alpha \sin \beta \qquad \text{• } \cos(-\beta) = \cos \beta$$
$$\qquad\qquad\qquad\qquad\qquad\qquad\qquad \sin(-\beta) = -\sin \beta$$

Thus

$$\sin(\alpha - \beta) = \sin \alpha \cos \beta - \cos \alpha \sin \beta$$

The identity for $\tan(\alpha + \beta)$ is a result of the identity $\tan \theta = \dfrac{\sin \theta}{\cos \theta}$ and the identities for $\sin(\alpha + \beta)$ and $\cos(\alpha + \beta)$.

$$\tan(\alpha + \beta) = \frac{\sin(\alpha + \beta)}{\cos(\alpha + \beta)} = \frac{\sin \alpha \cos \beta + \cos \alpha \sin \beta}{\cos \alpha \cos \beta - \sin \alpha \sin \beta}$$

$$= \frac{\dfrac{\sin \alpha \cos \beta}{\cos \alpha \cos \beta} + \dfrac{\cos \alpha \sin \beta}{\cos \alpha \cos \beta}}{\dfrac{\cos \alpha \cos \beta}{\cos \alpha \cos \beta} - \dfrac{\sin \alpha \sin \beta}{\cos \alpha \cos \beta}}$$

• Multiply both the numerator and the denominator by $\dfrac{1}{\cos \alpha \cos \beta}$ and simplify.

Therefore,

$$\tan(\alpha + \beta) = \frac{\tan \alpha + \tan \beta}{1 - \tan \alpha \tan \beta}$$

The tangent function is an odd function, so $\tan(-\theta) = -\tan \theta$. Rewriting $(\alpha - \beta)$ as $[\alpha + (-\beta)]$ enables us to derive an identity for $\tan(\alpha - \beta)$.

$$\tan(\alpha - \beta) = \tan[\alpha + (-\beta)] = \frac{\tan \alpha + \tan(-\beta)}{1 - \tan \alpha \tan(-\beta)}$$

Therefore,

$$\tan(\alpha - \beta) = \frac{\tan \alpha - \tan \beta}{1 + \tan \alpha \tan \beta}$$

The sum and difference identities can be used to simplify some trigonometric expressions.

EXAMPLE 3   Simplify Trigonometric Expressions

Write each expression in terms of a single trigonometric function.

a. $\sin 5x \cos 3x - \cos 5x \sin 3x$ b. $\dfrac{\tan 4\alpha + \tan \alpha}{1 - \tan 4\alpha \tan \alpha}$

Solution

a. $\sin 5x \cos 3x - \cos 5x \sin 3x = \sin(5x - 3x) = \sin 2x$

b. $\dfrac{\tan 4\alpha + \tan \alpha}{1 - \tan 4\alpha \tan \alpha} = \tan(4\alpha + \alpha) = \tan 5\alpha$

▶ **TRY EXERCISE 26, PAGE 222**

EXAMPLE 4   Evaluate a Trigonometric Function

Given $\tan \alpha = -\dfrac{4}{3}$ for α in Quadrant II and $\tan \beta = -\dfrac{5}{12}$ for β in Quadrant IV, find $\sin(\alpha + \beta)$.

Solution

See **Figure 3.5.** Because $\tan \alpha = \dfrac{y}{x} = -\dfrac{4}{3}$ and the terminal side of α is in Quadrant II, $P_1(-3, 4)$ is a point on the terminal side of α. Similarly, $P_2(12, -5)$ is a point on the terminal side of β.

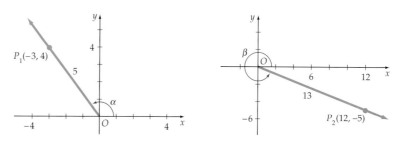

FIGURE 3.5

Using the Pythagorean Theorem, we find that the length of the line segment OP_1 is 5 and the length of OP_2 is 13.

$$\sin(\alpha + \beta) = \sin \alpha \cos \beta + \cos \alpha \sin \beta$$

$$= \frac{4}{5} \cdot \frac{12}{13} + \frac{-3}{5} \cdot \frac{-5}{13} = \frac{48}{65} + \frac{15}{65} = \frac{63}{65}$$

▶ **TRY EXERCISE 38, PAGE 222**

EXAMPLE 5   Verify an Identity

Verify the identity $\cos(\pi - \theta) = -\cos\theta$.

Solution

Use the identity for $\cos(\alpha - \beta)$.

$$\cos(\pi - \theta) = \cos\pi\cos\theta + \sin\pi\sin\theta = -1\cdot\cos\theta + 0\cdot\sin\theta = -\cos\theta$$

▶ **TRY EXERCISE 50, PAGE 223**

Figure 3.6 shows the graphs of $f(\theta) = \cos(\pi - \theta)$ and $g(\theta) = -\cos\theta$ on the same coordinate axes. The fact that the graphs appear to be identical supports the verification in Example 5.

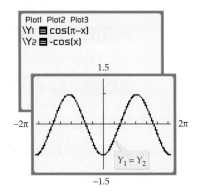

FIGURE 3.6

EXAMPLE 6   Verify an Identity

Verify the identity $\dfrac{\cos 4\theta}{\sin\theta} - \dfrac{\sin 4\theta}{\cos\theta} = \dfrac{\cos 5\theta}{\sin\theta\cos\theta}$.

Solution

Subtract the fractions on the left side of the equation.

$$\frac{\cos 4\theta}{\sin\theta} - \frac{\sin 4\theta}{\cos\theta} = \frac{\cos 4\theta\cos\theta - \sin 4\theta\sin\theta}{\sin\theta\cos\theta}$$

$$= \frac{\cos(4\theta + \theta)}{\sin\theta\cos\theta} = \frac{\cos 5\theta}{\sin\theta\cos\theta}$$ • **Use the identity for $\cos(\alpha + \beta)$.**

▶ **TRY EXERCISE 62, PAGE 223**

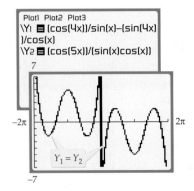

FIGURE 3.7

Figure 3.7 shows the graph of $f(\theta) = \dfrac{\cos 4\theta}{\sin\theta} - \dfrac{\sin 4\theta}{\cos\theta}$ and the graph of $g(\theta) = \dfrac{\cos 5\theta}{\sin\theta\cos\theta}$ on the same coordinate axes. The fact that the graphs appear to be identical supports the verification in Example 6.

• REDUCTION FORMULAS

The sum or difference identities can be used to write expressions such as

$$\sin(\theta + k\pi) \qquad \sin(\theta + 2k\pi) \qquad \text{and} \qquad \cos[\theta + (2k + 1)\pi]$$

where k is an integer, as expressions involving only $\sin\theta$ or $\cos\theta$. The resulting formulas are called reduction formulas.

EXAMPLE 7   **Find Reduction Formulas**

Write as a function involving only $\sin\theta$.

$$\sin[\theta + (2k + 1)\pi], \quad \text{where } k \text{ is an integer}$$

Solution

Applying the identity $\sin(\alpha + \beta) = \sin\alpha\cos\beta + \cos\alpha\sin\beta$ yields

$$\sin[\theta + (2k + 1)\pi] = \sin\theta\cos[(2k + 1)\pi] + \cos\theta\sin[(2k + 1)\pi]$$

If k is an integer, then $2k + 1$ is an odd integer. The cosine of any odd multiple of π equals -1, and the sine of any odd multiple of π is 0. This gives us

$$\sin[\theta + (2k + 1)\pi] = (\sin\theta)(-1) + (\cos\theta)(0) = -\sin\theta$$

Thus $\sin[\theta + (2k + 1)\pi] = -\sin\theta$ for any integer k.

▶ **TRY EXERCISE 74, PAGE 223**

❓ QUESTION   Is $\sin(\theta + 2k\pi) = \sin\theta$ a reduction formula?

 TOPICS FOR DISCUSSION

1. Does $\sin(\alpha + \beta) = \sin\alpha + \sin\beta$ for all values of α and β? If not, find nonzero values of α and β for which $\sin(\alpha + \beta) \neq \sin\alpha + \sin\beta$.

2. If k is an integer, then $2k + 1$ is an odd integer. Do you agree? Explain.

3. What are the trigonometric cofunction identities? Explain.

4. Is $\tan(\theta + k\pi) = \tan\theta$, where k is an integer, a reduction formula? Explain.

❓ ANSWER   Yes.

EXERCISE SET 3.2

In Exercises 1 to 18, find the exact value of the expression.

1. $\sin(45° + 30°)$

2. $\sin(330° + 45°)$

3. $\cos(45° - 30°)$

▶ **4.** $\cos(120° - 45°)$

5. $\tan(45° - 30°)$

6. $\tan(240° - 45°)$

7. $\sin\left(\dfrac{5\pi}{4} - \dfrac{\pi}{6}\right)$

8. $\sin\left(\dfrac{4\pi}{3} + \dfrac{\pi}{4}\right)$

9. $\cos\left(\dfrac{3\pi}{4} + \dfrac{\pi}{6}\right)$

10. $\cos\left(\dfrac{\pi}{4} - \dfrac{\pi}{3}\right)$

11. $\tan\left(\dfrac{\pi}{6} + \dfrac{\pi}{4}\right)$

12. $\tan\left(\dfrac{11\pi}{6} - \dfrac{\pi}{4}\right)$

13. $\cos 212° \cos 122° + \sin 212° \sin 122°$

14. $\sin 167° \cos 107° - \cos 167° \sin 107°$

15. $\sin\dfrac{5\pi}{12}\cos\dfrac{\pi}{4} - \cos\dfrac{5\pi}{12}\sin\dfrac{\pi}{4}$

16. $\cos\dfrac{\pi}{12}\cos\dfrac{\pi}{4} - \sin\dfrac{\pi}{12}\sin\dfrac{\pi}{4}$

17. $\dfrac{\tan\dfrac{7\pi}{12} - \tan\dfrac{\pi}{4}}{1 + \tan\dfrac{7\pi}{12}\tan\dfrac{\pi}{4}}$

18. $\dfrac{\tan\dfrac{\pi}{6} + \tan\dfrac{\pi}{3}}{1 - \tan\dfrac{\pi}{6}\tan\dfrac{\pi}{3}}$

In Exercises 19 to 24, use a cofunction identity to write an equivalent expression for the given value.

19. $\sin 42°$

▶ **20.** $\cos 80°$

21. $\tan 15°$

22. $\cot 2°$

23. $\sec 25°$

24. $\csc 84°$

In Exercises 25 to 36, write each expression in terms of a single trigonometric function.

25. $\sin 7x \cos 2x - \cos 7x \sin 2x$

▶ **26.** $\sin x \cos 3x + \cos x \sin 3x$

27. $\cos x \cos 2x + \sin x \sin 2x$

28. $\cos 4x \cos 2x - \sin 4x \sin 2x$

29. $\sin 7x \cos 3x - \cos 7x \sin 3x$

30. $\cos x \cos 5x - \sin x \sin 5x$

31. $\cos 4x \cos(-2x) - \sin 4x \sin(-2x)$

32. $\sin(-x)\cos 3x - \cos(-x)\sin 3x$

33. $\sin\dfrac{x}{3}\cos\dfrac{2x}{3} + \cos\dfrac{x}{3}\sin\dfrac{2x}{3}$

34. $\cos\dfrac{3x}{4}\cos\dfrac{x}{4} + \sin\dfrac{3x}{4}\sin\dfrac{x}{4}$

35. $\dfrac{\tan 3x + \tan 4x}{1 - \tan 3x \tan 4x}$

36. $\dfrac{\tan 2x - \tan 3x}{1 + \tan 2x \tan 3x}$

In Exercises 37 to 48, find the exact value of the given functions.

37. Given $\tan\alpha = -\dfrac{4}{3}$, α in Quadrant II, and $\tan\beta = \dfrac{15}{8}$, β in Quadrant III, find

    a. $\sin(\alpha - \beta)$　　**b.** $\cos(\alpha + \beta)$　　**c.** $\tan(\alpha - \beta)$

▶ **38.** Given $\tan\alpha = \dfrac{24}{7}$, α in Quadrant I, and $\sin\beta = -\dfrac{8}{17}$, β in Quadrant III, find

    a. $\sin(\alpha + \beta)$　　**b.** $\cos(\alpha + \beta)$　　**c.** $\tan(\alpha - \beta)$

39. Given $\sin\alpha = \dfrac{3}{5}$, α in Quadrant I, and $\cos\beta = -\dfrac{5}{13}$, β in Quadrant II, find

    a. $\sin(\alpha - \beta)$　　**b.** $\cos(\alpha + \beta)$　　**c.** $\tan(\alpha - \beta)$

40. Given $\sin\alpha = \dfrac{24}{25}$, α in Quadrant II, and $\cos\beta = -\dfrac{4}{5}$, β in Quadrant III, find

    a. $\cos(\beta - \alpha)$　　**b.** $\sin(\alpha + \beta)$　　**c.** $\tan(\alpha + \beta)$

41. Given $\sin \alpha = -\dfrac{4}{5}$, α in Quadrant III, and $\cos \beta = -\dfrac{12}{13}$, β in Quadrant II, find

    a. $\sin(\alpha - \beta)$ **b.** $\cos(\alpha + \beta)$ **c.** $\tan(\alpha + \beta)$

42. Given $\sin \alpha = -\dfrac{7}{25}$, α in Quadrant IV, and $\cos \beta = \dfrac{8}{17}$, β in Quadrant IV, find

    a. $\sin(\alpha + \beta)$ **b.** $\cos(\alpha - \beta)$ **c.** $\tan(\alpha + \beta)$

43. Given $\cos \alpha = \dfrac{15}{17}$, α in Quadrant I, and $\sin \beta = -\dfrac{3}{5}$, β in Quadrant III, find

    a. $\sin(\alpha + \beta)$ **b.** $\cos(\alpha - \beta)$ **c.** $\tan(\alpha - \beta)$

44. Given $\cos \alpha = -\dfrac{7}{25}$, α in Quadrant II, and $\sin \beta = -\dfrac{12}{13}$, β in Quadrant IV, find

    a. $\sin(\alpha + \beta)$ **b.** $\cos(\alpha + \beta)$ **c.** $\tan(\alpha - \beta)$

45. Given $\cos \alpha = -\dfrac{3}{5}$, α in Quadrant III, and $\sin \beta = \dfrac{5}{13}$, β in Quadrant I, find

    a. $\sin(\alpha - \beta)$ **b.** $\cos(\alpha + \beta)$ **c.** $\tan(\alpha + \beta)$

46. Given $\cos \alpha = \dfrac{8}{17}$, α in Quadrant IV, and $\sin \beta = -\dfrac{24}{25}$, β in Quadrant III, find

    a. $\sin(\alpha - \beta)$ **b.** $\cos(\alpha + \beta)$ **c.** $\tan(\alpha + \beta)$

47. Given $\sin \alpha = \dfrac{3}{5}$, α in Quadrant I, and $\tan \beta = \dfrac{5}{12}$, β in Quadrant III, find

    a. $\sin(\alpha + \beta)$ **b.** $\cos(\alpha - \beta)$ **c.** $\tan(\alpha - \beta)$

48. Given $\tan \alpha = \dfrac{15}{8}$, α in Quadrant I, and $\tan \beta = -\dfrac{7}{24}$, β in Quadrant IV, find

    a. $\sin(\alpha - \beta)$ **b.** $\cos(\alpha - \beta)$ **c.** $\tan(\alpha + \beta)$

In Exercises 49 to 72, verify the identity.

49. $\cos\left(\dfrac{\pi}{2} - \theta\right) = \sin \theta$ ▶ **50.** $\cos(\theta + \pi) = -\cos \theta$

51. $\sin\left(\theta + \dfrac{\pi}{2}\right) = \cos \theta$ **52.** $\sin(\theta + \pi) = -\sin \theta$

53. $\tan\left(\theta + \dfrac{\pi}{4}\right) = \dfrac{\tan \theta + 1}{1 - \tan \theta}$

54. $\tan 2\theta = \dfrac{2 \tan \theta}{1 - \tan^2 \theta}$ **55.** $\cos\left(\dfrac{3\pi}{2} - \theta\right) = -\sin \theta$

56. $\sin\left(\dfrac{3\pi}{2} + \theta\right) = -\cos \theta$ **57.** $\cot\left(\dfrac{\pi}{2} - \theta\right) = \tan \theta$

58. $\cot(\pi + \theta) = \cot \theta$ **59.** $\csc(\pi - \theta) = \csc \theta$

60. $\sec\left(\dfrac{\pi}{2} - \theta\right) = \csc \theta$

61. $\sin 6x \cos 2x - \cos 6x \sin 2x = 2 \sin 2x \cos 2x$

▶ **62.** $\cos 5x \cos 3x + \sin 5x \sin 3x = \cos^2 x - \sin^2 x$

63. $\cos(\alpha + \beta) + \cos(\alpha - \beta) = 2 \cos \alpha \cos \beta$

64. $\cos(\alpha - \beta) - \cos(\alpha + \beta) = 2 \sin \alpha \sin \beta$

65. $\sin(\alpha + \beta) + \sin(\alpha - \beta) = 2 \sin \alpha \cos \beta$

66. $\sin(\alpha - \beta) - \sin(\alpha + \beta) = -2 \cos \alpha \sin \beta$

67. $\dfrac{\cos(\alpha - \beta)}{\sin(\alpha + \beta)} = \dfrac{\cot \alpha + \tan \beta}{1 + \cot \alpha \tan \beta}$

68. $\dfrac{\sin(\alpha + \beta)}{\sin(\alpha - \beta)} = \dfrac{1 + \cot \alpha \tan \beta}{1 - \cot \alpha \tan \beta}$

69. $\sin\left(\dfrac{\pi}{2} + \alpha - \beta\right) = \cos \alpha \cos \beta + \sin \alpha \sin \beta$

70. $\cos\left(\dfrac{\pi}{2} + \alpha + \beta\right) = -(\sin \alpha \cos \beta + \cos \alpha \sin \beta)$

71. $\sin 3x = 3 \sin x - 4 \sin^3 x$

72. $\cos 3x = 4 \cos^3 x - 3 \cos x$

In Exercises 73 to 78, write the given expression as a function that involves only $\sin \theta$, $\cos \theta$, or $\tan \theta$. (In Exercises 76, 77, and 78, assume k is an integer.)

73. $\cos(\theta + 3\pi)$ ▶ **74.** $\sin(\theta + 2\pi)$

75. $\tan(\theta + \pi)$ **76.** $\cos[\theta + (2k + 1)\pi]$

77. $\sin(\theta + 2k\pi)$ **78.** $\sin(\theta - k\pi)$

In Exercises 79 to 82, compare the graphs of each side of the equation to predict whether the equation is an identity.

79. $\sin\left(\dfrac{\pi}{2} - x\right) = \cos x$ **80.** $\cos(x + \pi) = -\cos x$

81. $\sin 7x \cos 2x - \cos 7x \sin 2x = \sin 5x$

82. $\sin 3x = 3 \sin x - 4 \sin^3 x$

CONNECTING CONCEPTS

In Exercises 83 to 89, verify the identity.

83. $\sin(x - y) \cdot \sin(x + y) = \sin^2 x \cos^2 y - \cos^2 x \sin^2 y$

84. $\sin(x + y + z) = \sin x \cos y \cos z + \cos x \sin y \cos z + \cos x \cos y \sin z - \sin x \sin y \sin z$

85. $\cos(x + y + z) = \cos x \cos y \cos z - \sin x \sin y \cos z - \sin x \cos y \sin z - \cos x \sin y \sin z$

86. $\dfrac{\sin(x + y)}{\sin x \sin y} = \cot x + \cot y$

87. $\dfrac{\cos(x - y)}{\cos x \sin y} = \cot y + \tan x$

88. $\dfrac{\sin(x + h) - \sin x}{h} = \cos x \dfrac{\sin h}{h} + \sin x \dfrac{(\cos h - 1)}{h}$

89. $\dfrac{\cos(x + h) - \cos x}{h} = \cos x \dfrac{(\cos h - 1)}{h} - \sin x \dfrac{\sin h}{h}$

90. **MODEL RESISTANCE** The drag (resistance) on a fish when it is swimming is two to three times the drag when it is gliding. To compensate for this, some fish swim in a saw-tooth pattern, as shown in the accompanying figure. The ratio of the amount of energy the fish expends when swimming upward at angle β and then gliding down at angle α to the energy it expends swimming horizontally is given by

$$E_R = \frac{k \sin \alpha + \sin \beta}{k \sin(\alpha + \beta)}$$

where k is a value such that $2 \le k \le 3$, and k depends on the assumptions we make about the amount of drag experienced by the fish. Find E_R for $k = 2$, $\alpha = 10°$, and $\beta = 20°$.

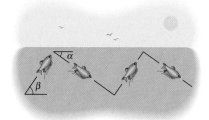

PREPARE FOR SECTION 3.3

91. Use the identity for $\sin(\alpha + \beta)$ to rewrite $\sin 2\alpha$. [3.2]

92. Use the identity for $\cos(\alpha + \beta)$ to rewrite $\cos 2\alpha$. [3.2]

93. Use the identity for $\tan(\alpha + \beta)$ to rewrite $\tan 2\alpha$. [3.2]

94. Compare $\tan \dfrac{\alpha}{2}$ and $\dfrac{\sin \alpha}{1 + \cos \alpha}$ for $\alpha = 60°$, $\alpha = 90°$, and $\alpha = 120°$. [2.2]

95. Verify that $\sin 2\alpha = 2 \sin \alpha$ is *not* an identity. [2.2]

96. Verify that $\cos \dfrac{\alpha}{2} = \dfrac{1}{2} \cos \alpha$ is *not* an identity. [2.2]

PROJECTS

1. **INTERSECTING LINES** In the figure shown at the right, two nonvertical lines intersect in a plane. The slope of line l_1 is m_1 and the slope of line l_2 is m_2.

a. Show that the tangent of the smallest positive angle γ from l_1 to l_2 is given by

$$\tan \gamma = \frac{m_2 - m_1}{1 + m_1 m_2}$$

b. Two nonvertical lines intersect at the point $(1, 5)$. The measure of the smallest positive angle between the lines is $\gamma = 60°$. The first line is given by $y = 0.5x + 4.5$. What is the equation (in slope-intercept form) of the second line?

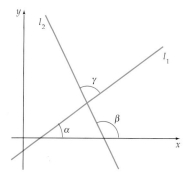

| SECTION 3.3 | # DOUBLE- AND HALF-ANGLE IDENTITIES |

- **DOUBLE-ANGLE IDENTITIES**
- **HALF-ANGLE IDENTITIES**

● DOUBLE-ANGLE IDENTITIES

By using the sum identities, we can derive identities for $f(2\alpha)$, where f is a trigonometric function. These are called the *double-angle identities*. To find the sine of a double angle, substitute α for β in the identity for $\sin(\alpha + \beta)$.

$$\sin (\alpha + \beta) = \sin \alpha \cos \beta + \cos \alpha \sin \beta$$

$$\sin (\alpha + \alpha) = \sin \alpha \cos \alpha + \cos \alpha \sin \alpha \qquad \text{• Let } \beta = \alpha.$$

$$\sin 2\alpha = 2 \sin \alpha \cos \alpha$$

A double-angle identity for cosine is derived in a similar manner.

$$\cos(\alpha + \beta) = \cos \alpha \cos \beta - \sin \alpha \sin \beta$$

$$\cos(\alpha + \alpha) = \cos \alpha \cos \alpha - \sin \alpha \sin \alpha \qquad \text{• Let } \beta = \alpha.$$

$$\cos 2\alpha = \cos^2 \alpha - \sin^2 \alpha$$

There are two alternative forms of the double-angle identity for $\cos 2\alpha$. Using $\cos^2 \alpha = 1 - \sin^2 \alpha$, we can rewrite the identity for $\cos 2\alpha$ as follows:

$$\cos 2\alpha = \cos^2 \alpha - \sin^2 \alpha$$

$$\cos 2\alpha = (1 - \sin^2 \alpha) - \sin^2 \alpha \qquad \text{• } \cos^2 \alpha = 1 - \sin^2 \alpha$$

$$\cos 2\alpha = 1 - 2 \sin^2 \alpha$$

We also can rewrite $\cos 2\alpha$ as

$$\cos 2\alpha = \cos^2 \alpha - \sin^2 \alpha$$

$$\cos 2\alpha = \cos^2 \alpha - (1 - \cos^2 \alpha) \qquad \text{• } \sin^2 \alpha = 1 - \cos^2 \alpha$$

$$\cos 2\alpha = 2 \cos^2 \alpha - 1$$

INTEGRATING TECHNOLOGY

One way of showing that $\sin 2x \neq 2 \sin x$ is by graphing $y = \sin 2x$ and $y = 2 \sin x$ and observing that the graphs are not the same.

The double-angle identity for the tangent function is derived from the identity for $\tan(\alpha + \beta)$ with $\beta = \alpha$.

$$\tan(\alpha + \beta) = \frac{\tan \alpha + \tan \beta}{1 - \tan \alpha \tan \beta}$$

$$\tan(\alpha + \alpha) = \frac{\tan \alpha + \tan \alpha}{1 - \tan \alpha \tan \alpha} \qquad \text{• Let } \beta = \alpha.$$

$$\tan 2\alpha = \frac{2 \tan \alpha}{1 - \tan^2 \alpha}$$

The double-angle identities are often used to write a trigonometric expression in terms of a single trigonometric function.

EXAMPLE 1   Simplify a Trigonometric Expression

Write $4 \sin 5\theta \cos 5\theta$ as a single trigonometric function.

Solution

$$4 \sin 5\theta \cos 5\theta = 2(2 \sin 5\theta \cos 5\theta) \qquad \text{• Use } 2 \sin \alpha \cos \alpha = \sin 2\alpha,$$
$$= 2(\sin 10\theta) = 2 \sin 10\theta \qquad \text{with } \alpha = 5\theta.$$

▶ **TRY EXERCISE 2, PAGE 230**

? QUESTION Does $\sin \theta \cos \theta = \dfrac{1}{2} \sin 2\theta$?

The double-angle identities can also be used to evaluate some trigonometric expressions.

EXAMPLE 2   Evaluate a Trigonometric Function

For an angle α in Quadrant I, $\sin \alpha = \dfrac{4}{5}$. Find $\sin 2\alpha$.

Solution

Use the identity $\sin 2\alpha = 2 \sin \alpha \cos \alpha$. Find $\cos \alpha$ by substituting for $\sin \alpha$ in $\sin^2 \alpha + \cos^2 \alpha = 1$ and solving for $\cos \alpha$.

$$\cos \alpha = \sqrt{1 - \sin^2 \alpha} = \sqrt{1 - \left(\frac{4}{5}\right)^2} = \frac{3}{5} \qquad \text{• } \cos \alpha > 0 \text{ if } \alpha \text{ is in Quadrant I.}$$

Substitute the values of $\sin \alpha$ and $\cos \alpha$ in the double-angle formula for $\sin 2\alpha$.

$$\sin 2\alpha = 2 \sin \alpha \cos \alpha = 2\left(\frac{4}{5}\right)\left(\frac{3}{5}\right) = \frac{24}{25}$$

▶ **TRY EXERCISE 26, PAGE 230**

? ANSWER Yes. $\sin \theta \cos \theta = \dfrac{2 \sin \theta \cos \theta}{2} = \dfrac{\sin 2\theta}{2} = \dfrac{1}{2} \sin 2\theta.$

EXAMPLE 3 **Verify a Double-Angle Identity**

Verify the identity $\csc 2\alpha = \dfrac{1}{2}(\tan \alpha + \cot \alpha)$.

Solution

Work on the right-hand side of the equation.

$$\frac{1}{2}(\tan \alpha + \cot \alpha) = \frac{1}{2}\left(\frac{\sin \alpha}{\cos \alpha} + \frac{\cos \alpha}{\sin \alpha}\right) = \frac{1}{2}\left(\frac{\sin^2 \alpha + \cos^2 \alpha}{\cos \alpha \sin \alpha}\right)$$

$$= \frac{1}{2 \cos \alpha \sin \alpha} = \frac{1}{\sin 2\alpha} = \csc 2\alpha$$

▶ **TRY EXERCISE 54, PAGE 231**

● HALF-ANGLE IDENTITIES

An identity for one-half an angle, $\dfrac{\alpha}{2}$, is called a *half-angle identity*. To derive a half-angle identity for $\sin \dfrac{\alpha}{2}$, we solve for $\sin^2 \theta$ in the following double-angle identity for $\cos 2\theta$.

$$\cos 2\theta = 1 - 2\sin^2 \theta$$

$$\sin^2 \theta = \frac{1 - \cos 2\theta}{2}$$

Substitute $\dfrac{\alpha}{2}$ for θ and take the square root of each side of the equation.

$$\sin^2 \frac{\alpha}{2} = \frac{1 - \cos 2\left(\dfrac{\alpha}{2}\right)}{2}$$

$$\sin \frac{\alpha}{2} = \pm \sqrt{\frac{1 - \cos \alpha}{2}}$$

The sign of the radical is determined by the quadrant in which the terminal side of angle $\dfrac{\alpha}{2}$ lies.

In a similar manner, we derive an identity for $\cos \dfrac{\alpha}{2}$.

$$\cos 2\theta = 2\cos^2 \theta - 1$$

$$\cos^2 \theta = \frac{1 + \cos 2\theta}{2}$$

take note

The sign of $\sqrt{\dfrac{1-\cos\alpha}{2}}$ depends on the quadrant in which the terminal side of $\dfrac{\alpha}{2}$ lies, not the terminal side of α.

Substitute $\dfrac{\alpha}{2}$ for θ and take the square root of each side of the equation.

$$\cos^2\frac{\alpha}{2} = \frac{1+\cos 2\left(\dfrac{\alpha}{2}\right)}{2}$$

$$\cos\frac{\alpha}{2} = \pm\sqrt{\frac{1+\cos\alpha}{2}}$$

Two different identities for $\tan\dfrac{\alpha}{2}$ are possible.

$$\tan\frac{\alpha}{2} = \frac{\sin\dfrac{\alpha}{2}}{\cos\dfrac{\alpha}{2}} = \frac{\sin\dfrac{\alpha}{2}}{\cos\dfrac{\alpha}{2}}\cdot\frac{2\cos\dfrac{\alpha}{2}}{2\cos\dfrac{\alpha}{2}}$$

$$= \frac{2\sin\dfrac{\alpha}{2}\cos\dfrac{\alpha}{2}}{2\cos^2\dfrac{\alpha}{2}} = \frac{\sin 2\left(\dfrac{\alpha}{2}\right)}{2\left(\pm\sqrt{\dfrac{1+\cos\alpha}{2}}\right)^2}$$ • $\cos\dfrac{\alpha}{2} = \pm\sqrt{\dfrac{1+\cos\alpha}{2}}$

$$\tan\frac{\alpha}{2} = \frac{\sin\alpha}{1+\cos\alpha}$$ • See **Figure 3.8.**

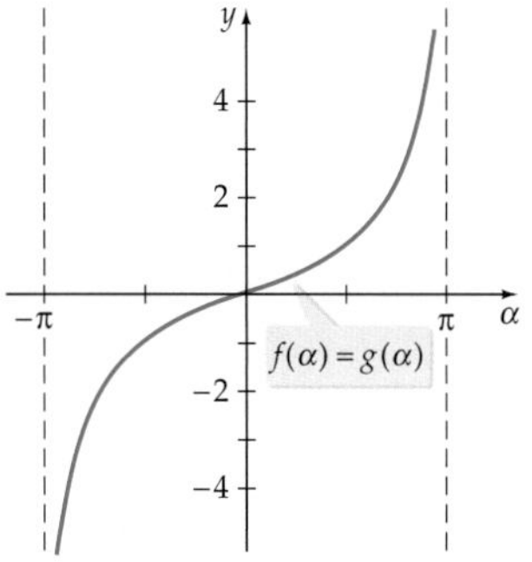

$f(\alpha) = \tan\dfrac{\alpha}{2}$

$g(\alpha) = \dfrac{\sin\alpha}{1+\cos\alpha}$

FIGURE 3.8

To obtain another identity for $\tan\dfrac{\alpha}{2}$, multiply by the conjugate of the denominator.

$$\tan\frac{\alpha}{2} = \frac{\sin\alpha}{1+\cos\alpha}\cdot\frac{1-\cos\alpha}{1-\cos\alpha}$$ • $\alpha \neq 2k\pi$, where k is an integer

$$= \frac{\sin\alpha(1-\cos\alpha)}{1-\cos^2\alpha}$$

$$= \frac{\sin\alpha(1-\cos\alpha)}{\sin^2\alpha}$$

$$\tan\frac{\alpha}{2} = \frac{1-\cos\alpha}{\sin\alpha}$$

EXAMPLE 4 Verify a Half-Angle Identity

Verify the identity $2\csc x\cos^2\dfrac{x}{2} = \dfrac{\sin x}{1-\cos x}$.

Solution

Work on the left side of the identity.

$$2\csc x\cos^2\frac{x}{2} = 2\csc x\left(\frac{1+\cos x}{2}\right)$$ • $\cos^2\dfrac{x}{2} = \dfrac{1+\cos x}{2}$

$$= \frac{1+\cos x}{\sin x}$$ • $\csc x = \dfrac{1}{\sin x}$

$$= \frac{1 + \cos x}{\sin x} \cdot \frac{1 - \cos x}{1 - \cos x}$$

• Multiply the numerator and denominator by the conjugate of the numerator.

$$= \frac{1 - \cos^2 x}{\sin x(1 - \cos x)}$$

$$= \frac{\sin^2 x}{\sin x(1 - \cos x)}$$

• $1 - \cos^2 x = \sin^2 x$

$$= \frac{\sin x}{1 - \cos x}$$

▶ **TRY EXERCISE 72, PAGE 232**

EXAMPLE 5   **Verify a Half-Angle Identity**

Verify the identity $\tan \dfrac{\alpha}{2} = \sin \alpha + \cos \alpha \cot \alpha - \cot \alpha$.

Solution

Work on the left side of the identity.

$$\tan \frac{\alpha}{2} = \frac{1 - \cos \alpha}{\sin \alpha} = \frac{\sin^2 \alpha + \cos^2 \alpha - \cos \alpha}{\sin \alpha}$$

• $1 = \sin^2 \alpha + \cos^2 \alpha$

$$= \frac{\sin^2 \alpha}{\sin \alpha} + \frac{\cos^2 \alpha}{\sin \alpha} - \frac{\cos \alpha}{\sin \alpha}$$

• Write each numerator over the common denominator.

$$= \sin \alpha + \cos \alpha \cot \alpha - \cot \alpha$$

▶ **TRY EXERCISE 78, PAGE 232**

Here is a summary of the double-angle and half-angle identities.

Double-Angle Identities

$$\sin 2\alpha = 2 \sin \alpha \cos \alpha$$

$$\cos 2\alpha = \cos^2 \alpha - \sin^2 \alpha = 1 - 2 \sin^2 \alpha = 2 \cos^2 \alpha - 1$$

$$\tan 2\alpha = \frac{2 \tan \alpha}{1 - \tan^2 \alpha}$$

Half-Angle Identities

$$\sin \frac{\alpha}{2} = \pm \sqrt{\frac{1 - \cos \alpha}{2}}$$

$$\cos \frac{\alpha}{2} = \pm \sqrt{\frac{1 + \cos \alpha}{2}}$$

$$\tan \frac{\alpha}{2} = \frac{\sin \alpha}{1 + \cos \alpha} = \frac{1 - \cos \alpha}{\sin \alpha}$$

TOPICS FOR DISCUSSION

1. True or false: If $\sin \alpha = \sin \beta$, then $\alpha = \beta$. Why?

2. Does $\sin 2\alpha = 2 \sin \alpha$ for all values of α? If not, find a value of α for which $\sin 2\alpha \neq 2 \sin \alpha$.

3. Because

$$\tan \frac{\alpha}{2} = \frac{\sin \alpha}{1 + \cos \alpha} \quad \text{and} \quad \tan \frac{\alpha}{2} = \frac{1 - \cos \alpha}{\sin \alpha}$$

are both identities, it follows that

$$\frac{\sin \alpha}{1 + \cos \alpha} = \frac{1 - \cos \alpha}{\sin \alpha}$$

is also an identity. Do you agree? Explain.

4. Is $\sin 10x = 2 \sin 5x \cos 5x$ an identity? Explain.

5. Is $\sin \dfrac{\alpha}{2} = \cos \dfrac{\alpha}{2}$ an identity? Explain.

EXERCISE SET 3.3

In Exercises 1 to 8, write each trigonometric expression in terms of a single trigonometric function.

1. $2 \sin 2\alpha \cos 2\alpha$

▶ **2.** $2 \sin 3\theta \cos 3\theta$

3. $1 - 2 \sin^2 5\beta$

4. $2 \cos^2 2\beta - 1$

5. $\cos^2 3\alpha - \sin^2 3\alpha$

6. $\cos^2 6\alpha - \sin^2 6\alpha$

7. $\dfrac{2 \tan 3\alpha}{1 - \tan^2 3\alpha}$

8. $\dfrac{2 \tan 4\theta}{1 - \tan^2 4\theta}$

In Exercises 9 to 24, use the half-angle identities to find the exact value of each trigonometric expression.

9. $\sin 75°$

10. $\cos 105°$

11. $\tan 67.5°$

12. $\tan 165°$

13. $\cos 157.5°$

14. $\sin 112.5°$

15. $\sin 22.5°$

16. $\cos 67.5°$

17. $\sin \dfrac{7\pi}{8}$

18. $\cos \dfrac{5\pi}{8}$

19. $\cos \dfrac{5\pi}{12}$

20. $\sin \dfrac{3\pi}{8}$

21. $\tan \dfrac{7\pi}{12}$

22. $\tan \dfrac{3\pi}{8}$

23. $\cos \dfrac{\pi}{12}$

24. $\sin \dfrac{\pi}{8}$

In Exercises 25 to 36, find the exact values of $\sin 2\theta$, $\cos 2\theta$, and $\tan 2\theta$ given the following information.

25. $\cos \theta = -\dfrac{4}{5}$ θ is in Quadrant II.

▶ **26.** $\cos \theta = \dfrac{24}{25}$ θ is in Quadrant IV.

27. $\sin \theta = \dfrac{8}{17}$ θ is in Quadrant II.

28. $\sin \theta = -\dfrac{9}{41}$ θ is in Quadrant III.

29. $\tan \theta = -\dfrac{24}{7}$ θ is in Quadrant IV.

30. $\tan \theta = \dfrac{4}{3}$ θ is in Quadrant I.

31. $\sin \theta = \dfrac{15}{17}$ θ is in Quadrant I.

32. $\sin \theta = -\dfrac{3}{5}$ θ is in Quadrant III.

33. $\cos \theta = \dfrac{40}{41}$ θ is in Quadrant IV.

34. $\cos \theta = \dfrac{4}{5}$ θ is in Quadrant IV.

35. $\tan \theta = \dfrac{15}{8}$ θ is in Quadrant III.

36. $\tan \theta = -\dfrac{40}{9}$ θ is in Quadrant II.

In Exercises 37 to 48, find the exact values of the sine, cosine, and tangent of $\dfrac{\alpha}{2}$ given the following information.

37. $\sin \alpha = \dfrac{5}{13}$ α is in Quadrant II.

38. $\sin \alpha = -\dfrac{7}{25}$ α is in Quadrant III.

39. $\cos \alpha = -\dfrac{8}{17}$ α is in Quadrant III.

40. $\cos \alpha = \dfrac{12}{13}$ α is in Quadrant I.

41. $\tan \alpha = \dfrac{4}{3}$ α is in Quadrant I.

42. $\tan \alpha = -\dfrac{8}{15}$ α is in Quadrant II.

43. $\cos \alpha = \dfrac{24}{25}$ α is in Quadrant IV.

44. $\sin \alpha = -\dfrac{9}{41}$ α is in Quadrant IV.

45. $\sec \alpha = \dfrac{17}{15}$ α is in Quadrant I.

46. $\csc \alpha = -\dfrac{5}{3}$ α is in Quadrant IV.

47. $\cot \alpha = \dfrac{8}{15}$ α is in Quadrant III.

48. $\sec \alpha = -\dfrac{13}{5}$ α is in Quadrant II.

In Exercises 49 to 94, verify the given identity.

49. $\sin 3x \cos 3x = \dfrac{1}{2} \sin 6x$

50. $\cos 8x = \cos^2 4x - \sin^2 4x$

51. $\sin^2 x + \cos 2x = \cos^2 x$ **52.** $\dfrac{\cos 2x}{\sin^2 x} = \cot^2 x - 1$

53. $\dfrac{1 + \cos 2x}{\sin 2x} = \cot x$ ▶ **54.** $\dfrac{1}{1 - \cos 2x} = \dfrac{1}{2} \csc^2 x$

55. $\dfrac{\sin 2x}{1 - \sin^2 x} = 2 \tan x$

56. $\dfrac{\cos^2 x - \sin^2 x}{2 \sin x \cos x} = \cot 2x$

57. $1 - \tan^2 x = \dfrac{\cos 2x}{\cos^2 x}$

58. $\tan 2x = \dfrac{2 \sin x \cos x}{\cos^2 x - \sin^2 x}$

59. $\sin 2x - \tan x = \tan x \cos 2x$

60. $\sin 2x - \cot x = -\cot x \cos 2x$

61. $\cos^4 x - \sin^4 x = \cos 2x$

62. $\sin 4x = 4 \sin x \cos^3 x - 4 \cos x \sin^3 x$

63. $\cos^2 x - 2 \sin^2 x \cos^2 x - \sin^2 x + 2 \sin^4 x = \cos^2 2x$

64. $2 \cos^4 x - \cos^2 x - 2 \sin^2 x \cos^2 x + \sin^2 x = \cos^2 2x$

65. $\cos 4x = 1 - 8 \cos^2 x + 8 \cos^4 x$

66. $\sin 4x = 4 \sin x \cos x - 8 \cos x \sin^3 x$

67. $\cos 3x - \cos x = 4 \cos^3 x - 4 \cos x$

68. $\sin 3x + \sin x = 4 \sin x - 4 \sin^3 x$

69. $\sin^3 x + \cos^3 x = (\sin x + \cos x)\left(1 - \dfrac{1}{2} \sin 2x\right)$

70. $\cos^3 x - \sin^3 x = (\cos x - \sin x)\left(1 + \dfrac{1}{2} \sin 2x\right)$

71. $\sin^2 \dfrac{x}{2} = \dfrac{\sec x - 1}{2 \sec x}$

▶ **72.** $\cos^2 \dfrac{x}{2} = \dfrac{\sec x + 1}{2 \sec x}$

73. $\tan \dfrac{x}{2} = \csc x - \cot x$

74. $\tan \dfrac{x}{2} = \dfrac{\tan x}{\sec x + 1}$

75. $2 \sin \dfrac{x}{2} \cos \dfrac{x}{2} = \sin x$

76. $\cos^2 \dfrac{x}{2} - \sin^2 \dfrac{x}{2} = \cos x$

77. $\left(\cos \dfrac{x}{2} + \sin \dfrac{x}{2} \right)^2 = 1 + \sin x$

▶ **78.** $\tan^2 \dfrac{x}{2} = \dfrac{\sec x - 1}{\sec x + 1}$

79. $\sin^2 \dfrac{x}{2} \sec x = \dfrac{1}{2}(\sec x - 1)$

80. $\cos^2 \dfrac{x}{2} \sec x = \dfrac{1}{2}(\sec x + 1)$

81. $\cos^2 \dfrac{x}{2} - \cos x = \sin^2 \dfrac{x}{2}$

82. $\sin^2 \dfrac{x}{2} + \cos x = \cos^2 \dfrac{x}{2}$

83. $\sin^2 \dfrac{x}{2} - \cos^2 \dfrac{x}{2} = -\cos x$

84. $\cos^2 \dfrac{x}{2} - \sin^2 \dfrac{x}{2} = \dfrac{1}{2} \csc x \sin 2x$

85. $\sin 2x - \cos x = (\cos x)(2 \sin x - 1)$

86. $\dfrac{\cos 2x}{\sin^2 x} = \csc^2 x - 2$

87. $\tan 2x = \dfrac{2}{\cot x - \tan x}$

88. $\dfrac{2 \cos 2x}{\sin 2x} = \cot x - \tan x$

89. $2 \tan \dfrac{x}{2} = \dfrac{\sin^2 x + 1 - \cos^2 x}{(\sin x)(1 + \cos x)}$

90. $\dfrac{1}{2} \csc^2 \dfrac{x}{2} = \csc^2 x + \cot x \csc x$

91. $\csc 2x = \dfrac{1}{2} \csc x \sec x$

92. $\sec 2x = \dfrac{\sec^2 x}{2 - \sec^2 x}$

93. $\cos \dfrac{x}{5} = 1 - 2 \sin^2 \dfrac{x}{10}$

94. $\sec^2 \dfrac{x}{2} = \dfrac{2}{1 + \cos x}$

CONNECTING CONCEPTS

 In Exercises 95 to 98, compare the graphs of each side of the equation to predict whether the equation is an identity.

95. $\sin^2 x + \cos 2x = \cos^2 x$ **96.** $\dfrac{\sin 2x}{1 - \sin^2 x} = 2 \tan x$

97. $\sin \dfrac{x}{2} \cos \dfrac{x}{2} = \sin x$

98. $\left(\cos \dfrac{x}{2} + \sin \dfrac{x}{2} \right)^2 = 1 + \sin x$

In Exercises 99 to 102, verify the identity.

99. $\dfrac{\sin^3 x + \cos^3 x}{\sin x + \cos x} = 1 - \dfrac{1}{2} \sin 2x$

100. $\cos^4 x = \dfrac{1}{8} \cos 4x + \dfrac{1}{2} \cos 2x + \dfrac{3}{8}$

101. $\sin \dfrac{x}{2} - \cos \dfrac{x}{2} = \sqrt{1 - \sin x}, \ 0° \le x \le 90°$

102. $\dfrac{\sin x - \sin 2x}{\cos x + \cos 2x} = -\tan \dfrac{x}{2}$

103. If $x + y = 90°$, verify that $\sin(x - y) = -\cos 2x$.

104. If $x + y = 90°$, verify that $\sin(x - y) = \cos 2y$.

105. If $x + y = 180°$, verify that $\sin(x - y) = -\sin 2x$.

106. If $x + y = 180°$, verify that $\cos(x - y) = -\cos 2x$.

PREPARE FOR SECTION 3.4

107. Use sum and difference identities to rewrite $\frac{1}{2}[\sin(\alpha + \beta) + \sin(\alpha - \beta)]$. [3.2]

108. Use sum and difference identities to rewrite $\frac{1}{2}[\cos(\alpha + \beta) + \cos(\alpha - \beta)]$. [3.2]

109. Compare $\sin x - \sin y$ and $2\cos\dfrac{x+y}{2}\sin\dfrac{x-y}{2}$ for $x = \pi$ and $y = \dfrac{\pi}{6}$. [2.4]

110. Use a sum identity to rewrite $\sqrt{2}\sin\left(x + \dfrac{\pi}{4}\right)$. [3.2]

111. Find a real number x and a real number y to verify that $\sin x - \sin y = \sin(x - y)$ is *not* an identity. [2.4]

112. Evaluate $\sqrt{a^2 + b^2}$ for $a = -1$ and $b = \sqrt{3}$.

PROJECTS

1. MACH NUMBERS   Earnst Mach (1838–1916) was an Austrian physicist. He made a study of the motion of objects at high speeds. Today we often state the speed of aircraft in terms of a *Mach number*. A **Mach number** is the speed of an object divided by the speed of sound. For example, a plane flying at the speed of sound is said to have a speed M of Mach 1. Mach 2 is twice the speed of sound. An airplane that travels faster than the speed of sound creates a sonic boom. This sonic boom emanates from the airplane in the shape of a cone.

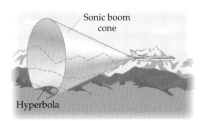

The following equation shows the relationship between the measure of the cone's vertex angle α and the Mach speed M of an aircraft that is flying faster than the speed of sound.

$$M \sin\frac{\alpha}{2} = 1$$

a. If $\alpha = \dfrac{\pi}{4}$, determine the Mach speed M of the airplane.

State your answer as an *exact* value and as a decimal accurate to the nearest hundredth.

b. How fast is Mach 1 in miles per hour?

c.   The speed of sound varies according to the altitude and the temperature of the atmosphere. Write an essay that describes this relationship.

2. VISUAL INSIGHT

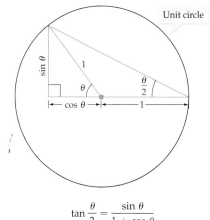

$$\tan\frac{\theta}{2} = \frac{\sin\theta}{1 + \cos\theta}$$

Explain how the figure above can be used to verify the half-angle identity shown.

SECTION 3.4

IDENTITIES INVOLVING THE SUM OF TRIGONOMETRIC FUNCTIONS

Some applications require that a product of trigonometric functions be written as a sum or difference of these functions. Other applications require that the sum or difference of trigonometric functions be represented as a product of these functions. The *product-to-sum identities* are particularly useful in these types of problems.

• THE PRODUCT-TO-SUM IDENTITIES

The product-to-sum identities can be derived by using the sum or difference identities. Adding the identities for $\sin(\alpha + \beta)$ and $\sin(\alpha - \beta)$, we have

$$\sin(\alpha + \beta) = \sin \alpha \cos \beta + \cos \alpha \sin \beta$$

$$\underline{\sin(\alpha - \beta) = \sin \alpha \cos \beta - \cos \alpha \sin \beta}$$

$$\sin(\alpha + \beta) + \sin(\alpha - \beta) = 2 \sin \alpha \cos \beta \qquad \text{• Add the identities.}$$

Solving for $\sin \alpha \cos \beta$, we obtain the first product-to-sum identity:

$$\sin \alpha \cos \beta = \frac{1}{2}[\sin(\alpha + \beta) + \sin(\alpha - \beta)]$$

The identity for $\cos \alpha \sin \beta$ is obtained when $\sin(\alpha - \beta)$ is subtracted from $\sin(\alpha + \beta)$. The result is

$$\cos \alpha \sin \beta = \frac{1}{2}[\sin(\alpha + \beta) - \sin(\alpha - \beta)]$$

In like manner, the identities for $\cos(\alpha + \beta)$ and $\cos(\alpha - \beta)$ are used to derive the identities for $\cos \alpha \cos \beta$ and $\sin \alpha \sin \beta$.

$$\cos \alpha \cos \beta = \frac{1}{2}[\cos(\alpha + \beta) + \cos(\alpha - \beta)]$$

$$\sin \alpha \sin \beta = \frac{1}{2}[\cos(\alpha - \beta) - \cos(\alpha + \beta)]$$

The product-to-sum identities can be used to verify some identities.

> **EXAMPLE I** **Verify an Identity**
>
> Verify the identity $\cos 2x \sin 5x = \dfrac{1}{2}(\sin 7x + \sin 3x)$.
>
> **Solution**
>
> $$\cos 2x \sin 5x = \frac{1}{2}[\sin(2x + 5x) - \sin(2x - 5x)] \qquad \bullet \text{ Use the product-}$$
> $$\text{to-sum identity for}$$
> $$= \frac{1}{2}[\sin 7x - \sin(-3x)] \qquad\qquad \cos \alpha \sin \beta.$$
> $$= \frac{1}{2}(\sin 7x + \sin 3x) \qquad\qquad \bullet \sin(-3x) = -\sin 3x$$

▶ **TRY EXERCISE 36, PAGE 239**

● THE SUM-TO-PRODUCT IDENTITIES

The *sum-to-product identities* can be derived from the product-to-sum identities. To derive the sum-to-product identity for $\sin x + \sin y$, we first let $x = \alpha + \beta$ and $y = \alpha - \beta$. Then

$$x + y = \alpha + \beta + \alpha - \beta \quad \text{and} \quad x - y = \alpha + \beta - (\alpha - \beta)$$
$$x + y = 2\alpha \qquad\qquad\qquad x - y = 2\beta$$
$$\alpha = \frac{x + y}{2} \qquad\qquad\qquad \beta = \frac{x - y}{2}$$

Substituting these expressions for α and β into the product-to-sum identity

$$\frac{1}{2}[\sin(\alpha + \beta) + \sin(\alpha - \beta)] = \sin \alpha \cos \beta$$

yields

$$\sin\left(\frac{x + y}{2} + \frac{x - y}{2}\right) + \sin\left(\frac{x + y}{2} - \frac{x - y}{2}\right) = 2 \sin \frac{x + y}{2} \cos \frac{x - y}{2}$$

Simplifying the left side, we have a sum-to-product identity.

$$\sin x + \sin y = 2 \sin \frac{x + y}{2} \cos \frac{x - y}{2}$$

In like manner, three other sum-to-product identities can be derived from the other product-to-sum identities. The proofs of these identities are left as exercises.

$$\sin x - \sin y = 2 \cos \frac{x + y}{2} \sin \frac{x - y}{2}$$

$$\cos x + \cos y = 2 \cos \frac{x + y}{2} \cos \frac{x - y}{2}$$

$$\cos x - \cos y = -2 \sin \frac{x + y}{2} \sin \frac{x - y}{2}$$

EXAMPLE 2   **Write the Difference of Trigonometric Expressions as a Product**

Write $\sin 4\theta - \sin \theta$ as the product of two functions.

Solution

$$\sin 4\theta - \sin \theta = 2 \cos \frac{4\theta + \theta}{2} \sin \frac{4\theta - \theta}{2} = 2 \cos \frac{5\theta}{2} \sin \frac{3\theta}{2}$$

▶ **TRY EXERCISE 22, PAGE 239**

❓ **QUESTION** Does $\cos 4\theta + \cos 2\theta = 2 \cos 3\theta \cos \theta$?

EXAMPLE 3   **Verify a Sum-to-Product Identity**

Verify the identity $\dfrac{\sin 6x + \sin 2x}{\sin 6x - \sin 2x} = \tan 4x \cot 2x$.

Solution

$$\frac{\sin 6x + \sin 2x}{\sin 6x - \sin 2x} = \frac{2 \sin \dfrac{6x + 2x}{2} \cos \dfrac{6x - 2x}{2}}{2 \cos \dfrac{6x + 2x}{2} \sin \dfrac{6x - 2x}{2}} = \frac{\sin 4x \cos 2x}{\cos 4x \sin 2x}$$

$$= \tan 4x \cot 2x$$

▶ **TRY EXERCISE 44, PAGE 240**

● **FUNCTIONS OF THE FORM $f(x) = a \sin x + b \cos x$**

The function given by $f(x) = a \sin x + b \cos x$ can be written in the form $f(x) = k \sin(x + \alpha)$. This form of the function is useful in graphing and engineering applications because the amplitude, period, and phase shift can be readily calculated.

Let $P(a, b)$ be a point on a coordinate plane, and let α represent an angle in standard position. See **Figure 3.9.** To rewrite $y = a \sin x + b \cos x$, multiply and divide the expression $a \sin x + b \cos x$ by $\sqrt{a^2 + b^2}$.

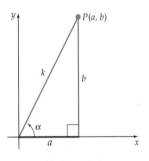

FIGURE 3.9

$$a \sin x + b \cos x = \frac{\sqrt{a^2 + b^2}}{\sqrt{a^2 + b^2}}(a \sin x + b \cos x)$$

$$= \sqrt{a^2 + b^2}\left(\frac{a}{\sqrt{a^2 + b^2}} \sin x + \frac{b}{\sqrt{a^2 + b^2}} \cos x\right) \qquad (1)$$

From the definition of the sine and cosine of an angle in standard position, let

$$k = \sqrt{a^2 + b^2}, \quad \cos \alpha = \frac{a}{\sqrt{a^2 + b^2}}, \quad \text{and} \quad \sin \alpha = \frac{b}{\sqrt{a^2 + b^2}}$$

❓ **ANSWER** Yes. $\cos 4\theta + \cos 2\theta = 2 \cos\left(\dfrac{4\theta + 2\theta}{2}\right) \cos\left(\dfrac{4\theta - 2\theta}{2}\right) = 2 \cos 3\theta \cos \theta$.

Substituting these expressions into Equation (1) yields

$$a \sin x + b \cos x = k(\cos \alpha \sin x + \sin \alpha \cos x)$$

Now, using the identity for the sine of the sum of two angles, we have

$$a \sin x + b \cos x = k \sin(x + \alpha)$$

Thus $a \sin x + b \cos x = k \sin(x + \alpha)$, where $k = \sqrt{a^2 + b^2}$ and α is the angle for which $\sin \alpha = \dfrac{b}{\sqrt{a^2 + b^2}}$ and $\cos \alpha = \dfrac{a}{\sqrt{a^2 + b^2}}$.

EXAMPLE 4   Rewrite $a \sin x + b \cos x$

Rewrite $\sin x + \cos x$ in the form $k \sin(x + \alpha)$.

Solution

Comparing $\sin x + \cos x$ to $a \sin x + b \cos x$, $a = 1$ and $b = 1$. Thus $k = \sqrt{1^2 + 1^2} = \sqrt{2}$, $\sin \alpha = \dfrac{1}{\sqrt{2}}$, and $\cos \alpha = \dfrac{1}{\sqrt{2}}$. Thus $\alpha = \dfrac{\pi}{4}$.

$$\sin x + \cos x = k \sin(x + \alpha) = \sqrt{2} \sin\left(x + \frac{\pi}{4}\right)$$

▶ **TRY EXERCISE 62, PAGE 240**

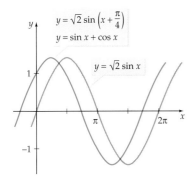

$y = \sqrt{2} \sin\left(x + \dfrac{\pi}{4}\right)$
$y = \sin x + \cos x$
$y = \sqrt{2} \sin x$

FIGURE 3.10

The graphs of $y = \sin x + \cos x$ and $y = \sqrt{2} \sin\left(x + \dfrac{\pi}{4}\right)$ are both the graph of $y = \sqrt{2} \sin x$ shifted $\dfrac{\pi}{4}$ units to the left. See **Figure 3.10.**

EXAMPLE 5   **Use an Identity to Graph a Trigonometric Function**

Graph $f(x) = -\sin x + \sqrt{3} \cos x$.

Solution

First, we write $f(x)$ as $k \sin(x + \alpha)$. Let $a = -1$ and $b = \sqrt{3}$; then $k = \sqrt{(-1)^2 + (\sqrt{3})^2} = 2$. The point $P\left(-1, \sqrt{3}\right)$ is in the second quadrant (see **Figure 3.11**). Let α be an angle in standard position with P on its terminal side. Let α' be the reference angle for α. Then

$$\sin \alpha' = \frac{\sqrt{3}}{2}$$

$$\alpha' = \frac{\pi}{3}$$

$$\alpha = \pi - \alpha' = \pi - \frac{\pi}{3} = \frac{2\pi}{3}$$

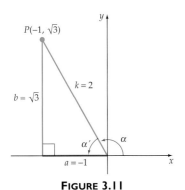

$P(-1, \sqrt{3})$
$b = \sqrt{3}$
$k = 2$
α'
α
$a = -1$

FIGURE 3.11

Continued ▶

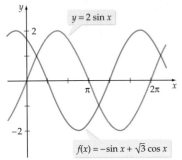

$y = 2 \sin x$

$f(x) = -\sin x + \sqrt{3} \cos x$

FIGURE 3.12

Substituting 2 for k and $\dfrac{2\pi}{3}$ for α in $y = k \sin(x + \alpha)$, we have

$$y = 2 \sin\left(x + \frac{2\pi}{3}\right)$$

The phase shift is $-\dfrac{c}{b} = -\dfrac{2\pi}{3}$. Thus the graph of the equation

$f(x) = -\sin x + \sqrt{3} \cos x$ is the graph of $y = 2 \sin x$ shifted $\dfrac{2\pi}{3}$ units to the

left. See **Figure 3.12.**

▶ **TRY EXERCISE 70, PAGE 240**

We now list the identities that have been discussed in this section.

Product-to-Sum Identities

$$\sin \alpha \cos \beta = \frac{1}{2}[\sin(\alpha + \beta) + \sin(\alpha - \beta)]$$

$$\cos \alpha \sin \beta = \frac{1}{2}[\sin(\alpha + \beta) - \sin(\alpha - \beta)]$$

$$\cos \alpha \cos \beta = \frac{1}{2}[\cos(\alpha + \beta) + \cos(\alpha - \beta)]$$

$$\sin \alpha \sin \beta = \frac{1}{2}[\cos(\alpha - \beta) - \cos(\alpha + \beta)]$$

Sum-to-Product Identities

$$\sin x + \sin y = 2 \sin \frac{x + y}{2} \cos \frac{x - y}{2}$$

$$\cos x + \cos y = 2 \cos \frac{x + y}{2} \cos \frac{x - y}{2}$$

$$\sin x - \sin y = 2 \cos \frac{x + y}{2} \sin \frac{x - y}{2}$$

$$\cos x - \cos y = -2 \sin \frac{x + y}{2} \sin \frac{x - y}{2}$$

Sums of the Form $a \sin x + b \cos x$

$$a \sin x + b \cos x = k \sin(x + \alpha)$$

where $k = \sqrt{a^2 + b^2}$, $\sin \alpha = \dfrac{b}{\sqrt{a^2 + b^2}}$, and $\cos \alpha = \dfrac{a}{\sqrt{a^2 + b^2}}$.

TOPICS FOR DISCUSSION

1. A student claims that the *exact* value of $\sin 75° \cos 15°$ is $\dfrac{2 + \sqrt{3}}{4}$. Do you agree? Explain.

2. Do you agree with the following work? Explain.

$$\cos 195° + \cos 105° = \cos(195° + 105°) = \cos 300° = -\frac{1}{2}$$

3. The graphs of $y_1 = \sin x + \cos x$ and $y_2 = \sqrt{2} \sin\left(x + \dfrac{\pi}{4}\right)$ are identical. Do you agree? Explain.

4. Explain to a classmate how to determine the amplitude of the graph of $y = a \sin x + b \cos x$.

EXERCISE SET 3.4

In Exercises 1 to 8, write each expression as the sum or difference of two functions.

1. $2 \sin x \cos 2x$

2. $2 \sin 4x \sin 2x$

3. $\cos 6x \sin 2x$

4. $\cos 3x \cos 5x$

5. $2 \sin 5x \cos 3x$

6. $2 \sin 2x \cos 6x$

7. $\sin x \sin 5x$

8. $\cos 3x \sin x$

In Exercises 9 to 16, find the exact value of each expression. Do not use a calculator.

9. $\cos 75° \cos 15°$

10. $\sin 105° \cos 15°$

11. $\cos 157.5° \sin 22.5°$

12. $\sin 195° \cos 15°$

13. $\sin \dfrac{13\pi}{12} \cos \dfrac{\pi}{12}$

14. $\sin \dfrac{11\pi}{12} \sin \dfrac{7\pi}{12}$

15. $\sin \dfrac{\pi}{12} \cos \dfrac{7\pi}{12}$

16. $\cos \dfrac{17\pi}{12} \sin \dfrac{7\pi}{12}$

In Exercises 17 to 32, write each expression as the product of two functions.

17. $\sin 4\theta + \sin 2\theta$

18. $\cos 5\theta - \cos 3\theta$

19. $\cos 3\theta + \cos \theta$

20. $\sin 7\theta - \sin 3\theta$

21. $\cos 6\theta - \cos 2\theta$

▶ 22. $\cos 3\theta + \cos 5\theta$

23. $\cos \theta + \cos 7\theta$

24. $\sin 3\theta + \sin 7\theta$

25. $\sin 5\theta + \sin 9\theta$

26. $\cos 5\theta - \cos \theta$

27. $\cos 2\theta - \cos \theta$

28. $\sin 2\theta + \sin 6\theta$

29. $\cos \dfrac{\theta}{2} - \cos \theta$

30. $\sin \dfrac{3\theta}{4} + \sin \dfrac{\theta}{2}$

31. $\sin \dfrac{\theta}{2} - \sin \dfrac{\theta}{3}$

32. $\cos \theta + \cos \dfrac{\theta}{2}$

In Exercises 33 to 48, verify the identity.

33. $2 \cos \alpha \cos \beta = \cos(\alpha + \beta) + \cos(\alpha - \beta)$

34. $2 \sin \alpha \sin \beta = \cos(\alpha - \beta) - \cos(\alpha + \beta)$

35. $2 \cos 3x \sin x = 2 \sin x \cos x - 8 \cos x \sin^3 x$

▶ 36. $\sin 5x \cos 3x = \sin 4x \cos 4x + \sin x \cos x$

37. $2 \cos 5x \cos 7x = \cos^2 6x - \sin^2 6x + 2 \cos^2 x - 1$

38. $\sin 3x \cos x = \sin x \cos x(3 - 4 \sin^2 x)$

39. $\sin 3x - \sin x = 2 \sin x - 4 \sin^3 x$

40. $\cos 5x - \cos 3x = -8 \sin^2 x(2 \cos^3 x - \cos x)$

41. $\sin 2x + \sin 4x = 2 \sin x \cos x(4 \cos^2 x - 1)$

42. $\cos 3x + \cos x = 4 \cos^3 x - 2 \cos x$

43. $\dfrac{\sin 3x - \sin x}{\cos 3x - \cos x} = -\cot 2x$

▶ **44.** $\dfrac{\cos 5x - \cos 3x}{\sin 5x + \sin 3x} = -\tan x$

45. $\dfrac{\sin 5x + \sin 3x}{4 \sin x \cos^3 x - 4 \sin^3 x \cos x} = 2 \cos x$

46. $\dfrac{\cos 4x - \cos 2x}{\sin 2x - \sin 4x} = \tan 3x$

47. $\sin(x + y) \cos(x - y) = \sin x \cos x + \sin y \cos y$

48. $\sin(x + y) \sin(x - y) = \sin^2 x - \sin^2 y$

In Exercises 49 to 58, write the given equation in the form $y = k \sin(x + \alpha)$, where the measure of α is in degrees.

49. $y = -\sin x - \cos x$

50. $y = \sqrt{3} \sin x - \cos x$

51. $y = \dfrac{1}{2} \sin x - \dfrac{\sqrt{3}}{2} \cos x$

52. $y = \dfrac{\sqrt{3}}{2} \sin x - \dfrac{1}{2} \cos x$

53. $y = \dfrac{1}{2} \sin x - \dfrac{1}{2} \cos x$

54. $y = -\dfrac{\sqrt{3}}{2} \sin x - \dfrac{1}{2} \cos x$

55. $y = -3 \sin x + 3 \cos x$

56. $y = \dfrac{\sqrt{2}}{2} \sin x + \dfrac{\sqrt{2}}{2} \cos x$

57. $y = \pi \sin x - \pi \cos x$

58. $y = -0.4 \sin x + 0.4 \cos x$

In Exercises 59 to 66, write the given equation in the form $y = k \sin(x + \alpha)$, where the measure of α is in radians.

59. $y = -\sin x + \cos x$

60. $y = -\sqrt{3} \sin x - \cos x$

61. $y = \dfrac{\sqrt{3}}{2} \sin x + \dfrac{1}{2} \cos x$ ▶ **62.** $y = \sin x + \sqrt{3} \cos x$

63. $y = -10 \sin x + 10\sqrt{3} \cos x$

64. $y = 3 \sin x - 3\sqrt{3} \cos x$

65. $y = -5 \sin x + 5 \cos x$ **66.** $y = 3 \sin x - 3 \cos x$

In Exercises 67 to 76, graph one cycle of each equation.

67. $y = -\sin x - \sqrt{3} \cos x$ **68.** $y = -\sqrt{3} \sin x + \cos x$

69. $y = 2 \sin x + 2 \cos x$ ▶ **70.** $y = \sin x + \sqrt{3} \cos x$

71. $y = -\sqrt{3} \sin x - \cos x$ **72.** $y = -\sin x + \cos x$

73. $y = -5 \sin x + 5\sqrt{3} \cos x$

74. $y = -\sqrt{2} \sin x + \sqrt{2} \cos x$

75. $y = 6\sqrt{3} \sin x - 6 \cos x$

76. $y = 5\sqrt{2} \sin x + 5\sqrt{2} \cos x$

77. **TONES ON A TOUCH-TONE PHONE**

   a. Write an equation of the form $p(t) = \cos(2\pi f_1 t) + \cos(2\pi f_2 t)$ that models the tone produced by pressing the 5 key on a touch-tone phone. (*Hint:* See the Chapter Opener, page 205.)

   b. Use a sum-to-product identity to write your equation from part **a.** in the form
   $$p(t) = A \cos(B\pi t) \cos(C\pi t)$$

   c. When a sound with frequency f_1 is combined with a sound with frequency f_2, the combined sound has a frequency of $\dfrac{f_1 + f_2}{2}$. What is the frequency of the tone produced when the 5 key is pressed?

78. **TONES ON A TOUCH-TONE PHONE**

   a. Write an equation of the form $p(t) = \cos(2\pi f_1 t) + \cos(2\pi f_2 t)$ that models the tone produced by pressing the 8 key on a touch-tone phone.

   b. Use a sum-to-product identity to write your equation from part **a.** in the form
   $$p(t) = A \cos(B\pi t) \cos(C\pi t)$$

   c. What is the frequency of the tone produced when the 8 key on a touch-tone phone is pressed? (*Hint:* See part **c.** of Exercise 77.)

In Exercises 79 to 84, compare the graphs of each side of the equation to predict whether the equation is an identity.

79. $\sin 3x - \sin x = 2 \sin x - 4 \sin^3 x$

80. $\dfrac{\sin 3x - \sin x}{\cos 3x - \cos x} = -\dfrac{1}{\tan 2x}$

81. $-\sqrt{3}\sin x - \cos x = 2\sin\left(x - \dfrac{5\pi}{6}\right)$

82. $-\sqrt{3}\sin x + \cos x = 2\sin\left(x + \dfrac{5\pi}{6}\right)$

83. $\dfrac{1}{2}\sin x - \dfrac{\sqrt{3}}{2}\cos x = \sin\left(x - \dfrac{\pi}{3}\right)$

84. $\dfrac{\sqrt{3}}{2}\sin x + \dfrac{1}{2}\cos x = \sin\left(x + \dfrac{\pi}{6}\right)$

CONNECTING CONCEPTS

85. Derive the sum-to-product identity
$$\cos x + \cos y = 2\cos\frac{x+y}{2}\cos\frac{x-y}{2}$$

86. Derive the product-to-sum identity
$$\sin x \sin y = \frac{1}{2}[\cos(x-y) - \cos(x+y)]$$

87. If $x + y = 180°$, show that $\sin x + \sin y = 2\sin x$.

88. If $x + y = 360°$, show that $\cos x + \cos y = 2\cos x$.

In Exercises 89 to 94, verify the identity.

89. $\sin 2x + \sin 4x + \sin 6x = 4\sin 3x \cos 2x \cos x$

90. $\sin 4x - \sin 2x + \sin 6x = 4\cos 3x \sin 2x \cos x$

91. $\dfrac{\cos 10x + \cos 8x}{\sin 10x - \sin 8x} = \cot x$

92. $\dfrac{\sin 10x + \sin 2x}{\cos 10x + \cos 2x} = \dfrac{2\tan 3x}{1 - \tan^2 3x}$

93. $\dfrac{\sin 2x + \sin 4x + \sin 6x}{\cos 2x + \cos 4x + \cos 6x} = \tan 4x$

94. $\dfrac{\sin 2x + \sin 6x}{\cos 6x - \cos 2x} = -\cot 2x$

95. Verify that $\cos^2 x - \sin^2 x = \cos 2x$ by using a product-to-sum identity.

96. Verify that $2\sin x \cos x = \sin 2x$ by using a product-to-sum identity.

97. Verify that $a\sin x + b\cos x = k\cos(x - \alpha)$, where
$$k = \sqrt{a^2 + b^2} \text{ and } \tan\alpha = \frac{a}{b}.$$

98. Verify that $a\sin cx + b\cos cx = k\sin(cx + \alpha)$, where
$$k = \sqrt{a^2 + b^2} \text{ and } \tan\alpha = \frac{b}{a}.$$

PREPARE FOR SECTION 3.5

99. What is a one-to-one function? [1.3]

100. State the horizontal line test. [1.3]

101. Find $f[g(x)]$ given that $f(x) = 2x + 4$ and $g(x) = \dfrac{1}{2}x - 2$. [1.5]

102. If f and f^{-1} are inverse functions, then determine $f[f^{-1}(x)]$ for any x in the domain of f^{-1}. [1.6]

103. If f and f^{-1} are inverse functions, then explain how the graph of f^{-1} is related to the graph of f. [1.6]

104. Use the horizontal line test to determine whether the graph of $y = \sin x$, where x is any real number, is a one-to-one function. [1.3/2.5]

PROJECTS

1. INTERFERENCE OF SOUND WAVES The following figure on page 242 shows the wave-forms of two tuning forks. The frequency of one of the tuning forks is 10 cycles per second, and the frequency of the other tuning fork is 8 cycles per second. Each of the sound waves can be modeled by an equation of the form
$$p(t) = A\cos 2\pi ft$$

where p is the pressure produced on the eardrum at time t, A is the amplitude of the sound wave, and f is the frequency of the sound.

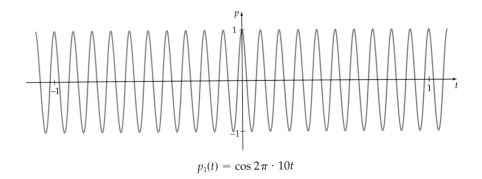

$$p_1(t) = \cos 2\pi \cdot 10t$$

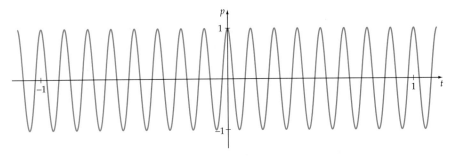

$$p_2(t) = \cos 2\pi \cdot 8t$$

If the two tuning forks are struck at the same time, with the same force, the sound that we hear fluctuates between a loud tone and silence. These regular fluctuations are called **beats.** The loud periods occur when the sound waves reinforce (interfere constructively with) one another, and the nearly silent periods occur when the waves interfere destructively with each other. The pressure p produced on the eardrum from the combined sound waves is given by

$$p(t) = p_1 + p_2 = \cos 2\pi \cdot 10t + \cos 2\pi \cdot 8t$$

The following figure shows the graph of p.

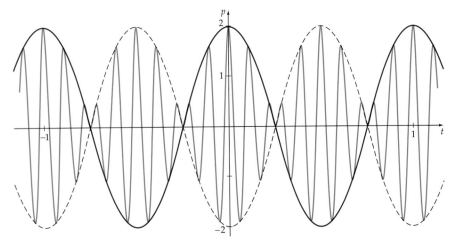

Graph of $p = p_1 + p_2 = \cos 2\pi \cdot 10t + \cos 2\pi \cdot 8t$
showing the beats in the combined sounds

a. Use the sum-to-product identity for $\cos x + \cos y$ to write p as a product.

b. Explain why the graph of $p = A \cos 2\pi f_1 t + A \cos 2\pi f_2 t$ can be thought of as a cosine curve with period $\dfrac{2}{f_1 + f_2}$ and a *variable* amplitude of

$$2A \cos\left[2\pi\left(\frac{f_1 - f_2}{2}\right)t\right]$$

c. The rate of the beats produced by two sounds, with the same intensity, equals the absolute value of the difference between their frequencies. Consider two tuning forks that are struck at the same time with the same force and held on a sounding board. The tuning forks have frequencies of 564 and 568 cycles per second, respectively. How many beats will be heard each second?

d. A piano tuner strikes a tuning fork and a key on a piano that is supposed to have the same frequency as the tuning fork. The piano tuner notices that the sound produced by the piano is lower than that produced by the tuning fork. The piano tuner also notes that the combined sound of the piano and the tuning fork has 2 beats per second. How much lower is the frequency of the piano than the frequency of the tuning fork?

2. The photo shows a matched set of tuning forks. The fork on the left has an adjustable weight that can be used to vary the frequency of its tone. Check with the physics department at your school to see if a matched set of tuning forks is available. Use the tuning forks to demonstrate to your classmates the phenomenon of beats.

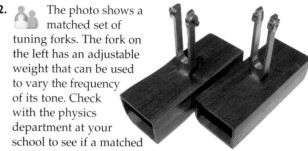

See Section 1.6 if you need to review the concept of an inverse function.

SECTION 3.5

INVERSE TRIGONOMETRIC FUNCTIONS

- **INVERSE TRIGONOMETRIC FUNCTIONS**
- **COMPOSITION OF TRIGONOMETRIC FUNCTIONS AND THEIR INVERSES**
- **GRAPHS OF INVERSE TRIGONOMETRIC FUNCTIONS**
- **AN APPLICATION INVOLVING AN INVERSE TRIGONOMETRIC FUNCTION**

● INVERSE TRIGONOMETRIC FUNCTIONS

Because the graph of $y = \sin x$ fails the horizontal line test, it is not the graph of a one-to-one function. Therefore, it does not have an inverse function. **Figure 3.13** shows the graph of $y = \sin x$ on the interval $-2\pi \le x \le 2\pi$ and the graph of the inverse relation $x = \sin y$. Note that the graph of $x = \sin y$ does not satisfy the vertical line test and therefore is not the graph of a function.

If the domain of $y = \sin x$ is restricted to $-\dfrac{\pi}{2} \le x \le \dfrac{\pi}{2}$, the graph of $y = \sin x$ satisfies the horizontal line test and therefore the function has an inverse function. The graphs of $y = \sin x$ for $-\dfrac{\pi}{2} \le x \le \dfrac{\pi}{2}$ and its inverse are shown in **Figure 3.14.**

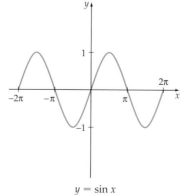

$y = \sin x$

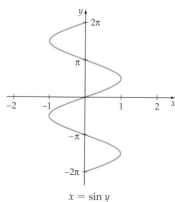

$x = \sin y$

FIGURE 3.13

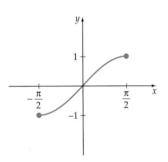

$$y = \sin x: -\frac{\pi}{2} \le x \le \frac{\pi}{2}$$

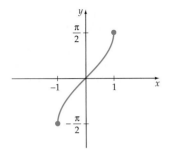

$$y = \sin^{-1} x: -1 \le x \le 1$$

FIGURE 3.14

> ### take note
>
> The -1 in $\sin^{-1} x$ is not an exponent.
> The -1 is used to denote the
> inverse function. To use -1 as an
> exponent for a sine function,
> enclose the function in
> parentheses.
>
> $$(\sin x)^{-1} = \frac{1}{\sin x} = \csc x$$
>
> $$\sin^{-1} x \ne \frac{1}{\sin x}$$

To find the inverse of the function defined by $y = \sin x$, with $-\frac{\pi}{2} \le x \le \frac{\pi}{2}$, interchange x and y. Then solve for y.

$$y = \sin x \qquad \bullet \; -\frac{\pi}{2} \le x \le \frac{\pi}{2}$$

$$x = \sin y \qquad \bullet \text{ Interchange } x \text{ and } y.$$

$$y = ? \qquad \bullet \textbf{ Solve for } y.$$

Unfortunately, there is no algebraic solution for y. Thus we establish new notation and write

$$y = \sin^{-1} x$$

which is read "y is the inverse sine of x." Some textbooks use the notation arcsin x instead of $\sin^{-1} x$.

Definition of $\sin^{-1} x$

$$y = \sin^{-1} x \quad \text{if and only if} \quad x = \sin y$$

$$\text{where } -1 \le x \le 1 \text{ and } -\frac{\pi}{2} \le y \le \frac{\pi}{2}.$$

It is convenient to think of the value of an inverse trigonometric function as an angle. For instance, if $y = \sin^{-1}\left(\frac{1}{2}\right)$, then y is the angle in the interval $\left[-\frac{\pi}{2}, \frac{\pi}{2}\right]$ whose sine is $\frac{1}{2}$. Thus $y = \frac{\pi}{6}$.

Because the graph of $y = \cos x$ fails the horizontal line test, it is not the graph of a one-to-one function. Therefore, it does not have an inverse function. **Figure 3.15** shows the graph of $y = \cos x$ on the interval $-2\pi \le x \le 2\pi$ and the graph of the inverse relation $x = \cos y$. Note that the graph of $x = \cos y$ does not satisfy the vertical line test and therefore is not the graph of a function.

If the domain of $y = \cos x$ is restricted to $0 \le x \le \pi$, the graph of $y = \cos x$ satisfies the horizontal line test and therefore is the graph of a one-to-one function. The graph of $y = \cos x$ for $0 \le x \le \pi$ and that of $x = \cos y$ are shown in **Figure 3.16.**

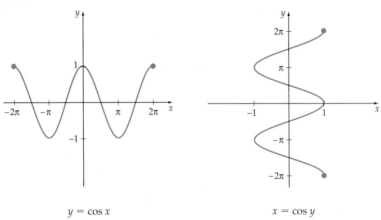

$$y = \cos x \qquad\qquad\qquad x = \cos y$$

FIGURE 3.15

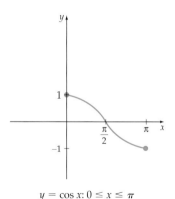

$y = \cos x : 0 \le x \le \pi$

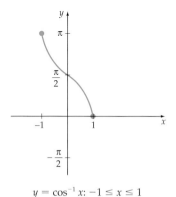

$y = \cos^{-1} x : -1 \le x \le 1$

FIGURE 3.16

To find the inverse of the function defined by $y = \cos x$, with $0 \le x \le \pi$, interchange x and y. Then solve for y.

$$y = \cos x \qquad \bullet\ 0 \le x \le \pi$$
$$x = \cos y \qquad \bullet\ \text{Interchange } x \text{ and } y.$$
$$y = ? \qquad \bullet\ \text{Solve for } y.$$

As in the case for the inverse sine function, there is no algebraic solution for y. Thus the notation for the inverse cosine function becomes $y = \cos^{-1} x$. We can write the following definition of the inverse cosine function.

Definition of $\cos^{-1} x$

$$y = \cos^{-1} x \quad \text{if and only if} \quad x = \cos y$$

where $-1 \le x \le 1$ and $0 \le y \le \pi$.

Because the graphs of $y = \tan x$, $y = \csc x$, $y = \sec x$, and $y = \cot x$ fail the horizontal line test, these functions are not one-to-one functions. Therefore, these functions do not have inverse functions. If the domains of all these functions are restricted in a certain way, however, the graphs satisfy the horizontal line test. Thus each of these functions has an inverse function over a restricted domain. **Table 3.2** on page 246 shows the restricted function and the inverse function for $\tan x$, $\csc x$, $\sec x$, and $\cot x$.

The choice of ranges for $y = \sec^{-1} x$ and $y = \csc^{-1} x$ is not universally accepted. For example, some calculus texts use $\left[0, \dfrac{\pi}{2}\right) \cup \left[\pi, \dfrac{3\pi}{2}\right)$ as the range of $y = \sec^{-1} x$. This definition has some advantages and some disadvantages that are explained in more advanced mathematics courses.

EXAMPLE 1 **Evaluate Inverse Functions**

Find the exact value of each inverse function.

a. $y = \tan^{-1} \dfrac{\sqrt{3}}{3}$ b. $y = \cos^{-1}\left(-\dfrac{\sqrt{2}}{2}\right)$

Solution

a.   Because $y = \tan^{-1} \dfrac{\sqrt{3}}{3}$, y is the angle whose measure is in the interval $\left(-\dfrac{\pi}{2}, \dfrac{\pi}{2}\right)$, and $\tan y = \dfrac{\sqrt{3}}{3}$. Therefore, $y = \dfrac{\pi}{6}$.

b.   Because $y = \cos^{-1}\left(-\dfrac{\sqrt{2}}{2}\right)$, y is the angle whose measure is in the interval $[0, \pi]$, and $\cos y = -\dfrac{\sqrt{2}}{2}$. Therefore, $y = \dfrac{3}{4}\pi$.

▶ **TRY EXERCISE 2, PAGE 253**

TABLE 3.2

	$y = \tan x$	$y = \tan^{-1} x$	$y = \csc x$	$y = \csc^{-1} x$
Domain	$-\dfrac{\pi}{2} < x < \dfrac{\pi}{2}$	$-\infty < x < \infty$	$-\dfrac{\pi}{2} \le x \le \dfrac{\pi}{2},\, x \ne 0$	$x \le -1 \text{ or } x \ge 1$
Range	$-\infty < y < \infty$	$-\dfrac{\pi}{2} < y < \dfrac{\pi}{2}$	$y \le -1 \text{ or } y \ge 1$	$-\dfrac{\pi}{2} \le y \le \dfrac{\pi}{2},\, y \ne 0$
Asymptotes	$x = -\dfrac{\pi}{2},\, x = \dfrac{\pi}{2}$	$y = -\dfrac{\pi}{2},\, y = \dfrac{\pi}{2}$	$x = 0$	$y = 0$
Graph				

	$y = \sec x$	$y = \sec^{-1} x$	$y = \cot x$	$y = \cot^{-1} x$
Domain	$0 \le x \le \pi,\, x \ne \dfrac{\pi}{2}$	$x \le -1 \text{ or } x \ge 1$	$0 < x < \pi$	$-\infty < x < \infty$
Range	$y \le -1 \text{ or } y \ge 1$	$0 \le y \le \pi,\, y \ne \dfrac{\pi}{2}$	$-\infty < y < \infty$	$0 < y < \pi$
Asymptotes	$x = \dfrac{\pi}{2}$	$y = \dfrac{\pi}{2}$	$x = 0,\, x = \pi$	$y = 0,\, y = \pi$
Graph				

A calculator may not have keys for the inverse secant, cosecant, and cotangent functions. The following procedure shows an identity for the inverse cosecant function in terms of the inverse sine function. If we need to determine y, which is the angle whose cosecant is x, we can rewrite $y = \csc^{-1} x$ as follows:

$$y = \csc^{-1} x$$

- Domain: $x \le -1$ or $x \ge 1$

  Range: $-\dfrac{\pi}{2} \le y \le \dfrac{\pi}{2},\, y \ne 0$

$$\csc y = x$$

- **Definition of inverse function**

$$\frac{1}{\sin y} = x$$

- **Substitute** $\dfrac{1}{\sin y}$ **for csc y.**

$$\sin y = \frac{1}{x}$$ • Solve for sin y.

$$y = \sin^{-1}\frac{1}{x}$$ • Write using inverse notation.

$$\csc^{-1}x = \sin^{-1}\frac{1}{x}$$ • Replace y with $\csc^{-1}x$.

Thus $\csc^{-1}x$ is the same as $\sin^{-1}\frac{1}{x}$. There is a similar identity for $\sec^{-1}x$.

Identities for $\csc^{-1}x$, $\sec^{-1}x$, and $\cot^{-1}x$

If $x \leq -1$ or $x \geq 1$, then

$$\csc^{-1}x = \sin^{-1}\frac{1}{x} \quad \text{and} \quad \sec^{-1}x = \cos^{-1}\frac{1}{x}$$

If x is a real number, then

$$\cot^{-1}x = \frac{\pi}{2} - \tan^{-1}x$$

• COMPOSITION OF TRIGONOMETRIC FUNCTIONS AND THEIR INVERSES

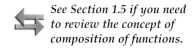

See Section 1.5 if you need to review the concept of composition of functions.

Recall that a function f and its inverse f^{-1} have the property that $f[f^{-1}(x)] = x$ for all x in the domain of f^{-1} and that $f^{-1}[f(x)] = x$ for all x in the domain of f. Applying this property to the functions $\sin x$, $\cos x$, and $\tan x$ and their inverse functions produces the following theorems.

Composition of Trigonometric Functions and Their Inverses

- If $-1 \leq x \leq 1$, then $\sin(\sin^{-1}x) = x$, and $\cos(\cos^{-1}x) = x$.

- If x is any real number, then $\tan(\tan^{-1}x) = x$.

- If $-\frac{\pi}{2} \leq x \leq \frac{\pi}{2}$, then $\sin^{-1}(\sin x) = x$.

- If $0 \leq x \leq \pi$, then $\cos^{-1}(\cos x) = x$.

- If $-\frac{\pi}{2} < x < \frac{\pi}{2}$, then $\tan^{-1}(\tan x) = x$.

In the next example we make use of some of the composition theorems to evaluate trigonometric expressions.

EXAMPLE 2   Evaluate the Composition of a Function and Its Inverse

Find the exact value of each composition of functions.

a. $\sin(\sin^{-1} 0.357)$ b. $\cos^{-1}(\cos 3)$ c. $\tan[\tan^{-1}(-11.27)]$

d. $\sin(\sin^{-1} \pi)$ e. $\cos(\cos^{-1} 0.277)$ f. $\tan^{-1}\left(\tan \dfrac{4\pi}{3}\right)$

Solution

a. Because 0.357 is in the interval $[-1, 1]$, $\sin(\sin^{-1} 0.357) = 0.357$.

b. Because 3 is in the interval $[0, \pi]$, $\cos^{-1}(\cos 3) = 3$.

c. Because -11.27 is a real number, $\tan[\tan^{-1}(-11.27)] = -11.27$.

d. Because π is not in the domain of the inverse sine function, $\sin(\sin^{-1} \pi)$ is undefined.

e. Because 0.277 is in the interval $[-1, 1]$, $\cos(\cos^{-1} 0.277) = 0.277$.

f. $\dfrac{4\pi}{3}$ is not in the interval $\left(-\dfrac{\pi}{2}, \dfrac{\pi}{2}\right)$; however, the reference angle for $\theta = \dfrac{4\pi}{3}$ is $\theta' = \dfrac{\pi}{3}$. Thus $\tan^{-1}\left(\tan \dfrac{4\pi}{3}\right) = \tan^{-1}\left(\tan \dfrac{\pi}{3}\right)$. Because $\dfrac{\pi}{3}$ is in the interval $\left(-\dfrac{\pi}{2}, \dfrac{\pi}{2}\right)$, $\tan^{-1}\left(\tan \dfrac{\pi}{3}\right) = \dfrac{\pi}{3}$. Hence $\tan^{-1}\left(\tan \dfrac{4\pi}{3}\right) = \dfrac{\pi}{3}$.

▶ **TRY EXERCISE 24, PAGE 253**

❓ **QUESTION** Is $\tan^{-1}(\tan x) = x$ an identity?

It is often easy to evaluate a trigonometric expression by referring to a sketch of a right triangle that satisfies given conditions. In Example 3 we make use of this technique.

EXAMPLE 3   Evaluate a Trigonometric Expression

Find the exact value of $\sin\left(\cos^{-1} \dfrac{2}{5}\right)$.

❓ **ANSWER** No. $\tan^{-1}(\tan x) = x$ only if $-\dfrac{\pi}{2} < x < \dfrac{\pi}{2}$.

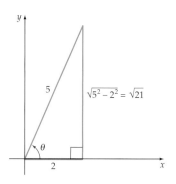

FIGURE 3.17

Solution

Let $\theta = \cos^{-1}\dfrac{2}{5}$, which implies $\cos\theta = \dfrac{2}{5}$. Because $\cos\theta$ is positive, θ is a

first-quadrant angle. We draw a right triangle with base 2 and hypotenuse 5

so that we can view θ, as shown in **Figure 3.17.** The height of the triangle is

$\sqrt{5^2 - 2^2} = \sqrt{21}$. Our goal is to find $\sin\theta$, which by definition is

$\dfrac{\text{opp}}{\text{hyp}} = \dfrac{\sqrt{21}}{5}$. Thus

$$\sin\left(\cos^{-1}\dfrac{2}{5}\right) = \sin(\theta) = \dfrac{\sqrt{21}}{5}$$

▶ **TRY EXERCISE 46, PAGE 253**

In Example 4, we sketch two right triangles to evaluate the given expression.

EXAMPLE 4   Evaluate a Trigonometric Expression

Find the exact value of $\sin\left[\sin^{-1}\dfrac{3}{5} + \cos^{-1}\left(-\dfrac{5}{13}\right)\right]$.

Solution

Let $\alpha = \sin^{-1}\dfrac{3}{5}$. Thus $\sin\alpha = \dfrac{3}{5}$. Let $\beta = \cos^{-1}\left(-\dfrac{5}{13}\right)$, which implies

that $\cos\beta = -\dfrac{5}{13}$. Sketch angles α and β as shown in **Figure 3.18.** We wish

to evaluate

$$\sin\left[\sin^{-1}\dfrac{3}{5} + \cos^{-1}\left(-\dfrac{5}{13}\right)\right] = \sin(\alpha + \beta)$$
$$= \sin\alpha\cos\beta + \cos\alpha\sin\beta \qquad (1)$$

A close look at the triangles in **Figure 3.18** shows us that

$$\cos\alpha = \dfrac{4}{5} \quad \text{and} \quad \sin\beta = \dfrac{12}{13}$$

Substituting in Equation (1) gives us our desired result.

$$\sin\left[\sin^{-1}\dfrac{3}{5} + \cos^{-1}\left(-\dfrac{5}{13}\right)\right] = \sin\alpha\cos\beta + \cos\alpha\sin\beta$$
$$= \left(\dfrac{3}{5}\right)\left(-\dfrac{5}{13}\right) + \left(\dfrac{4}{5}\right)\left(\dfrac{12}{13}\right) = \dfrac{33}{65}$$

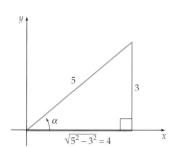

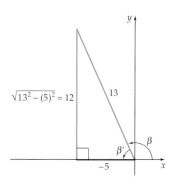

FIGURE 3.18

▶ **TRY EXERCISE 54, PAGE 254**

In Example 5 we make use of the identity $\cos(\cos^{-1} x) = x$, where $-1 \le x \le 1$,
to solve an equation.

EXAMPLE 5 **Solve an Inverse Trigonometric Equation**

Solve $\sin^{-1}\dfrac{3}{5} + \cos^{-1} x = \pi$.

Solution

Solve for $\cos^{-1} x$, and then take the cosine of both sides of the equation.

$$\sin^{-1}\frac{3}{5} + \cos^{-1} x = \pi$$

$$\cos^{-1} x = \pi - \sin^{-1}\frac{3}{5}$$

$$\cos(\cos^{-1} x) = \cos\left(\pi - \sin^{-1}\frac{3}{5}\right)$$

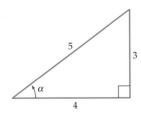

FIGURE 3.19

$$x = \cos(\pi - \alpha)$$

• Let $\alpha = \sin^{-1}\dfrac{3}{5}$. Note that α is the angle whose sine is $\dfrac{3}{5}$. (See **Figure 3.19.**)

$$= \cos\pi\cos\alpha + \sin\pi\sin\alpha$$ • Difference identity for cosine.

$$= (-1)\cos\alpha + (0)\sin\alpha$$

$$= -\cos\alpha$$

$$= -\frac{4}{5}$$ • $\cos\alpha = \dfrac{4}{5}$ (See **Figure 3.19.**)

▶ **TRY EXERCISE 64, PAGE 254**

EXAMPLE 6 **Verify a Trigonometric Identity That Involves Inverses**

Verify the identity $\sin^{-1} x + \cos^{-1} x = \dfrac{\pi}{2}$.

Solution

Let $\alpha = \sin^{-1} x$ and $\beta = \cos^{-1} x$. These equations imply that $\sin\alpha = x$ and $\cos\beta = x$. From the right triangles in **Figure 3.20,**

$$\cos\alpha = \sqrt{1 - x^2} \qquad \text{and} \qquad \sin\beta = \sqrt{1 - x^2}$$

Our goal is to show $\sin^{-1} x + \cos^{-1} x$ equals $\dfrac{\pi}{2}$.

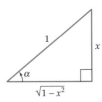

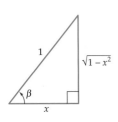

FIGURE 3.20

$$\sin^{-1} x + \cos^{-1} x = \alpha + \beta$$

$$= \cos^{-1}[\cos(\alpha + \beta)]$$ • Because $0 \le \alpha + \beta \le \pi$, we can apply $\alpha + \beta = \cos^{-1}[\cos(\alpha + \beta)]$.

$$= \cos^{-1}[\cos\alpha\cos\beta - \sin\alpha\sin\beta]$$ • Addition identity
for cosine

$$= \cos^{-1}\left[\left(\sqrt{1-x^2}\right)(x) - (x)\left(\sqrt{1-x^2}\right)\right]$$

$$= \cos^{-1}0 = \frac{\pi}{2}$$

▶ **TRY EXERCISE 72, PAGE 254**

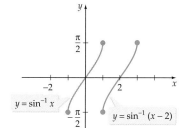

FIGURE 3.21

● GRAPHS OF INVERSE TRIGONOMETRIC FUNCTIONS

The inverse trigonometric functions can be graphed by using the procedures of stretching, shrinking, and translation that were discussed earlier in the text. For instance, the graph of $y = \sin^{-1}(x - 2)$ is a horizontal shift 2 units to the right of the graph of $y = \sin^{-1}x$, as shown in **Figure 3.21.**

EXAMPLE 7   **Graph an Inverse Function**

Graph: $y = \cos^{-1}x + 1$

Solution

Recall that the graph of $y = f(x) + c$ is a vertical translation of the graph of f. Because $c = 1$, a positive number, the graph of $y = \cos^{-1}x + 1$ is the graph of $y = \cos^{-1}x$ shifted 1 unit up. See **Figure 3.22.**

▶ **TRY EXERCISE 76, PAGE 254**

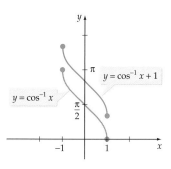

FIGURE 3.22

🖩 INTEGRATING TECHNOLOGY

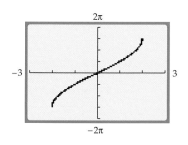

$y = 3\sin^{-1}0.5x$

FIGURE 3.23

When you use a graphing utility to draw the graph of an inverse trigonometric function, use the properties of these functions to verify the correctness of your graph. For instance, the graph of $y = 3\sin^{-1}0.5x$ is shown in **Figure 3.23.** The domain of $y = \sin^{-1}x$ is $-1 \le x \le 1$. Therefore, the domain of $y = 3\sin^{-1}0.5x$ is $-1 \le 0.5x \le 1$ or, multiplying the inequality by 2, $-2 \le x \le 2$. This is consistent with the graph in **Figure 3.23.**

The range of $y = \sin^{-1}x$ is $-\dfrac{\pi}{2} \le y \le \dfrac{\pi}{2}$. Thus the range of

$y = 3\sin^{-1}0.5x$ is $-\dfrac{3\pi}{2} \le y \le \dfrac{3\pi}{2}$. This is also consistent with the graph.

Verifying some of the properties of $y = \sin^{-1}x$ serves as a check that you have correctly entered the equation for the graph.

• AN APPLICATION INVOLVING AN INVERSE TRIGONOMETRIC FUNCTION

EXAMPLE 8   **Solve an Application**

 A camera is placed on a deck of a pool as shown in **Figure 3.24.** A diver is 18 feet above the camera lens. The extended length of the diver is 8 feet.

a. Show that the angle θ subtended at the lens by the diver is

$$\theta = \tan^{-1}\frac{26}{x} - \tan^{-1}\frac{18}{x}$$

b. For what values of x will $\theta = 9°$?

c. What value of x maximizes θ?

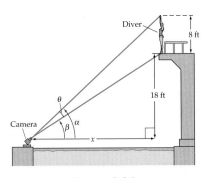

FIGURE 3.24

Solution

a. From **Figure 3.24** we see that $\alpha = \tan^{-1}\dfrac{26}{x}$ and $\beta = \tan^{-1}\dfrac{18}{x}$. Because

$\theta = \alpha - \beta$, we have $\theta = \tan^{-1}\dfrac{26}{x} - \tan^{-1}\dfrac{18}{x}$.

b. Use a graphing utility to graph $\theta = \tan^{-1}\dfrac{26}{x} - \tan^{-1}\dfrac{18}{x}$ and

$\theta = \dfrac{\pi}{20}\left(9° = \dfrac{\pi}{20} \text{ radians}\right)$. See **Figure 3.25.** Use the "intersect" command to show that θ is $9°$ for $x \approx 12.22$ feet and $x \approx 38.29$ feet.

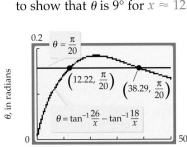

FIGURE 3.25

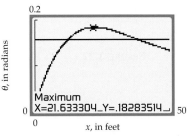

FIGURE 3.26

c. Use the "maximum" command to show that the maximum value of

$\theta = \tan^{-1}\dfrac{26}{x} - \tan^{-1}\dfrac{18}{x}$ occurs when $x \approx 21.63$ feet. See **Figure 3.26.**

▶ **TRY EXERCISE 84, PAGE 254**

TOPICS FOR DISCUSSION

1. Is the equation

$$\tan^{-1} x = \frac{1}{\tan x}$$

true for all values of x, true for some values of x, or false for all values of x?

2. Are there real numbers x for which the following is true? Explain.

$$\sin(\sin^{-1} x) \neq \sin^{-1}(\sin x)$$

3. Explain how to find the value of $\sec^{-1} 3$ by using a scientific calculator.

4. Explain how you can determine the range of $y = (2 \cos^{-1} x) - 1$ using

   a. algebra   **b.** a graph

EXERCISE SET 3.5

In Exercises 1 to 20, find the exact radian value.

1. $\sin^{-1} 1$ ▶ **2.** $\sin^{-1} \dfrac{\sqrt{2}}{2}$

3. $\cos^{-1}\left(-\dfrac{\sqrt{3}}{2}\right)$ **4.** $\cos^{-1}\left(-\dfrac{1}{2}\right)$ **5.** $\tan^{-1}(-1)$

6. $\tan^{-1} \sqrt{3}$ **7.** $\cot^{-1} \dfrac{\sqrt{3}}{3}$ **8.** $\cot^{-1} 1$

9. $\sec^{-1} 2$ **10.** $\sec^{-1} \dfrac{2\sqrt{3}}{3}$ **11.** $\csc^{-1}\left(-\sqrt{2}\right)$

12. $\csc^{-1}(-2)$ **13.** $\sin^{-1}\left(-\dfrac{\sqrt{3}}{2}\right)$ **14.** $\sin^{-1} \dfrac{1}{2}$

15. $\cos^{-1}\left(-\dfrac{1}{2}\right)$ **16.** $\cos^{-1} \dfrac{\sqrt{3}}{2}$ **17.** $\tan^{-1} \dfrac{\sqrt{3}}{3}$

18. $\tan^{-1} 1$ **19.** $\cot^{-1} \sqrt{3}$ **20.** $\cot^{-1}(-1)$

In Exercises 21 to 56, find the exact value of the given expression. If an exact value cannot be given, give the value to the nearest ten-thousandth.

21. $\cos\left(\cos^{-1} \dfrac{1}{2}\right)$ **22.** $\cos(\cos^{-1} 2)$

23. $\tan(\tan^{-1} 2)$ ▶ **24.** $\tan\left(\tan^{-1} \dfrac{1}{2}\right)$

25. $\sin\left(\tan^{-1} \dfrac{3}{4}\right)$ **26.** $\cos\left(\sin^{-1} \dfrac{5}{13}\right)$

27. $\tan\left(\sin^{-1} \dfrac{\sqrt{2}}{2}\right)$ **28.** $\sin\left[\cos^{-1}\left(-\dfrac{\sqrt{3}}{2}\right)\right]$

29. $\cos(\sec^{-1} 2)$ **30.** $\sin^{-1}(\sin 2)$

31. $\sin^{-1}\left(\sin \dfrac{\pi}{6}\right)$ **32.** $\sin^{-1}\left(\sin \dfrac{5\pi}{6}\right)$

33. $\cos^{-1}\left(\sin \dfrac{\pi}{4}\right)$ **34.** $\cos^{-1}\left(\cos \dfrac{5\pi}{4}\right)$

35. $\sin^{-1}\left(\tan \dfrac{\pi}{3}\right)$ **36.** $\cos^{-1}\left(\tan \dfrac{2\pi}{3}\right)$

37. $\tan^{-1}\left(\sin \dfrac{\pi}{6}\right)$ **38.** $\cot^{-1}\left(\cos \dfrac{2\pi}{3}\right)$

39. $\sin^{-1}\left[\cos\left(-\dfrac{2\pi}{3}\right)\right]$ **40.** $\cos^{-1}\left[\tan\left(-\dfrac{\pi}{3}\right)\right]$

41. $\tan\left(\sin^{-1} \dfrac{1}{2}\right)$ **42.** $\cot(\csc^{-1} 2)$

43. $\sec\left(\sin^{-1} \dfrac{1}{4}\right)$ **44.** $\csc\left(\cos^{-1} \dfrac{3}{4}\right)$

45. $\cos\left(\sin^{-1} \dfrac{7}{25}\right)$ ▶ **46.** $\tan\left(\cos^{-1} \dfrac{3}{5}\right)$

47. $\sec\left(\tan^{-1}\dfrac{12}{5}\right)$

48. $\csc\left(\sin^{-1}\dfrac{12}{13}\right)$

49. $\cos\left(2\sin^{-1}\dfrac{\sqrt{2}}{2}\right)$

50. $\tan\left(2\sin^{-1}\dfrac{\sqrt{3}}{2}\right)$

51. $\sin\left(2\sin^{-1}\dfrac{4}{5}\right)$

52. $\cos(2\tan^{-1}1)$

53. $\sin\left(\sin^{-1}\dfrac{2}{3}+\cos^{-1}\dfrac{1}{2}\right)$

▶ **54.** $\cos\left(\sin^{-1}\dfrac{3}{4}+\cos^{-1}\dfrac{5}{13}\right)$

55. $\tan\left(\cos^{-1}\dfrac{1}{2}-\sin^{-1}\dfrac{3}{4}\right)$

56. $\sec\left(\cos^{-1}\dfrac{2}{3}+\sin^{-1}\dfrac{2}{3}\right)$

In Exercises 57 to 66, solve the equation for x algebraically.

57. $\sin^{-1}x=\cos^{-1}\dfrac{5}{13}$

58. $\tan^{-1}x=\sin^{-1}\dfrac{24}{25}$

59. $\sin^{-1}(x-1)=\dfrac{\pi}{2}$

60. $\cos^{-1}\left(x-\dfrac{1}{2}\right)=\dfrac{\pi}{3}$

61. $\tan^{-1}\left(x+\dfrac{\sqrt{2}}{2}\right)=\dfrac{\pi}{4}$

62. $\sin^{-1}(x-2)=-\dfrac{\pi}{6}$

63. $\sin^{-1}\dfrac{3}{5}+\cos^{-1}x=\dfrac{\pi}{4}$

▶ **64.** $\sin^{-1}x+\cos^{-1}\dfrac{4}{5}=\dfrac{\pi}{6}$

65. $\sin^{-1}\dfrac{\sqrt{2}}{2}+\cos^{-1}x=\dfrac{2\pi}{3}$

66. $\cos^{-1}x+\sin^{-1}\dfrac{\sqrt{3}}{2}=\dfrac{\pi}{2}$

In Exercises 67 to 70, evaluate each expression.

67. $\cos(\sin^{-1}x)$

68. $\tan(\cos^{-1}x)$

69. $\sin(\sec^{-1}x)$

70. $\sec(\sin^{-1}x)$

In Exercises 71 to 74, verify the identity.

71. $\sin^{-1}x+\sin^{-1}(-x)=0$

▶ **72.** $\cos^{-1}x+\cos^{-1}(-x)=\pi$

73. $\tan^{-1}x+\tan^{-1}\dfrac{1}{x}=\dfrac{\pi}{2},\ x>0$

74. $\sec^{-1}\dfrac{1}{x}+\csc^{-1}\dfrac{1}{x}=\dfrac{\pi}{2}$

In Exercises 75 to 82, use stretching, shrinking, and translation procedures to graph each equation.

75. $y=\sin^{-1}x+2$

▶ **76.** $y=\cos^{-1}(x-1)$

77. $y=\sin^{-1}(x+1)-2$

78. $y=\tan^{-1}(x-1)+2$

79. $y=2\cos^{-1}x$

80. $y=-2\tan^{-1}x$

81. $y=\tan^{-1}(x+1)-2$

82. $y=\sin^{-1}(x-2)+1$

83. **DOT-MATRIX PRINTING** In dot-matrix printing, the *blank-area factor* is the ratio of the blank area (unprinted area) to the total area of the line. If circular dots are used to print, then the blank-area factor is given by

$$\frac{A}{(S)(D)}=1-\frac{1}{2}\left[1-\left(\frac{S}{D}\right)^2+\frac{D}{S}\sin^{-1}\left(\frac{S}{D}\right)\right]$$

where $A=A_1+A_2$, A_1 and A_2 are the areas of the regions shown in the figure, S is the distance between the centers of overlapping dots, and D is the diameter of a dot.

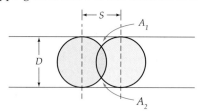

Calculate, to four decimal places, the blank-area factor where

a. $D=0.2$ millimeter and $S=0.1$ millimeter

b. $D=0.16$ millimeter and $S=0.1$ millimeter

▶ **84.** **VOLUME IN A WATER TANK** The volume V of water (measured in cubic feet) in a horizontal cylindrical tank of radius 5 feet and length 12 feet is given by

$$V(x)=12\left[25\cos^{-1}\left(\frac{5-x}{5}\right)-(5-x)\sqrt{10x-x^2}\right]$$

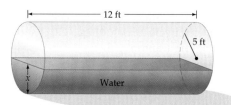

where x is the depth of the water in feet.

a. 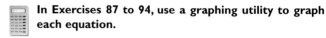 Graph V over its domain $0 \le x \le 10$.

b. Write a sentence that explains why the graph of V increases more rapidly when x increases from 4.9 feet to 5 feet than it does when x increases from 0.1 foot to 0.2 foot.

c. If $x = 4$ feet, find the volume (to the nearest 0.01 cubic foot) of the water in the tank.

d. Find the depth x (to the nearest 0.01 foot) if there are 288 cubic feet of water in the tank.

85. Graph $f(x) = \cos^{-1} x$ and $g(x) = \sin^{-1} \sqrt{1 - x^2}$ on the same coordinate axes. Does $f(x) = g(x)$ on the interval $[-1, 1]$?

86. Graph $y = \cos(\cos^{-1} x)$ on $[-1, 1]$. Graph $y = \cos^{-1} (\cos x)$ on $[-2\pi, 2\pi]$.

In Exercises 87 to 94, use a graphing utility to graph each equation.

87. $y = \csc^{-1} 2x$

88. $y = 0.5 \sec^{-1} \dfrac{x}{2}$

89. $y = \sec^{-1}(x - 1)$

90. $y = \sec^{-1}(x + \pi)$

91. $y = 2 \tan^{-1} 2x$

92. $y = \tan^{-1}(x - 1)$

93. $y = \cot^{-1} \dfrac{x}{3}$

94. $y = 2 \cot^{-1}(x - 1)$

CONNECTING CONCEPTS

In Exercises 95 to 98, verify the identity.

95. $\cos(\sin^{-1} x) = \sqrt{1 - x^2}$

96. $\sec(\sin^{-1} x) = \dfrac{\sqrt{1 - x^2}}{1 - x^2}$

97. $\tan(\csc^{-1} x) = \dfrac{\sqrt{x^2 - 1}}{x^2 - 1}, x > 1$

98. $\sin(\cot^{-1} x) = \dfrac{\sqrt{x^2 + 1}}{x^2 + 1}$

In Exercises 99 to 102, solve for y in terms of x.

99. $5x = \tan^{-1} 3y$

100. $2x = \dfrac{1}{2} \sin^{-1} 2y$

101. $x - \dfrac{\pi}{3} = \cos^{-1}(y - 3)$

102. $x + \dfrac{\pi}{2} = \tan^{-1}(2y - 1)$

PREPARE FOR SECTION 3.6

103. Use the quadratic formula to solve $3x^2 - 5x - 4 = 0$. [1.1]

104. Use a Pythagorean identity to write $\sin^2 x$ as a function involving $\cos^2 x$. [3.1]

105. Evaluate $\dfrac{\pi}{2} + 2k\pi$ for $k = 1, 2$, and 3.

106. Factor by grouping: $x^2 - \dfrac{\sqrt{3}}{2} x + x - \dfrac{\sqrt{3}}{2}$.

107. Use a graphing utility to construct a scatter plot for the data at the right. Use a viewing window with Xmin=0, Xmax=40, Ymin=0, and Ymax=100. [1.7]

x	y
3	14
7	55
11	90
15	99
19	80
23	44
27	8
31	4

108. Solve $2x^2 - 2x = 0$ by factoring. [1.1]

PROJECTS

I. VISUAL INSIGHT

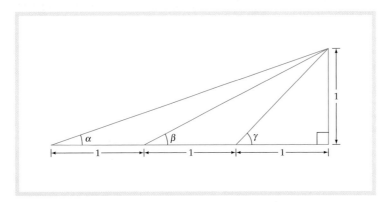

Explain how the figure above can be used to verify each identity.

a. $\tan^{-1}\dfrac{1}{3} + \tan^{-1}\dfrac{1}{2} = \dfrac{\pi}{4}$ (*Hint:* Start by using an identity to find the value of $\tan(\alpha + \beta)$.) **b.** $\alpha + \beta = \gamma$

| SECTION 3.6 | TRIGONOMETRIC EQUATIONS |

- SOLVE TRIGONOMETRIC EQUATIONS
- AN APPLICATION INVOLVING A TRIGONOMETRIC EQUATION
- MODEL SINUSOIDAL DATA

• SOLVE TRIGONOMETRIC EQUATIONS

Consider the equation $\sin x = \dfrac{1}{2}$. The graph of $y = \sin x$, along with the line $y = \dfrac{1}{2}$, is shown in **Figure 3.27.** The x values of the intersections of the two graphs are the solutions of $\sin x = \dfrac{1}{2}$. The solutions in the interval $0 \le x < 2\pi$ are $x = \dfrac{\pi}{6}$ and $\dfrac{5\pi}{6}$.

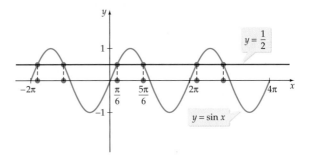

FIGURE 3.27

If we remove the restriction $0 \le x < 2\pi$, there are many more solutions. Because the sine function is periodic with a period of 2π, other solutions are obtained by adding $2k\pi$, k an integer, to either of the previous solutions. Thus the solutions of $\sin x = \dfrac{1}{2}$ are

$$x = \frac{\pi}{6} + 2k\pi, \quad k \text{ an integer}$$

$$x = \frac{5\pi}{6} + 2k\pi, \quad k \text{ an integer}$$

? QUESTION How many solutions does the equation $\cos x = \dfrac{\sqrt{3}}{2}$ have on the interval $0 \le x < 2\pi$?

Algebraic methods and trigonometric identities are used frequently to find the solutions of trigonometric equations. Algebraic methods that are often employed include solving by factoring, solving by using the quadratic formula, and squaring each side of the equation.

EXAMPLE 1   Solve a Trigonometric Equation by Factoring

Solve $2 \sin^2 x \cos x - \cos x = 0$, where $0 \le x < 2\pi$.

Algebraic Solution

$2 \sin^2 x \cos x - \cos x = 0$

$\quad \cos x(2 \sin^2 x - 1) = 0$ • **Factor cos x from each term.**

$\cos x = 0 \quad$ or $\quad 2 \sin^2 x - 1 = 0$ • **Use the Principle of Zero Products.**

$x = \dfrac{\pi}{2}, \dfrac{3\pi}{2} \qquad\qquad \sin^2 x = \dfrac{1}{2}$ • **Solve each equation for x with $0 \le x < 2\pi$.**

$$\sin x = \pm\frac{\sqrt{2}}{2}$$

$$x = \frac{\pi}{4}, \frac{3\pi}{4}, \frac{5\pi}{4}, \frac{7\pi}{4}$$

The solutions in the interval $0 \le x < 2\pi$ are $\dfrac{\pi}{4}, \dfrac{\pi}{2}, \dfrac{3\pi}{4}, \dfrac{5\pi}{4}, \dfrac{3\pi}{2}$, and $\dfrac{7\pi}{4}$.

Visualize the Solution

The solutions are the x-coordinates of the x-intercepts of $y = 2 \sin^2 x \cos x - \cos x$ on the interval $[0, 2\pi)$. See **Figure 3.28**.

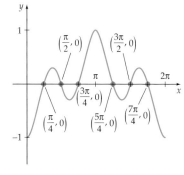

$y = 2 \sin^2 x \cos x - \cos x$

FIGURE 3.28

▶ TRY EXERCISE 14, PAGE 266

? ANSWER Two

Squaring both sides of an equation may not produce an equivalent equation. Thus, when this method is used, the proposed solutions must be checked to eliminate any extraneous solutions.

EXAMPLE 2 **Solve a Trigonometric Equation by Squaring Each Side of the Equation**

Solve $\sin x + \cos x = 1$, where $0 \le x < 2\pi$.

Algebraic Solution

$$\sin x + \cos x = 1 \qquad \text{• Solve for } \sin x.$$
$$\sin x = 1 - \cos x$$
$$\sin^2 x = (1 - \cos x)^2 \qquad \text{• Square each side.}$$
$$\sin^2 x = 1 - 2\cos x + \cos^2 x$$
$$1 - \cos^2 x = 1 - 2\cos x + \cos^2 x \qquad \text{• } \sin^2 x = 1 - \cos^2 x$$
$$2\cos^2 x - 2\cos x = 0$$
$$2\cos x(\cos x - 1) = 0 \qquad \text{• Factor.}$$
$$2\cos x = 0 \quad \text{or} \quad \cos x = 1$$
$$x = \frac{\pi}{2}, \frac{3\pi}{2} \qquad\qquad x = 0 \qquad \begin{array}{l}\text{• Solve each equation for } x \\ \text{with } 0 \le x < 2\pi.\end{array}$$

Squaring each side of an equation may introduce extraneous solutions. Therefore, we must check the solutions. A check will show that 0 and $\dfrac{\pi}{2}$ are solutions but $\dfrac{3\pi}{2}$ is not a solution.

Visualize the Solution

The solutions are the x-coordinates of the points of intersection of $y = \sin x + \cos x$ and $y = 1$ on the interval $[0, 2\pi)$. See **Figure 3.29.**

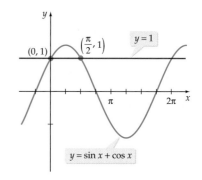

FIGURE 3.29

▶ **TRY EXERCISE 52, PAGE 266**

EXAMPLE 3 **Solve a Trigonometric Equation by Using the Quadratic Formula**

Solve $3\cos^2 x - 5\cos x - 4 = 0$, where $0 \le x < 2\pi$.

Algebraic Solution

The given equation is quadratic in form and cannot be factored easily. However, we can use the quadratic formula to solve for $\cos x$.

$$3\cos^2 x - 5\cos x - 4 = 0 \qquad \text{• } a = 3, b = -5, c = -4$$
$$\cos x = \frac{-(-5) \pm \sqrt{(-5)^2 - 4(3)(-4)}}{(2)(3)} = \frac{5 \pm \sqrt{73}}{6}$$

The equation $\cos x = \dfrac{5 + \sqrt{73}}{6}$ does not have a solution because $\dfrac{5 + \sqrt{73}}{6} > 2$ and for any x the maximum value of $\cos x$ is 1. Thus

Visualize the Solution

The solutions are the x-coordinates of the x-intercepts of $y = 3\cos^2 x - 5\cos x - 4$ on the interval $[0, 2\pi)$. See **Figure 3.30.**

$\cos x = \dfrac{5 - \sqrt{73}}{6}$, and because $\dfrac{5 - \sqrt{73}}{6}$ is a negative number (about -0.59), the equation $\cos x = \dfrac{5 - \sqrt{73}}{6}$ will have two solutions on the interval $[0, 2\pi)$. Thus

$$x = \cos^{-1}\left(\dfrac{5 - \sqrt{73}}{6}\right) \approx 2.2027 \qquad \text{or}$$

$$x = 2\pi - \cos^{-1}\left(\dfrac{5 - \sqrt{73}}{6}\right) \approx 4.0805$$

To the nearest 0.0001, the solutions on the interval $[0, 2\pi)$ are 2.2027 and 4.0805.

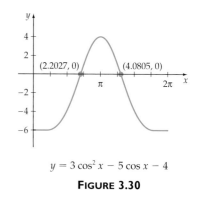

$y = 3\cos^2 x - 5\cos x - 4$

FIGURE 3.30

▶ **TRY EXERCISE 56, PAGE 266**

When solving equations that contain multiple angles, we must be sure we find all the solutions of the equation for the given interval. For example, to find all solutions of $\sin 2x = \dfrac{1}{2}$, where $0 \le x < 2\pi$, we first solve for $2x$.

$$\sin 2x = \dfrac{1}{2}$$

$$2x = \dfrac{\pi}{6} + 2k\pi \quad \text{or} \quad 2x = \dfrac{5\pi}{6} + 2k\pi \qquad \bullet \; k \text{ is an integer.}$$

Solving for x, we have $x = \dfrac{\pi}{12} + k\pi$ or $x = \dfrac{5\pi}{12} + k\pi$. Substituting integers for k, we obtain

$$k = 0: \qquad x = \dfrac{\pi}{12} \quad \text{or} \quad x = \dfrac{5\pi}{12}$$

$$k = 1: \qquad x = \dfrac{13\pi}{12} \quad \text{or} \quad x = \dfrac{17\pi}{12}$$

$$k = 2: \qquad x = \dfrac{25\pi}{12} \quad \text{or} \quad x = \dfrac{29\pi}{12}$$

Note that for $k \ge 2$, $x \ge 2\pi$, and the solutions to $\sin 2x = \dfrac{1}{2}$ are not in the interval $0 \le x < 2\pi$. Thus, for $0 \le x < 2\pi$, the solutions are $\dfrac{\pi}{12}, \dfrac{5\pi}{12}, \dfrac{13\pi}{12},$ and $\dfrac{17\pi}{12}$. See **Figure 3.31.**

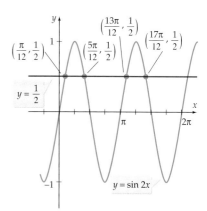

$y = \sin 2x$

FIGURE 3.31

EXAMPLE 4 **Solve a Trigonometric Equation**

Solve: $\sin 3x = 1$

Algebraic Solution

The equation $\sin 3x = 1$ implies

$$3x = \frac{\pi}{2} + 2k\pi, \quad k \text{ an integer}$$

$$x = \frac{\pi}{6} + \frac{2k\pi}{3}, \quad k \text{ an integer} \qquad \bullet \text{ Divide each side by 3.}$$

Because x is not restricted to a finite interval, the given equation has an infinite number of solutions. All of the solutions are represented by the equation

$$x = \frac{\pi}{6} + \frac{2k\pi}{3}, \quad \text{where } k \text{ is an integer}$$

Visualize the Solution

The solutions are the x-coordinates of the points of intersection of $y = \sin 3x$ and $y = 1$. **Figure 3.32** shows eight of the points of intersection.

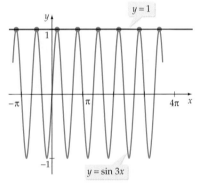

FIGURE 3.32

▶ **TRY EXERCISE 66, PAGE 267**

EXAMPLE 5 **Solve a Trigonometric Equation**

Solve $\sin^2 2x - \dfrac{\sqrt{3}}{2} \sin 2x + \sin 2x - \dfrac{\sqrt{3}}{2} = 0$, where $0° \le x < 360°$.

Algebraic Solution

Factor the left side of the equation by grouping, and then set each factor equal to zero.

$$\sin^2 2x - \frac{\sqrt{3}}{2} \sin 2x + \sin 2x - \frac{\sqrt{3}}{2} = 0$$

$$\sin 2x \left(\sin 2x - \frac{\sqrt{3}}{2} \right) + \left(\sin 2x - \frac{\sqrt{3}}{2} \right) = 0$$

$$(\sin 2x + 1)\left(\sin 2x - \frac{\sqrt{3}}{2} \right) = 0$$

$$\sin 2x + 1 = 0 \quad \text{or} \quad \sin 2x - \frac{\sqrt{3}}{2} = 0$$

$$\sin 2x = -1 \qquad\qquad \sin 2x = \frac{\sqrt{3}}{2}$$

Visualize the Solution

The solutions are the x-coordinates of the x-intercepts of

$$y = \sin^2 2x - \frac{\sqrt{3}}{2} \sin 2x$$
$$+ \sin 2x - \frac{\sqrt{3}}{2}$$

on the interval $[0, 2\pi)$. See **Figure 3.33.**

The equation $\sin 2x = -1$ implies that $2x = 270° + 360° \cdot k$, k an integer. Thus $x = 135° + 180° \cdot k$. The solutions of this equation with $0° \le x < 360°$ are $135°$ and $315°$. Similarly, the equation $\sin 2x = \dfrac{\sqrt{3}}{2}$ implies

$$2x = 60° + 360° \cdot k \quad \text{or} \quad 2x = 120° + 360° \cdot k$$
$$x = 30° + 180° \cdot k \qquad\qquad x = 60° + 180° \cdot k$$

The solutions with $0° \le x < 360°$ are $30°$, $60°$, $210°$, and $240°$. Combining the solutions from each equation, we have $30°$, $60°$, $135°$, $210°$, $240°$, and $315°$ as our solutions.

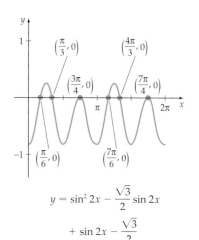

$$y = \sin^2 2x - \frac{\sqrt{3}}{2} \sin 2x$$
$$+ \sin 2x - \frac{\sqrt{3}}{2}$$

FIGURE 3.33

▶ **TRY EXERCISE 84, PAGE 267**

In Example 6, algebraic methods do not provide the solutions, so we rely on a graph.

EXAMPLE 6 **Approximate Solutions Graphically**

Use a graphing utility to approximate the solutions of $x + 3 \cos x = 0$.

Solution

The solutions are the x-intercepts of $y = x + 3 \cos x$. See **Figure 3.34.** A close-up view of the graph of $y = x + 3 \cos x$ shows that, to the nearest thousandth, the solutions are

$$x_1 = -1.170, \quad x_2 = 2.663, \quad \text{and} \quad x_3 = 2.938$$

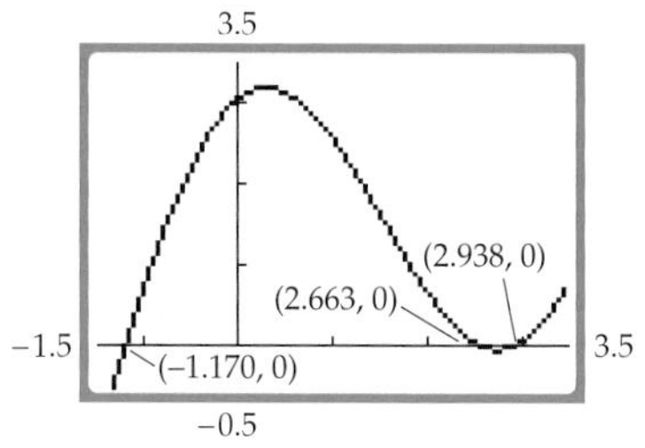

▶ **TRY EXERCISE 86, PAGE 267**

$y = x + 3 \cos x$

FIGURE 3.34

● AN APPLICATION INVOLVING A TRIGONOMETRIC EQUATION

EXAMPLE 7 **Solve a Projectile Application**

A projectile is fired at an angle of inclination θ from the horizon with an initial velocity v_0. Its range d (neglecting air resistance) is given by

$$d = \frac{v_0^2}{16} \sin \theta \cos \theta$$

where v_0 is measured in feet per second and d is measured in feet. See **Figure 3.35.**

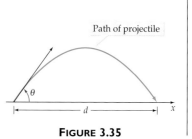

Path of projectile

FIGURE 3.35

Continued ▶

a. If $v_0 = 325$ feet per second, find the angles θ (in degrees) for which the projectile will hit a target 2295 feet downrange.

b. What is the maximum horizontal range for a projectile that has an initial velocity of 474 feet per second?

c. Determine the angle of inclination that produces the maximum range.

Solution

a. We need to solve

$$2295 = \frac{325^2}{16} \sin \theta \cos \theta \tag{1}$$

for θ, where $0° < \theta < 90°$.

Method 1   The following solutions were obtained by using a graphing utility to graph $d = 2295$ and $d = \dfrac{325^2}{16} \sin \theta \cos \theta$. See **Figure 3.36.** Thus there are two angles for which the projectile will hit the target. To the nearest thousandth of a degree, they are

$$\theta = 22.025° \quad \text{and} \quad \theta = 67.975°$$

It should be noted that the graph in **Figure 3.36** is *not* a graph of the path of the projectile. It is a graph of the distance d as a function of the angle θ.

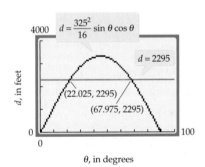

4000 $d = \dfrac{325^2}{16} \sin \theta \cos \theta$

d, in feet

$d = 2295$

(22.025, 2295)
(67.975, 2295)

0   100
0

θ, in degrees

FIGURE 3.36

Method 2   To solve algebraically, we proceed as follows. Multiply each side of Equation (1) by 16 and divide by 325^2 to produce

$$\sin \theta \cos \theta = \frac{(16)(2295)}{325^2}$$

The identity $2 \sin \theta \cos \theta = \sin 2\theta$ gives us $\sin \theta \cos \theta = \dfrac{\sin 2\theta}{2}$.

Hence

$$\frac{\sin 2\theta}{2} = \frac{(16)(2295)}{325^2}$$

$$\sin 2\theta = 2\frac{(16)(2295)}{325^2} \approx 0.69529$$

There are two angles in the interval $[0°, 180°]$ whose sines are 0.69529. One is $\sin^{-1} 0.69529$, and the other one is the *reference angle* for $\sin^{-1} 0.69529$. Therefore,

$$2\theta \approx \sin^{-1} 0.69529 \quad \text{or} \quad 2\theta \approx 180° - \sin^{-1} 0.69529$$

$$\theta \approx \frac{1}{2} \sin^{-1} 0.69529 \quad \text{or} \quad \theta \approx \frac{1}{2}(180° - \sin^{-1} 0.69529)$$

$$\theta \approx 22.025° \quad \text{or} \quad \theta \approx 67.975°$$

These are the same angles that we obtained using Method 1.

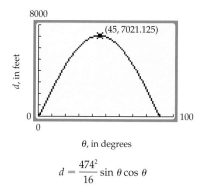

$$d = \frac{474^2}{16} \sin \theta \cos \theta$$

FIGURE 3.37

b. Use a graphing utility to find that the graph of $d = \frac{474^2}{16} \sin \theta \cos \theta$ has a maximum value of $d = 7021.125$ feet. See **Figure 3.37.**

c. In part **b.**, the maximum value is attained for $\theta = 45°$. To prove that this is true in general, we use $2 \sin \theta \cos \theta = \sin 2\theta$ to write

$$d = \frac{v_0^2}{16} \sin \theta \cos \theta \qquad \text{as} \qquad d = \frac{v_0^2}{32} \sin 2\theta$$

This equation enables us to determine that d will attain its maximum when $\sin 2\theta$ attains its maximum—that is, when $2\theta = 90°$, or $\theta = 45°$.

▶ **TRY EXERCISE 92, PAGE 267**

● MODEL SINUSOIDAL DATA

Data that can be closely modeled by a function of the form $y = a \sin (bx + c) + d$ are called **sinusoidal data.** Many graphing utilities are designed to perform a **sine regression** to find the sine function that provides the best least-squares fit for sinusoidal data. For instance, the TI-83 uses the command **SinReg** and the TI-86 uses **SinR** to perform a sine regression. The process generally works best for those sets of data for which we have a good estimate of the period.

In Example 8 we use a sine regression to model the percent of illumination of the moon, as seen from Earth. See **Figure 3.38.**

FIGURE 3.38

EXAMPLE 8   Use a Sine Regression to Model Data

Table 3.3 on the following page shows the percent of the moon illuminated, at midnight Central Standard Time, for selected days of January 2006. Find the regression function that models the percent of the moon illuminated as a function of the day, with January 1, 2006 represented by $x = 1$. Use the function to estimate the percent of the moon illuminated at midnight Central Standard Time on January 21, 2006. *Note:* The illumination cycle of the moon has a period of about 29.53 days.

Continued ▶

TABLE 3.3

Midnight of: Date in 2006 (CST)	Day Number	% of Moon Illuminated
Jan. 3	3	14
Jan. 7	7	55
Jan. 11	11	90
Jan. 15	15	99
Jan. 19	19	80
Jan. 23	23	44
Jan. 27	27	8
Jan. 31	31	4

The percent of the moon illuminated is nearly the same regardless of one's position on Earth.

Source: The Astronomical Applications Department of the U.S. Naval Observatory.

Solution

See Section 1.7 if you need to review how to construct a scatter plot.

1. **Construct a scatter plot of the data.** Enter the data from **Table 3.3** into your graphing utility. See **Figure 3.39.** The sinusoidal nature of the scatter plot in **Figure 3.40** indicates that the data can be effectively modeled by a sine function.

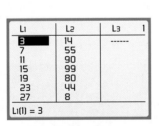

FIGURE 3.39

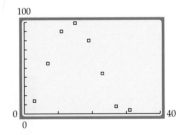

FIGURE 3.40

INTEGRATING TECHNOLOGY

A sine regression requires at least four data points. The iterations number 16 in **Figure 3.41** is the maximum number of times the SinReg command will iterate to find a regression equation. Any integer from 1 to 16 can be used; however, the integer 16 generally produces more accurate results than smaller integers.

2. **Find the regression equation.** On a TI-83 graphing calculator, the input shown in **Figure 3.41** produces the results in **Figure 3.42.**

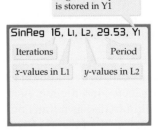

FIGURE 3.41

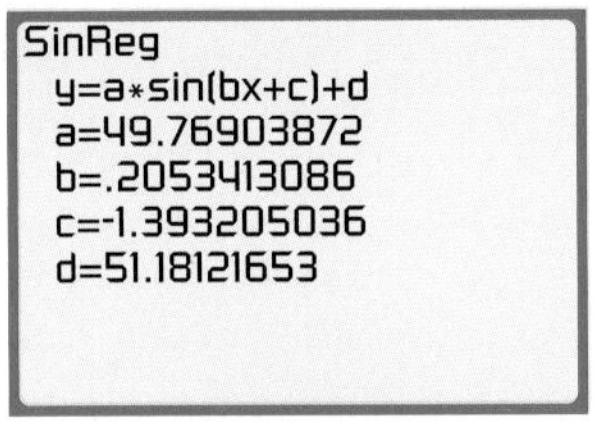

FIGURE 3.42

The regression equation is $y \approx 49.77 \sin(0.2053x - 1.393) + 51.18$.

3. **Examine the fit.** The SinReg command does not yield a correlation coefficient. However, a graph of the regression equation and the scatter plot of the data shows that the regression equation provides a good model. See **Figure 3.43.**

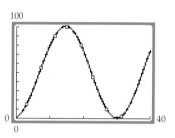

FIGURE 3.43

According to the following result, the percent of the moon illuminated at midnight (CST) on January 21, 2006 ($x = 21$) will be about 62%.

$$y \approx 49.77 \sin(0.2053(21) - 1.393) + 51.18 \approx 62$$

▶ **TRY EXERCISE 94, PAGE 268**

TOPICS FOR DISCUSSION

1. Explain why it is not necessary for a graphing utility to have a cosine regression.

2. A student finds that $x = 0$ is a solution of $\sin x = x$. Because the function $y = \sin x$ has a period of 2π, the student reasons that $\pm 2\pi$, $\pm 4\pi$, $\pm 6\pi, \ldots$ are also solutions. Explain why the student is not correct.

3. How many solutions does $2 \sin\left(x - \dfrac{\pi}{2}\right) = 5$ have on the interval $0 \le x < 2\pi$? Explain.

4. How many solutions does $\sin\left(\dfrac{1}{x}\right) = 0$ have on the interval $0 < x < \dfrac{\pi}{2}$? Explain.

5. On the interval $0 \le x < 2\pi$, the equation $\sin x = \dfrac{1}{2}$ has solutions of $x = \dfrac{\pi}{6}$ and $x = \dfrac{5\pi}{6}$. How would you write the solutions of $\sin x = \dfrac{1}{2}$ if the real number x were not restricted to the interval $[0, 2\pi)$?

EXERCISE SET 3.6

In Exercises 1 to 22, solve each equation for exact solutions in the interval $0 \le x < 2\pi$.

1. $\sec x - \sqrt{2} = 0$

2. $2 \sin x = \sqrt{3}$

3. $\tan x - \sqrt{3} = 0$

4. $\cos x - 1 = 0$

5. $2 \sin x \cos x = \sqrt{2} \cos x$

6. $2 \sin x \cos x = \sqrt{3} \sin x$

7. $\sin^2 x - 1 = 0$

8. $\cos^2 x - 1 = 0$

9. $4 \sin x \cos x - 2\sqrt{3} \sin x - 2\sqrt{2} \cos x + \sqrt{6} = 0$

10. $\sec^2 x + \sqrt{3} \sec x - \sqrt{2} \sec x - \sqrt{6} = 0$

11. $\csc x - \sqrt{2} = 0$

12. $3 \cot x + \sqrt{3} = 0$

13. $2 \sin^2 x + 1 = 3 \sin x$

▶ **14.** $2 \cos^2 x + 1 = -3 \cos x$

15. $4 \cos^2 x - 3 = 0$

16. $2 \sin^2 x - 1 = 0$

17. $2 \sin^3 x = \sin x$

18. $4 \cos^3 x = 3 \cos x$

19. $4 \sin^2 x + 2\sqrt{3} \sin x - \sqrt{3} = 2 \sin x$

20. $\tan^2 x + \tan x - \sqrt{3} = \sqrt{3} \tan x$

21. $\sin^4 x = \sin^2 x$

22. $\cos^4 x = \cos^2 x$

In Exercises 23 to 60, solve each equation, where $0° \le x < 360°$. Round approximate solutions to the nearest tenth of a degree.

23. $\cos x - 0.75 = 0$

24. $\sin x + 0.432 = 0$

25. $3 \sin x - 5 = 0$

26. $4 \cos x - 1 = 0$

27. $3 \sec x - 8 = 0$

28. $4 \csc x + 9 = 0$

29. $\cos x + 3 = 0$

30. $\sin x - 4 = 0$

31. $3 - 5 \sin x = 4 \sin x + 1$

32. $4 \cos x - 5 = \cos x - 3$

33. $\dfrac{1}{2} \sin x + \dfrac{2}{3} = \dfrac{3}{4} \sin x + \dfrac{3}{5}$

34. $\dfrac{2}{5} \cos x - \dfrac{1}{2} = \dfrac{1}{3} - \dfrac{1}{2} \cos x$

35. $3 \tan^2 x - 2 \tan x = 0$

36. $4 \cot^2 x + 3 \cot x = 0$

37. $3 \cos x + \sec x = 0$

38. $5 \sin x - \csc x = 0$

39. $\tan^2 x = 3 \sec^2 x - 2$

40. $\csc^2 x - 1 = 3 \cot^2 x + 2$

41. $2 \sin^2 x = 1 - \cos x$

42. $\cos^2 x + 4 = 2 \sin x - 3$

43. $3 \cos^2 x + 5 \cos x - 2 = 0$

44. $2 \sin^2 x + 5 \sin x + 3 = 0$

45. $2 \tan^2 x - \tan x - 10 = 0$

46. $2 \cot^2 x - 7 \cot x + 3 = 0$

47. $3 \sin x \cos x - \cos x = 0$

48. $\tan x \sin x - \sin x = 0$

49. $2 \sin x \cos x - \sin x - 2 \cos x + 1 = 0$

50. $6 \cos x \sin x - 3 \cos x - 4 \sin x + 2 = 0$

51. $2 \sin x - \cos x = 1$

▶ **52.** $\sin x + 2 \cos x = 1$

53. $2 \sin x - 3 \cos x = 1$

54. $\sqrt{3} \sin x + \cos x = 1$

55. $3 \sin^2 x - \sin x - 1 = 0$

▶ **56.** $2 \cos^2 x - 5 \cos x - 5 = 0$

57. $2 \cos x - 1 + 3 \sec x = 0$

58. $3 \sin x - 5 + \csc x = 0$

59. $\cos^2 x - 3 \sin x + 2 \sin^2 x = 0$

60. $\sin^2 x = 2 \cos x + 3 \cos^2 x$

In Exercises 61 to 70, find the exact solutions, in radians, of each trigonometric equation.

61. $\tan 2x - 1 = 0$

62. $\sec 3x - \dfrac{2\sqrt{3}}{3} = 0$

63. $\sin 5x = 1$

64. $\cos 4x = -\dfrac{\sqrt{2}}{2}$

65. $\sin 2x - \sin x = 0$

▶ **66.** $\cos 2x = -\dfrac{\sqrt{3}}{2}$

67. $\sin\left(2x + \dfrac{\pi}{6}\right) = -\dfrac{1}{2}$

68. $\cos\left(2x - \dfrac{\pi}{4}\right) = -\dfrac{\sqrt{2}}{2}$

69. $\sin^2\dfrac{x}{2} + \cos x = 1$

70. $\cos^2\dfrac{x}{2} - \cos x = 1$

In Exercises 71 to 84, find exact solutions, where $0 \le x < 2\pi$.

71. $\cos 2x = 1 - 3\sin x$

72. $\cos 2x = 2\cos x - 1$

73. $\sin 4x - \sin 2x = 0$

74. $\sin 4x - \cos 2x = 0$

75. $\tan\dfrac{x}{2} = \sin x$

76. $\tan\dfrac{x}{2} = 1 - \cos x$

77. $\sin 2x \cos x + \cos 2x \sin x = 0$

78. $\cos 2x \cos x - \sin 2x \sin x = 0$

79. $\sin x \cos 2x - \cos x \sin 2x = \dfrac{\sqrt{3}}{2}$

80. $\cos 2x \cos x + \sin 2x \sin x = -1$

81. $\sin 3x - \sin x = 0$

82. $\cos 3x + \cos x = 0$

83. $2\sin x \cos x + 2\sin x - \cos x - 1 = 0$

▶ **84.** $2\sin x \cos x - 2\sqrt{2}\sin x - \sqrt{3}\cos x + \sqrt{6} = 0$

In Exercises 85 to 88, use a graphing utility to solve the equation. State each solution accurate to the nearest ten-thousandth.

85. $\cos x = x$, where $0 \le x < 2\pi$

▶ **86.** $2\sin x = x$, where $0 \le x < 2\pi$

87. $\sin 2x = \dfrac{1}{x}$, where $-4 \le x \le 4$

88. $\cos x = \dfrac{1}{x}$, where $0 \le x \le 5$

89. Use a graphing utility to solve $\cos x = x^3 - x$ by graphing each side and finding the x-value of all points of intersection. Round to the nearest hundredth.

90. Approximate the largest value of k for which the equation $\sin x \cos x = k$ has a solution.

PROJECTILES Exercises 91 and 92 make use of the following. A projectile is fired at an angle of inclination θ from the horizon with an initial velocity v_0. Its range d (neglecting air resistance) is given by

$$d = \dfrac{v_0^2}{16} \sin\theta \cos\theta$$

where v_0 is measured in feet per second and d is measured in feet.

91. If $v_0 = 288$ feet per second, use a graphing utility to find the angles θ (to the nearest hundredth of a degree) for which the projectile will hit a target 1295 feet downrange.

▶ **92.** Use a graphing utility to find the maximum horizontal range, to the nearest tenth of a foot, for a projectile that has an initial velocity of 375 feet per second. What value of θ produces this maximum horizontal range?

93. **SUNRISE TIME** The table below shows the sunrise time for Atlanta, Georgia, for selected days in 2004.

Date	Day of the Year, x	Sunrise Time
Jan. 1	1	7:42
Feb. 1	32	7:35
Mar. 1	61	7:06
April 1	92	6:25
May 1	122	5:48
June 1	153	5:28
July 1	183	5:31
Aug. 1	214	5:50
Sept. 1	245	6:12
Oct. 1	275	6:32
Nov. 1	306	6:57
Dec. 1	336	7:25

Source: The U.S. Naval Observatory. *Note:* The times do not reflect daylight savings time.

a. Use a graphing utility to find the sine regression function that models the sunrise time, in hours, as a function of the day of the year. Let $x = 1$ represent January 1, 2004. Assume that the sunrise times have a period of 365.25 days.

b. Use the regression function to estimate the sunrise time (to the nearest minute) for March 11, 2004 ($x = 71$).

▶ **94.**   **SUNSET TIME** The table below shows the sunset time for Sioux City, Iowa, for selected days in 2006.

a. Use a graphing utility to find the sine regression function that models the sunset time, in hours, as a function of the day of the year. Let $x = 1$ represent January 1, 2006.

Date	Day of the Year, x	Sunset Time
Jan. 1	1	17:03
Feb. 1	32	17:39
Mar. 1	60	18:15
April 1	91	18:52
May 1	121	19:26
June 1	152	19:56
July 1	182	20:07
Aug. 1	213	19:45
Sept. 1	244	19:00
Oct. 1	274	18:07
Nov. 1	305	17:19
Dec. 1	335	16:54

Source: The U.S. Naval Observatory. *Note:* The times do not reflect daylight savings time.

Assume that the sunset times have a period of 365.25 days.

b. Use the regression function to estimate the sunset time (to the nearest minute) for May 21, 2006 ($x = 141$).

95.   **PERCENT OF THE MOON ILLUMINATED** The table below shows the percent of the moon illuminated at midnight, Central Standard Time, for selected days in October and November of 2005.

Midnight of: Date in 2005	Day Number	% of Moon Illuminated
Oct. 1	1	5
Oct. 5	5	3
Oct. 9	9	33
Oct. 13	13	77
Oct. 17	17	100
Oct. 21	21	84
Oct. 25	25	48
Oct. 29	29	14
Nov. 2	33	0
Nov. 6	37	20
Nov. 10	41	63
Nov. 14	45	96
Nov. 18	49	95
Nov. 22	53	66
Nov. 26	57	29
Nov. 30	61	2

Source: The U.S. Naval Observatory.

a. Use a graphing utility to find the sine regression function that models the percent of the moon illuminated as a function of the day of the year. Let $x = 1$ represent October 1, 2005. Use 29.53 days for the period of the data.

b. Use the regression function to estimate the percent of the moon illuminated (to the nearest 1 percent) at midnight Central Standard Time on October 31, 2005.

96. HOURS OF DAYLIGHT The table below shows the hours of daylight for Houston, Texas, for selected days in 2006.

Date	Day of the Year, x	Hours of Daylight (hours:minutes)
Jan. 1	1	10:17
Feb. 1	32	10:48
Mar. 1	60	11:34
April 1	91	12:29
May 1	121	13:20
June 1	152	13:56
July 1	182	14:01
Aug. 1	213	13:33
Sept. 1	244	12:45
Oct. 1	274	11:52
Nov. 1	305	11:00
Dec. 1	335	10:23

Source: Data extracted from sunrise-sunset times given on the World Wide Web by the U.S. Naval Observatory.

a. Use a graphing utility to find the sine regression function that models the hours of daylight as a function of the day of the year. Let $x = 1$ represent January 1, 2006. Use 365.25 for the period of the data.

b. Use the regression function to estimate the hours of daylight (stated in hours and minutes, with the minutes rounded to the nearest minute) for Houston on May 12, 2006.

97. ALTITUDE OF THE SUN The table at the top of the next column shows the altitude of the sun for Detroit, Michigan, at selected times during October 19, 2007.

a. Use a graphing utility to find the sine regression function that models the altitude, in degrees, of the sun as a function of the time of day. Use 24.03 hours (the time from sunrise October 19 to sunrise October 20) for the period.

b. Use the regression function to estimate the altitude of the sun (to the nearest 0.1 degree) on October 19, 2007, at 9:25.

Time of day	Altitude (degrees)
6:00	−9.9
7:00	1.4
8:00	11.5
9:00	21.0
10:00	29.0
11:00	34.7
12:00	37.5
13:00	36.7
14:00	32.6
15:00	25.7
16:00	17.0
17:00	7.2
18:00	−3.6

Source: The U.S. Naval Observatory.

98. RAINFALL TOTALS FOR SAN FRANCISCO The table below shows some average monthly rainfall totals for San Francisco, California.

Month	Month Number	Average Rainfall (inches)
January	1	4.48
March	3	2.58
May	5	0.35
July	7	0.04
September	9	0.24
November	11	2.49

a. Use a graphing utility to find the sine regression function that models the average monthly rainfall totals as a function of the month number. Use 12 months for the period.

b. Use the regression function to estimate San Francisco's average rainfall total (to the nearest 0.01 inch) for the month of April. How does this result compare with the recorded value of 1.48 inches?

c. Use the regression function to estimate San Francisco's average rainfall total for the month of June. Explain how you know that this result is incorrect.

In Exercises 99 and 100, use a graphing utility.

99. MODEL THE DAYLIGHT HOURS For a particular day of the year t, the number of daylight hours in Mexico City can be approximated by

$$d(t) = 1.208 \sin\left(\frac{2\pi(t - 80)}{365}\right) + 12.133$$

where t is an integer and $t = 1$ corresponds to January 1. According to d, how many days per year will Mexico City have at least 12 hours of daylight?

100. MODEL THE DAYLIGHT HOURS For a particular day of the year t, the number of daylight hours in New Orleans can be approximated by

$$d(t) = 1.792 \sin\left(\frac{2\pi(t - 80)}{365}\right) + 12.145$$

where t is an integer and $t = 1$ corresponds to January 1. According to d, how many days per year will New Orleans have at least 10.75 hours of daylight?

101. CROSS-SECTIONAL AREA A rain gutter is constructed from a long sheet of aluminum that measures 9 inches in width. The aluminum is to be folded as shown by the cross section in the following diagram.

a. Verify that the area of the cross section is $A = 9 \sin \theta(\cos \theta + 1)$, where $0° < \theta \leq 90°$.

b. What values of θ, to the nearest degree, produce a cross-sectional area of 10.5 square inches?

c. Determine the value of θ that produces the cross section with the maximum area.

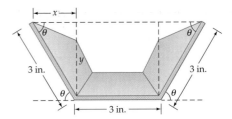

102. OBSERVATION ANGLE A person with an eye level of 5 feet 6 inches is standing in front of a painting, as shown in the following diagram. The bottom of the painting is 3 feet above floor level, and the painting is 6 feet in height. The angle θ shown in the figure is called the *observation angle* for the painting. The person is d feet from the painting.

a. Verify that $\theta = \tan^{-1}\left(\dfrac{6d}{d^2 - 8.75}\right)$.

b. Find the distance d, to the nearest tenth of a foot, for which $\theta = \dfrac{\pi}{6}$.

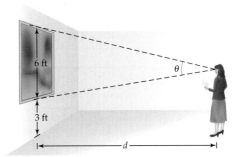

CONNECTING CONCEPTS

In Exercises 103 to 112, solve each equation for exact solutions in the interval $0 \leq x < 2\pi$.

103. $\sqrt{3} \sin x + \cos x = \sqrt{3}$

104. $\sin x - \cos x = 1$

105. $-\sin x + \sqrt{3} \cos x = \sqrt{3}$

106. $-\sqrt{3} \sin x - \cos x = 1$

107. $\cos 5x - \cos 3x = 0$

108. $\cos 5x - \cos x - \sin 3x = 0$

109. $\sin 3x + \sin x = 0$

110. $\sin 3x + \sin x - \sin 2x = 0$

111. $\cos 4x + \cos 2x = 0$

112. $\cos 4x + \cos 2x - \cos 3x = 0$

113. MODEL THE MOVEMENT OF A BUS As bus A_1 makes a left turn, the back B of the bus moves to the right. If bus A_2 were waiting at a stoplight while A_1 turned left, as shown in the figure, there is a chance the two buses would scrape against one another. For a bus 28 feet long and 8 feet wide, the movement of the back of the bus to the right can be approximated by

$$x = \sqrt{(4 + 18 \cot \theta)^2 + 100} - (4 + 18 \cot \theta)$$

where θ is the angle the bus driver has turned the front of the bus. Find the value of x for $\theta = 20°$ and $\theta = 30°$. Round to the nearest hundredth of a foot.

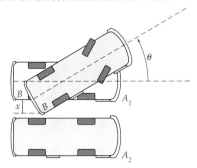

114. OPTIMAL BRANCHING OF BLOOD VESSELS It is hypothesized that the system of blood vessels in primates has evolved so that it has an optimal structure. In the case of a blood vessel splitting into two vessels, as shown in the accompanying figure, we assume that both new branches carry equal amounts of blood. A model of the angle θ is given by the equation $\cos \theta = 2^{(x-4)/(x+4)}$. The value of x is such that $1 \le x \le 2$ and depends on assumptions about the thickness of the blood vessels. Assuming this is an accurate model, find the values of the angle θ.

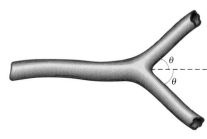

PROJECTS

1. THE MOONS OF SATURN The accompanying figure shows the east-west displacement of five moons of Saturn. The figure shows that the period of the sine curve that models the displacement of the moon Titan is about 15.95 Earth days. The period of the sine curve that models the displacement of the moon Rhea is about 4.52 Earth days.

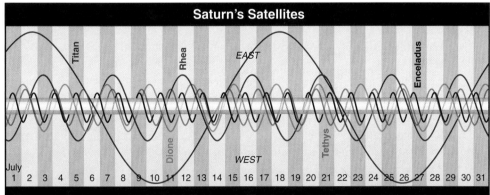

Source: "Saturn's Satellites" from *Sky & Telescope*, July 2003, p. 103. Copyright 2003 by Sky Publishing Company. Reprinted with permission.

a. Write an equation of the form

$$d(t) = A \sin (Bt + C)$$

that models the displacement of the moon Titan. Use $t = 0$ to represent the beginning of July 1, 2003, and think of "east" as a positive displacement and of "west" as a negative displacement.

b. Work part **a.** for the moon Rhea.

c. An astronomer viewed the moons Titan and Rhea at 10 P.M. on August 10, 2003. Were these moons on opposite sides of Saturn or on the same side at that time?

EXPLORING CONCEPTS WITH TECHNOLOGY

Approximate an Inverse Trigonometric Function with Polynomials

The function $y = \sin^{-1} x$ can be approximated by polynomials. For example, consider the following:

$$f_1(x) = x + \frac{x^3}{2 \cdot 3} \qquad \text{where } -1 \le x \le 1$$

$$f_2(x) = x + \frac{x^3}{2 \cdot 3} + \frac{1 \cdot 3 x^5}{2 \cdot 4 \cdot 5} \qquad \text{where } -1 \le x \le 1$$

$$f_3(x) = x + \frac{x^3}{2 \cdot 3} + \frac{1 \cdot 3 x^5}{2 \cdot 4 \cdot 5} + \frac{1 \cdot 3 \cdot 5 x^7}{2 \cdot 4 \cdot 6 \cdot 7} \qquad \text{where } -1 \le x \le 1$$

$$f_4(x) = x + \frac{x^3}{2 \cdot 3} + \frac{1 \cdot 3 x^5}{2 \cdot 4 \cdot 5} + \frac{1 \cdot 3 \cdot 5 x^7}{2 \cdot 4 \cdot 6 \cdot 7} + \frac{1 \cdot 3 \cdot 5 \cdot 7 x^9}{2 \cdot 4 \cdot 6 \cdot 8 \cdot 9} \qquad \text{where } -1 \le x \le 1$$

$$\vdots$$

$$f_n(x) = x + \frac{x^3}{2 \cdot 3} + \frac{1 \cdot 3 x^5}{2 \cdot 4 \cdot 5} + \frac{1 \cdot 3 \cdot 5 x^7}{2 \cdot 4 \cdot 6 \cdot 7} + \cdots + \frac{(2n)! \, x^{2n+1}}{(2^n n!)^2 (2n + 1)}$$

where $\quad -1 \le x \le 1, n! = 1 \cdot 2 \cdot 3 \cdots (n - 1)n$

and $\quad (2n)! = 1 \cdot 2 \cdot 3 \cdots (2n - 1)(2n)$

Use a graphing utility for the following exercises.

1. Graph $y = f_1(x)$, $y = f_2(x)$, $y = f_3(x)$, and $y = f_4(x)$ on the viewing window Xmin $= -1$, Xmax $= 1$, Ymin $= -1.5708$, Ymax $= 1.5708$.

2. Determine the values of x for which $f_3(x)$ and $\sin^{-1} x$ differ by less than 0.001. That is, determine the values of x for which

$$\left| f_3(x) - \sin^{-1} x \right| < 0.001$$

3. Determine the values of x for which

$$\left| f_4(x) - \sin^{-1} x \right| < 0.001$$

4. Write all seven terms of $f_6(x)$. Graph $y = f_6(x)$ and $y = \sin^{-1} x$ on the viewing window Xmin $= -1$, Xmax $= 1$, Ymin $= -\dfrac{\pi}{2}$, Ymax $= \dfrac{\pi}{2}$.

5. Write all seven terms of $f_6(1)$. What do you notice about the size of a term compared to that of the previous term?

6. What is the largest-degree term in $f_{10}(x)$?

CHAPTER **3** SUMMARY

3.1 Verification of Trigonometric Identities

- Trigonometric identities are verified by using algebraic methods and previously proved identities. Here are a few fundamental trigonometric identities.

$$\sin x = \frac{1}{\csc x} \qquad \cos x = \frac{1}{\sec x} \qquad \tan x = \frac{1}{\cot x}$$

$$\tan x = \frac{\sin x}{\cos x} \qquad \cot x = \frac{\cos x}{\sin x}$$

$$\sin^2 x + \cos^2 x = 1;\ \tan^2 x + 1 = \sec^2 x;$$
$$1 + \cot^2 x = \csc^2 x$$

3.2 Sum, Difference, and Cofunction Identities

- Sum and difference identities for the cosine function are

$$\cos(\alpha - \beta) = \cos \alpha \cos \beta + \sin \alpha \sin \beta$$
$$\cos(\alpha + \beta) = \cos \alpha \cos \beta - \sin \alpha \sin \beta$$

- Sum and difference identities for the sine function are

$$\sin(\alpha - \beta) = \sin \alpha \cos \beta - \cos \alpha \sin \beta$$
$$\sin(\alpha + \beta) = \sin \alpha \cos \beta + \cos \alpha \sin \beta$$

- Sum and difference identities for the tangent function are

$$\tan(\alpha + \beta) = \frac{\tan \alpha + \tan \beta}{1 - \tan \alpha \tan \beta}$$

$$\tan(\alpha - \beta) = \frac{\tan \alpha - \tan \beta}{1 + \tan \alpha \tan \beta}$$

- The cofunction identities are

$$\sin(90° - \theta) = \cos \theta \qquad \cos(90° - \theta) = \sin \theta$$
$$\tan(90° - \theta) = \cot \theta \qquad \cot(90° - \theta) = \tan \theta$$
$$\sec(90° - \theta) = \csc \theta \qquad \csc(90° - \theta) = \sec \theta$$

where θ is in degrees. If θ is in radian measure, replace 90° with $\frac{\pi}{2}$.

3.3 Double- and Half-Angle Identities

- The double-angle identities are

$$\sin 2\alpha = 2 \sin \alpha \cos \alpha$$
$$\cos 2\alpha = \cos^2 \alpha - \sin^2 \alpha$$
$$= 1 - 2 \sin^2 \alpha$$
$$= 2 \cos^2 \alpha - 1$$
$$\tan 2\alpha = \frac{2 \tan \alpha}{1 - \tan^2 \alpha}$$

- The half-angle identities are

$$\sin \frac{\alpha}{2} = \pm \sqrt{\frac{1 - \cos \alpha}{2}}$$

$$\cos \frac{\alpha}{2} = \pm \sqrt{\frac{1 + \cos \alpha}{2}}$$

$$\tan \frac{\alpha}{2} = \frac{\sin \alpha}{1 + \cos \alpha} = \frac{1 - \cos \alpha}{\sin \alpha}$$

3.4 Identities Involving the Sum of Trigonometric Functions

- The product-to-sum identities are

$$\sin \alpha \cos \beta = \frac{1}{2}[\sin(\alpha + \beta) + \sin(\alpha - \beta)]$$

$$\cos \alpha \sin \beta = \frac{1}{2}[\sin(\alpha + \beta) - \sin(\alpha - \beta)]$$

$$\cos \alpha \cos \beta = \frac{1}{2}[\cos(\alpha + \beta) + \cos(\alpha - \beta)]$$

$$\sin \alpha \sin \beta = \frac{1}{2}[\cos(\alpha - \beta) - \cos(\alpha + \beta)]$$

- The sum-to-product identities are

$$\sin x + \sin y = 2 \sin \frac{x + y}{2} \cos \frac{x - y}{2}$$

$$\cos x - \cos y = -2 \sin \frac{x + y}{2} \sin \frac{x - y}{2}$$

$$\sin x - \sin y = 2 \cos \frac{x + y}{2} \sin \frac{x - y}{2}$$

$$\cos x + \cos y = 2 \cos \frac{x + y}{2} \cos \frac{x - y}{2}$$

- For sums of the form $a \sin x + b \cos x$,

$$a \sin x + b \cos x = k \sin(x + \alpha)$$

where $k = \sqrt{a^2 + b^2}$, $\sin \alpha = \dfrac{b}{\sqrt{a^2 + b^2}}$, and

$$\cos \alpha = \frac{a}{\sqrt{a^2 + b^2}}.$$

3.5 Inverse Trigonometric Functions

- The inverse of $y = \sin x$ is $y = \sin^{-1} x$, with $-1 \leq x \leq 1$ and $-\dfrac{\pi}{2} \leq y \leq \dfrac{\pi}{2}$.

- The inverse of $y = \cos x$ is $y = \cos^{-1} x$, with $-1 \leq x \leq 1$ and $0 \leq y \leq \pi$.

- The inverse of $y = \tan x$ is $y = \tan^{-1} x$, with $-\infty < x < \infty$ and $-\dfrac{\pi}{2} < y < \dfrac{\pi}{2}$.

- The inverse of $y = \cot x$ is $y = \cot^{-1} x$, with $-\infty < x < \infty$ and $0 < y < \pi$.

- The inverse of $y = \csc x$ is $y = \csc^{-1} x$, with $x \leq -1$ or $x \geq 1$ and $-\dfrac{\pi}{2} \leq y \leq \dfrac{\pi}{2}$, $y \neq 0$.

- The inverse of $y = \sec x$ is $y = \sec^{-1} x$, with $x \leq -1$ or $x \geq 1$ and $0 \leq y \leq \pi$, $y \neq \dfrac{\pi}{2}$.

3.6 Trigonometric Equations

- Algebraic methods and identities are used to solve trigonometric equations. Because the trigonometric functions are periodic, there may be an infinite number of solutions. If solutions cannot be found by algebraic methods, then we often use a graphing utility to find approximate solutions.

CHAPTER 3 TRUE/FALSE EXERCISES

In Exercises 1 to 12, answer true or false. If the statement is false, give a reason or an example to show that the statement is false.

1. $\dfrac{\tan \alpha}{\tan \beta} = \dfrac{\alpha}{\beta}$

2. $\dfrac{\sin x}{\cos y} = \tan \dfrac{x}{y}$

3. $\sin^{-1} x = \csc x^{-1}$

4. $\sin 2\alpha = 2 \sin \alpha$ for all α

5. $\sin(\alpha + \beta) = \sin \alpha + \sin \beta$

6. An equation that has an infinite number of solutions is an identity.

7. If $\tan \alpha = \tan \beta$, then $\alpha = \beta$.

8. $\cos^{-1}(\cos x) = x$

9. $\cos(\cos^{-1} x) = x$

10. $\csc^{-1} \dfrac{1}{\alpha} = \dfrac{1}{\csc \alpha}$

11. If $0° \leq \theta \leq 90°$, then $\cos \theta = \sin(180° - \theta)$.

12. $\sin^2 \theta = \sin \theta^2$

CHAPTER 3 REVIEW EXERCISES

In Exercises 1 to 10, find the exact value.

1. $\cos (45° + 30°)$

2. $\tan (210° - 45°)$

3. $\sin\left(\dfrac{2\pi}{3} + \dfrac{\pi}{4}\right)$

4. $\sec\left(\dfrac{4\pi}{3} - \dfrac{\pi}{4}\right)$

5. $\sin(60° - 135°)$

6. $\cos\left(\dfrac{5\pi}{3} - \dfrac{7\pi}{4}\right)$

7. $\sin\left(22\dfrac{1}{2}\right)°$

8. $\cos 105°$

9. $\tan\left(67\dfrac{1}{2}\right)°$

10. $\sin 112.5°$

In Exercises 11 to 14, find the exact values of the given functions.

11. Given $\sin \alpha = \dfrac{1}{2}$, α in Quadrant I, and $\cos \beta = \dfrac{1}{2}$, β in Quadrant IV, find

    a. $\cos(\alpha - \beta)$ **b.** $\tan 2\alpha$ **c.** $\sin\left(\dfrac{\beta}{2}\right)$

12. Given $\sin \alpha = \dfrac{\sqrt{3}}{2}$, α in Quadrant II, and $\cos \beta = -\dfrac{1}{2}$, β in Quadrant III, find

    a. $\sin(\alpha + \beta)$ **b.** $\sec 2\beta$ **c.** $\cos\left(\dfrac{\alpha}{2}\right)$

13. Given $\sin \alpha = -\dfrac{1}{2}$, α in Quadrant IV, and $\cos \beta = -\dfrac{\sqrt{3}}{2}$, β in Quadrant III, find

    a. $\sin(\alpha - \beta)$ **b.** $\tan 2\alpha$ **c.** $\cos\left(\dfrac{\beta}{2}\right)$

14. Given $\sin \alpha = \dfrac{\sqrt{2}}{2}$, α in Quadrant I, and $\cos \beta = \dfrac{\sqrt{3}}{2}$, β in Quadrant IV, find

    a. $\cos(\alpha - \beta)$ **b.** $\tan 2\beta$ **c.** $\sin 2\alpha$

In Exercises 15 to 20, write the given expression as a single trigonometric function.

15. $2 \sin 3x \cos 3x$

16. $\dfrac{\tan 2x + \tan x}{1 - \tan 2x \tan x}$

17. $\sin 4x \cos x - \cos 4x \sin x$

18. $\cos^2 2\theta - \sin^2 2\theta$

19. $\dfrac{\sin 2\theta}{\cos 2\theta}$

20. $\dfrac{1 - \cos 2\theta}{\sin 2\theta}$

In Exercises 21 to 24, write each expression as the product of two functions.

21. $\cos 2\theta - \cos 4\theta$

22. $\sin 3\theta - \sin 5\theta$

23. $\sin 6\theta + \sin 2\theta$

24. $\sin 5\theta - \sin \theta$

In Exercises 25 to 42, verify the identity.

25. $\dfrac{1}{\sin x - 1} + \dfrac{1}{\sin x + 1} = -2 \tan x \sec x$

26. $\dfrac{\sin x}{1 - \cos x} = \csc x + \cot x, \quad 0 < x < \dfrac{\pi}{2}$

27. $\dfrac{1 + \sin x}{\cos^2 x} = \tan^2 x + 1 + \tan x \sec x$

28. $\cos^2 x - \sin^2 x - \sin 2x = \dfrac{\cos^2 2x - \sin^2 2x}{\cos 2x + \sin 2x}$

29. $\dfrac{1}{\cos x} - \cos x = \tan x \sin x$

30. $\sin(270° - \theta) - \cos(270° - \theta) = \sin \theta - \cos \theta$

31. $\sin\left(\dfrac{\pi}{4} - \alpha\right) = \dfrac{\sqrt{2}}{2}(\cos \alpha - \sin \alpha)$

32. $\sin(180° - \alpha + \beta) = \sin \alpha \cos \beta - \cos \alpha \sin \beta$

33. $\dfrac{\sin 4x - \sin 2x}{\cos 4x - \cos 2x} = -\cot 3x$

34. $2 \sin x \sin 3x = (1 - \cos 2x)(1 + 2 \cos 2x)$

35. $\sin x - \cos 2x = (2 \sin x - 1)(\sin x + 1)$

36. $\cos 4x = 1 - 8 \sin^2 x + 8 \sin^4 x$

37. $\tan 4x = \dfrac{4 \tan x - 4 \tan^3 x}{1 - 6 \tan^2 x + \tan^4 x}$

38. $\dfrac{\sin 2x - \sin x}{\cos 2x + \cos x} = \dfrac{1 - \cos x}{\sin x}$

39. $2 \cos 4x \sin 2x = 2 \sin 3x \cos 3x - 2 \sin x \cos x$

40. $2 \sin x \sin 2x = 4 \cos x \sin^2 x$

41. $\cos(x + y) \cos(x - y) = \cos^2 x + \cos^2 y - 1$

42. $\cos(x + y) \sin(x - y) = \sin x \cos x - \sin y \cos y$

In Exercises 43 to 46, evaluate each expression.

43. $\sec\left(\sin^{-1}\dfrac{12}{13}\right)$

44. $\cos\left(\sin^{-1}\dfrac{3}{5}\right)$

45. $\cos\left[\sin^{-1}\left(-\dfrac{3}{5}\right) + \cos^{-1}\dfrac{5}{13}\right]$

46. $\cos\left(2 \sin^{-1}\dfrac{3}{5}\right)$

In Exercises 47 and 48, solve each equation.

47. $2 \sin^{-1}(x - 1) = \dfrac{\pi}{3}$

48. $\sin^{-1} x + \cos^{-1}\dfrac{4}{5} = \dfrac{\pi}{2}$

In Exercises 49 and 50, solve each equation on $0° \le x < 360°$.

49. $4 \sin^2 x + 2\sqrt{3} \sin x - 2 \sin x - \sqrt{3} = 0$

50. $2 \sin x \cos x - \sqrt{2} \cos x - 2 \sin x + \sqrt{2} = 0$

In Exercises 51 and 52, solve the trigonometric equation where x is in radians. Round approximate solutions to four decimal places.

51. $3 \cos^2 x + \sin x = 1$

52. $\tan^2 x - 2 \tan x - 3 = 0$

In Exercises 53 and 54, solve each equation on $0 \le x < 2\pi$.

53. $\sin 3x \cos x - \cos 3x \sin x = \dfrac{1}{2}$

54. $\cos\left(2x - \dfrac{\pi}{3}\right) = -\dfrac{\sqrt{3}}{2}$

In Exercises 55 to 58, write the equation in the form $y = k \sin(x + \alpha)$, where the measure of α is in radians. Graph one period of each function.

55. $f(x) = \sqrt{3} \sin x + \cos x$

56. $f(x) = -2 \sin x - 2 \cos x$

57. $f(x) = -\sin x - \sqrt{3} \cos x$

58. $f(x) = \dfrac{\sqrt{3}}{2} \sin x - \dfrac{1}{2} \cos x$

In Exercises 59 to 62, graph each function.

59. $f(x) = 2 \cos^{-1} x$ **60.** $f(x) = \sin^{-1}(x - 1)$

61. $f(x) = \sin^{-1} \dfrac{x}{2}$ **62.** $f(x) = \sec^{-1} 2x$

63.   SUNRISE TIME The table below shows the sunrise time for Madison, Wisconsin, for selected days in 2006.

Date	Day of the Year, x	Sunrise Time (hours:minutes)
Jan. 1	1	7:29
Feb. 1	32	7:13
Mar. 1	60	6:34
April 1	91	5:40
May 1	121	4:51
June 1	152	4:21
July 1	182	4:22
Aug. 1	213	4:48
Sept. 1	244	5:22
Oct. 1	274	5:55
Nov. 1	305	6:32
Dec. 1	335	7:09

Source: The U.S. Naval Observatory. *Note:* The times are Central Standard Times. The times do not reflect daylight savings time.

a. Use a graphing utility to find the sine regression function that models the sunrise time, in hours, as a function of the day of the year. Let $x = 1$ represent January 1, 2006. Assume that the sunrise times have a period of 365.25 days.

b. Use the regression function to estimate the sunrise time (to the nearest minute) for April 14, 2006 ($x = 104$).

CHAPTER 3 TEST

1. Verify the identity $1 + \sin^2 x \sec^2 x = \sec^2 x$.

2. Verify the identity

$$\frac{1}{\sec x - \tan x} - \frac{1}{\sec x + \tan x} = 2 \tan x$$

3. Verify the identity $\cos^3 x + \cos x \sin^2 x = \cos x$.

4. Verify the identity $\csc x - \cot x = \dfrac{1 - \cos x}{\sin x}$.

5. Find the exact value of $\sin 195°$.

6. Given $\sin \alpha = -\dfrac{3}{5}$, α in Quadrant III, and $\cos \beta = -\dfrac{\sqrt{2}}{2}$, β in Quadrant II, find $\sin(\alpha + \beta)$.

7. Verify the identity $\sin\left(\theta - \dfrac{3\pi}{2}\right) = \cos \theta$.

8. Write $\cos 6x \sin 3x + \sin 6x \cos 3x$ in terms of a single trigonometric function.

9. Find the exact value of $\cos 2\theta$ given that $\sin \theta = \dfrac{4}{5}$ and θ is in Quadrant II.

10. Verify the identity $\tan \dfrac{\theta}{2} + \dfrac{\cos \theta}{\sin \theta} = \csc \theta$.

11. Verify the identity $\sin^2 2x + 4 \cos^4 x = 4 \cos^2 x$.

12. Find the exact value of $\sin 15° \cos 75°$.

13. Write $y = -\dfrac{\sqrt{3}}{2} \sin x + \dfrac{1}{2} \cos x$ in the form $y = k \sin(x + \alpha)$, where α is measured in radians.

14. Use a calculator to approximate the radian measure of $\cos^{-1} 0.7644$ to the nearest thousandth.

15. Find the exact value of $\sin\left(\cos^{-1} \dfrac{12}{13}\right)$.

16. Graph: $y = \sin^{-1}(x + 2)$

17. Solve $3 \sin x - 2 = 0$, where $0° \le x < 360°$. (State solutions to the nearest 0.1°.)

18. Solve $\sin x \cos x - \dfrac{\sqrt{3}}{2} \sin x = 0$, where $0 \le x < 2\pi$.

19. Find the exact solutions of $\sin 2x + \sin x - 2 \cos x - 1 = 0$, where $0 \le x < 2\pi$.

20. **ALTITUDE OF THE SUN** The table shows the altitude of the sun for Fort Lauderdale, Florida, at selected times during April 20, 2005.

Time of day	Altitude (degrees)
6:00	1.2
7:00	14.1
8:00	27.5
9:00	41.0
10:00	54.1
11:00	66.4
12:00	74.9
13:00	72.7
14:00	62.3
15:00	49.6
16:00	36.4
17:00	22.9
18:00	9.6
19:00	−3.7

Source: The U.S. Naval Observatory.

a. Use a graphing utility to find the sine regression function that models the altitude of the sun as a function of the time of day. Use 23.983 hours (the time from sunrise April 20 to sunrise April 21) for the period.

b. Use the regression function to estimate the altitude of the sun (to the nearest 0.1 degree) on April 20, 2005, at 10:40.

CUMULATIVE REVIEW EXERCISES

1. Solve: $-2x + 1 < 7$

2. Explain how to use the graph of $y = f(x)$ to produce the graph of $y = f(x + 1) + 2$.

3. Explain how to use the graph of $y = f(x)$ to produce the graph of $y = -f(x)$.

4. Determine whether $f(x) = x - \sin x$ is an even function or an odd function.

5. Find the inverse of $f(x) = \dfrac{5x}{x - 1}$.

6. Convert $240°$ to radians.

7. Convert $\dfrac{5\pi}{3}$ to degrees.

8. Evaluate: $\sin \dfrac{\pi}{3}$

9. Evaluate: $\csc 60°$

10. Find $\tan \theta$, given that θ is an acute angle and $\sin \theta = \dfrac{2}{3}$.

11. Determine the sign of $\cot \theta$ given that $\pi < \theta < \dfrac{3\pi}{2}$.

12. What is the measure of the reference angle for the angle $\theta = 310°$?

13. What is the measure of the reference angle for the angle $\theta = \dfrac{5\pi}{3}$?

14. Find the x- and y-coordinates of the point defined by $W\left(\dfrac{\pi}{3}\right)$, where W is the wrapping function.

15. Find the amplitude, the period, and the phase shift for the graph of $y = 0.43 \sin\left(2x - \dfrac{\pi}{6}\right)$.

16. Evaluate: $\sin^{-1} \dfrac{1}{2}$

17. Use a calculator to evaluate $\cos^{-1}(-0.8)$. Round to the nearest thousandth.

18. Use interval notation to state the domain of $f(x) = \cos^{-1} x$.

19. Use interval notation to state the range of $f(x) = \tan^{-1} x$.

20. Find the exact solutions of $2 \cos^2 x - 1 = -\sin x$, where $0 \le x < 2\pi$.

4.1 THE LAW OF SINES
4.2 THE LAW OF COSINES AND AREA
4.3 VECTORS

APPLICATIONS OF TRIGONOMETRY

Trigonometry and Indirect Measurement

In Chapter 2 we used trigonometric functions to find the unknown length of a side of a given *right triangle*. In this chapter we present theorems that can be used to find the length of a side or the measure of an angle of a given triangle even if it is not a right triangle. The theorems in this chapter are used often in the areas of navigation, surveying, and the design of structures. Architects often use trigonometry to find the unknown distance between two points. For instance, in the following diagram, the length a of the steel brace from C to B can be determined using known values and the Law of Sines, which is a major theorem presented in this chapter.

The Burji Al Arab hotel in Dubai, United Arab Emirates, is the world's tallest hotel. It is 321 meters in height and was designed in the shape of a billowing sail.

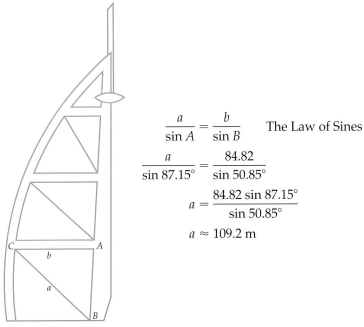

$$\frac{a}{\sin A} = \frac{b}{\sin B} \qquad \text{The Law of Sines}$$

$$\frac{a}{\sin 87.15°} = \frac{84.82}{\sin 50.85°}$$

$$a = \frac{84.82 \sin 87.15°}{\sin 50.85°}$$

$$a \approx 109.2 \text{ m}$$

In triangle ABC, $b = 84.82$ meters, $A = 87.15°$, $B = 50.85°$

VIDEO & DVD   SSG   WWW

See **Exercises 25 and 26, pages 286 and 287,** for additional application exercises that can be solved by applying the Law of Sines.

Devising a Plan

One of the most important aspects of problem solving is the process of devising a plan. In some applications there may be more than one plan (method) that can be used to obtain the solution. For instance, consider the following classic problem.

> You have eight coins. They all look identical, but one is a fake and is slightly lighter than the others. Explain how you can use a balance scale to determine which coin is the fake in exactly
> **a.** three weighings.
> **b.** two weighings.

   In this Chapter you will often need to solve a triangle, which means to determine all unknown measures of the triangle. In most cases you will solve a triangle by using one of two theorems, which are known as the Law of Sines and the Law of Cosines. The guidelines on page 292 provide the information needed to choose between these theorems.

THE LAW OF SINES

● THE LAW OF SINES

Solving a triangle involves finding the lengths of all sides and the measures of all angles in the triangle. In this section and the next we develop formulas for solving an **oblique triangle,** which is a triangle that does not contain a right angle. The *Law of Sines* can be used to solve oblique triangles in which either two angles and a side (AAS) or two sides and an angle opposite one of the sides (SSA) are known. In **Figure 4.1,** altitude CD is drawn from C. The length of the altitude is h. Triangles ACD and BCD are right triangles.

Using the definition of the sine of an angle of a right triangle, we have from **Figure 4.1**

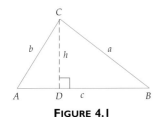

FIGURE 4.1

$$\sin B = \frac{h}{a} \qquad\qquad \sin A = \frac{h}{b}$$

$$h = a \sin B \quad (1) \qquad h = b \sin A \quad (2)$$

Equating the values of h in Equations (1) and (2), we obtain

$$a \sin B = b \sin A$$

Dividing each side of the equation by $\sin A \sin B$, we obtain

$$\frac{a}{\sin A} = \frac{b}{\sin B}$$

Similarly, when an altitude is drawn to a different side, the following formulas result:

$$\frac{c}{\sin C} = \frac{b}{\sin B} \quad \text{and} \quad \frac{c}{\sin C} = \frac{a}{\sin A}$$

---

take note

The Law of Sines may also be written as

$$\frac{\sin A}{a} = \frac{\sin B}{b} = \frac{\sin C}{c}$$

The Law of Sines

If A, B, and C are the measures of the angles of a triangle and a, b, and c are the lengths of the sides opposite these angles, then

$$\frac{a}{\sin A} = \frac{b}{\sin B} = \frac{c}{\sin C}$$

EXAMPLE 1   **Solve a Triangle Using the Law of Sines (AAS)**

Solve triangle ABC if $A = 42°$, $B = 63°$, and $c = 18$ centimeters.

Continued ▶

Solution

Find C by using the fact that the sum of the interior angles of a triangle is $180°$.

$$A + B + C = 180°$$
$$42° + 63° + C = 180°$$
$$C = 75°$$

Use the Law of Sines to find a.

$$\frac{a}{\sin A} = \frac{c}{\sin C}$$

$$\frac{a}{\sin 42°} = \frac{18}{\sin 75°} \qquad \bullet\ A = 42°, c = 18, C = 75°$$

$$a = \frac{18 \sin 42°}{\sin 75°} \approx 12 \text{ centimeters}$$

Use the Law of Sines again, this time to find b.

$$\frac{b}{\sin B} = \frac{c}{\sin C}$$

$$\frac{b}{\sin 63°} = \frac{18}{\sin 75°} \qquad \bullet\ B = 63°, c = 18, C = 75°$$

$$b = \frac{18 \sin 63°}{\sin 75°} \approx 17 \text{ centimeters}$$

The solution is $C = 75°$, $a \approx 12$ centimeters, and $b \approx 17$ centimeters. A scale drawing can be used to see if these results are reasonable. See **Figure 4.2.**

▶ **TRY EXERCISE 4, PAGE 286**

> **take note**
>
> We have used the rounding conventions stated on page 134 to determine the number of significant digits to be used for a and b.

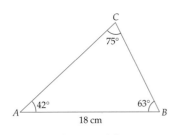

FIGURE 4.2

● THE AMBIGUOUS CASE (SSA)

When you are given two sides of a triangle and an angle opposite one of them, you may find that the triangle is not unique. Some information may result in two triangles, and some may result in no triangle at all. It is because of this that the case of knowing two sides and an angle opposite one of them (SSA) is called the *ambiguous case* of the Law of Sines.

Suppose that sides a and c and the nonincluded angle A of a triangle are known and we are then asked to solve triangle ABC. The relationships among h, the height of the triangle, a (the side opposite $\angle A$), and c determine whether there are no, one, or two triangles.

Case I   First consider the case in which $\angle A$ is an acute angle (see **Figure 4.3**). There are four possible situations.

1. $a < h$; there is no possible triangle.

2. $a = h$; there is one triangle, a right triangle.

3. $h < a < c$; there are two possible triangles. One has all acute angles, and the second has one obtuse angle.

4. $a \geq c$; there is one triangle, which is not a right triangle.

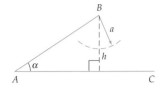

1. $a < h$; no triangle

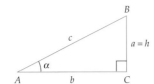

2. $a = h$; one triangle

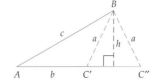

3. $h < a < c$; two triangles

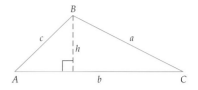

4. $a \geq c$; one triangle

FIGURE 4.3

Case 1: A is an acute angle.

Case 2   Now consider the case in which $\angle A$ is an obtuse angle (see **Figure 4.4**). Here, there are two possible situations.

1. $a \leq c$; there is no triangle.

2. $a > c$; there is one triangle.

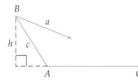

1. $a \leq c$; no triangle

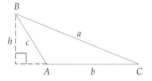

2. $a > c$; one triangle

FIGURE 4.4

Case 2: A is an obtuse angle.

take note

When you solve for an angle of a triangle by using the Law of Sines, be aware that the value of the inverse sine function will give the measure of an acute angle. If the situation is the ambiguous case (SSA), you must consider a second, obtuse angle by using the supplement of the angle. You can use a scale drawing to see whether your results are reasonable.

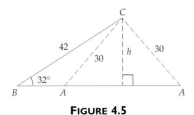

FIGURE 4.5

EXAMPLE 2   **Solve a Triangle Using the Law of Sines (SSA)**

a.   Find A, given triangle ABC with $B = 32°$, $a = 42$, and $b = 30$.

b.   Find C, given triangle ABC with $A = 57°$, $a = 15$ feet, and $c = 20$ feet.

Solution

a.
$$\frac{b}{\sin B} = \frac{a}{\sin A}$$

$$\frac{30}{\sin 32°} = \frac{42}{\sin A}$$

• $B = 32°, a = 42, b = 30$

$$\sin A = \frac{42 \sin 32°}{30} \approx 0.7419$$

$$A \approx 48° \text{ or } 132°$$

• The two angles with measure between 0° and 180° that have a sine of 0.7419 are approximately 48° and 132°.

To check that $A \approx 132°$ is a valid result, add $132°$ to the measure of the given angle B ($32°$). Because $132° + 32° < 180°$, we know that $A \approx 132°$ is a valid result. Thus angle $A \approx 48°$ or $A \approx 132°$ ($\angle BAC$ in **Figure 4.5**).

Continued ▶

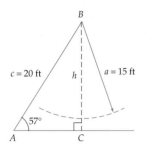

FIGURE 4.6

b.
$$\frac{a}{\sin A} = \frac{c}{\sin C}$$

$$\frac{15}{\sin 57°} = \frac{20}{\sin C} \qquad \bullet \, A = 57°, a = 15, c = 20$$

$$\sin C = \frac{20 \sin 57°}{15} \approx 1.1182$$

Because 1.1182 is not in the range of the sine function, there is no solution of the equation. Thus there is no triangle for these values of A, a, and c. See **Figure 4.6.**

▶ **TRY EXERCISE 18, PAGE 286**

● APPLICATIONS OF THE LAW OF SINES

EXAMPLE 3   **Solve an Application Using the Law of Sines**

A radio antenna 85 feet high is located on top of an office building. At a distance AD from the base of the building, the angle of elevation to the top of the antenna is 26°, and the angle of elevation to the bottom of the antenna is 16°. Find the height of the building.

Solution

Sketch the diagram. See **Figure 4.7.** Find B and β.

$$B = 90° - 26° = 64°$$
$$\beta = 26° - 16° = 10°$$

Because we know the length BC and the measure of β, we can use triangle ABC and the Law of Sines to find length AC.

$$\frac{BC}{\sin \beta} = \frac{AC}{\sin B}$$

$$\frac{85}{\sin 10°} = \frac{AC}{\sin 64°} \qquad \bullet \, BC = 85, \beta = 10°, B = 64°$$

$$AC = \frac{85 \sin 64°}{\sin 10°}$$

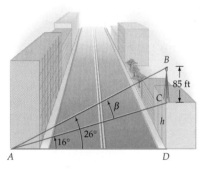

FIGURE 4.7

Having found AC, we can now find the height of the building.

$$\sin 16° = \frac{h}{AC}$$

$$h = AC \sin 16°$$

$$= \frac{85 \sin 64°}{\sin 10°} \sin 16° \approx 121 \text{ feet} \qquad \bullet \text{ Substitute for } AC.$$

The height of the building to two significant digits is 120 feet.

▶ **TRY EXERCISE 26, PAGE 287**

take note

In Example 3 we rounded the height of the building to two significant digits to comply with the rounding convention given on page 134.

In navigation and surveying problems, there are two commonly used methods for specifying direction. The angular direction in which a craft is pointed is called the heading. Heading is expressed in terms of an angle measured clockwise from north. **Figure 4.8** shows a heading of 65° and a heading of 285°.

The angular direction used to locate one object in relation to another object is called the bearing. Bearing is expressed in terms of the acute angle formed by a north–south line and the line of direction. **Figure 4.9** shows a bearing of N38°W and a bearing of S15°E.

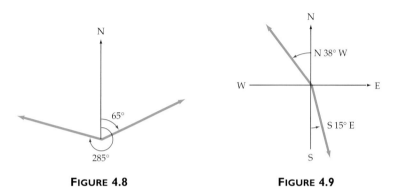

FIGURE 4.8   **FIGURE 4.9**

❓ QUESTION   Can a bearing of N50°E be written as N310°W?

EXAMPLE 4   **Solve an Application**

A ship with a heading of 330° first sighted a lighthouse (point *B*) at a bearing of N65°E. After traveling 8.5 miles, the ship observed the lighthouse at a bearing of S50°E. Find the distance from the ship to the lighthouse when the first sighting was made.

Solution

From **Figure 4.10** we see that the measure of $\angle CAB = 65° + 30° = 95°$, the measure of $\angle BCA = 50° - 30° = 20°$, and $B = 180° - 95° - 20° = 65°$. Use triangle *ABC* and the Law of Sines to find *c*.

$$\frac{b}{\sin B} = \frac{c}{\sin C}$$

$$\frac{8.5}{\sin 65°} = \frac{c}{\sin 20°} \qquad \bullet\, b = 8.5, B = 65°, C = 20°$$

$$c = \frac{8.5 \sin 20°}{\sin 65°} \approx 3.2$$

The lighthouse was 3.2 miles (to two significant digits) from the ship when the first sighting was made.

▶ TRY EXERCISE 36, PAGE 288

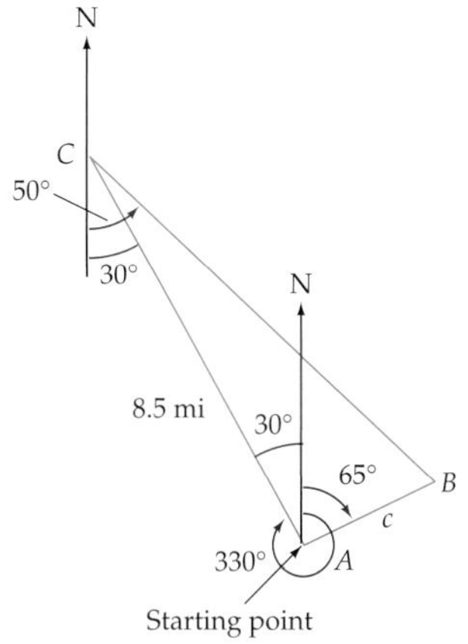

N
C
50°
30°
N
8.5 mi
30°
65°
B
c
330°
A
Starting point

FIGURE 4.10

❓ ANSWER   No. A bearing is always expressed using an acute angle.

TOPICS FOR DISCUSSION

1. Is it possible to solve a triangle if the only given information consists of the measures of the three angles of the triangle? Explain.

2. Explain why it is not possible (in general) to use the Law of Sines to solve a triangle for which we are given only the lengths of all the sides.

3. Draw a triangle with dimensions $A = 30°$, $c = 3$ inches, and $a = 2.5$ inches. Is your answer unique? That is, can more than one triangle with the given dimensions be drawn?

4. Argue for or against the following proposition: In a scalene triangle (a triangle with no congruent sides), the largest angle is always opposite the longest side and the smallest angle is always opposite the shortest side.

EXERCISE SET 4.1

In Exercises 1 to 40, round answers according to the rounding conventions on page 134.

In Exercises 1 to 12, solve the triangles.

1. $A = 42°$, $B = 61°$, $a = 12$

2. $B = 25°$, $C = 125°$, $b = 5.0$

3. $A = 110°$, $C = 32°$, $b = 12$

▶ **4.** $B = 28°$, $C = 78°$, $c = 44$

5. $A = 132°$, $a = 22$, $b = 16$

6. $B = 82.0°$, $b = 6.0$, $c = 3.0$

7. $A = 82.0°$, $B = 65.4°$, $b = 36.5$

8. $B = 54.8°$, $C = 72.6°$, $a = 14.4$

9. $A = 33.8°$, $C = 98.5°$, $c = 102$

10. $B = 36.9°$, $C = 69.2°$, $a = 166$

11. $C = 114.2°$, $c = 87.2$, $b = 12.1$

12. $A = 54.32°$, $a = 24.42$, $c = 16.92$

In Exercises 13 to 24, solve the triangles that exist.

13. $A = 37°$, $c = 40$, $a = 28$

14. $B = 32°$, $c = 14$, $b = 9.0$

15. $C = 65°$, $b = 10$, $c = 8.0$

16. $A = 42°$, $a = 12$, $c = 18$

17. $A = 30°$, $a = 1.0$, $b = 2.4$

▶ **18.** $B = 22.6°$, $b = 5.55$, $a = 13.8$

19. $A = 14.8°$, $c = 6.35$, $a = 4.80$

20. $C = 37.9°$, $b = 3.50$, $c = 2.84$

21. $C = 47.2°$, $a = 8.25$, $c = 5.80$

22. $B = 52.7°$, $b = 12.3$, $c = 16.3$

23. $B = 117.32°$, $b = 67.25$, $a = 15.05$

24. $A = 49.22°$, $a = 16.92$, $c = 24.62$

25. **HURRICANE WATCH** A satellite weather map shows a hurricane off the coast of North Carolina. Use the information in the map to find the distance from the hurricane to Nags Head.

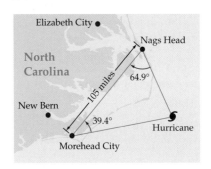

▶ **26.** NAVAL MANEUVERS The distance between an aircraft carrier and a Navy destroyer is 7620 feet. The angle of elevation from the destroyer to a helicopter is 77.2°, and the angle of elevation from the aircraft carrier to the helicopter is 59.0°. The helicopter is in the same vertical plane as the two ships, as shown in the following figure. Use this data to determine the distance x from the helicopter to the aircraft carrier.

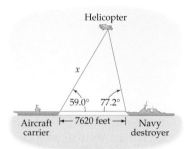

27. CHOOSING A GOLF STRATEGY The following diagram shows two ways to play a golf hole. One is to hit the ball down the fairway on your first shot and then hit an approach shot to the green on your second shot. A second way is to hit directly toward the pin. Due to the water hazard, this is a more risky strategy. The distance AB is 165 yards, BC is 155 yards, and angle $A = 42.0°$. Find the distance AC from the tee directly to the pin. Assume that angle B is an obtuse angle.

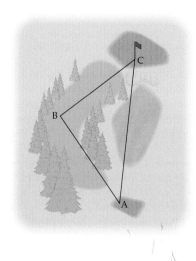

28. DRIVING DISTANCE A golfer drives a golf ball from the tee at point A to point B, as shown in the following diagram. The distance AC from the tee directly to the pin is 365 yards. Angle A measures 11.2°, and angle C measures 22.9°.

   a. Find the distance AB that the golfer drove the ball.

   b. Find the distance BC from the present position of the ball to the pin.

29. DISTANCE TO A HOT AIR BALLOON The angle of elevation to a balloon from one observer is 67°, and the angle of elevation from another observer, 220 feet away, is 31°. If the balloon is in the same vertical plane as the two observers and in between them, find the distance of the balloon from the first observer.

30. LENGTH OF A DIAGONAL The longer side of a parallelogram is 6.0 meters. The measure of $\angle BAD$ is 56° and α is 35°. Find the length of the longer diagonal.

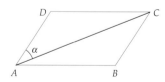

31. DISTANCE ACROSS A CANYON To find the distance across a canyon, a surveying team locates points A and B on one side of the canyon and point C on the other side of the canyon. The distance between A and B is 85 yards. The measure of $\angle CAB$ is 68°, and the measure of $\angle CBA$ is 75°. Find the distance across the canyon.

32. HEIGHT OF A KITE Two observers, in the same vertical plane as a kite and at a distance of 30 feet apart, observe the kite at angles 62° and 78°, as shown in the following diagram. Find the height of the kite.

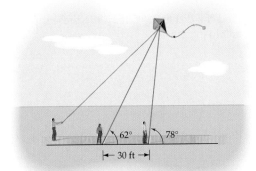

33. LENGTH OF A GUY WIRE A telephone pole 35 feet high is situated on an 11° slope from the horizontal. The measure of angle *CAB* is 21°. Find the length of the guy wire *AC*.

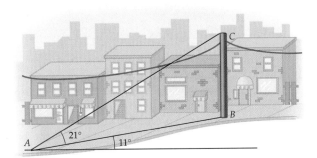

34. DIMENSIONS OF A PLOT OF LAND Three roads intersect in such a way as to form a triangular piece of land. See the accompanying figure. Find the lengths of the other two sides of the land.

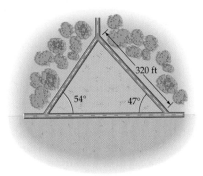

35. HEIGHT OF A HILL A surveying team determines the height of a hill by placing a 12-foot pole at the top of the hill and measuring the angles of elevation to the bottom and the top of the pole. They find the angles of elevation to be as shown in the following figure. Find the height of the hill.

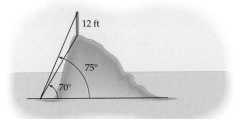

▶ 36. DISTANCE TO A FIRE Two fire lookouts are located on mountains 20 miles apart. Lookout *B* is at a bearing of S65°E from lookout *A*. A fire was sighted at a bearing of N50°E from *A* and at a bearing of N8°E from *B*. Find the distance of the fire from lookout *A*.

37. DISTANCE TO A LIGHTHOUSE A navigator on a ship sights a lighthouse at a bearing of N36°E. After traveling 8.0 miles at a heading of 332°, the ship sights the lighthouse at a bearing of S82°E. How far is the ship from the lighthouse at the second sighting?

38. MINIMUM DISTANCE The navigator on a ship traveling due east at 8 mph sights a lighthouse at a bearing of S55°E. One hour later it is sighted at a bearing of S25°W. Find the closest the ship came to the lighthouse.

39. DISTANCE BETWEEN AIRPORTS An airplane flew 450 miles at a bearing of N65°E from airport *A* to airport *B*. The plane then flew at a bearing of S38°E to airport *C*. Find the distance from *A* to *C* if the bearing from airport *A* to airport *C* is S60°E.

40. LENGTH OF A BRACE A 12-foot solar panel is to be installed on a roof with a 15° pitch. Find the length of the vertical brace *d* if the panel must be installed to make a 40° angle with the horizontal.

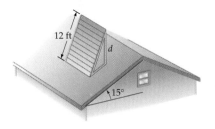

CONNECTING CONCEPTS
---

41. **DISTANCES BETWEEN HOUSES** House B is located at a bearing of N67°E from house A. House C is 300 meters from house A at a bearing of S68°E. House B is located at a bearing of N11°W from house C. Find the distance from house A to house B.

42. Show that for any triangle ABC, $\dfrac{a-b}{b} = \dfrac{\sin A - \sin B}{\sin B}$.

43. Show that for any triangle ABC, $\dfrac{a+b}{b} = \dfrac{\sin A + \sin B}{\sin B}$.

44. Show that for any triangle ABC, $\dfrac{a-b}{a+b} = \dfrac{\sin A - \sin B}{\sin A + \sin B}$.

45. **MAXIMUM LENGTH OF A ROD** The longest rod that can be carried horizontally around a corner from a hall 3 meters wide into one that is 5 meters wide is

the minimum of the length L of the dashed line shown in the figure below.

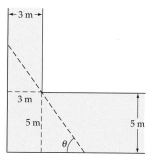

Use similar triangles to show that the length L is a function of the angle θ, given by

$$L(\theta) = \frac{5}{\sin \theta} + \frac{3}{\cos \theta}$$

Use a graphing utility to graph L and estimate the minimum value of L.

PREPARE FOR SECTION 4.2
---

46. Evaluate $\sqrt{a^2 + b^2 - 2ab \cos C}$ for $a = 10.0$, $b = 15.0$, and $C = 110.0°$. Round your result to the nearest tenth. [2.3]

47. Find the area of a triangle with a base of 6 inches and a height of 8.5 inches.

48. Solve $c^2 = a^2 + b^2 - 2ab \cos C$ for C. [3.5]

49. The **semiperimeter** of a triangle is defined as one-half the perimeter of the triangle. Find the semiperimeter of a triangle with sides of 6 meters, 9 meters, and 10 meters.

50. Evaluate $\sqrt{s(s-a)(s-b)(s-c)}$ for $a = 3$, $b = 4$, $c = 5$, and $s = \dfrac{a+b+c}{2}$.

51. State a relationship between the lengths a, b, and c in the triangle shown at the right. [1.2]

PROJECTS
---

1. **FERMAT'S PRINCIPLE AND SNELL'S LAW** State Fermat's Principle and Snell's Law. The refractive index of the glass in a ring is found to be 1.82. The refractive index of a particular diamond in a ring is 2.38. Use Snell's Law to explain what this means in terms of the reflective properties of the glass and the diamond.

THE LAW OF COSINES AND AREA

• THE LAW OF COSINES

The *Law of Cosines* can be used to solve triangles in which two sides and the included angle (SAS) are known or in which three sides (SSS) are known. Consider the triangle in **Figure 4.11**. The height BD is drawn from B perpendicular to the x-axis. The triangle BDA is a right triangle, and the coordinates of B are $(a \cos C, a \sin C)$. The coordinates of A are $(b, 0)$. Using the distance formula, we can find the distance c.

$$c = \sqrt{(a \cos C - b)^2 + (a \sin C - 0)^2}$$
$$c^2 = a^2 \cos^2 C - 2ab \cos C + b^2 + a^2 \sin^2 C$$
$$c^2 = a^2(\cos^2 C + \sin^2 C) + b^2 - 2ab \cos C$$
$$c^2 = a^2 + b^2 - 2ab \cos C$$

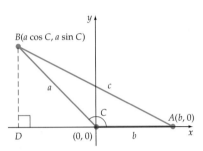

FIGURE 4.11

The Law of Cosines

If A, B, and C are the measures of the angles of a triangle and a, b, and c are the lengths of the sides opposite these angles, then

$$c^2 = a^2 + b^2 - 2ab \cos C$$
$$a^2 = b^2 + c^2 - 2bc \cos A$$
$$b^2 = a^2 + c^2 - 2ac \cos B$$

EXAMPLE 1   Use the Law of Cosines (SAS)

In triangle ABC, $B = 110.0°$, $a = 10.0$ centimeters, and $c = 15.0$ centimeters. See **Figure 4.12.** Find b.

Solution

The Law of Cosines can be used because two sides and the included angle are known.

$$b^2 = a^2 + c^2 - 2ac \cos B$$
$$= 10.0^2 + 15.0^2 - 2(10.0)(15.0) \cos 110.0°$$
$$b = \sqrt{10.0^2 + 15.0^2 - 2(10.0)(15.0) \cos 110.0°}$$
$$b \approx 20.7 \text{ centimeters}$$

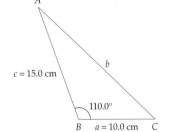

FIGURE 4.12

▶ TRY EXERCISE 12, PAGE 296

In the next example we know the length of each side, but we do not know the measure of any of the angles.

EXAMPLE 2   **Use the Law of Cosines (SSS)**

In triangle ABC, $a = 32$ feet, $b = 20$ feet, and $c = 40$ feet. Find B. This is the SSS case.

Solution

$$b^2 = a^2 + c^2 - 2ac \cos B$$

$$\cos B = \frac{a^2 + c^2 - b^2}{2ac} \qquad \text{• Solve for cos } B.$$

$$= \frac{32^2 + 40^2 - 20^2}{2(32)(40)} \qquad \text{• Substitute for } a, b, \text{ and } c.$$

$$B = \cos^{-1}\left(\frac{32^2 + 40^2 - 20^2}{2(32)(40)}\right) \qquad \text{• Solve for angle } B.$$

$$B \approx 30° \qquad \text{• To the nearest degree}$$

▶ **TRY EXERCISE 18, PAGE 296**

● **AN APPLICATION OF THE LAW OF COSINES**

EXAMPLE 3   **Solve an Application Using the Law of Cosines**

A car traveled 3.0 miles at a heading of 78°. The road turned, and the car traveled another 4.3 miles at a heading of 138°. Find the distance and the bearing of the car from the starting point.

Solution

Sketch a diagram (see **Figure 4.13**). First find B.

$$B = 78° + (180° - 138°) = 120°$$

Use the Law of Cosines to find b.

$$b^2 = a^2 + c^2 - 2ac \cos B$$

$$= 4.3^2 + 3.0^2 - 2(4.3)(3.0) \cos 120° \qquad \text{• Substitute for } a, c, \text{ and } B.$$

$$b = \sqrt{4.3^2 + 3.0^2 - 2(4.3)(3.0) \cos 120°}$$

$$b \approx 6.4 \text{ miles}$$

Find A.

$$\cos A = \frac{b^2 + c^2 - a^2}{2bc}$$

$$A = \cos^{-1}\left(\frac{b^2 + c^2 - a^2}{2bc}\right) \approx \cos^{-1}\left(\frac{6.4^2 + 3.0^2 - 4.3^2}{(2)(6.4)(3.0)}\right)$$

$$A \approx 35°$$

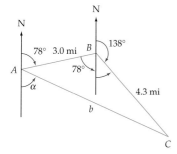

FIGURE 4.13

take note

The measure of A in Example 3 can also be determined by using the Law of Sines.

Continued ▶

The bearing of the present position of the car from the starting point A can be determined by calculating the measure of angle α in **Figure 4.13.**

$$\alpha \approx 180° - (78° + 35°) = 67°$$

The distance is approximately 6.4 miles, and the bearing (to the nearest degree) is S67°E.

▶ **TRY EXERCISE 52, PAGE 298**

There are five different cases that we may encounter when solving an oblique triangle. They are listed in the following guideline, along with the law that can be used to solve the triangle.

A Guideline for Choosing Between the Law of Sines and the Law of Cosines

Apply the Law of Sines to solve an oblique triangle for each of the following cases.

ASA The measures of two angles of the triangle and the length of the included side are known.

AAS The measures of two angles of the triangle and the length of a side opposite one of these angles are known.

SSA The lengths of two sides of the triangle and the measure of an angle opposite one of these sides are known. This case is called the ambiguous case. It may yield one solution, two solutions, or no solution.

Apply the Law of Cosines to solve an oblique triangle for each of the following cases.

SSS The lengths of all three sides of the triangle are known. After finding the measure of an angle, you can complete your solution by using the Law of Sines.

SAS The lengths of two sides of the triangle and the measure of the included angle are known. After finding the measure of the third side, you can complete your solution by using the Law of Sines.

❓ **QUESTION** In triangle ABC, $A = 40°$, $C = 60°$, and $b = 114$. Should you use the Law of Sines or the Law of Cosines to solve this triangle?

❓ **ANSWER** The Law of Sines.

● AREA OF A TRIANGLE

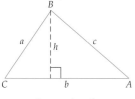

Acute triangle

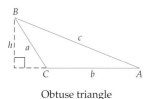

Obtuse triangle

FIGURE 4.14

The formula $A = \dfrac{1}{2}bh$ can be used to find the area of a triangle when the base and height are given. In this section we will find the areas of triangles when the height is not given. We will use K for the area of a triangle because the letter A is often used to represent the measure of an angle.

Consider the areas of the acute and obtuse triangles in **Figure 4.14.**

Height of each triangle: $h = c \sin A$

Area of each triangle: $K = \dfrac{1}{2}bh$

$K = \dfrac{1}{2}bc \sin A$ • **Substitute for h.**

Thus we have established the following theorem.

Area of a Triangle

The area K of triangle ABC is one-half the product of the lengths of any two sides and the sine of the included angle. Thus

$$K = \frac{1}{2}bc \sin A$$

$$K = \frac{1}{2}ab \sin C$$

$$K = \frac{1}{2}ac \sin B$$

> *take note*
>
> Because each formula requires two sides and the included angle, it is necessary to learn only one formula.

EXAMPLE 4   **Find the Area of a Triangle**

Given angle $A = 62°$, $b = 12$ meters, and $c = 5.0$ meters, find the area of triangle ABC.

Solution

In **Figure 4.15,** two sides and the included angle of the triangle are given. Using the formula for area, we have

$$K = \frac{1}{2}bc \sin A = \frac{1}{2}(12)(5.0)(\sin 62°) \approx 26 \text{ square meters}$$

FIGURE 4.15

▶ **TRY EXERCISE 26, PAGE 297**

When two angles and an included side are given, the Law of Sines is used to derive a formula for the area of a triangle. First, solve for c in the Law of Sines.

$$\frac{c}{\sin C} = \frac{b}{\sin B}$$

$$c = \frac{b \sin C}{\sin B}$$

Substitute for c in the formula $K = \dfrac{1}{2}bc \sin A$.

$$K = \frac{1}{2}bc \sin A = \frac{1}{2}b\left(\frac{b \sin C}{\sin B}\right)\sin A$$

$$K = \frac{b^2 \sin C \sin A}{2 \sin B}$$

In like manner, the following two alternative formulas can be derived for the area of a triangle.

$$K = \frac{a^2 \sin B \sin C}{2 \sin A} \quad \text{and} \quad K = \frac{c^2 \sin A \sin B}{2 \sin C}$$

EXAMPLE 5   Find the Area of a Triangle

Given $A = 32°$, $C = 77°$, and $a = 14$ inches, find the area of triangle ABC.

Solution

To use the area formula, we need to know two angles and the included side. Therefore, we need to determine the measure of angle B.

$$B = 180° - 32° - 77° = 71°$$

Thus

$$K = \frac{a^2 \sin B \sin C}{2 \sin A} = \frac{14^2 \sin 71° \sin 77°}{2 \sin 32°} \approx 170 \text{ square inches}$$

▶ **TRY EXERCISE 28, PAGE 297**

MATH MATTERS

Recent findings indicate that Heron's formula for finding the area of a triangle was first discovered by Archimedes. However, the formula is called Heron's formula in honor of the geometer Heron of Alexandria (A.D. 50), who gave an ingenious proof of the theorem in his work *Metrica*. Because Heron of Alexandria was also known as Hero, some texts refer to Heron's formula as Hero's formula.

● HERON'S FORMULA

The Law of Cosines can be used to derive *Heron's formula* for the area of a triangle in which three sides of the triangle are given.

Heron's Formula for Finding the Area of a Triangle

If a, b, and c are the lengths of the sides of a triangle, then the area K of the triangle is

$$K = \sqrt{s(s - a)(s - b)(s - c)}, \quad \text{where } s = \frac{1}{2}(a + b + c)$$

Because s is one-half the perimeter of the triangle, it is called the **semiperimeter**.

EXAMPLE 6   **Find an Area by Heron's Formula**

Find, to two significant digits, the area of the triangle with $a = 7.0$ meters, $b = 15$ meters, and $c = 12$ meters.

Solution

Calculate the semiperimeter s.

$$s = \frac{a + b + c}{2} = \frac{7.0 + 15 + 12}{2} = 17$$

Use Heron's formula.

$$K = \sqrt{s(s - a)(s - b)(s - c)}$$
$$= \sqrt{17(17 - 7.0)(17 - 15)(17 - 12)}$$
$$= \sqrt{1700} \approx 41 \text{ square meters}$$

▶ **TRY EXERCISE 36, PAGE 297**

EXAMPLE 7   **Use Heron's Formula to Solve an Application**

The original portion of the Luxor Hotel in Las Vegas has the shape of a square pyramid. Each face of the pyramid is an isosceles triangle with a base of 646 feet and sides of length 576 feet. Assuming that the glass on the exterior of the Luxor Hotel costs $35 per square foot, determine the cost of the glass, to the nearest $10,000, for one of the triangular faces of the hotel.

Solution

The lengths (in feet) of the sides of a triangular face are $a = 646$, $b = 576$, and $c = 576$.

$$s = \frac{a + b + c}{2} = \frac{646 + 576 + 576}{2} = 899 \text{ feet}$$

$$K = \sqrt{s(s - a)(s - b)(s - c)}$$
$$= \sqrt{899(899 - 646)(899 - 576)(899 - 576)}$$
$$= \sqrt{23,729,318,063}$$
$$\approx 154,043 \text{ square feet}$$

The cost C of the glass is the product of the cost per square foot and the area.

$$C \approx 35 \cdot 154,043 = 5,391,505$$

The approximate cost of the glass for one face of the Luxor Hotel is $5,390,000.

▶ **TRY EXERCISE 58, PAGE 298**

The pyramid portion of the Luxor Hotel, Las Vegas, Nevada

TOPICS FOR DISCUSSION

1. Explain why there is no triangle that has sides of lengths $a = 2$ inches, $b = 11$ inches, and $c = 3$ inches.

2. The Pythagorean Theorem is a special case of the Law of Cosines. Explain.

3. To solve a triangle in which the lengths of the three sides are given (SSS), a mathematics professor recommends the following procedure.

   (i) Use the Law of Cosines to find the measure of the largest angle.

   (ii) Use the Law of Sines to find the measure of a second angle.

   (iii) Find the measure of the third angle by using the formula $A + B + C = 180°$.

   Explain why this procedure is easier than using the Law of Cosines to find the measure of all three angles.

4. To solve a triangle in which the lengths of two sides and the measure of the included angle are given (SAS), a tutor recommends the following procedure.

   (i) Use the Law of Cosines to find the length of the third side.

   (ii) Use the Law of Sines to find the smaller of the unknown angles.

   (iii) Find the measure of the third angle by using the formula $A + B + C = 180°$.

   Explain why this procedure is easier than using the Law of Cosines three times to find each of the unknowns.

EXERCISE SET 4.2

In Exercises 1 to 52, round answers according to the rounding conventions on page 134.

In Exercises 1 to 14, find the third side of the triangle.

1. $a = 12, b = 18, C = 44°$

2. $b = 30, c = 24, A = 120°$

3. $a = 120, c = 180, B = 56°$

4. $a = 400, b = 620, C = 116°$

5. $b = 60, c = 84, A = 13°$

6. $a = 122, c = 144, B = 48°$

7. $a = 9.0, b = 7.0, C = 72°$

8. $b = 12, c = 22, A = 55°$

9. $a = 4.6, b = 7.2, C = 124°$

10. $b = 12.3, c = 14.5, A = 6.5°$

11. $a = 25.9, c = 33.4, B = 84.0°$

▶ **12.** $a = 14.2, b = 9.30, C = 9.20°$

13. $a = 122, c = 55.9, B = 44.2°$

14. $b = 444.8, c = 389.6, A = 78.44°$

In Exercises 15 to 24, given three sides of a triangle, find the specified angle.

15. $a = 25, b = 32, c = 40$; find A.

16. $a = 60, b = 88, c = 120$; find B.

17. $a = 8.0, b = 9.0, c = 12$; find C.

▶ **18.** $a = 108, b = 132, c = 160$; find A.

19. $a = 80.0, b = 92.0, c = 124$; find B.

20. $a = 166, b = 124, c = 139$; find B.

21. $a = 1025, b = 625.0, c = 1420$; find C.

22. $a = 4.7, b = 3.2, c = 5.9$; find A.

23. $a = 32.5, b = 40.1, c = 29.6$; find B.

24. $a = 112.4, b = 96.80, c = 129.2$; find C.

In Exercises 25 to 36, find the area of the given triangle. Round each area to the same number of significant digits given for each of the given sides.

25. $A = 105°, b = 12, c = 24$

▶ **26.** $B = 127°, a = 32, c = 25$

27. $A = 42°, B = 76°, c = 12$

▶ **28.** $B = 102°, C = 27°, a = 8.5$

29. $a = 16, b = 12, c = 14$

30. $a = 32, b = 24, c = 36$

31. $B = 54.3°, a = 22.4, b = 26.9$

32. $C = 18.2°, b = 13.4, a = 9.84$

33. $A = 116°, B = 34°, c = 8.5$

34. $B = 42.8°, C = 76.3°, c = 17.9$

35. $a = 3.6, b = 4.2, c = 4.8$

▶ **36.** $a = 13.3, b = 15.4, c = 10.2$

37. DISTANCE BETWEEN AIRPORTS A plane leaves airport A and travels 560 miles to airport B at a bearing of N32°E. The plane leaves airport B and travels to airport C 320 miles away at a bearing of S72°E. Find the distance from airport A to airport C.

38. LENGTH OF A STREET A developer has a triangular lot at the intersection of two streets. The streets meet at an angle of 72°, and the lot has 300 feet of frontage along one street and 416 feet of frontage along the other street. Find the length of the third side of the lot.

39. BASEBALL In a baseball game, a batter hits a ground ball 26 feet in the direction of the pitcher's mound. See the figure at the top of the next column. The pitcher runs forward and reaches for the ball. At that moment, how far is the ball from first base? (*Note:* A baseball infield is a square that measures 90 feet on each side.)

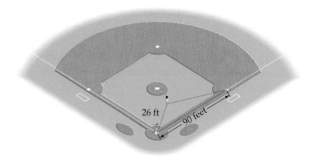

26 ft   90 feet

40. B-2 BOMBER The leading edge of each wing of the B-2 Stealth Bomber measures 105.6 feet in length. The angle between the wing's leading edges ($\angle ABC$) is 109.05°. What is the wing span (the distance from A to C) of the B-2 Bomber?

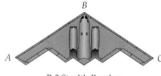

B-2 Stealth Bomber

41. ANGLE BETWEEN THE DIAGONALS OF A BOX The rectangular box in the figure measures 6.50 feet by 3.25 feet by 4.75 feet. Find the measure of the angle θ that is formed by the union of the diagonal shown on the front of the box and the diagonal shown on the right side of the box.

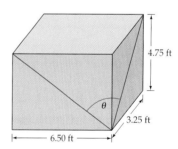

4.75 ft

θ

3.25 ft

6.50 ft

42. SUBMARINE RESCUE MISSION The surface ships shown in the figure below have determined the indicated distances. Use this data to determine the depth of the submarine below the surface of the water. Assume that the line segment between the surface ships is directly above the submarine.

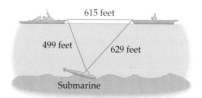

615 feet

499 feet   629 feet

Submarine

43. DISTANCE BETWEEN SHIPS Two ships left a port at the same time. One ship traveled at a speed of 18 mph at a heading of 318°. The other ship traveled at a speed of 22 mph at a heading of 198°. Find the distance between the two ships after 10 hours of travel.

44. DISTANCE ACROSS A LAKE Find the distance across a lake, using the measurements shown in the figure.

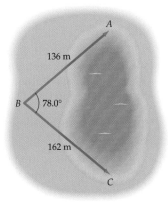

136 m

A

B ⟩ 78.0°

162 m

C

45. A regular hexagon is inscribed in a circle with a radius of 40 centimeters. Find the length of one side of the hexagon.

46. A regular pentagon is inscribed in a circle with a radius of 25 inches. Find the length of one side of the pentagon.

47. The lengths of the diagonals of a parallelogram are 20 inches and 32 inches. The diagonals intersect at an angle of 35°. Find the lengths of the sides of the parallelogram. (*Hint:* The diagonals of a parallelogram bisect one another.)

48. The sides of a parallelogram are 10 feet and 14 feet. The longer diagonal of the parallelogram is 18 feet. Find the length of the shorter diagonal of the parallelogram. (*Hint:* The diagonals of a parallelogram bisect one another.)

49. The sides of a parallelogram are 30 centimeters and 40 centimeters. The shorter diagonal of the parallelogram is 44 centimeters. Find the length of the longer diagonal of the parallelogram. (*Hint:* The diagonals of a parallelogram bisect one another.)

50. ANGLE BETWEEN BOUNDARIES OF A LOT A triangular city lot has sides of 224 feet, 182 feet, and 165 feet. Find the angle between the longer two sides of the lot.

51. DISTANCE TO A PLANE A plane traveling at 180 mph passes 400 feet directly over an observer. The plane is traveling along a straight path with an angle of elevation of 14°. Find the distance of the plane from the observer 10 seconds after the plane has passed directly overhead.

▶ **52. DISTANCE BETWEEN SHIPS** A ship leaves a port at a speed of 16 mph at a heading of 32°. One hour later another ship leaves the port at a speed of 22 mph at a heading of 254°. Find the distance between the ships 4 hours after the first ship leaves the port.

53. AREA OF A TRIANGULAR LOT Find the area of a triangular piece of land that is bounded by sides of 236 meters, 620 meters, and 814 meters.

54. Find the exact area of a parallelogram with sides of exactly 8 feet and 12 feet. The shorter diagonal is exactly 10 feet.

55. Find the exact area of a square inscribed in a circle with a radius of exactly 9 inches.

56. Find the exact area of a regular hexagon inscribed in a circle with a radius of exactly 24 centimeters.

57. COST OF A LOT A commercial piece of real estate is priced at $2.20 per square foot. Find, to the nearest $1000, the cost of a triangular lot measuring 212 feet by 185 feet by 240 feet.

▶ **58. COST OF A LOT** An industrial piece of real estate is priced at $4.15 per square foot. Find, to the nearest $1000, the cost of a triangular lot measuring 324 feet by 516 feet by 412 feet.

59. AREA OF A PASTURE Find the number of acres in a pasture whose shape is a triangle measuring 800 feet by 1020 feet by 680 feet. (An acre is 43,560 square feet.)

60. AREA OF A HOUSING TRACT Find the number of acres in a housing tract whose shape is a triangle measuring 420 yards by 540 yards by 500 yards. (An acre is 4840 square yards.)

The following identity is one of *Mollweide's formulas*. It applies to any triangle *ABC*.

$$\frac{a - b}{c} = \frac{\sin\left(\dfrac{A - B}{2}\right)}{\cos\left(\dfrac{C}{2}\right)}$$

The formula is intriguing because it contains the dimensions of all angles and all sides of △*ABC*. The formula can be used to check whether a triangle has been solved correctly. Substitute the dimensions of a given triangle in the formula and compare the value of the left side of the formula with the value of the right side. If the two results are not reasonably close, then you know that at least one dimension is incorrect. The results generally will not be identical because each dimension is an approximation. In Exercises 61 and 62, you can assume that a triangle has an incorrect dimension if the value of the left side and the value of the right side of the above formula differ by more than 0.02.

61. CHECK DIMENSIONS OF TRUSSES The following diagram shows some of the steel trusses in an airplane hangar.

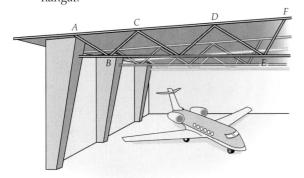

An architect has determined the following dimensions for △*ABC* and △*DEF*:

△*ABC*: $A = 53.5°$, $B = 86.5°$, $C = 40.0°$
    $a = 13.0$ feet, $b = 16.1$ feet, $c = 10.4$ feet

△*DEF*: $D = 52.1°$, $E = 59.9°$, $F = 68.0°$
    $d = 17.2$ feet, $e = 21.3$ feet, $f = 22.8$ feet

Use Mollweide's formula to determine if either △*ABC* or △*DEF* has an incorrect dimension.

62. CHECK DIMENSIONS OF TRUSSES The following diagram shows some of the steel trusses in a railroad bridge.

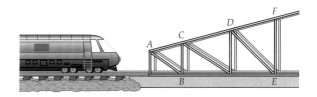

A structural engineer has determined the following dimensions for △*ABC* and △*DEF*:

△*ABC*: $A = 34.1°$, $B = 66.2°$, $C = 79.7°$
    $a = 9.23$ feet, $b = 15.1$ feet, $c = 16.2$ feet

△*DEF*: $D = 45.0°$, $E = 56.2°$, $F = 78.8°$
    $d = 13.6$ feet, $e = 16.0$ feet, $f = 18.9$ feet

Use Mollweide's formula to determine if either △*ABC* or △*DEF* has an incorrect dimension.

CONNECTING CONCEPTS

HERON'S FORMULA FOR QUADRILATERALS The following formula is a generalization of Heron's formula. It gives the area *K* of a convex quadrilateral with sides of lengths *a*, *b*, *c*, and *d*.

$$K = \sqrt{(s - a)(s - b)(s - c)(s - d)}$$

where

$$s = \frac{a + b + c + d}{2}$$

63. Verify that Heron's formula for quadrilaterals produces the correct area for each of the following:

  a. A square with sides of length 8.

  b. A rectangle with sides of lengths 5 and 11.

64. Use Heron's formula for quadrilaterals to find the area (to the nearest 0.01 square unit) of the following convex quadrilateral.

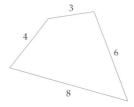

65. Find the measure of the angle formed by the sides P_1P_2 and P_1P_3 of a triangle with vertices at $P_1(-2, 4)$, $P_2(2, 1)$, and $P_3(4, -3)$.

66. A regular pentagon is inscribed in a circle with a radius of 4 inches. Find the perimeter of the pentagon.

67. An equilateral triangle is inscribed in a circle with a radius of 10 centimeters. Find the perimeter of the triangle.

68. Given a triangle ABC, prove that

$$a^2 = b^2 + c^2 - 2bc \cos A$$

69. Use the Law of Cosines to show that

$$\cos A = \frac{(b + c - a)(b + c + a)}{2bc} - 1$$

70. Prove that $K = xy \sin A$ for a parallelogram, where x and y are the lengths of adjacent sides, A is the measure of the angle between side x and side y, and K is the area of the parallelogram.

71. Show that the area of the parallelogram in the figure is $K = 2ab \sin C$.

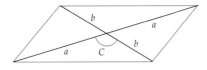

72. Given a regular hexagon inscribed in a circle with a radius of 10 inches, find the area of a segment of the circle (see the dark-shaded region in the following figure).

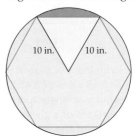

73. Find the volume of the triangular prism shown in the figure.

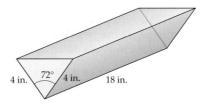

74. Show that the area of the circumscribed triangle in the figure is $K = rs$, where $s = \dfrac{a + b + c}{2}$.

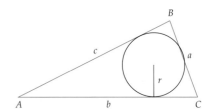

PREPARE FOR SECTION 4.3

75. Evaluate: $\sqrt{\left(\dfrac{3}{5}\right)^2 + \left(-\dfrac{4}{5}\right)^2}$

76. Use a calculator to evaluate $10 \cos 228°$. Round to the nearest thousandth. [2.3]

77. Solve $\tan \alpha = \left|\dfrac{-\sqrt{3}}{3}\right|$ for α, where α is an acute angle measured in degrees. [3.5]

78. Solve $\cos \alpha = \dfrac{-17}{\sqrt{338}}$ for α, where α is an obtuse angle measured in degrees. Round to the nearest tenth of a degree. [3.5]

79. Rationalize the denominator of $\dfrac{1}{\sqrt{5}}$.

80. Rationalize the denominator of $\dfrac{28}{\sqrt{68}}$.

PROJECTS
---

1. The following identity is known as the **Law of Tangents.** Given a triangle ABC with sides a, b, and c, then

$$\frac{a - b}{a + b} = \frac{\tan\left(\dfrac{A - B}{2}\right)}{\tan\left(\dfrac{A + B}{2}\right)}$$

The Law of Tangents can be used to solve triangles in which you are given the measures of two angles and the length of a side or you are given the lengths of two sides and the measure of the included angle (SAS).

a. Use the Law of Tangents to find the length b in triangle ABC with $A = 14.4°$, $B = 25.8°$, and $a = 123$.

b. Use the Law of Tangents to find the measures of angles A and B in triangle ABC with $C = 25.2°$, $a = 18.9$, and $b = 15.2$. (*Hint:* In triangle ABC, $A + B = 180° - C$.)

2. VISUAL INSIGHT

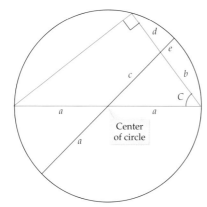

$$bd = (a + c)e$$
$$bd = (a + c)(a - c)$$
$$b(2a \cos C - b) = (a + c)(a - c)$$
$$2ab \cos C - b^2 = a^2 - c^2$$
$$c^2 = a^2 + b^2 - 2ab \cos C$$

Give the rule or reason that justifies each step in the above proof of the Law of Cosines.

SECTION 4.3

VECTORS

● VECTORS
● UNIT VECTORS
● APPLICATIONS OF VECTORS
● DOT PRODUCT
● SCALAR PROJECTION
● PARALLEL AND PERPENDICULAR VECTORS
● WORK: AN APPLICATION OF THE DOT PRODUCT

● VECTORS

In scientific applications, some measurements, such as area, mass, distance, speed, and time, are completely described by a real number and a unit. Examples include 30 square feet (area), 25 meters/second (speed), and 5 hours (time). These measurements are **scalar quantities,** and the number used to indicate the magnitude of the measurement is called a **scalar.** Two other examples of scalar quantities are volume and temperature.

For other quantities, besides the numerical and unit description, it is also necessary to include a *direction* to describe the quantity completely. For example, applying a force of 25 pounds at various angles to a small metal box will influence how the box moves. In **Figure 4.16,** applying the 25-pound force

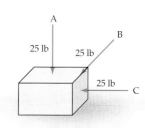

FIGURE 4.16

straight down (A) will not move the box to the left. However, applying the 25-pound force (C) parallel to the floor will move the box along the floor.

Vector quantities have a *magnitude* (numerical and unit description) and a *direction*. Force is a vector quantity. Velocity is another. Velocity includes the speed (magnitude) and a direction. A velocity of 40 mph east is different from a velocity of 40 mph north. Displacement is another vector quantity; it consists of distance (a scalar) moved in a certain direction; for example, we might speak of a displacement of 13 centimeters at an angle of 15° from the positive *x*-axis.

Definition of a Vector

A **vector** is a directed line segment. The length of the line segment is the magnitude of the vector, and the direction of the vector is measured by an angle.

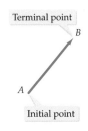

FIGURE 4.17

The point *A* for the vector in **Figure 4.17** is called the **initial point** (or tail) of the vector, and the point *B* is the **terminal point** (or head) of the vector. An arrow over the letters ($\overrightarrow{AB}$), an arrow over a single letter ($\vec{V}$), or boldface type (**AB** or **V**) is used to denote a vector. The magnitude of the vector is the length of the line segment and is denoted by $\|\overrightarrow{AB}\|$, $\|\vec{V}\|$, $\|\mathbf{AB}\|$, or $\|\mathbf{V}\|$.

Equivalent vectors have the same magnitude and the same direction. The vectors in **Figure 4.18** are equivalent. They have the same magnitude and direction.

Multiplying a vector by a positive real number (other than 1) changes the magnitude of the vector but not its direction. If **v** is any vector, then 2**v** is the vector that has the same direction as **v** but is twice the magnitude of **v**. The multiplication of 2 and **v** is called the **scalar multiplication** of the vector **v** and the scalar 2. Multiplying a vector by a negative number *a* reverses the direction of the vector and multiplies the magnitude of the vector by |*a*|. See **Figure 4.19.**

FIGURE 4.18

The sum of two vectors, called the **resultant vector** or the **resultant,** is the single equivalent vector that will have the same effect as the application of those two vectors. For example, a displacement of 40 meters along the positive *x*-axis and then 30 meters in the positive *y* direction is equivalent to a vector of magnitude 50 meters at an angle of approximately 37° to the positive *x*-axis. See **Figure 4.20.**

Vectors can be added graphically by using the *triangle method* or the *parallelogram method*. In the triangle method, shown in **Figure 4.21,** the tail of **V** is placed at the head of **U**. The vector connecting the tail of **U** with the head of **V** is the sum **U** + **V**.

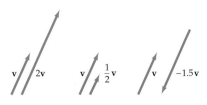

FIGURE 4.19

The parallelogram method of adding two vectors graphically places the tails of the two vectors **U** and **V** together, as in **Figure 4.22.** Complete the parallelogram so that **U** and **V** are sides of the parallelogram. The diagonal beginning at the tails of the two vectors is **U** + **V**.

To find the difference between two vectors, first rewrite the expression as **V** − **U** = **V** + (−**U**). The difference is shown geometrically in **Figure 4.23.**

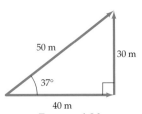

FIGURE 4.20

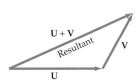

FIGURE 4.21

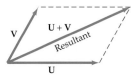

FIGURE 4.22

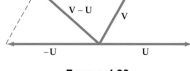

FIGURE 4.23

By introducing a coordinate plane, it is possible to develop an analytic approach to vectors. Recall from our discussion about equivalent vectors that a vector can be moved in the plane as long as *the magnitude and direction* are not changed.

With this in mind, consider **AB**, whose initial point is $A(2, -1)$ and whose terminal point is $B(-3, 4)$. If this vector is moved so that the initial point is at the origin O, the terminal point becomes $P(-5, 5)$, as shown in **Figure 4.24.** The vector **OP** is equivalent to the vector **AB**.

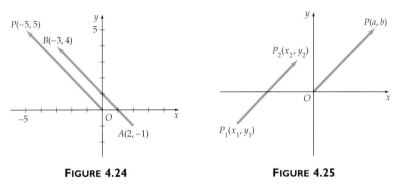

FIGURE 4.24   **FIGURE 4.25**

In **Figure 4.25,** let $P_1(x_1, y_1)$ be the initial point of a vector and $P_2(x_2, y_2)$ its terminal point. Then an equivalent vector **OP** has its initial point at the origin and its terminal point at $P(a, b)$, where $a = x_2 - x_1$ and $b = y_2 - y_1$. The vector **OP** can be denoted by $\mathbf{v} = \langle a, b \rangle$; a and b are called the **components** of the vector.

EXAMPLE 1   **Find the Components of a Vector**

Find the components of the vector **AB** whose tail is the point $A(2, -1)$ and whose head is the point $B(-2, 6)$. Determine a vector **v** that is equivalent to **AB** and has an initial point at the origin.

Algebraic Solution

The components of **AB** are $\langle a, b \rangle$, where

$$a = x_2 - x_1 = -2 - 2 = -4 \quad \text{and} \quad b = y_2 - y_1 = 6 - (-1) = 7$$

Thus $\mathbf{v} = \langle -4, 7 \rangle$.

Visualize the Solution

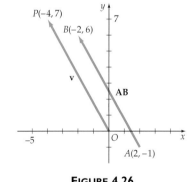

FIGURE 4.26

▶ **TRY EXERCISE 6, PAGE 313**

The magnitude and direction of a vector can be found from its components. For instance, the head of vector **v** sketched in **Figure 4.26** is the ordered pair $(-4, 7)$. Applying the Pythagorean Theorem, we find

$$\|\mathbf{v}\| = \sqrt{(-4)^2 + 7^2} = \sqrt{16 + 49} = \sqrt{65}$$

Let θ be the angle made by the positive x-axis and $\mathbf{v}$. Let α be the reference angle for θ. Then

$$\tan \alpha = \left| \frac{b}{a} \right| = \left| \frac{7}{-4} \right| = \frac{7}{4}$$

$$\alpha = \tan^{-1} \frac{7}{4} \approx 60° \qquad \bullet \ \alpha \text{ is the reference angle.}$$

$$\theta = 180° - 60° = 120° \qquad \bullet \ \theta \text{ is the angle made by the vector and the positive } x\text{-axis.}$$

The magnitude of $\mathbf{v}$ is $\sqrt{65}$, and its direction is $120°$ as measured from the positive x-axis. The angle between a vector and the positive x-axis is called the **direction angle** of the vector. Because $\mathbf{AB}$ in **Figure 4.26** is equivalent to $\mathbf{v}$, $\|\mathbf{AB}\| = \sqrt{65}$ and the direction angle of $\mathbf{AB}$ is also $120°$.

Expressing vectors in terms of components provides a convenient method for performing operations on vectors.

Fundamental Vector Operations

If $\mathbf{v} = \langle a, b \rangle$ and $\mathbf{w} = \langle c, d \rangle$ are two vectors and k is a real number, then

1. $\|\mathbf{v}\| = \sqrt{a^2 + b^2}$

2. $\mathbf{v} + \mathbf{w} = \langle a, b \rangle + \langle c, d \rangle = \langle a + c, b + d \rangle$

3. $k\mathbf{v} = k\langle a, b \rangle = \langle ka, kb \rangle$

In terms of components, the zero vector $\mathbf{0} = \langle 0, 0 \rangle$. The additive inverse of a vector $\mathbf{v} = \langle a, b \rangle$ is given by $-\mathbf{v} = \langle -a, -b \rangle$.

EXAMPLE 2 **Perform Operations on Vectors**

Given $\mathbf{v} = \langle -2, 3 \rangle$ and $\mathbf{w} = \langle 4, -1 \rangle$, find

a. $\|\mathbf{w}\|$ b. $\mathbf{v} + \mathbf{w}$ c. $-3\mathbf{v}$ d. $2\mathbf{v} - 3\mathbf{w}$

Solution

a. $\|\mathbf{w}\| = \sqrt{4^2 + (-1)^2} = \sqrt{17}$ c. $-3\mathbf{v} = -3\langle -2, 3 \rangle = \langle 6, -9 \rangle$

b. $\mathbf{v} + \mathbf{w} = \langle -2, 3 \rangle + \langle 4, -1 \rangle$ d. $2\mathbf{v} - 3\mathbf{w} = 2\langle -2, 3 \rangle - 3\langle 4, -1 \rangle$

$\phantom{\mathbf{v} + \mathbf{w}} = \langle -2 + 4, 3 + (-1) \rangle$ $= \langle -4, 6 \rangle - \langle 12, -3 \rangle$

$\phantom{\mathbf{v} + \mathbf{w}} = \langle 2, 2 \rangle$ $= \langle -16, 9 \rangle$

▶ **TRY EXERCISE 20, PAGE 313**

● UNIT VECTORS

A **unit vector** is a vector whose magnitude is 1. For example, the vector $\mathbf{v} = \left\langle \frac{3}{5}, -\frac{4}{5} \right\rangle$ is a unit vector because

$$\|\mathbf{v}\| = \sqrt{\left(\frac{3}{5}\right)^2 + \left(-\frac{4}{5}\right)^2} = \sqrt{\frac{9}{25} + \frac{16}{25}} = \sqrt{\frac{25}{25}} = 1$$

Given any nonzero vector **v**, we can obtain a unit vector in the direction of **v** by dividing each component of **v** by the magnitude of **v**, $\|\mathbf{v}\|$.

EXAMPLE 3   Find a Unit Vector

Find a unit vector **u** in the direction of $\mathbf{v} = \langle -4, 2 \rangle$.

Solution

Find the magnitude of **v**.

$$\|\mathbf{v}\| = \sqrt{(-4)^2 + 2^2} = \sqrt{16 + 4} = \sqrt{20} = 2\sqrt{5}$$

Divide each component of **v** by $\|\mathbf{v}\|$.

$$\mathbf{u} = \left\langle \frac{-4}{2\sqrt{5}}, \frac{2}{2\sqrt{5}} \right\rangle = \left\langle \frac{-2}{\sqrt{5}}, \frac{1}{\sqrt{5}} \right\rangle = \left\langle -\frac{2\sqrt{5}}{5}, \frac{\sqrt{5}}{5} \right\rangle.$$

A unit vector in the direction of **v** is **u**.

▶ **TRY EXERCISE 8, PAGE 313**

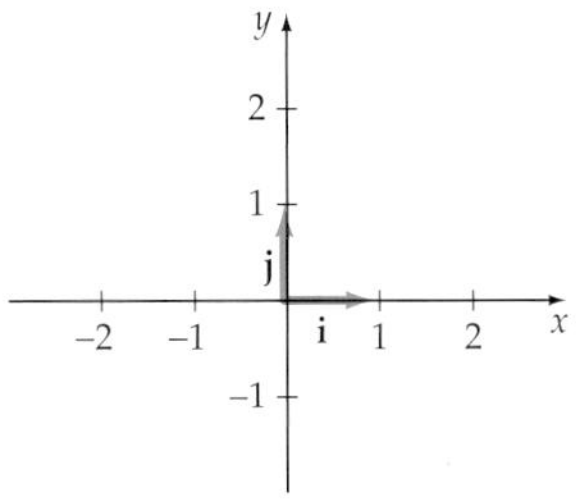

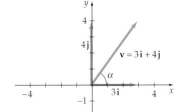

FIGURE 4.27

Two unit vectors, one parallel to the x-axis and one parallel to the y-axis, are of special importance. See **Figure 4.27**.

Definition of Unit Vectors i and j

$$\mathbf{i} = \langle 1, 0 \rangle \qquad \mathbf{j} = \langle 0, 1 \rangle$$

The vector $\mathbf{v} = \langle 3, 4 \rangle$, can be written in terms of the unit vectors **i** and **j** as shown in **Figure 4.28**.

$$\langle 3, 4 \rangle = \langle 3, 0 \rangle + \langle 0, 4 \rangle \qquad \text{• Vector Addition Property}$$
$$= 3\langle 1, 0 \rangle + 4\langle 0, 1 \rangle \qquad \text{• Scalar multiplication of a vector}$$
$$= 3\mathbf{i} + 4\mathbf{j} \qquad \text{• Definition of i and j}$$

FIGURE 4.28

By means of scalar multiplication and addition of vectors, any vector can be expressed in terms of the unit vectors **i** and **j**. Let $\mathbf{v} = \langle a_1, a_2 \rangle$. Then

$$\mathbf{v} = \langle a_1, a_2 \rangle = a_1\langle 1, 0 \rangle + a_2\langle 0, 1 \rangle = a_1\mathbf{i} + a_2\mathbf{j}$$

This gives the following result.

Representation of a Vector in Terms of i and j

If **v** is a vector and $\mathbf{v} = \langle a_1, a_2 \rangle$, then $\mathbf{v} = a_1\mathbf{i} + a_2\mathbf{j}$.

The rules for addition and scalar multiplication of vectors can be restated in terms of **i** and **j**. If $\mathbf{v} = a_1\mathbf{i} + a_2\mathbf{j}$ and $\mathbf{w} = b_1\mathbf{i} + b_2\mathbf{j}$, then

$$\mathbf{v} + \mathbf{w} = (a_1\mathbf{i} + a_2\mathbf{j}) + (b_1\mathbf{i} + b_2\mathbf{j}) = (a_1 + b_1)\mathbf{i} + (a_2 + b_2)\mathbf{j}$$
$$k\mathbf{v} = k(a_1\mathbf{i} + a_2\mathbf{j}) = ka_1\mathbf{i} + ka_2\mathbf{j}$$

| EXAMPLE 4 | Operate on Vectors Written in Terms of i and j |

Given $\mathbf{v} = 3\mathbf{i} - 4\mathbf{j}$ and $\mathbf{w} = 5\mathbf{i} + 3\mathbf{j}$, find $3\mathbf{v} - 2\mathbf{w}$.

Solution

$$
\begin{aligned}
3\mathbf{v} - 2\mathbf{w} &= 3(3\mathbf{i} - 4\mathbf{j}) - 2(5\mathbf{i} + 3\mathbf{j}) \\
&= (9\mathbf{i} - 12\mathbf{j}) - (10\mathbf{i} + 6\mathbf{j}) \\
&= (9 - 10)\mathbf{i} + (-12 - 6)\mathbf{j} \\
&= -\mathbf{i} - 18\mathbf{j}
\end{aligned}
$$

▶ **TRY EXERCISE 26, PAGE 313**

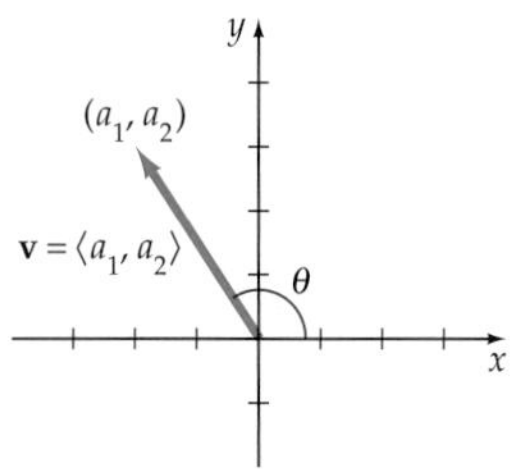

FIGURE 4.29

The components a_1 and a_2 of the vector $\mathbf{v} = \langle a_1, a_2 \rangle$ can be expressed in terms of the magnitude of $\mathbf{v}$ and the direction angle of $\mathbf{v}$ (the angle that $\mathbf{v}$ makes with the positive x-axis). Consider the vector $\mathbf{v}$ in **Figure 4.29**. Then

$$\|\mathbf{v}\| = \sqrt{(a_1)^2 + (a_2)^2}$$

From the definitions of sine and cosine, we have

$$\cos \theta = \frac{a_1}{\|\mathbf{v}\|} \quad \text{and} \quad \sin \theta = \frac{a_2}{\|\mathbf{v}\|}$$

Rewriting the last two equations, we find that the components of $\mathbf{v}$ are

$$a_1 = \|\mathbf{v}\| \cos \theta \quad \text{and} \quad a_2 = \|\mathbf{v}\| \sin \theta$$

Horizontal and Vertical Components of a Vector

Let $\mathbf{v} = \langle a_1, a_2 \rangle$, where $\mathbf{v} \neq \mathbf{0}$, the zero vector. Then

$$a_1 = \|\mathbf{v}\| \cos \theta \quad \text{and} \quad a_2 = \|\mathbf{v}\| \sin \theta$$

where θ is the angle between the positive x-axis and $\mathbf{v}$.

The **horizontal component** of $\mathbf{v}$ is $\|\mathbf{v}\| \cos \theta$. The **vertical component** of $\mathbf{v}$ is $\|\mathbf{v}\| \sin \theta$.

❓ **QUESTION** Is $\mathbf{u} = \cos \theta \mathbf{i} + \sin \theta \mathbf{j}$ a unit vector?

Any nonzero vector can be written in terms of its horizontal and vertical components. Let $\mathbf{v} = a_1 \mathbf{i} + a_2 \mathbf{j}$. Then

$$
\begin{aligned}
\mathbf{v} &= a_1 \mathbf{i} + a_2 \mathbf{j} \\
&= \left(\|\mathbf{v}\| \cos \theta \right) \mathbf{i} + \left(\|\mathbf{v}\| \sin \theta \right) \mathbf{j} \\
&= \|\mathbf{v}\| (\cos \theta \mathbf{i} + \sin \theta \mathbf{j})
\end{aligned}
$$

$\|\mathbf{v}\|$ is the magnitude of $\mathbf{v}$, and the vector $\cos \theta \mathbf{i} + \sin \theta \mathbf{j}$ is a unit vector. The last equation shows that any vector $\mathbf{v}$ can be written as the product of its magnitude and a unit vector in the direction of $\mathbf{v}$.

❓ **ANSWER** Yes, because $\|\cos \theta \mathbf{i} + \sin \theta \mathbf{j}\| = \sqrt{\cos^2 \theta + \sin^2 \theta} = \sqrt{1} = 1$.

EXAMPLE 5   **Find the Horizontal and Vertical Components of a Vector**

Find the approximate horizontal and vertical components of a vector **v** of magnitude 10 meters with direction angle 228°. Write the vector in the form $\mathbf{v} = a_1\mathbf{i} + a_2\mathbf{j}$.

Solution

$a_1 = 10 \cos 228° \approx -6.7$

$a_2 = 10 \sin 228° \approx -7.4$

The approximate horizontal and vertical components are -6.7 and -7.4, respectively.

$$\mathbf{v} \approx -6.7\mathbf{i} - 7.4\mathbf{j}$$

▶ **TRY EXERCISE 36, PAGE 313**

take note

Ground speed is the magnitude of the resultant of the plane's velocity vector and the wind velocity vector.

● **APPLICATIONS OF VECTORS**

Consider an object on which two vectors are acting simultaneously. This occurs when a boat is moving in a current or an airplane is flying in a wind. The **airspeed** of a plane is the speed at which the plane would be moving if there were no wind. The actual velocity of a plane is the velocity relative to the ground. The magnitude of the actual velocity is the **ground speed.**

EXAMPLE 6   **Solve an Application Involving Airspeed**

An airplane is traveling with an airspeed of 320 mph and a heading of 62°. A wind of 42 mph is blowing at a heading of 125°. Find the ground speed and the course of the airplane.

Solution

Sketch a diagram similar to **Figure 4.30** showing the relevant vectors. **AB** represents the heading and the airspeed, **AD** represents the wind velocity, and **AC** represents the course and the ground speed. By vector addition, **AC** = **AB** + **AD**. From the figure,

$$\mathbf{AB} = 320(\cos 28°\mathbf{i} + \sin 28°\mathbf{j})$$
$$\mathbf{AD} = 42[\cos (-35°)\mathbf{i} + \sin (-35°)\mathbf{j}]$$
$$\mathbf{AC} = 320(\cos 28°\mathbf{i} + \sin 28°\mathbf{j}) + 42[\cos (-35°)\mathbf{i} + \sin (-35°)\mathbf{j}]$$
$$\approx (282.5\mathbf{i} + 150.2\mathbf{j}) + (34.4\mathbf{i} - 24.1\mathbf{j})$$
$$= 316.9\mathbf{i} + 126.1\mathbf{j}$$

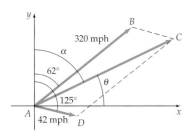

FIGURE 4.30

AC is the course of the plane. The ground speed is $\|\mathbf{AC}\|$. The heading is $\alpha = 90° - \theta$.

$$\|\mathbf{AC}\| = \sqrt{(316.9)^2 + (126.1)^2} \approx 340$$

$$\alpha = 90° - \theta = 90° - \tan^{-1}\left(\frac{126.1}{316.9}\right) \approx 68°$$

The ground speed is approximately 340 mph at a heading of 68°.

▶ **TRY EXERCISE 40, PAGE 314**

There are numerous problems involving force that can be solved by using vectors. One type involves objects that are resting on a ramp. For these problems, we frequently try to find the components of a force vector relative to the ramp rather than to the x-axis.

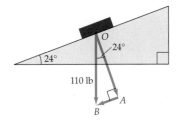

FIGURE 4.31

EXAMPLE 7   Solve an Application Involving Force

A 110-pound box is on a 24° ramp. Find the component of the force that is parallel to the ramp.

Solution

The force-of-gravity vector (**OB**) is the sum of two components, one parallel to the ramp, **AB**, and the other (called the *normal component*) perpendicular to the ramp, **OA**. (See **Figure 4.31**). **AB** is the vector that represents the force tending to move the box down the ramp. Because triangle OAB is a right triangle and $\angle AOB$ is 24°,

$$\sin 24° = \frac{\|\mathbf{AB}\|}{110}$$

$$\|\mathbf{AB}\| = 110 \sin 24° \approx 45$$

The component of the force parallel to the ramp is approximately 45 pounds.

▶ **TRY EXERCISE 42, PAGE 314**

● DOT PRODUCT

We have considered the product of a real number (scalar) and a vector. We now turn our attention to the product of two vectors. Finding the *dot product* of two vectors is one way to multiply a vector by a vector. The dot product of two vectors is a real number and *not* a vector. The dot product is also called the *inner product* or the *scalar product*. This product is useful in engineering and physics.

Definition of Dot Product

Given $\mathbf{v} = \langle a, b \rangle$ and $\mathbf{w} = \langle c, d \rangle$, the **dot product** of $\mathbf{v}$ and $\mathbf{w}$ is given by

$$\mathbf{v} \cdot \mathbf{w} = ac + bd$$

EXAMPLE 8   Find the Dot Product of Two Vectors

Find the dot product of $\mathbf{v} = \langle 6, -2 \rangle$ and $\mathbf{w} = \langle -2, 4 \rangle$.

Solution

$$\mathbf{v} \cdot \mathbf{w} = 6(-2) + (-2)4 = -12 - 8 = -20$$

▶ **TRY EXERCISE 50, PAGE 314**

? QUESTION Is the dot product of two vectors a vector or a real number?

If the vectors in Example 8 were given in terms of the vectors $\mathbf{i}$ and $\mathbf{j}$, then $\mathbf{v} = 6\mathbf{i} - 2\mathbf{j}$ and $\mathbf{w} = -2\mathbf{i} + 4\mathbf{j}$. In this case,

$$\mathbf{v} \cdot \mathbf{w} = (6\mathbf{i} - 2\mathbf{j}) \cdot (-2\mathbf{i} + 4\mathbf{j}) = 6(-2) + (-2)4 = -20$$

Properties of the Dot Product

In the following properties, $\mathbf{u}$, $\mathbf{v}$, and $\mathbf{w}$ are vectors and a is a scalar.

1. $\mathbf{v} \cdot \mathbf{w} = \mathbf{w} \cdot \mathbf{v}$

2. $\mathbf{u} \cdot (\mathbf{v} + \mathbf{w}) = \mathbf{u} \cdot \mathbf{v} + \mathbf{u} \cdot \mathbf{w}$

3. $a(\mathbf{u} \cdot \mathbf{v}) = (a\mathbf{u}) \cdot \mathbf{v} = \mathbf{u} \cdot (a\mathbf{v})$

4. $\mathbf{v} \cdot \mathbf{v} = \|\mathbf{v}\|^2$

5. $\mathbf{0} \cdot \mathbf{v} = 0$

6. $\mathbf{i} \cdot \mathbf{i} = \mathbf{j} \cdot \mathbf{j} = 1$

7. $\mathbf{i} \cdot \mathbf{j} = \mathbf{j} \cdot \mathbf{i} = 0$

The proofs of these properties follow from the definition of dot product. Here is the proof of the fourth property. Let $\mathbf{v} = a\mathbf{i} + b\mathbf{j}$.

$$\mathbf{v} \cdot \mathbf{v} = (a\mathbf{i} + b\mathbf{j}) \cdot (a\mathbf{i} + b\mathbf{j}) = a^2 + b^2 = \|\mathbf{v}\|^2$$

Rewriting the fourth property of the dot product yields an alternative way of expressing the magnitude of a vector.

Magnitude of a Vector in Terms of the Dot Product

If $\mathbf{v} = \langle a, b \rangle$, then $\|\mathbf{v}\| = \sqrt{\mathbf{v} \cdot \mathbf{v}}$.

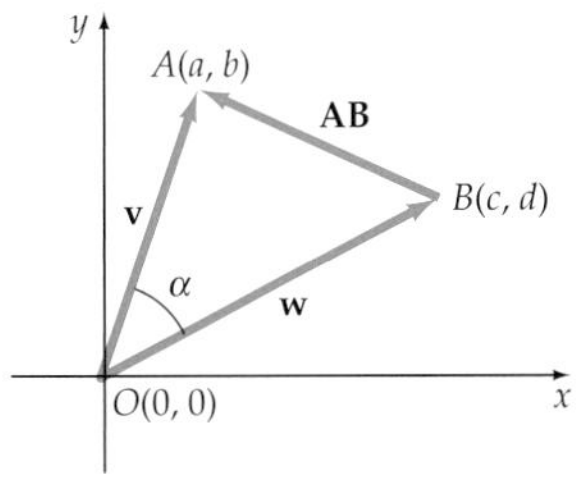

The Law of Cosines can be used to derive an alternative formula for the dot product. Consider the vectors $\mathbf{v} = \langle a, b \rangle$ and $\mathbf{w} = \langle c, d \rangle$ as shown in **Figure 4.32**. Using the Law of Cosines for triangle OAB, we have

$$\|\mathbf{AB}\|^2 = \|\mathbf{v}\|^2 + \|\mathbf{w}\|^2 - 2\|\mathbf{v}\|\|\mathbf{w}\| \cos \alpha$$

By the distance formula, $\|\mathbf{AB}\|^2 = (a - c)^2 + (b - d)^2$, $\|\mathbf{v}\|^2 = a^2 + b^2$, and $\|\mathbf{w}\|^2 = c^2 + d^2$. Thus

$$(a - c)^2 + (b - d)^2 = (a^2 + b^2) + (c^2 + d^2) - 2\|\mathbf{v}\|\|\mathbf{w}\|\cos \alpha$$

$$a^2 - 2ac + c^2 + b^2 - 2bd + d^2 = a^2 + b^2 + c^2 + d^2 - 2\|\mathbf{v}\|\|\mathbf{w}\| \cos \alpha$$

$$-2ac - 2bd = -2\|\mathbf{v}\|\|\mathbf{w}\| \cos \alpha$$

$$ac + bd = \|\mathbf{v}\|\|\mathbf{w}\| \cos \alpha$$

$$\mathbf{v} \cdot \mathbf{w} = \|\mathbf{v}\|\|\mathbf{w}\| \cos \alpha \qquad \bullet\ \mathbf{v} \cdot \mathbf{w} = ac + bd$$

FIGURE 4.32

? ANSWER A real number

Alternative Formula for the Dot Product

If **v** and **w** are two nonzero vectors and α is the smallest nonnegative angle between **v** and **w**, then $\mathbf{v} \cdot \mathbf{w} = \|\mathbf{v}\| \|\mathbf{w}\| \cos \alpha$.

Solving the alternative formula for the dot product for $\cos \alpha$, we have a formula for the cosine of the angle between two vectors.

Angle Between Two Vectors

If **v** and **w** are two nonzero vectors and α is the smallest nonnegative angle between **v** and **w**, then $\cos \alpha = \dfrac{\mathbf{v} \cdot \mathbf{w}}{\|\mathbf{v}\| \|\mathbf{w}\|}$ and $\alpha = \cos^{-1}\left(\dfrac{\mathbf{v} \cdot \mathbf{w}}{\|\mathbf{v}\| \|\mathbf{w}\|}\right)$.

EXAMPLE 9   Find the Angle Between Two Vectors

Find the measure of the smallest positive angle between the vectors $\mathbf{v} = 2\mathbf{i} - 3\mathbf{j}$ and $\mathbf{w} = -\mathbf{i} + 5\mathbf{j}$, as shown in **Figure 4.33**.

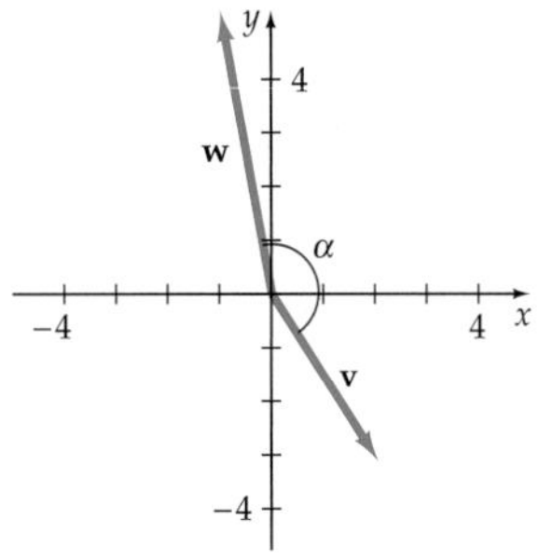

Solution

Use the equation for the angle between two vectors.

$$\cos \alpha = \frac{\mathbf{v} \cdot \mathbf{w}}{\|\mathbf{v}\|\|\mathbf{w}\|} = \frac{(2\mathbf{i} - 3\mathbf{j}) \cdot (-\mathbf{i} + 5\mathbf{j})}{(\sqrt{2^2 + (-3)^2})(\sqrt{(-1)^2 + 5^2})}$$

$$= \frac{-2 - 15}{\sqrt{13}\sqrt{26}} = \frac{-17}{\sqrt{338}}$$

$$\alpha = \cos^{-1}\left(\frac{-17}{\sqrt{338}}\right) \approx 157.6°$$

The angle between the two vectors is approximately 157.6°.

▶ **TRY EXERCISE 60, PAGE 314**

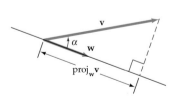

FIGURE 4.33

● SCALAR PROJECTION

Let $\mathbf{v} = \langle a_1, a_2 \rangle$ and $\mathbf{w} = \langle b_1, b_2 \rangle$ be two nonzero vectors, and let α be the angle between the vectors. Two possible configurations, one for which α is an acute angle and one for which α is an obtuse angle, are shown in **Figure 4.34.** In each case, a right triangle is formed by drawing a line segment from the head of **v** to a line through **w**.

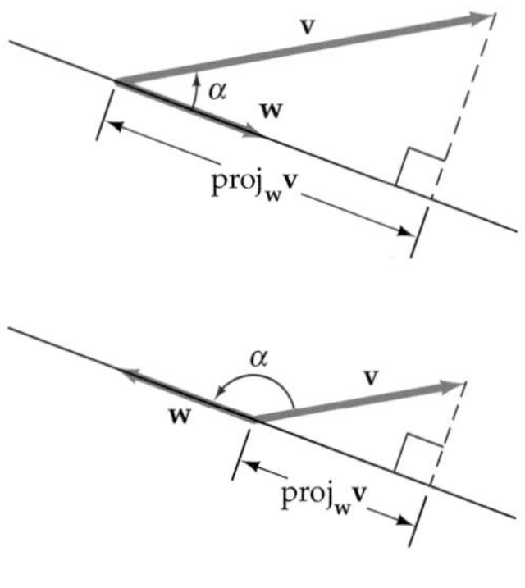

Definition of the Scalar Projection of v on w

If **v** and **w** are two nonzero vectors and α is the smallest positive angle between **v** and **w**, then the scalar projection of **v** on **w**, $\text{proj}_{\mathbf{w}}\mathbf{v}$, is given by

$$\text{proj}_{\mathbf{w}}\mathbf{v} = \|\mathbf{v}\| \cos \alpha$$

FIGURE 4.34

To derive an alternate formula for $\text{proj}_w\mathbf{v}$, consider the dot product $\mathbf{v} \cdot \mathbf{w} = \|\mathbf{v}\| \|\mathbf{w}\| \cos \alpha$. Solving for $\|\mathbf{v}\| \cos \alpha$, which is $\text{proj}_w\mathbf{v}$, we have

$$\text{proj}_w\mathbf{v} = \frac{\mathbf{v} \cdot \mathbf{w}}{\|\mathbf{w}\|}$$

When the angle α between the two vectors is an acute angle, $\text{proj}_w\mathbf{v}$ is positive. When α is an obtuse angle, $\text{proj}_w\mathbf{v}$ is negative.

EXAMPLE 10   Find the Projection of v on w

Given $\mathbf{v} = 2\mathbf{i} + 4\mathbf{j}$ and $\mathbf{w} = -2\mathbf{i} + 8\mathbf{j}$ as shown in **Figure 4.35,** find $\text{proj}_w\mathbf{v}$.

Solution

Use the equation $\text{proj}_w\mathbf{v} = \dfrac{\mathbf{v} \cdot \mathbf{w}}{\|\mathbf{w}\|}$.

$$\text{proj}_w \mathbf{v} = \frac{(2\mathbf{i} + 4\mathbf{j}) \cdot (-2\mathbf{i} + 8\mathbf{j})}{\sqrt{(-2)^2 + 8^2}} = \frac{28}{\sqrt{68}} = \frac{14\sqrt{17}}{17} \approx 3.4$$

▶ **TRY EXERCISE 62, PAGE 314**

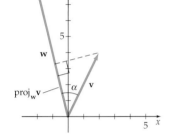

FIGURE 4.35

● PARALLEL AND PERPENDICULAR VECTORS

Two vectors are *parallel* when the angle α between the vectors is $0°$ or $180°$, as shown in **Figure 4.36.** When the angle α is $0°$, the vectors point in the same direction; the vectors point in opposite directions when α is $180°$.

Let $\mathbf{v} = a_1\mathbf{i} + b_1\mathbf{j}$, let c be a real number, and let $\mathbf{w} = c\mathbf{v}$. Because $\mathbf{w}$ is a constant multiple of $\mathbf{v}$, $\mathbf{w}$ and $\mathbf{v}$ are parallel vectors. When $c > 0$, the vectors point in the same direction. When $c < 0$, the vectors point in opposite directions.

Two vectors are *perpendicular* when the angle between the vectors is $90°$. See **Figure 4.37.** Perpendicular vectors are referred to as **orthogonal vectors**. If $\mathbf{v}$ and $\mathbf{w}$ are two nonzero orthogonal vectors, then from the formula for the angle between two vectors and the fact that $\cos \alpha = 0$, we have

$$0 = \frac{\mathbf{v} \cdot \mathbf{w}}{\|\mathbf{v}\| \|\mathbf{w}\|}$$

If a fraction equals zero, the numerator must be zero. Thus, for orthogonal vectors $\mathbf{v}$ and $\mathbf{w}$, $\mathbf{v} \cdot \mathbf{w} = 0$. This gives the following result.

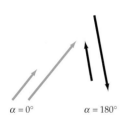

$\alpha = 0°$ $\alpha = 180°$

FIGURE 4.36

$90°$

FIGURE 4.37

Condition for Perpendicular Vectors

Two nonzero vectors $\mathbf{v}$ and $\mathbf{w}$ are orthogonal if and only if $\mathbf{v} \cdot \mathbf{w} = 0$.

● WORK: AN APPLICATION OF THE DOT PRODUCT

When a 5-pound force is used to lift a box from the ground a distance of 4 feet, *work* is done. The amount of **work** is the product of the force on the box and the distance the box is moved. In this case the work is 20 foot-pounds. When the box

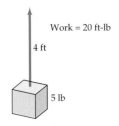

Work = 20 ft-lb

4 ft

5 lb

FIGURE 4.38

25 lb

37°

FIGURE 4.39

is lifted, the force and the displacement vector (the direction in which and the distance the box was moved) are in the same direction. (See **Figure 4.38**.)

Now consider a sled being pulled by a child along the ground by a rope attached to the sled, as shown in **Figure 4.39**. The force vector (along the rope) is *not* in the same direction as the displacement vector (parallel to the ground). In this case the dot product is used to determine the work done by the force.

Definition of Work

The work W done by a force $\mathbf{F}$ applied along a displacement $\mathbf{s}$ is

$$W = \mathbf{F} \cdot \mathbf{s}$$

In the case of the child pulling the sled 7 feet, the work done is

$$W = \mathbf{F} \cdot \mathbf{s}$$
$$= \|\mathbf{F}\| \, \|\mathbf{s}\| \cos \alpha \qquad \bullet \; \alpha \text{ is the angle between F and s.}$$
$$= (25)(7) \cos 37° \approx 140 \text{ foot-pounds}$$

EXAMPLE 11   Solve a Work Problem

A force of 50 pounds on a rope is used to drag a box up a ramp that is inclined 10°. If the rope makes an angle of 37° with the ground, find the work done in moving the box 15 feet along the ramp. See **Figure 4.40**.

Solution

We will provide two solutions to this example.

Method 1   From the last section, $\mathbf{u} = \cos 10° \mathbf{i} + \sin 10° \mathbf{j}$ is a unit vector parallel to the ramp. Multiplying $\mathbf{u}$ by 15 (the magnitude of the displacement vector) gives $\mathbf{s} = 15(\cos 10° \mathbf{i} + \sin 10° \mathbf{j})$. Similarly, the force vector is $\mathbf{F} = 50(\cos 37° \mathbf{i} + \sin 37° \mathbf{j})$. The work done is given by the dot product.

$$W = \mathbf{F} \cdot \mathbf{s} = 50(\cos 37° \mathbf{i} + \sin 37° \mathbf{j}) \cdot 15(\cos 10° \mathbf{i} + \sin 10° \mathbf{j})$$
$$= [50 \cdot 15](\cos 37° \cos 10° + \sin 37° \sin 10°) \approx 668.3 \text{ foot-pounds}$$

Method 2   When we write the work equation as $W = \|\mathbf{F}\|\|\mathbf{s}\| \cos \alpha$, α is the angle between the force and the displacement. Thus $\alpha = 37° - 10° = 27°$. The work done is

$$W = \|\mathbf{F}\|\|\mathbf{s}\| \cos \alpha = 50 \cdot 15 \cdot \cos 27° \approx 668.3 \text{ foot-pounds}$$

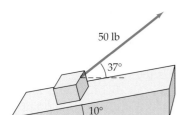

50 lb

37°

10°

FIGURE 4.40

▶ **TRY EXERCISE 70, PAGE 314**

TOPICS FOR DISCUSSION

1. Is the dot product of two vectors a vector or a scalar? Explain.

2. Is the projection of $\mathbf{v}$ on $\mathbf{w}$ a vector or a scalar? Explain.

3. Is the nonzero vector $\langle a, b \rangle$ perpendicular to the vector $\langle -b, a \rangle$? Explain.

4. Explain how to determine the angle between the vector $\langle 3, 4 \rangle$ and the vector $\langle 5, -1 \rangle$.

5. Consider the nonzero vector $\mathbf{u} = \langle a, b \rangle$ and the vector
$$\mathbf{v} = \left\langle \frac{a}{\sqrt{a^2 + b^2}}, \frac{b}{\sqrt{a^2 + b^2}} \right\rangle.$$

   a. Are the vectors parallel? Explain.

   b. Which one of the vectors is a unit vector?

   c. Which vector has the larger magnitude? Explain.

EXERCISE SET 4.3

In Exercises 1 to 6, find the components of a vector with the given initial and terminal points. Write an equivalent vector in terms of its components.

1. $P_1(-3, 0); P_2(4, -1)$

2. $P_1(5, -1); P_2(3, 1)$

3. $P_1(4, 2); P_2(-3, -3)$

4. $P_1(0, -3); P_2(0, 4)$

5. $P_1(2, -5); P_2(2, 3)$

▶ 6. $P_1(3, -2); P_2(3, 0)$

In Exercises 7 to 14, find the magnitude and direction of each vector. Find the unit vector in the direction of the given vector.

7. $\mathbf{v} = \langle -3, 4 \rangle$

▶ 8. $\mathbf{v} = \langle 6, 10 \rangle$

9. $\mathbf{v} = \langle 20, -40 \rangle$

10. $\mathbf{v} = \langle -50, 30 \rangle$

11. $\mathbf{v} = 2\mathbf{i} - 4\mathbf{j}$

12. $\mathbf{v} = -5\mathbf{i} + 6\mathbf{j}$

13. $\mathbf{v} = 42\mathbf{i} - 18\mathbf{j}$

14. $\mathbf{v} = -22\mathbf{i} - 32\mathbf{j}$

In Exercises 15 to 23, perform the indicated operations where $\mathbf{u} = \langle -2, 4 \rangle$ and $\mathbf{v} = \langle -3, -2 \rangle$.

15. $3\mathbf{u}$

16. $-4\mathbf{v}$

17. $2\mathbf{u} - \mathbf{v}$

18. $4\mathbf{v} - 2\mathbf{u}$

19. $\frac{2}{3}\mathbf{u} + \frac{1}{6}\mathbf{v}$

▶ 20. $\frac{3}{4}\mathbf{u} - 2\mathbf{v}$

21. $\|\mathbf{u}\|$

22. $\|\mathbf{v} + 2\mathbf{u}\|$

23. $\|3\mathbf{u} - 4\mathbf{v}\|$

In Exercises 24 to 32, perform the indicated operations where $\mathbf{u} = 3\mathbf{i} - 2\mathbf{j}$ and $\mathbf{v} = -2\mathbf{i} + 3\mathbf{j}$.

24. $-2\mathbf{u}$

25. $4\mathbf{v}$

▶ 26. $3\mathbf{u} + 2\mathbf{v}$

27. $6\mathbf{u} + 2\mathbf{v}$

28. $\frac{1}{2}\mathbf{u} - \frac{3}{4}\mathbf{v}$

29. $\frac{2}{3}\mathbf{v} + \frac{3}{4}\mathbf{u}$

30. $\|\mathbf{v}\|$

31. $\|\mathbf{u} - 2\mathbf{v}\|$

32. $\|2\mathbf{v} + 3\mathbf{u}\|$

In Exercises 33 to 36, find the horizontal and vertical components of each vector. Write an equivalent vector in the form $\mathbf{v} = a_1\mathbf{i} + a_2\mathbf{j}$.

33. Magnitude = 5, direction angle = $27°$

34. Magnitude = 4, direction angle = $127°$

35. Magnitude = 4, direction angle = $\dfrac{\pi}{4}$

▶ 36. Magnitude = 2, direction angle = $\dfrac{8\pi}{7}$

37. **GROUND SPEED OF A PLANE** A plane is flying at an airspeed of 340 mph at a heading of 124°. A wind of 45 mph is blowing from the west. Find the ground speed of the plane.

38. **HEADING OF A BOAT** A person who can row 2.6 mph in still water wants to row due east across a river. The river is flowing from the north at a rate of 0.8 mph. Determine the heading of the boat that will be required for it to travel due east across the river.

39. **GROUND SPEED AND COURSE OF A PLANE** A pilot is flying at a heading of 96° at 225 mph. A 50-mph wind is blowing from the southwest at a heading of 37°. Find the ground speed and course of the plane.

▶ **40.** COURSE OF A BOAT The captain of a boat is steering at a heading of 327° at 18 mph. The current is flowing at 4 mph at a heading of 60°. Find the course (to the nearest degree) of the boat.

41. MAGNITUDE OF A FORCE Find the magnitude of the force necessary to keep a 3000-pound car from sliding down a ramp inclined at an angle of 5.6°.

▶ **42.** ANGLE OF A RAMP A 120-pound force keeps an 800-pound object from sliding down an inclined ramp. Find the angle of the ramp.

43. MAGNITUDE OF THE NORMAL COMPONENT A 25-pound box is resting on a ramp that is inclined 9.0°. Find the magnitude of the normal component of force.

44. MAGNITUDE OF THE NORMAL COMPONENT Find the magnitude of the normal component of force for a 50-pound crate that is resting on a ramp that is inclined 12°.

In Exercises 45 to 52, find the dot product of the vectors.

45. $\mathbf{v} = \langle 3, -2 \rangle; \mathbf{w} = \langle 1, 3 \rangle$ **46.** $\mathbf{v} = \langle 2, 4 \rangle; \mathbf{w} = \langle 0, 2 \rangle$

47. $\mathbf{v} = \langle 4, 1 \rangle; \mathbf{w} = \langle -1, 4 \rangle$ **48.** $\mathbf{v} = \langle 2, -3 \rangle; \mathbf{w} = \langle 3, 2 \rangle$

49. $\mathbf{v} = \mathbf{i} + 2\mathbf{j}; \mathbf{w} = -\mathbf{i} + \mathbf{j}$

▶ **50.** $\mathbf{v} = 5\mathbf{i} + 3\mathbf{j}; \mathbf{w} = 4\mathbf{i} - 2\mathbf{j}$

51. $\mathbf{v} = 6\mathbf{i} - 4\mathbf{j}; \mathbf{w} = -2\mathbf{i} - 3\mathbf{j}$

52. $\mathbf{v} = -4\mathbf{i} + 2\mathbf{j}; \mathbf{w} = -2\mathbf{i} - 4\mathbf{j}$

In Exercises 53 to 60, find the angle between the two vectors. State which pairs of vectors are orthogonal.

53. $\mathbf{v} = \langle 2, -1 \rangle; \mathbf{w} = \langle 3, 4 \rangle$ **54.** $\mathbf{v} = \langle 1, -5 \rangle; \mathbf{w} = \langle -2, 3 \rangle$

55. $\mathbf{v} = \langle 0, 3 \rangle; \mathbf{w} = \langle 2, 2 \rangle$ **56.** $\mathbf{v} = \langle -1, 7 \rangle; \mathbf{w} = \langle 3, -2 \rangle$

57. $\mathbf{v} = 5\mathbf{i} - 2\mathbf{j}; \mathbf{w} = 2\mathbf{i} + 5\mathbf{j}$

58. $\mathbf{v} = 8\mathbf{i} + \mathbf{j}; \mathbf{w} = -\mathbf{i} + 8\mathbf{j}$

59. $\mathbf{v} = 5\mathbf{i} + 2\mathbf{j}; \mathbf{w} = -5\mathbf{i} - 2\mathbf{j}$

▶ **60.** $\mathbf{v} = 3\mathbf{i} - 4\mathbf{j}; \mathbf{w} = 6\mathbf{i} - 12\mathbf{j}$

In Exercises 61 to 68, find $\text{proj}_{\mathbf{w}}\mathbf{v}$.

61. $\mathbf{v} = \langle 6, 7 \rangle; \mathbf{w} = \langle 3, 4 \rangle$ ▶ **62.** $\mathbf{v} = \langle -7, 5 \rangle; \mathbf{w} = \langle -4, 1 \rangle$

63. $\mathbf{v} = \langle -3, 4 \rangle; \mathbf{w} = \langle 2, 5 \rangle$ **64.** $\mathbf{v} = \langle 2, 4 \rangle; \mathbf{w} = \langle -1, 5 \rangle$

65. $\mathbf{v} = 2\mathbf{i} + \mathbf{j}; \mathbf{w} = 6\mathbf{i} + 3\mathbf{j}$

66. $\mathbf{v} = 5\mathbf{i} + 2\mathbf{j}; \mathbf{w} = -5\mathbf{i} - 2\mathbf{j}$

67. $\mathbf{v} = 3\mathbf{i} - 4\mathbf{j}; \mathbf{w} = -6\mathbf{i} + 12\mathbf{j}$

68. $\mathbf{v} = 2\mathbf{i} + 2\mathbf{j}; \mathbf{w} = -4\mathbf{i} - 2\mathbf{j}$

69. WORK A 150-pound box is dragged 15 feet along a level floor. Find the work done if a force of 75 pounds at an angle of 32° is used.

▶ **70.** WORK A 100-pound force is pulling a sled loaded with bricks that weighs 400 pounds. The force is at an angle of 42° with the displacement. Find the work done in moving the sled 25 feet.

71. WORK A rope is being used to pull a box up a ramp that is inclined at 15°. The rope exerts a force of 75 pounds on the box, and it makes an angle of 30° with the plane of the ramp. Find the work done in moving the box 12 feet.

72. WORK A dock worker exerts a force on a box sliding down the ramp of a truck. The ramp makes an angle of 48° with the road, and the worker exerts a 50-pound force parallel to the road. Find the work done in sliding the box 6 feet.

─────── *CONNECTING CONCEPTS* ───────

73. For $\mathbf{u} = \langle -1, 1 \rangle$, $\mathbf{v} = \langle 2, 3 \rangle$, and $\mathbf{w} = \langle 5, 5 \rangle$, find the sum of the three vectors geometrically by using the triangle method of adding vectors.

74. For $\mathbf{u} = \langle 1, 2 \rangle$, $\mathbf{v} = \langle 3, -2 \rangle$, and $\mathbf{w} = \langle -1, 4 \rangle$, find $\mathbf{u} + \mathbf{v} - \mathbf{w}$ geometrically by using the triangle method of adding vectors.

75. Find a vector that has the initial point $(3, -1)$ and is equivalent to $\mathbf{v} = 2\mathbf{i} - 3\mathbf{j}$.

76. Find a vector that has the initial point $(-2, 4)$ and is equivalent to $\mathbf{v} = \langle -1, 3 \rangle$.

77. If $\mathbf{v} = 2\mathbf{i} - 5\mathbf{j}$ and $\mathbf{w} = 5\mathbf{i} + 2\mathbf{j}$ have the same initial point, is $\mathbf{v}$ perpendicular to $\mathbf{w}$? Why or why not?

78. If $\mathbf{v} = \langle 5, 6 \rangle$ and $\mathbf{w} = \langle 6, 5 \rangle$ have the same initial point, is $\mathbf{v}$ perpendicular to $\mathbf{w}$? Why or why not?

79. Let $\mathbf{v} = \langle -2, 7 \rangle$. Find a vector perpendicular to $\mathbf{v}$.

80. Let $\mathbf{w} = 4\mathbf{i} + \mathbf{j}$. Find a vector perpendicular to $\mathbf{w}$.

In Example 7 of this section, if the box were to be kept from sliding down the ramp, it would be necessary to provide a force of 45 pounds parallel to the ramp but pointed _up_ the ramp. Some of this force would be provided by a frictional force between the box and the ramp. The force of friction is $\mathbf{F}_\mu = \mu \mathbf{N}$, where μ is a constant called the coefficient of friction, and $\mathbf{N}$ is the normal component of the force of gravity. In Exercises 81 and 82, find the frictional force.

81. FRICTIONAL FORCE A 50-pound box is resting on a ramp inclined at $12°$. Find the force of friction if the coefficient of friction, μ, is 0.13.

82. FRICTIONAL FORCE A car weighing 2500 pounds is resting on a ramp inclined at $15°$. Find the frictional force if the coefficient of friction, μ, is 0.21.

83. Is the dot product an associative operation? That is, given any nonzero vectors $\mathbf{u}$, $\mathbf{v}$, and $\mathbf{w}$, does

$$(\mathbf{u} \cdot \mathbf{v}) \cdot \mathbf{w} = \mathbf{u} \cdot (\mathbf{v} \cdot \mathbf{w})?$$

84. Prove that $\mathbf{v} \cdot \mathbf{w} = \mathbf{w} \cdot \mathbf{v}$.

85. Prove that $c(\mathbf{v} \cdot \mathbf{w}) = (c\mathbf{v}) \cdot \mathbf{w}$.

86. Show that the dot product of two nonzero vectors is positive if the angle between the vectors is an acute angle and negative if the angle between the two vectors is an obtuse angle.

87. COMPARISON OF WORK DONE Consider the following two situations. (1) A rope is being used to pull a box up a ramp inclined at an angle α. The rope exerts a force $\mathbf{F}$ on the box, and the rope makes an angle θ with the ramp. The box is pulled s feet. (2) A rope is being used to pull a box along a level floor. The rope exerts the same force $\mathbf{F}$ on the box. The box is pulled the same s feet. In which case is more work done?

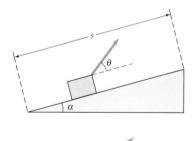

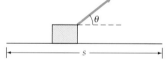

PROJECTS

1. SAME DIRECTION OR OPPOSITE DIRECTIONS Let $\mathbf{v} = c\mathbf{w}$, where c is a nonzero real number and $\mathbf{w}$ is a nonzero vector. Show that $\dfrac{\mathbf{v} \cdot \mathbf{w}}{\|\mathbf{v}\| \|\mathbf{w}\|} = \pm 1$ and that the result is 1 when $c > 0$ and -1 when $c < 0$.

2. THE LAW OF COSINES AND VECTORS Prove that $\|\mathbf{v} - \mathbf{w}\|^2 = \|\mathbf{v}\|^2 + \|\mathbf{w}\|^2 - 2\mathbf{v} \cdot \mathbf{w}$.

3. PROJECTION RELATIONSHIPS What is the relationship between the nonzero vectors $\mathbf{v}$ and $\mathbf{w}$ if

a. $\text{proj}_\mathbf{w}\mathbf{v} = 0$?   **b.** $\text{proj}_\mathbf{w}\mathbf{v} = \|\mathbf{v}\|$?

Optimal Branching of Arteries

The physiologist Jean Louis Poiseuille (1799–1869) developed several laws concerning the flow of blood. One of his laws states that the resistance R of a blood vessel of length l and radius r is given by

$$R = k\frac{l}{r^4} \tag{1}$$

The number k is a variation constant that depends on the viscosity of the blood. **Figure 4.41** shows a large artery with radius r_1 and a smaller artery with radius r_2. The branching angle between the arteries is θ. Make use of Poiseuille's Law, Equation (1), to show that the resistance R of the blood along the path $P_1P_2P_3$ is

$$R = k\left(\frac{a - b\cot\theta}{r_1^4} + \frac{b\csc\theta}{r_2^4}\right) \tag{2}$$

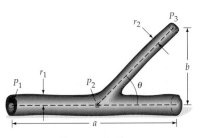

FIGURE 4.41

Use a graphing utility to graph R with $k = 0.0563$, $a = 8$ centimeters, $b = 4$ centimeters, $r_1 = 0.4$ centimeter, and $r_2 = \dfrac{3}{4}r_1 = 0.3$ centimeter. Then estimate (to the nearest degree) the angle θ that minimizes R. By using calculus, it can be demonstrated that R is minimized when

$$\cos\theta = \left(\frac{r_2}{r_1}\right)^4 \tag{3}$$

This equation is remarkable because it is much simpler than Equation (2) and because it does not involve the distance a or b. Solve Equation (3) for θ, with $r_2 = \dfrac{3}{4}r_1$. How does this value of θ compare with the value of θ you obtained by graphing?

CHAPTER 4 SUMMARY

4.1 The Law of Sines

- The Law of Sines is used to solve triangles when two angles and a side are given (AAS) or when two sides and an angle opposite one of them are given (SSA).

$$\frac{a}{\sin A} = \frac{b}{\sin B} = \frac{c}{\sin C}$$

4.2 The Law of Cosines and Area

- The Law of Cosines, $a^2 = b^2 + c^2 - 2bc\cos A$, is used to solve general triangles when two sides and the included angle (SAS) or three sides (SSS) of the triangle are given.

- The area K of triangle ABC is

$$K = \frac{1}{2}bc\sin A = \frac{b^2\sin C\sin A}{2\sin B}$$

- The area of a triangle for which three sides are given (Heron's formula) is:

$$K = \sqrt{s(s-a)(s-b)(s-c)}, \quad \text{where } s = \frac{1}{2}(a + b + c)$$

4.3 Vectors

- A vector is a quantity with magnitude and direction. Two vectors are equivalent if they have the same magnitude and the same direction. The resultant of two or more vectors is the sum of the vectors.

- Vectors can be added by the parallelogram method, the triangle method, or addition of the x- and y-components.

- If $\mathbf{v} = \langle a, b \rangle$ and k is a real number, then $k\mathbf{v} = \langle ka, kb \rangle$.

- The dot product of $\mathbf{v} = \langle a, b \rangle$ and $\mathbf{w} = \langle c, d \rangle$ is given by

$$\mathbf{v} \cdot \mathbf{w} = ac + bd$$

- If $\mathbf{v}$ and $\mathbf{w}$ are two nonzero vectors and α is the smallest positive angle between $\mathbf{v}$ and $\mathbf{w}$, then $\cos \alpha = \dfrac{\mathbf{v} \cdot \mathbf{w}}{\|\mathbf{v}\| \, \|\mathbf{w}\|}$.

CHAPTER 4 TRUE/FALSE EXERCISES

For Exercises 1 to 10, answer true or false. If the statement is false, state a reason or give an example to show that the statement is false.

1. The Law of Cosines can be used to solve any triangle given two sides and an angle.

2. The Law of Sines can be used to solve any triangle given two angles and any side.

3. In any triangle, the longest side is opposite the largest angle.

4. If two vectors have the same magnitude, then they are equal.

5. It is possible for the sum of two nonzero vectors to equal zero.

6. The expression $a^2 = b^2 + c^2 + 2bc \cos D$ is true for triangle ABC, in which angle D is the supplement of angle A.

7. The measure of angle α formed by two vectors is greater than or equal to $0°$ and less than or equal to $180°$.

8. If A, B, and C are the angles of a triangle, then

$$\sin(A + B + C) = 0$$

9. Let $\mathbf{v} = a\mathbf{i} + b\mathbf{j}$. Then $\mathbf{v} \cdot \mathbf{v} = a^2\mathbf{i} + b^2\mathbf{j}$.

10. If $\mathbf{v}$ and $\mathbf{w}$ are vectors with $\mathbf{v} \cdot \mathbf{w} = 0$, then $\mathbf{v} = \mathbf{0}$ or $\mathbf{w} = \mathbf{0}$.

CHAPTER 4 REVIEW EXERCISES

In Exercises 1 to 10, solve each triangle.

1. $A = 37°, b = 14, C = 92°$

2. $B = 77.4°, c = 11.8, C = 94.0°$

3. $a = 12, b = 15, c = 20$

4. $a = 24, b = 32, c = 28$

5. $a = 18, b = 22, C = 35°$

6. $b = 102, c = 150, A = 82°$

7. $A = 105°, a = 8, c = 10$

8. $C = 55°, c = 80, b = 110$

9. $A = 55°, B = 80°, c = 25$

10. $B = 25°, C = 40°, c = 40$

In Exercises 11 to 18, find the area of each triangle. Round each area accurate to two significant digits.

11. $a = 24, b = 30, c = 36$

12. $a = 9.0, b = 7.0, c = 12$

13. $a = 60, b = 44, C = 44°$

14. $b = 8.0, c = 12, A = 75°$

15. $b = 50, c = 75, C = 15°$

16. $b = 18, a = 25, A = 68°$

17. $A = 110°, a = 32, b = 15$

18. $A = 45°, c = 22, b = 18$

In Exercises 19 and 20, find the components of each vector with the given initial and terminal points. Write an equivalent vector in terms of its components.

19. $P_1(-2, 4)$; $P_2(3, 7)$ **20.** $P_1(-4, 0)$; $P_2(-3, 6)$

In Exercises 21 to 24, find the magnitude and the direction angle of each vector.

21. $\mathbf{v} = \langle -4, 2 \rangle$ **22.** $\mathbf{v} = \langle 6, -3 \rangle$

23. $\mathbf{u} = -2\mathbf{i} + 3\mathbf{j}$ **24.** $\mathbf{u} = -4\mathbf{i} - 7\mathbf{j}$

In Exercises 25 to 28, find a unit vector in the direction of the given vector.

25. $\mathbf{w} = \langle -8, 5 \rangle$ **26.** $\mathbf{w} = \langle 7, -12 \rangle$

27. $\mathbf{v} = 5\mathbf{i} + \mathbf{j}$ **28.** $\mathbf{v} = 3\mathbf{i} - 5\mathbf{j}$

In Exercises 29 and 30, perform the indicated operation where $\mathbf{u} = \langle 3, 2 \rangle$ and $\mathbf{v} = \langle -4, -1 \rangle$.

29. $\mathbf{v} - \mathbf{u}$ **30.** $2\mathbf{u} - 3\mathbf{v}$

In Exercises 31 and 32, perform the indicated operation where $\mathbf{u} = 10\mathbf{i} + 6\mathbf{j}$ and $\mathbf{v} = 8\mathbf{i} - 5\mathbf{j}$.

31. $-\mathbf{u} + \dfrac{1}{2}\mathbf{v}$ **32.** $\dfrac{2}{3}\mathbf{v} - \dfrac{3}{4}\mathbf{u}$

33. GROUND SPEED OF A PLANE A plane is flying at an airspeed of 400 mph at a heading of 204°. A wind of 45 mph is blowing from the east. Find the ground speed of the plane.

34. ANGLE OF A RAMP A 40-pound force keeps a 320-pound object from sliding down an inclined ramp. Find the angle of the ramp.

In Exercises 35 to 38, find the dot product of the vectors.

35. $\mathbf{u} = \langle 3, 7 \rangle$; $\mathbf{v} = \langle -1, 3 \rangle$

36. $\mathbf{v} = \langle -8, 5 \rangle$; $\mathbf{u} = \langle 2, -1 \rangle$

37. $\mathbf{v} = -4\mathbf{i} - \mathbf{j}$; $\mathbf{u} = 2\mathbf{i} + \mathbf{j}$

38. $\mathbf{u} = -3\mathbf{i} + 7\mathbf{j}$; $\mathbf{v} = -2\mathbf{i} + 2\mathbf{j}$

In Exercises 39 to 42, find the angle between the vectors.

39. $\mathbf{u} = \langle 7, -4 \rangle$; $\mathbf{v} = \langle 2, 3 \rangle$

40. $\mathbf{v} = \langle -5, 2 \rangle$; $\mathbf{u} = \langle 2, -4 \rangle$

41. $\mathbf{v} = 6\mathbf{i} - 11\mathbf{j}$; $\mathbf{u} = 2\mathbf{i} + 4\mathbf{j}$

42. $\mathbf{u} = \mathbf{i} - 5\mathbf{j}$; $\mathbf{v} = \mathbf{i} + 5\mathbf{j}$

In Exercises 43 and 44, find $\text{proj}_w\mathbf{v}$.

43. $\mathbf{v} = \langle -2, 5 \rangle$; $\mathbf{w} = \langle 5, 4 \rangle$

44. $\mathbf{v} = 4\mathbf{i} - 7\mathbf{j}$; $\mathbf{w} = -2\mathbf{i} - 5\mathbf{j}$

45. WORK A 120-pound box is dragged 14 feet along a level floor. Find the work done if a force of 60 pounds at an angle of 38° is used.

CHAPTER 4 TEST

1. Solve triangle ABC if $A = 70°$, $C = 16°$, and $c = 14$.

2. Find B in triangle ABC if $A = 140°$, $b = 13$, and $a = 45$.

3. In triangle ABC, $C = 42°$, $a = 20$, and $b = 12$. Find side c.

4. In triangle ABC, $a = 32$, $b = 24$, and $c = 18$. Find angle B.

In Exercises 5 to 7, round your answers to two significant digits.

5. Given angle $C = 110°$, side $a = 7.0$, and side $b = 12$, find the area of triangle ABC.

6. Given angle $B = 42°$, angle $C = 75°$, and side $b = 12$, find the area of triangle ABC.

7. Given side $a = 17$, side $b = 55$, and side $c = 42$, find the area of triangle ABC.

8. Given $\mathbf{v} = -2\mathbf{i} + 3\mathbf{j}$, find $|\mathbf{v}|$.

9. A vector has a magnitude of 12 and direction 220°. Write an equivalent vector in the form $\mathbf{v} = a_1\mathbf{i} + a_2\mathbf{j}$. Round a_1 and a_2 to four significant digits.

10. Find $3\mathbf{u} - 5\mathbf{v}$ given the vectors $\mathbf{u} = 2\mathbf{i} - 3\mathbf{j}$ and $\mathbf{v} = 5\mathbf{i} + 4\mathbf{j}$.

11. Find the dot product of $\mathbf{u} = -2\mathbf{i} + 3\mathbf{j}$ and $\mathbf{v} = 5\mathbf{i} + 3\mathbf{j}$.

12. Find the smallest positive angle, to the nearest degree, between the vectors $\mathbf{u} = \langle 3, 5 \rangle$ and $\mathbf{v} = \langle -6, 2 \rangle$.

13. One ship leaves a port at 1:00 P.M. traveling at 12 mph at a heading of 65°. At 2:00 P.M. another ship leaves the port traveling at 18 mph at a heading of 142°. Find the distance between the ships at 3:00 P.M.

14. Two fire lookouts are located 12 miles apart. Lookout A is at a bearing of N32°W from lookout B. A fire was sighted at a bearing of S82°E from A and N72°E from B. Find the distance of the fire from lookout B.

15. A triangular commercial piece of real estate is priced at $8.50 per square foot. Find the cost of the lot, to the nearest $100, that measures 112 feet by 165 feet by 140 feet.

CUMULATIVE REVIEW EXERCISES

1. Find the distance between $P_1(-3, 4)$ and $P_2(4, -1)$.

2. Given $f(x) = \cos x$ and $g(x) = \sin x$, find $(f + g)(x)$.

3. Given $f(x) = \sec x$ and $g(x) = \cos x$, find $(f \circ g)(x)$.

4. Given $f(x) = \dfrac{1}{2}x - 3$, find $f^{-1}(x)$.

5. How is the graph of $F(x) = f(x - 2) + 3$ related to the graph of $y = f(x)$?

6. For the right triangle shown at the right, find a.

7. Graph $y = 3 \sin 2\pi x$.

8. Graph $y = \dfrac{1}{4} \tan 2x$.

9. Graph $y = 2 \sin\left(\dfrac{\pi x}{2}\right) + 1$.

10. What are the amplitude, period, and phase shift of the graph of $y = 3 \sin\left(\dfrac{1}{3}x - \dfrac{\pi}{2}\right)$.

11. What are the amplitude, period, and phase shift of the graph of $y = \sin x + \cos x$.

12. Find c for the triangle at the right.

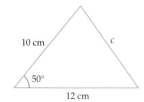

13. Verify the identity $\dfrac{1}{\cos x} - \cos x = \sin x \tan x$.

14. Evaluate $\sin^{-1} \sin\left(\dfrac{2\pi}{3}\right)$.

15. Evaluate $\tan\left(\cos^{-1}\left(\dfrac{12}{13}\right)\right)$.

16. Solve $\sin x \tan x - \dfrac{1}{2} \tan x = 0$ for $0 \le x < 2\pi$.

17. Find the magnitude and direction angle for the vector $\langle 4, -3 \rangle$. Round the angle to the nearest tenth of a degree.

18. Find the angle between the vectors $\mathbf{v} = \langle 1, 2 \rangle$ and $\mathbf{w} = \langle -2, 3 \rangle$. Round to the nearest tenth of a degree.

19. A person who can row at 3 miles per hour in still water wants to row due west across a river. This river is flowing north to south at a rate of 1 mile per hour. Determine the heading of the boat that is required to travel due west across the river.

20. An airplane is traveling with an airspeed of 515 mph at a heading of 54°. A wind of 150 mph is blowing at a heading of 120°. Find the ground speed and the course of the plane.

CHAPTER 5

COMPLEX NUMBERS

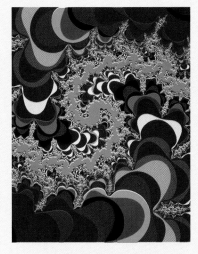

Fractals

Look at the sequence of figures below. Each succeeding figure is created by constructing right triangles using the line segments in the preceding figure as the hypotenuses.

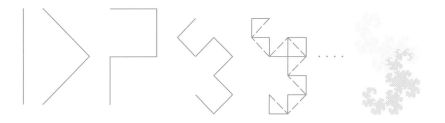

If this process is repeated over and over, the sequence of figures becomes like the last figure on the right, which is called a *fractal*.

One method of creating fractals is to use complex numbers, the topic of this chapter. The Exploring Concepts with Technology at the end of this chapter shows how a fractal called the Mandelbrot set is created.

Verifying Results

One important aspect of problem solving involves the process of checking to see if your results satisfy the conditions of the original problem. This process will be especially important in this chapter when you solve an equation or an inequality.

Here is an example that illustrates the importance of checking your results. The problem seems easy, but many students fail to get the correct answer on their first attempt.

Two volumes of the series *Mathematics: Its Content, Methods, and Meaning* are on a shelf, with no space between the volumes. Each volume is 1 inch thick without its covers. Each cover is $\frac{1}{8}$ inch thick. A bookworm bores horizontally from the first page of Volume I to the last page of Volume II. How far does the bookworm travel?

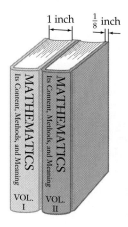

Once you have obtained your solution, try to check it by closely examining two books placed as shown above. Check to make sure you have the proper starting and ending positions. The correct answer is $\frac{1}{4}$ inch.

COMPLEX NUMBERS

● INTRODUCTION TO COMPLEX NUMBERS

Recall that $\sqrt{9} = 3$ because $3^2 = 9$. Now consider the expression $\sqrt{-9}$. To find $\sqrt{-9}$, we need to find a number c such that $c^2 = -9$. However, the square of any real number c (except zero) is a *positive* number. Consequently, we must expand our concept of number to include numbers whose squares are negative numbers.

Around the 17th century, a new number, called an *imaginary number*, was defined so that a negative number would have a square root. The letter i was chosen to represent the number whose square is -1.

Definition of i

The number i, called the **imaginary unit,** is the number such that $i^2 = -1$.

The principal square root of a negative number is defined in terms of i.

Principal Square Root of a Negative Number

If a is a positive real number, then $\sqrt{-a} = i\sqrt{a}$. The number $i\sqrt{a}$ is called an **imaginary number.**

Here are some examples of imaginary numbers.

$$\sqrt{-36} = i\sqrt{36} = 6i \qquad \sqrt{-18} = i\sqrt{18} = 3i\sqrt{2}$$
$$\sqrt{-23} = i\sqrt{23} \qquad\qquad \sqrt{-1} = i\sqrt{1} = i$$

It is customary to write i in front of a radical sign, as we did for $i\sqrt{23}$, to avoid confusing $\sqrt{a}\,i$ with $\sqrt{ai}$.

Complex Numbers

A **complex number** is a number of the form $a + bi$, where a and b are real numbers and $i = \sqrt{-1}$. The number a is the **real part** of $a + bi$, and b is the **imaginary part.**

Here are some examples of complex numbers.

$-3 + 5i$	• Real part: -3; imaginary part: 5
$2 - 6i$	• Real part: 2; imaginary part: -6
5	• Real part: 5; imaginary part: 0
$7i$	• Real part: 0; imaginary part: 7

Note from these examples that a real number is a complex number whose imaginary part is zero and that an imaginary number is a complex number whose real part is zero.

MATH MATTERS

It may seem strange to just invent new numbers, but that is how mathematics evolves. For instance, negative numbers were not an accepted part of mathematics until well into the 13th century. In fact, these numbers often were referred to as "fictitious numbers."

In the 17th century, Rene Descartes called square roots of negative numbers "imaginary numbers," an unfortunate choice of words, and started using the letter i to denote these numbers. These numbers were subjected to the same skepticism as negative numbers.

It is important to understand that these numbers are not *imaginary* in the dictionary sense of the word. It is similar to the situation of negative numbers being called fictitious.

If you think of a number line, then the numbers to the right of zero are positive numbers and the numbers to the left of zero are negative numbers. One way to think of an imaginary number is to visualize it as *up* or *down* from zero. See the Project on page 330 for more information on this topic.

? **QUESTION** What is the real part and the imaginary part of $3 - 5i$?

Note from the following diagram that the real numbers are a subset of the complex numbers, and imaginary numbers are a subset of the complex numbers. The real numbers and imaginary numbers are disjoint sets.

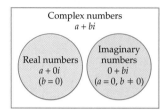

Example 1 illustrates writing a complex number in the standard form $a + bi$.

EXAMPLE 1   **Write a Complex Number in Standard Form**

Write $7 + \sqrt{-45}$ in the form $a + bi$.

Solution

$$7 + \sqrt{-45} = 7 + i\sqrt{45}$$
$$= 7 + i\sqrt{9} \cdot \sqrt{5}$$
$$= 7 + 3i\sqrt{5}$$

▶ **TRY EXERCISE 8, PAGE 328**

● **ADDITION AND SUBTRACTION OF COMPLEX NUMBERS**

All the standard arithmetic operations that are applied to real numbers can be applied to complex numbers.

Definition of Addition and Subtraction of Complex Numbers

If $a + bi$ and $c + di$ are complex numbers, then

Addition $(a + bi) + (c + di) = (a + c) + (b + d)i$

Subtraction $(a + bi) - (c + di) = (a - c) + (b - d)i$

Basically, these rules say that to add two complex numbers, add the real parts and add the imaginary parts. To subtract two complex numbers, subtract the real parts and subtract the imaginary parts.

? **ANSWER**   Real part: 3; imaginary part: -5

| EXAMPLE 2 | Add or Subtract Complex Numbers |

Simplify.

a. $(7 - 2i) + (-2 + 4i)$ b. $(-9 + 4i) - (2 - 6i)$

Solution

a. $(7 - 2i) + (-2 + 4i) = (7 + (-2)) + (-2 + 4)i = 5 + 2i$

b. $(-9 + 4i) - (2 - 6i) = (-9 - 2) + (4 - (-6))i = -11 + 10i$

▶ TRY EXERCISE 18, PAGE 329

● MULTIPLICATION OF COMPLEX NUMBERS

When multiplying complex numbers, the term i^2 is frequently a part of the product. Recall that $i^2 = -1$. Therefore,

$$3i(5i) = 15i^2 = 15(-1) = -15$$
$$-2i(6i) = -12i^2 = -12(-1) = 12$$
$$4i(3 - 2i) = 12i - 8i^2 = 12i - 8(-1) = 8 + 12i$$

take note

Recall that the definition of the product of radical expressions required that the radicand be a positive number. Therefore, when multiplying expressions containing negative radicands, we must first rewrite the expression using i and a positive radicand.

When multiplying square roots of negative numbers, first rewrite the radical expressions using i. For instance,

$$\sqrt{-6} \cdot \sqrt{-24} = i\sqrt{6} \cdot i\sqrt{24} \qquad \bullet\ \sqrt{-6} = i\sqrt{6},\ \sqrt{-24} = i\sqrt{24}$$
$$= i^2\sqrt{144} = -1 \cdot 12$$
$$= -12$$

Note from this example that it would have been incorrect to multiply the radicands of the two radical expressions. To illustrate:

$$\sqrt{-6} \cdot \sqrt{-24} \neq \sqrt{(-6)(-24)}$$

❷ QUESTION What is the product of $\sqrt{-2}$ and $\sqrt{-8}$?

To multiply two complex numbers, we use the following definition.

Definition of Multiplication of Complex Numbers

If $a + bi$ and $c + di$ are complex numbers, then

$$(a + bi)(c + di) = (ac - bd) + (ad + bc)i$$

Because every complex number can be written as a sum of two terms, it is natural to perform multiplication on complex numbers in a manner consistent with

---

❷ ANSWER $\sqrt{-2} \cdot \sqrt{-8} = i\sqrt{2} \cdot i\sqrt{8} = i^2\sqrt{16} = -1 \cdot 4 = -4$

the operation defined on binomials and the definition $i^2 = -1$. By using this analogy, you can multiply complex numbers without memorizing the definition.

EXAMPLE 3   Multiply Complex Numbers

Simplify. **a.** $(3 - 4i)(2 + 5i)$ **b.** $\left(2 + \sqrt{-3}\right)\left(4 - 5\sqrt{-3}\right)$

Solution

a. $(3 - 4i)(2 + 5i) = 6 + 15i - 8i - 20i^2$

$\qquad\qquad\qquad\quad = 6 + 15i - 8i - 20(-1)$ • **Replace i^2 by -1.**

$\qquad\qquad\qquad\quad = 6 + 15i - 8i + 20$ • **Simplify.**

$\qquad\qquad\qquad\quad = 26 + 7i$

b. $\left(2 + \sqrt{-3}\right)\left(4 - 5\sqrt{-3}\right) = \left(2 + i\sqrt{3}\right)\left(4 - 5i\sqrt{3}\right)$

$\qquad\qquad\qquad\qquad\qquad\quad = 8 - 10i\sqrt{3} + 4i\sqrt{3} - 5i^2\sqrt{9}$

$\qquad\qquad\qquad\qquad\qquad\quad = 8 - 10i\sqrt{3} + 4i\sqrt{3} - 5(-1)(3)$

$\qquad\qquad\qquad\qquad\qquad\quad = 8 - 10i\sqrt{3} + 4i\sqrt{3} + 15 = 23 - 6i\sqrt{3}$

▶ **TRY EXERCISE 34, PAGE 329**

● **DIVISION OF COMPLEX NUMBERS**

Recall that the number $\dfrac{3}{\sqrt{2}}$ is not in simplest form because there is a radical expression in the denominator. Similarly, $\dfrac{3}{i}$ is not in simplest form because $i = \sqrt{-1}$. To write this expression in simplest form, multiply the numerator and denominator by i.

$$\frac{3}{i} \cdot \frac{i}{i} = \frac{3i}{i^2} = \frac{3i}{-1} = -3i$$

Here is another example.

$$\frac{3 - 6i}{2i} = \frac{3 - 6i}{2i} \cdot \frac{i}{i} = \frac{3i - 6i^2}{2i^2} = \frac{3i - 6(-1)}{2(-1)}$$

$$= \frac{3i + 6}{-2} = -3 - \frac{3}{2}i$$

Recall that to simplify the quotient $\dfrac{2 + \sqrt{3}}{5 + 2\sqrt{3}}$, we multiply the numerator and denominator by the conjugate of $5 + 2\sqrt{3}$, which is $5 - 2\sqrt{3}$. In a similar manner, to find the quotient of two complex numbers, we multiply the numerator and denominator by the conjugate of the denominator.

The complex numbers $a + bi$ and $a - bi$ are called **complex conjugates** or **conjugates** of each other. The conjugate of the complex number z is denoted by $\bar{z}$. For instance,

$$\overline{2 + 5i} = 2 - 5i \qquad \text{and} \qquad \overline{3 - 4i} = 3 + 4i$$

INTEGRATING TECHNOLOGY

Some graphing calculators can be used to perform operations on complex numbers. Here are some typical screens for a TI-83 Plus.

Press ⏺MODE⏺. Use the down arrow key to highlight $a + bi$.

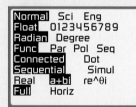

```
Normal  Sci  Eng
Float   0123456789
Radian  Degree
Func    Par  Pol  Seq
Connected    Dot
Sequential   Simul
Real   a+bi  re^θi
Full   Horiz
```

Press ⏺ENTER⏺ ⏺2nd⏺ [QUIT].
Here are two examples of computations on complex numbers. To enter an i, use ⏺2nd⏺ [i], which is above the decimal point key.

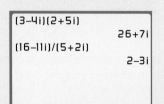

```
(3−4i)(2+5i)
                26+7i
(16−11i)/(5+2i)
                2−3i
```

Consider the product of a complex number and its conjugate. For instance,

$$(2 + 5i)(2 - 5i) = 4 - 10i + 10i - 25i^2$$
$$= 4 - 25(-1) = 4 + 25$$
$$= 29$$

Note that the product is a *real* number. This is always true.

Product of Complex Conjugates

The product of a complex number and its conjugate is a real number. That is, $(a + bi)(a - bi) = a^2 + b^2$.

For instance, $(5 + 3i)(5 - 3i) = 5^2 + 3^2 = 25 + 9 = 34$.

The next example shows how the quotient of two complex numbers is determined by using conjugates.

EXAMPLE 4   Divide Complex Numbers

Simplify: $\dfrac{16 - 11i}{5 + 2i}$

Solution

$$\frac{16 - 11i}{5 + 2i} = \frac{16 - 11i}{5 + 2i} \cdot \frac{5 - 2i}{5 - 2i}$$

$$= \frac{80 - 32i - 55i + 22i^2}{5^2 + 2^2}$$

$$= \frac{80 - 32i - 55i + 22(-1)}{25 + 4}$$

$$= \frac{80 - 87i - 22}{29}$$

$$= \frac{58 - 87i}{29}$$

$$= \frac{29(2 - 3i)}{29} = 2 - 3i$$

• Multiply numerator and denominator by the conjugates of the denominator.

▶ **TRY EXERCISE 48, PAGE 329**

● POWERS OF *i*

The following powers of *i* illustrate a pattern:

$$i^1 = i \qquad\qquad i^5 = i^4 \cdot i = 1 \cdot i = i$$
$$i^2 = -1 \qquad\qquad i^6 = i^4 \cdot i^2 = 1(-1) = -1$$
$$i^3 = i^2 \cdot i = (-1)i = -i \qquad i^7 = i^4 \cdot i^3 = 1(-i) = -i$$
$$i^4 = i^2 \cdot i^2 = (-1)(-1) = 1 \qquad i^8 = (i^4)^2 = 1^2 = 1$$

Because $i^4 = 1$, $(i^4)^n = 1^n = 1$ for any integer n. Thus it is possible to evaluate powers of i by factoring out powers of i^4, as shown in the following:

$$i^{27} = (i^4)^6 \cdot i^3 = 1^6 \cdot i^3 = 1 \cdot (-i) = -i$$

The following theorem can be used to evaluate powers of i.

Powers of *i*

If n is a positive integer, then $i^n = i^r$, where r is the remainder of the division of n by 4.

EXAMPLE 5 — Evaluate a Power of *i*

Evaluate: i^{153}

Solution

Use the powers of i theorem.

$$i^{153} = i^1 = i \qquad \text{• Remainder of } 153 \div 4 \text{ is } 1.$$

▶ **TRY EXERCISE 60, PAGE 329**

TOPICS FOR DISCUSSION

1. What is an imaginary number? What is a complex number?

2. How are the real numbers related to the complex numbers?

3. Is zero a complex number?

4. What is the conjugate of a complex number?

5. If a and b are real numbers and $ab = 0$, then $a = 0$ or $b = 0$. Is the same true for complex numbers? That is, if u and v are complex numbers and $uv = 0$, must one of the numbers be zero?

EXERCISE SET 5.1

In Exercises 1 to 10, write the complex number in standard form.

1. $\sqrt{-81}$

2. $\sqrt{-64}$

3. $\sqrt{-98}$

4. $\sqrt{-27}$

5. $\sqrt{16} + \sqrt{-81}$

6. $\sqrt{25} + \sqrt{-9}$

7. $5 + \sqrt{-49}$

▶ **8.** $6 - \sqrt{-1}$

9. $8 - \sqrt{-18}$

10. $11 + \sqrt{-48}$

In Exercises 11 to 36, simplify and write the complex number in standard form.

11. $(5 + 2i) + (6 - 7i)$

12. $(4 - 8i) + (5 + 3i)$

13. $(-2 - 4i) - (5 - 8i)$

14. $(3 - 5i) - (8 - 2i)$

15. $(1 - 3i) + (7 - 2i)$

16. $(2 - 6i) + (4 - 7i)$

17. $(-3 - 5i) - (7 - 5i)$

▶ **18.** $(5 - 3i) - (2 + 9i)$

19. $8i - (2 - 8i)$

20. $3 - (4 - 5i)$

21. $5i \cdot 8i$

22. $(-3i)(2i)$

23. $\sqrt{-50} \cdot \sqrt{-2}$

24. $\sqrt{-12} \cdot \sqrt{-27}$

25. $3(2 + 5i) - 2(3 - 2i)$

26. $3i(2 + 5i) + 2i(3 - 4i)$

27. $(4 + 2i)(3 - 4i)$

28. $(6 + 5i)(2 - 5i)$

29. $(-3 - 4i)(2 + 7i)$

30. $(-5 - i)(2 + 3i)$

31. $(4 - 5i)(4 + 5i)$

32. $(3 + 7i)(3 - 7i)$

33. $\left(3 + \sqrt{-4}\right)\left(2 - \sqrt{-9}\right)$

▶ **34.** $\left(5 + 2\sqrt{-16}\right)\left(1 - \sqrt{-25}\right)$

35. $\left(3 + 2\sqrt{-18}\right)\left(2 + 2\sqrt{-50}\right)$

36. $\left(5 - 3\sqrt{-48}\right)\left(2 - 4\sqrt{-27}\right)$

In Exercises 37 to 54, write each expression as a complex number in standard form.

37. $\dfrac{6}{i}$

38. $\dfrac{-8}{2i}$

39. $\dfrac{6 + 3i}{i}$

40. $\dfrac{4 - 8i}{4i}$

41. $\dfrac{1}{7 + 2i}$

42. $\dfrac{5}{3 + 4i}$

43. $\dfrac{2i}{1 + i}$

44. $\dfrac{5i}{2 - 3i}$

45. $\dfrac{5 - i}{4 + 5i}$

46. $\dfrac{4 + i}{3 + 5i}$

47. $\dfrac{3 + 2i}{3 - 2i}$

▶ **48.** $\dfrac{8 - i}{2 + 3i}$

49. $\dfrac{-7 + 26i}{4 + 3i}$

50. $\dfrac{-4 - 39i}{5 - 2i}$

51. $(3 - 5i)^2$

52. $(2 + 4i)^2$

53. $(1 + 2i)^3$

54. $(2 - i)^3$

In Exercises 55 to 62, evaluate the power of i.

55. i^{15} **56.** i^{66} **57.** $-i^{40}$ **58.** $-i^{51}$

59. $\dfrac{1}{i^{25}}$ ▶ **60.** $\dfrac{1}{i^{83}}$ **61.** i^{-34} **62.** i^{-52}

In Exercises 63 to 68, evaluate $\dfrac{-b + \sqrt{b^2 - 4ac}}{2a}$ for the given values of a, b, and c. Write your answer as a complex number in standard form.

63. $a = 3, b = -3, c = 3$

64. $a = 2, b = 4, c = 4$

65. $a = 2, b = 6, c = 6$

66. $a = 2, b = 1, c = 3$

67. $a = 4, b = -4, c = 2$

68. $a = 3, b = -2, c = 4$

CONNECTING CONCEPTS

The property that the product of conjugates of the form $(a + bi)(a - bi)$ is equal to $a^2 + b^2$ can be used to factor the sum of two perfect squares over the set of complex numbers. For example, $x^2 + y^2 = (x + yi)(x - yi)$. In Exercises 69 to 74, factor the binomial over the set of complex numbers.

69. $x^2 + 16$

70. $x^2 + 9$

71. $z^2 + 25$

72. $z^2 + 64$

73. $4x^2 + 81$

74. $9x^2 + 1$

75. Show that if $x = 1 + 2i$, then $x^2 - 2x + 5 = 0$.

76. Show that if $x = 1 - 2i$, then $x^2 - 2x + 5 = 0$.

77. When we think of the cube root of 8, $\sqrt[3]{8}$, we normally mean the *real* cube root of 8 and write $\sqrt[3]{8} = 2$. However, there are two other cube roots of 8 that are complex numbers. Verify that $-1 + i\sqrt{3}$ and $-1 - i\sqrt{3}$ are cube roots of 8 by showing that $\left(-1 + i\sqrt{3}\right)^3 = 8$ and $\left(-1 - i\sqrt{3}\right)^3 = 8$.

78. It is possible to find the square root of a complex number.

Verify that $\sqrt{i} = \dfrac{\sqrt{2}}{2}(1 + i)$ by showing that

$$\left[\frac{\sqrt{2}}{2}(1+i)\right]^2 = i.$$

79. Simplify $i + i^2 + i^3 + i^4 + \cdots + i^{28}$.

80. Simplify $i + i^2 + i^3 + i^4 + \cdots + i^{100}$.

PREPARE FOR SECTION 5.2

81. Simplify: $(1 + i)(2 + i)$ [5.1]

82. Simplify: $\dfrac{2 + i}{3 - i}$ [5.1]

83. What is the conjugate of $2 + 3i$? [5.1]

84. What is the conjugate of $3 - 5i$? [5.1]

85. Use the quadratic formula to find the solutions of $x^2 + x = -1$. [1.1/5.1]

86. Solve: $x^2 + 9 = 0$ [1.1/5.1]

PROJECTS

ARGAND DIAGRAM Just as we can graph a real number on a real number line, we can graph a complex number. This is accomplished by using one number line for the real part of the complex number and one number line for the imaginary part of the complex number. These two number lines are drawn perpendicular to each other and pass through their respective origins, as shown below.

The result is called the *complex plane* or an *Argand diagram* after Jean-Robert Argand (1768–1822), an accountant and amateur mathematician. Although he is given credit for this representation of complex numbers, Caspar Wessel (1745–1818) actually conceived the idea before Argand.

To graph the complex number $3 + 4i$, start at 3 on the real axis. Now move 4 units up (for positive numbers move up, for negative numbers move down) and place a dot at that point, as shown in the diagram. Graphs of several other complex numbers are also shown.

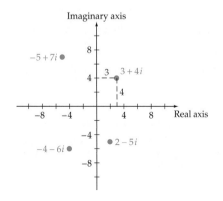

In Exercises 1 to 8, graph the complex number.

1. $2 + 5i$ **2.** $4 - 3i$ **3.** $-2 + 6i$ **4.** $-3 - 5i$

5. 4 **6.** $-2i$ **7.** $3i$ **8.** -5

The absolute value of a complex number is given by $|a + bi| = \sqrt{a^2 + b^2}$. **In Exercises 9 to 12, find the absolute value of the complex number.**

9. $2 + 5i$ **10.** $4 - 3i$ **11.** $-2 + 6i$ **12.** $-3 - 5i$

13. The additive inverse of $a + bi$ is $-a - bi$. Show that the absolute value of a complex number and the absolute value of its additive inverse are equal.

14. A *real* number and its additive inverse are the same distance from zero but on opposite sides of zero on a real number line. Describe the relationship between the graphs of a complex number and its additive inverse.

TRIGONOMETRIC FORM OF COMPLEX NUMBERS

● GRAPHICAL REPRESENTATION OF A COMPLEX NUMBER

Real numbers are graphed as points on a number line. Complex numbers can be graphed in a coordinate plane called the **complex plane.** The horizontal axis of the complex plane is called the **real axis;** the vertical axis is called the **imaginary axis.**

A complex number written in the form $z = a + bi$ is written in **standard form** or **rectangular form.** The graph of $a + bi$ is associated with the point $P(a, b)$ in the complex plane. **Figure 5.1** shows the graphs of several complex numbers.

● ABSOLUTE VALUE OF A COMPLEX NUMBER

The length of the line segment from the origin to the point $(-3, 4)$ in the complex plane is the *absolute value* of $z = -3 + 4i$. See **Figure 5.2.** From the Pythagorean

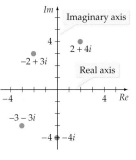

FIGURE 5.1

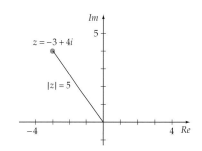

FIGURE 5.2

Theorem, the absolute value of $z = -3 + 4i$ is

$$\sqrt{(-3)^2 + 4^2} = \sqrt{25} = 5$$

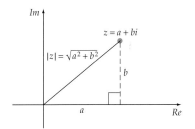

FIGURE 5.3

Definition of the Absolute Value of a Complex Number

The absolute value of the complex number $z = a + bi$, denoted by $|z|$, is

$$|z| = |a + bi| = \sqrt{a^2 + b^2}$$

Thus $|z|$ is the distance from the origin to z (see **Figure 5.3**).

❓ QUESTION The conjugate of $a + bi$ is $a - bi$. Does $|a + bi| = |a - bi|$?

❓ ANSWER Yes.

● TRIGONOMETRIC FORM OF A COMPLEX NUMBER

A complex number $z = a + bi$ can be written in terms of trigonometric functions. Consider the complex number graphed in **Figure 5.4.** We can write a and b in terms of the sine and the cosine.

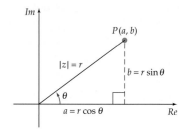

$$\cos \theta = \frac{a}{r} \qquad \sin \theta = \frac{b}{r}$$

$$a = r \cos \theta \qquad b = r \sin \theta$$

FIGURE 5.4

where $r = |z| = \sqrt{a^2 + b^2}$. Substituting for a and b in $z = a + bi$, we obtain

$$z = r \cos \theta + ir \sin \theta = r(\cos \theta + i \sin \theta)$$

The expression $z = r(\cos \theta + i \sin \theta)$ is known as the **trigonometric form** of a complex number. The trigonometric form of a complex number is also called the **polar form** of the complex number. The notation $\cos \theta + i \sin \theta$ is often abbreviated as cis θ using the c from $\cos \theta$, the imaginary unit i, and the s from $\sin \theta$.

Trigonometric Form of a Complex Number

The complex number $z = a + bi$ can be written in trigonometric form as

$$z = r(\cos \theta + i \sin \theta) = r \text{ cis } \theta$$

where $a = r \cos \theta$, $b = r \sin \theta$, $r = \sqrt{a^2 + b^2}$, and $\tan \theta = \dfrac{b}{a}$.

In this text we will often write the trigonometric form of a complex number in its abbreviated form $z = r$ cis θ. The value of r is called the **modulus** of the complex number z, and the angle θ is called the **argument** of the complex number z. The modulus r and the argument θ of a complex number $z = a + bi$ are given by

$$r = \sqrt{a^2 + b^2} \qquad \text{and} \qquad \cos \theta = \frac{a}{r}, \quad \sin \theta = \frac{b}{r}$$

We also can write $\alpha = \tan^{-1} \left| \dfrac{b}{a} \right|$, where α is the reference angle for θ. As a result of the periodic nature of the sine and cosine functions, the trigonometric form of a complex number is not unique. Because $\cos \theta = \cos(\theta + 2k\pi)$ and $\sin \theta = \sin(\theta + 2k\pi)$, where k is an integer, the following complex numbers are equal.

$$r \text{ cis } \theta = r \text{ cis}(\theta + 2k\pi) \quad \text{for } k \text{ an integer}$$

For example, $2 \text{ cis } \dfrac{\pi}{6} = 2 \text{ cis}\left(\dfrac{\pi}{6} + 2\pi\right)$.

EXAMPLE 1   Write a Complex Number in Trigonometric Form

Write $z = -2 - 2i$ in trigonometric form.

Solution

Find the modulus and the argument of z. Then substitute these values in the trigonometric form of z.

$$r = \sqrt{(-2)^2 + (-2)^2} = \sqrt{8} = 2\sqrt{2}$$

To determine θ, we first determine α. See **Figure 5.5.**

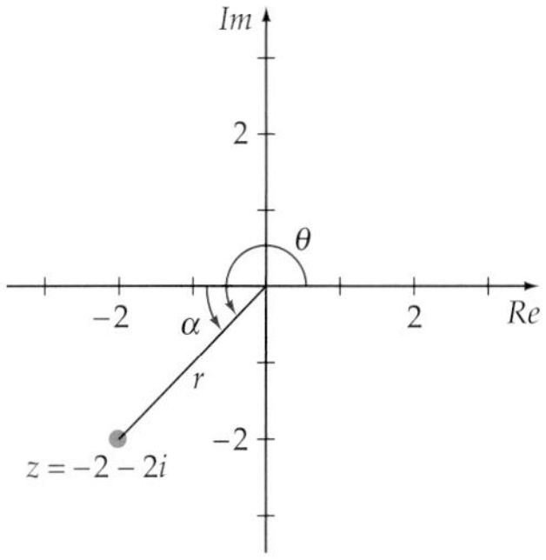

$$\alpha = \tan^{-1}\left|\frac{b}{a}\right|$$
        • α is the reference angle of angle θ.

$$\alpha = \tan^{-1}\left|\frac{-2}{-2}\right| = \tan^{-1}1 = 45°$$

$$\theta = 180° + 45° = 225°$$
        • Because z is in the third quadrant, $180° < \theta < 270°$.

The trigonometric form is

$$z = r\ \text{cis}\ \theta = 2\sqrt{2}\ \text{cis}\ 225°$$
        • $r = 2\sqrt{2},\ \theta = 225°$

▶ **TRY EXERCISE 12, PAGE 336**

Im

FIGURE 5.5

EXAMPLE 2   Write a Complex Number in Standard Form

Write $z = 2\ \text{cis}\ 120°$ in standard form.

Solution

Write z in the form $r(\cos\theta + i\sin\theta)$ and then evaluate $\cos\theta$ and $\sin\theta$. See **Figure 5.6.**

$$z = 2\ \text{cis}\ 120° = 2(\cos 120° + i\sin 120°) = 2\left(-\frac{1}{2} + \frac{\sqrt{3}}{2}i\right) = -1 + i\sqrt{3}$$

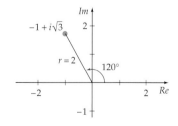

FIGURE 5.6

▶ **TRY EXERCISE 26, PAGE 337**

• THE PRODUCT AND THE QUOTIENT OF COMPLEX NUMBERS WRITTEN IN TRIGONOMETRIC FORM

Let z_1 and z_2 be two complex numbers written in trigonometric form. The product of z_1 and z_2 can be found by using trigonometric identities. If $z_1 = r_1(\cos \theta_1 + i \sin \theta_1)$ and $z_2 = r_2(\cos \theta_2 + i \sin \theta_2)$, then

$$z_1 z_2 = r_1(\cos \theta_1 + i \sin \theta_1) \cdot r_2(\cos \theta_2 + i \sin \theta_2)$$

$$= r_1 r_2(\cos \theta_1 \cos \theta_2 + i \cos \theta_1 \sin \theta_2 + i \sin \theta_1 \cos \theta_2 + i^2 \sin \theta_1 \sin \theta_2)$$

$$= r_1 r_2[(\cos \theta_1 \cos \theta_2 - \sin \theta_1 \sin \theta_2) + i(\sin \theta_1 \cos \theta_2 + \cos \theta_1 \sin \theta_2)]$$

$$= r_1 r_2[\cos(\theta_1 + \theta_2) + i \sin(\theta_1 + \theta_2)] \qquad \bullet \text{ Identities for } \cos(\theta_1 + \theta_2)$$
$$\text{and } \sin(\theta_1 + \theta_2)$$

$$z_1 z_2 = r_1 r_2 \operatorname{cis}(\theta_1 + \theta_2) \qquad \bullet \textbf{ The Product Property of Complex Numbers}$$

Thus the modulus for the product of two complex numbers in trigonometric form is the product of the moduli of the two numbers, and the argument of the product is the sum of the arguments of the two numbers.

EXAMPLE 3 **Find the Product of Two Complex Numbers**

Find the product of $z_1 = -1 + i\sqrt{3}$ and $z_2 = -\sqrt{3} + i$ by using the trigonometric form of the complex numbers. Write the answer in standard form.

Solution

Write each complex number in trigonometric form. Then use the Product Property of Complex Numbers. See **Figure 5.7.**

$$z_1 = -1 + i\sqrt{3} = 2 \operatorname{cis} \frac{2\pi}{3} \qquad \bullet \, r_1 = 2, \, \theta_1 = \frac{2\pi}{3}$$

$$z_2 = -\sqrt{3} + i = 2 \operatorname{cis} \frac{5\pi}{6} \qquad \bullet \, r_2 = 2, \, \theta_2 = \frac{5\pi}{6}$$

$$z_1 z_2 = 2 \operatorname{cis} \frac{2\pi}{3} \cdot 2 \operatorname{cis} \frac{5\pi}{6}$$

$$= 4 \operatorname{cis}\left(\frac{2\pi}{3} + \frac{5\pi}{6}\right)$$

$$= 4 \operatorname{cis} \frac{3\pi}{2}$$

$$= 4\left(\cos \frac{3\pi}{2} + i \sin \frac{3\pi}{2}\right)$$

$$= 4(0 - i)$$

$$= -4i$$

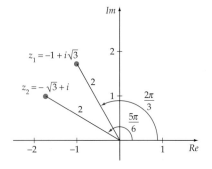

$z_1 = -1 + i\sqrt{3}$

$z_2 = -\sqrt{3} + i$

FIGURE 5.7

▶ TRY EXERCISE 52, PAGE 337

Similarly, the quotient of z_1 and z_2 can be found by using trigonometric identities. If $z_1 = r_1(\cos \theta_1 + i \sin \theta_1)$ and $z_2 = r_2(\cos \theta_2 + i \sin \theta_2)$, then

$$\frac{z_1}{z_2} = \frac{r_1(\cos \theta_1 + i \sin \theta_1)}{r_2(\cos \theta_2 + i \sin \theta_2)}$$

$$= \frac{r_1(\cos \theta_1 + i \sin \theta_1)(\cos \theta_2 - i \sin \theta_2)}{r_2(\cos \theta_2 + i \sin \theta_2)(\cos \theta_2 - i \sin \theta_2)}$$

$$= \frac{r_1(\cos \theta_1 \cos \theta_2 - i \cos \theta_1 \sin \theta_2 + i \sin \theta_1 \cos \theta_2 - i^2 \sin \theta_1 \sin \theta_2)}{r_2(\cos^2 \theta_2 - i^2 \sin^2 \theta_2)}$$

$$= \frac{r_1[(\cos \theta_1 \cos \theta_2 + \sin \theta_1 \sin \theta_2) + i(\sin \theta_1 \cos \theta_2 - \cos \theta_1 \sin \theta_2)]}{r_2(\cos^2 \theta_2 + \sin^2 \theta_2)}$$

$$= \frac{r_1}{r_2}[\cos(\theta_1 - \theta_2) + i \sin(\theta_1 - \theta_2)] \qquad \bullet \text{ Identities for } \cos(\theta_1 - \theta_2),$$
$$\sin(\theta_1 - \theta_2), \text{ and}$$
$$\cos^2 \theta_2 + \sin^2 \theta_2$$

$$\frac{z_1}{z_2} = \frac{r_1}{r_2} \text{cis}(\theta_1 - \theta_2) \qquad \bullet \textbf{The Quotient Property of}$$
$$\textbf{Complex Numbers}$$

Thus the modulus for the quotient of two complex numbers in trigonometric form is the quotient of the moduli of the two numbers, and the argument of the quotient is the difference of the arguments of the two numbers.

EXAMPLE 4 **Find the Quotient of Two Complex Numbers**

Find the quotient of $z_1 = -1 + i$ and $z_2 = \sqrt{3} - i$ by using the trigonometric form of the complex numbers. Write the answer in standard form.

Solution

Write the numbers in trigonometric form. Then use the Quotient Property of Complex Numbers. See **Figure 5.8.**

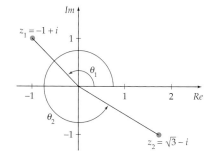

$$z_1 = -1 + i = \sqrt{2} \text{ cis } 135° \qquad \bullet r_1 = \sqrt{2}, \theta_1 = 135°$$

$$z_2 = \sqrt{3} - i = 2 \text{ cis } 330° \qquad \bullet r_2 = 2, \theta_2 = 330°$$

$$\frac{z_1}{z_2} = \frac{-1 + i}{\sqrt{3} - i} = \frac{\sqrt{2} \text{ cis } 135°}{2 \text{ cis } 330°}$$

$$= \frac{\sqrt{2}}{2} \text{ cis}(135° - 330°)$$

$$= \frac{\sqrt{2}}{2} \text{ cis}(-195°)$$

$$= \frac{\sqrt{2}}{2}[\cos(-195°) + i \sin(-195°)]$$

$$= \frac{\sqrt{2}}{2}(\cos 195° - i \sin 195°) \qquad \bullet \cos(-x) = \cos x,$$
$$\sin(-x) = -\sin x$$

$$\approx -0.6830 + 0.1830i$$

FIGURE 5.8

▶ **TRY EXERCISE 56, PAGE 337**

Here is a summary of the product and quotient theorems.

Product and Quotient of Complex Numbers Written in Trigonometric Form

Let $z_1 = r_1(\cos \theta_1 + i \sin \theta_1)$ and $z_2 = r_2(\cos \theta_2 + i \sin \theta_2)$. Then

$$z_1 z_2 = r_1 r_2 [\cos(\theta_1 + \theta_2) + i \sin(\theta_1 + \theta_2)]$$

$$z_1 z_2 = r_1 r_2 \operatorname{cis}(\theta_1 + \theta_2) \qquad \bullet \text{ Using cis notation}$$

and

$$\frac{z_1}{z_2} = \frac{r_1}{r_2}[\cos(\theta_1 - \theta_2) + i \sin(\theta_1 - \theta_2)]$$

$$\frac{z_1}{z_2} = \frac{r_1}{r_2} \operatorname{cis}(\theta_1 - \theta_2) \qquad \bullet \text{ Using cis notation}$$

 TOPICS FOR DISCUSSION

1. Explain why the absolute value of $a + bi$ is equal to the absolute value of $-a + bi$.

2. Describe the graph of all complex numbers with an absolute value of 5.

3. Explain two different ways to find $\dfrac{4}{i}$.

4. The complex numbers z_1 and z_2 both have an absolute value of 1. What is the absolute value of the product $z_1 z_2$? Explain.

5. Explain how to use the Product Property to prove that the product of a complex number and its conjugate is a real number.

EXERCISE SET 5.2

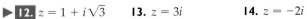

In Exercises 1 to 8, graph each complex number. Find the absolute value of each complex number.

1. $z = -2 - 2i$ **2.** $z = 4 - 4i$ **3.** $z = \sqrt{3} - i$

4. $z = 1 + i\sqrt{3}$ **5.** $z = -2i$ **6.** $z = -5$

7. $z = 3 - 5i$ **8.** $z = -5 - 4i$

In Exercises 9 to 16, write each complex number in trigonometric form.

9. $z = 1 - i$ **10.** $z = -4 - 4i$ **11.** $z = \sqrt{3} - i$

▶ **12.** $z = 1 + i\sqrt{3}$ **13.** $z = 3i$ **14.** $z = -2i$

15. $z = -5$ **16.** $z = 3$

In Exercises 17 to 34, write each complex number in standard form.

17. $z = 2(\cos 45° + i \sin 45°)$

18. $z = 3(\cos 240° + i \sin 240°)$

19. $z = (\cos 315° + i \sin 315°)$

20. $z = 5(\cos 120° + i \sin 120°)$

21. $z = 6 \operatorname{cis} 135°$ **22.** $z = \operatorname{cis} 315°$

23. $z = 8 \operatorname{cis} 0°$

24. $z = 5 \operatorname{cis} 90°$

25. $z = 2\left(\cos \dfrac{5\pi}{6} + i \sin \dfrac{5\pi}{6}\right)$

▶ **26.** $z = 4\left(\cos \dfrac{5\pi}{3} + i \sin \dfrac{5\pi}{3}\right)$

27. $z = 3\left(\cos \dfrac{3\pi}{2} + i \sin \dfrac{3\pi}{2}\right)$

28. $z = 5(\cos \pi + i \sin \pi)$

29. $z = 8 \operatorname{cis} \dfrac{3\pi}{4}$

30. $z = 9 \operatorname{cis} \dfrac{4\pi}{3}$

31. $z = 9 \operatorname{cis} \dfrac{11\pi}{6}$

32. $z = \operatorname{cis} \dfrac{3\pi}{2}$

33. $z = 2 \operatorname{cis} 2$

34. $z = 5 \operatorname{cis} 4$

In Exercises 35 to 42, multiply the complex numbers. Write the answer in trigonometric form.

35. $2 \operatorname{cis} 30° \cdot 3 \operatorname{cis} 225°$

36. $4 \operatorname{cis} 120° \cdot 6 \operatorname{cis} 315°$

37. $3(\cos 122° + i \sin 122°) \cdot 4(\cos 213° + i \sin 213°)$

38. $8(\cos 88° + i \sin 88°) \cdot 12(\cos 112° + i \sin 112°)$

39. $5\left(\cos \dfrac{2\pi}{3} + i \sin \dfrac{2\pi}{3}\right) \cdot 2\left(\cos \dfrac{2\pi}{5} + i \sin \dfrac{2\pi}{5}\right)$

40. $5 \operatorname{cis} \dfrac{11\pi}{12} \cdot 3 \operatorname{cis} \dfrac{4\pi}{3}$

41. $4 \operatorname{cis} 2.4 \cdot 6 \operatorname{cis} 4.1$

42. $7 \operatorname{cis} 0.88 \cdot 5 \operatorname{cis} 1.32$

In Exercises 43 to 50, divide the complex numbers. Write the answer in standard form.

43. $\dfrac{32 \operatorname{cis} 30°}{4 \operatorname{cis} 150°}$

44. $\dfrac{15 \operatorname{cis} 240°}{3 \operatorname{cis} 135°}$

45. $\dfrac{27(\cos 315° + i \sin 315°)}{9(\cos 225° + i \sin 225°)}$

46. $\dfrac{9(\cos 25° + i \sin 25°)}{3(\cos 175° + i \sin 175°)}$

47. $\dfrac{12 \operatorname{cis} \dfrac{2\pi}{3}}{4 \operatorname{cis} \dfrac{11\pi}{6}}$

48. $\dfrac{10 \operatorname{cis} \dfrac{\pi}{3}}{5 \operatorname{cis} \dfrac{\pi}{4}}$

49. $\dfrac{25(\cos 3.5 + i \sin 3.5)}{5(\cos 1.5 + i \sin 1.5)}$

50. $\dfrac{18(\cos 0.56 + i \sin 0.56)}{6(\cos 1.22 + i \sin 1.22)}$

In Exercises 51 to 58, perform the indicated operation in trigonometric form. Write the solution in standard form.

51. $\left(1 - i\sqrt{3}\right)(1 + i)$

▶ **52.** $\left(\sqrt{3} - i\right)\left(1 + i\sqrt{3}\right)$

53. $(3 - 3i)(1 + i)$

54. $(2 + 2i)\left(\sqrt{3} - i\right)$

55. $\dfrac{1 + i\sqrt{3}}{1 - i\sqrt{3}}$

▶ **56.** $\dfrac{1 + i}{1 - i}$

57. $\dfrac{\sqrt{2} - i\sqrt{2}}{1 + i}$

58. $\dfrac{1 + i\sqrt{3}}{4 - 4i}$

CONNECTING CONCEPTS

In Exercises 59 to 64, perform the indicated operation in trigonometric form. Write the solution in standard form.

59. $\left(\sqrt{3} - i\right)(2 + 2i)\left(2 - 2i\sqrt{3}\right)$

60. $(1 - i)\left(1 + i\sqrt{3}\right)\left(\sqrt{3} - i\right)$

61. $\dfrac{\sqrt{3} + i\sqrt{3}}{\left(1 - i\sqrt{3}\right)(2 - 2i)}$

62. $\dfrac{\left(2 - 2i\sqrt{3}\right)\left(1 - i\sqrt{3}\right)}{4\sqrt{3} + 4i}$

63. $(1 - 3i)(2 + 3i)(4 + 5i)$

64. $\dfrac{(2 - 5i)(1 - 6i)}{3 + 4i}$

65. Use the trigonometric forms of the complex numbers z and $\bar{z}$ to find $z \cdot \bar{z}$. (Note that $\bar{z}$ is the conjugate of z.)

66. Use the trigonometric forms of z and $\bar{z}$ to find $\dfrac{z}{\bar{z}}$.

PREPARE FOR SECTION 5.3

67. Find $\left(\dfrac{\sqrt{2}}{2} + \dfrac{\sqrt{2}}{2}i\right)^2$. [5.1]

68. Find the real root of $x^3 - 8 = 0$.

69. Find the real root of $x^5 - 243 = 0$.

70. Write $2 + 2i$ in trigonometric form. [5.2]

71. Write $2(\cos 150° + i \sin 150°)$ in standard form. [5.2]

72. Find the absolute value of $\dfrac{\sqrt{2}}{2} - \dfrac{\sqrt{2}}{2}i$. [5.2]

PROJECTS

1. A GEOMETRIC INTERPRETATION Multiplying a real number by a number greater than 1 increases the magnitude of that number. For example, multiplying 2 by 3 triples the magnitude of 2. Explain the effect of multiplying a real number by i, the imaginary unit. Now multiply a complex number by i and note the effect. The use of a complex plane may be helpful in your explanation.

2. **HISTORICAL PERSPECTIVE** The complex number system was not widely accepted in the mathematical community until the 1600s. Other kinds of numbers, such as zero and negative numbers, also did not gain immediate acceptance when they were first introduced. Write a brief history of negative numbers. Be sure to include the contributions of early Chinese and Middle-Eastern mathematicians.

SECTION 5.3   DE MOIVRE'S THEOREM

- DE MOIVRE'S THEOREM
- DE MOIVRE'S THEOREM FOR FINDING ROOTS

• DE MOIVRE'S THEOREM

De Moivre's Theorem is a procedure for finding powers and roots of complex numbers when the complex numbers are expressed in trigonometric form. This theorem can be illustrated by repeated multiplication of a complex number.

Let $z = r \operatorname{cis} \theta$. Then z^2 can be written as

$$z \cdot z = r \operatorname{cis} \theta \cdot r \operatorname{cis} \theta$$

$$z^2 = r^2 \operatorname{cis} 2\theta$$

The product $z^2 \cdot z$ is

$$z^2 \cdot z = r^2 \operatorname{cis} 2\theta \cdot r \operatorname{cis} \theta$$

$$z^3 = r^3 \operatorname{cis} 3\theta$$

If we continue this process, the results suggest a formula known as De Moivre's Theorem for the nth power of a complex number.

Abraham de Moivre (1667–1754) was a French mathematician who fled to England during the expulsion of the Huguenots in 1685.

De Moivre made important contributions in probability theory and analytic geometry. In 1718 he published *The Doctrine of Chance*, in which he developed the theory of annuities, mortality statistics, and the concept of statistical independence. In 1730 de Moivre stated the theorem at the top of the page, which we now call De Moivre's Theorem. This theorem is significant because it provides a connection between trigonometry and mathematical analysis.

Although de Moivre was well respected in the mathematical community and was elected to the Royal Society of England, he never was able to secure a university teaching position. His income came mainly from tutoring, and he died in poverty.

De Moivre is often remembered for predicting the day of his own death. At one point in his life, he noticed that he was sleeping a few minutes longer each night. Thus he calculated that he would die when he needed 24 hours of sleep. As it turned out, his calculation was correct.

De Moivre's Theorem

If $z = r \operatorname{cis} \theta$ and n is a positive integer, then

$$z^n = r^n \operatorname{cis} n\theta$$

EXAMPLE 1   Find the Power of a Complex Number

Find $(2 \operatorname{cis} 30°)^5$. Write the answer in standard form.

Solution

By De Moivre's Theorem,

$$\begin{aligned}
(2 \operatorname{cis} 30°)^5 &= 2^5 \operatorname{cis}(5 \cdot 30°) \\
&= 2^5[\cos(5 \cdot 30°) + i \sin(5 \cdot 30°)] \\
&= 32(\cos 150° + i \sin 150°) \\
&= 32\left(-\frac{\sqrt{3}}{2} + \frac{1}{2}i\right) = -16\sqrt{3} + 16i
\end{aligned}$$

▶ **TRY EXERCISE 6, PAGE 342**

EXAMPLE 2   Use De Moivre's Theorem

Find $(1 + i)^8$ using De Moivre's Theorem. Write the answer in standard form.

Solution

Convert $1 + i$ to trigonometric form and then use De Moivre's Theorem.

$$(1 + i)^8 = \left(\sqrt{2} \operatorname{cis} 45°\right)^8 = \left(\sqrt{2}\right)^8 \operatorname{cis} 8(45°) = 16 \operatorname{cis} 360°$$
$$= 16(\cos 360° + i \sin 360°) = 16(1 + 0i) = 16$$

▶ **TRY EXERCISE 14, PAGE 342**

❷ QUESTION Is $(1 + i)^4$ a real number?

---

❷ ANSWER Yes. $(1 + i)^4 = \left(\sqrt{2} \operatorname{cis} 45°\right)^4 = \left(\sqrt{2}\right)^4 \operatorname{cis} 180° = -4.$

● DE MOIVRE'S THEOREM FOR FINDING ROOTS

De Moivre's Theorem can be used to find the nth roots of any number.

De Moivre's Theorem for Finding Roots

If $z = r \text{ cis } \theta$ is a complex number, then there are n distinct nth roots of z given by

$$w_k = r^{1/n} \text{ cis } \frac{\theta + 360°k}{n} \quad \text{for } k = 0, 1, 2, \ldots, n - 1, \text{ and } n \geq 1.$$

EXAMPLE 3   **Find Cube Roots by De Moivre's Theorem**

Find the three cube roots of 27.

Algebraic Solution

Write 27 in trigonometric form: $27 = 27 \text{ cis } 0°$. Then, from De Moivre's Theorem for finding roots, the cube roots of 27 are

$$w_k = 27^{1/3} \text{ cis } \frac{0° + 360°k}{3} \quad \text{for } k = 0, 1, 2$$

Substitute for k to find the three cube roots of 27.

$w_0 = 27^{1/3} \text{ cis } 0°$ • $k = 0; \dfrac{0° + 360°(0)}{3} = 0°$

$\quad = 3(\cos 0° + i \sin 0°)$

$\quad = 3$

$w_1 = 27^{1/3} \text{ cis } 120°$ • $k = 1; \dfrac{0° + 360°(1)}{3} = 120°$

$\quad = 3(\cos 120° + i \sin 120°)$

$\quad = -\dfrac{3}{2} + \dfrac{3\sqrt{3}}{2}i$

$w_2 = 27^{1/3} \text{ cis } 240°$ • $k = 2; \dfrac{0° + 360°(2)}{3} = 240°$

$\quad = 3(\cos 240° + i \sin 240°)$

$\quad = -\dfrac{3}{2} - \dfrac{3\sqrt{3}}{2}i$

For $k = 3$, $\dfrac{0° + 1080°}{3} = 360°$. The angles start repeating; thus there are only three cube roots of 27. The three cube roots are graphed in **Figure 5.9**.

Visualize the Solution

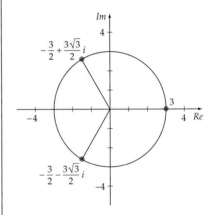

FIGURE 5.9

Note that the arguments of the three cube roots of 27 are 0°, 120°, and 240° and that $|w_0| = |w_1| = |w_2| = 3$. In geometric terms, this means that the three cube roots of 27 are equally spaced on a circle centered at the origin with a radius of 3.

▶ **TRY EXERCISE 26, PAGE 343**

EXAMPLE 4   **Find the Fifth Roots of a Complex Number**

Find the fifth roots of $z = 1 + i\sqrt{3}$.

Algebraic Solution

Write z in trigonometric form: $z = r \text{ cis } \theta$.

$$r = \sqrt{1^2 + \left(\sqrt{3}\right)^2} = 2$$

$$z = 2 \text{ cis } 60° \qquad \bullet\ \theta = \tan^{-1}\frac{\sqrt{3}}{1} = 60°$$

From De Moivre's Theorem, the modulus of each root is $\sqrt[5]{2}$, and the arguments are determined by $\dfrac{60° + 360°k}{5}$, $k = 0, 1, 2, 3, 4$.

$$w_k = \sqrt[5]{2} \text{ cis } \frac{60° + 360°k}{5} \qquad \bullet\ k = 0, 1, 2, 3, 4$$

Substitute for k to find the five fifth roots of z.

$w_0 = \sqrt[5]{2} \text{ cis } 12°$ $\qquad \bullet\ k = 0;\ \dfrac{60° + 360°(0)}{5} = 12°$

$w_1 = \sqrt[5]{2} \text{ cis } 84°$ $\qquad \bullet\ k = 1;\ \dfrac{60° + 360°(1)}{5} = 84°$

$w_2 = \sqrt[5]{2} \text{ cis } 156°$ $\qquad \bullet\ k = 2;\ \dfrac{60° + 360°(2)}{5} = 156°$

$w_3 = \sqrt[5]{2} \text{ cis } 228°$ $\qquad \bullet\ k = 3;\ \dfrac{60° + 360°(3)}{5} = 228°$

$w_4 = \sqrt[5]{2} \text{ cis } 300°$ $\qquad \bullet\ k = 4;\ \dfrac{60° + 360°(4)}{5} = 300°$

▶ **TRY EXERCISE 28, PAGE 343**

Visualize the Solution

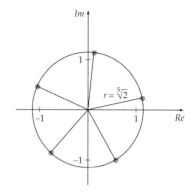

FIGURE 5.10

The five fifth roots of $1 + i\sqrt{3}$ are graphed in **Figure 5.10.** Note that the roots are equally spaced on a circle with center $(0, 0)$ and a radius of $\sqrt[5]{2} \approx 1.15$.

Keep the following properties in mind as you compute the n distinct nth roots of the complex number z.

Properties of the nth Roots of z

Geometric Property
All nth roots of z are equally spaced on a circle with center $(0, 0)$ and a radius of $|z|^{1/n}$.

Absolute Value Properties
1. If $|z| = 1$, then each nth root of z has an absolute value of 1.

2. If $|z| > 1$, then each nth root of z has an absolute value of $|z|^{1/n}$, and $|z|^{1/n}$ is greater than 1 but less than $|z|$.

3. If $|z| < 1$, then each nth root of z has an absolute value of $|z|^{1/n}$, and $|z|^{1/n}$ is less than 1 but greater than $|z|$.

Argument Property

Given that the argument of z is θ, then the argument of w_0 is $\dfrac{\theta}{n}$ and the arguments of the remaining nth roots can be determined by adding multiples of $\dfrac{360°}{n}$ (or $\dfrac{2\pi}{n}$ if you are using radians) to $\dfrac{\theta}{n}$.

❓ QUESTION Are all fourth roots of 1 equally spaced on a circle with center $(0, 0)$ and a radius of 1?

TOPICS FOR DISCUSSION

1. How many solutions are there for $z = (a + bi)^8$? How many solutions are there for $z^8 = a + bi$? Explain.

2. A student claims that $z^3 = 8$ has solutions of

$$2, \quad \frac{-1 + i\sqrt{3}}{2}, \quad \text{and} \quad \frac{-1 - i\sqrt{3}}{2}$$

   Cube each of these to see whether the student is correct.

3. To solve $z^4 = 16$, a student first observes that $z = 2$ is one solution. The other solutions are equally spaced on a circle with center $(0, 0)$ and radius 2, so the student reasons that $z = 2i$, $z = -2$, and $z = -2i$ are the other three solutions. Do you agree? Explain.

4. If $|z| = 1$, then the n solutions of $w^n = z$ all have an absolute value of 1. Explain.

5. If z is a solution of $z^2 = c + di$, then the conjugate of z is also a solution. Do you agree? Explain.

❓ ANSWER Yes.

EXERCISE SET 5.3

In Exercises 1 to 14, find the indicated power. Write the answer in standard form.

1. $[2(\cos 30° + i \sin 30°)]^8$

2. $(\cos 240° + i \sin 240°)^{12}$

3. $[2(\cos 240° + i \sin 240°)]^5$

4. $[2(\cos 45° + i \sin 45°)]^{10}$

5. $(2 \operatorname{cis} 225°)^5$

6. $(2 \operatorname{cis} 330°)^4$

7. $\left(2 \operatorname{cis} \dfrac{2\pi}{3}\right)^6$

8. $\left(4 \operatorname{cis} \dfrac{5\pi}{6}\right)^3$

9. $(1 - i)^{10}$

10. $\left(1 + i\sqrt{3}\right)^8$

11. $(2 + 2i)^7$

12. $\left(2\sqrt{3} - 2i\right)^5$

13. $\left(\dfrac{\sqrt{2}}{2} + i\dfrac{\sqrt{2}}{2}\right)^6$

14. $\left(-\dfrac{\sqrt{2}}{2} + i\dfrac{\sqrt{2}}{2}\right)^{12}$

In Exercises 15 to 28, find all of the indicated roots. Write all answers in standard form.

15. The two square roots of 9

16. The two square roots of 16

17. The six sixth roots of 64

18. The five fifth roots of 32

19. The five fifth roots of -1

20. The four fourth roots of -16

21. The three cube roots of 1

22. The three cube roots of i

23. The four fourth roots of $1 + i$

24. The five fifth roots of $-1 + i$

25. The three cube roots of $2 - 2i\sqrt{3}$

▶ **26.** The three cube roots of $-2 + 2i\sqrt{3}$

27. The two square roots of $-16 + 16i\sqrt{3}$

▶ **28.** The two square roots of $-1 + i\sqrt{3}$

In Exercises 29 to 40, find all roots of the equation. Write the answers in trigonometric form.

29. $x^3 + 8 = 0$ **30.** $x^5 - 32 = 0$

31. $x^4 + i = 0$ **32.** $x^3 - 2i = 0$

33. $x^3 - 27 = 0$ **34.** $x^5 + 32i = 0$

35. $x^4 + 81 = 0$ **36.** $x^3 - 64i = 0$

37. $x^4 - \left(1 - i\sqrt{3}\right) = 0$ **38.** $x^3 + \left(2\sqrt{3} - 2i\right) = 0$

39. $x^3 + \left(1 + i\sqrt{3}\right) = 0$ **40.** $x^6 - \left(4 - 4i\right) = 0$

CONNECTING CONCEPTS

41. Show that the conjugate of $z = r(\cos\theta + i\sin\theta)$ is equal to $\bar{z} = r(\cos\theta - i\sin\theta)$.

42. Show that if $z = r(\cos\theta + i\sin\theta)$, then
$$z^{-1} = r^{-1}(\cos\theta - i\sin\theta)$$

43. Show that if $z = r(\cos\theta + i\sin\theta)$, then
$$z^{-2} = r^{-2}(\cos 2\theta - i\sin 2\theta)$$

Note that Exercises 42 and 43 suggest that if
$z = r(\cos\theta + i\sin\theta)$, then for positive integers n,

$$z^{-n} = r^{-n}(\cos n\theta - i\sin n\theta)$$

44. Use the above equation to find z^{-4} for $z = 1 - i\sqrt{3}$.

PROJECTS

1. VERIFY IDENTITIES Raise $(\cos\theta + i\sin\theta)$ to the second power by using De Moivre's Theorem. Now square $(\cos\theta + i\sin\theta)$ as a binomial. Equate the real and imaginary parts of the two complex numbers and show that

a. $\cos 2\theta = \cos^2\theta - \sin^2\theta$ **b.** $\sin 2\theta = 2\sin\theta\cos\theta$

2. DISCOVER IDENTITIES Raise $(\cos\theta + i\sin\theta)$ to the fourth power by using De Moivre's Theorem. Now find the fourth power of the binomial $(\cos\theta + i\sin\theta)$ by multiplying. Equate the real and imaginary parts of the two complex numbers.

a. What identity have you discovered for $\cos 4\theta$?

b. What identity have you discovered for $\sin 4\theta$?

The Mandelbrot Replacement Procedure

The following procedure is called the **Mandelbrot replacement procedure.**

> **Mandelbrot Replacement Procedure**
>
> Pick a complex number s.
>
> **1.** Square s and add the result to s.
>
> **2.** Square the last result and add it to s.
>
> **3.** Repeat step 2.

The number s is referred to as the seed of the procedure. The number s is a seed in the sense that each seed produces a different sequence of numbers. Some seeds produce sequences that grow without bound. Some seeds produce sequences that grow toward some constant. Still other seeds yield sequences that are cyclic. Consider the following illustrations.

- Let the seed $s = 1$.

$$1^2 + 1 = 2, \qquad 2^2 + 1 = 5, \qquad 5^2 + 1 = 26, \qquad 26^2 + 1 = 677$$

As the replacement procedure continues, we get larger and larger numbers.

- Let the seed $s = -1$.

$$(-1)^2 + (-1) = 0, \qquad 0^2 + (-1) = -1, \qquad (-1)^2 + (-1) = 0, \ldots$$

As the replacement procedure continues, the results *cycle*: $0, -1, 0, -1, 0, \ldots$.

- Let the seed $s = 0.25$.

$$(0.25)^2 + 0.25 = 0.3125, \qquad (0.3125)^2 + 0.25 = 0.34765625,$$

$$(0.34765625)^2 + 0.25 \approx 0.3708648682, \ldots$$

1. Use your calculator to continue the above Mandelbrot replacement procedure with seed $s = 0.25$. What number do you have after

    a. 25 applications of step 2?

    b. 50 applications of step 2?

    c. 75 applications of step 2?

    d. What constant do you think the sequence of numbers is approaching?

- Let the seed $s = i$.

$$i^2 + 1 = -1 + i, \qquad (-1 + i)^2 + i = -i, \ldots$$

2. **a.** What is the next number produced by the Mandelbrot replacement procedure?

    b. What happens as the procedure is continued?

The Mandelbrot replacement procedure can be used to determine special kinds of numbers called **attractors.** The attractors produced by the Mandelbrot replacement procedure are an essential part of the Mandelbrot set, which is shown in black in **Figure 5.11.**

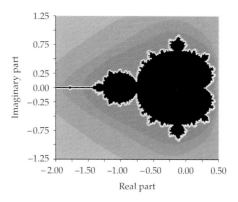

FIGURE 5.11

CHAPTER 5 SUMMARY

5.1 Complex Numbers

- The number i, called the *imaginary unit*, is the number such that $i^2 = -1$.

- If a is a positive real number, then $\sqrt{-a} = i\sqrt{a}$. The number $i\sqrt{a}$ is called an *imaginary number*.

- A *complex number* is a number of the form $a + bi$, where a and b are real numbers and $i = \sqrt{-1}$. The number a is the *real part* of $a + bi$, and b is the *imaginary part*.

- The complex numbers $a + bi$ and $a - bi$ are called *complex conjugates* or *conjugates* of each other.

- **Operations on Complex Numbers**

  $(a + bi) + (c + di) = (a + c) + (b + d)i$

  $(a + bi) - (c + di) = (a - c) + (b - d)i$

  $(a + bi)(c + di) = (ac - bd) + (ad + bc)i$

  $\dfrac{a + bi}{c + di} = \dfrac{a + bi}{c + di} \cdot \dfrac{c - di}{c - di}$ • Multiply numerator and denominator by the conjugate of the denominator.

5.2 Trigonometric Form of Complex Numbers

- The complex number $z = a + bi$ can be written in trigonometric form as

  $$z = r(\cos\theta + i\sin\theta) = r\operatorname{cis}\theta$$

  where $a = r\cos\theta$, $b = r\sin\theta$, $r = \sqrt{a^2 + b^2}$, and $\tan\theta = \dfrac{b}{a}$.

- If $z_1 = r_1(\cos\theta_1 + i\sin\theta_1)$ and $z_2 = r_2(\cos\theta_2 + i\sin\theta_2)$, then

  $$z_1 z_2 = r_1 r_2 \operatorname{cis}(\theta_1 + \theta_2)$$

  and

  $$\frac{z_1}{z_2} = \frac{r_1}{r_2}\operatorname{cis}(\theta_1 - \theta_2)$$

5.3 De Moivre's Theorem

- **De Moivre's Theorem**
  If $z = r\operatorname{cis}\theta$ and n is a positive integer, then

  $$z^n = r^n \operatorname{cis} n\theta$$

- If $z = r\operatorname{cis}\theta$, then the n distinct roots of z are given by

  $$w_k = r^{1/n}\operatorname{cis}\frac{\theta + 360°k}{n} \quad \text{for } k = 0, 1, 2, \ldots, n - 1$$

CHAPTER 5 TRUE/FALSE EXERCISES

In Exercises 1 to 12, answer true or false. If the statement is false, give an example to show that the statement is false.

1. If n is a negative number, then $\sqrt{n^2} = ni$.

2. Every real number a is a complex number.

3. The sum of a complex number z and its conjugate $\bar{z}$ is a real number.

4. The product of a complex number z and its conjugate $\bar{z}$ is a real number.

5. The quotient of a complex number z and its conjugate $\bar{z}$ is a real number

6. Multiplication of two complex numbers is commutative.

7. The n roots of a complex number can be graphed on a circle centered at $(0, 0)$ and are equally spaced around the circle.

8. If $z = r(\cos \theta + i \sin \theta)$, then $z^2 = r^2(\cos^2 \theta + i \sin^2 \theta)$.

9. $|a + bi| = \sqrt{a^2 + b^2}$

10. $i = \cos \pi + i \sin \pi$

11. $1 = \cos \pi + i \sin \pi$

12. $z = \cos 45° + i \sin 45°$ is a square root of i.

CHAPTER 5 REVIEW EXERCISES

In Exercises 1 to 4, write the complex number in standard form and give its conjugate.

1. $3 - \sqrt{-64}$

2. $\sqrt{-4} + 6$

3. $-2 + \sqrt{-5}$

4. $-5 - \sqrt{-27}$

In Exercises 5 to 20, simplify and write the complex number in standard form.

5. $(\sqrt{-4})(\sqrt{-4})$

6. $(-\sqrt{-27})(\sqrt{-3})$

7. $(3 + 7i) + (2 - 5i)$

8. $(3 - 4i) + (-6 + 8i)$

9. $(6 - 8i) - (9 - 11i)$

10. $(-3 - 5i) - (2 + 10i)$

11. $(5 + 3i)(2 - 5i)$

12. $(-2 - 3i)(-4 + 7i)$

13. $\dfrac{-2i}{3 - 4i}$

14. $\dfrac{4 + i}{7 - 2i}$

15. $i(2i) - (1 + i)^2$

16. $(2 - i)^3$

17. $(3 + \sqrt{-4}) - (-3 - \sqrt{-16})$

18. $(-2 + \sqrt{-9}) + (-3 - \sqrt{-81})$

19. $(2 - \sqrt{-3})(2 + \sqrt{-3})$

20. $(3 - \sqrt{-5})(2 + \sqrt{-5})$

In Exercises 21 to 24, simplify and write each complex number as i, $-i$, 1, or -1.

21. i^{27} **22.** i^{105} **23.** $\dfrac{i}{i^{17}}$ **24.** i^{62}

In Exercises 25 to 28, find the indicated absolute value of each complex number.

25. $|-8i|$

26. $|2 - 3i|$

27. $|-4 + 5i|$

28. $|-1 - i|$

In Exercises 29 to 32, write the complex numbers in trigonometric form.

29. $z = 2 - 2i$

30. $z = -\sqrt{3} + i$

31. $z = -3 + 2i$

32. $z = 4 - i$

In Exercises 33 to 36, write the complex numbers in standard form.

33. $z = 5(\cos 315° + i \sin 315°)$

34. $z = 6\left(\cos \dfrac{4\pi}{3} + i \sin \dfrac{4\pi}{3}\right)$

35. $z = 2(\cos 2 + i \sin 2)$

36. $z = 3(\cos 115° + i \sin 115°)$

In Exercises 37 to 42, multiply the complex numbers. Write the answers in standard form.

37. $3(\cos 225° + i \sin 225°) \cdot 10(\cos 45° + i \sin 45°)$

38. $5(\cos 162° + i \sin 162°) \cdot 2(\cos 63° + i \sin 63°)$

39. $3(\cos 12° + i \sin 12°) \cdot 4(\cos 126° + i \sin 126°)$

40. $(\cos 23° + i \sin 23°) \cdot 4(\cos 233° + i \sin 233°)$

41. $3(\cos 1.8 + i \sin 1.8) \cdot 5(\cos 2.5 + i \sin 2.5)$

42. $6(\cos 3.1 + i \sin 3.1) \cdot 5(\cos 4.3 + i \sin 4.3)$

In Exercises 43 to 48, divide the complex numbers. Write the answers in trigonometric form.

43. $\dfrac{6(\cos 50° + i \sin 50°)}{2(\cos 150° + i \sin 150°)}$

44. $\dfrac{30(\cos 165° + i \sin 165°)}{10(\cos 55° + i \sin 55°)}$

45. $\dfrac{40(\cos 66° + i \sin 66°)}{8(\cos 125° + i \sin 125°)}$

46. $\dfrac{2(\cos 150° + i \sin 150°)}{\sqrt{2}\,(\cos 200° + i \sin 200°)}$

47. $\dfrac{10(\cos 3.7 + i \sin 3.7)}{6(\cos 1.8 + i \sin 1.8)}$

48. $\dfrac{4(\cos 1.2 + i \sin 1.2)}{8(\cos 5.2 + i \sin 5.2)}$

In Exercises 49 to 54, find the indicated power. Write the answers in standard form.

49. $[3(\cos 45° + i \sin 45°)]^5$

50. $\left[\cos\left(\dfrac{11\pi}{8} \right) + i \sin\left(\dfrac{11\pi}{8} \right) \right]^8$

51. $(1 - i\sqrt{3})^7$

52. $(-2 - 2i)^{10}$

53. $(\sqrt{2} - i\sqrt{2})^5$

54. $(3 - 4i)^5$

In Exercises 55 to 60, find the indicated roots. Write the answers in trigonometric form.

55. The three cube roots of $27i$

56. The four fourth roots of $8i$

57. The four fourth roots of 256

58. The five fifth roots of $-16\sqrt{2} - 16\sqrt{2}i$

59. The four fourth roots of 81

60. The three cube roots of -125

CHAPTER 5 TEST

1. Write $6 + \sqrt{-9}$ in the form $a + bi$.

2. Simplify: $\sqrt{-18}$

3. Simplify $(3 + \sqrt{-4}) + (7 - \sqrt{-9})$. Write the answer in standard form.

4. Simplify $(-1 + \sqrt{-25}) - (8 - \sqrt{-16})$. Write the answer in standard form.

5. Simplify: $(\sqrt{-12})(\sqrt{-3})$

6. Simplify: i^{263}

7. Simplify: $(3 + 7i) - (-2 - 9i)$

8. Simplify: $(-6 - 9i)(4 + 3i)$

9. Simplify: $(3 - 5i)(-3 + 5i)$

10. Simplify: $\dfrac{4 - 5i}{i}$

11. Simplify: $\dfrac{2 - 7i}{4 + 3i}$

12. Simplify: $\dfrac{6 + 2i}{1 - i}$

13. Find the absolute value of $3 - 5i$.

14. Write $3 - 3i$ in trigonometric form.

15. Write $-6i$ in trigonometric form.

16. Write $4(\cos 120° + i \sin 120°)$ in standard form.

17. Write $5(\cos 225° + i \sin 225°)$ in standard form.

18. Simplify $3(\cos 28° + i \sin 28°) \cdot 4(\cos 17° + i \sin 17°)$. Write the answer in standard form.

19. Simplify $5(\cos 115° + i \sin 115°) \cdot 4(\cos 10° + i \sin 10°)$. Write the answer in standard form.

20. Simplify $\dfrac{24(\cos 258° + i \sin 258°)}{6(\cos 78° + i \sin 78°)}$. Write the answer in standard form.

21. Simplify $\dfrac{18(\cos 50° + i \sin 50°)}{3(\cos 140° + i \sin 140°)}$. Write the answer in standard form.

22. Simplify $(2 - 2i\sqrt{3})^{12}$. Write the answer in standard form.

In Exercises 23 to 25, write the indicated roots and solutions in trigonometric form.

23. Find the six sixth roots of 64.

24. Find the three cube roots of $-1 + i\sqrt{3}$.

25. Find the five solutions of $z^5 + 32 = 0$.

CUMULATIVE REVIEW EXERCISES

1. Solve $x^2 - x - 6 \leq 0$. Write the answer using interval notation.

2. What is the domain of $f(x) = \dfrac{x^2}{x^2 - 4}$?

3. Find c in the domain of $f(x) = \dfrac{x}{x + 1}$ so that $f(c) = 2$.

4. Given $f(x) = \sin 3x$ and $g(x) = \dfrac{x^2 - 1}{3}$, find $(f \circ g)(x)$.

5. Given $f(x) = \dfrac{x}{x - 1}$, find $f^{-1}(3)$.

6. Convert $\dfrac{3\pi}{2}$ radians to degrees.

7. Find the length of the hypotenuse a for the right triangle shown at the right.

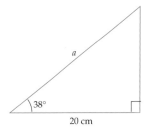

38°

20 cm

8. If t is any real number, what are the values of a and b for the inequality $a \leq \sin t \leq b$?

9. Graph $y = 3 \sin \pi x$.

10. Graph $y = \dfrac{1}{2} \tan \dfrac{\pi x}{4}$.

11. Verify the identity $\dfrac{\sin x}{1 + \cos x} = \sec x - \tan x$.

12. Express $\sin 2x \cos 3x - \sin 3x \cos 2x$ in terms of the sine function.

13. Given $\sin \alpha = \dfrac{4}{5}$ in Quadrant I and $\cos \beta = \dfrac{12}{13}$ in Quadrant IV, find $\cos(\alpha + \beta)$.

14. Find the exact value of $\sin\left[\sin^{-1}\left(\dfrac{3}{5}\right) + \cos^{-1}\left(-\dfrac{5}{13}\right)\right]$.

15. Solve $\sin 2x = \sqrt{3} \sin x$ for $0 \leq x < 2\pi$.

16. For the triangle at the right, find the length of side c. Round to the nearest ten.

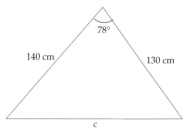

78°

140 cm   130 cm

c

17. Find the angle between the vectors $\mathbf{v} = 3\mathbf{i} + 2\mathbf{j}$ and $\mathbf{w} = 5\mathbf{i} - 3\mathbf{j}$. Round to the nearest tenth of a degree.

18. A force of 100 pounds on a rope is used to drag a box up a ramp that is inclined 15°. If the rope makes an angle of 30° with the ground, find the work done in moving the box 15 feet along the ramp. Round to the nearest foot-pound.

19. Write the complex number $2 + 2i$ in trigonometric form.

20. Find the three cube roots of -3.

CHAPTER 6

TOPICS IN ANALYTIC GEOMETRY

Sonic Booms

When the speed of a boat exceeds the speed at which waves can propagate through the water, a wake is created. In a similar way, when a plane travels at speeds that exceed the speed of sound, a "sound wake" is created. This sound wake results in a sonic boom.

The sound wake is in the form of a cone with the airplane at the vertex. When the wave-cone intersects flat land, a hyperbola is formed. People standing along the hyperbola hear the sonic boom at the same time. Hyperbolas are one of the topics of this chapter. See Exercise 55 on page 387.

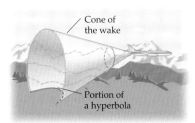

Cone of the wake

Portion of a hyperbola

The photo at the left, by George Standsbery, shows a B1-B Lancer as it accelerates through the speed of sound. The white cloud is condensation from the shock wave created when the plane exceeded the speed of sound. In what may seem like a totally unrelated phenomenon, the sound produced when a bullwhip is snapped occurs for exactly the same reason as a sonic boom: The tip of the whip exceeds the speed of sound, thereby producing the characteristic popping sound.

  VIDEO & DVD SSG WWW

Proof by Contradiction

In a detective television show, a suspect might be asked, "Where were you Friday night?", to which the suspect replies, "I was in New York." However, several eyewitnesses agree that they saw the suspect in California on that Friday night. Their testimony *contradicts* what the suspect claimed to be a factual statement.

A similar strategy is used by mathematicians to prove some theorems. The mathematician assumes that the theorem is false and then shows that the assumption contradicts a statement that is known to be true. This method of proof is called a *proof by contradiction*. To illustrate this method, consider the following theorem.

The $\sqrt{3}$ is an irrational number.

To prove this theorem, we begin by assuming that $\sqrt{3}$ is not an irrational number—that is, it is a rational number. This is the opposite of what we want to prove. If $\sqrt{3}$ is a rational number, then $\sqrt{3}$ can be represented as the ratio of two integers. That is, $\sqrt{3} = \dfrac{a}{b}$, where a and b are integers with no common factors and $b \neq 0$. From this assumption, we have

$$\sqrt{3} = \frac{a}{b}$$

$$3 = \frac{a^2}{b^2} \qquad \text{• Square both sides of the equation.}$$

$$3b^2 = a^2 \qquad \text{• Multiply each side by } b^2.$$

The last equation implies that a^2 is divisible by 3. Because 3 is prime, a is divisible by 3. Thus $a = 3k$ for some integer k and $a^2 = (3k)^2 = 9k^2$.

Replacing a^2 by $9k^2$, we have

$$3b^2 = 9k^2$$

$$b^2 = 3k^2 \qquad \text{• Divide each side by 3.}$$

The equation $b^2 = 3k^2$ implies that b is divisible by 3. Thus we have shown that both a and b are divisible by 3. This, however, contradicts our statement that a and b have no common factors. This contradiction means that our assumption that $\sqrt{3}$ can be represented as the quotient of integers is not possible, and therefore, $\sqrt{3}$ must be an irrational number.

PARABOLAS

take note

If the intersection of a plane and a cone is a point, a line, or two intersecting lines, then the intersection is called a *degenerate conic section.*

The graph of a parabola, a circle, an ellipse, or a hyperbola can be formed by the intersection of a plane and a cone. Hence these figures are referred to as conic sections. See **Figure 6.1.**

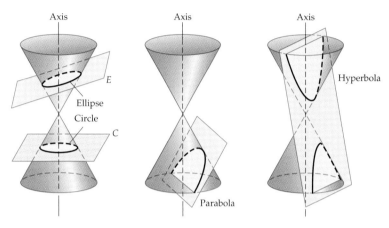

FIGURE 6.1
Cones intersected by planes

A plane perpendicular to the axis of the cone intersects the cone in a circle (plane *C*). The plane *E*, tilted so that it is not perpendicular to the axis, intersects the cone in an ellipse. When the plane is parallel to a line on the surface of the cone, the plane intersects the cone in a parabola. When the plane intersects both portions of the cone, a hyperbola is formed.

● PARABOLAS WITH VERTEX AT $(0, 0)$

Besides the geometric description of a conic section just given, a conic section can be defined as a set of points. This method uses some specified conditions about the curve to determine which points in a coordinate system are points of the graph. For example, a parabola can be defined by the following set of points.

Definition of a Parabola

A **parabola** is the set of points in the plane that are equidistant from a fixed line (the **directrix**) and a fixed point (the **focus**) not on the directrix.

The line that passes through the focus and is perpendicular to the directrix is called the **axis of symmetry** of the parabola. The midpoint of the line segment between the focus and directrix on the axis of symmetry is the **vertex** of the parabola, as shown in **Figure 6.2.**

Using this definition of a parabola, we can determine an equation of a parabola. Suppose that the coordinates of the vertex of a parabola are $V(0, 0)$ and

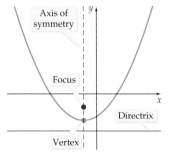

FIGURE 6.2

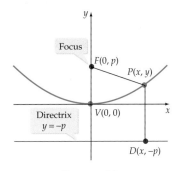

FIGURE 6.3

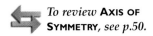

To review **AXIS OF SYMMETRY,** *see p.50.*

the axis of symmetry is the y-axis. The equation of the directrix is $y = -p, p > 0$. The focus lies on the axis of symmetry and is the same distance from the vertex as the vertex is from the directrix. Thus the coordinates of the focus are $F(0, p)$, as shown in **Figure 6.3.**

Let $P(x, y)$ be any point P on the parabola. Then, using the distance formula and the fact that the distance between any point P on the parabola and the focus is equal to the distance from the point P to the directrix, we can write the equation

$$d(P, F) = d(P, D)$$

By the distance formula,

$$\sqrt{(x - 0)^2 + (y - p)^2} = y + p$$

Now, squaring each side and simplifying, we get

$$\left(\sqrt{(x - 0)^2 + (y - p)^2}\right)^2 = (y + p)^2$$
$$x^2 + y^2 - 2py + p^2 = y^2 + 2py + p^2$$
$$x^2 = 4py$$

This is an equation of a parabola with vertex at the origin and the y-axis as its axis of symmetry. The equation of a parabola with vertex at the origin and the x-axis as its axis of symmetry is derived in a similar manner.

Standard Forms of the Equation of a Parabola with Vertex at the Origin

Axis of Symmetry Is the y-Axis

The standard form of the equation of a parabola with vertex $(0, 0)$ and the y-axis as its axis of symmetry is $x^2 = 4py$. The focus is $(0, p)$, and the equation of the directrix is $y = -p$.

Axis of Symmetry Is the x-Axis

The standard form of the equation of a parabola with vertex $(0, 0)$ and the x-axis as its axis of symmetry is $y^2 = 4px$. The focus is $(p, 0)$ and the equation of the directrix is $x = -p$.

take note

The tests for y-axis and x-axis symmetry can be used to verify these statements and provide connections to earlier topics on symmetry.

In the equation $x^2 = 4py$, $x^2 \geq 0$. Therefore, $4py \geq 0$. Thus if $p > 0$, then $y \geq 0$, and the parabola opens up. If $p < 0$, then $y \leq 0$, and the parabola opens down. A similar analysis shows that for $y^2 = 4px$, the parabola opens to the right when $p > 0$ and opens to the left when $p < 0$.

❓ QUESTION Does the graph of $y^2 = -4x$ open up, down, to the left, or to the right?

EXAMPLE 1   Find the Focus and Directrix of a Parabola

Find the focus and directrix of the parabola given by the equation

$$y = -\frac{1}{2}x^2.$$

❓ ANSWER To the left.

Solution

Because the x term is squared, the standard form of the equation is $x^2 = 4py$.

$$y = -\frac{1}{2}x^2$$

$$x^2 = -2y \qquad \bullet \textbf{Write the given equation in standard form.}$$

Comparing this equation with $x^2 = 4py$ gives

$$4p = -2$$

$$p = -\frac{1}{2}$$

Because p is negative, the parabola opens down, and the focus is below the vertex $(0, 0)$, as shown in **Figure 6.4.** The coordinates of the focus are $\left(0, -\frac{1}{2}\right)$. The equation of the directrix is $y = \frac{1}{2}$.

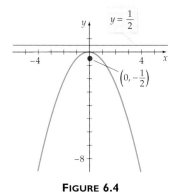

FIGURE 6.4

▶ TRY EXERCISE 4, PAGE 358

EXAMPLE 2   **Find the Equation of a Parabola in Standard Form**

Find the equation of the parabola in standard form with vertex at the origin and focus at $(-2, 0)$.

Solution

Because the vertex is $(0, 0)$ and the focus is at $(-2, 0)$, $p = -2$. The graph of the parabola opens toward the focus, so in this case the parabola opens to the left. The equation of the parabola in standard form that opens to the left is $y^2 = 4px$. Substitute -2 for p in this equation and simplify.

$$y^2 = 4(-2)x = -8x$$

The equation of the parabola is $y^2 = -8x$.

▶ TRY EXERCISE 28, PAGE 358

INTEGRATING TECHNOLOGY

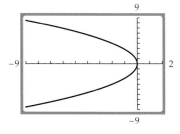

FIGURE 6.5

The graph of $y^2 = -8x$ is shown in **Figure 6.5.** Note that the graph is not the graph of a function. To graph $y^2 = -8x$ with a graphing utility, we first solve for y to produce $y = \pm\sqrt{-8x}$. From this equation we can see that for any $x < 0$, there are two values of y. For example, when $x = -2$,

$$y = \pm\sqrt{(-8)(-2)} = \pm\sqrt{16} = \pm 4$$

The graph of $y^2 = -8x$ in **Figure 6.5** was drawn by graphing both $y_1 = \sqrt{-8x}$ and $y_2 = -\sqrt{-8x}$ in the same window.

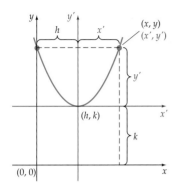

FIGURE 6.6

• PARABOLAS WITH VERTEX AT (h, k)

The equation of a parabola with a vertical or horizontal axis of symmetry and with the vertex at a point (h, k) can be found by using the translations discussed previously. Consider a coordinate system with coordinate axes labeled x' and y' placed so that its origin is at (h, k) of the xy-coordinate system.

The relationship between an ordered pair in the $x'y'$-coordinate system and in the xy-coordinate system is given by the transformation equations

$$x' = x - h$$
$$y' = y - k \tag{1}$$

Now consider a parabola with vertex at (h, k), as shown in **Figure 6.6.** Create a new coordinate system with axes labeled x' and y' and with its origin at (h, k). The equation of a parabola in the $x'y'$-coordinate system is

$$(x')^2 = 4py' \tag{2}$$

Using the transformation Equations (1), we can substitute the expressions for x' and y' into Equation (2). The standard form of the equation of the parabola with vertex (h, k) and a vertical axis of symmetry is

$$(x - h)^2 = 4p(y - k)$$

Similarly, we can derive the standard form of the equation of the parabola with vertex (h, k) and a horizontal axis of symmetry.

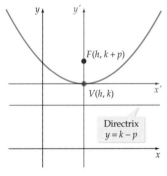

FIGURE 6.7

Standard Forms of the Equation of a Parabola with Vertex at (h, k)

Vertical Axis of Symmetry

The standard form of the equation of the parabola with vertex $V(h, k)$ and a vertical axis of symmetry is

$$(x - h)^2 = 4p(y - k)$$

The focus is $(h, k + p)$, and the equation of the directrix is $y = k - p$. See **Figure 6.7.**

Horizontal Axis of Symmetry

The standard form of the equation of the parabola with vertex (h, k) and a horizontal axis of symmetry is

$$(y - k)^2 = 4p(x - h)$$

The focus is $(h + p, k)$, and the equation of the directrix is $x = h - p$.

EXAMPLE 3 Find the Focus and Directrix of a Parabola

Find the equation of the directrix and the coordinates of the vertex and focus of the parabola given by the equation $3x + 2y^2 + 8y - 4 = 0$.

Solution

Rewrite the equation so that the y terms are on one side of the equation, and then complete the square on y.

$$3x + 2y^2 + 8y - 4 = 0$$
$$2y^2 + 8y = -3x + 4$$
$$2(y^2 + 4y) = -3x + 4$$
$$2(y^2 + 4y + 4) = -3x + 4 + 8 \qquad \bullet \text{ Complete the square. Note that } 2 \cdot 4 = 8 \text{ is added to each side.}$$
$$2(y + 2)^2 = -3(x - 4) \qquad \bullet \text{ Simplify and then factor.}$$
$$(y + 2)^2 = -\frac{3}{2}(x - 4) \qquad \bullet \text{ Write the equation in standard form.}$$

Comparing this equation to $(y - k)^2 = 4p(x - h)$, we have a parabola that opens to the left with vertex $(4, -2)$ and $4p = -\frac{3}{2}$. Thus $p = -\frac{3}{8}$.

The coordinates of the focus are

$$\left(4 + \left(-\frac{3}{8}\right), -2\right) = \left(\frac{29}{8}, -2\right)$$

The equation of the directrix is

$$x = 4 - \left(-\frac{3}{8}\right) = \frac{35}{8}$$

Choosing some values for y and finding the corresponding values for x, we plot a few points. Because the line $y = -2$ is the axis of symmetry, for each point on one side of the axis of symmetry there is a corresponding point on the other side. Two points are $(-2, 1)$ and $(-2, -5)$. See **Figure 6.8.**

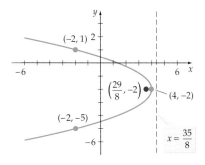

FIGURE 6.8

▶ TRY EXERCISE 20, PAGE 358

EXAMPLE 4   **Find the Equation in Standard Form of a Parabola**

Find the equation in standard form of the parabola with directrix $x = -1$ and focus $(3, 2)$.

Solution

The vertex is the midpoint of the line segment joining the focus $(3, 2)$ and the point $(-1, 2)$ on the directrix.

$$(h, k) = \left(\frac{-1 + 3}{2}, \frac{2 + 2}{2}\right) = (1, 2)$$

The standard form of the equation is $(y - k)^2 = 4p(x - h)$. The distance from the vertex to the focus is 2. Thus $4p = 4(2) = 8$, and the equation of the parabola in standard form is $(y - 2)^2 = 8(x - 1)$. See **Figure 6.9.**

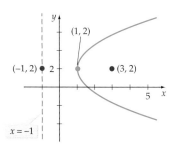

FIGURE 6.9

▶ TRY EXERCISE 30, PAGE 358

● APPLICATIONS

A principle of physics states that when light is reflected from a point P on a surface, the angle of incidence (that of the incoming ray) equals the angle of reflection (that of the outgoing ray). See **Figure 6.10.** This principle applied to parabolas has some useful consequences.

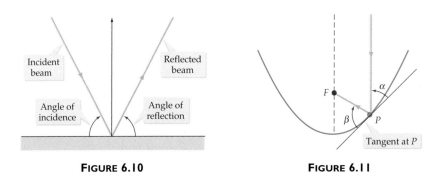

FIGURE 6.10

FIGURE 6.11

Optical Property of a Parabola

The line tangent to a parabola at a point P makes equal angles with the line through P and parallel to the axis of symmetry and the line through P and the focus of the parabola (see **Figure 6.11**).

A cross section of the reflecting mirror of a telescope has the shape of a parabola. The incoming parallel rays of light are reflected from the surface of the mirror and to the focus. See **Figure 6.12.**

Flashlights and car headlights also make use of this property. The light bulb is positioned at the focus of the parabolic reflector, which causes the reflected light to be reflected outward in parallel rays. See **Figure 6.13.**

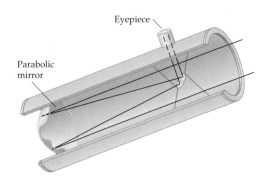

FIGURE 6.12

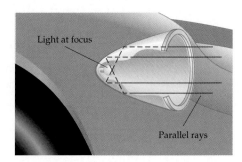

FIGURE 6.13

EXAMPLE 5   **Find the Focus of a Satellite Dish**

A satellite dish has the shape of a paraboloid. The signals that it receives are reflected to a receiver that is located at the focus of the paraboloid. If the dish is 8 feet across at its opening and 1.25 feet deep at its center, determine the location of its focus.

Solution

Figure 6.14 shows that a cross section of the paraboloid along its axis of symmetry is a parabola. **Figure 6.15** shows this cross section placed in a rectangular coordinate system with the vertex of the parabola at $(0, 0)$ and the axis of symmetry of the parabola on the y-axis. The parabola has an equation of the form

$$4py = x^2$$

Because the parabola contains the point $(4, 1.25)$, this equation is satisfied by the substitutions $x = 4$ and $y = 1.25$. Thus we have

$$4p(1.25) = 4^2$$
$$5p = 16$$
$$p = \frac{16}{5}$$

The focus of the satellite dish is on the axis of symmetry of the dish, and it is $3\frac{1}{5}$ feet above the vertex of the dish. See **Figure 6.15.**

▶ **TRY EXERCISE 38, PAGE 358**

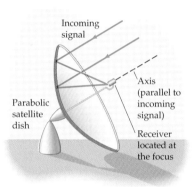

FIGURE 6.14

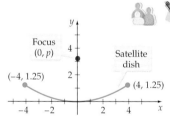

FIGURE 6.15

TOPICS FOR DISCUSSION

1. Do the graphs of the parabola given by $y = x^2$ and the vertical line given by $x = 10,000$ intersect? Explain.

2. A student claims that the focus of the parabola given by $y = 8x^2$ is at $(0, 2)$ because $4p = 8$ implies that $p = 2$. Explain the error in the student's reasoning.

3. "The vertex of a parabola is always halfway between its focus and its directrix." Do you agree? Explain.

4. A tutor claims that the graph of $(x - h)^2 = 4p(y - k)$ has a y-intercept of $\left(0, \frac{h^2}{4p} + k\right)$. Explain why the tutor is correct.

EXERCISE SET 6.1

In Exercises I to 26, find the vertex, focus, and directrix of the parabola given by each equation. Sketch the graph.

1. $x^2 = -4y$

2. $2y^2 = x$

3. $y^2 = \dfrac{1}{3}x$

▶ **4.** $x^2 = -\dfrac{1}{4}y$

5. $(x - 2)^2 = 8(y + 3)$

6. $(y + 1)^2 = 6(x - 1)$

7. $(y + 4)^2 = -4(x - 2)$

8. $(x - 3)^2 = -(y + 2)$

9. $(y - 1)^2 = 2x + 8$

10. $(x + 2)^2 = 3y - 6$

11. $(2x - 4)^2 = 8y - 16$

12. $(3x + 6)^2 = 18y - 36$

13. $x^2 + 8x - y + 6 = 0$

14. $x^2 - 6x + y + 10 = 0$

15. $x + y^2 - 3y + 4 = 0$

16. $x - y^2 - 4y + 9 = 0$

17. $2x - y^2 - 6y + 1 = 0$

18. $3x + y^2 + 8y + 4 = 0$

19. $x^2 + 3x + 3y - 1 = 0$

▶ **20.** $x^2 + 5x - 4y - 1 = 0$

21. $2x^2 - 8x - 4y + 3 = 0$

22. $6x - 3y^2 - 12y + 4 = 0$

23. $2x + 4y^2 + 8y - 5 = 0$

24. $4x^2 - 12x + 12y + 7 = 0$

25. $3x^2 - 6x - 9y + 4 = 0$

26. $2x - 3y^2 + 9y + 5 = 0$

27. Find the equation in standard form of the parabola with vertex at the origin and focus $(0, -4)$.

▶ **28.** Find the equation in standard form of the parabola with vertex at the origin and focus $(5, 0)$.

29. Find the equation in standard form of the parabola with vertex at $(-1, 2)$ and focus $(-1, 3)$.

▶ **30.** Find the equation in standard form of the parabola with vertex at $(2, -3)$ and focus $(0, -3)$.

31. Find the equation in standard form of the parabola with focus $(3, -3)$ and directrix $y = -5$.

32. Find the equation in standard form of the parabola with focus $(-2, 4)$ and directrix $x = 4$.

33. Find the equation in standard form of the parabola that has vertex $(-4, 1)$, has its axis of symmetry parallel to the y-axis, and passes through the point $(-2, 2)$.

34. Find the equation in standard form of the parabola that has vertex $(3, -5)$, has its axis of symmetry parallel to the x-axis, and passes through the point $(4, 3)$.

35. **STRUCTURAL DEFECTS** Ultra-sound is used as a nondestructive method of determining whether a support beam for a structure has an internal fracture. In one scanning procedure, if the resulting image is a parabola, engineers know that there is a structural defect. Suppose that a scan produced an image whose equation is

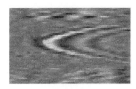

$$x = -0.325y^2 + 13y + 120$$

Determine the vertex and focus of the graph of this parabola.

36. **FOUNTAIN DESIGN** A fountain in a shopping mall has two parabolic arcs of water intersecting as shown below. If the equation of one parabola is $y = -0.25x^2 + 2x$ and the equation of the second parabola is $y = -0.25x^2 + 4.5x - 16.25$, how high above the base of the fountain do the parabolas intersect? All dimensions are in feet.

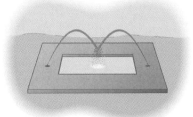

37. **SATELLITE DISH** A satellite dish has the shape of a paraboloid. The signals that it receives are reflected to a receiver that is located at the focus of the paraboloid. If the dish is 8 feet across at its opening and 1 foot deep at its vertex, determine the location (distance above the vertex of the dish) of its focus.

▶ **38.** **RADIO TELESCOPES** The antenna of a radio telescope is a paraboloid measuring 81 feet across with a depth of 16 feet. Determine, to the nearest 0.1 of a foot, the distance from the vertex to the focus of this antenna.

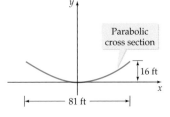

39. CAPTURING THE SOUND During televised football games, a parabolic microphone is used to capture sounds. The shield of the microphone is a paraboloid with a diameter of 18.75 inches and a depth of 3.66 inches. To pick up the sounds, a microphone is placed at the focus of the paraboloid. How far (to the nearest 0.1 of an inch) from the vertex of the paraboloid should the microphone be placed?

40. THE LOVELL TELESCOPE The Lovell Telescope is a radio telescope located at the Jodrell Bank Observatory in Cheshire, England. The dish of the telescope has the shape of a paraboloid with a diameter of 250 feet and a focal length of 75 feet.

a. Find an equation of a cross section of the paraboloid that passes through the vertex of the paraboloid. Assume that the dish has its vertex at (0, 0) and a vertical axis of symmetry.

b. Find the depth of the dish. Round to the nearest foot.

41. The surface area of a paraboloid with radius r and depth d is given by $S = \dfrac{\pi r}{6d^2}[(r^2 + 4d^2)^{3/2} - r^3]$.

Approximate (to the nearest 100 square feet) the surface area of:

a. The radio telescope in Exercise 38.

b. The Lovell Telescope (see Exercise 40).

42. THE HALE TELESCOPE The parabolic mirror in the Hale telescope at the Palomar Observatory in southern Califor-

nia has a diameter of 200 inches, and it has a concave depth of 3.75375 inches. Determine the location of its focus (to the nearest inch).

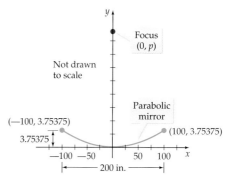

Mirror in the Hale Telescope

43. THE LICK TELESCOPE The parabolic mirror in the Lick telescope at the Lick Observatory on Mount Hamilton has a diameter of 120 inches, and it has a focal length of 600 inches. In the construction of the mirror, workers ground the mirror as shown in the following diagram. Determine the dimension a, which is the concave depth of the mirror.

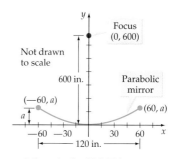

Mirror in the Lick Telescope

44. HEADLIGHT DESIGN A light source is to be placed on the axis of symmetry of the parabolic reflector shown in the figure below. How far to the right of the vertex point should the light source be located if the designer wishes the reflected light rays to form a beam of parallel rays?

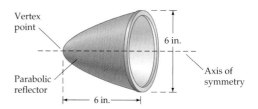

In Exercises 45 to 48, graph each equation, and find the coordinates of the points of intersection of the two graphs to the nearest ten-thousandth.

45. $y = 2x^2 - x - 1$
$y = x$

46. $y = x^2 + 2x - 4$
$y = x - 1$

47. $y = 2x^2 - 1$
$y = x^2 + x + 3$

48. $y = 2x^2 - x - 1$
$y = x^2 - 4$

CONNECTING CONCEPTS

In Exercises 49 to 51, use the following definition of latus rectum: The line segment that has endpoints on a parabola, passes through the focus of the parabola, and is perpendicular to the axis of symmetry is called the *latus rectum* **of the parabola.**

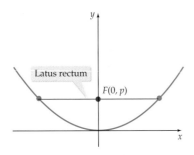

49. Find the length of the latus rectum for the parabola $x^2 = 4y$.

50. Find the length of the latus rectum for the parabola $y^2 = -8x$.

51. Find the length of the latus rectum for any parabola in terms of $|p|$, the distance from the vertex of the parabola to the focus.

The result of Exercise 51 can be stated as the following theorem: Two points on a parabola will be 2|p| units on each side of the axis of symmetry on the line through the focus and perpendicular to that axis.

52. Use the theorem to sketch a graph of the parabola given by the equation $(x - 3)^2 = 2(y + 1)$.

53. Use the theorem to sketch a graph of the parabola given by the equation $(y + 4)^2 = -(x - 1)$.

54. By using the definition of a parabola, find the equation in standard form of the parabola with $V(0, 0)$, $F(-c, 0)$, and directrix $x = c$.

55. Sketch a graph of $4(y - 2) = x|x| - 1$.

56. Find the equation of the directrix of the parabola with vertex at the origin and focus at the point $(1, 1)$.

57. Find the equation of the parabola with vertex at the origin and focus at the point $(1, 1)$. (*Hint:* You will need the answer to Exercise 56 and the definition of a parabola.)

PREPARE FOR SECTION 6.2

58. Find the midpoint and the length of the line segment between $P_1(5, 1)$ and $P_2(-1, 5)$. [1.2]

59. Solve: $x^2 + 6x - 16 = 0$ [1.1]

60. Solve: $x^2 - 2x = 2$ [1.1]

61. Complete the square of $x^2 - 8x$ and write the result as the square of a binomial. [1.2]

62. Find the center and radius of the circle whose equation is $x^2 + y^2 + 2x + 4y - 4 = 0$ [1.2]

63. Graph: $(x - 2)^2 + (y + 3)^2 = 16$ [1.2]

PROJECTS

1. **PARABOLAS AND TANGENTS** Calculus procedures can be used to show that the equation of a tangent line to the parabola $4py = x^2$ at the point (x_0, y_0) is given by

$$y - y_0 = \left(\frac{1}{2p} x_0\right)(x - x_0)$$

Use this equation to verify each of the following statements.

a. If two tangent lines to a parabola intersect at right angles, then the point of intersection of the tangent lines is on the directrix of the parabola.

b. If two tangent lines to a parabola intersect at right angles, then the focus of the parabola is located on the line segment that connects the two points of tangency.

c. The tangent line to the parabola $4py = x^2$ at the point (x_0, y_0) intersects the y-axis at the point $(0, -y_0)$.

SECTION 6.2

ELLIPSES

- **ELLIPSES WITH CENTER AT $(0,0)$**
- **ELLIPSES WITH CENTER AT (h, k)**
- **ECCENTRICITY OF AN ELLIPSE**
- **APPLICATIONS**
- **ACOUSTIC PROPERTY OF AN ELLIPSE**

An ellipse is another of the conic sections formed when a plane intersects a right circular cone. If β is the angle at which the plane intersects the axis of the cone and α is the angle shown in **Figure 6.16,** an ellipse is formed when $\alpha < \beta < 90°$. If $\beta = 90°$, then a circle is formed.

take note

If the plane intersects the cone at the vertex of the cone so that the resulting figure is a point, the point is a *degenerate ellipse*. See the accompanying figure.

Degenerate ellipse

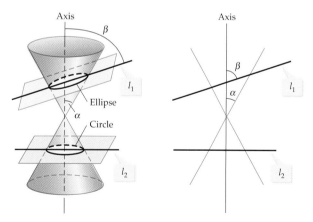

FIGURE 6.16

As was the case for a parabola, there is a definition for an ellipse in terms of a certain set of points in the plane.

Definition of an Ellipse

An **ellipse** is the set of all points in the plane, the sum of whose distances from two fixed points (**foci**) is a positive constant.

We can use this definition to draw an ellipse, equipped only with a piece of string and two tacks (see **Figure 6.17**). Tack the ends of the string to the foci, and trace a curve with a pencil held tight against the string. The resulting curve is an ellipse. The positive constant mentioned in the definition of an ellipse is the length of the string.

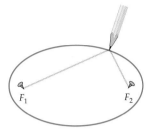

FIGURE 6.17

● ELLIPSES WITH CENTER AT (0, 0)

The graph of an ellipse has two axes of symmetry (see **Figure 6.18**). The longer axis is called the **major axis.** The foci of the ellipse are on the major axis. The shorter axis is called the **minor axis.** It is customary to denote the length of the major axis as $2a$ and the length of the minor axis as $2b$. The **semiaxes** are one-half the axes in length. Thus the length of the semimajor axis is denoted by a and the length of the semiminor axis by b. The **center** of the ellipse is the midpoint of the major axis. The endpoints of the major axis are the **vertices** (plural of *vertex*) of the ellipse.

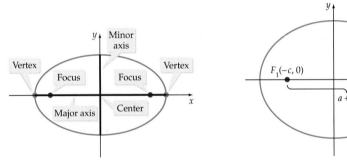

FIGURE 6.18   **FIGURE 6.19**

Consider the point $V_2(a, 0)$, which is one vertex of an ellipse, and the points $F_2(c, 0)$ and $F_1(-c, 0)$, which are the foci of the ellipse shown in **Figure 6.19.** The distance from V_2 to F_1 is $a + c$. Similarly, the distance from V_2 to F_2 is $a - c$. From the definition of an ellipse, the sum of the distances from any point on the ellipse to the foci is a positive constant. By adding the expressions $a + c$ and $a - c$, we have

$$(a + c) + (a - c) = 2a$$

Thus the positive constant referred to in the definition of an ellipse is $2a$, the length of the major axis.

Now let $P(x, y)$ be any point on the ellipse (see **Figure 6.20**). By using the definition of an ellipse, we have

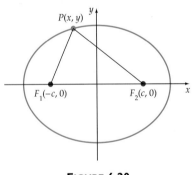

FIGURE 6.20

$$d(P, F_1) + d(P, F_2) = 2a$$
$$\sqrt{(x + c)^2 + y^2} + \sqrt{(x - c)^2 + y^2} = 2a$$

Subtract the second radical from each side of the equation, and then square each side.

$$\left[\sqrt{(x + c)^2 + y^2}\right]^2 = \left[2a - \sqrt{(x - c)^2 + y^2}\right]^2$$
$$(x + c)^2 + y^2 = 4a^2 - 4a\sqrt{(x - c)^2 + y^2} + (x - c)^2 + y^2$$

$$x^2 + 2cx + c^2 + y^2 = 4a^2 - 4a\sqrt{(x-c)^2 + y^2} + x^2 - 2cx + c^2 + y^2$$

$$4cx - 4a^2 = -4a\sqrt{(x-c)^2 + y^2}$$

$$[-cx + a^2]^2 = \left[a\sqrt{(x-c)^2 + y^2}\right]^2$$ • Divide by -4, and then square each side.

$$c^2x^2 - 2cxa^2 + a^4 = a^2x^2 - 2cxa^2 + a^2c^2 + a^2y^2$$

$$-a^2x^2 + c^2x^2 - a^2y^2 = -a^4 + a^2c^2$$ • Rewrite with x and y terms on the left side.

$$-(a^2 - c^2)x^2 - a^2y^2 = -a^2(a^2 - c^2)$$ • Factor and let $b^2 = a^2 - c^2$.

$$-b^2x^2 - a^2y^2 = -a^2b^2$$ • Divide each side by $-a^2b^2$.

$$\frac{x^2}{a^2} + \frac{y^2}{b^2} = 1$$ • An equation of an ellipse with center at $(0, 0)$

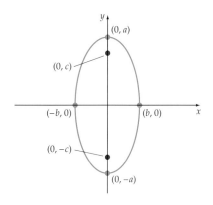

a. Major axis on x-axis

b. Major axis on y-axis

FIGURE 6.21

Standard Forms of the Equation of an Ellipse with Center at the Origin

Major Axis on the x-Axis

The standard form of the equation of an ellipse with the center at the origin and major axis on the x-axis (see **Figure 6.21a**) is given by

$$\frac{x^2}{a^2} + \frac{y^2}{b^2} = 1, \quad a > b$$

The length of the major axis is $2a$. The length of the minor axis is $2b$. The coordinates of the vertices are $(a, 0)$ and $(-a, 0)$, and the coordinates of the foci are $(c, 0)$ and $(-c, 0)$, where $c^2 = a^2 - b^2$.

Major Axis on the y-Axis

The standard form of the equation of an ellipse with the center at the origin and major axis on the y-axis (see **Figure 6.21b**) is given by

$$\frac{x^2}{b^2} + \frac{y^2}{a^2} = 1, \quad a > b$$

The length of the major axis is $2a$. The length of the minor axis is $2b$. The coordinates of the vertices are $(0, a)$ and $(0, -a)$, and the coordinates of the foci are $(0, c)$ and $(0, -c)$, where $c^2 = a^2 - b^2$.

❓ **QUESTION** For the graph of $\dfrac{x^2}{16} + \dfrac{y^2}{25} = 1$, is the major axis on the x-axis or the y-axis?

❓ **ANSWER** Because $25 > 16$, the major axis is on the y-axis.

EXAMPLE I   **Find the Vertices and Foci of an Ellipse**

Find the vertices and foci of the ellipse given by the equation $\dfrac{x^2}{25} + \dfrac{y^2}{49} = 1$.
Sketch the graph.

Solution

Because the y^2 term has the larger denominator, the major axis is on the y-axis.

$$a^2 = 49 \qquad b^2 = 25 \qquad c^2 = a^2 - b^2$$
$$a = 7 \qquad b = 5 \qquad = 49 - 25 = 24$$
$$c = \sqrt{24} = 2\sqrt{6}$$

The vertices are $(0, 7)$ and $(0, -7)$. The foci are $\left(0, 2\sqrt{6}\right)$ and $\left(0, -2\sqrt{6}\right)$. See **Figure 6.22.**

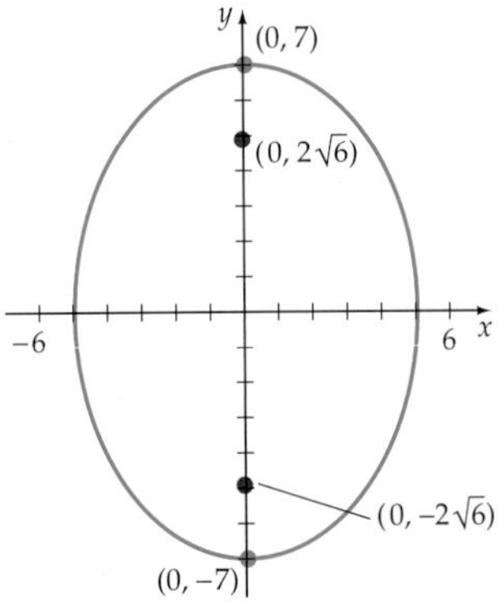

▶ **TRY EXERCISE 20, PAGE 371**

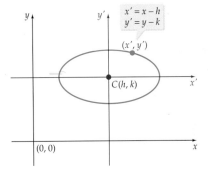

$$\frac{x^2}{25} + \frac{y^2}{49} = 1$$

FIGURE 6.22

An ellipse with foci $(3, 0)$ and $(-3, 0)$ and major axis of length 10 is shown in **Figure 6.23.** To find the equation of the ellipse in standard form, we must find a^2 and b^2. Because the foci are on the major axis, the major axis is on the x-axis. The length of the major axis is $2a$. Thus $2a = 10$. Solving for a, we have $a = 5$ and $a^2 = 25$.

Because the foci are $(3, 0)$ and $(-3, 0)$ and the center of the ellipse is the midpoint between the two foci, the distance from the center of the ellipse to a focus is 3. Therefore, $c = 3$. To find b^2, use the equation

$$c^2 = a^2 - b^2$$
$$9 = 25 - b^2$$
$$b^2 = 16$$

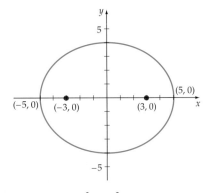

$$\frac{x^2}{25} + \frac{y^2}{16} = 1$$

FIGURE 6.23

The equation of the ellipse in standard form is $\dfrac{x^2}{25} + \dfrac{y^2}{16} = 1$.

● **ELLIPSES WITH CENTER AT (h, k)**

The equation of an ellipse with center (h, k) and with horizontal or vertical major axis can be found by using a translation of coordinates. On a coordinate system with axes labeled x' and y', the standard form of the equation of an ellipse with center at the origin of the $x'y'$-coordinate system is

$$\frac{(x')^2}{a^2} + \frac{(y')^2}{b^2} = 1$$

Now place the origin of the $x'y'$-coordinate system at (h, k) in an xy-coordinate system. See **Figure 6.24.**

FIGURE 6.24

The relationship between an ordered pair in the $x'y'$-coordinate system and one in the xy-coordinate system is given by the transformation equations

$$x' = x - h$$
$$y' = y - k$$

Substitute the expressions for x' and y' into the equation of an ellipse. The equation of the ellipse with center at (h, k) is

$$\frac{(x - h)^2}{a^2} + \frac{(y - k)^2}{b^2} = 1$$

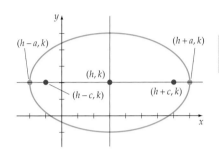

a. Major axis parallel to x-axis

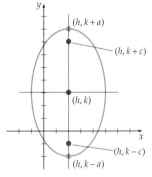

b. Major axis parallel to y-axis

FIGURE 6.25

Standard Forms of the Equation of an Ellipse with Center at (h, k)

Major Axis Parallel to the x-Axis

The standard form of the equation of an ellipse with the center at (h, k) and major axis parallel to the x-axis (see **Figure 6.25a**) is given by

$$\frac{(x - h)^2}{a^2} + \frac{(y - k)^2}{b^2} = 1, \quad a > b$$

The length of the major axis is $2a$. The length of the minor axis is $2b$. The coordinates of the vertices are $(h + a, k)$ and $(h - a, k)$, and the coordinates of the foci are $(h + c, k)$ and $(h - c, k)$, where $c^2 = a^2 - b^2$.

Major Axis Parallel to the y-Axis

The standard form of the equation of an ellipse with the center at (h, k) and major axis parallel to the y-axis (see **Figure 6.25b**) is given by

$$\frac{(x - h)^2}{b^2} + \frac{(y - k)^2}{a^2} = 1, \quad a > b$$

The length of the major axis is $2a$. The length of the minor axis is $2b$. The coordinates of the vertices are $(h, k + a)$ and $(h, k - a)$, and the coordinates of the foci are $(h, k + c)$ and $(h, k - c)$, where $c^2 = a^2 - b^2$.

EXAMPLE 2   Find the Vertices and Foci of an Ellipse

Find the vertices and foci of the ellipse $4x^2 + 9y^2 - 8x + 36y + 4 = 0$. Sketch the graph.

Solution

Write the equation of the ellipse in standard form by completing the square.

$$4x^2 + 9y^2 - 8x + 36y + 4 = 0$$
$$4x^2 - 8x + 9y^2 + 36y = -4 \qquad \text{• Rearrange terms.}$$
$$4(x^2 - 2x) + 9(y^2 + 4y) = -4 \qquad \text{• Factor.}$$
$$4(x^2 - 2x + 1) + 9(y^2 + 4y + 4) = -4 + 4 + 36 \qquad \text{• Complete the square.}$$
$$4(x - 1)^2 + 9(y + 2)^2 = 36 \qquad \text{• Factor.}$$
$$\frac{(x - 1)^2}{9} + \frac{(y + 2)^2}{4} = 1 \qquad \text{• Divide by 36.}$$

Continued ▶

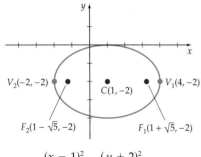

$$\frac{(x-1)^2}{9} + \frac{(y+2)^2}{4} = 1$$

FIGURE 6.26

From the equation of the ellipse in standard form, the coordinates of the center of the ellipse are (1, −2). Because the larger denominator is 9, the major axis is parallel to the *x*-axis and $a^2 = 9$. Thus $a = 3$. The vertices are (4, −2) and (−2, −2).

To find the coordinates of the foci, we find *c*.

$$c^2 = a^2 - b^2 = 9 - 4 = 5$$
$$c = \sqrt{5}$$

The foci are $\left(1 + \sqrt{5}, -2\right)$ and $\left(1 - \sqrt{5}, -2\right)$. See **Figure 6.26.**

▶ **TRY EXERCISE 26, PAGE 371**

INTEGRATING TECHNOLOGY

A graphing utility can be used to graph an ellipse. For instance, consider the equation $4x^2 + 9y^2 - 8x + 36y + 4 = 0$ from Example 2. Rewrite the equation as

$$9y^2 + 36y + (4x^2 - 8x + 4) = 0$$

In this form, the equation is a quadratic equation in terms of the variable *y* with

$$A = 9, B = 36, \text{ and } C = 4x^2 - 8x + 4$$

Apply the quadratic formula to produce

$$y = \frac{-36 \pm \sqrt{1296 - 36(4x^2 - 8x + 4)}}{18}$$

To review **QUADRATIC FORMULA**, *see p. 6.*

The graph of $Y1 = \dfrac{-36 + \sqrt{1296 - 36(4x^2 - 8x + 4)}}{18}$ is the part of the ellipse on or above the line $y = -2$ (see **Figure 6.27**).

The graph of $Y2 = \dfrac{-36 - \sqrt{1296 - 36(4x^2 - 8x + 4)}}{18}$ is the part of the ellipse on or below the line $y = -2$, as shown in **Figure 6.27.**

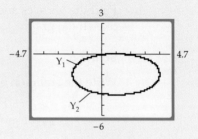

FIGURE 6.27

One advantage of this graphing procedure is that it does not require us to write the given equation in standard form. A disadvantage of the graphing procedure is that it does not indicate where the foci of the ellipse are located.

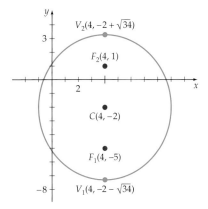

FIGURE 6.28

EXAMPLE 3   **Find the Equation of an Ellipse**

Find the standard form of the equation of the ellipse with center at $(4, -2)$, foci $F_2(4, 1)$ and $F_1(4, -5)$, and minor axis of length 10, as shown in **Figure 6.28.**

Solution

Because the foci are on the major axis, the major axis is parallel to the y-axis. The distance from the center of the ellipse to a focus is c. The distance between the center $(4, -2)$ and the focus $(4, 1)$ is 3. Therefore, $c = 3$.

The length of the minor axis is $2b$. Thus $2b = 10$ and $b = 5$.
To find a^2, use the equation $c^2 = a^2 - b^2$.

$$9 = a^2 - 25$$
$$a^2 = 34$$

Thus the equation in standard form is

$$\frac{(x - 4)^2}{25} + \frac{(y + 2)^2}{34} = 1$$

▶ **TRY EXERCISE 42, PAGE 371**

● ECCENTRICITY OF AN ELLIPSE

The graph of an ellipse can be very long and thin, or it can be much like a circle. The **eccentricity** of an ellipse is a measure of its "roundness."

Eccentricity (e) of an Ellipse

The eccentricity e of an ellipse is the ratio of c to a, where c is the distance from the center to a focus and a is one-half the length of the major axis. (See **Figure 6.29.**) That is,

$$e = \frac{c}{a}$$

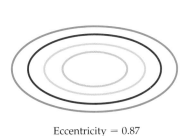

Eccentricity = 0.87

FIGURE 6.29

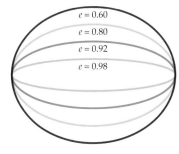

FIGURE 6.30

Because $c < a$, for an ellipse, $0 < e < 1$. When $e \approx 0$, the graph is almost a circle. When $e \approx 1$, the graph is long and thin. See **Figure 6.30.**

EXAMPLE 4 **Find the Eccentricity of an Ellipse**

Find the eccentricity of the ellipse given by $8x^2 + 9y^2 = 18$.

Solution

First, write the equation of the ellipse in standard form. Divide each side of the equation by 18.

$$\frac{8x^2}{18} + \frac{9y^2}{18} = 1$$

$$\frac{4x^2}{9} + \frac{y^2}{2} = 1$$

$$\frac{x^2}{9/4} + \frac{y^2}{2} = 1 \qquad \bullet \frac{4}{9} = \frac{1}{9/4}$$

The last step is necessary because the standard form of the equation has coefficients of 1 in the numerator. Thus

$$a^2 = \frac{9}{4} \qquad \text{and} \qquad a = \frac{3}{2}$$

Use the equation $c^2 = a^2 - b^2$ to find c.

$$c^2 = \frac{9}{4} - 2 = \frac{1}{4} \qquad \text{and} \qquad c = \sqrt{\frac{1}{4}} = \frac{1}{2}$$

Now find the eccentricity.

$$e = \frac{c}{a} = \frac{1/2}{3/2} = \frac{1}{3}$$

The eccentricity of the ellipse is $\frac{1}{3}$.

▶ **TRY EXERCISE 48, PAGE 371**

TABLE 6.1

Planet	Eccentricity
Mercury	0.206
Venus	0.007
Earth	0.017
Mars	0.093
Jupiter	0.049
Saturn	0.051
Uranus	0.046
Neptune	0.005
Pluto	0.250

● **APPLICATIONS**

The planets travel around the sun in elliptical orbits. The sun is located at a focus of the orbit. The eccentricities of the orbits for the planets in our solar system are given in **Table 6.1.**

❓ QUESTION Which planet has the most nearly circular orbit?

The terms *perihelion* and *aphelion* are used to denote the position of a planet in its orbit around the sun. The perihelion is the point nearest the sun; the aphelion is the point farthest from the sun. See **Figure 6.31.** The length of the semimajor axis of a planet's elliptical orbit is called the *mean distance* of the planet from the sun.

❓ ANSWER Neptune has the smallest eccentricity, so it is the planet with the most nearly circular orbit.

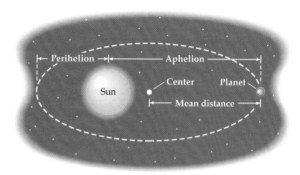

FIGURE 6.31

EXAMPLE 5   Determine an Equation for the Orbit of Earth

 Earth has a mean distance of 93 million miles and a perihelion distance of 91.5 million miles. Find an equation for Earth's orbit.

Solution

A mean distance of 93 million miles implies that the length of the semi-major axis of the orbit is $a = 93$ million miles. Earth's aphelion distance is the length of the major axis less the length of the perihelion distance. Thus

$$\text{Aphelion distance} = 2(93) - 91.5 = 94.5 \text{ million miles}$$

The distance c from the sun to the center of Earth's orbit is

$$c = \text{aphelion distance} - 93 = 94.5 - 93 = 1.5 \text{ million miles}$$

The length b of the semiminor axis of the orbit is

$$b = \sqrt{a^2 - c^2} = \sqrt{93^2 - 1.5^2} = \sqrt{8646.75}$$

An equation of Earth's orbit is

$$\frac{x^2}{93^2} + \frac{y^2}{8646.75} = 1$$

▶ **TRY EXERCISE 56, PAGE 372**

● ACOUSTIC PROPERTY OF AN ELLIPSE

Sound waves, although different from light waves, have a similar reflective property. When sound is reflected from a point P on a surface, the angle of incidence equals the angle of reflection. Applying this principle to a room with an elliptical ceiling results in what are called whispering galleries. These galleries are based on the following theorem.

The Reflective Property of an Ellipse

The lines from the foci to a point on an ellipse make equal angles with the tangent line at that point. See **Figure 6.32**.

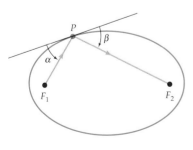

$\alpha = \beta$

FIGURE 6.32

The Statuary Hall in the Capital Building in Washington, D.C., is a whispering gallery. Two people standing at the foci of the elliptical ceiling can whisper and yet hear each other even though they are a considerable distance apart. The whisper from one person is reflected to the person standing at the other focus.

EXAMPLE 6   Locate the Foci of a Whispering Gallery

A room 88 feet long is constructed to be a whispering gallery. The room has an elliptical ceiling, as shown in **Figure 6.33.** If the maximum height of the ceiling is 22 feet, determine where the foci are located.

Solution

The length a of the semimajor axis of the elliptical ceiling is 44 feet. The height b of the semiminor axis is 22 feet. Thus

$$c^2 = a^2 - b^2$$
$$c^2 = 44^2 - 22^2$$
$$c = \sqrt{44^2 - 22^2} \approx 38.1 \text{ feet}$$

The foci are located about 38.1 feet from the center of the elliptical ceiling along its major axis.

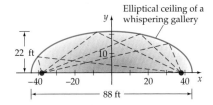

Elliptical ceiling of a whispering gallery

FIGURE 6.33

▶ **TRY EXERCISE 58, PAGE 372**

TOPICS FOR DISCUSSION

1. In every ellipse, the length of the semimajor axis a is greater than the length of the semiminor axis b and greater than the distance c from a focus to the center of the ellipse. Do you agree? Explain.

2. How many vertices does an ellipse have?

3. Every ellipse has two y-intercepts. Do you agree? Explain.

4. Explain why the eccentricity of every ellipse is a number between 0 and 1.

EXERCISE SET 6.2

In Exercises 1 to 32, find the center, vertices, and foci of the ellipse given by each equation. Sketch the graph.

1. $\dfrac{x^2}{16} + \dfrac{y^2}{25} = 1$

2. $\dfrac{x^2}{49} + \dfrac{y^2}{36} = 1$

3. $\dfrac{x^2}{9} + \dfrac{y^2}{4} = 1$

4. $\dfrac{x^2}{64} + \dfrac{y^2}{25} = 1$

5. $\dfrac{x^2}{7} + \dfrac{y^2}{9} = 1$

6. $\dfrac{x^2}{5} + \dfrac{y^2}{4} = 1$

7. $\dfrac{4x^2}{9} + \dfrac{y^2}{16} = 1$

8. $\dfrac{x^2}{9} + \dfrac{9y^2}{16} = 1$

9. $\dfrac{(x-3)^2}{25} + \dfrac{(y+2)^2}{16} = 1$

10. $\dfrac{(x+3)^2}{9} + \dfrac{(y+1)^2}{16} = 1$

11. $\dfrac{(x+2)^2}{9} + \dfrac{y^2}{25} = 1$

12. $\dfrac{x^2}{25} + \dfrac{(y-2)^2}{81} = 1$

13. $\dfrac{(x-1)^2}{21} + \dfrac{(y-3)^2}{4} = 1$ 14. $\dfrac{(x+5)^2}{9} + \dfrac{(y-3)^2}{7} = 1$

15. $\dfrac{9(x-1)^2}{16} + \dfrac{(y+1)^2}{9} = 1$ 16. $\dfrac{(x+6)^2}{25} + \dfrac{25y^2}{144} = 1$

17. $3x^2 + 4y^2 = 12$ 18. $5x^2 + 4y^2 = 20$

19. $25x^2 + 16y^2 = 400$ ▶ 20. $25x^2 + 12y^2 = 300$

21. $64x^2 + 25y^2 = 400$ 22. $9x^2 + 64y^2 = 144$

23. $4x^2 + y^2 - 24x - 8y + 48 = 0$

24. $x^2 + 9y^2 + 6x - 36y + 36 = 0$

25. $5x^2 + 9y^2 - 20x + 54y + 56 = 0$

▶ 26. $9x^2 + 16y^2 + 36x - 16y - 104 = 0$

27. $16x^2 + 9y^2 - 64x - 80 = 0$

28. $16x^2 + 9y^2 + 36y - 108 = 0$

29. $25x^2 + 16y^2 + 50x - 32y - 359 = 0$

30. $16x^2 + 9y^2 - 64x - 54y + 1 = 0$

31. $8x^2 + 25y^2 - 48x + 50y + 47 = 0$

32. $4x^2 + 9y^2 + 24x + 18y + 44 = 0$

In Exercises 33 to 44, find the equation in standard form of each ellipse, given the information provided.

33. Center $(0, 0)$, major axis of length 10, foci at $(4, 0)$ and $(-4, 0)$

34. Center $(0, 0)$, minor axis of length 6, foci at $(0, 4)$ and $(0, -4)$

35. Vertices $(6, 0)$, $(-6, 0)$; ellipse passes through $(0, -4)$ and $(0, 4)$

36. Vertices $(7, 0)$, $(-7, 0)$; ellipse passes through $(0, 5)$ and $(0, -5)$

37. Major axis of length 12 on the x-axis, center at $(0, 0)$; ellipse passes through $(2, -3)$

38. Major axis of length 8, center at $(0, 0)$; ellipse passes through $(-2, 2)$

39. Center $(-2, 4)$, vertices $(-6, 4)$ and $(2, 4)$, foci at $(-5, 4)$ and $(1, 4)$

40. Center $(0, 3)$, minor axis of length 4, foci at $(0, 0)$ and $(0, 6)$

41. Center $(2, 4)$, major axis parallel to the y-axis and of length 10; ellipse passes through the point $(3, 3)$

▶ 42. Center $(-4, 1)$, minor axis parallel to the y-axis and of length 8; ellipse passes through the point $(0, 4)$

43. Vertices $(5, 6)$ and $(5, -4)$, foci at $(5, 4)$ and $(5, -2)$

44. Vertices $(-7, -1)$ and $(5, -1)$, foci at $(-5, -1)$ and $(3, -1)$

In Exercises 45 to 52, use the eccentricity of each ellipse to find its equation in standard form.

45. Eccentricity $\dfrac{2}{5}$, major axis on the x-axis and of length 10, center at $(0, 0)$

46. Eccentricity $\dfrac{3}{4}$, foci at $(9, 0)$ and $(-9, 0)$

47. Foci at $(0, -4)$ and $(0, 4)$, eccentricity $\dfrac{2}{3}$

▶ 48. Foci at $(0, -3)$ and $(0, 3)$, eccentricity $\dfrac{1}{4}$

49. Eccentricity $\dfrac{2}{5}$, foci at $(-1, 3)$ and $(3, 3)$

50. Eccentricity $\dfrac{1}{4}$, foci at $(-2, 4)$ and $(-2, -2)$

51. Eccentricity $\dfrac{2}{3}$, major axis of length 24 on the y-axis, center at $(0, 0)$

52. Eccentricity $\dfrac{3}{5}$, major axis of length 15 on the x-axis, center at $(0, 0)$

53. **MEDICINES** A *lithotripter* is an instrument used to remove a kidney stone in a patient without having to do surgery. A high-frequency sound wave is emitted from a source that is located at the focus of an ellipse. The patient is placed so that the kidney stone is located at the other focus of the ellipse. If the equation of the ellipse is

$$\frac{(x-11)^2}{484} + \frac{y^2}{64} = 1 \ (x \text{ and } y \text{ are measured in centimeters}),$$

where, to the nearest centimeter, should the patient's kidney stone be placed so that the reflected sound hits the kidney stone?

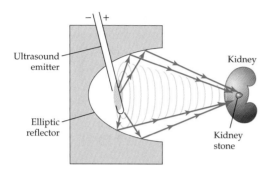

54. **CONSTRUCTION** A circular vent pipe is placed on a roof that has a slope of $\frac{4}{5}$, as shown in the figure at the right.

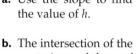

   a. Use the slope to find the value of h.

   b. The intersection of the vent pipe and the roof is an ellipse. To the nearest thousandth of an inch, what are the lengths of the major and minor axes?

   c. Find an equation of the ellipse that should be cut from the roof so that the pipe will fit.

55. **THE ORBIT OF SATURN** The distance from Saturn to the sun at Saturn's aphelion is 934.34 million miles, and the distance from Saturn to the sun at its perihelion is 835.14 million miles. Find an equation for the orbit of Saturn.

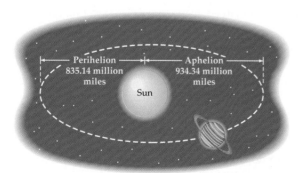

56. **THE ORBIT OF VENUS** Venus has a mean distance from the sun of 67.08 million miles, and the distance

from Venus to the sun at its aphelion is 67.58 million miles. Find an equation for the orbit of Venus.

57. **WHISPERING GALLERY** An architect wishes to design a large room that will be a whispering gallery. See Example 6. The ceiling of the room has a cross section that is an ellipse, as shown in the following figure.

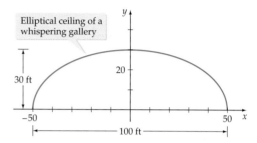

How far to the right and to the left of center are the foci located?

▶ **58.** **WHISPERING GALLERY** An architect wishes to design a large room 100 feet long that will be a whispering gallery. The ceiling of the room has a cross section that is an ellipse, as shown in the following figure.

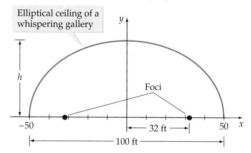

If the foci are to be located 32 feet to the right and to the left of center, find the height h of the elliptical ceiling (to the nearest 0.1 foot).

59. **HALLEY'S COMET** Find the equation of the path of Halley's comet in astronomical units by letting the sun (one focus) be at the origin and letting the other focus be on the positive x-axis. The length of the major axis of the orbit of Halley's comet is approximately 36 astronomical units (36 AU), and the length of the minor axis is 9 AU (1 AU = 92,960,000 miles).

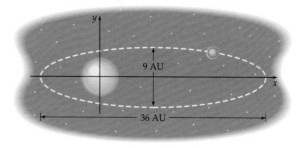

60. ELLIPTICAL RECEIVERS Some satellite receivers are made in an elliptical shape that enables the receiver to pick up signals from two satellites. The receiver shown below has a major axis of 24 inches and a minor axis of 18 inches.

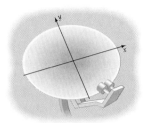

Determine, to the nearest 0.1 inch, the coordinates in the *xy*-plane of the foci of the ellipse. (*Note:* Because the receiver has only a slight curvature, we can estimate the location of the foci by assuming the receiver is flat.)

In Exercises 61 and 62, use the following formula for the perimeter *p* of an ellipse with semimajor axis *a* and semiminor axis *b*.

$$p \approx \pi \sqrt{2(a^2 + b^2)}$$

61. ELLIPTICAL EXERCISE EQUIPMENT Many exercise clubs have installed elliptical trainers. These machines are similar to step machines except that the motion of the feet follows an elliptical path. On one elliptical trainer, the path of a person's foot is elliptical with a major axis of 16 inches and a

minor axis of 10 inches. How many revolutions must the left foot make to complete a distance of 1 mile on this elliptical trainer?

62. ORBIT OF MARS Mars travels around the sun in an elliptical orbit. The orbit has a major axis of 3.04 AU and a minor axis of 2.99 AU. (1 AU is 1 astronomical unit, or approximately 92,960,000 miles, the average distance of Earth from the sun.) Estimate, to the nearest million miles, the perimeter of the orbit of Mars.

63. THE COLOSSEUM The base of the Colosseum in Rome has an elliptical shape.

a. Find an equation in standard form for the base of the Colosseum, which has a major axis of 615 feet and a minor axis of 510 feet.

b. The area of an ellipse with a semimajor axis of length *a* and a semiminor axis of length *b* is given by $A = \pi ab$. Find, to the nearest 100 square feet, the area of the base of the Colosseum.

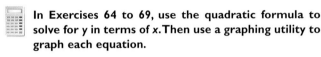

In Exercises 64 to 69, use the quadratic formula to solve for *y* in terms of *x*. Then use a graphing utility to graph each equation.

64. $16x^2 + 9y^2 - 64x - 80 = 0$

65. $16x^2 + 9y^2 + 36y - 108 = 0$

66. $25x^2 + 16y^2 + 50x - 32y - 359 = 0$

67. $16x^2 + 9y^2 - 64x - 54y + 1 = 0$

68. $8x^2 + 25y^2 - 48x + 50y + 47 = 0$

69. $4x^2 + 9y^2 + 24x + 18y + 44 = 0$

CONNECTING CONCEPTS

70. Explain why the graph of $4x^2 + 9y - 16x - 2 = 0$ is or is not an ellipse. Sketch the graph of this equation.

In Exercises 71 to 74, find the equation in standard form of each ellipse by using the definition of an ellipse.

71. Find the equation of the ellipse with foci at $(-3, 0)$ and $(3, 0)$ that passes through the point $\left(3, \dfrac{9}{2}\right)$.

72. Find the equation of the ellipse with foci at $(0, 4)$ and $(0, -4)$ that passes through the point $\left(\dfrac{9}{5}, 4\right)$.

73. Find the equation of the ellipse with foci at $(-1, 2)$ and $(3, 2)$ that passes through the point $(3, 5)$.

74. Find the equation of the ellipse with foci at $(-1, 1)$ and $(-1, 7)$ that passes through the point $\left(\dfrac{3}{4}, 1\right)$.

In Exercises 75 and 76, find the latus rectum of the given ellipse. A line segment with endpoints on the ellipse that is perpendicular to the major axis and passes through a focus is a *latus rectum* of the ellipse.

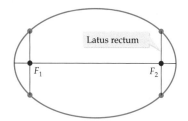

Latus rectum

F_1 F_2

75. Find the length of a latus rectum of the ellipse given by

$$\frac{(x-1)^2}{9} + \frac{(y+1)^2}{16} = 1$$

76. Find the length of a latus rectum of the ellipse given by

$$9x^2 + 16y^2 - 36x + 96y + 36 = 0$$

77. Show that for any ellipse, the length of a latus rectum is $\dfrac{2b^2}{a}$.

78. Use the definition of an ellipse to find the equation of an ellipse with center at $(0, 0)$ and foci at $(0, c)$ and $(0, -c)$.

Recall that a parabola has a directrix that is a line perpendicular to the axis of symmetry. An ellipse has two directrices, both of which are perpendicular to the major axis and outside the ellipse. For an ellipse with center at the

origin and whose major axis is the x-axis, the equations of the directrices are $x = \dfrac{a^2}{c}$ and $x = -\dfrac{a^2}{c}$.

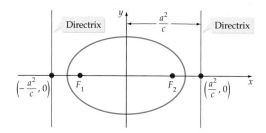

Directrix $\dfrac{a^2}{c}$ Directrix

$\left(-\dfrac{a^2}{c}, 0\right)$ F_1 F_2 $\left(\dfrac{a^2}{c}, 0\right)$

79. Find the directrices of the ellipse in Exercise 3.

80. Find the directrices of the ellipse in Exercise 4.

81. Let $P(x, y)$ be a point on the ellipse $\dfrac{x^2}{12} + \dfrac{y^2}{8} = 1$. Show that the distance from the point P to the focus $(2, 0)$ divided by the distance from the point P to the directrix $x = 6$ equals the eccentricity. (*Hint:* Solve the equation of the ellipse for y^2. Substitute this value for y^2 after applying the distance formula.)

82. Generalize the results of Exercise 81. That is, show that if $P(x, y)$ is a point on the ellipse $\dfrac{x^2}{a^2} + \dfrac{y^2}{b^2} = 1$, where $F(c, 0)$ is a focus and $x = \dfrac{a^2}{c}$ is a directrix, then the following equation is true: $e = \dfrac{d(P, F)}{d(P, D)}$. (*Hint:* Solve the equation of the ellipse for y^2. Substitute this value for y^2 after applying the distance formula.)

PREPARE FOR SECTION 6.3

83. Find the midpoint and the length of the line segment between $P_1(4, -3)$ and $P_2(-2, 1)$. [1.2]

84. Solve: $(x - 1)(x + 3) = 5$ [1.1]

85. Simplify: $\dfrac{4}{\sqrt{8}}$

86. Complete the square of $4x^2 + 24x$ and write the result as the square of a binomial. [1.1]

87. What is the length of the major axis for the ellipse whose equation is $\dfrac{x^2}{9} + \dfrac{y^2}{16} = 1$? [6.2]

88. Graph: $\dfrac{(x-2)^2}{16} + \dfrac{(y+3)^2}{9} = 1$ [6.2]

PROJECTS

1. I. M. PEI'S OVAL The poet and architect I. M. Pei suggested that the oval with the most appeal to the eye is given by the equation

$$\left(\frac{x}{a}\right)^{3/2} + \left(\frac{y}{b}\right)^{3/2} = 1$$

Use a graphing utility to graph this equation with $a = 5$ and $b = 3$. Then compare your graph with the graph of

$$\left(\frac{x}{5}\right)^2 + \left(\frac{y}{3}\right)^2 = 1$$

2. KEPLER'S LAWS The German astronomer Johannes Kepler (1571–1630) derived three laws that describe how the planets orbit the sun. Write an essay that includes biographical information about Kepler and a statement of Kepler's Laws. In addition, use Kepler's Laws to answer the following questions.

a. Where is a planet located in its orbit around the sun when it achieves its greatest velocity?

b. What is the period of Mars if it has a mean distance from the sun of 1.52 astronomical units? (*Hint:* Use Earth as a reference with a period of 1 year and a mean distance from the sun of 1 astronomical unit.)

3. NEPTUNE The position of the planet Neptune was discovered by using celestial mechanics and mathematics. Write an essay that tells how, when, and by whom Neptune was discovered.

4. GRAPH THE COLOSSEUM Some of the Colosseum scenes in the movie *Gladiator* (Universal Studios, 2000) were computer-generated.

a. You can create a simple but accurate scale image of the exterior of the Colosseum by using a computer and the mathematics software program *Maple*. Open a new *Maple* worksheet and enter the following two commands.

with(plots);
plots[implicitplot3d]((x^2)/(307.5^2)+(y^2)/(255^2)= 1, x= –310..310, y= –260..260, z=0..157, scaling=CONSTRAINED, style=PATCHNOGRID, axes=FRAMED);

Execute each of the commands by placing the cursor in a command and pressing the $\boxed{\text{ENTER}}$ key. After execution of the second command a three-dimensional graph will appear. Click and drag on the graph to rotate the image.

b. A graphing calculator can also be used to generate a simple "graph" of the exterior of the Colosseum. Here is a procedure for the TI-83 graphing calculator.

Enter the following in the WINDOW menu.

Xmin=-4.7 Xmax=4.7 Xscl=1 Ymin=-4 Ymax=9 Yscl=1

Enter the following formulas in the Y= menu.

$Y_1 = \sqrt{(9-X^2)}$ $Y_2 = Y_1+4$ $Y_3 = -Y_1$ $Y_4 = Y_3+4$

Press: QUIT (2nd MODE)

Enter: Shade(Y₃,Y₄) and press ENTER.
        Note: "Shade(" is in the DRAW menu.

Explain why this "Colosseum graph" appears to be constructed with ellipses even though the functions entered in the Y= menu are the equations of semicircles.

HYPERBOLAS

The hyperbola is a conic section formed when a plane intersects a right circular cone at a certain angle. If β is the angle at which the plane intersects the axis of the cone and α is the angle shown in **Figure 6.34,** a hyperbola is formed when $0° < \beta < \alpha$ or when the plane is parallel to the axis of the cone.

As with the other conic sections, there is a definition of a hyperbola in terms of a certain set of points in the plane.

> ***take note***
>
> If the plane intersects the cone along the axis of the cone, the resulting curve is two intersecting straight lines. This is the *degenerate* form of a hyperbola. See the accompanying figure.
>
> Degenerate hyperbola

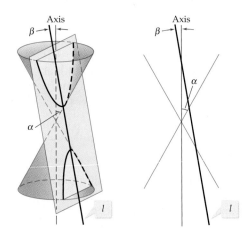

FIGURE 6.34

Definition of a Hyperbola

A **hyperbola** is the set of all points in the plane, the difference between whose distances from two fixed points (foci) is a positive constant.

This definition differs from that of an ellipse in that the ellipse was defined in terms of the *sum* of two distances, whereas the hyperbola is defined in terms of the *difference* of two distances.

● HYPERBOLAS WITH CENTER AT (0, 0)

The **transverse axis** of a hyperbola is the line segment joining the intercepts (see **Figure 6.35**). The midpoint of the transverse axis is called the **center** of the hyperbola. The **conjugate axis** passes through the center of the hyperbola and is perpendicular to the transverse axis.

The length of the transverse axis is customarily represented as $2a$, and the distance between the two foci is represented as $2c$. The length of the conjugate axis is represented as $2b$.

The **vertices** of a hyperbola are the points where the hyperbola intersects the transverse axis.

To determine the positive constant stated in the definition of a hyperbola, consider the point $V_1(a, 0)$, which is one vertex of a hyperbola, and the points $F_1(c, 0)$ and $F_2(-c, 0)$, which are the foci of the hyperbola (see **Figure 6.36**). The difference between the distance from $V_1(a, 0)$ to $F_1(c, 0)$, $c - a$, and the distance from $V_1(a, 0)$ to $F_2(-c, 0)$, $c + a$, must be a constant. By subtracting these distances, we find

$$|(c - a) - (c + a)| = |-2a| = 2a$$

Thus the constant is $2a$ and is the length of the transverse axis. The absolute value is used to ensure that the distance is a positive number.

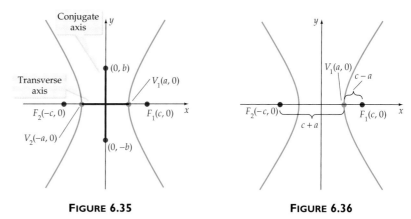

FIGURE 6.35   FIGURE 6.36

Standard Forms of the Equation of a Hyperbola with Center at the Origin

Transverse Axis on the x-Axis

The standard form of the equation of a hyperbola with the center at the origin and transverse axis on the x-axis (see **Figure 6.37a**) is given by

$$\frac{x^2}{a^2} - \frac{y^2}{b^2} = 1$$

The coordinates of the vertices are $(a, 0)$ and $(-a, 0)$, and the coordinates of the foci are $(c, 0)$ and $(-c, 0)$, where $c^2 = a^2 + b^2$.

Transverse Axis on the y-Axis

The standard form of the equation of a hyperbola with the center at the origin and transverse axis on the y-axis (see **Figure 6.37b**) is given by

$$\frac{y^2}{a^2} - \frac{x^2}{b^2} = 1$$

The coordinates of the vertices are $(0, a)$ and $(0, -a)$, and the coordinates of the foci are $(0, c)$ and $(0, -c)$, where $c^2 = a^2 + b^2$.

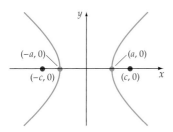

a. Transverse axis on the x-axis

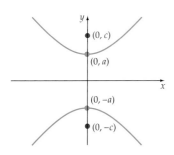

b. Transverse axis on the y-axis

FIGURE 6.37

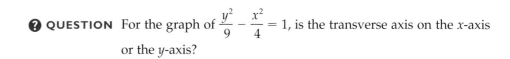

❓ **QUESTION** For the graph of $\dfrac{y^2}{9} - \dfrac{x^2}{4} = 1$, is the transverse axis on the x-axis or the y-axis?

❓ **ANSWER** Because the y-term is positive, the transverse axis is on the y-axis.

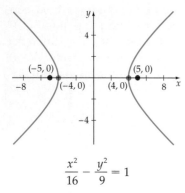

$$\frac{x^2}{16} - \frac{y^2}{9} = 1$$

FIGURE 6.38

By looking at the equations, it is possible to determine the location of the transverse axis by finding which term in the equation is positive. When the x^2 term is positive, the transverse axis is on the x-axis. When the y^2 term is positive, the transverse axis is on the y-axis.

Consider the hyperbola given by the equation $\frac{x^2}{16} - \frac{y^2}{9} = 1$. Because the x^2 term is positive, the transverse axis is on the x-axis, $a^2 = 16$, and thus $a = 4$. The vertices are $(4, 0)$ and $(-4, 0)$. To find the foci, we determine c.

$$c^2 = a^2 + b^2 = 16 + 9 = 25$$
$$c = \sqrt{25} = 5$$

The foci are $(5, 0)$ and $(-5, 0)$. The graph is shown in **Figure 6.38**.

Each hyperbola has two asymptotes that pass through the center of the hyperbola. The asymptotes of the hyperbola are a useful guide to sketching the graph of the hyperbola.

Asymptotes of a Hyperbola with Center at the Origin

The **asymptotes** of the hyperbola defined by $\frac{x^2}{a^2} - \frac{y^2}{b^2} = 1$ are given by the equations $y = \frac{b}{a}x$ and $y = -\frac{b}{a}x$ (see **Figure 6.39a**).

The asymptotes of the hyperbola defined by $\frac{y^2}{a^2} - \frac{x^2}{b^2} = 1$ are given by the equations $y = \frac{a}{b}x$ and $y = -\frac{a}{b}x$ (see **Figure 6.39b**).

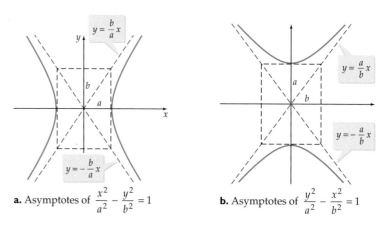

a. Asymptotes of $\frac{x^2}{a^2} - \frac{y^2}{b^2} = 1$ **b.** Asymptotes of $\frac{y^2}{a^2} - \frac{x^2}{b^2} = 1$

FIGURE 6.39

One method for remembering the equations of the asymptotes is to write the equation of a hyperbola in standard form but to replace 1 by 0 and then solve for y.

$$\frac{x^2}{a^2} - \frac{y^2}{b^2} = 0 \quad \text{so} \quad y^2 = \frac{b^2}{a^2}x^2, \text{ or } y = \pm\frac{b}{a}x$$

$$\frac{y^2}{a^2} - \frac{x^2}{b^2} = 0 \quad \text{so} \quad y^2 = \frac{a^2}{b^2}x^2, \text{ or } y = \pm\frac{a}{b}x$$

EXAMPLE 1   Find the Vertices, Foci, and Asymptotes of a Hyperbola

Find the vertices, foci, and asymptotes of the hyperbola given by the equation $\dfrac{y^2}{9} - \dfrac{x^2}{4} = 1$. Sketch the graph.

Solution

Because the y^2 term is positive, the transverse axis is on the y-axis. We know $a^2 = 9$; thus $a = 3$. The vertices are $V_1(0, 3)$ and $V_2(0, -3)$.

$$c^2 = a^2 + b^2 = 9 + 4$$
$$c = \sqrt{13}$$

The foci are $F_1\left(0, \sqrt{13}\right)$ and $F_2\left(0, -\sqrt{13}\right)$.

Because $a = 3$ and $b = 2$ ($b^2 = 4$), the equations of the asymptotes are $y = \dfrac{3}{2}x$ and $y = -\dfrac{3}{2}x$.

To sketch the graph, we draw a rectangle that has its center at the origin and has dimensions equal to the lengths of the transverse and conjugate axes. The asymptotes are extensions of the diagonals of the rectangle. See **Figure 6.40**.

▶ **TRY EXERCISE 4, PAGE 385**

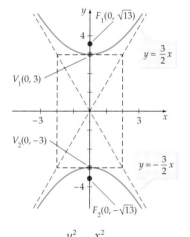

$$\dfrac{y^2}{9} - \dfrac{x^2}{4} = 1$$

FIGURE 6.40

● HYPERBOLAS WITH CENTER AT (h, k)

Using a translation of coordinates similar to that used for ellipses, we can write the equation of a hyperbola with its center at the point (h, k). Given coordinate axes labeled x' and y', an equation of a hyperbola with center at the origin is

$$\dfrac{(x')^2}{a^2} - \dfrac{(y')^2}{b^2} = 1 \qquad (1)$$

Now place the origin of this coordinate system at the point (h, k) of the xy-coordinate system, as shown in **Figure 6.41**. The relationship between an ordered pair in the $x'y'$-coordinate system and one in the xy-coordinate system is given by the transformation equations

$$x' = x - h$$
$$y' = y - k$$

Substitute the expressions for x' and y' into Equation (1). The equation of the hyperbola with center at (h, k) is

$$\dfrac{(x - h)^2}{a^2} - \dfrac{(y - k)^2}{b^2} = 1$$

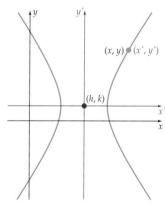

FIGURE 6.41

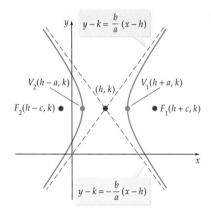

a. Transverse axis parallel to the x-axis

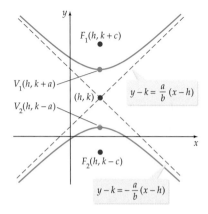

b. Transverse axis parallel to the y-axis

FIGURE 6.42

Standard Forms of the Equation of a Hyperbola with Center at (h, k)

Transverse Axis Parallel to the x-Axis

The standard form of the equation of a hyperbola with center at (h, k) and transverse axis parallel to the x-axis (see **Figure 6.42a**) is given by

$$\frac{(x - h)^2}{a^2} - \frac{(y - k)^2}{b^2} = 1$$

The coordinates of the vertices are $V_1(h + a, k)$ and $V_2(h - a, k)$. The coordinates of the foci are $F_1(h + c, k)$ and $F_2(h - c, k)$, where $c^2 = a^2 + b^2$. The equations of the asymptotes are $y - k = \pm\dfrac{b}{a}(x - h)$.

Transverse Axis Parallel to the y-Axis

The standard form of the equation of a hyperbola with center at (h, k) and transverse axis parallel to the y-axis (see **Figure 6.42b**) is given by

$$\frac{(y - k)^2}{a^2} - \frac{(x - h)^2}{b^2} = 1$$

The coordinates of the vertices are $V_1(h, k + a)$ and $V_2(h, k - a)$. The coordinates of the foci are $F_1(h, k + c)$ and $F_2(h, k - c)$, where $c^2 = a^2 + b^2$.

The equations of the asymptotes are $y - k = \pm\dfrac{a}{b}(x - h)$.

EXAMPLE 2   Find the Vertices, Foci, and Asymptotes of a Hyperbola

Find the vertices, foci, and asymptotes of the hyperbola given by the equation $4x^2 - 9y^2 - 16x + 54y - 29 = 0$. Sketch the graph.

Solution

Write the equation of the hyperbola in standard form by completing the square.

$$4x^2 - 9y^2 - 16x + 54y - 29 = 0$$
$$4x^2 - 16x - 9y^2 + 54y = 29 \qquad \text{• Rearrange terms.}$$
$$4(x^2 - 4x) - 9(y^2 - 6y) = 29 \qquad \text{• Factor.}$$
$$4(x^2 - 4x + 4) - 9(y^2 - 6y + 9) = 29 + 16 - 81 \qquad \text{• Complete the square.}$$
$$4(x - 2)^2 - 9(y - 3)^2 = -36 \qquad \text{• Factor.}$$
$$\frac{(y - 3)^2}{4} - \frac{(x - 2)^2}{9} = 1 \qquad \text{• Divide by } -36.$$

The coordinates of the center are $(2, 3)$. Because the term containing $(y - 3)^2$ is positive, the transverse axis is parallel to the y-axis. We know $a^2 = 4$; thus

$a = 2$. The vertices are $(2, 5)$ and $(2, 1)$. See **Figure 6.43.**

$$c^2 = a^2 + b^2 = 4 + 9$$
$$c = \sqrt{13}$$

The foci are $\left(2, 3 + \sqrt{13}\right)$ and $\left(2, 3 - \sqrt{13}\right)$. We know $b^2 = 9$; thus $b = 3$. The equations of the asymptotes are $y - 3 = \pm\left(\dfrac{2}{3}\right)(x - 2)$, which simplifies to

$$y = \frac{2}{3}x + \frac{5}{3} \qquad \text{and} \qquad y = -\frac{2}{3}x + \frac{13}{3}$$

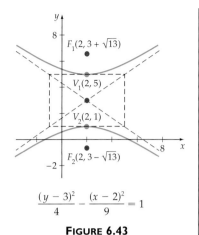

$$\frac{(y - 3)^2}{4} - \frac{(x - 2)^2}{9} = 1$$

FIGURE 6.43

▶ **TRY EXERCISE 26, PAGE 386**

🖩 INTEGRATING TECHNOLOGY

A graphing utility can be used to graph a hyperbola. For instance, consider the equation $4x^2 - 9y^2 - 16x + 54y - 29 = 0$ from Example 2. Rewrite the equation as

$$-9y^2 + 54y + (4x^2 - 16x - 29) = 0$$

In this form, the equation is a quadratic equation in terms of the variable y with

$$A = -9, B = 54, \text{ and } C = 4x^2 - 16x - 29$$

Apply the quadratic formula to produce

$$y = \frac{-54 \pm \sqrt{2916 + 36(4x^2 - 16x - 29)}}{-18}$$

The graph of $Y1 = \dfrac{-54 + \sqrt{2916 + 36(4x^2 - 16x - 29)}}{-18}$ is the upper branch of the hyperbola (see **Figure 6.44**).

The graph of $Y2 = \dfrac{-54 - \sqrt{2916 + 36(4x^2 - 16x - 29)}}{-18}$ is the lower branch of the hyperbola, as shown in **Figure 6.44.**

One advantage of this graphing procedure is that it does not require us to write the given equation in standard form. A disadvantage of the graphing procedure is that it does not indicate where the foci of the hyperbola are located.

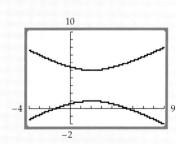

FIGURE 6.44

• ECCENTRICITY OF A HYPERBOLA

The graph of a hyperbola can be very wide or very narrow. The **eccentricity** of a hyperbola is a measure of its "wideness."

Eccentricity (e) of a Hyperbola

The eccentricity e of a hyperbola is the ratio of c to a, where c is the distance from the center to a focus and a is the length of the semitransverse axis.

$$e = \frac{c}{a}$$

For a hyperbola, $c > a$ and therefore $e > 1$. As the eccentricity of the hyperbola increases, the graph becomes wider and wider, as shown in **Figure 6.45.**

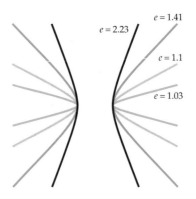

FIGURE 6.45

EXAMPLE 3 **Find the Equation of a Hyperbola Given Its Eccentricity**

Find the standard form of the equation of the hyperbola that has eccentricity $\frac{3}{2}$, center at the origin, and a focus $(6, 0)$.

Solution

Because the focus is located at $(6, 0)$ and the center is at the origin, $c = 6$. An extension of the transverse axis contains the foci, so the transverse axis is on the x-axis.

$$e = \frac{3}{2} = \frac{c}{a}$$

$$\frac{3}{2} = \frac{6}{a} \qquad \text{• Substitute } \mathbf{6} \text{ for } c.$$

$$a = 4 \qquad \text{• Solve for } a.$$

To find b^2, use the equation $c^2 = a^2 + b^2$ and the values for c and a.

$$c^2 = a^2 + b^2$$
$$36 = 16 + b^2$$
$$b^2 = 20$$

The equation of the hyperbola is $\dfrac{x^2}{16} - \dfrac{y^2}{20} = 1$. See **Figure 6.46.**

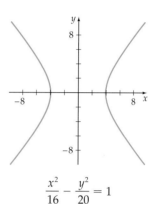

$$\dfrac{x^2}{16} - \dfrac{y^2}{20} = 1$$

FIGURE 6.46

▶ **TRY EXERCISE 48, PAGE 386**

Caroline Herschel (1750–1848) became interested in mathematics and astronomy after her brother William discovered the planet Uranus. She was the first woman to receive credit for the discovery of a comet. In fact, between 1786 and 1797 she discovered eight comets. In 1828 she completed a catalog of over 2000 nebulae for which the Royal Astronomical Society of England presented her with its prestigious gold medal.

● **APPLICATIONS**

Orbits of Comets   In Section 6.2 we noted that the orbits of the planets are elliptical. Some comets have elliptical orbits also, the most notable being Halley's comet, whose eccentricity is 0.97.

   Other comets have hyperbolic orbits with the sun at a focus. These comets pass by the sun only once. The velocity of a comet determines whether its orbit is elliptical or hyperbolic. See **Figure 6.47.**

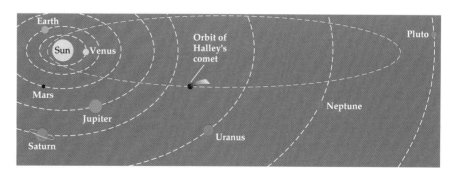

FIGURE 6.47

Hyperbolas as an Aid to Navigation   Consider two radio transmitters, T_1 and T_2, placed some distance apart. A ship with electronic equipment measures the difference between the times it takes signals from the transmitters to reach the ship.

Because the difference between the times is proportional to the difference between the distances of the ship from the transmitters, the ship must be located on the hyperbola with foci at the two transmitters.

Using a third transmitter, T_3, we can find a second hyperbola with foci T_2 and T_3. The ship lies on the intersection of the two hyperbolas, as shown in **Figure 6.48**.

EXAMPLE 4   Determine the Position of a Ship

Two radio transmitters are positioned along a coastline, 500 miles apart. See **Figure 6.49.** Using a LORAN (LOng RAnge Navigation) system, a ship determines that a radio signal from transmitter T_1 reaches the ship 1600 microseconds before it receives a simultaneous signal from transmitter T_2.

a. Find an equation of a hyperbola (with foci located at T_1 and T_2) on which the ship lies. See **Figure 6.49.** (Assume the radio signals travel at 0.186 mile per microsecond.)

b. If the ship is directly north of transmitter T_1, determine how far (to the nearest mile) the ship is from the transmitter.

Solution

a. The ship lies on a hyperbola at point B, with foci at T_1 and T_2. The difference of the distances $d(T_2, B)$ and $d(T_1, B)$ is given by

$$\text{Distance} = \text{rate} \times \text{time}$$
$$= 0.186 \text{ mile/microsecond} \times 1600 \text{ microseconds}$$
$$= 297.6 \text{ mile}$$

Thus the ship is located on a hyperbola with transverse axis of 297.6 miles and semitransverse axis $a = 148.8$ miles. **Figure 6.49** shows that the foci are located at $(250, 0)$ and $(-250, 0)$. Thus $c = 250$ miles, and

$$b = \sqrt{c^2 - a^2} = \sqrt{250^2 - 148.8^2} \approx 200.9 \text{ miles}$$

The ship is located on the hyperbola given by

$$\frac{x^2}{148.8^2} - \frac{y^2}{200.9^2} = 1$$

b. If the ship is directly north of T_1, then $x = 250$, and the distance from the ship to the transmitter T_1 is y, where

$$-\frac{y^2}{200.9^2} = 1 - \frac{250^2}{148.8^2}$$

$$y = \frac{200.9}{148.8}\sqrt{250^2 - 148.8^2} \approx 271 \text{ miles}$$

The ship is about 271 miles north of transmitter T_1.

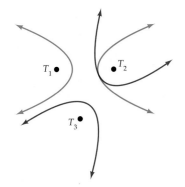

FIGURE 6.48

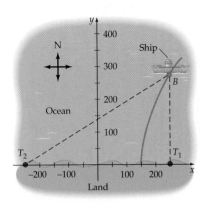

FIGURE 6.49

▶ TRY EXERCISE 54, PAGE 387

Hyperbolas also have a reflective property that makes them useful in many applications.

Reflective Property of a Hyperbola

A ray of light directed toward one focus of a hyperbolic mirror is reflected toward the other focus. See **Figures 6.50** and **6.51**.

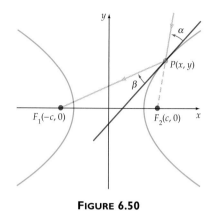

FIGURE 6.50

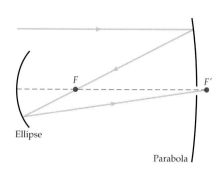

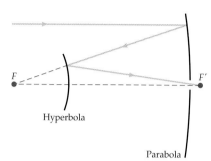

FIGURE 6.51

TOPICS FOR DISCUSSION

1. In every hyperbola, the distance c from a focus to the center of the hyperbola is greater than the length of the semitransverse axis a. Do you agree? Explain.

2. How many vertices does a hyperbola have?

3. Explain why the eccentricity of every hyperbola is a number greater than 1.

4. Is the conjugate axis of a hyperbola perpendicular to the transverse axis of the hyperbola?

EXERCISE SET 6.3

In Exercises 1 to 26, find the center, vertices, foci, and asymptotes for the hyperbola given by each equation. Graph each equation.

1. $\dfrac{x^2}{16} - \dfrac{y^2}{25} = 1$

2. $\dfrac{x^2}{16} - \dfrac{y^2}{9} = 1$

3. $\dfrac{y^2}{4} - \dfrac{x^2}{25} = 1$

▶ 4. $\dfrac{y^2}{25} - \dfrac{x^2}{36} = 1$

5. $\dfrac{x^2}{7} - \dfrac{y^2}{9} = 1$

6. $\dfrac{x^2}{5} - \dfrac{y^2}{4} = 1$

7. $\dfrac{4x^2}{9} - \dfrac{y^2}{16} = 1$

8. $\dfrac{x^2}{9} - \dfrac{9y^2}{16} = 1$

9. $\dfrac{(x-3)^2}{16} - \dfrac{(y+4)^2}{9} = 1$

10. $\dfrac{(x+3)^2}{25} - \dfrac{y^2}{4} = 1$

11. $\dfrac{(y+2)^2}{4} - \dfrac{(x-1)^2}{16} = 1$

12. $\dfrac{(y-2)^2}{36} - \dfrac{(x+1)^2}{49} = 1$

13. $\dfrac{(x+2)^2}{9} - \dfrac{y^2}{25} = 1$

14. $\dfrac{x^2}{25} - \dfrac{(y-2)^2}{81} = 1$

15. $\dfrac{9(x-1)^2}{16} - \dfrac{(y+1)^2}{9} = 1$

16. $\dfrac{(x+6)^2}{25} - \dfrac{25y^2}{144} = 1$

17. $x^2 - y^2 = 9$

18. $4x^2 - y^2 = 16$

19. $16y^2 - 9x^2 = 144$

20. $9y^2 - 25x^2 = 225$

21. $9y^2 - 36x^2 = 4$

22. $16x^2 - 25y^2 = 9$

23. $x^2 - y^2 - 6x + 8y - 3 = 0$

24. $4x^2 - 25y^2 + 16x + 50y - 109 = 0$

25. $9x^2 - 4y^2 + 36x - 8y + 68 = 0$

▶ **26.** $16x^2 - 9y^2 - 32x - 54y + 79 = 0$

In Exercises 27 to 32, use the quadratic formula to solve for *y* in terms of *x*. Then use a graphing utility to graph each equation.

27. $4x^2 - y^2 + 32x + 6y + 39 = 0$

28. $x^2 - 16y^2 + 8x - 64y + 16 = 0$

29. $9x^2 - 16y^2 - 36x - 64y + 116 = 0$

30. $2x^2 - 9y^2 + 12x - 18y + 18 = 0$

31. $4x^2 - 9y^2 + 8x - 18y - 6 = 0$

32. $2x^2 - 9y^2 - 8x + 36y - 46 = 0$

In Exercises 33 to 46, find the equation in standard form of the hyperbola that satisfies the stated conditions.

33. Vertices $(3, 0)$ and $(-3, 0)$, foci $(4, 0)$ and $(-4, 0)$

34. Vertices $(0, 2)$ and $(0, -2)$, foci $(0, 3)$ and $(0, -3)$

35. Foci $(0, 5)$ and $(0, -5)$, asymptotes $y = 2x$ and $y = -2x$

36. Foci $(4, 0)$ and $(-4, 0)$, asymptotes $y = x$ and $y = -x$

37. Vertices $(0, 3)$ and $(0, -3)$, passing through $(2, 4)$

38. Vertices $(5, 0)$ and $(-5, 0)$, passing through $(-1, 3)$

39. Asymptotes $y = \frac{1}{2}x$ and $y = -\frac{1}{2}x$, vertices $(0, 4)$ and $(0, -4)$

40. Asymptotes $y = \frac{2}{3}x$ and $y = -\frac{2}{3}x$, vertices $(6, 0)$ and $(-6, 0)$

41. Vertices $(6, 3)$ and $(2, 3)$, foci $(7, 3)$ and $(1, 3)$

42. Vertices $(-1, 5)$ and $(-1, -1)$, foci $(-1, 7)$ and $(-1, -3)$

43. Foci $(1, -2)$ and $(7, -2)$, slope of an asymptote $\frac{5}{4}$

44. Foci $(-3, -6)$ and $(-3, -2)$, slope of an asymptote 1

45. Passing through $(9, 4)$, slope of an asymptote $\frac{1}{2}$, center $(7, 2)$, transverse axis parallel to the *y*-axis

46. Passing through $(6, 1)$, slope of an asymptote 2, center $(3, 3)$, transverse axis parallel to the *x*-axis

In Exercises 47 to 52, use the eccentricity to find the equation in standard form of each hyperbola.

47. Vertices $(1, 6)$ and $(1, 8)$, eccentricity 2

▶ **48.** Vertices $(2, 3)$ and $(-2, 3)$, eccentricity $\frac{5}{2}$

49. Eccentricity 2, foci $(4, 0)$ and $(-4, 0)$

50. Eccentricity $\frac{4}{3}$, foci $(0, 6)$ and $(0, -6)$

51. Center $(4, 1)$, conjugate axis of length 4, eccentricity $\frac{4}{3}$ (*Hint:* There are two answers.)

52. Center $(-3, -3)$, conjugate axis of length 6, eccentricity 2 (*Hint:* There are two answers.)

53. LORAN Two radio transmitters are positioned along the coast, 250 miles apart. A signal is sent simultaneously from each transmitter. The signal from transmitter T_2 is received by a ship's LORAN 500 microseconds after it receives the signal from T_1. The radio signals travel 0.186 mile per microsecond.

    a. Find an equation of a hyperbola, with foci at T_1 and T_2, on which the ship is located.

    b. If the ship is 100 miles east of the *y*-axis, determine its distance from the coastline (to the nearest mile).

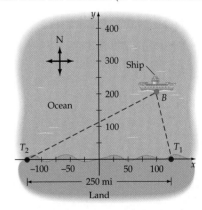

▶ **54. LORAN** Two radio transmitters are positioned along the coast, 300 miles apart. A signal is sent simultaneously from each transmitter. The signal from transmitter T_1 is received by a ship's LORAN 800 microseconds after it receives the signal from T_2. The radio signals travel 0.186 mile per microsecond.

    a. Find an equation of a hyperbola, with foci at T_1 and T_2, on which the ship is located.

b. If the ship continues to travel so that the difference of 800 microseconds is maintained, determine the point at which the ship will reach the coastline.

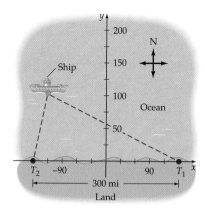

55. SONIC BOOMS When a plane exceeds the speed of sound, a sonic boom is produced by the wake of the sound waves. (See the chapter opener on page 349.) For a plane flying at 10,000 feet, the circular wave front can be given by

$$y^2 = x^2 + (z - 10{,}000)^2$$

where z is the height of the wave front above Earth. See the diagram below. Note that the xy-plane is Earth's surface, which is approximately flat over small distances.

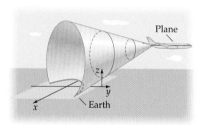

Find and name the equation formed when the wave front hits Earth.

56. WATER WAVES If two pebbles are dropped into a pond at different places $F_1(-2, 0)$ and $F_2(2, 0)$, circular waves are

propagated with F_1 as the center of one set of circular waves (in green) and F_2 as the center of the other set of circular waves (in red).

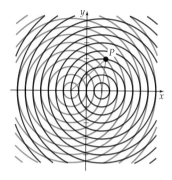

Let P be a point at which the waves intersect. In the diagram above, $|F_1P - F_2P| = 2$.

a. What curve is generated by connecting all points P for which $|F_1P - F_2P| = 2$?

b. What is the equation of the curve in part **a.**?

In Exercises 57 to 64, identify the graph of each equation as a parabola, an ellipse, or a hyperbola. Graph each equation.

57. $4x^2 + 9y^2 - 16x - 36y + 16 = 0$

58. $2x^2 + 3y - 8x + 2 = 0$

59. $5x - 4y^2 + 24y - 11 = 0$

60. $9x^2 - 25y^2 - 18x + 50y = 0$

61. $x^2 + 2y - 8x = 0$

62. $9x^2 + 16y^2 + 36x - 64y - 44 = 0$

63. $25x^2 + 9y^2 - 50x - 72y - 56 = 0$

64. $(x - 3)^2 + (y - 4)^2 = (x + 1)^2$

CONNECTING CONCEPTS

In Exercises 65 to 68, use the definition of a hyperbola to find the equation of the hyperbola in standard form.

65. Foci $(2, 0)$ and $(-2, 0)$; passes through the point $(2, 3)$

66. Foci $(0, 3)$ and $(0, -3)$; passes through the point $\left(\dfrac{5}{2}, 3\right)$

67. Foci $(0, 4)$ and $(0, -4)$; passes through the point $\left(\dfrac{7}{3}, 4\right)$

68. Foci $(5, 0)$ and $(-5, 0)$; passes through the point $\left(5, \dfrac{9}{4}\right)$

Recall that an ellipse has two directrices that are lines perpendicular to the line containing the foci. A hyperbola also has two directrices; they are perpendicular to the transverse axis and outside the hyperbola. For a hyperbola with center at the origin and transverse axis on the *x*-axis, the equations of the directrices are $x = \dfrac{a^2}{c}$ and $x = -\dfrac{a^2}{c}$. In Exercises 69 to 72, use this information to solve each exercise.

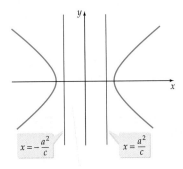

$x = -\dfrac{a^2}{c}$ $x = \dfrac{a^2}{c}$

69. Find the directrices for the hyperbola in Exercise 1.

70. Find the directrices for the hyperbola in Exercise 2.

71. Let $P(x, y)$ be a point on the hyperbola $\dfrac{x^2}{9} - \dfrac{y^2}{16} = 1$. Show that the distance from the point P to the focus $(5, 0)$ divided by the distance from the point P to the directrix $x = \dfrac{9}{5}$ equals the eccentricity.

72. Generalize the results of Exercise 71. That is, show that if $P(x, y)$ is a point on the hyperbola $\dfrac{x^2}{a^2} - \dfrac{y^2}{b^2} = 1$, $F(c, 0)$ is a focus, and $x = \dfrac{a^2}{c}$ is a directrix, then the following equation is true:

$$e = \frac{d(P, F)}{d(P, D)}$$

73. Sketch a graph of $\dfrac{x|x|}{16} - \dfrac{y|y|}{9} = 1$.

74. Sketch a graph of $\dfrac{x|x|}{16} + \dfrac{y|y|}{9} = 1$.

PREPARE FOR SECTION 6.4

75. Expand $\cos(\alpha + \beta)$. [3.2]

76. Expand $\sin(\alpha + \beta)$. [3.2]

77. Solve $\cot 2\alpha = \dfrac{\sqrt{3}}{3}$ for $0 < \alpha < \dfrac{\pi}{2}$. [3.6]

78. If $\sin \alpha = \dfrac{1}{2}$ and $\cos \alpha = -\dfrac{\sqrt{3}}{2}$, find α. Write the answer in degrees. [3.6]

79. Identify the graph of $4x^2 - 6y^2 + 9x + 16y - 8 = 0$. [6.3]

80. Graph: $4x - y^2 - 2y + 3 = 0$ [6.1]

PROJECTS

1. **A HYPERBOLIC PARABOLOID** A *hyperbolic paraboloid* is a three-dimensional figure. Some of its cross sections are parabolas and some are hyperbolas. Make a drawing of a hyperbolic paraboloid. Explain the relationship that exists between the equations of the parabolic cross sections and the relationship that exists between the equations of the hyperbolic cross sections.

2. **A HYPERBOLOID OF ONE SHEET** Make a sketch of a *hyperboloid of one sheet*. Explain the different cross sections of the hyperboloid of one sheet. Do some research on nuclear power plants, and explain why nuclear cooling towers are designed in the shape of hyperboloids of one sheet.

SECTION 6.4

ROTATION OF AXES

- THE ROTATION THEOREM FOR CONICS
- THE CONIC IDENTIFICATION THEOREM
- USE A GRAPHING UTILITY TO GRAPH SECOND-DEGREE EQUATIONS IN TWO VARIABLES

take note

Some choices of the constants A, B, C, D, E, and F may result in a degenerate conic or an equation that has no solutions.

● THE ROTATION THEOREM FOR CONICS

The equation of a conic with axes parallel to the coordinate axes can be written in a general form.

General Equation of a Conic with Axes Parallel to Coordinate Axes

The **general equation of a conic** with axes parallel to the coordinate axes and not both A and C equal to zero is

$$Ax^2 + Cy^2 + Dx + Ey + F = 0$$

The graph of the equation is a parabola when $AC = 0$, an ellipse when $AC > 0$, and a hyperbola when $AC < 0$.

The terms Dx, Ey, and F determine the translation of the conic from the origin. The general equation of a conic is a *second-degree equation* in two variables. A more general second-degree equation can be written that contains a Bxy term.

General Second-Degree Equation in Two Variables

The **general second-degree equation in two variables** is

$$Ax^2 + Bxy + Cy^2 + Dx + Ey + F = 0$$

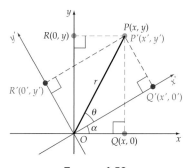

FIGURE 6.52

The Bxy term ($B \neq 0$) determines a rotation of the conic so that its axes are no longer parallel to the coordinate axes.

A **rotation of axes** is a rotation of the x- and y-axes about the origin to another position denoted by x' and y'. We denote the measure of the **angle of rotation** by α.

Let P be some point in the plane, and let r represent the distance of P from the origin. The coordinates of P relative to the xy-coordinate system and the $x'y'$-coordinate system are $P(x, y)$ and $P(x', y')$, respectively.

Let $Q(x, 0)$ and $R(0, y)$ be the projections of P onto the x- and the y-axis and let $Q'(x', 0')$ and $R'(0', y')$ be the projections of P onto the x'- and the y'-axis. (See **Figure 6.52**.) The angle between the x'-axis and OP is denoted by θ. We can express the coordinates of P in each coordinate system in terms of α and θ.

$$x = r\cos(\theta + \alpha) \qquad x' = r\cos\theta$$

$$y = r\sin(\theta + \alpha) \qquad y' = r\sin\theta$$

Applying the addition formulas for $\cos(\theta + \alpha)$ and $\sin(\theta + \alpha)$, we get

$$x = r\cos(\theta + \alpha) = r\cos\theta\cos\alpha - r\sin\theta\sin\alpha$$

$$y = r\sin(\theta + \alpha) = r\sin\theta\cos\alpha + r\cos\theta\sin\alpha$$

Now, substituting x' for $r\cos\theta$ and y' for $r\sin\theta$ into these equations yields

$$x = x'\cos\alpha - y'\sin\alpha \qquad \bullet\ x' = r\cos\theta,\ y' = r\sin\alpha$$

$$y = y'\cos\alpha + x'\sin\alpha$$

This proves the equations labeled (1) of the following theorem.

Rotation-of-Axes Formulas

Suppose that an xy-coordinate system and an $x'y'$-coordinate system have the same origin and that α is the angle between the positive x-axis and the positive x'-axis. If the coordinates of a point P are (x, y) in one system and (x', y') in the rotated system, then

$$\left. \begin{array}{l} x = x'\cos\alpha - y'\sin\alpha \\ y = y'\cos\alpha + x'\sin\alpha \end{array} \right\} \quad (1) \qquad \left. \begin{array}{l} x' = x\cos\alpha + y\sin\alpha \\ y' = y\cos\alpha - x\sin\alpha \end{array} \right\} \quad (2)$$

The derivations of the formulas for x' and y' are left as an exercise.

As we have noted, the appearance of the Bxy $(B \neq 0)$ term in the general second-degree equation indicates that the graph of the conic has been rotated. The angle through which the axes have been rotated can be determined from the following theorem.

Rotation Theorem for Conics

Let $Ax^2 + Bxy + Cy^2 + Dx + Ey + F = 0$, $B \neq 0$, be the equation of a conic in an xy-coordinate system, and let α be an angle of rotation such that

$$\cot 2\alpha = \frac{A - C}{B}, \quad 0° < 2\alpha < 180°$$

Then the equation of the conic in the rotated coordinate system will be

$$A'x'^2 + C'y'^2 + D'x' + E'y' + F' = 0$$

where $0° < 2\alpha < 180°$ and

$$A' = A\cos^2\alpha + B\cos\alpha\sin\alpha + C\sin^2\alpha \tag{3}$$

$$C' = A\sin^2\alpha - B\cos\alpha\sin\alpha + C\cos^2\alpha \tag{4}$$

$$D' = D\cos\alpha + E\sin\alpha \tag{5}$$

$$E' = -D\sin\alpha + E\cos\alpha \tag{6}$$

$$F' = F \tag{7}$$

❓ QUESTION For $x^2 - 4xy + 9y^2 + 6x - 8y - 20 = 0$, what is the value of $\cot 2\alpha$?

EXAMPLE 1   **Use the Rotation Theorem to Sketch a Conic**

Sketch the graph of $7x^2 - 6\sqrt{3}xy + 13y^2 - 16 = 0$.

Solution

We are given

$$A = 7, \quad B = -6\sqrt{3}, \quad C = 13, \quad D = 0, \quad E = 0, \quad \text{and} \quad F = -16$$

The angle of rotation α can be determined by solving

$$\cot 2\alpha = \frac{A - C}{B} = \frac{7 - 13}{-6\sqrt{3}} = \frac{-6}{-6\sqrt{3}} = \frac{1}{\sqrt{3}} = \frac{\sqrt{3}}{3}$$

This gives us $2\alpha = 60°$, or $\alpha = 30°$. Because $\alpha = 30°$, we have

$$\sin \alpha = \frac{1}{2} \quad \text{and} \quad \cos \alpha = \frac{\sqrt{3}}{2}$$

We determine the coefficients A', C', D', E', and F' by using Equations (3) to (7).

$$A' = 7\left(\frac{\sqrt{3}}{2}\right)^2 + (-6\sqrt{3})\left(\frac{\sqrt{3}}{2}\right)\left(\frac{1}{2}\right) + 13\left(\frac{1}{2}\right)^2 = 4$$

$$C' = 7\left(\frac{1}{2}\right)^2 - (-6\sqrt{3})\left(\frac{\sqrt{3}}{2}\right)\left(\frac{1}{2}\right) + 13\left(\frac{\sqrt{3}}{2}\right)^2 = 16$$

$$D' = 0\left(\frac{\sqrt{3}}{2}\right)^2 + 0\left(\frac{1}{2}\right) = 0$$

$$E' = -0\left(\frac{1}{2}\right) + 0\left(\frac{\sqrt{3}}{2}\right) = 0$$

$$F' = F = -16$$

The equation of the conic in the $x'y'$-plane is $4(x')^2 + 16(y')^2 - 16 = 0$ or

$$\frac{(x')^2}{2^2} + \frac{(y')^2}{1^2} = 1$$

This is the equation of an ellipse that is centered at the origin of an $x'y'$-coordinate system. The ellipse has a semimajor axis $a = 2$ and a semiminor axis $b = 1$. See **Figure 6.53.**

▶ **TRY EXERCISE 8, PAGE 396**

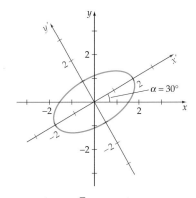

$7x^2 - 6\sqrt{3}xy + 13y^2 - 16 = 0$

FIGURE 6.53

❓ ANSWER $\cot 2\alpha = \dfrac{1 - 9}{-4} = 2$

In Example 1, the angle of rotation α was 30°, which is a special angle. In the next example, we demonstrate a technique that is often used when the angle of rotation is not a special angle.

EXAMPLE 2   Use the Rotation Theorem to Sketch a Conic

Sketch the graph of $32x^2 - 48xy + 18y^2 - 15x - 20y = 0$.

Solution

We are given

$$A = 32, \quad B = -48, \quad C = 18, \quad D = -15, \quad E = -20, \quad \text{and} \quad F = 0$$

Therefore,

$$\cot 2\alpha = \frac{A - C}{B} = \frac{32 - 18}{-48} = -\frac{7}{24}$$

Figure 6.54 shows an angle 2α for which $\cot 2\alpha = -\dfrac{7}{24}$. From **Figure 6.54** we conclude that $\cos 2\alpha = -\dfrac{7}{25}$. The half-angle identities can be used to determine $\sin \alpha$ and $\cos \alpha$.

$$\sin \alpha = \sqrt{\frac{1 - (-7/25)}{2}} = \frac{4}{5} \quad \text{and} \quad \cos \alpha = \sqrt{\frac{1 + (-7/25)}{2}} = \frac{3}{5}$$

A calculator can be used to determine that $\alpha \approx 53.1°$.

Equations (3) to (7) give us

$$A' = 32\left(\frac{3}{5}\right)^2 + (-48)\left(\frac{3}{5}\right)\left(\frac{4}{5}\right) + 18\left(\frac{4}{5}\right)^2 = 0$$

$$C' = 32\left(\frac{4}{5}\right)^2 - (-48)\left(\frac{3}{5}\right)\left(\frac{4}{5}\right) + 18\left(\frac{3}{5}\right)^2 = 50$$

$$D' = (-15)\left(\frac{3}{5}\right) + (-20)\left(\frac{4}{5}\right) = -25$$

$$E' = -(-15)\left(\frac{4}{5}\right) + (-20)\left(\frac{3}{5}\right) = 0$$

$$F' = F = 0$$

The equation of the conic in the $x'y'$-plane is $50(y')^2 - 25x' = 0$, or

$$(y')^2 = \frac{1}{2}x'$$

This is the equation of a parabola. Because $4p = \dfrac{1}{2}$, we know $p = \dfrac{1}{8}$, and the focus of the parabola is at $\left(\dfrac{1}{8}, 0\right)$ on the x'-axis. See **Figure 6.55.**

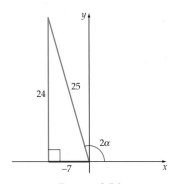

FIGURE 6.54

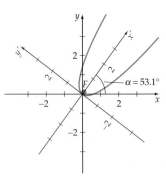

$32x^2 - 48xy + 18y^2 - 15x - 20y = 0$

FIGURE 6.55

▶ **TRY EXERCISE 18, PAGE 396**

• THE CONIC IDENTIFICATION THEOREM

The following theorem provides us with a procedure that can be used to identify the type of conic that will be produced by graphing an equation that is in the form of the general second-degree equation in two variables.

> **Conic Identification Theorem**
>
> The graph of
>
> $$Ax^2 + Bxy + Cy^2 + Dx + Ey + F = 0$$
>
> is either a conic or a degenerate conic. If the graph is a conic, then the graph can be identified by its *discriminant* $B^2 - 4AC$. The graph is
>
> - an ellipse or a circle, provided $B^2 - 4AC < 0$.
>
> - a parabola, provided $B^2 - 4AC = 0$.
>
> - a hyperbola, provided $B^2 - 4AC > 0$.

EXAMPLE 3   **Identify Conic Sections**

Each of the following equations has a graph that is a nondegenerate conic. Compute $B^2 - 4AC$ to identify the type of conic given by each equation.

a. $2x^2 - 4xy + 2y^2 - 6x - 10 = 0$ b. $-2xy + 11 = 0$

c. $3x^2 + 5xy + 4y^2 - 8x + 10y + 6 = 0$ d. $xy - 3y^2 + 2 = 0$

Solution

a.   Because $B^2 - 4AC = (-4)^2 - 4(2)(2) = 0$, the graph is a parabola.

b.   Because $B^2 - 4AC = (-2)^2 - 4(0)(0) > 0$, the graph is a hyperbola.

c.   Because $B^2 - 4AC = 5^2 - 4(3)(4) < 0$, the graph is an ellipse or a circle.

d.   Because $B^2 - 4AC = 1^2 - 4(0)(-3) > 0$, the graph is a hyperbola.

▶ **TRY EXERCISE 30, PAGE 396**

● USE A GRAPHING UTILITY TO GRAPH SECOND-DEGREE EQUATIONS IN TWO VARIABLES

INTEGRATING TECHNOLOGY

To graph a general second-degree equation in two variables with a graphing utility requires that we first solve the general equation for y. Consider the general second-degree equation in two variables

$$Ax^2 + Bxy + Cy^2 + Dx + Ey + F = 0 \qquad (8)$$

where A, B, C, D, E, and F are real constants, and $C \neq 0$. To solve Equation (8) for y, we first rewrite the equation as

$$Cy^2 + (Bx + E)y + (Ax^2 + Dx + F) = 0 \qquad (9)$$

Applying the quadratic formula to Equation (9) yields

$$y = \frac{-(Bx + E) \pm \sqrt{(Bx + E)^2 - 4C(Ax^2 + Dx + F)}}{2C} \qquad (10)$$

Thus the graph of Equation (8) can be constructed by graphing both

$$y_1 = \frac{-(Bx + E) + \sqrt{(Bx + E)^2 - 4C(Ax^2 + Dx + F)}}{2C} \qquad (11)$$

and

$$y_2 = \frac{-(Bx + E) - \sqrt{(Bx + E)^2 - 4C(Ax^2 + Dx + F)}}{2C} \qquad (12)$$

on the same grid.

INTEGRATING TECHNOLOGY

A TI-82/83 graphing calculator program is available to graph a rotated conic section by entering its coefficients. This program, ROTATE, can be found on our website at

http://college.hmco.com

EXAMPLE 4   **Use a Graphing Utility to Graph a Conic**

 Use a graphing utility to graph each conic.

a. $7x^2 + 6xy + 2.5y^2 - 14x + 4y + 9 = 0$

b. $x^2 + 5xy + 3y^2 - 25x - 84y + 375 = 0$

c. $3x^2 - 6xy + 3y^2 - 15x - 12y - 8 = 0$

Solution

Enter y_1 (Equation 11) and y_2 (Equation 12) into the function editing menu of a graphing utility.

a. Store the following constants in place of the indicated variables.

$$A = 7, \quad B = 6, \quad C = 2.5, \quad D = -14, \quad E = 4, \quad \text{and} \quad F = 9$$

Graph y_1 and y_2 on the same screen. The union of the two graphs is an ellipse. See **Figure 6.56.**

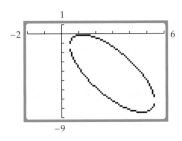

$7x^2 + 6xy + 2.5y^2 - 14x + 4y + 9 = 0$

FIGURE 6.56

b. Store the following constants in place of the indicated variables.

$$A = 1, \quad B = 5, \quad C = 3, \quad D = -25, \quad E = -84, \quad \text{and} \quad F = 375$$

Graph y_1 and y_2 on the same screen. The union of the two graphs is a hyperbola. See **Figure 6.57.**

c. Store the following constants in place of the indicated variables.

$$A = 3, \quad B = -6, \quad C = 3, \quad D = -15, \quad E = -12, \quad \text{and} \quad F = -8$$

Graph y_1 and y_2 on the same screen. The union of the two graphs is a parabola. See **Figure 6.58.**

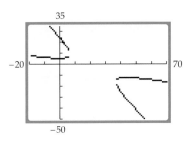

$x^2 + 5xy + 3y^2 - 25x - 84y + 375 = 0$

FIGURE 6.57

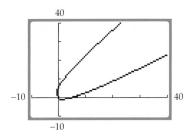

$3x^2 - 6xy + 3y^2 - 15x - 12y - 8 = 0$

FIGURE 6.58

 TRY EXERCISE 24, PAGE 396

TOPICS FOR DISCUSSION

1. Two students disagree about the graph of $x^2 + y^2 = -4$. One student states that the graph of the equation is an ellipse. The other student claims that the equation does not have a graph. Which student is correct? Explain.

2. The graph of $4x^2 - y^2 = 0$ consists of two intersecting lines. What are the equations of the lines?

3. The graph of $xy = 12$ is a hyperbola. What are the equations of the asymptotes of the hyperbola?

4. Explain why the graph of the ellipse in **Figure 6.56** shows a gap at the left side and the right side of the ellipse.

5. Explain why any conic that is given by a general second-degree equation in which $D = 0$ and $E = 0$ will have a quadratic equation in the rotated coordinate system with $D' = 0$ and $E' = 0$.

EXERCISE SET 6.4

In Exercises 1 to 6, find an angle of rotation α that elimi-nates the xy term. State approximate solutions to the nearest 0.1°.

1. $xy = 3$

2. $5x^2 - 3xy - 5y^2 - 1 = 0$

3. $9x^2 - 24xy + 16y^2 - 320x - 240y = 0$

4. $x^2 + 4xy + 4y^2 - 6x - 5 = 0$

5. $5x^2 - 6\sqrt{3}xy - 11y^2 + 4x - 3y + 2 = 0$

6. $5x^2 + 4xy + 8y^2 - 6x + 3y - 12 = 0$

In Exercises 7 to 18, find an angle of rotation α that eliminates the xy term. Then find an equation in $x'y'$-coordinates. Graph the equation.

7. $xy = 4$

▶ **8.** $xy = -10$

9. $6x^2 - 6xy + 14y^2 - 45 = 0$

10. $11x^2 - 10\sqrt{3}xy + y^2 - 20 = 0$

11. $x^2 + 4xy - 2y^2 - 1 = 0$

12. $9x^2 - 24xy + 16y^2 + 100 = 0$

13. $3x^2 + 2\sqrt{3}xy + y^2 + 2x - 2\sqrt{3}y + 16 = 0$

14. $x^2 + 2xy + y^2 + 2\sqrt{2}x - 2\sqrt{2}y = 0$

15. $9x^2 - 24xy + 16y^2 - 40x - 30y + 100 = 0$

16. $24x^2 + 16\sqrt{3}xy + 8y^2 - x + \sqrt{3}y - 8 = 0$

17. $6x^2 + 24xy - y^2 - 12x + 26y + 11 = 0$

▶ **18.** $x^2 + 4xy + 4y^2 - 2\sqrt{5}x + \sqrt{5}y = 0$

In Exercises 19 to 24, use a graphing utility to graph each equation.

19. $6x^2 - xy + 2y^2 + 4x - 12y + 7 = 0$

20. $5x^2 - 2xy + 10y^2 - 6x - 9y - 20 = 0$

21. $x^2 - 6xy + y^2 - 2x - 5y + 4 = 0$

22. $2x^2 - 10xy + 3y^2 - x - 8y - 7 = 0$

23. $3x^2 - 6xy + 3y^2 + 10x - 8y - 2 = 0$

▶ **24.** $2x^2 - 8xy + 8y^2 + 20x - 24y - 3 = 0$

25. Find the equations of the asymptotes, relative to an xy-coordinate system, for the hyperbola defined by the equation in Exercise 11. Assume that the xy-coordinate system has the same origin as the $x'y'$-coordinate system.

26. Find the coordinates of the foci and the equation of the directrix, relative to an xy-coordinate system, for the parabola defined by the equation in Exercise 14. Assume that the xy-coordinate system has the same origin as the $x'y'$-coordinate system.

27. Find the coordinates of the foci, relative to an xy-coordinate system, for the ellipse defined by the equation in Exercise 9. Assume that the xy-coordinate system has the same origin as the $x'y'$-coordinate system.

In Exercises 28 to 36, identify the graph of each equation as a parabola, an ellipse (or a circle), or a hyperbola.

28. $xy = 4$

29. $x^2 + xy - y^2 - 40 = 0$

▶ **30.** $11x^2 - 10\sqrt{3}xy + y^2 - 20 = 0$

31. $3x^2 + 2\sqrt{3}xy + y^2 - 3x + 2y + 20 = 0$

32. $9x^2 - 24xy + 16y^2 + 8x - 12y - 20 = 0$

33. $4x^2 - 4xy + y^2 - 12y + 20 = 0$

34. $5x^2 + 4xy + 8y^2 - 6x + 3y - 12 = 0$

35. $5x^2 - 6\sqrt{3}xy - 11y^2 + 4x - 3y + 2 = 0$

36. $6x^2 - 6xy + 14y^2 - 14x + 12y - 60 = 0$

CONNECTING CONCEPTS

---

37. By using the rotation-of-axes equations, show that for every choice of α, the equation $x^2 + y^2 = r^2$ becomes $x'^2 + y'^2 = r^2$.

38. The vertices of a hyperbola are $(1, 1)$ and $(-1, -1)$. The foci are $(\sqrt{2}, \sqrt{2})$ and $(-\sqrt{2}, -\sqrt{2})$. Find an equation of the hyperbola.

39. The vertices on the major axis of an ellipse are the points $(2, 4)$ and $(-2, -4)$. The foci are the points $(\sqrt{2}, 2\sqrt{2})$ and $(-\sqrt{2}, -2\sqrt{2})$. Find an equation of the ellipse.

40. The vertex of a parabola is the origin, and the focus is the point $(1, 3)$. Find an equation of the parabola.

41. **AN INVARIANT THEOREM** Let $Ax^2 + Bxy + Cy^2 + Dx + Ey + F = 0$ be an equation of a conic in an xy-coordinate system. Let the equation of the conic in the rotated $x'y'$-coordinate system be $A'(x')^2 + B'x'y' + C'(y')^2 + D'x' + E'y' + F' = 0$. Show that

$$A' + C' = A + C$$

42. **AN INVARIANT THEOREM** Let $Ax^2 + Bxy + Cy^2 + Dx + Ey + F = 0$ be an equation of a conic in an xy-coordinate system. Let the equation of the conic in the rotated $x'y'$-coordinate system be $A'x'^2 + B'x'y' + C'y'^2 + D'x' + E'y' + F' = 0$. Show that

$$B'^2 - 4A'C' = B^2 - 4AC$$

43. Using the result of Exercise 42, show that, except in degenerate cases,

$$B^2 - 4AC \begin{cases} < 0 \text{ for ellipses} \\ = 0 \text{ for parabolas} \\ > 0 \text{ for hyperbolas} \end{cases}$$

44. Derive Equation (2) of the rotation-of-axes formulas.

PREPARE FOR SECTION 6.5

---

45. Is $\sin x$ an even or odd function? [2.5]

46. Is $\cos x$ an even or odd function? [2.5]

47. Solve $\tan \alpha = -\sqrt{3}$ for $0 < \alpha < 2\pi$. [3.6]

48. If $\sin \alpha = -\dfrac{\sqrt{3}}{2}$ and $\cos \alpha = -\dfrac{1}{2}$, find α. Write the answer in degrees. [3.6]

49. Write $(r \cos \theta)^2 + (r \sin \theta)^2$ in simplest form. [3.1]

50. For the graph at the right, find the coordinates of point A. Round each coordinate to the nearest tenth. [2.2]

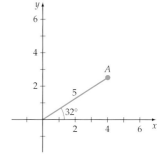

PROJECTS

---

1. **USE THE INVARIANT THEOREMS** The results of Exercises 41 and 42 illustrate that when the rotation theorem is used to transform equations of the form

$$Ax^2 + Bxy + Cy^2 - F = 0$$

to the form

$$A'(x')^2 + C'(y')^2 - F = 0$$

the following relationships hold:

$$A + C = A' + C' \quad \text{and} \quad B^2 - 4AC = (B')^2 - 4A'C'$$

Use these equations to transform $10x^2 + 24xy + 17y^2 - 26 = 0$ to the form

$$A'(x')^2 + C'(y')^2 - 26 = 0$$

without applying the rotation theorem.

INTRODUCTION TO POLAR COORDINATES

Until now, we have used a *rectangular coordinate system* to locate a point in the coordinate plane. An alternative method is to use a *polar coordinate system,* wherein a point is located by giving a distance from a fixed point and an angle from some fixed direction.

● THE POLAR COORDINATE SYSTEM

A **polar coordinate system** is formed by drawing a horizontal ray. The ray is called the **polar axis,** and the endpoint of the ray is called the **pole.** A point $P(r, \theta)$ in the plane is located by specifying a distance r from the pole and an angle θ measured from the polar axis to the line segment OP. The angle can be measured in degrees or radians. See **Figure 6.59.**

The coordinates of the pole are $(0, \theta)$, where θ is an arbitrary angle. Positive angles are measured counterclockwise from the polar axis. Negative angles are measured clockwise from the axis. Positive values of r are measured along the ray that makes an angle of θ from the polar axis. Negative values of r are measured along the ray that makes an angle of $\theta + 180°$ from the polar axis. See **Figures 6.60** and **6.61.**

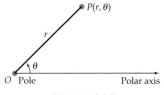

FIGURE 6.59

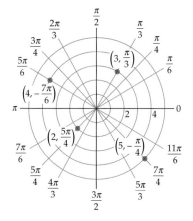

FIGURE 6.60

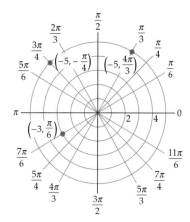

FIGURE 6.61

In a rectangular coordinate system, there is a one-to-one correspondence between the points in the plane and the ordered pairs (x, y). This is not true for a polar coordinate system. For polar coordinates, the relationship is one-to-many. Infinitely many ordered-pair descriptions correspond to each point $P(r, \theta)$ in a polar coordinate system.

For example, consider a point whose coordinates are $P(3, 45°)$. Because there are 360° in one complete revolution around a circle, the point P also could be written as $(3, 405°)$, as $(3, 765°)$, as $(3, 1125°)$, and generally as $(3, 45° + n \cdot 360°)$, where n is an integer. It is also possible to describe the point $P(3, 45°)$ by $(-3, 225°)$, $(-3, -135°)$, and $(3, -315°)$, to name just a few options.

The relationship between an ordered pair and a point is not one-to-many. That is, given an ordered pair (r, θ), there is exactly one point in the plane that corresponds to that ordered pair.

● GRAPHS OF EQUATIONS IN A POLAR COORDINATE SYSTEM

A **polar equation** is an equation in r and θ. A **solution** to a polar equation is an ordered pair (r, θ) that satisfies the equation. The **graph** of a polar equation is the set of all points whose ordered pairs are solutions of the equation.

The graph of the polar equation $\theta = \dfrac{\pi}{6}$ is a line. Because θ is independent of r, θ is $\dfrac{\pi}{6}$ radian from the polar axis for all values of r. The graph is a line that makes an angle of $\dfrac{\pi}{6}$ radian (30°) from the polar axis. See **Figure 6.62.**

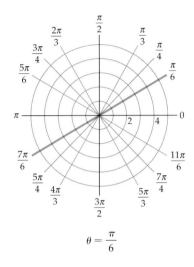

$\theta = \dfrac{\pi}{6}$

FIGURE 6.62

Polar Equations of a Line

The graph of $\theta = \alpha$ is a line through the pole at an angle of α from the polar axis. See **Figure 6.63a.**

The graph of $r \sin \theta = a$ is a horizontal line passing through the point $\left(a, \dfrac{\pi}{2}\right)$. See **Figure 6.63b.**

The graph of $r \cos \theta = a$ is a vertical line passing through the point $(a, 0)$. See **Figure 6.63c.**

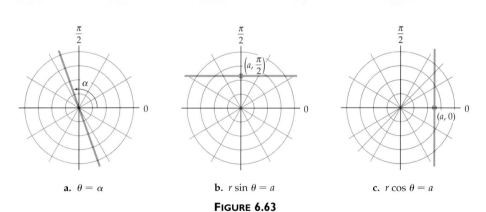

a. $\theta = \alpha$ **b.** $r \sin \theta = a$ **c.** $r \cos \theta = a$

FIGURE 6.63

Figure 6.64 is the graph of the polar equation $r = 2$. Because r is independent of θ, r is 2 units from the pole for all values of θ. The graph is a circle of radius 2 with center at the pole.

The Graph of $r = a$

The graph of $r = a$ is a circle with center at the pole and radius a.

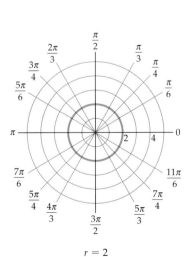

$r = 2$

FIGURE 6.64

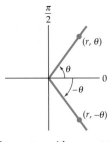

Symmetry with respect to
the line $\theta = 0$

FIGURE 6.65

Suppose that whenever the ordered pair (r, θ) lies on the graph of a polar equation, $(r, -\theta)$ also lies on the graph. From **Figure 6.65,** the graph will have symmetry with respect to the polar axis $\theta = 0$. Thus one test for symmetry is to replace θ by $-\theta$ in the polar equation. If the resulting equation is equivalent to the original equation, the graph is symmetric with respect to the polar axis.

Table 6.2 shows the types of symmetry and their associated tests. For each type, if the recommended substitution results in an equivalent equation, the graph will have the indicated symmetry. **Figure 6.66** illustrates the tests for symmetry with respect to the line $\theta = \dfrac{\pi}{2}$ and for symmetry with respect to the pole.

TABLE 6.2   Tests for Symmetry

Substitution	Symmetry with respect to
$-\theta$ for θ	The line $\theta = 0$
$\pi - \theta$ for θ, $-r$ for r	The line $\theta = 0$
$\pi - \theta$ for θ	The line $\theta = \dfrac{\pi}{2}$
$-\theta$ for θ, $-r$ for r	The line $\theta = \dfrac{\pi}{2}$
$-r$ for r	The pole
$\pi + \theta$ for θ	The pole

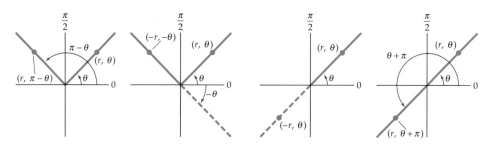

Symmetry with respect to the line $\theta = \dfrac{\pi}{2}$ Symmetry with respect to the pole

FIGURE 6.66

The graph of a polar equation may have a type of symmetry even though a test for that symmetry fails. For example, as we will see later, the graph of $r = \sin 2\theta$ is symmetric with respect to the line $\theta = 0$. However, using the symmetry test of substituting $-\theta$ for θ, we have

$$\sin 2(-\theta) = -\sin 2\theta = -r \neq r$$

Thus this test fails to show symmetry with respect to the line $\theta = 0$. The symmetry test of substituting $\pi - \theta$ for θ and $-r$ for r establishes symmetry with respect to the line $\theta = 0$.

EXAMPLE 1   **Graph a Polar Equation**

Show that the graph of $r = 4 \cos \theta$ is symmetric with respect to the polar axis. Graph the equation.

Solution

Test for symmetry with respect to the polar axis. Replace θ by $-\theta$.

$$r = 4 \cos(-\theta) = 4 \cos \theta \qquad \bullet \cos(-\theta) = \cos \theta$$

Because replacing θ by $-\theta$ results in the original equation $r = 4 \cos \theta$, the graph is symmetric with respect to the polar axis.

To graph the equation, begin choosing various values of θ and finding the corresponding values of r. However, before doing so, consider two further observations that will reduce the number of points you must choose.

First, because the cosine function is a periodic function with period 2π, it is only necessary to choose points between 0 and 2π (0° and 360°). Second, when $\dfrac{\pi}{2} < \theta < \dfrac{3\pi}{2}$, $\cos \theta$ is negative, which means that any θ between these values will produce a negative r. Thus the point will be in the first or fourth quadrant. That is, we need consider only angles θ in the first or fourth quadrants. However, because the graph is symmetric with respect to the polar axis, it is only necessary to choose values of θ between 0 and $\dfrac{\pi}{2}$.

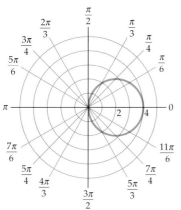

$r = 4 \cos \theta$

FIGURE 6.67

						By symmetry			
θ	0	$\dfrac{\pi}{6}$	$\dfrac{\pi}{4}$	$\dfrac{\pi}{3}$	$\dfrac{\pi}{2}$	$-\dfrac{\pi}{6}$	$-\dfrac{\pi}{4}$	$-\dfrac{\pi}{3}$	$-\dfrac{\pi}{2}$
r	4.0	3.5	2.8	2.0	0.0	3.5	2.8	2.0	0.0

The graph of $r = 4 \cos \theta$ is a circle with center at (2, 0). See **Figure 6.67.**

▶ **TRY EXERCISE 14, PAGE 409**

Polar Equations of a Circle

The graph of the equation $r = a$ is a circle with center at the pole and radius a. See **Figure 6.68a.**

The graph of the equation $r = a \cos \theta$ is a circle that is symmetric with respect to the line $\theta = 0$. See **Figure 6.68b.**

The graph of $r = a \sin \theta$ is a circle that is symmetric with respect to the line $\theta = \dfrac{\pi}{2}$. See **Figure 6.68c.**

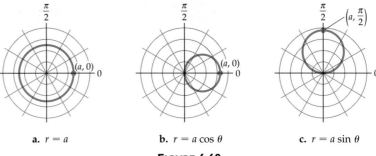

a. $r = a$ **b.** $r = a \cos \theta$ **c.** $r = a \sin \theta$

FIGURE 6.68

Just as there are specifically named curves in an xy-coordinate system (such as parabola and ellipse), there are named curves in an $r\theta$-coordinate system. Two of the many types are the *limaçon* and the *rose curve*.

Polar Equations of a Limaçon

The graph of the equation $r = a + b \cos \theta$ is a **limaçon** that is symmetric with respect to the line $\theta = 0$.

The graph of the equation $r = a + b \sin \theta$ is a limaçon that is symmetric with respect to the line $\theta = \dfrac{\pi}{2}$.

In the special case where $|a| = |b|$, the graph is called a **cardioid.**

The graph of $r = a + b \cos \theta$ is shown in **Figure 6.69** for various values of a and b.

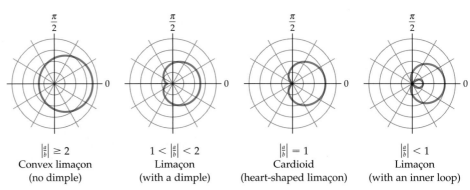

$\left|\dfrac{a}{b}\right| \geq 2$
Convex limaçon
(no dimple)

$1 < \left|\dfrac{a}{b}\right| < 2$
Limaçon
(with a dimple)

$\left|\dfrac{a}{b}\right| = 1$
Cardioid
(heart-shaped limaçon)

$\left|\dfrac{a}{b}\right| < 1$
Limaçon
(with an inner loop)

FIGURE 6.69

EXAMPLE 2   **Sketch the Graph of a Limaçon**

Sketch the graph of $r = 2 - 2 \sin \theta$.

Solution

From the general equation of a limaçon $r = a + b \sin \theta$ with $|a| = |b|$ ($|2| = |-2|$), the graph of $r = 2 - 2 \sin \theta$ is a cardioid that is symmetric with respect to the line $\theta = \dfrac{\pi}{2}$.

Because we know that the graph is heart-shaped, we can sketch the graph by finding r for a few values of θ. When $\theta = 0$, $r = 2$. When $\theta = \dfrac{\pi}{2}$, $r = 0$. When $\theta = \pi$, $r = 2$. When $\theta = \dfrac{3\pi}{2}$, $r = 4$. Sketching a heart-shaped curve through the four points

$$(2, 0), \quad \left(0, \frac{\pi}{2}\right), \quad (2, \pi), \quad \text{and} \quad \left(4, \frac{3\pi}{2}\right)$$

produces the cardioid in **Figure 6.70**.

▶ TRY EXERCISE 20, PAGE 409

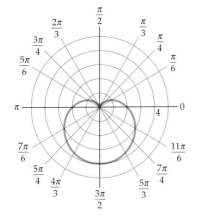

$r = 2 - 2 \sin \theta$

FIGURE 6.70

Example 3 gives the details necessary for using a graphing utility to construct a polar graph.

EXAMPLE 3   **Use a Graphing Utility to Sketch the Graph of a Limaçon**

 Use a graphing utility to graph $r = 3 - 2 \cos \theta$.

Solution

From the general equation of a limaçon $r = a + b \cos \theta$, with $a = 3$ and $b = -2$, we know that the graph will be a limaçon with a dimple. The graph will be symmetric with respect to the line $\theta = 0$.

Use polar mode with angle measure in radians. Enter the equation $r = 3 - 2 \cos \theta$ in the polar function editing menu. The graph in **Figure 6.71** was produced with a *TI-83* by using a window defined by the following:

θmin=0	Xmin=-6	Ymin=-4
θmax=2π	Xmax=6	Ymax=4
θstep=0.1	Xscl=1	Yscl=1

▶ TRY EXERCISE 26, PAGE 409

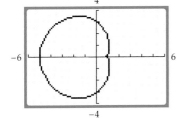

$r = 3 - 2 \cos \theta$

FIGURE 6.71

Polar Equations of Rose Curves

The graphs of the equations $r = a \cos n\theta$ and $r = a \sin n\theta$ are **rose curves.** When n is an even number, the number of petals is $2n$. See **Figure 6.72a.** When n is an odd number, the number of petals is n. See **Figure 6.72b.**

INTEGRATING TECHNOLOGY

When using a graphing utility in polar mode, choose the value of θstep carefully. If θstep is set too small, the graphing utility may require an excessively long period of time to complete the graph. If θstep is set too large, the resulting graph may give only a very rough approximation of the actual graph.

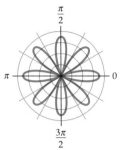
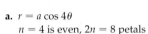
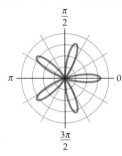

a. $r = a \cos 4\theta$
$n = 4$ is even, $2n = 8$ petals

b. $r = a \cos 5\theta$
$n = 5$ is odd, 5 petals

FIGURE 6.72

❓ QUESTION How many petals are in the graph of   **a.** $r = 4 \cos 3\theta$?
   b. $r = 5 \sin 2\theta$?

EXAMPLE 4   Sketch the Graph of a Rose Curve

Sketch the graph of $r = 2 \sin 3\theta$.

Solution

From the general equation of a rose curve $r = a \sin n\theta$, with $a = 2$ and $n = 3$, the graph of $r = 2 \sin 3\theta$ is a rose curve that is symmetric with respect to the line $\theta = \dfrac{\pi}{2}$. Because n is an odd number ($n = 3$), there will be three petals in the graph.

Choose some values for θ and find the corresponding values of r. Use symmetry to sketch the graph. See **Figure 6.73.**

θ	0	$\dfrac{\pi}{18}$	$\dfrac{\pi}{6}$	$\dfrac{5\pi}{18}$	$\dfrac{\pi}{3}$	$\dfrac{7\pi}{18}$	$\dfrac{\pi}{2}$
r	0.0	1.0	2.0	1.0	0.0	−1.0	−2.0

▶ TRY EXERCISE 16, PAGE 409

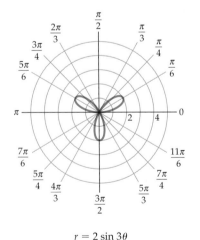

$r = 2 \sin 3\theta$

FIGURE 6.73

❓ ANSWER   **a.** Because 3 is an odd number, there are three petals in the graph. **b.** Because 2 is an even number, there are 2(2) = 4 petals in the graph.

EXAMPLE 5   Use a Graphing Utility to Sketch the Graph of a Rose Curve

    Use a graphing utility to graph $r = 4 \cos 2\theta$.

Solution

From the general equation of a rose curve $r = a \cos n\theta$, with $a = 4$ and $n = 2$, we know that the graph will be a rose curve with $2n = 4$ petals. The very tip of each petal will be $a = 4$ units away from the pole. Our symmetry tests also indicate that the graph is symmetric with respect to the line $\theta = 0$, the line $\theta = \dfrac{\pi}{2}$, and the pole.

Use polar mode with angle measure in radians. Enter the equation $r = 4 \cos 2\theta$ in the polar function editing menu. The graph in **Figure 6.74** was produced with a *TI-83* by using a window defined by the following:

θmin=0	Xmin=-6	Ymin=-4
θmax=2π	Xmax=6	Ymax=4
θstep=0.1	Xscl=1	Yscl=1

▶ **TRY EXERCISE 32, PAGE 409**

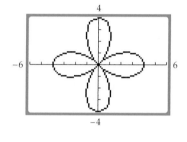

$r = 4 \cos 2\theta$

FIGURE 6.74

● **TRANSFORMATIONS BETWEEN RECTANGULAR AND POLAR COORDINATES**

A transformation between coordinate systems is a set of equations that relate the coordinates of a point in one system with the coordinates of the point in a second system. By superimposing a rectangular coordinate system on a polar system, we can derive the set of transformation equations.

Construct a polar coordinate system and a rectangular system such that the pole coincides with the origin and the polar axis coincides with the positive x-axis. Let a point P have coordinates (x, y) in one system and (r, θ) in the other ($r > 0$).

From the definitions of $\sin \theta$ and $\cos \theta$, we have

$$\frac{x}{r} = \cos \theta \quad \text{or} \quad x = r \cos \theta$$

$$\frac{y}{r} = \sin \theta \quad \text{or} \quad y = r \sin \theta$$

It can be shown that these equations are also true when $r < 0$.

Thus, given the point (r, θ) in a polar coordinate system (see **Figure 6.75**), the coordinates of the point in the xy-coordinate system are given by

$$x = r \cos \theta \qquad y = r \sin \theta$$

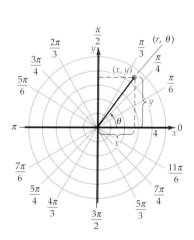

FIGURE 6.75

For example, to find the point in the xy-coordinate system that corresponds to the point $\left(4, \dfrac{2\pi}{3}\right)$ in the $r\theta$-coordinate system, substitute 4 for r and $\dfrac{2\pi}{3}$ for θ into the equations and simplify.

$$x = 4\cos\frac{2\pi}{3} = 4\left(-\frac{1}{2}\right) = -2$$

$$y = 4\sin\frac{2\pi}{3} = 4\left(\frac{\sqrt{3}}{2}\right) = 2\sqrt{3}$$

The point $\left(4, \dfrac{2\pi}{3}\right)$ in the $r\theta$-coordinate system is $\left(-2, 2\sqrt{3}\right)$ in the xy-coordinate system.

To find the polar coordinates of a given point in the xy-coordinate system, use the Pythagorean Theorem and the definition of the tangent function. Let $P(x, y)$ be a point in the plane, and let r be the distance from the origin to the point P. Then $r = \sqrt{x^2 + y^2}$.

From the definition of the tangent function of an angle in a right triangle,

$$\tan\theta = \frac{y}{x}$$

Thus θ is the angle whose tangent is $\dfrac{y}{x}$. The quadrant for θ depends on the sign of x and the sign of y.

The equations of transformation between a polar and a rectangular coordinate system are summarized as follows:

Transformations between Polar and Rectangular Coordinates

Given the point (r, θ) in the polar coordinate system, the transformation equations to change from polar to rectangular coordinates are

$$x = r\cos\theta \qquad y = r\sin\theta$$

Given the point (x, y) in the rectangular coordinate system, the transformation equations to change from rectangular to polar coordinates are

$$r = \sqrt{x^2 + y^2} \qquad \tan\theta = \frac{y}{x}, \quad x \neq 0$$

where $r \geq 0, 0 \leq \theta < 2\pi$, and θ is chosen so that the point lies in the appropriate quadrant. If $x = 0$, then $\theta = \dfrac{\pi}{2}$ or $\theta = \dfrac{3\pi}{2}$.

EXAMPLE 6 Transform from Polar to Rectangular Coordinates

Find the rectangular coordinates of the points whose polar coordinates

are: a. $\left(6, \dfrac{3\pi}{4}\right)$ b. $(-4, 30°)$

Solution

Use the equations $x = r \cos \theta$ and $y = r \sin \theta$.

a. $x = 6 \cos \dfrac{3\pi}{4} = -3\sqrt{2}$ $y = 6 \sin \dfrac{3\pi}{4} = 3\sqrt{2}$

The rectangular coordinates of $\left(6, \dfrac{3\pi}{4}\right)$ are $\left(-3\sqrt{2}, 3\sqrt{2}\right)$.

b. $x = -4 \cos 30° = -2\sqrt{3}$ $y = -4 \sin 30° = -2$

The rectangular coordinates of $(-4, 30°)$ are $\left(-2\sqrt{3}, -2\right)$.

▶ **TRY EXERCISE 44, PAGE 409**

EXAMPLE 7 Transform from Rectangular to Polar Coordinates

Find the polar coordinates of the point whose rectangular coordinates are $\left(-2, -2\sqrt{3}\right)$.

Solution

Use the equations $r = \sqrt{x^2 + y^2}$ and $\tan \theta = \dfrac{y}{x}$.

$$r = \sqrt{(-2)^2 + (-2\sqrt{3})^2} = \sqrt{4 + 12} = \sqrt{16} = 4$$

$$\tan \theta = \frac{-2\sqrt{3}}{-2} = \sqrt{3}$$

From this and the fact that $\left(-2, -2\sqrt{3}\right)$ lies in the third quadrant, $\theta = \dfrac{4\pi}{3}$.

The polar coordinates of $\left(-2, -2\sqrt{3}\right)$ are $\left(4, \dfrac{4\pi}{3}\right)$.

▶ **TRY EXERCISE 48, PAGE 409**

● **WRITE POLAR COORDINATE EQUATIONS AS RECTANGULAR EQUATIONS AND RECTANGULAR COORDINATE EQUATIONS AS POLAR EQUATIONS**

Using the transformation equations, it is possible to write a polar coordinate equation in rectangular form or a rectangular coordinate equation in polar form.

EXAMPLE 8   **Write a Polar Coordinate Equation in Rectangular Form**

Find a rectangular form of the equation $r^2 \cos 2\theta = 3$.

Solution

$$r^2 \cos 2\theta = 3$$

$$r^2(1 - 2\sin^2\theta) = 3 \qquad \bullet \cos 2\theta = 1 - 2\sin^2\theta$$

$$r^2 - 2r^2\sin^2\theta = 3$$

$$r^2 - 2(r\sin\theta)^2 = 3$$

$$x^2 + y^2 - 2y^2 = 3 \qquad \bullet r^2 = x^2 + y^2; \sin\theta = \frac{y}{r}$$

$$x^2 - y^2 = 3$$

A rectangular form of $r^2 \cos 2\theta = 3$ is $x^2 - y^2 = 3$.

▶ **TRY EXERCISE 56, PAGE 409**

EXAMPLE 9   **Write a Rectangular Coordinate Equation in Polar Form**

Find a polar form of the equation $x^2 + y^2 - 2x = 3$.

Solution

$$x^2 + y^2 - 2x = 3$$

$$(r\cos\theta)^2 + (r\sin\theta)^2 - 2r\cos\theta = 3 \qquad \bullet \text{ Use the transformation equations } x = r\cos\theta \text{ and } y = r\sin\theta.$$

$$r^2(\cos^2\theta + \sin^2\theta) - 2r\cos\theta = 3 \qquad \bullet \text{ Simplify.}$$

$$r^2 - 2r\cos\theta = 3$$

A polar form of $x^2 + y^2 - 2x = 3$ is $r^2 - 2r\cos\theta = 3$.

▶ **TRY EXERCISE 64, PAGE 409**

TOPICS FOR DISCUSSION

1. In what quadrant is the point $(-2, 150°)$ located?

2. To explain why the graph of $\theta = \dfrac{\pi}{6}$ is a line, a tutor rewrites the equation in the form $\theta = \dfrac{\pi}{6} + 0 \cdot r$. In this form, regardless of the value of r, $\theta = \dfrac{\pi}{6}$. Use an analogous approach to explain why the graph of $r = a$ is a circle.

3. Two students use a graphing calculator to graph the polar equation $r = 2\sin\theta$. One graph appears to be a circle, and the other graph appears to be an ellipse. Both graphs are correct. Explain.

4. Is the graph of $r^2 = 6\cos 2\theta$ a rose curve with four petals? Explain your answer.

EXERCISE SET 6.5

In Exercises 1 to 8, plot the point on a polar coordinate system.

1. $(2, 60°)$

2. $(3, -90°)$

3. $(1, 315°)$

4. $(2, 400°)$

5. $\left(-2, \dfrac{\pi}{4}\right)$

6. $\left(4, \dfrac{7\pi}{6}\right)$

7. $\left(-3, \dfrac{5\pi}{3}\right)$

8. $(-3, \pi)$

In Exercises 9 to 24, sketch the graph of each polar equation.

9. $r = 3$

10. $r = 5$

11. $\theta = 2$

12. $\theta = -\dfrac{\pi}{3}$

13. $r = 6\cos\theta$

▶ 14. $r = 4\sin\theta$

15. $r = 4\cos 2\theta$

▶ 16. $r = 5\cos 3\theta$

17. $r = 2\sin 5\theta$

18. $r = 3\cos 5\theta$

19. $r = 2 - 3\sin\theta$

▶ 20. $r = 2 - 2\cos\theta$

21. $r = 4 + 3\sin\theta$

22. $r = 2 + 4\sin\theta$

23. $r = 2(1 - 2\sin\theta)$

24. $r = 4(1 - \sin\theta)$

In Exercises 25 to 40, use a graphing utility to graph each equation.

25. $r = 3 + 3\cos\theta$

▶ 26. $r = 4 - 4\sin\theta$

27. $r = 4\cos 3\theta$

28. $r = 2\sin 4\theta$

29. $r = 3\sec\theta$

30. $r = 4\csc\theta$

31. $r = -5\csc\theta$

▶ 32. $r = -4\sec\theta$

33. $r = 4\sin(3.5\theta)$

34. $r = 6\cos(2.25\theta)$

35. $r = \theta, 0 \le \theta \le 6\pi$

36. $r = -\theta, 0 \le \theta \le 6\pi$

37. $r = 2^\theta, 0 \le \theta \le 2\pi$

38. $r = \dfrac{1}{\theta}, 0 \le \theta \le 4\pi$

39. $r = \dfrac{6\cos 7\theta + 2\cos 3\theta}{\cos\theta}$

40. $r = \dfrac{4\cos 3\theta + \cos 5\theta}{\cos\theta}$

In Exercises 41 to 48, transform the given coordinates to the indicated ordered pair.

41. $\left(1, -\sqrt{3}\right)$ to (r, θ)

42. $\left(-2\sqrt{3}, 2\right)$ to (r, θ)

43. $\left(-3, \dfrac{2\pi}{3}\right)$ to (x, y)

▶ 44. $\left(2, -\dfrac{\pi}{3}\right)$ to (x, y)

45. $\left(0, -\dfrac{\pi}{2}\right)$ to (x, y)

46. $\left(3, \dfrac{5\pi}{6}\right)$ to (x, y)

47. $(3, 4)$ to (r, θ)

▶ 48. $(12, -5)$ to (r, θ)

In Exercises 49 to 60, find a rectangular form of each of the equations.

49. $r = 3\cos\theta$

50. $r = 2\sin\theta$

51. $r = 3\sec\theta$

52. $r = 4\csc\theta$

53. $r = 4$

54. $\theta = \dfrac{\pi}{4}$

55. $r = \tan\theta$

▶ 56. $r = \cot\theta$

57. $r = \dfrac{2}{1 + \cos\theta}$

58. $r = \dfrac{2}{1 - \sin\theta}$

59. $r(\sin\theta - 2\cos\theta) = 6$

60. $r(2\cos\theta + \sin\theta) = 3$

In Exercises 61 to 68, find a polar form of each of the equations.

61. $y = 2$

62. $x = -4$

63. $x^2 + y^2 = 4$

▶ 64. $2x - 3y = 6$

65. $x^2 = 8y$

66. $y^2 = 4y$

67. $x^2 - y^2 = 25$

68. $x^2 + 4y^2 = 16$

In Exercises 69 to 76, use a graphing utility to graph each equation.

69. $r = 3\cos\left(\theta + \dfrac{\pi}{4}\right)$

70. $r = 2\sin\left(\theta - \dfrac{\pi}{6}\right)$

71. $r = 2\sin\left(2\theta - \dfrac{\pi}{3}\right)$

72. $r = 3\cos\left(2\theta + \dfrac{\pi}{4}\right)$

73. $r = 2 + 2 \sin\left(\theta - \dfrac{\pi}{6}\right)$ **74.** $r = 3 - 2 \cos\left(\theta + \dfrac{\pi}{3}\right)$ **75.** $r = 1 + 3 \cos\left(\theta + \dfrac{\pi}{3}\right)$ **76.** $r = 2 - 4 \sin\left(\theta - \dfrac{\pi}{4}\right)$

CONNECTING CONCEPTS

77. Explain why the graph of $r^2 = \cos^2 \theta$ and the graph of $r = \cos \theta$ are not the same.

78. Explain why the graph of $r = \cos 2\theta$ and the graph of $r = 2 \cos^2 \theta - 1$ are identical.

In Exercises 79 to 86, use a graphing utility to graph each equation.

79. $r^2 = 4 \cos 2\theta$ (lemniscate)

80. $r^2 = -2 \sin 2\theta$ (lemniscate)

81. $r = 2(1 + \sec \theta)$ (conchoid)

82. $r = 2 \cos 2\theta \sec \theta$ (strophoid)

83. $r\theta = 2$ (spiral)

84. $r = 2 \sin \theta \cos^2 2\theta$ (bifolium)

85. $r = |\theta|$

86. $r = \ln \theta$

87. The graph of

$$r = 1.5^{\sin \theta} - 2.5 \cos 4\theta + \sin^7 \dfrac{\theta}{15}$$

is a *butterfly curve* similar to the one shown below.

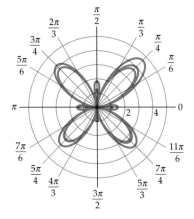

Use a graphing utility to graph the butterfly curve for

a. $0 \le \theta \le 5\pi$ **b.** $0 \le \theta \le 20\pi$

For additional information on butterfly curves, read "The Butterfly Curve" by Temple H. Fay, *The American Mathematical Monthly*, vol. 96, no. 5 (May 1989), p. 442.

PREPARE FOR SECTION 6.6

88. Find the eccentricity of the graph of $\dfrac{x^2}{25} + \dfrac{y^2}{16} = 1$. [6.2]

89. What is the equation of the directrix of the graph of $y^2 = 4x$? [6.1]

90. Solve $y = 2(1 + yx)$ for y. [1.1]

91. For the function $f(x) = \dfrac{3}{1 + \sin x}$, what is the smallest positive value of x that must be excluded from the domain of f? [2.2]

92. Let e be the eccentricity of a hyperbola. Which of the following statements is true: $e = 0$, $0 < e < 1$, $e = 1$, or $e > 1$? [6.3]

93. Write $\dfrac{4 \sec x}{2 \sec x - 1}$ in terms of $\cos x$. [3.1]

PROJECTS

1. A POLAR DISTANCE FORMULA Let $P_1(r_1, \theta_1)$ and $P_2(r_2, \theta_2)$ be two distinct points in the $r\theta$-plane.

a. Verify that the distance d between the points is
$$d = \sqrt{r_1^2 + r_2^2 - 2r_1r_2 \cos(\theta_2 - \theta_1)}$$

b. Use the above formula to find the distance (to the nearest hundredth) between $(3, 60°)$ and $(5, 170°)$.

c. Does the formula $d = \sqrt{r_1^2 + r_2^2 - 2r_1r_2 \cos(\theta_1 - \theta_2)}$ also produce the correct distance between P_1 and P_2? Explain.

2. ANOTHER POLAR FORM FOR A CIRCLE

a. Verify that the graph of the polar equation $r = a \sin \theta + b \cos \theta$ is a circle. Assume that a and b are not both 0.

b. What are the center (in rectangular coordinates) and the radius of the circle?

SECTION 6.6

POLAR EQUATIONS OF THE CONICS

● POLAR EQUATIONS OF THE CONICS

The definition of a parabola was given in terms of a point (the focus) and a line (the directrix). The definitions of both the ellipse and the hyperbola were given in terms of two points (the foci). It is possible to define each conic in terms of a point and a line.

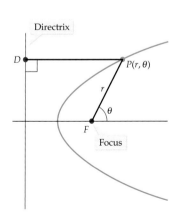

FIGURE 6.76

Focus-Directrix Definitions of the Conics

Let F be a fixed point and D a fixed line in a plane. Consider the set of all points P such that $\dfrac{d(P, F)}{d(P, D)} = e$, where e is a constant. The graph is a parabola for $e = 1$, an ellipse for $0 < e < 1$, and a hyperbola for $e > 1$. See **Figure 6.76.**

The fixed point is a focus of the conic, and the fixed line is a directrix. The constant e is the eccentricity of the conic. Using this definition, we can derive the polar equations of the conics.

Standard Form of the Polar Equations of the Conics

Let the pole be a focus of a conic section of eccentricity e with directrix d units from the focus. Then the equation of the conic is given by one of the following:

$$r = \frac{ed}{1 + e \cos \theta} \quad (1) \qquad\qquad r = \frac{ed}{1 - e \cos \theta} \quad (2)$$

Vertical directrix to the right of the pole

Vertical directrix to the left of the pole

$$r = \frac{ed}{1 + e \sin \theta} \quad (3) \qquad\qquad r = \frac{ed}{1 - e \sin \theta} \quad (4)$$

Horizontal directrix above the pole

Horizontal directrix below the pole

When the equation involves $\cos \theta$, the polar axis is an axis of symmetry.

When the equation involves $\sin \theta$, the line $\theta = \dfrac{\pi}{2}$ is an axis of symmetry.

Graphs of examples are shown in **Figure 6.77.**

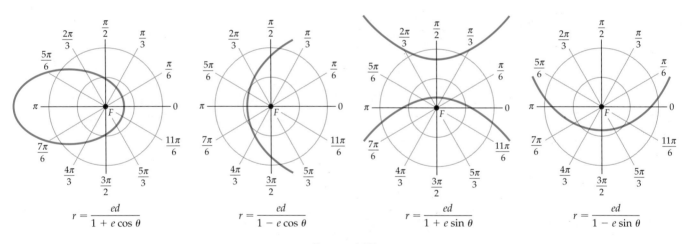

$$r = \frac{ed}{1 + e \cos \theta} \qquad r = \frac{ed}{1 - e \cos \theta} \qquad r = \frac{ed}{1 + e \sin \theta} \qquad r = \frac{ed}{1 - e \sin \theta}$$

FIGURE 6.77

? QUESTION Is the graph of $r = \dfrac{4}{5 - 3 \sin \theta}$ a parabola, an ellipse, or a hyperbola?

We will derive Equation (2). Let $P(r, \theta)$ be any point on a conic section. Then, by definition,

$$\frac{d(P, F)}{d(P, D)} = e \quad \text{or} \quad d(P, F) = e \cdot d(P, D)$$

? ANSWER The eccentricity is $\dfrac{3}{5}$, which is less than 1. The graph is an ellipse.

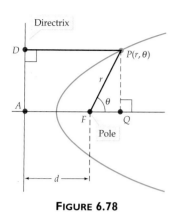

FIGURE 6.78

From **Figure 6.78**, $d(P, F) = r$ and $d(P, D) = d(A, Q)$. But note that

$$d(A, Q) = d(A, F) + d(F, Q) = d + r \cos \theta$$

Thus

$$r = e(d + r \cos \theta) \qquad \cdot \; d(P, F) = e \cdot d(P, D)$$

$$= ed + er \cos \theta$$

$$r - er \cos \theta = ed \qquad \cdot \; \text{Subtract } er \cos \theta.$$

$$r = \frac{ed}{1 - e \cos \theta} \qquad \cdot \; \text{Solve for } r.$$

The remaining equations can be derived in a similar manner.

• GRAPH A CONIC GIVEN IN POLAR FORM

EXAMPLE 1   **Sketch the Graph of a Hyperbola Given in Polar Form**

Describe and sketch the graph of $r = \dfrac{8}{2 - 3 \sin \theta}$.

Solution

Write the equation in standard form by dividing the numerator and denominator by 2, the constant term in the denominator.

$$r = \frac{4}{1 - \dfrac{3}{2} \sin \theta}$$

Because e is the coefficient of $\sin \theta$ and $e = \dfrac{3}{2} > 1$, the graph is a hyperbola with a focus at the pole. Because the equation contains the expression $\sin \theta$, the transverse axis is on the line $\theta = \dfrac{\pi}{2}$.

To find the vertices, choose θ equal to $\dfrac{\pi}{2}$ and $\dfrac{3\pi}{2}$. The corresponding values of r are -8 and $\dfrac{8}{5}$. The vertices are $\left(-8, \dfrac{\pi}{2} \right)$ and $\left(\dfrac{8}{5}, \dfrac{3\pi}{2} \right)$. By choosing θ equal to 0 and π, we can determine the points $(4, 0)$ and $(4, \pi)$ on the upper branch of the hyperbola. The lower branch can be determined by symmetry.

Plot some points (r, θ) for additional values of θ and corresponding values of r. See **Figure 6.79**.

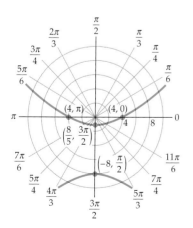

FIGURE 6.79

▶ **TRY EXERCISE 2, PAGE 416**

EXAMPLE 2   Sketch the Graph of an Ellipse Given in Polar Form

Describe and sketch the graph of $r = \dfrac{4}{2 + \cos\theta}$.

Solution

Write the equation in standard form by dividing the numerator and denominator by 2, which is the constant term in the denominator.

$$r = \dfrac{2}{1 + \dfrac{1}{2}\cos\theta}$$

Thus $e = \dfrac{1}{2}$ and the graph is an ellipse with a focus at the pole. Because the equation contains the expression $\cos\theta$, the major axis is on the polar axis.

To find the vertices, choose θ equal to 0 and π. The corresponding values for r are $\dfrac{4}{3}$ and 4. The vertices on the major axis are $\left(\dfrac{4}{3}, 0\right)$ and $(4, \pi)$. Plot some points (r, θ) for additional values of θ and the corresponding values of r. Two possible points are $\left(2, \dfrac{\pi}{2}\right)$ and $\left(2, \dfrac{3\pi}{2}\right)$. See the graph of the ellipse in **Figure 6.80.**

▶ TRY EXERCISE 4, PAGE 416

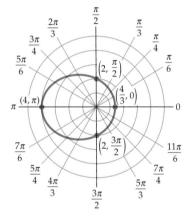

FIGURE 6.80

• WRITE THE POLAR EQUATION OF A CONIC

EXAMPLE 3   Find the Equation of a Conic in Polar Form

Find the equation of the parabola, shown in **Figure 6.81,** with vertex at $\left(2, \dfrac{\pi}{2}\right)$ and focus at the pole.

Solution

Because the vertex is on the line $\theta = \dfrac{\pi}{2}$ and the focus is at the pole, the axis of symmetry is the line $\theta = \dfrac{\pi}{2}$. Thus the equation of the parabola must involve $\sin\theta$. The parabola has a horizontal directrix above the pole, so the equation has the form

$$r = \dfrac{ed}{1 + e\sin\theta}$$

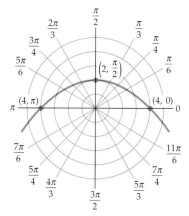

The distance from the vertex to the focus is 2, so the distance from the focus to the directrix is 4. Because the graph of the equation is a parabola, the eccentricity is 1. The equation is

$$r = \frac{(1)(4)}{1 + (1)\sin\theta} \qquad \bullet\ e = 1, d = 4$$

$$r = \frac{4}{1 + \sin\theta}$$

▶ **TRY EXERCISE 24, PAGE 416**

$(2, \frac{\pi}{2})$

$(4, \pi)$

$(4, 0)$

FIGURE 6.81

❓ **QUESTION** In Example 3, why is there no point on the parabola that corresponds to $\theta = \dfrac{3\pi}{2}$?

TOPICS FOR DISCUSSION

1. Is the graph of $r = \dfrac{12}{2 + \cos\theta}$ a parabola? Explain.

2. The graph of $r = \dfrac{2\sec\theta}{2\sec\theta + 1}$ is an ellipse except for the fact that it has two holes. Where are the holes located?

3. Are there two different ellipses that have a focus at the pole and a vertex at $\left(1, \dfrac{\pi}{2}\right)$? Explain.

4. Does the parabola given by $r = \dfrac{6}{1 + \sin\theta}$ have a horizontal axis of symmetry or a vertical axis of symmetry? Explain.

❓ **ANSWER** When $\theta = \dfrac{3\pi}{2}$, $\sin\theta = -1$. Thus $1 + \sin\theta = 0$, and $r = \dfrac{4}{1 + \sin\theta}$ is undefined.

EXERCISE SET 6.6

In Exercises 1 to 14, describe and sketch the graph of each equation.

1. $r = \dfrac{12}{3 - 6\cos\theta}$

▶ **2.** $r = \dfrac{8}{2 - 4\cos\theta}$

3. $r = \dfrac{8}{4 + 3\sin\theta}$

▶ **4.** $r = \dfrac{6}{3 + 2\cos\theta}$

5. $r = \dfrac{9}{3 - 3\sin\theta}$

6. $r = \dfrac{5}{2 - 2\sin\theta}$

7. $r = \dfrac{10}{5 + 6\cos\theta}$

8. $r = \dfrac{8}{2 + 4\cos\theta}$

9. $r = \dfrac{4\sec\theta}{2\sec\theta - 1}$

10. $r = \dfrac{3\sec\theta}{2\sec\theta + 2}$

11. $r = \dfrac{12\csc\theta}{6\csc\theta - 2}$

12. $r = \dfrac{3\csc\theta}{2\csc\theta + 2}$

13. $r = \dfrac{3}{\cos\theta - 1}$

14. $r = \dfrac{2}{\sin\theta + 2}$

In Exercises 15 to 20, find a rectangular equation for the graphs in Exercises 1 to 6.

In Exercises 21 to 28, find a polar equation of the conic with focus at the pole and the given eccentricity and directrix.

21. $e = 2, r\cos\theta = -1$

22. $e = \dfrac{3}{2}, r\sin\theta = 1$

23. $e = 1, r\sin\theta = 2$

▶ **24.** $e = 1, r\cos\theta = -2$

25. $e = \dfrac{2}{3}, r\sin\theta = -4$

26. $e = \dfrac{1}{2}, r\cos\theta = 2$

27. $e = \dfrac{3}{2}, r = 2\sec\theta$

28. $e = \dfrac{3}{4}, r = 2\csc\theta$

29. Find the polar equation of the parabola with a focus at the pole and vertex $(2, \pi)$.

30. Find the polar equation of the ellipse with a focus at the pole, vertex at $(4, 0)$, and eccentricity $\dfrac{1}{2}$.

31. Find the polar equation of the hyperbola with a focus at the pole, vertex at $\left(1, \dfrac{3\pi}{2}\right)$, and eccentricity 2.

32. Find the polar equation of the ellipse with a focus at the pole, vertex at $\left(2, \dfrac{3\pi}{2}\right)$, and eccentricity $\dfrac{2}{3}$.

In Exercises 33 to 40, use a graphing utility to graph each equation. Write a sentence that explains how to obtain the graph from the graph of r as given in the exercise listed to the right of each equation.

33. $r = \dfrac{12}{3 - 6\cos\left(\theta - \dfrac{\pi}{6}\right)}$ (Compare with Exercise 1.)

34. $r = \dfrac{8}{2 - 4\cos\left(\theta - \dfrac{\pi}{2}\right)}$ (Compare with Exercise 2.)

35. $r = \dfrac{8}{4 + 3\sin(\theta - \pi)}$ (Compare with Exercise 3.)

36. $r = \dfrac{6}{3 + 2\cos\left(\theta - \dfrac{\pi}{3}\right)}$ (Compare with Exercise 4.)

37. $r = \dfrac{9}{3 - 3\sin\left(\theta + \dfrac{\pi}{6}\right)}$ (Compare with Exercise 5.)

38. $r = \dfrac{5}{2 - 2\sin\left(\theta + \dfrac{\pi}{2}\right)}$ (Compare with Exercise 6.)

39. $r = \dfrac{10}{5 + 6\cos(\theta + \pi)}$ (Compare with Exercise 7.)

40. $r = \dfrac{8}{2 + 4\cos\left(\theta + \dfrac{\pi}{3}\right)}$ (Compare with Exercise 8.)

CONNECTING CONCEPTS

 In Exercises 41 to 46, use a graphing utility to graph each equation.

41. $r = \dfrac{3}{3 - \sec \theta}$

42. $r = \dfrac{5}{4 - 2 \csc \theta}$

43. $r = \dfrac{3}{1 + 2 \csc \theta}$

44. $r = \dfrac{4}{1 + 3 \sec \theta}$

45. $r = 4 \sin \sqrt{2}\theta, 0 \le \theta \le 12\pi$

46. $r = 4 \cos \sqrt{3}\theta, 0 \le \theta \le 8\pi$

47. Let $P(r, \theta)$ satisfy the equation $r = \dfrac{ed}{1 - e \cos \theta}$. Show that $\dfrac{d(P, F)}{d(P, D)} = e$.

48. Show that the equation of a conic with a focus at the pole and directrix $r \sin \theta = d$ is given by $r = \dfrac{ed}{1 + e \sin \theta}$.

PREPARE FOR SECTION 6.7

49. Complete the square of $y^2 + 3y$ and write the result as the square of a binomial. [1.1]

50. If $x = 2t + 1$ and $y = x^2$, write y in terms of t. [1.3]

51. Identify the graph of $\left(\dfrac{x - 2}{3}\right)^2 + \left(\dfrac{y - 3}{2}\right)^2 = 1$. [6.2]

52. Let $x = \sin t$ and $y = \cos t$. What is the value of $x^2 + y^2$? [3.1]

53. Solve $y = \ln t$ for t. [5.5]

54. What are the domain and range of $f(t) = 3 \cos 2t$? Write the answer using interval notation. [2.5]

PROJECTS

1. **POLAR EQUATION OF A LINE** Verify that the polar equation of a line that is d units from the pole is given by

$$r = \dfrac{d}{\cos(\theta - \theta_p)}$$

where θ_p is the angle from the polar axis to a line segment that passes through the pole and is perpendicular to the line.

2. **POLAR EQUATION OF A CIRCLE THAT PASSES THROUGH THE POLE** Verify that the polar equation of a circle with center (a, θ_c) that passes through the pole is given by

$$r = 2a \cos(\theta - \theta_c)$$

PARAMETRIC EQUATIONS

• PARAMETRIC EQUATIONS

The graph of a function is a graph for which no vertical line can intersect the graph more than once. For a graph that is not the graph of a function (an ellipse or a hyperbola, for example), it is frequently useful to describe the graph by *parametric equations.*

Curve and Parametric Equations

Let t be a number in an interval I. A **curve** is the set of ordered pairs (x, y), where

$$x = f(t), \qquad y = g(t) \quad \text{for } t \in I$$

The variable t is called a **parameter,** and the equations $x = f(t)$ and $y = g(t)$ are **parametric equations.**

For instance,

$$x = 2t - 1, \qquad y = 4t + 1 \quad \text{for } t \in (-\infty, \infty)$$

is an example of a set of parametric equations. By choosing arbitrary values of t, ordered pairs (x, y) can be created, as shown in the table below.

t	$x = 2t - 1$	$y = 4t + 1$	(x, y)
-2	-5	-7	$(-5, -7)$
0	-1	1	$(-1, 1)$
$\dfrac{1}{2}$	0	3	$(0, 3)$
2	3	9	$(3, 9)$

By plotting the points and drawing a curve through the points, a graph of the parametric equations is produced. See **Figure 6.82.**

❓ QUESTION If $x = t^2 + 1$ and $y = 3 - t$, what ordered pair corresponds to $t = -3$?

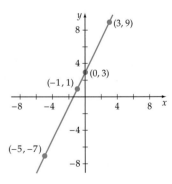

FIGURE 6.82

❓ ANSWER $(10, 6)$

● GRAPH A CURVE GIVEN BY PARAMETRIC EQUATIONS

EXAMPLE 1   Sketch the Graph of a Curve Given in Parametric Form

Sketch the graph of the curve given by the parametric equations

$$x = t^2 + t, \quad y = t - 1 \quad \text{for } t \in (-\infty, \infty)$$

Solution

Begin by making a table of values of t and the corresponding values of x and y. Five values of t were arbitrarily chosen for the table that follows. Many more values might be necessary to determine an accurate graph.

t	$x = t^2 + t$	$y = t - 1$	(x, y)
-2	2	-3	$(2, -3)$
-1	0	-2	$(0, -2)$
0	0	-1	$(0, -1)$
1	2	0	$(2, 0)$
2	6	1	$(6, 1)$

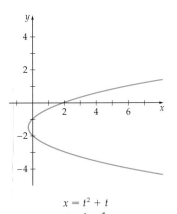

$$x = t^2 + t$$
$$y = t - 1$$

FIGURE 6.83

Graph the ordered pairs (x, y) and then draw a smooth curve through the points. See **Figure 6.83**.

▶ TRY EXERCISE 6, PAGE 424

● ELIMINATE THE PARAMETER OF A PARAMETRIC EQUATION

It may not be clear from Example 1 and the corresponding graph that the curve is a parabola. By **eliminating the parameter**, we can write one equation in x and y that is equivalent to the two parametric equations.

To eliminate the parameter, solve $y = t - 1$ for t.

$$y = t - 1 \quad \text{or} \quad t = y + 1$$

Substitute $y + 1$ for t in $x = t^2 + t$ and then simplify.

$$x = (y + 1)^2 + (y + 1)$$
$$= y^2 + 3y + 2 \qquad \bullet \text{The equation of a parabola}$$

Complete the square and write the equation in standard form.

$$\left(x + \frac{1}{4}\right) = \left(y + \frac{3}{2}\right)^2 \qquad \bullet \text{This is the equation of a parabola with vertex at } \left(-\frac{1}{4}, -\frac{3}{2}\right).$$

EXAMPLE 2 **Eliminate the Parameter and Sketch the Graph of a Curve**

Eliminate the parameter and sketch the curve of the parametric equations

$$x = \sin t, \qquad y = \cos t \quad \text{for } 0 \le t < 2\pi$$

Solution

The process of eliminating the parameter sometimes involves trigonometric identities. To eliminate the parameter for the equations, square each side of each equation and then add.

$$x^2 = \sin^2 t$$

$$y^2 = \cos^2 t$$

$$x^2 + y^2 = \sin^2 t + \cos^2 t$$

Thus, using the trigonometric identity $\sin^2 t + \cos^2 t = 1$, we get

$$x^2 + y^2 = 1$$

This is the equation of a circle with center $(0, 0)$ and radius equal to 1. See **Figure 6.84.**

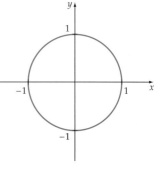

FIGURE 6.84

▶ **TRY EXERCISE 12, PAGE 424**

A parametric representation of a curve is not unique. That is, it is possible that a curve may be given by many different pairs of parametric equations. We will demonstrate this by using the equation of a line and providing two different parametric representations of the line.

Consider a line with slope m passing through the point (x_1, y_1). By the point-slope formula, the equation of the line is

$$y - y_1 = m(x - x_1)$$

Let $t = x - x_1$. Then $y - y_1 = mt$. A parametric representation is

$$x = x_1 + t, \qquad y = y_1 + mt \quad \text{for } t \text{ a real number} \tag{1}$$

Let $x - x_1 = \cot t$. Then $y - y_1 = m \cot t$. A parametric representation is

$$x = x_1 + \cot t, \qquad y = y_1 + m \cot t \quad \text{for } 0 < t < \pi \tag{2}$$

It can be verified that Equations (1) and (2) represent the original line.

Example 3 illustrates that the domain of the parameter t can be used to determine the domain and range of the function.

| EXAMPLE 3 | **Sketch the Graph of a Curve Given by Parametric Equations** |

Eliminate the parameter and sketch the graph of the curve that is given by the parametric equations

$$x = 2 + 3\cos t, \qquad y = 3 + 2\sin t \quad \text{for } 0 \le t \le \pi$$

Solution

Rewrite each equation in terms of the trigonometric function.

$$\frac{x-2}{3} = \cos t \qquad \frac{y-3}{2} = \sin t \tag{3}$$

Using the trigonometric identity $\cos^2 t + \sin^2 t = 1$, we have

$$\cos^2 t + \sin^2 t = \left(\frac{x-2}{3}\right)^2 + \left(\frac{y-3}{2}\right)^2 = 1$$

$$\frac{(x-2)^2}{9} + \frac{(y-3)^2}{4} = 1$$

This is the equation of an ellipse with center at (2, 3) and major axis parallel to the x-axis. However, because $0 \le t \le \pi$, it follows that $-1 \le \cos t \le 1$ and $0 \le \sin t \le 1$. Therefore, we have

$$-1 \le \frac{x-2}{3} \le 1 \qquad 0 \le \frac{y-3}{2} \le 1 \qquad \bullet \text{ **Using Equations (3)**}$$

Solving these inequalities for x and y yields

$$-1 \le x \le 5 \qquad \text{and} \qquad 3 \le y \le 5$$

Because the values of y are between 3 and 5, the graph of the parametric equations is only the top half of the ellipse. See **Figure 6.85**.

▶ **TRY EXERCISE 14, PAGE 424**

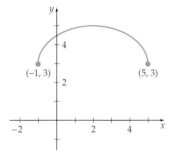

FIGURE 6.85

● THE BRACHISTOCHRONE PROBLEM

Parametric equations are useful in writing the equation of a moving point. One famous problem, involving a bead traveling down a frictionless wire, was posed in 1696 by the mathematician Johann Bernoulli. The problem was to determine the shape of a wire a bead could slide down such that the distance between two points was traveled in the shortest time. Problems that involve "shortest time" are called *brachistochrone problems*. They are very important in physics and form the basis for much of the classical theory of light propagation.

The answer to Bernoulli's problem is an arc of an inverted cycloid. See **Figure 6.86**. A **cycloid** is formed by letting a circle of radius a roll on a straight line L without slipping. See **Figure 6.87**. The curve traced by a point on the circumference of the circle is a cycloid. To find an equation for this curve, begin by placing a circle tangent to the x-axis with a point P on the circle and at the origin of a rectangular coordinate system.

FIGURE 6.86

Roll the circle along the *x*-axis. After the radius of the circle has rotated through an angle θ, the coordinates of the point $P(x, y)$ can be given by

$$x = h - a \sin \theta, \qquad y = k - a \cos \theta \qquad (4)$$

where $C(h, k)$ is the current center of the circle.

Because the radius of the circle is a, $k = a$. See **Figure 6.87.** Because the circle rolls without slipping, the arc length subtended by θ equals h. Thus $h = a\theta$. Substituting for h and k in Equations (4), we have, after factoring,

$$x = a(\theta - \sin \theta), \qquad y = a(1 - \cos \theta) \quad \text{for } \theta \geq 0$$

See **Figure 6.88.**

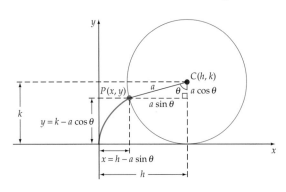

FIGURE 6.87

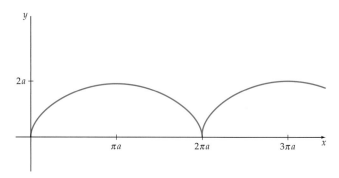

$$x = a(\theta - \sin \theta), y = a(1 - \cos \theta)$$

FIGURE 6.88

A cycloid

EXAMPLE 4   **Graph a Cycloid**

 Use a graphing utility to graph the cycloid given by

$$x = 4(\theta - \sin \theta), \qquad y = 4(1 - \cos \theta) \quad \text{for } 0 \leq \theta \leq 4\pi$$

Solution

Although θ is the parameter in the above equations, many graphing utilities, such as the TI-83, use T as the parameter for parametric equations. Thus to graph the equations for $0 \leq \theta \leq 4\pi$, we use Tmin = 0 and Tmax = 4π, as shown below. Use radian mode and parametric mode to produce the graph in **Figure 6.89.**

Tmin=0	Xmin=-6	Ymin=-4
Tmax=4π	Xmax=16π	Ymax=10
Tstep=0.5	Xscl=2π	Yscl=1

▶ **TRY EXERCISE 26, PAGE 424**

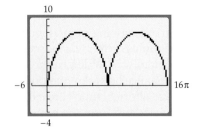

$$x = 4(\theta - \sin \theta)$$
$$y = 4(1 - \cos \theta)$$

FIGURE 6.89

● PARAMETRIC EQUATIONS AND PROJECTILE MOTION

The path of a projectile (assume air resistance is negligible) that is launched at an angle θ from the horizon with an initial velocity of v_0 feet per second is given by

the parametric equations

$$x = (v_0 \cos \theta)t, \qquad y = -16t^2 + (v_0 \sin \theta)t$$

where t is the time in seconds since the projectile was launched.

EXAMPLE 5   Sketch the Path of a Projectile

Use a graphing utility to sketch the path of a projectile that is launched at an angle of $\theta = 32°$ with an initial velocity of 144 feet per second. Use the graph to determine (to the nearest foot) the maximum height of the projectile and the range of the projectile. Assume the ground is level.

Solution

Use degree mode and parametric mode. Graph the parametric equations

$$x = (144 \cos 32°)t, \qquad y = -16t^2 + (144 \sin 32°)t \quad \text{for } 0 \le t \le 5$$

to produce the graph in **Figure 6.90.** Use the TRACE feature to determine that the maximum height of 91 feet is attained when $t \approx 2.38$ seconds and that the projectile strikes the ground about 581 feet downrange when $t \approx 4.76$ seconds.

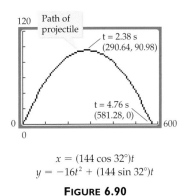

$$x = (144 \cos 32°)t$$
$$y = -16t^2 + (144 \sin 32°)t$$

FIGURE 6.90

In **Figure 6.90,** the angle of launch does not appear to be 32° because 1 foot on the x-axis is smaller than 1 foot on the y-axis.

 TRY EXERCISE 32, PAGE 425

TOPICS FOR DISCUSSION

1. It is always possible to eliminate the parameter of a pair of parametric equations. Do you agree? Explain.

2. The line $y = 3x + 5$ has more than one parametric representation. Do you agree? Explain.

3. Parametric equations are used only to graph functions. Do you agree? Explain.

4. Every function $y = f(x)$ can be written in parametric form by letting $x = t$ and $y = f(t)$. Do you agree? Explain.

EXERCISE SET 6.7

In Exercises 1 to 10, graph the parametric equations by plotting several points.

1. $x = 2t, y = -t$, for $t \in R$

2. $x = -3t, y = 6t$, for $t \in R$

3. $x = -t, y = t^2 - 1$, for $t \in R$

4. $x = 2t, y = 2t^2 - t + 1$, for $t \in R$

5. $x = t^2, y = t^3$, for $t \in R$

▶**6.** $x = t^2 + 1, y = t^2 - 1$, for $t \in R$

7. $x = 2 \cos t, y = 3 \sin t$, for $0 \le t < 2\pi$

8. $x = 1 - \sin t, y = 1 + \cos t$, for $0 \le t < 2\pi$

9. $x = 2^t, y = 2^{t+1}$, for $t \in R$

10. $x = t^2, y = 2 \log_2 t$, for $t \ge 1$

In Exercises 11 to 20, eliminate the parameter and graph the equation.

11. $x = \sec t, y = \tan t$, for $-\dfrac{\pi}{2} < t < \dfrac{\pi}{2}$

▶**12.** $x = 3 + 2 \cos t, y = -1 - 3 \sin t$, for $0 \le t < 2\pi$

13. $x = 2 - t^2, y = 3 + 2t^2$, for $t \in R$

▶**14.** $x = 1 + t^2, y = 2 - t^2$, for $t \in R$

15. $x = \cos^3 t, y = \sin^3 t$, for $0 \le t < 2\pi$

16. $x = e^{-t}, y = e^t$, for $t \in R$

17. $x = \sqrt{t + 1}, y = t$, for $t \ge -1$

18. $x = \sqrt{t}, y = 2t - 1$, for $t \ge 0$

19. $x = t^3, y = 3 \ln t$, for $t > 0$

20. $x = e^t, y = e^{2t}$, for $t \in R$

21. Eliminate the parameter for the curves

$$C_1: \quad x = 2 + t^2, \quad y = 1 - 2t^2$$

and $\qquad C_2: \quad x = 2 + t, \quad y = 1 - 2t$

and then discuss the differences between their graphs.

22. Eliminate the parameter for the curves

$$C_1: \quad x = \sec^2 t, \quad y = \tan^2 t$$

and $\qquad C_2: \quad x = 1 + t^2, \quad y = t^2$

for $0 \le t < \dfrac{\pi}{2}$, and then discuss the differences between their graphs.

23. Sketch the graph of

$$x = \sin t, \qquad y = \csc t \quad \text{for } 0 < t \le \dfrac{\pi}{2}$$

Sketch another graph for the same pair of equations but choose the domain of t to be $\pi < t \le \dfrac{3\pi}{2}$.

24. Discuss the differences between

$$C_1: \quad x = \cos t, \quad y = \cos^2 t$$

and $\qquad C_2: \quad x = \sin t, \quad y = \sin^2 t$

for $0 \le t \le \pi$.

25. Use a graphing utility to graph the cycloid $x = 2(t - \sin t), y = 2(1 - \cos t)$ for $0 \le t < 2\pi$.

▶**26.** Use a graphing utility to graph the cycloid $x = 3(t - \sin t), y = 3(1 - \cos t)$ for $0 \le t \le 12\pi$.

Parametric equations of the form $x = a \sin \alpha t$, $y = b \cos \beta t$, **for** $t \ge 0$, **are encountered in electrical circuit theory. The graphs of these equations are called** *Lissajous figures.* **Use Tstep = 0.5 and** $0 \le T \le 2\pi$.

27. Graph: $x = 5 \sin 2t, y = 5 \cos t$

28. Graph: $x = 5 \sin 3t, y = 5 \cos 2t$

29. Graph: $x = 4 \sin 2t, y = 4 \cos 3t$

30. Graph: $x = 5 \sin 10t, y = 5 \cos 9t$

In Exercises 31 to 34, graph the path of the projectile that is launched at an angle of θ with the horizon with an initial velocity of v_0. In each exercise, use the graph to determine the maximum height and the range of the projectile (to the nearest foot). Also state the time t at which the projectile reaches its maximum height and the time it hits the ground. Assume the ground is level and the only force acting on the projectile is gravity.

31. $\theta = 55°$, $v_0 = 210$ feet per second

▶ **32.** $\theta = 35°$, $v_0 = 195$ feet per second

33. $\theta = 42°$, $v_0 = 315$ feet per second

34. $\theta = 52°$, $v_0 = 315$ feet per second

CONNECTING CONCEPTS

35. Let $P_1(x_1, y_1)$ and $P_2(x_2, y_2)$ be two distinct points in the plane, and consider the line L passing through those points. Choose a point $P(x, y)$ on the line L. Show that

$$\frac{x - x_1}{x_2 - x_1} = \frac{y - y_1}{y_2 - y_1}$$

Use this result to demonstrate that $x = (x_2 - x_1)t + x_1$, $y = (y_2 - y_1)t + y_1$ is a parametric representation of the line through the two points.

36. Show that $x = h + a \sin t$, $y = k + b \cos t$, for $a > 0$, $b > 0$, and $0 \leq t < 2\pi$, are parametric equations for an ellipse with center at (h, k).

37. Suppose a string, held taut, is unwound from the circumference of a circle of radius a. The path traced by the end of the string is called the *involute* of a circle. Find parametric equations for the involute of a circle.

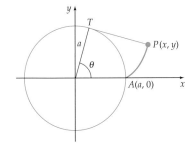

38. A circle of radius a rolls without slipping on the outside of a circle of radius $b > a$. Find the parametric equations of a point P on the smaller circle. The curve is called an *epicycloid*.

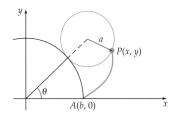

39. A circle of radius a rolls without slipping on the inside of a circle of radius $b > a$. Find the parametric equations of a point P on the smaller circle. The curve is called a *hypocycloid*.

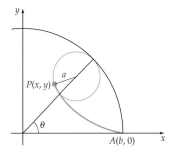

PROJECTS

1. PARAMETRIC EQUATIONS IN AN *XYZ*-COORDINATE SYSTEM

a. Graph the three-dimensional curve given by

$$x = 3 \cos t, \qquad y = 3 \sin t, \qquad z = 0.5t$$

b. Graph the three-dimensional curve given by

$$x = 3 \cos t, \qquad y = 6 \sin t, \qquad z = 0.5t$$

c. What is the main difference between these curves?

d. What name is given to curves of this type?

EXPLORING CONCEPTS
WITH TECHNOLOGY

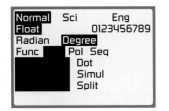

FIGURE 6.91

Using a Graphing Calculator to Find the *n*th Roots of *z*

In Chapter 5 we used De Moivre's Theorem to find the *n*th roots of a number. The parametric feature of a graphing calculator can also be used to find and display the *n*th roots of $z = r(\cos \theta + i \sin \theta)$. Here is the procedure for a TI-83 graphing calculator. Put the calculator in parametric and degree mode. See **Figure 6.91**. To find the *n*th roots of $z = r(\cos \theta + i \sin \theta)$, enter in the Y= menu

$$X_{1T} = r\textasciicircum(1/n)\cos(\theta/n+T) \quad \text{and} \quad Y_{1T} = r\textasciicircum(1/n)\sin(\theta/n+T)$$

In the WINDOW menu, set Tmin=0, Tmax=360, and Tstep=360/n. Set Xmin, Xmax, Ymin, and Ymax to appropriate values that will allow the roots to be seen in the graph window. Press GRAPH to display a polygon. The *x*- and *y*-coordinates of each vertex of the polygon represent a root of *z* in the rectangular form $x + yi$. Here is a specific example that illustrates this procedure.

Example Find the fourth roots of $z = 16i$.

In trigonometric form, $z = 16(\cos 90° + i \sin 90°)$. Thus in this example, $r = 16$, $\theta = 90°$, and $n = 4$. In the Y= menu, enter

$$X_{1T} = 16\textasciicircum(1/4)\cos(90/4+T) \quad \text{and} \quad Y_{1T} = 16\textasciicircum(1/4)\sin(90/4+T)$$

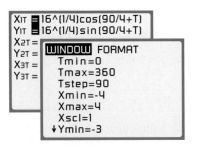

FIGURE 6.92

In the WINDOW menu, set

Tmin=0	Xmin=-4	Ymin=-3
Tmax=360	Xmax=4	Ymax=3
Tstep=360/4	Xscl=1	Yscl=1

See **Figure 6.92**. Press GRAPH to produce the quadrilateral in **Figure 6.93**. Use TRACE and the arrow key ▷ to move to each of the vertices of the quadrilateral. **Figure 6.93** shows that one of the roots of $z = 16i$ is $1.8477591 + 0.76536686i$. Continue to press the arrow key ▷ to find the other three roots, which are

$$-0.7653669 + 1.8477591i, -1.847759 - 0.7653669i, \text{ and}$$
$$0.76536686 - 1.847759i$$

Use a graphing calculator to estimate, in rectangular form, each of the following.

1. The cube roots of -27.

2. The fifth roots of $32i$.

3. The fourth roots of $\sqrt{8} + \sqrt{8}i$.

4. The sixth roots of $-64i$.

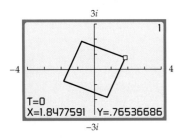

The vertices of the quadrilateral represent the fourth roots of 16*i* in the complex plane.

FIGURE 6.93

CHAPTER 6 SUMMARY

6.1 Parabolas

- The equations of a parabola with vertex at (h, k) and axis of symmetry parallel to a coordinate axis are given by

$(x - h)^2 = 4p(y - k)$; focus $(h, k + p)$; directrix $y = k - p$

$(y - k)^2 = 4p(x - h)$; focus $(h + p, k)$; directrix $x = h - p$

6.2 Ellipses

- The equations of an ellipse with center at (h, k) and major axis parallel to a coordinate axis are given by

$$\frac{(x - h)^2}{a^2} + \frac{(y - k)^2}{b^2} = 1; \text{ foci } (h \pm c, k); \text{ vertices } (h \pm a, k)$$

$$\frac{(x - h)^2}{b^2} + \frac{(y - k)^2}{a^2} = 1; \text{ foci } (h, k \pm c); \text{ vertices } (h, k \pm a)$$

For each equation, $a > b$ and $c^2 = a^2 - b^2$.

- The eccentricity e of an ellipse is given by $e = \dfrac{c}{a}$.

6.3 Hyperbolas

- The equations of a hyperbola with center at (h, k) and transverse axis parallel to a coordinate axis are given by

$$\frac{(x - h)^2}{a^2} - \frac{(y - k)^2}{b^2} = 1; \text{ foci } (h \pm c, k); \text{ vertices } (h \pm a, k)$$

$$\frac{(y - k)^2}{a^2} - \frac{(x - h)^2}{b^2} = 1; \text{ foci } (h, k \pm c); \text{ vertices } (h, k \pm a)$$

For each equation, $c^2 = a^2 + b^2$.

- The eccentricity e of a hyperbola is given by $e = \dfrac{c}{a}$.

6.4 Rotation of Axes

- The rotation-of-axes formulas are

$$\begin{cases} x = x' \cos \alpha - y' \sin \alpha \\ y = y' \cos \alpha + x' \sin \alpha \end{cases} \quad \begin{cases} x' = x \cos \alpha + y \sin \alpha \\ y' = y \cos \alpha - x \sin \alpha \end{cases}$$

- To eliminate the xy term from the general quadratic equation, rotate the coordinate axes through an angle α, where

$$\cot 2\alpha = \frac{A - C}{B}, \quad B \neq 0, \quad 0° < 2\alpha < 180°$$

- The graph of $Ax^2 + Bxy + Cy^2 + Dx + Ey + F = 0$ is either a conic or a degenerate conic. If the graph is a conic, then the graph can be identified by its *discriminant* $B^2 - 4AC$. The graph is

    an ellipse or a circle, provided $B^2 - 4AC < 0$.

    a parabola, provided $B^2 - 4AC = 0$.

    a hyperbola, provided $B^2 - 4AC > 0$.

- The graph of $Ax^2 + Bxy + Cy^2 + Dx + Ey + F = 0$ can be constructed by using a graphing utility to graph both

$$y_1 = \frac{-(Bx + E) + \sqrt{(Bx + E)^2 - 4(C)(Ax^2 + Dx + F)}}{2(C)}$$

and

$$y_2 = \frac{-(Bx + E) - \sqrt{(Bx + E)^2 - 4(C)(Ax^2 + Dx + F)}}{2(C)}$$

6.5 Introduction to Polar Coordinates

- A polar coordinate system is formed by drawing a horizontal ray (*polar axis*). The *pole* is the origin of a polar coordinate system.

- A point is specified by coordinates (r, θ), where r is a directed distance from the pole and θ is an angle measured from the polar axis.

    The transformation equations between a polar coordinate system and a rectangular coordinate system are

Polar to rectangular: $x = r \cos \theta$ $y = r \sin \theta$

Rectangular to polar: $r = \sqrt{x^2 + y^2}$ $\tan \theta = \dfrac{y}{x}$

6.6 Polar Equations of the Conics

- The polar equations of the conics are given by

$$r = \frac{ed}{1 \pm e \cos \theta} \quad \text{or} \quad r = \frac{ed}{1 \pm e \sin \theta}$$

where e is the eccentricity and d is the distance of the directrix from the focus.

When

$0 < e < 1$, the graph is an ellipse.

$e = 1$, the graph is a parabola.

$e > 1$, the graph is a hyperbola.

6.7 Parametric Equations

- Let t be a number in an interval I. A *curve* is a set of ordered pairs (x, y), where

$$x = f(t), \qquad y = g(t) \quad \text{for } t \in I$$

  The variable t is called a *parameter*, and the pair of equations are *parametric equations*.

- To *eliminate the parameter* is to find an equation in x and y that has the same graph as the given parametric equations.

- The path of a projectile (assume air resistance is negligible) that is launched at an angle θ from the horizon with an initial velocity of v_0 feet per second is given by

$$x = (v_0 \cos \theta)t, \qquad y = -16t^2 + (v_0 \sin \theta)t$$

  where t is the time in seconds since the projectile was launched.

CHAPTER 6 TRUE/FALSE EXERCISES

In Exercises 1 to 10, answer true or false. If the statement is false, give an example or a reason to show that the statement is false.

1. The graph of a parabola is the same shape as that of one branch of a hyperbola.

2. For the two axes of an ellipse (which is not a circle), the major axis and the minor axis, the major axis is always the longer axis.

3. For the two axes of a hyperbola, the transverse axis and the conjugate axis, the transverse axis is always the longer axis.

4. If two ellipses have the same foci, they have the same graph.

5. A hyperbola is similar to a parabola in that both curves have asymptotes.

6. If a hyperbola with center at the origin and a parabola with vertex at the origin have the same focus, $(0, c)$, then the two graphs always intersect.

7. The graphs of all the conic sections are not the graphs of functions.

8. Only the graph of a function can be written using parametric equations.

9. The graph of $x = \sin t$, $y = \cos t$, for $0 \le t < 2\pi$, and the graph of $x = \cos t$, $y = \sin t$, for $0 \le t < 2\pi$, are each the graph of a circle of radius 1 centered at the origin.

10. Each ordered pair (r, θ) in a polar coordinate system specifies exactly one point.

CHAPTER 6 REVIEW EXERCISES

In Exercises 1 to 12, if the equation is that of an ellipse or a hyperbola, find the center, vertices, and foci. For hyperbolas, find the equations of the asymptotes. If the equation is that of a parabola, find the vertex, the focus, and the equation of the directrix. Graph each equation.

1. $x^2 - y^2 = 4$

2. $y^2 = 16x$

3. $x^2 + 4y^2 - 6x + 8y - 3 = 0$

4. $3x^2 - 4y^2 + 12x - 24y - 36 = 0$

5. $3x - 4y^2 + 8y + 2 = 0$

6. $3x + 2y^2 - 4y - 7 = 0$

7. $9x^2 + 4y^2 + 36x - 8y + 4 = 0$

8. $11x^2 - 25y^2 - 44x - 50y - 256 = 0$

9. $4x^2 - 9y^2 - 8x + 12y - 144 = 0$

10. $9x^2 + 16y^2 + 36x - 16y - 104 = 0$

11. $4x^2 + 28x + 32y + 81 = 0$

12. $x^2 - 6x - 9y + 27 = 0$

In Exercises 13 to 20, find the equation of the conic that satisfies the given conditions.

13. Ellipse with vertices at $(7, 3)$ and $(-3, 3)$; length of minor axis is 8.

14. Hyperbola with vertices at $(4, 1)$ and $(-2, 1)$; eccentricity $\dfrac{4}{3}$.

15. Hyperbola with foci $(-5, 2)$ and $(1, 2)$; length of transverse axis is 4.

16. Parabola with focus $(2, -3)$ and directrix $x = 6$.

17. Parabola with vertex $(0, -2)$ and passing through the point $(3, 4)$.

18. Ellipse with eccentricity $\dfrac{2}{3}$ and foci $(-4, -1)$ and $(0, -1)$.

19. Hyperbola with vertices $(\pm 6, 0)$ and asymptotes whose equations are $y = \pm \dfrac{1}{9}x$.

20. Parabola passing through the points $(1, 0)$, $(2, 1)$, and $(0, 1)$ with axis of symmetry parallel to the y-axis.

In Exercises 21 to 24, write an equation without an xy term. Name the graph of the equation.

21. $11x^2 - 6xy + 19y^2 - 40 = 0$

22. $3x^2 + 6xy + 3y^2 - 4x + 5y - 12 = 0$

23. $x^2 + 2\sqrt{3}xy + 3y^2 + 8\sqrt{3}x - 8y + 32 = 0$

24. $xy - x - y - 1 = 0$

In Exercises 25 to 34, graph each polar equation.

25. $r = 4 \cos 3\theta$ **26.** $r = 1 + \cos \theta$

27. $r = 2(1 - 2 \sin \theta)$ **28.** $r = 4 \sin 4\theta$

29. $r = 5 \sin \theta$ **30.** $r = 3 \sec \theta$

31. $r = 4 \csc \theta$ **32.** $r = 4 \cos \theta$

33. $r = 3 + 2 \cos \theta$ **34.** $r = 4 + 2 \sin \theta$

In Exercises 35 to 38, change each equation to a polar equation.

35. $y^2 = 16x$ **36.** $x^2 + y^2 + 4x + 3y = 0$

37. $3x - 2y = 6$ **38.** $xy = 4$

In Exercises 39 to 42, change each equation to a rectangular equation.

39. $r = \dfrac{4}{1 - \cos \theta}$ **40.** $r = 3 \cos \theta - 4 \sin \theta$

41. $r^2 = \cos 2\theta$ **42.** $\theta = 1$

In Exercises 43 to 46, graph the conic given by each polar equation.

43. $r = \dfrac{4}{3 - 6 \sin \theta}$ **44.** $r = \dfrac{2}{1 + \cos \theta}$

45. $r = \dfrac{2}{2 - \cos \theta}$ **46.** $r = \dfrac{6}{4 + 3 \sin \theta}$

In Exercises 47 to 53, eliminate the parameter and graph the curve given by the parametric equations.

47. $x = 4t - 2$, $y = 3t + 1$, for $t \in R$

48. $x = 1 - t^2$, $y = 3 - 2t^2$, for $t \in R$

49. $x = 4 \sin t$, $y = 3 \cos t$, for $0 \le t < 2\pi$

50. $x = \sec t$, $y = 4 \tan t$, for $-\dfrac{\pi}{2} < t < \dfrac{\pi}{2}$

51. $x = \dfrac{1}{t}$, $y = -\dfrac{2}{t}$, for $t > 0$

52. $x = 1 + \cos t$, $y = 2 - \sin t$, for $0 \le t < 2\pi$

53. $x = \sqrt{t}$, $y = 2^{-t}$, for $t \ge 0$

54. Use a graphing utility to graph the cycloid given by
$$x = 3(t - \sin t), \qquad y = 3(1 - \cos t)$$
for $0 \le t \le 18\pi$.

55. Use a graphing utility to graph the conic given by
$$x^2 + 4xy + 2y^2 - 2x + 5y + 1 = 0$$

56. Use a graphing utility to graph

$$r = \dfrac{6}{3 + \sin\left(\theta + \dfrac{\pi}{4}\right)}$$

57. PHYSICS The path of a projectile (assume air resistance is negligible) that is launched at an angle θ from the horizon with an initial velocity of v_0 feet per second is given by the parametric equations

$$x = (v_0 \cos \theta)t, \qquad y = -16t^2 + (v_0 \sin \theta)t$$

where t is the time in seconds since the projectile was launched. Use a graphing utility to graph the path of a projectile that is launched at an angle of 33° with an initial velocity of 245 feet per second. Use the graph to determine the maximum height of the projectile to the nearest foot.

CHAPTER 6 TEST

1. Find the vertex, focus, and directrix of the parabola given by the equation $y = \dfrac{1}{8} x^2$.

2. Graph: $\dfrac{x^2}{16} + \dfrac{y^2}{1} = 1$

3. Find the vertices and foci of the ellipse given by the equation $25x^2 - 150x + 9y^2 + 18y + 9 = 0$.

4. Find the equation in standard form of the ellipse with center $(0, -3)$, foci $(-6, -3)$ and $(6, -3)$, and minor axis of length 6.

5. Graph: $\dfrac{y^2}{25} - \dfrac{x^2}{16} = 1$

6. Find the vertices, foci, and asymptotes of the hyperbola given by the equation $\dfrac{x^2}{36} - \dfrac{y^2}{64} = 1$.

7. Graph: $16y^2 + 32y - 4x^2 - 24x = 84$

8. For the equation $x^2 - 4xy - 5y^2 + 3x - 5y - 20 = 0$, determine what acute angle of rotation (to the nearest 0.01°) would eliminate the xy term.

9. Determine whether the graph of the following equation is the graph of a parabola, an ellipse, or a hyperbola.

$$8x^2 + 5xy + 2y^2 - 10x + 5y + 4 = 0$$

10. $P\left(1, -\sqrt{3}\right)$ are the coordinates of a point in an xy-coordinate system. Find the polar coordinates of P.

11. Graph: $r = 4 \cos \theta$

12. Graph: $r = 3(1 - \sin \theta)$

13. Graph: $r = 2 \sin 4\theta$

14. Find the rectangular coordinates of the point whose polar coordinates are $\left(5, \dfrac{7\pi}{3}\right)$.

15. Find the rectangular form of $r - r \cos \theta = 4$.

16. Write $r = \dfrac{4}{1 + \sin \theta}$ as an equation in rectangular coordinates.

17. Eliminate the parameter and graph the curve given by the parametric equations $x = t - 3$, $y = 2t^2$.

18. Eliminate the parameter and graph the curve given by the parametric equations $x = 4 \sin \theta$, $y = \cos \theta + 2$, where $0 \le \theta < 2\pi$.

19. Use a graphing utility to graph the cycloid given by

$$x = 2(t - \sin t), \qquad y = 2(1 - \cos t)$$

for $0 \le t \le 12\pi$.

20. The path of a projectile that is launched at an angle of 30° from the horizon with an initial velocity of 128 feet per second is given by

$$x = (128 \cos 30°)t, \qquad y = -16t^2 + (128 \sin 30°)t$$

where t is the time in seconds after the projectile is launched. Use a graphing utility to determine how far (to the nearest foot) the projectile will travel downrange if the ground is level.

CUMULATIVE REVIEW EXERCISES

1. Solve: $x^2 + 4x + 6 = 0$

2. Is the graph of $f(x) = x^3 - 4x$ symmetric with respect to the x-axis, the y-axis, or the origin, or does it exhibit none of these symmetries?

3. Given $f(x) = \sin(x)$ and $g(x) = 3x - 2$, find $(g \circ f)(x)$.

4. Convert $240°$ to radians.

5. An electric cart has 10-inch-radius wheels. What is the linear speed in miles per hour of this cart when the wheels are rotating at 3 radians per second? Round to the nearest mile per hour.

6. Given $\sin(t) = -\dfrac{\sqrt{3}}{2}, \dfrac{3\pi}{2} < t < 2\pi$, find $\tan(t)$.

7. Find the measure of a for the right triangle shown at the right. Round to the nearest tenth.

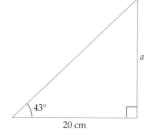

43°

20 cm

8. What are the amplitude and period of $y = \dfrac{1}{2}\cos\left(\dfrac{\pi x}{3}\right)$?

9. What is the period of $y = 2\tan\left(\dfrac{\pi x}{3}\right)$?

10. Verify the identity $\dfrac{\sin(x)}{1 - \cos(x)} = \csc(x) + \cot(x)$.

11. Given $\sin(\alpha) = \dfrac{3}{5}$ in Quadrant II and $\cos(\beta) = -\dfrac{5}{13}$ in Quadrant III, find $\sin(\alpha + \beta)$.

12. Find the exact value of $\sin\left(\cos^{-1}\dfrac{1}{5}\right)$.

13. Solve: $\cos^{-1}x = \sin^{-1}\dfrac{12}{13}$

14. To find the distance across a ravine, a surveying team locates points A and B on one side of the ravine and point C on the other side of the ravine. The distance between A and B is 155 feet. The measure of angle CAB is $71°$, and the measure of angle CBA is $80°$. Find the distance across the ravine. Round to the nearest foot.

15. The lengths of the sides of a triangular piece of wood are 2.5 feet, 4 feet, and 3.6 feet. Find the angle between the two longer sides of the triangle. Round to the nearest degree.

16. The magnitude of vector $\mathbf{v}$ is 30 and the direction angle is $145°$. Write the vector in the form $\mathbf{v} = a\mathbf{i} + b\mathbf{j}$, where a and b are rounded to the nearest tenth.

17. Are the vectors $\mathbf{v} = 3\mathbf{i} + 2\mathbf{j}$ and $\mathbf{w} = 5\mathbf{i} - 7\mathbf{j}$ perpendicular?

18. Write the complex number $z = -2 + 2i\sqrt{3}$ in polar form.

19. Sketch the graph of $r = 3\sin(2\theta)$.

20. Write x in terms of y by eliminating the parameter from the parametric equations $x = 2t - 1, y = 4t^2 + 1$.

Modeling Data with an Exponential Function

The following table shows the time, in hours, before the body of a scuba diver, wearing a 5-millimeter-thick wet suit, reaches hypothermia (95°F) for various water temperatures.

Water Temperature, °F	Time, hours
36	1.5
41	1.8
46	2.6
50	3.1
55	4.9

Source: Data extracted from the *American Journal of Physics,* vol. 71, no. 4 (April 2003), Fig. 3, p. 336.

The following function, which is an example of an exponential function, closely models the data in the table:

$$T(F) = 0.1509(1.0639)^F$$

In this function F represents the Fahrenheit temperature of the water, and T represents the time in hours. A diver can use the function to determine the time it takes to reach hypothermia for water temperatures that are not included in the table. See **Exercise 21, page 513.** The function $T(F)$ was determined by using exponential regression, which is one of the topics in Section 7.6.

VIDEO & DVD

SSG

WWW

FOCUS ON PROBLEM SOLVING

Use Two Methods to Solve and Compare Results

Sometimes it is possible to solve a problem in two or more ways. In such situations it is recommended that you use at least two methods to solve the problem, and compare your results. Here is an example of an application that can be solved in more than one way.

Example

In a league of eight basketball teams, each team plays every other team in the league exactly once. How many league games will take place?

Solution

Method 1: *Use an analytic approach.* Each of the eight teams must play the other seven teams. Using this information, you might be tempted to conclude that there will be $8 \cdot 7 = 56$ games, but this result is too large because it counts each game between two individual teams as two different games. Thus the number of league games will be

$$\frac{8 \cdot 7}{2} = \frac{56}{2} = 28$$

Method 2: *Make an organized list.* Use the letters A, B, C, D, E, F, G, and H to represent the eight teams. Use the notation AB to represent the game between team A and team B. Do not include BA in your list because it represents the same game between team A and team B.

AB	AC	AD	AE	AF	AG	AH
	BC	BD	BE	BF	BG	BH
		CD	CE	CF	CG	CH
			DE	DF	DG	DH
				EF	EG	EH
					FG	FH
						GH

The list shows that there will be 28 league games.

The procedure of using two different solution methods and comparing results is employed often in this chapter. For instance, see **Example 2, page 490.** In this example, a solution is found by applying algebraic procedures and also by graphing. Notice that both methods produce the same result.

EXPONENTIAL FUNCTIONS AND THEIR APPLICATIONS

SECTION 7.1

- **EXPONENTIAL FUNCTIONS**
- **GRAPHS OF EXPONENTIAL FUNCTIONS**
- **THE NATURAL EXPONENTIAL FUNCTION**
- **APPLICATIONS OF EXPONENTIAL FUNCTIONS**

● EXPONENTIAL FUNCTIONS

In 1965, Gordon Moore, one of the cofounders of Intel Corporation, observed that the maximum number of transistors that could be placed on a microprocessor seemed to be doubling every 18 to 24 months. **Table 7.1** below shows how the maximum number of transistors on various Intel processors has changed over time. (*Source:* Intel Museum home page.)

TABLE 7.1

Year	1971	1979	1983	1985	1990	1993	1995	1998	2000
Number of transistors per microprocessor (in thousands)	2.3	31	110	280	1200	3100	5500	14,000	42,000

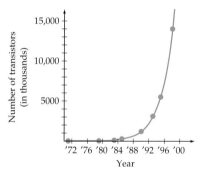

FIGURE 7.1

Moore's Law

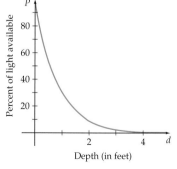

FIGURE 7.2

The curve that approximately passes through the points is a mathematical model of the data. See **Figure 7.1.** The model is based on an *exponential* function.

When light enters water, the intensity of the light decreases with the depth of the water. The graph in **Figure 7.2** shows a model, for Lake Michigan, of the decrease in the percentage of available light as the depth of the water increases. This model is also based on an exponential function.

Definition of an Exponential Function

The **exponential function with base b** is defined by

$$f(x) = b^x$$

where $b > 0$, $b \neq 1$, and x is a real number.

The base b of $f(x) = b^x$ is required to be positive. If the base were a negative number, the value of the function would be a complex number for some values of x. For instance, if $b = -4$ and $x = \dfrac{1}{2}$, then $f\left(\dfrac{1}{2}\right) = (-4)^{1/2} = 2i$. To avoid complex number values of a function, the base of any exponential function must be a nonnegative number. Also, b is defined such that $b \neq 1$ because $f(x) = 1^x = 1$ is a constant function.

In the following examples we evaluate $f(x) = 2^x$ at $x = 3$ and $x = -2$.

$$f(3) = 2^3 = 8 \qquad f(-2) = 2^{-2} = \frac{1}{2^2} = \frac{1}{4}$$

To evaluate the exponential function $f(x) = 2^x$ at an irrational number such as $x = \sqrt{2}$, we use a rational approximation of $\sqrt{2}$, such as 1.4142, and a calculator to obtain an approximation of the function. For instance, if $f(x) = 2^x$, then $f\left(\sqrt{2}\right) = 2^{\sqrt{2}} \approx 2^{1.4142} \approx 2.6651$.

EXAMPLE I **Evaluate an Exponential Function**

Evaluate $f(x) = 3^x$ at $x = 2$, $x = -4$, and $x = \pi$.

Solution

$$f(2) = 3^2 = 9$$

$$f(-4) = 3^{-4} = \frac{1}{3^4} = \frac{1}{81}$$

$$f(\pi) = 3^\pi \approx 3^{3.1415927} \approx 31.54428 \qquad \bullet \text{ Evaluate with the aid of a calculator.}$$

▶ **TRY EXERCISE 2, PAGE 444**

● GRAPHS OF EXPONENTIAL FUNCTIONS

The graph of $f(x) = 2^x$ is shown in **Figure 7.3.** The coordinates of some of the points on the curve are given in **Table 7.2.**

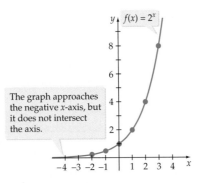

The graph approaches the negative x-axis, but it does not intersect the axis.

FIGURE 7.3

TABLE 7.2

x	$y = f(x) = 2^x$	(x, y)
−2	$f(-2) = 2^{-2} = \dfrac{1}{4}$	$\left(-2, \dfrac{1}{4}\right)$
−1	$f(-1) = 2^{-1} = \dfrac{1}{2}$	$\left(-1, \dfrac{1}{2}\right)$
0	$f(0) = 2^0 = 1$	$(0, 1)$
1	$f(1) = 2^1 = 2$	$(1, 2)$
2	$f(2) = 2^2 = 4$	$(2, 4)$
3	$f(3) = 2^3 = 8$	$(3, 8)$

Note the following properties of the graph of the exponential function $f(x) = 2^x$.

- The y-intercept is $(0, 1)$.
- The graph passes through $(1, 2)$.
- As x decreases without bound (that is, as $x \to -\infty$), $f(x) \to 0$.
- The graph is a smooth continuous increasing curve.

Now consider the graph of an exponential function for which the base is between 0 and 1. The graph of $f(x) = \left(\dfrac{1}{2}\right)^x$ is shown in **Figure 7.4.** The coordinates of some of the points on the curve are given in **Table 7.3.**

TABLE 7.3

x	$y = f(x) = \left(\dfrac{1}{2}\right)^x$	(x, y)
-3	$f(-3) = \left(\dfrac{1}{2}\right)^{-3} = 8$	$(-3, 8)$
-2	$f(-2) = \left(\dfrac{1}{2}\right)^{-2} = 4$	$(-2, 4)$
-1	$f(-1) = \left(\dfrac{1}{2}\right)^{-1} = 2$	$(-1, 2)$
0	$f(0) = \left(\dfrac{1}{2}\right)^{0} = 1$	$(0, 1)$
1	$f(1) = \left(\dfrac{1}{2}\right)^{1} = \dfrac{1}{2}$	$\left(1, \dfrac{1}{2}\right)$
2	$f(2) = \left(\dfrac{1}{2}\right)^{2} = \dfrac{1}{4}$	$\left(2, \dfrac{1}{4}\right)$

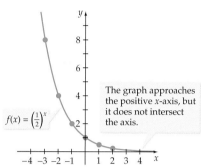

$f(x) = \left(\frac{1}{2}\right)^x$

The graph approaches the positive x-axis, but it does not intersect the axis.

FIGURE 7.4

Note the following properties of the graph of $f(x) = \left(\dfrac{1}{2}\right)^x$ in **Figure 7.4.**

- The y-intercept is $(0, 1)$.

- The graph passes through $\left(1, \dfrac{1}{2}\right)$.

- As x increases without bound, the y-values decrease toward 0. That is, as $x \to \infty$, $f(x) \to 0$.

- The graph is a smooth continuous decreasing curve.

The basic properties of exponential functions are provided in the following summary.

Properties of $f(x) = b^x$

For positive real numbers b, $b \neq 1$, the exponential function defined by $f(x) = b^x$ has the following properties:

1. The function f is a one-to-one function. It has the set of real numbers as its domain and the set of positive real numbers as its range.

2. The graph of f is a smooth continuous curve with a y-intercept of $(0, 1)$, and the graph passes through $(1, b)$.

3. If $b > 1$, f is an increasing function and the graph of f is asymptotic to the negative x-axis. [As $x \to \infty$, $f(x) \to \infty$, and as $x \to -\infty$, $f(x) \to 0$.] See **Figure 7.5a.**

4. If $0 < b < 1$, f is a decreasing function and the graph of f is asymptotic to the positive x-axis. [As $x \to -\infty$, $f(x) \to \infty$, and as $x \to \infty$, $f(x) \to 0$.] See **Figure 7.5b.**

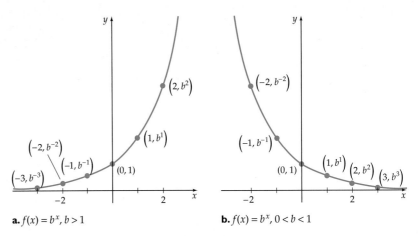

a. $f(x) = b^x, b > 1$ **b.** $f(x) = b^x, 0 < b < 1$

FIGURE 7.5

❓ QUESTION What is the x-intercept of the graph of $f(x) = \left(\dfrac{1}{3}\right)^x$?

EXAMPLE 2   **Graph an Exponential Function**

Graph $g(x) = \left(\dfrac{3}{4}\right)^x$.

Solution

Because the base $\dfrac{3}{4}$ is less than 1, we know that the graph of g is a decreasing function that is asymptotic to the positive x-axis. The y-intercept of the graph is the point $(0, 1)$, and the graph also passes through $\left(1, \dfrac{3}{4}\right)$.

Plot a few additional points (see **Table 7.4**), and then draw a smooth curve through the points as in **Figure 7.6.**

❓ ANSWER The graph does not have an x-intercept. As x increases, the graph approaches the x-axis, but it does not intersect the x-axis.

TABLE 7.4

x	$y = g(x) = \left(\dfrac{3}{4}\right)^x$	(x, y)
−3	$\left(\dfrac{3}{4}\right)^{-3} = \dfrac{64}{27}$	$\left(-3, \dfrac{64}{27}\right)$
−2	$\left(\dfrac{3}{4}\right)^{-2} = \dfrac{16}{9}$	$\left(-2, \dfrac{16}{9}\right)$
−1	$\left(\dfrac{3}{4}\right)^{-1} = \dfrac{4}{3}$	$\left(-1, \dfrac{4}{3}\right)$
2	$\left(\dfrac{3}{4}\right)^{2} = \dfrac{9}{16}$	$\left(2, \dfrac{9}{16}\right)$
3	$\left(\dfrac{3}{4}\right)^{3} = \dfrac{27}{64}$	$\left(3, \dfrac{27}{64}\right)$

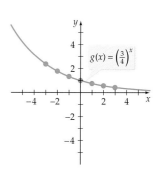

FIGURE 7.6

▶ **TRY EXERCISE 22, PAGE 445**

Consider the functions $F(x) = 2^x - 3$ and $G(x) = 2^{x-3}$. You can construct the graphs of these functions by plotting points; however, it is easier to construct their graphs by using translations of the graph of $f(x) = 2^x$, as shown in Example 3.

EXAMPLE 3   **Use a Translation to Produce a Graph**

a. Explain how to use the graph of $f(x) = 2^x$ to produce the graph of $F(x) = 2^x - 3$.

b. Explain how to use the graph of $f(x) = 2^x$ to produce the graph of $G(x) = 2^{x-3}$.

Solution

a. $F(x) = 2^x - 3 = f(x) - 3$. The graph of F is a vertical translation of f down 3 units, as shown in **Figure 7.7.**

b. $G(x) = 2^{x-3} = f(x - 3)$. The graph of G is a horizontal translation of f to the right 3 units, as shown in **Figure 7.8.**

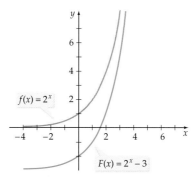

FIGURE 7.7

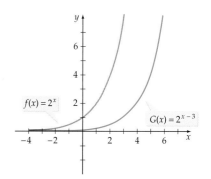

FIGURE 7.8

▶ **TRY EXERCISE 28, PAGE 445**

The graphs of some functions can be constructed by stretching, compressing, or reflecting the graph of an exponential function.

EXAMPLE 4 **Use Stretching or Reflecting Procedures to Produce a Graph**

a. Explain how to use the graph of $f(x) = 2^x$ to produce the graph of $M(x) = 2(2^x)$.

b. Explain how to use the graph of $f(x) = 2^x$ to produce the graph of $N(x) = 2^{-x}$.

Solution

a. $M(x) = 2(2^x) = 2f(x)$. The graph of M is a vertical stretching of f, as shown in **Figure 7.9**. If (x, y) is a point on the graph of $f(x) = 2^x$, then $(x, 2y)$ is a point on the graph of M.

b. $N(x) = 2^{-x} = f(-x)$. The graph of N is the graph of f reflected across the y-axis, as shown in **Figure 7.10**. If (x, y) is a point on the graph of $f(x) = 2^x$, then $(-x, y)$ is a point on the graph of N.

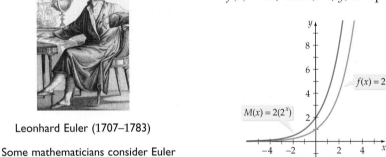

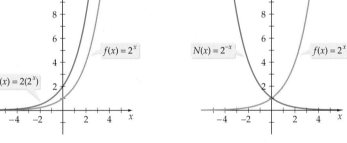

FIGURE 7.9   **FIGURE 7.10**

 TRY EXERCISE 30, PAGE 445

● THE NATURAL EXPONENTIAL FUNCTION

The irrational number π is often used in applications that involve circles. Another irrational number, denoted by the letter e, is useful in applications that involve growth or decay.

Definition of e

The **number e** is defined as the number that

$$\left(1 + \frac{1}{n}\right)^n$$

approaches as n increases without bound.

The letter e was chosen in honor of the Swiss mathematician Leonhard Euler. He was able to compute the value of e to several decimal places by evaluating $\left(1 + \dfrac{1}{n}\right)^n$ for large values of n, as shown in **Table 7.5.**

TABLE 7.5

Value of n	Value of $\left(1 + \dfrac{1}{n}\right)^n$
1	2
10	2.59374246
100	2.704813829
1000	2.716923932
10,000	2.718145927
100,000	2.718268237
1,000,000	2.718280469
10,000,000	2.718281693

The value of e accurate to eight decimal places is 2.71828183.

The Natural Exponential Function

For all real numbers x, the function defined by

$$f(x) = e^x$$

is called the **natural exponential function.**

A calculator can be used to evaluate e^x for specific values of x. For instance,

$$e^2 \approx 7.389056, \quad e^{3.5} \approx 33.115452, \quad \text{and} \quad e^{-1.4} \approx 0.246597$$

On a TI-83 calculator the e^x function is located above the $\boxed{\text{LN}}$ key.

To graph $f(x) = e^x$, use a calculator to find the range values for a few domain values. The range values in **Table 7.6** have been rounded to the nearest tenth.

TABLE 7.6

x	-2	-1	0	1	2
$f(x) = e^x$	0.1	0.4	1.0	2.7	7.4

INTEGRATING TECHNOLOGY

The graph below was produced on a TI-83 graphing calculator by entering e^x in the Y= menu.

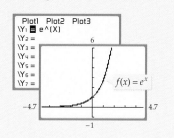

Plot the points given in **Table 7.6,** and then connect the points with a smooth curve. Because $e > 1$, we know that the graph is an increasing function. To the far left, the graph will approach the x-axis. The y-intercept is $(0, 1)$. See **Figure 7.11.** Note in **Figure 7.12** how the graph of $f(x) = e^x$ compares with the graphs of $g(x) = 2^x$ and $h(x) = 3^x$. You may have anticipated that the graph of $f(x) = e^x$ would lie between the two other graphs because e is between 2 and 3.

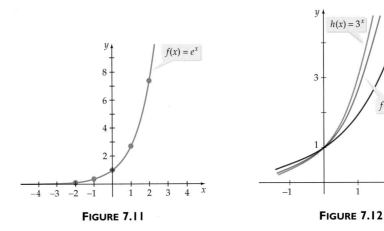

FIGURE 7.11   **FIGURE 7.12**

● APPLICATIONS OF EXPONENTIAL FUNCTIONS

Many applications can be effectively modeled by functions that involve an exponential function. For instance, in Example 5 we make use of a function that involves an exponential function to model the temperature of a cup of coffee.

EXAMPLE 5   Use a Mathematical Model

A cup of coffee is heated to 160°F and placed in a room that maintains a temperature of 70°F. The temperature T of the coffee, in degrees Fahrenheit, after t minutes is given by

$$T = 70 + 90e^{-0.0485t}$$

a.   Find the temperature of the coffee, to the nearest degree, 20 minutes after it is placed in the room.

b.   Use a graphing utility to determine when the temperature of the coffee will reach 90°F.

Solution

a. $T = 70 + 90e^{-0.0485t}$

$= 70 + 90e^{-0.0485 \cdot (20)}$ • **Substitute 20 for t.**

$\approx 70 + 34.1$

≈ 104.1

After 20 minutes the temperature of the coffee is about 104°F.

take note

In Example 5**b.**, we use a graphing utility to solve the equation $90 = 70 + 90e^{-0.0485t}$. Analytic methods of solving this type of equation without the use of a graphing utility will be developed in Section 7.4.

b. Graph $T = 70 + 90e^{-0.0485t}$ and $T = 90$. See the following figure.

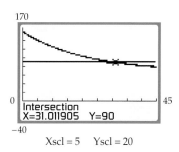

The graphs intersect at about $(31.01, 90)$. It takes the coffee about 31 minutes to cool to 90°F.

▶ TRY EXERCISE 44, PAGE 445

EXAMPLE 6   **Use a Mathematical Model**

 The weekly revenue R, in dollars, from the sale of a product varies with time according to the function

$$R(x) = \frac{1760}{8 + 14e^{-0.03x}}$$

where x is the number of weeks that have passed since the product was put on the market. What will the weekly revenue approach as time goes by?

Solution

Method 1   Use a graphing utility to graph $R(x)$, and use the TRACE feature to see what happens to the revenue as the time increases. The graph at the left shows that as the weeks go by, the weekly revenue will increase and approach $220.00 per week.

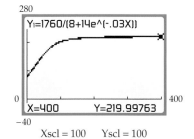

Method 2   Write the revenue function in the following form.

$$R(x) = \frac{1760}{8 + \dfrac{14}{e^{0.03x}}} \qquad \bullet\ 14e^{-0.03x} = \frac{14}{e^{0.03x}}$$

As x increases without bound, $e^{0.03x}$ increases without bound, and the fraction $\dfrac{14}{e^{0.03x}}$ approaches 0. Therefore, as $x \to \infty$, $R(x) \to \dfrac{1760}{8 + 0} = 220$. Both methods indicate that as the number of weeks increases, the revenue approaches $220 per week.

▶ TRY EXERCISE 54, PAGE 447

TOPICS FOR DISCUSSION

1. Explain how to use the graph of $f(x) = 2^x$ to produce the graph of $g(x) = 2^{(x-3)} + 4$.

2. At what point does the function $g(x) = e^{-x^2/2}$ take on its maximum value?

3. Without using a graphing utility, determine whether the revenue function $R(t) = 10 + e^{-0.05t}$ is an increasing function or a decreasing function.

4. Discuss the properties of the graph of $f(x) = b^x$ when $b > 1$.

5. What is the base of the natural exponential function? How is it calculated? What is its approximate value?

EXERCISE SET 7.1

In Exercises 1 to 8, evaluate the exponential function for the given x-values.

1. $f(x) = 3^x$; $x = 0$ and $x = 4$

▶ **2.** $f(x) = 5^x$; $x = 3$ and $x = -2$

3. $g(x) = 10^x$; $x = -2$ and $x = 3$

4. $g(x) = 4^x$; $x = 0$ and $x = -1$

5. $h(x) = \left(\dfrac{3}{2}\right)^x$; $x = 2$ and $x = -3$

6. $h(x) = \left(\dfrac{2}{5}\right)^x$; $x = -1$ and $x = 3$

7. $j(x) = \left(\dfrac{1}{2}\right)^x$; $x = -2$ and $x = 4$

8. $j(x) = \left(\dfrac{1}{4}\right)^x$; $x = -1$ and $x = 5$

In Exercises 9 to 14, use a calculator to evaluate the exponential function for the given x-value. Round to the nearest hundredth.

9. $f(x) = 2^x$, $x = 3.2$

10. $f(x) = 3^x$, $x = -1.5$

11. $g(x) = e^x$, $x = 2.2$

12. $g(x) = e^x$, $x = -1.3$

13. $h(x) = 5^x$, $x = \sqrt{2}$

14. $h(x) = 0.5^x$, $x = \pi$

15. Examine the following four functions and the graphs labeled **a, b, c,** and **d.** For each graph, determine which function has been graphed.

$$f(x) = 5^x \qquad g(x) = 1 + 5^{-x}$$
$$h(x) = 5^{x+3} \qquad k(x) = 5^x + 3$$

a.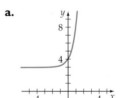

b.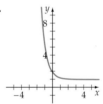

c.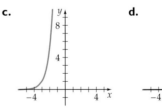

d.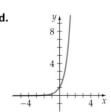

16. Examine the following four functions and the graphs labeled **a, b, c,** and **d.** For each graph, determine which function has been graphed.

$$f(x) = \left(\frac{1}{4}\right)^x \qquad g(x) = \left(\frac{1}{4}\right)^{-x}$$
$$h(x) = \left(\frac{1}{4}\right)^{x-2} \qquad k(x) = 3\left(\frac{1}{4}\right)^x$$

a.

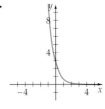

b.

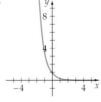

c.

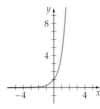

d.

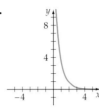

In Exercises 17 to 24, sketch the graph of each function.

17. $f(x) = 3^x$

18. $f(x) = 4^x$

19. $f(x) = 10^x$

20. $f(x) = 6^x$

21. $f(x) = \left(\dfrac{3}{2}\right)^x$

▶ 22. $f(x) = \left(\dfrac{5}{2}\right)^x$

23. $f(x) = \left(\dfrac{1}{3}\right)^x$

24. $f(x) = \left(\dfrac{2}{3}\right)^x$

In Exercises 25 to 34, explain how to use the graph of the first function f to produce the graph of the second function F.

25. $f(x) = 3^x,\ F(x) = 3^x + 2$

26. $f(x) = 4^x,\ F(x) = 4^x - 3$

27. $f(x) = 10^x,\ F(x) = 10^{x-2}$

▶ 28. $f(x) = 6^x,\ F(x) = 6^{x+5}$

29. $f(x) = \left(\dfrac{3}{2}\right)^x,\ F(x) = \left(\dfrac{3}{2}\right)^{-x}$

▶ 30. $f(x) = \left(\dfrac{5}{2}\right)^x,\ F(x) = -\left[\left(\dfrac{5}{2}\right)^x\right]$

31. $f(x) = \left(\dfrac{1}{3}\right)^x,\ F(x) = 2\left[\left(\dfrac{1}{3}\right)^x\right]$

32. $f(x) = \left(\dfrac{2}{3}\right)^x,\ F(x) = \dfrac{1}{2}\left[\left(\dfrac{2}{3}\right)^x\right]$

33. $f(x) = e^x,\ F(x) = e^{-x} + 2$

34. $f(x) = e^x,\ F(x) = e^{x-3} + 1$

In Exercises 35 to 42, use a graphing utility to graph each function. If the function has a horizontal asymptote, state the equation of the horizontal asymptote.

35. $f(x) = \dfrac{3^x + 3^{-x}}{2}$

36. $f(x) = 4 \cdot 3^{-x^2}$

37. $f(x) = \dfrac{e^x - e^{-x}}{2}$

38. $f(x) = \dfrac{e^x + e^{-x}}{2}$

39. $f(x) = -e^{(x-4)}$

40. $f(x) = 0.5e^{-x}$

41. $f(x) = \dfrac{10}{1 + 0.4e^{-0.5x}},$ $x \geq 0$

42. $f(x) = \dfrac{10}{1 + 1.5e^{-0.5x}},$ $x \geq 0$

43. **INTERNET CONNECTIONS** Data from Forrester Research suggest that the number of broadband [cable and digital subscriber line (DSL)] connections to the Internet can be modeled by $f(x) = 1.353(1.9025)^x$, where x is the number of years after January 1, 1998, and $f(x)$ is the number of connections in millions.

a. How many broadband Internet connections, to the nearest million, does this model predict will exist on January 1, 2005?

b. According to the model, in what year will the number of broadband connections first reach 300 million? [*Hint:* Use the intersect feature of a graphing utility to determine the x-coordinate of the point of intersection of the graphs of $f(x)$ and $y = 300$.]

▶ 44. **MEDICATION IN BLOODSTREAM** The function $A(t) = 200e^{-0.014t}$ gives the amount of medication, in milligrams, in a patient's bloodstream t minutes after the medication has been injected into the patient's bloodstream.

a. Find the amount of medication, to the nearest milligram, in the patient's bloodstream after 45 minutes.

b. Use a graphing utility to determine how long it will take, to the nearest minute, for the amount of medication in the patient's bloodstream to reach 50 milligrams.

45. **DEMAND FOR A PRODUCT** The demand d for a specific product, in items per month, is given by

$$d(p) = 25 + 880e^{-0.18p}$$

where p is the price, in dollars, of the product.

a. What will be the monthly demand, to the nearest unit, when the price of the product is $8 and when the price is $18?

b. What will happen to the demand as the price increases without bound?

46. **SALES** The monthly income I, in dollars, from a new product is given by

$$I(t) = 24{,}000 - 22{,}000e^{-0.005t}$$

where t is the time, in months, since the product was first put on the market.

a. What was the monthly income after the 10th month and after the 100th month?

b. What will the monthly income from the product approach as the time increases without bound?

47. **A PROBABILITY FUNCTION** The manager of a home improvement store finds that between 10 A.M. and 11 A.M., customers enter the store at the average rate of 45 customers per hour. The following function gives the probability that a customer will arrive within t minutes of 10 A.M. (*Note:* A probability of 0.6 means there is a 60% chance that a customer will arrive during a given time period.)

$$P(t) = 1 - e^{-0.75t}$$

a. Find the probability, to the nearest hundredth, that a customer will arrive within 1 minute of 10 A.M.

b. Find the probability, to the nearest hundredth, that a customer will arrive within 3 minutes of 10 A.M.

c. Use a graph of $P(t)$ to determine how many minutes, to the nearest tenth of a minute, it takes for $P(t)$ to equal 98%.

d. Write a sentence that explains the meaning of the answer in part **c.**

48. **A PROBABILITY FUNCTION** The owner of a sporting goods store finds that between 9 A.M. and 10 A.M., customers enter the store at the average rate of 12 customers per hour. The following function gives the probability that a customer will arrive within t minutes of 9 A.M.

$$P(t) = 1 - e^{-0.2t}$$

a. Find the probability, to the nearest hundredth, that a customer will arrive within 5 minutes of 9 A.M.

b. Find the probability, to the nearest hundredth, that a customer will arrive within 15 minutes of 9 A.M.

c. Use a graph of $P(t)$ to determine how many minutes, to the nearest 0.1 minute, it takes for $P(t)$ to equal 90%.

d. Write a sentence that explains the meaning of the answer in part **c.**

Exercises 49 and 50 involve the factorial function $x!$, which is defined for whole numbers x as

$$x! = \begin{cases} 1, & \text{if } x = 0 \\ x \cdot (x-1) \cdot (x-2) \cdots 3 \cdot 2 \cdot 1, & \text{if } x \geq 1 \end{cases}$$

For example, $3! = 3 \cdot 2 \cdot 1 = 6$ and $5! = 5 \cdot 4 \cdot 3 \cdot 2 \cdot 1 = 120$.

49. **QUEUING THEORY** During the 30-minute period before a Broadway play begins, the members of the audience arrive at the theater at the average rate of 12 people per minute. The probability that x people will arrive during a particular minute is given by $P(x) = \dfrac{12^x e^{-12}}{x!}$. Find the probability, to the nearest 0.1%, that

a. 9 people will arrive during a given minute.

b. 18 people will arrive during a given minute.

50. **QUEUING THEORY** During the period from 2:00 P.M. to 3:00 P.M., a bank finds that an average of seven people enter the bank every minute. The probability that x people will enter the bank during a particular minute is given by $P(x) = \dfrac{7^x e^{-7}}{x!}$. Find the probability, to the nearest 0.1%, that

a. only two people will enter the bank during a given minute.

b. 11 people will enter the bank during a given minute.

51. **E. COLI INFEC-TION** *Escherichia coli (E. coli)* is a bacterium that can reproduce at an exponential rate. The *E. coli* reproduce by dividing. A small number of *E. coli* bacteria in the large intestine of a human can trigger a serious infection within a few hours. Consider a particular *E. coli* infection that starts with 100 *E. coli* bacteria. Each bacterium splits into two parts every half hour. Assuming none of the bacteria die, the size of the *E. coli* population after t hours is given by $P(t) = 100 \cdot 2^{2t}$, where $0 \leq t \leq 16$.

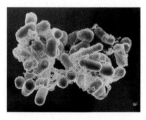

a. Find $P(3)$ and $P(6)$.

b. Use a graphing utility to find the time, to the nearest tenth of an hour, it takes for the *E. coli* population to number 1 billion.

52. RADIATION Lead shielding is used to contain radiation. The percentage of a certain radiation that can penetrate x millimeters of lead shielding is given by $I(x) = 100e^{-1.5x}$.

a. What percentage of radiation, to the nearest tenth of a percent, will penetrate a lead shield that is 1 millimeter thick?

b. How many millimeters of lead shielding are required so that less than 0.05% of the radiation penetrates the shielding? Round to the nearest millimeter.

53. AIDS An exponential function that approximates the number of people in the United States who have been infected with AIDS is given by $N(t) = 138,000(1.39)^t$, where t is the number of years after January 1, 1990.

a. According to this function, how many people had been infected with AIDS as of January 1, 1994? Round to the nearest thousand.

b. Use a graph to estimate during what year the number of people in the United States who had been infected with AIDS first reached 1.5 million.

▶ **54.** FISH POPULATION The number of bass in a lake is given by

$$P(t) = \frac{3600}{1 + 7e^{-0.05t}},$$

where t is the number of months that have passed since the lake was stocked with bass.

a. How many bass were in the lake immediately after it was stocked?

b. How many bass were in the lake 1 year after the lake was stocked?

c. What will happen to the bass population as t increases without bound?

55. THE PAY IT FORWARD MODEL In the movie *Pay It Forward*, Trevor McKinney, played by Haley Joel Osment, is given a school assignment to "think of an idea to change the world—and then put it into action." In response to this assignment, Trevor develops a *pay it forward* project. In this project, anyone who benefits from another person's good deed must do a good deed for three additional people. Each of these three people is then obligated to do a good deed for another three people, and so on.

The following diagram shows the number of people who have been a beneficiary of a good deed after 1 round and after 2 rounds of this project.

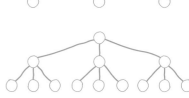

Three beneficiaries after one round.

A total of 12 beneficiaries after two rounds (3 + 9 = 12).

A mathematical model for the number of pay it forward beneficiaries after n rounds is given by $B(n) = \dfrac{3^{n+1} - 3}{2}$. Use this model to determine

a. the number of beneficiaries after 5 rounds and after 10 rounds. Assume that no person is a beneficiary of more than one good deed.

b. how many rounds are required to produce at least 2 million beneficiaries.

56. INTENSITY OF LIGHT The percent $I(x)$ of the original intensity of light striking the surface of a lake that is available x feet below the surface of the lake is given by $I(x) = 100e^{-0.95x}$.

a. What percentage of the light, to the nearest tenth of a percent, is available 2 feet below the surface of the lake?

b. At what depth, to the nearest hundredth of a foot, is the intensity of the light one-half the intensity at the surface?

57. 🖩 A TEMPERATURE MODEL A cup of coffee is heated to 180°F and placed in a room that maintains a temperature of 65°F. The temperature of the coffee after t minutes is given by $T(t) = 65 + 115e^{-0.042t}$.

a. Find the temperature, to the nearest degree, of the coffee 10 minutes after it is placed in the room.

b. Use a graphing utility to determine when, to the nearest tenth of a minute, the temperature of the coffee will reach 100°F.

58. 🖩 A TEMPERATURE MODEL Soup that is at a temperature of 170°F is poured into a bowl in a room that maintains a constant temperature. The temperature of the soup decreases according to the model given by $T(t) = 75 + 95e^{-0.12t}$, where t is time in minutes after the soup is poured.

a. What is the temperature, to the nearest tenth of a degree, of the soup after 2 minutes?

b. A certain customer prefers soup at a temperature of 110°F. How many minutes, to the nearest 0.1 minute, after the soup is poured does the soup reach that temperature?

c. What is the temperature of the room?

59. 🖩 MUSICAL SCALES Starting on the left side of a standard 88-key piano, the frequency, in vibrations per second, of the nth note is given by $f(n) = (27.5)2^{(n-1)/12}$.

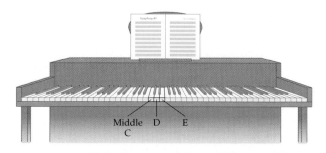

a. Using this formula, determine the frequency, to the nearest hundredth of a vibration per second, of middle C, key number 40 on an 88-key piano.

b. Is the difference in frequency between middle C (key number 40) and D (key number 42) the same as the difference in frequency between D (key number 42) and E (key number 44)? Explain.

CONNECTING CONCEPTS

60. Verify that the hyperbolic cosine function

$$\cosh(x) = \frac{e^x + e^{-x}}{2}$$ is an even function.

61. Verify that the hyperbolic sine function $\sinh(x) = \dfrac{e^x - e^{-x}}{2}$

is an odd function.

62. Graph $g(x) = 10^x$, and then sketch the graph of g reflected across the line given by $y = x$.

63. Graph $f(x) = e^x$, and then sketch the graph of f reflected across the line given by $y = x$.

In Exercises 64 to 67, determine the domain of the given function. Write the domain using interval notation.

64. $f(x) = \dfrac{e^x - e^{-x}}{e^x + e^{-x}}$

65. $f(x) = \dfrac{e^{|x|}}{1 + e^x}$

66. $f(x) = \sqrt{1 - e^x}$

67. $f(x) = \sqrt{e^x - e^{-x}}$

PREPARE FOR SECTION 7.2

68. If $2^x = 16$, determine the value of x. [7.1]

69. If $3^{-x} = \dfrac{1}{27}$, determine the value of x. [7.1]

70. If $x^4 = 625$, determine the value of x. [7.1]

71. Find the inverse of $f(x) = \dfrac{2x}{x + 3}$. [1.6]

72. State the domain of $g(x) = \sqrt{x - 2}$. [1.3]

73. If the range of $h(x)$ is the set of all positive real numbers, then what is the domain of $h^{-1}(x)$? [1.6]

PROJECTS

1. **THE SAINT LOUIS GATEWAY ARCH** The Gateway Arch in Saint Louis was designed in the shape of an inverted **catenary,** as shown by the red curve in the drawing below. The Gateway Arch is one of the largest optical illusions ever created. As you look at the arch (and its basic shape defined by the catenary curve), it appears to be much taller than it is wide. However, this is not the case. The height of the catenary is given by

$$h(x) = 693.8597 - 68.7672\left(\frac{e^{0.0100333x} + e^{-0.0100333x}}{2}\right)$$

where x and $h(x)$ are measured in feet and $x = 0$ represents the position at ground level that is directly below the highest point of the catenary.

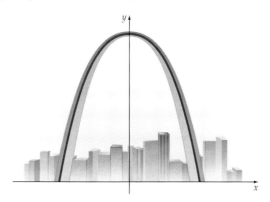

a. Use a graphing utility to graph $h(x)$.

b. Use your graph to find the height of the catenary for $x = 0$, 100, 200, and 299 feet. Round each result to the nearest tenth of a foot.

c. What is the width of the catenary at ground level and what is the maximum height of the catenary? Round each result to the nearest tenth of a foot.

d. By how much does the maximum height of the catenary exceed its width at ground level? Round to the nearest tenth of a foot.

2. **AN EXPONENTIAL REWARD** According to legend, when Sissa Ben Dahir of India invented the game of chess, King Shirham was so impressed with the game that he summoned the game's inventor and offered him the reward of his choosing. The inventor pointed to the chessboard and requested, for his reward, one grain of wheat on the first square, two grains of wheat on the second square, four grains on the third square, eight grains on the fourth square, and so on for all 64 squares on the chessboard. The King considered this a very modest reward and said he would grant the inventor's wish. The following table shows how many grains of rice are on each of the first six squares and the total number of grains of wheat needed to cover squares 1 to n for $n \le 6$.

Square number, n	Number of grains of wheat on square n	Total number of grains of wheat on squares 1 through n
1	1	1
2	2	3
3	4	7
4	8	15
5	16	31
6	32	63

a. If all 64 squares of the chessboard are piled with wheat as requested by Sissa Ben Dahir, how many grains of wheat are on the board?

b. A grain of wheat weighs approximately 0.000008 kilogram. Find the total weight of the wheat requested by Sissa Ben Dahir.

c. In a recent year, a total of 6.5×10^8 metric tons of wheat were produced in the world. At this level, how many years, to the nearest year, of wheat production would be required to fill the request of Sissa Ben Dahir? One metric ton equals 1000 kilograms.

LOGARITHMIC FUNCTIONS AND THEIR APPLICATIONS

SECTION **7.2**

- **LOGARITHMIC FUNCTIONS**
- **GRAPHS OF LOGARITHMIC FUNCTIONS**
- **DOMAINS OF LOGARITHMIC FUNCTIONS**
- **COMMON AND NATURAL LOGARITHMS**
- **APPLICATIONS OF LOGARITHMIC FUNCTIONS**

● LOGARITHMIC FUNCTIONS

Every exponential function of the form $g(x) = b^x$ is a one-to-one function and therefore has an inverse function. Sometimes we can determine the inverse of a function represented by an equation by interchanging the variables of its equation and then solving for the dependent variable. If we attempt to use this procedure for $g(x) = b^x$, we obtain

$$g(x) = b^x$$
$$y = b^x$$
$$x = b^y \qquad \text{• Interchange the variables.}$$

None of our previous methods can be used to solve the equation $x = b^y$ for the exponent y. Thus we need to develop a new procedure. One method would be to merely write

$$y = \text{the power of } b \text{ that produces } x$$

Although this would work, it is not very concise. We need a compact notation to represent "y is the power of b that produces x." This more compact notation is given in the following definition.

Logarithms were developed by John Napier (1550–1617) as a means of simplifying the calculations of astronomers. One of his ideas was to devise a method by which the product of two numbers could be determined by performing an addition.

Definition of a Logarithm and a Logarithmic Function

If $x > 0$ and b is a positive constant ($b \neq 1$), then

$$y = \log_b x \qquad \text{if and only if} \qquad b^y = x$$

The notation $\log_b x$ is read "the **logarithm** (or log) base b of x." The function defined by $f(x) = \log_b x$ is a **logarithmic function** with base b. This function is the inverse of the exponential function $g(x) = b^x$.

It is essential to remember that $f(x) = \log_b x$ is the inverse function of $g(x) = b^x$. Because these functions are inverses and because functions that are inverses have the property that $f(g(x)) = x$ and $g(f(x)) = x$, we have the following important relationships.

Composition of Logarithmic and Exponential Functions

Let $g(x) = b^x$ and $f(x) = \log_b x$ $(x > 0, b > 0, b \neq 1)$. Then

$$g(f(x)) = b^{\log_b x} = x \qquad \text{and} \qquad f(g(x)) = \log_b b^x = x$$

take note

The notation $\log_b x$ replaces the phrase "the power of b that produces x." For instance, "3 is the power of 2 that produces 8" is abbreviated $3 = \log_2 8$. In your work with logarithms, remember that a logarithm is an *exponent*.

As an example of these relationships, let $g(x) = 2^x$ and $f(x) = \log_2 x$. Then

$$2^{\log_2 x} = x \qquad \text{and} \qquad \log_2 2^x = x$$

The equations

$$y = \log_b x \qquad \text{and} \qquad b^y = x$$

are different ways of expressing the same concept.

Exponential Form and Logarithmic Form

The **exponential form** of $y = \log_b x$ is $b^y = x$.

The **logarithmic form** of $b^y = x$ is $y = \log_b x$.

These concepts are illustrated in the next two examples.

EXAMPLE 1 Change from Logarithmic to Exponential Form

Write each equation in its exponential form.

a. $3 = \log_2 8$ b. $2 = \log_{10}(x + 5)$ c. $\log_e x = 4$ d. $\log_b b^3 = 3$

Solution

Use the definition $y = \log_b x$ if and only if $b^y = x$.

┌── Logarithms are exponents. ──┐
a. $3 = \log_2 8$ if and only if $2^3 = 8$
└──────── Base ────────┘

b. $2 = \log_{10}(x + 5)$ if and only if $10^2 = x + 5$.

c. $\log_e x = 4$ if and only if $e^4 = x$.

d. $\log_b b^3 = 3$ if and only if $b^3 = b^3$.

▶ **TRY EXERCISE 4, PAGE 458**

EXAMPLE 2   **Change from Exponential to Logarithmic Form**

Write each equation in its logarithmic form.

a. $3^2 = 9$ **b.** $5^3 = x$ **c.** $a^b = c$ **d.** $b^{\log_b 5} = 5$

Solution

The logarithmic form of $b^y = x$ is $y = \log_b x$.

a. $3^2 = 9$ if and only if $2 = \log_3 9$

$$\overbrace{\qquad}^{\text{Exponent}} \qquad \underbrace{\qquad}_{\text{Base}}$$

b. $5^3 = x$ if and only if $3 = \log_5 x$.

c. $a^b = c$ if and only if $b = \log_a c$.

d. $b^{\log_b 5} = 5$ if and only if $\log_b 5 = \log_b 5$.

▶ **TRY EXERCISE 12, PAGE 458**

The definition of a logarithm and the definition of inverse functions can be used to establish many properties of logarithms. For instance:

- $\log_b b = 1$ because $b = b^1$.
- $\log_b 1 = 0$ because $1 = b^0$.
- $\log_b(b^x) = x$ because $b^x = b^x$.
- $b^{\log_b x} = x$ because $f(x) = \log_b x$ and $g(x) = b^x$ are inverse functions. Thus $g[f(x)] = x$.

We will refer to the preceding properties as the *basic logarithmic properties*.

Basic Logarithmic Properties

1. $\log_b b = 1$ **2.** $\log_b 1 = 0$ **3.** $\log_b(b^x) = x$ **4.** $b^{\log_b x} = x$

EXAMPLE 3   **Apply the Basic Logarithmic Properties**

Evaluate each of the following logarithms.

a. $\log_8 1$ **b.** $\log_5 5$ **c.** $\log_2(2^4)$ **d.** $3^{\log_3 7}$

Solution

a. By Property 2, $\log_8 1 = 0$.

b. By Property 1, $\log_5 5 = 1$.

c. By Property 3, $\log_2(2^4) = 4$.

d. By Property 4, $3^{\log_3 7} = 7$.

▶ **TRY EXERCISE 28, PAGE 458**

Some logarithms can be evaluated just by remembering that a logarithm is an exponent. For instance, $\log_5 25$ equals 2 because the base 5 raised to the second power equals 25.

- $\log_{10} 100 = 2$ because $10^2 = 100$.

- $\log_4 64 = 3$ because $4^3 = 64$.

- $\log_7 \dfrac{1}{49} = -2$ because $7^{-2} = \dfrac{1}{7^2} = \dfrac{1}{49}$.

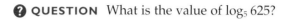 **QUESTION** What is the value of $\log_5 625$?

• GRAPHS OF LOGARITHMIC FUNCTIONS

Because $f(x) = \log_b x$ is the inverse function of $g(x) = b^x$, the graph of f is a reflection of the graph of g across the line given by $y = x$. The graph of $g(x) = 2^x$ is shown in **Figure 7.13**. **Table 7.7** below shows some of the ordered pairs on the graph of g.

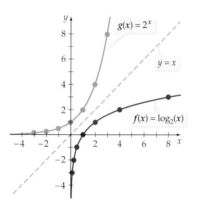

FIGURE 7.13

TABLE 7.7

x	-3	-2	-1	0	1	2	3
$g(x) = 2^x$	$\dfrac{1}{8}$	$\dfrac{1}{4}$	$\dfrac{1}{2}$	1	2	4	8

The graph of the inverse of g, which is $f(x) = \log_2 x$, is also shown in **Figure 7.13**. Some of the ordered pairs of f are shown in **Table 7.8**. Note that if (x, y) is a point on the graph of g, then (y, x) is a point on the graph of f. Also notice that the graph of f is a reflection of the graph of g across the line given by $y = x$.

TABLE 7.8

x	$\dfrac{1}{8}$	$\dfrac{1}{4}$	$\dfrac{1}{2}$	1	2	4	8
$f(x) = \log_2 x$	-3	-2	-1	0	1	2	3

The graph of a logarithmic function can be drawn by first rewriting the function in its exponential form. This procedure is illustrated in Example 4.

EXAMPLE 4   Graph a Logarithmic Function

Graph $f(x) = \log_3 x$.

Solution

To graph $f(x) = \log_3 x$, consider the equivalent exponential equation $x = 3^y$. Because this equation is solved for x, choose values of y and calculate the corresponding values of x, as shown in **Table 7.9**.

Continued ▶

ANSWER $\log_5 625 = 4$ because $5^4 = 625$.

TABLE 7.9

$x = 3^y$	$\frac{1}{9}$	$\frac{1}{3}$	1	3	9
y	-2	-1	0	1	2

Now plot the ordered pairs and connect the points with a smooth curve, as shown in **Figure 7.14.**

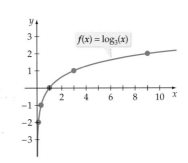

FIGURE 7.14

▶ **TRY EXERCISE 32, PAGE 459**

We can use a similar procedure to draw the graph of a logarithmic function with a fractional base. For instance, consider $y = \log_{2/3} x$. Rewriting this in exponential form gives us $\left(\dfrac{2}{3}\right)^y = x$. Choose values of y and calculate the corresponding x-values. See **Table 7.10.** Plot the points corresponding to the ordered pairs (x, y), and then draw a smooth curve through the points, as shown in **Figure 7.15.**

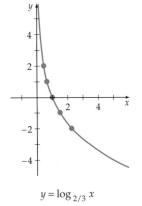

$y = \log_{2/3} x$

FIGURE 7.15

TABLE 7.10

$x = \left(\dfrac{2}{3}\right)^y$	$\left(\dfrac{2}{3}\right)^{-2} = \dfrac{9}{4}$	$\left(\dfrac{2}{3}\right)^{-1} = \dfrac{3}{2}$	$\left(\dfrac{2}{3}\right)^{0} = 1$	$\left(\dfrac{2}{3}\right)^{1} = \dfrac{2}{3}$	$\left(\dfrac{2}{3}\right)^{2} = \dfrac{4}{9}$
y	-2	-1	0	1	2

Properties of $f(x) = \log_b x$

For all positive real numbers b, $b \neq 1$, the function $f(x) = \log_b x$ has the following properties:

1. The domain of f consists of the set of positive real numbers and its range consists of the set of all real numbers.

2. The graph of f has an x-intercept of $(1, 0)$ and passes through $(b, 1)$.

3. If $b > 1$, f is an increasing function and its graph is asymptotic to the negative y-axis. [As $x \to \infty$, $f(x) \to \infty$, and as $x \to 0$ from the right, $f(x) \to -\infty$.] See **Figure 7.16a.**

4. If $0 < b < 1$, f is a decreasing function and its graph is asymptotic to the positive y-axis. [As $x \to \infty$, $f(x) \to -\infty$, and as $x \to 0$ from the right, $f(x) \to \infty$.] See **Figure 7.16b.**

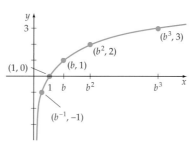
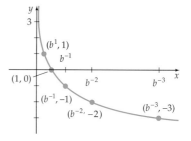

a. $f(x) = \log_b x,\ b > 1$ **b.** $f(x) = \log_b x,\ 0 < b < 1$

FIGURE 7.16

● DOMAINS OF LOGARITHMIC FUNCTIONS

The function $f(x) = \log_b x$ has as its domain the set of positive real numbers. The function $f(x) = \log_b(g(x))$ has as its domain the set of all x for which $g(x) > 0$. To determine the domain of a function such as $f(x) = \log_b(g(x))$, we must determine the values of x that make $g(x)$ positive. This process is illustrated in Example 5.

EXAMPLE 5   **Find the Domain of a Logarithmic Function**

Find the domain of each of the following logarithmic functions.

a. $f(x) = \log_6(x - 3)$ b. $F(x) = \log_2|x + 2|$ c. $R(x) = \log_5\left(\dfrac{x}{8 - x}\right)$

Solution

a.   Solving $(x - 3) > 0$ for x gives us $x > 3$. The domain of f consists of all real numbers greater than 3. In interval notation the domain is $(3, \infty)$.

b.   The solution set of $|x + 2| > 0$ consists of all real numbers x except $x = -2$. The domain of F consists of all real numbers $x \neq -2$. In interval notation the domain is $(-\infty, -2) \cup (-2, \infty)$.

c.   Solving $\left(\dfrac{x}{8 - x}\right) > 0$ yields the set of all real numbers x between 0 and 8. The domain of R is all real numbers x such that $0 < x < 8$. In interval notation the domain is $(0, 8)$.

▶ **TRY EXERCISE 40, PAGE 459**

Some logarithmic functions can be graphed by using horizontal and/or vertical translations of a previously drawn graph.

EXAMPLE 6 Use Translations to Graph Logarithmic Functions

Graph: **a.** $f(x) = \log_4(x + 3)$ **b.** $f(x) = \log_4 x + 3$

Solution

a. The graph of $f(x) = \log_4(x + 3)$ can be obtained by shifting the graph of $g(x) = \log_4 x$ to the left 3 units. See **Figure 7.17**. Note that the domain of f consists of all real numbers x greater than -3 because $x + 3 > 0$ for $x > -3$. The graph of f is asymptotic to the vertical line $x = -3$.

b. The graph of $f(x) = \log_4 x + 3$ can be obtained by shifting the graph of $g(x) = \log_4 x$ upward 3 units. See **Figure 7.18**.

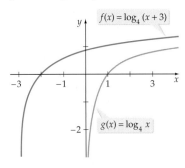

FIGURE 7.17

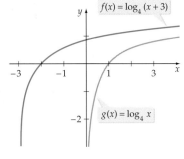

FIGURE 7.18

▶ **TRY EXERCISE 50, PAGE 459**

• COMMON AND NATURAL LOGARITHMS

Two of the most frequently used logarithmic functions are *common logarithms*, which have base 10, and *natural logarithms*, which have base e (the base of the natural exponential function).

Definition of Common and Natural Logarithms

The function defined by $f(x) = \log_{10} x$ is called the **common logarithmic function**. It is customarily written without stating the base as $f(x) = \log x$.

The function defined by $f(x) = \log_e x$ is called the **natural logarithmic function**. It is customarily written as $f(x) = \ln x$.

Most scientific or graphing calculators have a $\boxed{\text{LOG}}$ key for evaluating common logarithms and an $\boxed{\text{LN}}$ key to evaluate natural logarithms. For instance, using a graphing calculator,

$$\log 24 \approx 1.3802112 \qquad \text{and} \qquad \ln 81 \approx 4.3944492$$

The graphs of $f(x) = \log x$ and $f(x) = \ln x$ can be drawn using the same techniques we used to draw the graphs in the preceding examples. However, these graphs also can be produced with a graphing calculator by entering $\log x$ and $\ln x$ into the Y= menu. See **Figure 7.19** and **Figure 7.20.**

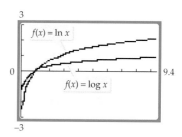

FIGURE 7.19   **FIGURE 7.20**

Observe that each graph passes through $(1, 0)$. Also note that as $x \to 0$ from the right, the functional values $f(x) \to -\infty$. Thus the y-axis is a vertical asymptote for each of the graphs. The domain of both $f(x) = \log x$ and $f(x) = \ln x$ is the set of positive real numbers. Each of these functions has a range consisting of the set of real numbers.

● APPLICATIONS OF LOGARITHMIC FUNCTIONS

Many applications can be modeled by logarithmic functions.

EXAMPLE 7   **Average Time of a Major League Baseball Game**

From 1981 to 1999, the average time of a major league baseball game tended to increase each year. If the year 1981 is represented by $x = 1$, then the function

$$T(x) = 149.57 + 7.63 \ln x$$

approximates the average time T, in minutes, of a major league baseball game for the years 1981 to 1999—that is, for $x = 1$ to $x = 19$.

a. Use the function T to determine the average time of a major league baseball game during the 1981 season and during the 1999 season.

b. By how much did the average time of a major league baseball game increase during the years 1981 to 1999?

Solution

a. The year 1981 is represented by $x = 1$ and the year 1999 by $x = 19$.

$$T(1) = 149.57 + 7.63 \ln(1) = 149.57$$

In 1981 the average time of a baseball game was about 149.57 minutes.

$$T(19) = 149.57 + 7.63 \ln(19) \approx 172.04$$

In 1999 the average time of a baseball game was about 172.04 minutes.

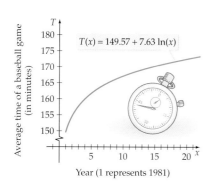

Continued ▶

MATH MATTERS

Although logarithms were originally developed to assist with computations, logarithmic functions have a much broader use today. They are often used in such disciplines as geology, acoustics, chemistry, physics, and economics, to name a few.

b. $T(19) - T(1) \approx 172.04 - 149.57 = 22.47.$ During the years 1981 to 1999, the average time of a baseball game increased by about 22.47 minutes.

▶ **Try Exercise 70, page 460**

Topics for Discussion

1. If $m > n$, must $\log_b m > \log_b n$?

2. For what values of x is $\ln x > \log x$?

3. What is the domain of $f(x) = \log(x^2 + 1)$? Explain why the graph of f does not have a vertical asymptote.

4. The subtraction $3 - 5$ does not have an answer if we require that the answer be positive. Keep this idea in mind as you work the rest of this exercise.

    Press the MODE key of a TI-83 graphing calculator, and choose "Real" from the menu. Now use the calculator to evaluate $\log(-2)$. What output is given by the calculator? Press the MODE key, and choose "a + bi" from the menu. Now use the calculator to evaluate $\log(-2)$. What output is given by the calculator? Write a sentence or two that explain why the output is different for these two evaluations.

Exercise Set 7.2

In Exercises 1 to 10, change each equation to its exponential form.

1. $\log 10 = 1$

2. $\log 10{,}000 = 4$

3. $\log_8 64 = 2$

▶ 4. $\log_4 64 = 3$

5. $\log_7 x = 0$

6. $\log_3 \dfrac{1}{81} = -4$

7. $\ln x = 4$

8. $\ln e^2 = 2$

9. $\ln 1 = 0$

10. $\ln x = -3$

In Exercises 11 to 20, change each equation to its logarithmic form. Assume $y > 0$ and $b > 0$.

11. $3^2 = 9$

▶ 12. $5^3 = 125$

13. $4^{-2} = \dfrac{1}{16}$

14. $10^0 = 1$

15. $b^x = y$

16. $2^x = y$

17. $y = e^x$

18. $5^1 = 5$

19. $100 = 10^2$

20. $2^{-4} = \dfrac{1}{16}$

In Exercises 21 to 30, evaluate each logarithm. Do not use a calculator.

21. $\log_4 16$

22. $\log_{3/2} \dfrac{8}{27}$

23. $\log_3 \dfrac{1}{243}$

24. $\log_b 1$

25. $\ln e^3$

26. $\log_b b$

27. $\log \dfrac{1}{100}$

▶ 28. $\log 1{,}000{,}000$

29. $\log_{0.5} 16$

30. $\log_{0.3} \dfrac{100}{9}$

In Exercises 31 to 38, graph each function by using its exponential form.

31. $f(x) = \log_4 x$

▶ 32. $f(x) = \log_6 x$

33. $f(x) = \log_{12} x$

34. $f(x) = \log_8 x$

35. $f(x) = \log_{1/2} x$

36. $f(x) = \log_{1/4} x$

37. $f(x) = \log_{5/2} x$

38. $f(x) = \log_{7/3} x$

In Exercises 39 to 48, find the domain of the function. Write the domains using interval notation.

39. $f(x) = \log_5(x - 3)$

▶ 40. $k(x) = \log_4(5 - x)$

41. $k(x) = \log_{2/3}(11 - x)$

42. $H(x) = \log_{1/4}(x^2 + 1)$

43. $P(x) = \ln(x^2 - 4)$

44. $J(x) = \ln\left(\dfrac{x - 3}{x}\right)$

45. $h(x) = \ln\left(\dfrac{x^2}{x - 4}\right)$

46. $R(x) = \ln(x^4 - x^2)$

47. $N(x) = \log_2(x^3 - x)$

48. $s(x) = \log_7(x^2 + 7x + 10)$

In Exercises 49 to 56, use translations of the graphs in Exercises 31 to 38 to produce the graph of the given function.

49. $f(x) = \log_4(x - 3)$

▶ 50. $f(x) = \log_6(x + 3)$

51. $f(x) = \log_{12} x + 2$

52. $f(x) = \log_8 x - 4$

53. $f(x) = 3 + \log_{1/2} x$

54. $f(x) = 2 + \log_{1/4} x$

55. $f(x) = 1 + \log_{5/2}(x - 4)$

56. $f(x) = \log_{7/3}(x - 3) - 1$

57. Examine the following four functions and the graphs labeled **a, b, c,** and **d.** Determine which graph is the graph of each function.

$$f(x) = \log_5(x - 2) \qquad g(x) = 2 + \log_5 x$$
$$h(x) = \log_5(-x) \qquad k(x) = -\log_5(x + 3)$$

a.

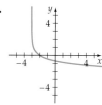

b.

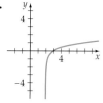

c.

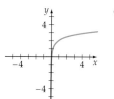

d.

58. Examine the following four functions and the graphs labeled **a, b, c,** and **d.** Determine which graph is the graph of each function.

$$f(x) = \ln x + 3 \qquad g(x) = \ln(x - 3)$$
$$h(x) = \ln(3 - x) \qquad k(x) = -\ln(-x)$$

a.

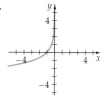

b.

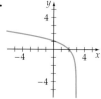

c.

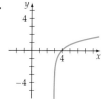

d.

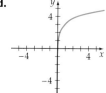

In Exercises 59 to 68, use a graphing utility to graph the function.

59. $f(x) = -2 \ln x$

60. $f(x) = -\log x$

61. $f(x) = |\ln x|$

62. $f(x) = \ln |x|$

63. $f(x) = \log \sqrt[3]{x}$

64. $f(x) = \ln \sqrt{x}$

65. $f(x) = \log(x + 10)$

66. $f(x) = \ln(x + 3)$

67. $f(x) = 3 \log |2x + 10|$

68. $f(x) = \dfrac{1}{2} \ln |x - 4|$

69. **MONEY MARKET RATES** The function
$$r(t) = 0.69607 + 0.60781 \ln t$$
gives the annual interest rate r, as a percent, a bank will pay on its money market accounts, where t is the term (the time the money is invested) in months.

a. What interest rate, to the nearest tenth of a percent, will the bank pay on a money market account with a term of 9 months?

b. What is the minimum number of complete months during which a person must invest to receive an interest rate of at least 3%?

▶ **70.** AVERAGE TYPING SPEED The following function models the average typing speed S, in words per minute, of a student who has been typing for t months.

$$S(t) = 5 + 29 \ln(t + 1), \quad 0 \leq t \leq 16$$

a. What was the student's average typing speed, to the nearest word per minute, when the student first started to type? What was the student's average typing speed, to the nearest word per minute, after 3 months?

b. Use a graph of S to determine how long, to the nearest tenth of a month, it will take the student to achieve an average typing speed of 65 words per minute.

71. ADVERTISING COSTS AND SALES The function

$$N(x) = 2750 + 180 \ln\left(\frac{x}{1000} + 1\right)$$

models the relationship between the dollar amount x spent on advertising a product and the number of units N that a company can sell.

a. Find the number of units that will be sold with advertising expenditures of $20,000, $40,000, and $60,000.

b. How many units will be sold if the company does not pay to advertise the product?

In anesthesiology it is necessary to accurately estimate the body surface area of a patient. One formula for estimating body surface area (*BSA*) was developed by Edith Boyd (University of Minnesota Press, 1935). Her formula for the *BSA* (in square meters) of a patient of height H (in centimeters) and weight W (in grams) is

$$BSA = 0.0003207 \cdot H^{0.3} \cdot W^{(0.7285 - 0.0188 \log W)}$$

MEDICINE **In Exercises 72 and 73, use Boyd's formula to estimate the body surface area of a patient with the given weight and height. Round to the nearest hundredth of a square meter.**

72. $W = 110$ pounds (49,895.2 grams); $H = 5$ feet 4 inches (162.56 centimeters)

73. $W = 180$ pounds (81,646.6 grams); $H = 6$ feet 1 inch (185.42 centimeters)

74. ASTRONOMY Astronomers measure the apparent brightness of a star by a unit called the **apparent magnitude.** This unit was created in the second century B.C. when the Greek astronomer Hipparchus classified the relative brightness of several stars. In his list he assigned the number 1 to the stars that appeared to be the brightest (Sirius, Vega, and Deneb). They are first-magnitude stars. Hipparchus assigned the number 2 to all the stars in the Big Dipper. They are second-magnitude stars. The following table shows the relationship between a star's brightness relative to a first-magnitude star and the star's apparent magnitude. Notice from the table that a first-magnitude star appears in the sky to be about 2.51 times as bright as a second-magnitude star.

Brightness relative to a first-magnitude star x	Apparent magnitude $M(x)$
1	1
$\dfrac{1}{2.51}$	2
$\dfrac{1}{6.31} \approx \dfrac{1}{2.51^2}$	3
$\dfrac{1}{15.85} \approx \dfrac{1}{2.51^3}$	4
$\dfrac{1}{39.82} \approx \dfrac{1}{2.51^4}$	5
$\dfrac{1}{100} \approx \dfrac{1}{2.51^5}$	6

The following logarithmic function gives the apparent magnitude $M(x)$ of a star as a function of its brightness x.

$$M(x) = -2.51 \log x + 1, \quad 0 < x \leq 1$$

a. Use $M(x)$ to find the apparent magnitude of a star that is $\dfrac{1}{10}$ as bright as a first-magnitude star. Round to the nearest hundredth.

b. Find the approximate apparent magnitude of a star that is $\dfrac{1}{400}$ as bright as a first-magnitude star. Round to the nearest hundredth.

c. Which star appears brighter: a star with an apparent magnitude of 12 or a star with an apparent magnitude of 15?

d. Is $M(x)$ an increasing function or a decreasing function?

75. NUMBER OF DIGITS IN b^x An engineer has determined that the number of digits N in the expansion of b^x, where both b and x are positive integers, is $N = \text{int}(x \log b) + 1$, where $\text{int}(x \log b)$ denotes the greatest integer of $x \log b$. (*Note:* The greatest integer of the real number x is x if x is an integer and is the largest integer less than x if x is not an integer. For example, the greatest integer of 5 is 5 and the greatest integer of 7.8 is 7.)

a. Because $2^{10} = 1024$, we know that 2^{10} has four digits. Use the equation $N = \text{int}(x \log b) + 1$ to verify this result.

b. Find the number of digits in 3^{200}.

c. Find the number of digits in 7^{4005}.

d. The largest known prime number as of November 17, 2003 was $2^{20996011} - 1$. Find the number of digits in this prime number. (*Hint:* Because $2^{20996011}$ is not a power of 10, both $2^{20996011}$ and $2^{20996011} - 1$ have the same number of digits.)

76. NUMBER OF DIGITS IN $9^{(9^9)}$ A science teacher has offered 10 points extra credit to any student who will write out all the digits in the expansion of $9^{(9^9)}$.

a. Use the formula from Exercise 75 to determine the number of digits in this number.

b. Assume that you can write 1000 digits per page and that 500 pages of paper are in a ream of paper. How many reams of paper, to the nearest tenth of a ream, are required to write out the expansion of $9^{(9^9)}$? Assume that you write on only one side of each page.

CONNECTING CONCEPTS

77. Use a graphing utility to graph $f(x) = \dfrac{e^x - e^{-x}}{2}$ and $g(x) = \ln(x + \sqrt{x^2 + 1})$ on the same screen. Use a square viewing window. What appears to be the relationship between f and g?

78. Use a graphing utility to graph $f(x) = \dfrac{e^x + e^{-x}}{2}$, for $x \geq 0$, and $g(x) = \ln(x + \sqrt{x^2 - 1})$, for $x \geq 1$, on the same screen. Use a square viewing window. What appears to be the relationship between f and g?

79. The functions $f(x) = \dfrac{e^x - e^{-x}}{e^x + e^{-x}}$ and $g(x) = \dfrac{1}{2} \ln \dfrac{1 + x}{1 - x}$ are inverse functions. The domain of f is the set of all real numbers. The domain of g is $\{x \,|\, -1 < x < 1\}$. Use this information to determine the range of f and the range of g.

80. Use a graph of $f(x) = \dfrac{2}{e^x + e^{-x}}$ to determine the domain and the range of f.

PREPARE FOR SECTION 7.3

In Exercises 81 to 86, use a calculator to compare each of the given expressions.

81. $\log 3 + \log 2$; $\log 6$ [7.2]

82. $\ln 8 - \ln 3$; $\ln\left(\dfrac{8}{3}\right)$ [7.2]

83. $3 \log 4$; $\log(4^3)$ [7.2]

84. $2 \ln 5$; $\ln(5^2)$ [7.2]

85. $\ln 5$; $\dfrac{\log 5}{\log e}$ [7.2]

86. $\log 8$; $\dfrac{\ln 8}{\ln 10}$ [7.2]

PROJECTS

1. **BENFORD'S LAW** The authors of this text know some interesting details about your finances. For instance, of the last 150 checks you have written, about 30% are for amounts that start with the number 1. Also, you have written about 3 times as many checks for amounts that start with the number 2 than you have for amounts that start with the number 7.

We are sure of these results because of a mathematical formula known as **Benford's Law.** This law was first discovered by the mathematician Simon Newcomb in 1881 and then rediscovered by the physicist Frank Benford in 1938. Benford's Law states that the probability P that the first digit of a number selected from a wide range of numbers is d is given by

$$P(d) = \log\left(1 + \frac{1}{d}\right)$$

a. Use Benford's Law to complete the table and the bar graph below.

d	$P(d) = \log\left(1 + \dfrac{1}{d}\right)$
1	0.301
2	0.176
3	0.125
4	
5	
6	
7	
8	
9	

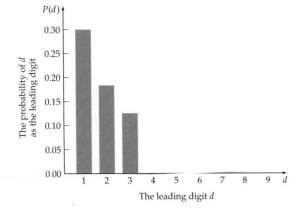

Benford's Law applies to most data with a wide range. For instance, it applies to

- the populations of the cities in the U.S.
- the numbers of dollars in the savings accounts at your local bank.
- the number of miles driven during a month by each person in a state.

b. Use the table in part **a.** to find the probability that in a U.S. city selected at random, the number of telephones in that city will be a number starting with 6.

c. Use the table in part **a.** to estimate how many times as many purchases you have made for dollar amounts that start with a 1 than for dollar amounts that start with a 9.

d. Explain why Benford's Law would not apply to the set of telephone numbers of the people living in a small city such as Le Mars, Iowa.

e. Explain why Benford's Law would not apply to the set of all the ages, in years, of students at a local high school.

AN APPLICATION OF BENFORD'S LAW Benford's Law has been used to identify fraudulent accountants. In most cases these accountants are unaware of Benford's Law and have replaced valid numbers with numbers selected at random. Their numbers do not conform to Benford's Law. Hence an audit is warranted.

SECTION 7.3 # LOGARITHMS AND LOGARITHMIC SCALES

- ● **PROPERTIES OF LOGARITHMS**
- ● **CHANGE-OF-BASE FORMULA**
- ● **LOGARITHMIC SCALES**

● PROPERTIES OF LOGARITHMS

In Section 7.2 we introduced the following basic properties of logarithms.

$$\log_b b = 1 \qquad \text{and} \qquad \log_b 1 = 0$$

Also, because exponential functions and logarithmic functions are inverses of each other, we observed the relationships

$$\log_b(b^x) = x \qquad \text{and} \qquad b^{\log_b x} = x$$

We can use the properties of exponents to establish the following additional logarithmic properties.

Properties of Logarithms

In the following properties, b, M, and N are positive real numbers ($b \neq 1$).

Product property	$\log_b(MN) = \log_b M + \log_b N$
Quotient property	$\log_b \dfrac{M}{N} = \log_b M - \log_b N$
Power property	$\log_b(M^p) = p \log_b M$
Logarithm-of-each-side property	$M = N$ implies $\log_b M = \log_b N$
One-to-one property	$\log_b M = \log_b N$ implies $M = N$

---

take note

Pay close attention to these properties. Note that

$$\log_b(MN) \neq \log_b M \cdot \log_b N$$

and

$$\log_b \frac{M}{N} \neq \frac{\log_b M}{\log_b N}$$

Also,

$$\log_b(M + N) \neq \log_b M + \log_b N$$

In fact, the expression $\log_b(M + N)$ cannot be expanded at all.

❓ QUESTION Is it true that $\ln 5 + \ln 10 = \ln 50$?

The above properties of logarithms are often used to rewrite logarithmic expressions in an equivalent form.

EXAMPLE 1   **Rewrite Logarithmic Expressions**

Use the properties of logarithms to express the following logarithms in terms of logarithms of x, y, and z.

a. $\log_5(xy^2)$ b. $\log_b \dfrac{2\sqrt{y}}{z^5}$

Solution

a. $\log_5(xy^2) = \log_5 x + \log_5 y^2$ • Product property

$\qquad\qquad\;\; = \log_5 x + 2\log_5 y$ • Power property

Continued ▶

❓ ANSWER Yes. By the product property, $\ln 5 + \ln 10 = \ln(5 \cdot 10)$.

b. $\log_b \dfrac{2\sqrt{y}}{z^5} = \log_b(2\sqrt{y}) - \log_b z^5$ • **Quotient property**

$\qquad = \log_b 2 + \log_b \sqrt{y} - \log_b z^5$ • **Product property**

$\qquad = \log_b 2 + \log_b y^{1/2} - \log_b z^5$ • **Replace $\sqrt{y}$ with $y^{1/2}$.**

$\qquad = \log_b 2 + \dfrac{1}{2}\log_b y - 5\log_b z$ • **Power property**

▶ **TRY EXERCISE 2, PAGE 471**

The properties of logarithms are also used to rewrite expressions that involve several logarithms as a single logarithm.

EXAMPLE 2   **Rewrite Logarithmic Expressions**

Use the properties of logarithms to rewrite each expression as a single logarithm with a coefficient of 1.

a. $2\log_b x + \dfrac{1}{2}\log_b(x + 4)$ b. $4\log_3(x + 2) - 3\log_3(x - 5)$

Solution

a. $2\log_b x + \dfrac{1}{2}\log_b(x + 4)$

$\qquad = \log_b x^2 + \log_b(x + 4)^{1/2}$ • **Power property**

$\qquad = \log_b[x^2(x + 4)^{1/2}]$ • **Product property**

$\qquad = \log_b(x^2\sqrt{x + 4})$

b. $4\log_3(x + 2) - 3\log_3(x - 5)$

$\qquad = \log_3(x + 2)^4 - \log_3(x - 5)^3$ • **Power property**

$\qquad = \log_3 \dfrac{(x + 2)^4}{(x - 5)^3}$ • **Quotient property**

▶ **TRY EXERCISE 10, PAGE 471**

● **CHANGE-OF-BASE FORMULA**

Recall that to determine the value of y in $\log_3 81 = y$, we are basically asking, "What power of 3 is equal to 81?" Because $3^4 = 81$, we have $\log_3 81 = 4$. Now suppose that we need to determine the value of $\log_3 50$. In this case we need to find the power of 3 that produces 50. Because $3^3 = 27$ and $3^4 = 81$, the value we are seeking is somewhere between 3 and 4. The following procedure can be used to produce an estimate of $\log_3 50$.

The exponential form of $\log_3 50 = y$ is $3^y = 50$. Applying logarithmic properties gives us

$$3^y = 50$$

$$\ln 3^y = \ln 50 \qquad \text{• Logarithm-of-each-side property}$$

$$y \ln 3 = \ln 50 \qquad \text{• Power property}$$

$$y = \dfrac{\ln 50}{\ln 3} \approx 3.56088 \qquad \text{• Solve for } y.$$

Thus $\log_3 50 \approx 3.56088$. In the preceding procedure we could just as well have used logarithms of any base and arrived at the same value. Thus any logarithm can be expressed in terms of logarithms of any base we wish. This general result is summarized in the following formula.

Change-of-Base Formula

If x, a, and b are positive real numbers with $a \neq 1$ and $b \neq 1$, then

$$\log_b x = \frac{\log_a x}{\log_a b}$$

Because most calculators use only common logarithms ($a = 10$) or natural logarithms ($a = e$), the change-of-base formula is used most often in the following form.

If x and b are positive real numbers and $b \neq 1$, then

$$\log_b x = \frac{\log x}{\log b} = \frac{\ln x}{\ln b}$$

EXAMPLE 3   Use the Change-of-Base Formula

Evaluate each logarithm. Round to the nearest hundred thousandth.

a. $\log_3 18$ b. $\log_{12} 400$

Solution

To approximate these logarithms, we may use the change-of-base formula with $a = 10$ or $a = e$. For this example we choose to use the change-of-base formula with $a = e$. That is, we will evaluate these logarithms by using the $\boxed{\text{LN}}$ key on a scientific or graphing calculator.

a. $\log_3 18 = \dfrac{\ln 18}{\ln 3} \approx 2.63093$ b. $\log_{12} 400 = \dfrac{\ln 400}{\ln 12} \approx 2.41114$

▶ **TRY EXERCISE 16, PAGE 472**

The change-of-base formula and a graphing calculator can be used to graph logarithmic functions that have a base other than 10 or e. For instance, to graph $f(x) = \log_3(2x + 3)$, we rewrite the function in terms of base 10 or base e. Using base 10 logarithms, we have $f(x) = \log_3(2x + 3) = \dfrac{\log(2x + 3)}{\log 3}$. The graph is shown in **Figure 7.21**.

take note

If common logarithms had been used for the calculation in Example 3**a.**, the final result would be the same.

$$\log_3 18 = \frac{\log 18}{\log 3} \approx 2.63093$$

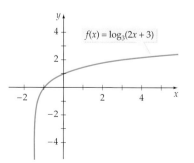

$f(x) = \log_3(2x + 3)$

FIGURE 7.21

EXAMPLE 4

Use the Change-of-Base Formula to Graph a Logarithmic Function

Graph $f(x) = \log_2|x - 3|$.

Solution

Rewrite f using the change-of-base formula. We will use the natural logarithm function; however, the common logarithm function could be used instead.

$$f(x) = \log_2|x - 3| = \frac{\ln|x - 3|}{\ln 2}$$

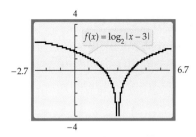

- Enter $\dfrac{\ln|x - 3|}{\ln 2}$ into **Y1**. Note that the domain of $f(x) = \log_2|x - 3|$ is all real numbers except 3, because $|x - 3| = 0$ when $x = 3$ and $|x - 3|$ is positive for all other values of x.

▶ **TRY EXERCISE 24, PAGE 472**

● LOGARITHMIC SCALES

Logarithmic functions are often used to scale very large (or very small) numbers into numbers that are easier to comprehend. For instance, the *Richter scale* magnitude of an earthquake uses a logarithmic function to convert the intensity of the earthquake's shock waves I into a number M, which for most earthquakes is in the range of 0 to 10. The intensity I of an earthquake is often given in terms of the constant I_0, where I_0 is the intensity of the smallest earthquake (called a **zero-level earthquake**) that can be measured on a seismograph near the earthquake's epicenter. The following formula is used to compute the Richter scale magnitude of an earthquake.

The Richter Scale Magnitude of an Earthquake

An earthquake with an intensity of I has a **Richter scale magnitude** of

$$M = \log\left(\frac{I}{I_0}\right)$$

where I_0 is the measure of the intensity of a zero-level earthquake.

EXAMPLE 5   Determine the Magnitude of an Earthquake

 Find the Richter scale magnitude (to the nearest 0.1) of the 1999 Joshua Tree, California earthquake that had an intensity of $I = 12{,}589{,}254I_0$.

Solution

$$M = \log\left(\frac{I}{I_0}\right) = \log\left(\frac{12{,}589{,}254I_0}{I_0}\right) = \log(12{,}589{,}254) \approx 7.1$$

The 1999 Joshua Tree earthquake had a Richter scale magnitude of 7.1.

▶ **TRY EXERCISE 56, PAGE 473**

> *take note*
>
> Notice in Example 5 that we didn't need to know the value of I_0 to determine the Richter scale magnitude of the quake.

If you know the Richter scale magnitude of an earthquake, you can determine the intensity of the earthquake.

EXAMPLE 6   Determine the Intensity of an Earthquake

 Find the intensity of the 1999 Taiwan earthquake, which measured 7.6 on the Richter scale.

Solution

$$\log\left(\frac{I}{I_0}\right) = 7.6$$

$$\frac{I}{I_0} = 10^{7.6}$$ • Write in exponential form.

$$I = 10^{7.6}I_0$$ • Solve for I.

$$I \approx 39{,}810{,}717I_0$$

The 1999 Taiwan earthquake had an intensity that was approximately 39,811,000 times the intensity of a zero-level earthquake.

▶ **TRY EXERCISE 58, PAGE 473**

In Example 7 we make use of the Richter scale magnitudes of two earthquakes to compare the intensities of the earthquakes.

EXAMPLE 7   Compare Earthquakes

The 1960 Chile earthquake had a Richter scale magnitude of 9.5. The 1989 San Francisco earthquake had a Richter scale magnitude of 7.1. Compare the intensities of the earthquakes.

Continued ▶

take note

The results of Example 7 show that if an earthquake has a Richter scale magnitude of M_1 and a smaller earthquake has a Richter scale magnitude of M_2, then the larger earthquake is $10^{M_1-M_2}$ times as intense as the smaller earthquake.

Solution

Let I_1 be the intensity of the Chilean earthquake and I_2 the intensity of the San Francisco earthquake. Then

$$\log\left(\frac{I_1}{I_0}\right) = 9.5 \qquad \text{and} \qquad \log\left(\frac{I_2}{I_0}\right) = 7.1$$

$$\frac{I_1}{I_0} = 10^{9.5} \qquad\qquad \frac{I_2}{I_0} = 10^{7.1}$$

$$I_1 = 10^{9.5}I_0 \qquad\qquad I_2 = 10^{7.1}I_0$$

To compare the intensities of the earthquakes, we compute the ratio I_1/I_2.

$$\frac{I_1}{I_2} = \frac{10^{9.5}I_0}{10^{7.1}I_0} = \frac{10^{9.5}}{10^{7.1}} = 10^{9.5-7.1} = 10^{2.4} \approx 251$$

The earthquake in Chile was approximately 251 times as intense as the San Francisco earthquake.

▶ **TRY EXERCISE 60, PAGE 473**

Seismologists generally determine the Richter scale magnitude of an earthquake by examining a *seismogram*. See **Figure 7.22.**

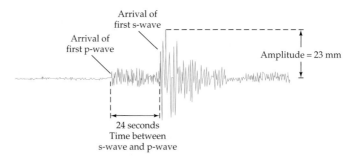

FIGURE 7.22

The magnitude of an earthquake cannot be determined just by examining the amplitude of a seismogram because this amplitude decreases as the distance between the epicenter of the earthquake and the observation station increases. To account for the distance between the epicenter and the observation station, a seismologist examines a seismogram for both small waves called **p-waves** and larger waves called **s-waves.** The Richter scale magnitude M of the earthquake is a function of both the amplitude A of the s-waves and the difference in time t between the occurrence of the s-waves and the p-waves. In the 1950s, Charles Richter developed the following formula to determine the magnitude of an earthquake from the data in a seismogram.

Amplitude-Time-Difference Formula

The Richter scale magnitude M of an earthquake is given by

$$M = \log A + 3 \log 8t - 2.92$$

where A is the amplitude, in millimeters, of the s-waves on a seismogram and t is the difference in time, in seconds, between the s-waves and the p-waves.

EXAMPLE 8 **Determine the Magnitude of an Earthquake from Its Seismogram**

Find the Richter scale magnitude of the earthquake that produced the seismogram in **Figure 7.22**.

Solution

$$
\begin{aligned}
M &= \log A + 3 \log 8t - 2.92 \\
&= \log 23 + 3 \log[8 \cdot 24] - 2.92 \qquad \text{• Substitute 23 for } A \text{ and 24 for } t. \\
&\approx 1.36173 + 6.84990 - 2.92 \\
&\approx 5.3
\end{aligned}
$$

The earthquake had a magnitude of about 5.3 on the Richter scale.

take note

The Richter scale magnitude is usually rounded to the nearest tenth.

▶ **TRY EXERCISE 64, PAGE 473**

Logarithmic scales are also used in chemistry. One example concerns the pH of a liquid, which is a measure of the liquid's **acidity** or **alkalinity.** (You may have tested the pH of a swimming pool or an aquarium.) Pure water, which is considered neutral, has a pH of 7.0. The pH scale ranges from 0 to 14, with 0 corresponding to the most acidic solutions and 14 to the most alkaline. Lemon juice has a pH of about 2, whereas household ammonia measures about 11.

Specifically, the pH of a solution is a function of the hydronium-ion concentration of the solution. Because the hydronium-ion concentration of a solution can be very small (with values such as 0.00000001), pH uses a logarithmic scale.

take note

One mole is equivalent to 6.022×10^{23} ions.

The pH of a Solution

The **pH of a solution** with a hydronium-ion concentration of H^+ moles per liter is given by

$$\text{pH} = -\log[H^+]$$

EXAMPLE 9   **Find the pH of a Solution**

Find the pH of each liquid. Round to the nearest tenth.

a. Orange juice with $H^+ = 2.8 \times 10^{-4}$ mole per liter

b. Milk with $H^+ = 3.97 \times 10^{-7}$ mole per liter

c. Rainwater with $H^+ = 6.31 \times 10^{-5}$ mole per liter

d. A baking soda solution with $H^+ = 3.98 \times 10^{-9}$ mole per liter

Solution

a. $pH = -\log[H^+] = -\log(2.8 \times 10^{-4}) \approx 3.6$
The orange juice has a pH of 3.6.

b. $pH = -\log[H^+] = -\log(3.97 \times 10^{-7}) \approx 6.4$
The milk has a pH of 6.4.

c. $pH = -\log[H^+] = -\log(6.31 \times 10^{-5}) \approx 4.2$
The rainwater has a pH of 4.2.

d. $pH = -\log[H^+] = -\log(3.98 \times 10^{-9}) \approx 8.4$
The baking soda solution has a pH of 8.4.

▶ **TRY EXERCISE 48, PAGE 472**

MATH MATTERS

The pH scale was created by the Danish biochemist Søren Sørensen in 1909 to measure the acidity of water used in the brewing of beer. pH is an abbreviation for *pondus hydrogenii*, which translates as "potential hydrogen."

Figure 7.23 illustrates the pH scale, along with the corresponding hydronium-ion concentrations. A solution on the left half of the scale, with a pH of less than 7, is an **acid**, and a solution on the right half of the scale is an **alkaline solution** or a **base**. Because the scale is logarithmic, a solution with a pH of 5 is 10 times more acidic than a solution with a pH of 6. From Example 9 we see that the orange juice, rainwater, and milk are acids, whereas the baking soda solution is a base.

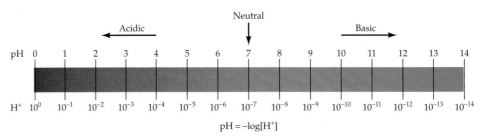

FIGURE 7.23

EXAMPLE 10   **Find the Hydronium-Ion Concentration**

A sample of blood has a pH of 7.3. Find the hydronium-ion concentration of the blood.

Solution

$$\text{pH} = -\log[\text{H}^+]$$
$$7.3 = -\log[\text{H}^+] \qquad \bullet \text{ Substitute 7.3 for pH.}$$
$$-7.3 = \log[\text{H}^+] \qquad \bullet \text{ Multiply both sides by } -1.$$
$$10^{-7.3} = \text{H}^+ \qquad \bullet \text{ Change to exponential form.}$$
$$5.0 \times 10^{-8} \approx \text{H}^+$$

The hydronium-ion concentration of the blood is about 5.0×10^{-8} mole per liter.

▶ TRY EXERCISE 50, PAGE 473

TOPICS FOR DISCUSSION

1. The function $f(x) = \log_b x$ is defined only for $x > 0$. Explain why this condition is imposed.

2. If p and q are positive numbers, explain why $\ln(p + q)$ isn't normally equal to $\ln p + \ln q$.

3. If $f(x) = \log_b x$ and $f(c) = f(d)$, can we conclude that $c = d$?

4. Give examples of situations in which it is advantageous to use logarithmic scales.

EXERCISE SET 7.3

In Exercises 1 to 8, write the given logarithm in terms of logarithms of x, y, and z.

1. $\log_b(xyz)$

▶ **2.** $\ln \dfrac{z^3}{\sqrt{xy}}$

3. $\ln \dfrac{x}{z^4}$

4. $\log_5 \dfrac{xy^2}{z^4}$

5. $\log_2 \dfrac{\sqrt{x}}{y^3}$

6. $\log_b\left(x\sqrt[3]{y}\right)$

7. $\log_7 \dfrac{\sqrt{xz}}{y^2}$

8. $\ln \sqrt[3]{x^2\sqrt{y}}$

In Exercises 9 to 14, write each logarithmic expression as a single logarithm with a coefficient of 1. Simplify when possible.

9. $\log(x + 5) + 2 \log x$

▶ **10.** $3 \log_2 t - \dfrac{1}{3} \log_2 u + 4 \log_2 v$

11. $\ln(x^2 - y^2) - \ln(x - y)$

12. $\dfrac{1}{2} \log_8(x + 5) - 3 \log_8 y$

13. $3 \log x + \dfrac{1}{3} \log y + \log(x + 1)$

14. $\ln(xz) - \ln\left(x\sqrt{y}\right) + 2 \ln \dfrac{y}{z}$

In Exercises 15 to 22, use the change-of-base formula to approximate the logarithm accurate to the nearest ten thousandth.

15. $\log_7 20$

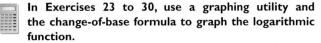 **16.** $\log_5 37$

17. $\log_{11} 8$

18. $\log_{50} 22$

19. $\log_6 \dfrac{1}{3}$

20. $\log_3 \dfrac{7}{8}$

21. $\log_9 \sqrt{17}$

22. $\log_4 \sqrt{7}$

In Exercises 23 to 30, use a graphing utility and the change-of-base formula to graph the logarithmic function.

23. $f(x) = \log_4 x$

▶ **24.** $g(x) = \log_8(5 - x)$

25. $g(x) = \log_8(x - 3)$

26. $t(x) = \log_9(5 - x)$

27. $h(x) = \log_3(x - 3)^2$

28. $J(x) = \log_{12}(-x)$

29. $F(x) = -\log_5|x - 2|$

30. $n(x) = \log_2\sqrt{x - 8}$

In Exercises 31 to 40, determine if the statement is true or false for all $x > 0$, $y > 0$. If it is false, write an example that disproves the statement.

31. $\log_b(x + y) = \log_b x + \log_b y$

32. $\log_b(xy) = \log_b x \cdot \log_b y$

33. $\log_b(xy) = \log_b x + \log_b y$

34. $\log_b x \cdot \log_b y = \log_b x + \log_b y$

35. $\log_b x - \log_b y = \log_b(x - y), \quad x > y$

36. $\log_b \dfrac{x}{y} = \dfrac{\log_b x}{\log_b y}$

37. $\dfrac{\log_b x}{\log_b y} = \log_b x - \log_b y$

38. $\log_b(x^n) = n \log_b x$

39. $(\log_b x)^n = n \log_b x$

40. $\log_b \sqrt{x} = \dfrac{1}{2} \log_b x$

41. Evaluate the following *without* using a calculator.

$$\log_3 5 \cdot \log_5 7 \cdot \log_7 9$$

42. Evaluate the following *without* using a calculator.

$$\log_5 20 \cdot \log_{20} 60 \cdot \log_{60} 100 \cdot \log_{100} 125$$

43. Which is larger, 500^{501} or 506^{500}? These numbers are too large for most calculators to handle. (They each have 1353 digits!) (*Hint:* Let $x = 500^{501}$ and $y = 506^{500}$ and then compare $\ln x$ with $\ln y$.)

44. Which number is smaller, $\dfrac{1}{50^{300}}$ or $\dfrac{1}{151^{233}}$?

45. **ANIMATED MAPS** A software company that creates interactive maps for Web sites has designed an animated zooming feature so that when a user selects the zoom-in option, the map appears to expand on a location. This is accomplished by displaying several intermediate maps to give the illusion of motion. The company has determined that zooming in on a location is more informative and pleasing to observe when the scale of each step of the animation is determined using the equation

$$S_n = S_0 \cdot 10^{\frac{n}{N}(\log S_f - \log S_0)}$$

where S_n represents the scale of the current step n ($n = 0$ corresponds to the initial scale), S_0 is the starting scale of the map, S_f is the final scale, and N is the number of steps in the animation following the initial scale. (If the initial scale of the map is 1:200, then $S_0 = 200$.) Determine the scales to be used at each intermediate step if a map is to start with a scale of 1:1,000,000 and proceed through five intermediate steps to end with a scale of 1:500,000.

46. **ANIMATED MAPS** Use the equation in Exercise 45 to determine the scales for each stage of an animated map zoom that goes from a scale of 1:250,000 to a scale of 1:100,000 in four steps (following the initial scale).

47. **pH** Milk of magnesia has a hydronium-ion concentration of about 3.97×10^{-11} mole per liter. Determine the pH of milk of magnesia and state whether it is an acid or a base.

▶ **48.** **pH** Vinegar has a hydronium-ion concentration of 1.26×10^{-3} mole per liter. Determine the pH of vinegar and state whether it is an acid or a base.

49. **HYDRONIUM-ION CONCENTRATION** A morphine solution has a pH of 9.5. Determine the hydronium-ion concentration of the morphine solution.

▶ **50.** HYDRONIUM-ION CONCENTRATION A rainstorm in New York City produced rainwater with a pH of 5.6. Determine the hydronium-ion concentration of the rainwater.

51. DECIBEL LEVEL The range of sound intensities that the human ear can detect is so large that a special decibel scale (named after Alexander Graham Bell) is used to measure and compare sound intensities. The **decibel level** dB of a sound is given by

$$dB(I) = 10 \log \left(\frac{I}{I_0} \right)$$

where I_0 is the intensity of sound that is barely audible to the human ear. Find the decibel level for the following sounds. Round to the nearest tenth of a decibel.

Sound	Intensity
a. Automobile traffic	$I = 1.58 \times 10^8 \cdot I_0$
b. Quiet conversation	$I = 10{,}800 \cdot I_0$
c. Fender guitar	$I = 3.16 \times 10^{11} \cdot I_0$
d. Jet engine	$I = 1.58 \times 10^{15} \cdot I_0$

52. COMPARISON OF SOUND INTENSITIES A team in Arizona installed a 48,000-watt sound system in a Ford Bronco that it claims can output 175-decibel sound. The human pain threshold for sound is 125 decibels. How many times more intense is the sound from the Bronco than the human pain threshold?

53. COMPARISON OF SOUND INTENSITIES How many times more intense is a sound that measures 120 decibels than a sound that measures 110 decibels?

54. DECIBEL LEVEL If the intensity of a sound is doubled, what is the increase in the decibel level? (*Hint:* Find $dB(2I) - dB(I)$.)

55. EARTHQUAKE MAGNITUDE What is the Richter scale magnitude of an earthquake with an intensity of $I = 100{,}000 I_0$?

▶ **56.** EARTHQUAKE MAGNITUDE The Colombia earthquake of 1906 had an intensity of $I = 398{,}107{,}000 I_0$. What did it measure on the Richter scale?

57. EARTHQUAKE INTENSITY The Coalinga, California, earthquake of 1983 had a Richter scale magnitude of 6.5. Find the intensity of this earthquake.

▶ **58.** EARTHQUAKE INTENSITY The earthquake that occurred just south of Concepción, Chile, in 1960 had a Richter scale magnitude of 9.5. Find the intensity of this earthquake.

59. COMPARISON OF EARTHQUAKES Compare the intensity of an earthquake that measures 5.0 on the Richter scale to the intensity of an earthquake that measures 3.0 on the Richter scale by finding the ratio of the larger intensity to the smaller intensity.

▶ **60.** COMPARISON OF EARTHQUAKES How many times more intense was the 1960 earthquake in Chile, which measured 9.5 on the Richter scale, than the San Francisco earthquake of 1906, which measured 8.3 on the Richter scale?

61. COMPARISON OF EARTHQUAKES On March 2, 1933, an earthquake of magnitude 8.9 on the Richter scale struck Japan. In October 1989, an earthquake of magnitude 7.1 on the Richter scale struck San Francisco, California. Compare the intensity of the larger earthquake to the intensity of the smaller earthquake by finding the ratio of the larger intensity to the smaller intensity.

62. COMPARISON OF EARTHQUAKES An earthquake that occurred in China in 1978 measured 8.2 on the Richter scale. In 1988, an earthquake in California measured 6.9 on the Richter scale. Compare the intensity of the larger earthquake to the intensity of the smaller earthquake by finding the ratio of the larger intensity to the smaller intensity.

63. EARTHQUAKE MAGNITUDE Find the Richter scale magnitude of the earthquake that produced the seismogram in the following figure.

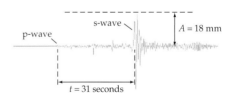

▶ **64.** EARTHQUAKE MAGNITUDE Find the Richter scale magnitude of the earthquake that produced the seismogram in the following figure.

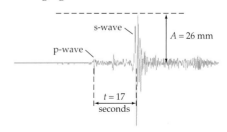

CONNECTING CONCEPTS

65. NOMOGRAMS AND LOGARITHMIC SCALES A **nomogram** is a diagram used to determine a numerical result by drawing a line across numerical scales. The following nomogram, used by Richter, determines the magnitude of an earthquake from its seismogram. To use the nomogram, mark the amplitude of a seismogram on the amplitude scale and mark the time between the s-wave and the p-wave on the S-P scale. Draw a line between these marks. The Richter scale magnitude of the earthquake that produced the seismogram is shown by the intersection of the line and the center scale. The example below shows that an earthquake with a seismogram amplitude of 23 millimeters and an S-P time of 24 seconds has a Richter scale magnitude of about 5.

The amplitude and the S-P time are shown on logarithmic scales. On the amplitude scale, the distance from 1 to 10 is the same as the distance from 10 to 100, because $\log 100 - \log 10 = \log 10 - \log 1$.

Use the nomogram at the left to determine the Richter scale magnitude of an earthquake with a seismogram

a. amplitude of 50 millimeters and S-P time of 40 seconds.

b. amplitude of 1 millimeter and S-P time of 30 seconds.

c. How do the results in parts **a.** and **b.** compare with the Richter scale magnitude produced by using the amplitude-time-difference formula?

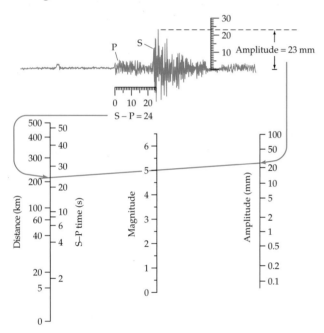

Richter's earthquake nomogram

PREPARE FOR SECTION 7.4

66. Use the definition of a logarithm to write the exponential equation $3^6 = 729$ in logarithmic form. [7.2]

67. Use the definition of a logarithm to write the logarithmic equation $\log_5 625 = 4$ in exponential form. [7.2]

68. Use the definition of a logarithm to write the exponential equation $a^{x+2} = b$ in logarithmic form. [7.2]

69. Solve for x: $4a = 7bx + 2cx$.

70. Solve for x: $165 = \dfrac{300}{1 + 12x}$.

71. Solve for x: $A = \dfrac{100 + x}{100 - x}$.

PROJECTS

1. LOGARITHMIC SCALES Sometimes **logarithmic scales** are used to better view a collection of data that span a wide range of values. For instance, consider the table below, which lists the approximate masses of various marine creatures in grams. Next we have attempted to plot the masses on a number line.

Animal	Mass (g)
Rotifer	0.000000006
Dwarf goby	0.30
Lobster	15,900
Leatherback turtle	851,000
Giant squid	1,820,000
Whale shark	4,700,000
Blue whale	120,000,000

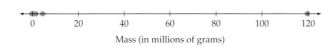

Mass (in millions of grams)

As you can see, we had to use such a large span of numbers that the data for most of the animals are bunched up at the left. Visually, this number line isn't very helpful for any comparisons.

a. Make a new number line, this time plotting the logarithm (base 10) of each of the masses.

b. Which number line is more helpful to compare the masses of the different animals?

c. If the data points for two animals on the logarithmic number line are 1 unit apart, how do the animals' masses compare? What if the points are 2 units apart?

2. LOGARITHMIC SCALES The distances of the planets in our solar system from the sun are given in the table at the top of the next column.

a. Draw a number line with an appropriate scale to plot the distances.

b. Draw a second number line, this time plotting the logarithm (base 10) of each distance.

c. Which number line do you find more helpful to compare the different distances?

d. If two distances are 3 units apart on the logarithmic number line, how do the distances of the corresponding planets compare?

Planet	Distance (million km)
Mercury	58
Venus	108
Earth	150
Mars	228
Jupiter	778
Saturn	1427
Uranus	2871
Neptune	4497
Pluto	5913

3. BIOLOGIC DIVERSITY To discuss the variety of species that live in a certain environment, a biologist needs a precise definition of *diversity*. Let $p_1, p_2, \ldots, p_n$ be the proportions of n species that live in an environment. The biologic diversity D of this system is

$$D = -(p_1 \log_2 p_1 + p_2 \log_2 p_2 + \cdots + p_n \log_2 p_n)$$

Suppose that an ecosystem has exactly five different varieties of grass: rye (R), bermuda (B), blue (L), fescue (F), and St. Augustine (A).

a. Calculate the diversity of this ecosystem if the proportions of these grasses are as shown in Table 7.1. Round to the nearest hundredth.

TABLE 7.1

R	B	L	F	A
$\dfrac{1}{5}$	$\dfrac{1}{5}$	$\dfrac{1}{5}$	$\dfrac{1}{5}$	$\dfrac{1}{5}$

b. Because bermuda and St. Augustine are virulent grasses, after a time the proportions will be as shown in Table 7.2. Calculate the diversity of this system. Does this system have more or less diversity than the system given in Table 7.1?

TABLE 7.2

R	B	L	F	A
$\dfrac{1}{8}$	$\dfrac{3}{8}$	$\dfrac{1}{16}$	$\dfrac{1}{8}$	$\dfrac{5}{16}$

c. After an even longer time period, the bermuda and St. Augustine grasses completely overrun the environment and the proportions are as shown in Table 7.3. Calculate the diversity of this system. (*Note:* Although the equation is not technically correct, for purposes of the diversity definition, we may say that $0 \log_2 0 = 0$. By using very small values of p_i, we can demonstrate that this definition makes sense.) Does this system have more or less diversity than the system given in Table 7.2?

d. Finally, the St. Augustine grasses overrun the bermuda grasses and the proportions are as shown in Table 7.4. Calculate the diversity of this system. Write a sentence that explains the meaning of the value you obtained.

TABLE 7.3

R	B	L	F	A
0	$\dfrac{1}{4}$	0	0	$\dfrac{3}{4}$

TABLE 7.4

R	B	L	F	A
0	0	0	0	1

SECTION 7.4

EXPONENTIAL AND LOGARITHMIC EQUATIONS

- SOLVE EXPONENTIAL EQUATIONS
- SOLVE LOGARITHMIC EQUATIONS
- APPLICATION

• SOLVE EXPONENTIAL EQUATIONS

If a variable appears in an exponent of a term of an equation, such as $2^{x+1} = 32$, then the equation is called an **exponential equation**. Example 1 uses the following equality-of-exponents theorem to solve $2^{x+1} = 32$.

Equality of Exponents Theorem

If $b^x = b^y$, then $x = y$, provided that $b > 0$ and $b \neq 1$.

EXAMPLE 1   **Solve an Exponential Equation**

Use the Equality of Exponents Theorem to solve $2^{x+1} = 32$.

Solution

$$2^{x+1} = 32$$
$$2^{x+1} = 2^5 \qquad \bullet \text{Write each side as a power of 2.}$$
$$x + 1 = 5 \qquad \bullet \text{Equate the exponents.}$$
$$x = 4$$

Check: Let $x = 4$, then $2^{x+1} = 2^{4+1}$
$$= 2^5$$
$$= 32$$

▶ **TRY EXERCISE 2, PAGE 483**

A graphing utility can also be used to find solutions of an equation of the form $f(x) = g(x)$. Either of the following two methods can be employed.

Using a Graphing Utility to Find the Solutions of $f(x) = g(x)$

Intersection Method   Graph $y_1 = f(x)$ and $y_2 = g(x)$ on the same screen. The solutions of $f(x) = g(x)$ are the x-coordinates of the points of intersection of the graphs.

Intercept Method   The solutions of $f(x) = g(x)$ are the x-coordinates of the x-intercepts of the graph of $y = f(x) - g(x)$.

Figures 7.24 and **7.25** illustrate the graphical methods for solving $2^{x+1} = 32$.

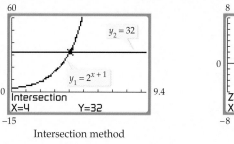

Intersection method

FIGURE 7.24

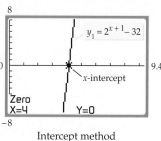

Intercept method

FIGURE 7.25

In Example 1 we were able to write both sides of the equation as a power of the same base. If you find it difficult to write both sides of an exponential equation in terms of the same base, then try the procedure of taking the logarithm of each side of the equation. This procedure is used in Example 2.

EXAMPLE 2   Solve an Exponential Equation

Solve: $5^x = 40$

Algebraic Solution

$$5^x = 40$$
$$\log(5^x) = \log 40 \qquad \text{• Take the logarithm of each side.}$$
$$x \log 5 = \log 40 \qquad \text{• Power property}$$
$$x = \frac{\log 40}{\log 5} \qquad \text{• Exact solution}$$
$$x \approx 2.3 \qquad \text{• Decimal approximation}$$

To the nearest tenth, the solution is 2.3.

Visualize the Solution

Intersection Method   The solution of $5^x = 40$ is the x-coordinate of the point of intersection of $y = 5^x$ and $y = 40$ (see **Figure 7.26**).

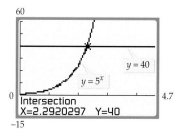

FIGURE 7.26

▶ **TRY EXERCISE 10, PAGE 483**

An alternative approach to solving the equation in Example 2 is to rewrite the exponential equation in logarithmic form: $5^x = 40$ is equivalent to the logarithmic equation $\log_5 40 = x$. Using the change-of-base formula, we find that $x = \log_5 40 = \dfrac{\log 40}{\log 5}$. In the following example, however, we must take logarithms of both sides to reach a solution.

EXAMPLE 3   **Solve an Exponential Equation**

Solve: $3^{2x-1} = 5^{x+2}$

Algebraic Solution

$$3^{2x-1} = 5^{x+2}$$

$$\ln 3^{2x-1} = \ln 5^{x+2}$$ • Take the natural logarithm of each side.

$$(2x - 1)\ln 3 = (x + 2)\ln 5$$ • Power property

$$2x \ln 3 - \ln 3 = x \ln 5 + 2 \ln 5$$ • Distributive property

$$2x \ln 3 - x \ln 5 = 2 \ln 5 + \ln 3$$ • Solve for x.

$$x(2 \ln 3 - \ln 5) = 2 \ln 5 + \ln 3$$

$$x = \frac{2 \ln 5 + \ln 3}{2 \ln 3 - \ln 5}$$ • Exact solution

$$x \approx 7.3$$ • Decimal approximation

To the nearest tenth, the solution is 7.3.

Visualize the Solution

Intercept Method   The solution of $3^{2x-1} = 5^{x+2}$ is the x-coordinate of the x-intercept of $y = 3^{2x-1} - 5^{x+2}$ (see **Figure 7.27**).

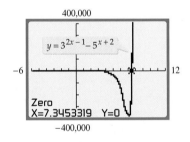

FIGURE 7.27

▶ **TRY EXERCISE 18, PAGE 483**

In Example 4 we solve an exponential equation that has two solutions.

EXAMPLE 4   **Solve an Exponential Equation Involving $b^x + b^{-x}$**

Solve: $\dfrac{2^x + 2^{-x}}{2} = 3$

Algebraic Solution

Multiplying each side by 2 produces

$$2^x + 2^{-x} = 6$$

$$2^{2x} + 2^0 = 6(2^x)$$ • Multiply each side by 2^x to clear negative exponents.

$$(2^x)^2 - 6(2^x) + 1 = 0$$ • Write in quadratic form.

$$(u)^2 - 6(u) + 1 = 0$$ • Substitute u for 2^x.

Visualize the Solution

Intersection Method   The solutions of $\dfrac{2^x + 2^{-x}}{2} = 3$ are the x-coordinates of the points of intersection of

By the quadratic formula,

$$u = \frac{6 \pm \sqrt{36 - 4}}{2} = \frac{6 \pm 4\sqrt{2}}{2} = 3 \pm 2\sqrt{2}$$

$$2^x = 3 \pm 2\sqrt{2}$$

$$\log 2^x = \log(3 \pm 2\sqrt{2})$$

$$x \log 2 = \log(3 \pm 2\sqrt{2})$$

$$x = \frac{\log(3 \pm 2\sqrt{2})}{\log 2} \approx \pm 2.54$$

- Replace u with 2^x.
- Take the common logarithm of each side.
- Power property

The approximate solutions are -2.54 and 2.54.

▶ **TRY EXERCISE 40, PAGE 483**

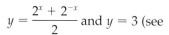

$y = \dfrac{2^x + 2^{-x}}{2}$ and $y = 3$ (see **Figure 7.28**).

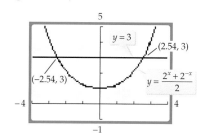

FIGURE 7.28

● SOLVE LOGARITHMIC EQUATIONS

Equations that involve logarithms are called logarithmic equations. The properties of logarithms, along with the definition of a logarithm, are often used to find the solutions of a logarithmic equation.

EXAMPLE 5   Solve a Logarithmic Equation

Solve: $\log(3x - 5) = 2$

Solution

$$\log(3x - 5) = 2$$
$$3x - 5 = 10^2 \qquad \text{• Definition of a logarithm}$$
$$3x = 105 \qquad \text{• Solve for } x.$$
$$x = 35$$

Check: $\log[3(35) - 5] = \log 100 = 2$

▶ **TRY EXERCISE 22, PAGE 483**

❓ **QUESTION** Can a negative number be a solution of a logarithmic equation?

❓ **ANSWER** Yes. For instance, -10 is a solution of $\log(-x) = 1$.

> **EXAMPLE 6**   **Solve a Logarithmic Equation**
>
> Solve: $\log 2x - \log(x - 3) = 1$
>
> **Solution**
>
> $$\log 2x - \log(x - 3) = 1$$
>
> $$\log \frac{2x}{x - 3} = 1 \qquad \bullet \textbf{ Quotient property}$$
>
> $$\frac{2x}{x - 3} = 10^1 \qquad \bullet \textbf{ Definition of logarithm}$$
>
> $$2x = 10x - 30 \qquad \bullet \textbf{ Solve for x.}$$
>
> $$-8x = -30$$
>
> $$x = \frac{15}{4}$$
>
> Check the solution by substituting $\frac{15}{4}$ into the original equation.
>
> ▶ **TRY EXERCISE 26, PAGE 483**

In Example 7 we make use of the one-to-one property of logarithms to find the solution of a logarithmic equation. This example illustrates that the process of solving a logarithmic equation by using logarithmic properties may introduce an extraneous solution.

> **EXAMPLE 7**   **Solve a Logarithmic Equation**
>
> Solve: $\ln(3x + 8) = \ln(2x + 2) + \ln(x - 2)$
>
> **Algebraic Solution**
>
> $$\ln(3x + 8) = \ln(2x + 2) + \ln(x - 2)$$
> $$\ln(3x + 8) = \ln[(2x + 2)(x - 2)] \qquad \bullet \textbf{ Product property}$$
> $$\ln(3x + 8) = \ln(2x^2 - 2x - 4)$$
> $$3x + 8 = 2x^2 - 2x - 4 \qquad \bullet \textbf{ One-to-one property of logarithms}$$
> $$0 = 2x^2 - 5x - 12$$
> $$0 = (2x + 3)(x - 4) \qquad \bullet \textbf{ Solve for x.}$$
> $$x = -\frac{3}{2} \quad \text{or} \quad x = 4$$
>
> Thus $-\frac{3}{2}$ and 4 are possible solutions. A check will show that 4 is a solution, but $-\frac{3}{2}$ is not a solution.
>
> **Visualize the Solution**
>
> The graph of $y = \ln(3x + 8) - \ln(2x + 2) - \ln(x - 2)$ has only one x-intercept (see **Figure 7.29**). Thus there is only one real solution.
>
>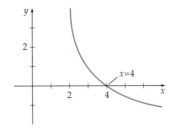
>
> $y = \ln(3x + 8) - \ln(2x + 2) - \ln(x - 2)$
>
> **FIGURE 7.29**

▶ **TRY EXERCISE 36, PAGE 483**

? **QUESTION** Why does $x = -\dfrac{3}{2}$ not check in Example 7?

● **APPLICATION**

| EXAMPLE 8 | **Velocity of a Sky Diver Experiencing Air Resistance** |

During the free-fall portion of a jump, the time t in seconds required for a sky diver to reach a velocity v in feet per second is given by

$$t = -\frac{175}{32} \ln\left(1 - \frac{v}{175}\right)$$

a. Determine the velocity of the diver after 5 seconds.

b. The graph of the above function has a vertical asymptote at $v = 175$. Explain the meaning of the vertical asymptote in the context of this example.

Solution

a. Substitute 5 for t and solve for v.

$$t = -\frac{175}{32} \ln\left(1 - \frac{v}{175}\right)$$

$$5 = -\frac{175}{32} \ln\left(1 - \frac{v}{175}\right) \qquad \text{• Replace } t \text{ with 5.}$$

$$\left(-\frac{32}{175}\right)5 = \ln\left(1 - \frac{v}{175}\right) \qquad \text{• Solve for } v.$$

$$-\frac{32}{35} = \ln\left(1 - \frac{v}{175}\right)$$

$$e^{-32/35} = 1 - \frac{v}{175} \qquad \text{• Write in exponential form.}$$

$$e^{-32/35} - 1 = -\frac{v}{175}$$

$$v = 175(1 - e^{-32/35})$$

$$v \approx 104.86$$

Continued ▶

take note

If air resistance is not considered, then the time in seconds required for a sky diver to reach a given velocity (in feet per second) is $t = \dfrac{v}{32}$. The function in Example 8 is a more realistic model of the time required to reach a given velocity during the free-fall of a sky diver who is experiencing air resistance.

? **ANSWER** If $x = -\dfrac{3}{2}$, the original equation becomes $\ln\left(\dfrac{7}{2}\right) = \ln(-1) + \ln\left(-\dfrac{7}{2}\right)$. This cannot be true, because the function $f(x) = \ln x$ is not defined for negative values of x.

After 5 seconds the velocity of the sky diver will be about 104.9 feet per second. See **Figure 7.30.**

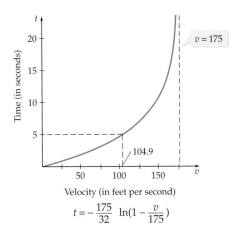

$$t = -\frac{175}{32} \ln\left(1 - \frac{v}{175}\right)$$

FIGURE 7.30

b. The vertical asymptote $v = 175$ indicates that the sky diver will not attain a velocity greater than 175 feet per second. In **Figure 7.30,** note that as $v \to 175$ from the left, $t \to \infty$.

▶ **TRY EXERCISE 68, PAGE 485**

 TOPICS FOR DISCUSSION

1. Discuss how to solve the equation $a = \log_b x$ for x.

2. What is the domain of $y = \log_4(2x - 5)$? Explain why this means that the equation $\log_4(x - 3) = \log_4(2x - 5)$ has no real number solution.

3. -8 is not a solution of the equation $\log_2 x + \log_2(x + 6) = 4$. Discuss at which step in the following solution the extraneous solution -8 was introduced.

$$\log_2 x + \log_2(x + 6) = 4$$
$$\log_2 x(x + 6) = 4$$
$$x(x + 6) = 2^4$$
$$x^2 + 6x = 16$$
$$x^2 + 6x - 16 = 0$$
$$(x + 8)(x - 2) = 0$$
$$x = -8 \quad \text{or} \quad x = 2$$

EXERCISE SET 7.4

In Exercises 1 to 46, solve for x algebraically.

1. $2^x = 64$

▶ 2. $3^x = 243$

3. $49^x = \dfrac{1}{343}$

4. $9^x = \dfrac{1}{243}$

5. $2^{5x+3} = \dfrac{1}{8}$

6. $3^{4x-7} = \dfrac{1}{9}$

7. $\left(\dfrac{2}{5}\right)^x = \dfrac{8}{125}$

8. $\left(\dfrac{2}{5}\right)^x = \dfrac{25}{4}$

9. $5^x = 70$

▶ 10. $6^x = 50$

11. $3^{-x} = 120$

12. $7^{-x} = 63$

13. $10^{2x+3} = 315$

14. $10^{6-x} = 550$

15. $e^x = 10$

16. $e^{x+1} = 20$

17. $2^{1-x} = 3^{x+1}$

▶ 18. $3^{x-2} = 4^{2x+1}$

19. $2^{2x-3} = 5^{-x-1}$

20. $5^{3x} = 3^{x+4}$

21. $\log(4x - 18) = 1$

▶ 22. $\log(x^2 + 19) = 2$

23. $\ln(x^2 - 12) = \ln x$

24. $\log(2x^2 + 3x) = \log(10x + 30)$

25. $\log_2 x + \log_2(x - 4) = 2$

▶ 26. $\log_3 x + \log_3(x + 6) = 3$

27. $\log(5x - 1) = 2 + \log(x - 2)$

28. $1 + \log(3x - 1) = \log(2x + 1)$

29. $\ln(1 - x) + \ln(3 - x) = \ln 8$

30. $\log(4 - x) = \log(x + 8) + \log(2x + 13)$

31. $\log \sqrt{x^3 - 17} = \dfrac{1}{2}$

32. $\log(x^3) = (\log x)^2$

33. $\log(\log x) = 1$

34. $\ln(\ln x) = 2$

35. $\ln(e^{3x}) = 6$

▶ 36. $\ln x = \dfrac{1}{2}\ln\left(2x + \dfrac{5}{2}\right) + \dfrac{1}{2}\ln 2$

37. $e^{\ln(x-1)} = 4$

38. $10^{\log(2x+7)} = 8$

39. $\dfrac{10^x - 10^{-x}}{2} = 20$

▶ 40. $\dfrac{10^x + 10^{-x}}{2} = 8$

41. $\dfrac{10^x + 10^{-x}}{10^x - 10^{-x}} = 5$

42. $\dfrac{10^x - 10^{-x}}{10^x + 10^{-x}} = \dfrac{1}{2}$

43. $\dfrac{e^x + e^{-x}}{2} = 15$

44. $\dfrac{e^x - e^{-x}}{2} = 15$

45. $\dfrac{1}{e^x - e^{-x}} = 4$

46. $\dfrac{e^x + e^{-x}}{e^x - e^{-x}} = 3$

In Exercises 47 to 56, use a graphing utility to approximate the solutions of the equation to the nearest hundredth.

47. $2^{-x+3} = x + 1$

48. $3^{x-2} = -2x - 1$

49. $e^{3-2x} - 2x = 1$

50. $2e^{x+2} + 3x = 2$

51. $3\log_2(x - 1) = -x + 3$

52. $2\log_3(2 - 3x) = 2x - 1$

53. $\ln(2x + 4) + \dfrac{1}{2}x = -3$

54. $2\ln(3 - x) + 3x = 4$

55. $2^{x+1} = x^2 - 1$

56. $\ln x = -x^2 + 4$

57. **POPULATION GROWTH** The population P of a city grows exponentially according to the function

$$P(t) = 8500(1.1)^t, \quad 0 \le t \le 8$$

where t is measured in years.

a. Find the population at time $t = 0$ and also at time $t = 2$.

b. When, to the nearest year, will the population reach 15,000?

58. **PHYSICAL FITNESS** After a race, a runner's pulse rate R in beats per minute decreases according to the function

$$R(t) = 145e^{-0.092t}, \quad 0 \le t \le 15$$

where t is measured in minutes.

a. Find the runner's pulse rate at the end of the race and also 1 minute after the end of the race.

b. How long, to the nearest minute, after the end of the race will the runner's pulse rate be 80 beats per minute?

59. RATE OF COOLING A can of soda at 79°F is placed in a refrigerator that maintains a constant temperature of 36°F. The temperature T of the soda t minutes after it is placed in the refrigerator is given by

$$T(t) = 36 + 43e^{-0.058t}$$

a. Find the temperature, to the nearest degree, of the soda 10 minutes after it is placed in the refrigerator.

b. When, to the nearest minute, will the temperature of the soda be 45°F?

60. MEDICINE During surgery, a patient's circulatory system requires at least 50 milligrams of an anesthetic. The amount of anesthetic present t hours after 80 milligrams of anesthetic is administered is given by

$$T(t) = 80(0.727)^t$$

a. How much, to the nearest milligram, of the anesthetic is present in the patient's circulatory system 30 minutes after the anesthetic is administered?

b. How long, to the nearest minute, can the operation last if the patient does not receive additional anesthetic?

61. PSYCHOLOGY Industrial psychologists study employee training programs to assess the effectiveness of the instruction. In one study, the percent score P on a test for a person who had completed t hours of training was given by

$$P = \frac{100}{1 + 30e^{-0.088t}}$$

a. Use a graphing utility to graph the equation for $t \geq 0$.

b. Use the graph to estimate (to the nearest hour) the number of hours of training necessary to achieve a 70% score on the test.

c. From the graph, determine the horizontal asymptote.

d. Write a sentence that explains the meaning of the horizontal asymptote.

62. PSYCHOLOGY An industrial psychologist has determined that the average percent score for an employee on a test of the employee's knowledge of the company's product is given by

$$P = \frac{100}{1 + 40e^{-0.1t}}$$

where t is the number of weeks on the job and P is the percent score.

a. Use a graphing utility to graph the equation for $t \geq 0$.

b. Use the graph to estimate (to the nearest week) the number of weeks of employment that are necessary for the average employee to earn a 70% score on the test.

c. Determine the horizontal asymptote of the graph.

d. Write a sentence that explains the meaning of the horizontal asymptote.

63. ECOLOGY A herd of bison was placed in a wildlife preserve that can support a maximum of 1000 bison. A population model for the bison is given by

$$B = \frac{1000}{1 + 30e^{-0.127t}}$$

where B is the number of bison in the preserve and t is time in years, with the year 1999 represented by $t = 0$.

a. Use a graphing utility to graph the equation for $t \geq 0$.

b. Use the graph to estimate (to the nearest year) the number of years before the bison population reaches 500.

c. Determine the horizontal asymptote of the graph.

d. Write a sentence that explains the meaning of the horizontal asymptote.

64. POPULATION GROWTH A yeast culture grows according to the equation

$$Y = \frac{50,000}{1 + 250e^{-0.305t}}$$

where Y is the number of yeast and t is time in hours.

a. Use a graphing utility to graph the equation for $t \geq 0$.

b. Use the graph to estimate (to the nearest hour) the number of hours before the yeast population reaches 35,000.

c. From the graph, estimate the horizontal asymptote.

d. Write a sentence that explains the meaning of the horizontal asymptote.

65. CONSUMPTION OF NATURAL RESOURCES A model for how long our coal resources will last is given by

$$T = \frac{\ln(300r + 1)}{\ln(r + 1)}$$

where r is the percent increase in consumption from current levels of use and T is the time (in years) before the resource is depleted.

a. Use a graphing utility to graph this equation.

b. If our consumption of coal increases by 3% per year, in how many years will we deplete our coal resources?

c. What percent increase in consumption of coal will deplete the resource in 100 years? Round to the nearest tenth of a percent.

66. CONSUMPTION OF NATURAL RESOURCES A model for how long our aluminum resources will last is given by

$$T = \frac{\ln(20{,}500r + 1)}{\ln(r + 1)}$$

where r is the percent increase in consumption from current levels of use and T is the time (in years) before the resource is depleted.

a. Use a graphing utility to graph this equation.

b. If our consumption of aluminum increases by 5% per year, in how many years (to the nearest year) will we deplete our aluminum resources?

c. What percent increase in consumption of aluminum will deplete the resource in 100 years? Round to the nearest tenth of a percent.

67. VELOCITY OF A MEDICAL CARE PACKAGE A medical care package is air lifted and dropped to a disaster area. During the free-fall portion of the drop, the time, in seconds, required for the package to obtain a velocity of v feet per second is given by the function

$$t = 2.43 \ln \frac{150 + v}{150 - v}, \quad 0 \le v < 150$$

a. Determine the velocity of the package 5 seconds after it is dropped. Round to the nearest foot per second.

b. Determine the vertical asymptote of the function.

c. Write a sentence that explains the meaning of the vertical asymptote in the context of this application.

▶ **68.** EFFECTS OF AIR RESISTANCE ON VELOCITY If we assume that air resistance is proportional to the square of the velocity, then the time t in seconds required for an object to reach a velocity v in feet per second is given by

$$t = \frac{9}{24} \ln \frac{24 + v}{24 - v}, 0 \le v < 24$$

a. Determine the velocity, to the nearest hundredth foot per second of the object after 1.5 seconds.

b. Determine the vertical asymptote for the graph of this function.

c. Write a sentence that describes the meaning of the vertical asymptote in the context of this problem.

69. TERMINAL VELOCITY WITH AIR RESISTANCE The velocity v of an object t seconds after it has been dropped from a height above the surface of the earth is given by the equation $v = 32t$ feet per second, assuming no air resistance. If we assume that air resistance is proportional to the square of the velocity, then the velocity after t seconds is given by

$$v = 100 \left(\frac{e^{0.64t} - 1}{e^{0.64t} + 1} \right)$$

a. In how many seconds will the velocity be 50 feet per second?

b. Determine the horizontal asymptote for the graph of this function.

c. Write a sentence that describes the meaning of the horizontal asymptote in the context of this problem.

70. TERMINAL VELOCITY WITH AIR RESISTANCE If we assume that air resistance is proportional to the square of the velocity, then the velocity v in feet per second of an object t seconds after it has been dropped is given by

$$v = 50 \left(\frac{e^{1.6t} - 1}{e^{1.6t} + 1} \right)$$

(See Exercise 69. The reason for the difference in the equations is that the proportionality constants are different.)

a. In how many seconds will the velocity be 20 feet per second?

b. Determine the horizontal asymptote for the graph of this function.

c. Write a sentence that describes the meaning of the horizontal asymptote in the context of this problem.

71. EFFECTS OF AIR RESISTANCE ON DISTANCE The distance s, in feet, that the object in Exercise 69 will fall in t seconds is given by

$$s = \frac{100^2}{32} \ln \left(\frac{e^{0.32t} + e^{-0.32t}}{2} \right)$$

a. Use a graphing utility to graph this equation for $t \ge 0$.

b. How long does it take for the object to fall 100 feet? Round to the nearest tenth of a second.

72. ▦ **EFFECTS OF AIR RESISTANCE ON DISTANCE** The distance s, in feet, that the object in Exercise 70 will fall in t seconds is given by

$$s = \frac{50^2}{40} \ln\left(\frac{e^{0.8t} + e^{-0.8t}}{2}\right)$$

a. Use a graphing utility to graph this equation for $t \geq 0$.

b. How long does it take for the object to fall 100 feet? Round to the nearest tenth of a second.

73. ▦ **RETIREMENT PLANNING** The retirement account for a graphic designer contains $250,000 on January 1, 2002, and earns interest at a rate of 0.5% per month. On February 1, 2002, the designer withdraws $2000 and plans to continue these withdrawals as retirement income each month. The value V of the account after x months is

$$V = 400,000 - 150,000(1.005)^x$$

If the designer wishes to leave $100,000 to a scholarship foundation, what is the maximum number of withdrawals (to the nearest month) the designer can make from this account and still have $100,000 to donate?

74. **HANGING CABLE** The height h, in feet, of any point P on the cable shown is given by

$$h(x) = 10(e^{x/20} + e^{-x/20}), \quad -15 \leq x \leq 15$$

where $|x|$ is the horizontal distance in feet between P and the y-axis.

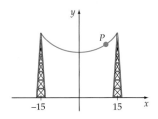

a. What is the lowest height of the cable?

b. What is the height of the cable 10 feet to the right of the y-axis? Round to the nearest tenth of a foot.

c. How far to the right of the y-axis is the cable 24 feet in height? Round to the nearest tenth of a foot.

CONNECTING CONCEPTS

75. The following argument seems to indicate that $0.125 > 0.25$. Find the first incorrect statement in the argument.

$$3 > 2$$
$$3(\log 0.5) > 2(\log 0.5)$$
$$\log 0.5^3 > \log 0.5^2$$
$$0.5^3 > 0.5^2$$
$$0.125 > 0.25$$

76. The following argument seems to indicate that $4 = 6$. Find the first incorrect statement in the argument.

$$4 = \log_2 16$$
$$4 = \log_2(8 + 8)$$
$$4 = \log_2 8 + \log_2 8$$
$$4 = 3 + 3$$
$$4 = 6$$

77. A common mistake that students make is to write $\log(x + y)$ as $\log x + \log y$. For what values of x and y does $\log(x + y) = \log x + \log y$? (*Hint:* Solve for x in terms of y.)

78. Let $f(x) = 2 \ln x$ and $g(x) = \ln x^2$. Does $f(x) = g(x)$ for all real numbers x?

79. ✎ Explain why the functions $F(x) = 1.4^x$ and $G(x) = e^{0.336x}$ represent essentially the same function.

80. Find the constant k that will make $f(t) = 2.2^t$ and $g(t) = e^{-kt}$ represent essentially the same function.

PREPARE FOR SECTION 7.5

81. Evaluate $A = 1000\left(1 + \dfrac{0.1}{12}\right)^{12t}$ for $t = 2$. Round to the nearest hundredth. [7.1]

82. Evaluate $A = 600\left(1 + \dfrac{0.04}{4}\right)^{4t}$ for $t = 8$. Round to the nearest hundredth. [7.1]

83. Solve $0.5 = e^{14k}$ for k. Round to the nearest ten-thousandth. [7.4]

84. Solve $0.85 = 0.5^{t/5730}$ for t. Round to the nearest ten. [7.4]

85. Solve $6 = \dfrac{70}{5 + 9e^{-k\cdot 12}}$ for k. Round to the nearest thousandth. [7.4]

86. Solve $2{,}000{,}000 = \dfrac{3^{n+1} - 3}{2}$ for n. Round to the nearest tenth. [7.4]

PROJECTS

1. NAVIGATING The pilot of a boat is trying to cross a river to a point O two miles due west of the boat's starting position by always pointing the nose of the boat toward O. Suppose the speed of the current is w miles per hour and the speed of the boat is v miles per hour. If point O is the origin and the boat's starting position is $(2, 0)$ (see the diagram at the right), then the equation of the boat's path is given by

$$y = \left(\frac{x}{2}\right)^{1-(w/v)} - \left(\frac{x}{2}\right)^{1+(w/v)}$$

a. If the speed of the current and the speed of the boat are the same, can the pilot reach point O by always having the nose of the boat pointed toward O? If not, at what point will the pilot arrive? Explain your answer.

b. If the speed of the current is greater than the speed of the boat, can the pilot reach point O by always pointing the nose of the boat toward point O? If not, where will the pilot arrive? Explain.

c. If the speed of the current is less than the speed of the boat, can the pilot reach point O by always pointing the nose of the boat toward point O? If not, where will the pilot arrive? Explain.

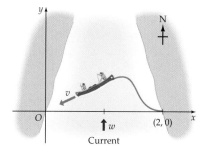

SECTION 7.5

EXPONENTIAL GROWTH AND DECAY

In many applications, a quantity N grows or decays according to the function $N(t) = N_0 e^{kt}$. In this function, N is a function of time t, and N_0 is the value of N at time $t = 0$. If k is a *positive* constant, then $N(t) = N_0 e^{kt}$ is called an **exponential growth function**. If k is a *negative* constant, then $N(t) = N_0 e^{kt}$ is called an **exponential decay function**. The following examples illustrate how growth and decay functions arise naturally in the investigation of certain phenomena.

Interest is money paid for the use of money. The interest I is called **simple interest** if it is a fixed percent r, per time period t, of the amount of money invested. The amount of money invested is called the **principal** P. Simple interest is computed using the formula $I = Prt$. For example, if \$1000 is invested at 12% for 3 years, the simple interest is

$$I = Prt = \$1000(0.12)(3) = \$360$$

The balance after t years is $A = P + I = P + Prt$. In the previous example, the \$1000 invested for 3 years produced \$360 interest. Thus the balance after 3 years is \$1000 + \$360 = \$1360.

COMPOUND INTEREST

In many financial transactions, interest is added to the principal at regular intervals so that interest is paid on interest as well as on the principal. Interest earned in this manner is called **compound interest**. For example, if \$1000 is invested at 12% annual interest compounded annually for 3 years, then the total interest after 3 years is

First-year interest	$\$1000(0.12) = \$120.00$
Second-year interest	$\$1120(0.12) = \$134.40$
Third-year interest	$\$1254.40(0.12) \approx \$150.53$
	$\$404.93$ • Total interest

This method of computing the balance can be tedious and time-consuming. A *compound interest formula* that can be used to determine the balance due after t years of compounding can be developed as follows.

Note that if P dollars is invested at an interest rate of r per year, then the balance after one year is $A_1 = P + Pr = P(1 + r)$, where Pr represents the interest earned for the year. Observe that A_1 is the product of the original principal P and $(1 + r)$. If the amount A_1 is reinvested for another year, then the balance after the second year is

$$A_2 = (A_1)(1 + r) = P(1 + r)(1 + r) = P(1 + r)^2$$

Successive reinvestments lead to the results shown in **Table 7.11**. The equation $A_t = P(1 + r)^t$ is valid if r is the annual interest rate paid during each of the t years.

TABLE 7.11

Number of Years	Balance
3	$A_3 = P(1 + r)^3$
4	$A_4 = P(1 + r)^4$
⋮	⋮
t	$A_t = P(1 + r)^t$

If r is an annual interest rate and n is the number of compounding periods per year, then the interest rate each period is r/n and the number of compounding periods after t years is nt. Thus the compound interest formula is expressed as follows:

The Compound Interest Formula

A principal P invested at an annual interest rate r, expressed as a decimal and compounded n times per year for t years, produces the balance

$$A = P\left(1 + \frac{r}{n}\right)^{nt}$$

EXAMPLE 1   Solve a Compound Interest Application

Find the balance if $1000 is invested at an annual interest rate of 10% for 2 years compounded

a. annually   b. monthly   c. daily

Solution

a. Use the compound interest formula with $P = 1000$, $r = 0.1$, $t = 2$, and $n = 1$.

$$A = \$1000\left(1 + \frac{0.1}{1}\right)^{1 \cdot 2} = \$1000(1.1)^2 = \$1210.00$$

b. Because there are 12 months in a year, use $n = 12$.

$$A = \$1000\left(1 + \frac{0.1}{12}\right)^{12 \cdot 2} \approx \$1000(1.008333333)^{24} \approx \$1220.39$$

c. Because there are 365 days in a year, use $n = 365$.

$$A = \$1000\left(1 + \frac{0.1}{365}\right)^{365 \cdot 2} \approx \$1000(1.000273973)^{730} \approx \$1221.37$$

▶ **TRY EXERCISE 4, PAGE 498**

To **compound continuously** means to increase the number of compounding periods without bound.

To derive a continuous compounding interest formula, substitute $\dfrac{1}{m}$ for $\dfrac{r}{n}$ in the compound interest formula

$$A = P\left(1 + \frac{r}{n}\right)^{nt} \tag{1}$$

to produce

$$A = P\left(1 + \frac{1}{m}\right)^{nt} \tag{2}$$

This substitution is motivated by the desire to express $\left(1 + \dfrac{r}{n}\right)^n$ as $\left[\left(1 + \dfrac{1}{m}\right)^m\right]^r$, which approaches e^r as m gets larger without bound.

Solving the equation $\dfrac{1}{m} = \dfrac{r}{n}$ for n yields $n = mr$, so the exponent nt can be written as mrt. Therefore Equation (2) can be expressed as

$$A = P\left(1 + \frac{1}{m}\right)^{mrt} = P\left[\left(1 + \frac{1}{m}\right)^m\right]^{rt} \tag{3}$$

By the definition of e, we know that as m increases without bound,

$$\left(1 + \frac{1}{m}\right)^m \qquad \text{approaches} \qquad e$$

Thus, using continuous compounding, Equation (3) simplifies to $A = Pe^{rt}$.

Continuous Compounding Interest Formula

If an account with principal P and annual interest rate r is compounded continuously for t years, then the balance is $A = Pe^{rt}$.

EXAMPLE 2   Solve a Continuous Compound Interest Application

Find the balance after 4 years on $800 invested at an annual rate of 6% compounded continuously.

Algebraic Solution

Use the continuous compounding formula with $P = 800$, $r = 0.06$, and $t = 4$.

$$\begin{aligned}
A &= Pe^{rt} \\
&= 800e^{0.06(4)} \\
&= 800e^{0.24} \\
&\approx 800(1.27124915) \\
&\approx 1017.00 \qquad \bullet \text{To the nearest cent}
\end{aligned}$$

The balance after 4 years will be $1017.00.

Visualize the Solution

Figure 7.31, a graph of $A = 800e^{0.06t}$, shows that the balance is about $1017.00 when $t = 4$.

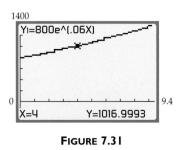

FIGURE 7.31

▶ **TRY EXERCISE 6, PAGE 498**

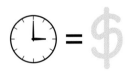

You have probably heard it said that time is money. In fact, many investors ask the question "How long will it take to double my money?" The following example answers this question for two different investments.

| EXAMPLE 3 | **Double Your Money** |

Find the time required for money invested at an annual rate of 6% to double in value if the investment is compounded

a. semiannually

b. continuously

Solution

a. Use $A = P\left(1 + \dfrac{r}{n}\right)^{nt}$ with $r = 0.06$, $n = 2$, and the balance A equal to twice the principal ($A = 2P$).

$$2P = P\left(1 + \frac{0.06}{2}\right)^{2t}$$

$$2 = \left(1 + \frac{0.06}{2}\right)^{2t} \qquad \text{• Divide each side by } P.$$

$$\ln 2 = \ln\left(1 + \frac{0.06}{2}\right)^{2t} \qquad \text{• Take the natural logarithm of each side.}$$

$$\ln 2 = 2t \ln\left(1 + \frac{0.06}{2}\right) \qquad \text{• Apply the power property.}$$

$$2t = \frac{\ln 2}{\ln\left(1 + \dfrac{0.06}{2}\right)} \qquad \text{• Solve for } t.$$

$$t = \frac{1}{2} \cdot \frac{\ln 2}{\ln\left(1 + \dfrac{0.06}{2}\right)}$$

$$t \approx 11.72$$

If the investment is compounded semiannually, it will double in value in about 11.72 years.

b. Use $A = Pe^{rt}$ with $r = 0.06$ and $A = 2P$.

$$2P = Pe^{0.06t}$$

$$2 = e^{0.06t} \qquad \text{• Divide each side by } P.$$

$$\ln 2 = 0.06t \qquad \text{• Write in logarithm form.}$$

$$t = \frac{\ln 2}{0.06} \qquad \text{• Solve for } t.$$

$$t \approx 11.55$$

If the investment is compounded continuously, it will double in value in about 11.55 years.

▶ **TRY EXERCISE 10, PAGE 498**

• EXPONENTIAL GROWTH

Given any two points on the graph of $N(t) = N_0e^{kt}$, you can use the given data to solve for the constants N_0 and k.

EXAMPLE 4 | **Find the Exponential Growth Function That Models Given Data**

a. Find the exponential growth function for a town whose population was 16,400 in 1990 and 20,200 in 2000.

b. Use the function from part **a.** to predict, to the nearest 100, the population of the town in 2005.

Solution

a. We need to determine N_0 and k in $N(t) = N_0e^{kt}$. If we represent the year 1990 by $t = 0$, then our given data are $N(0) = 16{,}400$ and $N(10) = 20{,}200$. Because N_0 is defined to be $N(0)$, we know that $N_0 = 16{,}400$. To determine k, substitute $t = 10$ and $N_0 = 16{,}400$ into $N(t) = N_0e^{kt}$ to produce

$$N(10) = 16{,}400e^{k \cdot 10}$$

$$20{,}200 = 16{,}400e^{10k} \qquad \text{• Substitute 20,200 for } N(10).$$

$$\frac{20{,}200}{16{,}400} = e^{10k} \qquad \text{• Solve for } e^{10k}.$$

$$\ln \frac{20{,}200}{16{,}400} = 10k \qquad \text{• Write in logarithmic form.}$$

$$\frac{1}{10} \ln \frac{20{,}200}{16{,}400} = k \qquad \text{• Solve for } k.$$

$$0.0208 \approx k$$

The exponential growth function is $N(t) \approx 16{,}400e^{0.0208t}$.

b. The year 1990 was represented by $t = 0$, so we will use $t = 15$ to represent the year 2005.

$$N(t) \approx 16{,}400e^{0.0208t}$$

$$N(15) \approx 16{,}400e^{0.0208 \cdot 15}$$

$$\approx 22{,}400 \quad \text{(nearest 100)}$$

The exponential growth function yields 22,400 as the approximate population of the town in 2005.

take note

Because $e^{0.0208} \approx 1.021$, the growth equation can also be written as
$$N(t) \approx 16{,}400(1.021)^t$$
In this form we see that the population is growing by 2.1% ($1.021 - 1 = 0.021 = 2.1\%$) per year.

▶ TRY EXERCISE 18, PAGE 499

● EXPONENTIAL DECAY

Many radioactive materials *decrease* in mass exponentially over time. This decrease, called radioactive decay, is measured in terms of half-life, which is defined as the time required for the disintegration of half the atoms in a sample of a radioactive substance. **Table 7.12** shows the half-lives of selected radioactive isotopes.

Table 7.12

Isotope	Half-Life
Carbon (^{14}C)	5730 years
Radium (^{226}Ra)	1660 years
Polonium (^{210}Po)	138 days
Phosphorus (^{32}P)	14 days
Polonium (^{214}Po)	1/10,000th of a second

EXAMPLE 5   **Find the Exponential Decay Function That Models Given Data**

Find the exponential decay function for the amount of phosphorus (^{32}P) that remains in a sample after t days.

Solution

When $t = 0$, $N(0) = N_0 e^{k(0)} = N_0$. Thus $N(0) = N_0$. Also, because the phosphorus has a half-life of 14 days (from **Table 7.12**), $N(14) = 0.5N_0$. To find k, substitute $t = 14$ into $N(t) = N_0 e^{kt}$ and solve for k.

$$N(14) = N_0 \cdot e^{k \cdot 14}$$

$$0.5N_0 = N_0 e^{14k} \qquad \text{• Substitute } 0.5N_0 \text{ for } N(14).$$

$$0.5 = e^{14k} \qquad \text{• Divide each side by } N_0.$$

$$\ln 0.5 = 14k \qquad \text{• Write in logarithmic form.}$$

$$\frac{1}{14} \ln 0.5 = k \qquad \text{• Solve for } k.$$

$$-0.0495 \approx k$$

The exponential decay function is $N(t) = N_0 e^{-0.0495t}$.

▶ **TRY EXERCISE 20, PAGE 499**

> *take note*
>
> Because $e^{-0.0495} \approx (0.5)^{1/14}$, the decay function $N(t) = N_0 e^{-0.0495t}$ can also be written as $N(t) = N_0(0.5)^{t/14}$. In this form it is easy to see that if t is increased by 14, then N will decrease by a factor of 0.5.

EXAMPLE 6   **Application to Air Resistance**

Assuming that air resistance is proportional to the velocity of a falling object, the velocity (in feet per second) of the object t seconds after it has been dropped is given by $v = 82(1 - e^{-0.39t})$.

a. Determine when the velocity will be 70 feet per second.

b. Write a sentence that explains the meaning of the horizontal asymptote, which is $v = 82$, in the context of this example.

Algebraic Solution

a.
$$v = 82(1 - e^{-0.39t})$$
$$70 = 82(1 - e^{-0.39t}) \quad \text{• Replace } v \text{ by 70.}$$
$$\frac{70}{82} = 1 - e^{-0.39t} \quad \text{• Divide each side by 82.}$$
$$e^{-0.39t} = 1 - \frac{70}{82} \quad \text{• Solve for } e^{-0.39t}.$$
$$-0.39t = \ln\frac{6}{41} \quad \text{• Write in logarithmic form.}$$
$$t = \frac{\ln(6/41)}{-0.39} \approx 4.9277246 \quad \text{• Solve for } t.$$

The time is approximately 4.9 seconds.

b. The horizontal asymptote $v = 82$ means that as time increases, the velocity of the object will approach but never reach or exceed 82 feet per second.

▶ TRY EXERCISE 32, PAGE 500

Visualize the Solution

a. A graph of $y = 82(1 - e^{-0.39x})$ and $y = 70$ shows that the x-coordinate of the point of intersection is about 4.9.

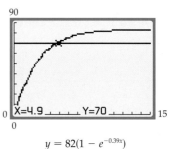

$$y = 82(1 - e^{-0.39x})$$

FIGURE 7.32

Note: The x value shown is rounded to the nearest tenth.

● **CARBON DATING**

The bone tissue in all living animals contains both carbon-12, which is nonradioactive, and carbon-14, which is radioactive with a half-life of approximately 5730 years. See **Figure 7.33**. As long as the animal is alive, the ratio of carbon-14 to carbon-12 remains constant. When the animal dies ($t = 0$), the carbon-14 begins to decay. Thus a bone that has a smaller ratio of carbon-14 to carbon-12 is older than a bone that has a larger ratio. The percent of carbon-14 present at time t is

$$P(t) = 0.5^{t/5730}$$

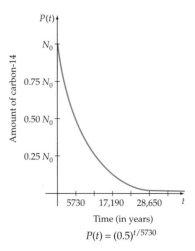

$$P(t) = (0.5)^{t/5730}$$

FIGURE 7.33

EXAMPLE 7   **Application to Archeology**

Find the age of a bone if it now has 85% of the carbon-14 it had when $t = 0$.

Solution

Let t be the time at which $P(t) = 0.85$.

$$0.85 = 0.5^{t/5730}$$

$$\ln 0.85 = \ln 0.5^{t/5730}$$ • Take the natural logarithm of each side.

$$\ln 0.85 = \frac{t}{5730} \ln 0.5$$ • Power property

$$5730\left(\frac{\ln 0.85}{\ln 0.5}\right) = t$$ • Solve for t.

$$1340 \approx t$$

The bone is about 1340 years old.

▶ TRY EXERCISE 24, PAGE 499

MATH MATTERS

The chemist Willard Frank Libby developed the carbon-14 dating technique in 1947. In 1960 he was awarded the Nobel Prize in chemistry for this achievement.

• THE LOGISTIC MODEL

The population growth function $P(t) = P_0 e^{kt}$ is called the **Malthusian growth model**. It was developed by Robert Malthus (1766–1834) in *An Essay on the Principle of Population Growth*, which was published in 1798. The Malthusian growth model is an unrestricted growth model that does not consider any limited resources that eventually will curb population growth.

The **logistic model** is a restricted growth model that takes into consideration the effects of limited resources. The logistic model was developed by Pierre Verhulst in 1836.

The Logistic Model (A Restricted Growth Model)

The magnitude of a population at time $t \geq 0$ is given by

$$P(t) = \frac{c}{1 + ae^{-bt}}$$

where c is the **carrying capacity** (the maximum population that can be supported by available resources as $t \to \infty$) and b is a positive constant called the **growth rate constant**.

The **initial population** is $P_0 = P(0)$. The constant a is related to the initial population P_0 and the carrying capacity c by the formula

$$a = \frac{c - P_0}{P_0}$$

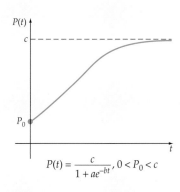

$$P(t) = \frac{c}{1 + ae^{-bt}}, \, 0 < P_0 < c$$

In the following example we determine a logistic growth model for a coyote population.

EXAMPLE 8   **Find and Use a Logistic Model**

At the beginning of 2002, the coyote population in a wilderness area was estimated at 200. By the beginning of 2004, the coyote population had increased to 250. A park ranger estimates that the carrying capacity of the wilderness area is 500 coyotes.

a. Use the given data to determine the growth rate constant for the logistic model of this coyote population.

b. Use the logistic model determined in part **a.** to predict the year in which the coyote population will first reach 400.

Solution

a. If we represent the beginning of the year 2002 by $t = 0$, then the beginning of the year 2004 will be represented by $t = 2$. In the logistic model, make the following substitutions: $P(2) = 250$, $c = 500$, and

$$a = \frac{c - P_0}{P_0} = \frac{500 - 200}{200} = 1.5.$$

$$P(t) = \frac{c}{1 + ae^{-bt}}$$

$$P(2) = \frac{500}{1 + 1.5e^{-b \cdot 2}}$$ • **Substitute the given values.**

$$250 = \frac{500}{1 + 1.5e^{-b \cdot 2}}$$

$$250(1 + 1.5e^{-b \cdot 2}) = 500$$ • **Solve for the growth rate constant b.**

$$1 + 1.5e^{-b \cdot 2} = \frac{500}{250}$$

$$1.5e^{-b \cdot 2} = 2 - 1$$

$$e^{-b \cdot 2} = \frac{1}{1.5}$$

$$-2b = \ln\left(\frac{1}{1.5}\right)$$

$$b = -\frac{1}{2}\ln\left(\frac{1}{1.5}\right)$$

$$b \approx 0.20273255$$

Using $a = 1.5$, $b = 0.20273255$, and $c = 500$ gives us the following logistic model.

$$P(t) = \frac{500}{1 + 1.5e^{-0.20273255t}}$$

b. To determine during what year the logistic model predicts the coyote population will first reach 400, replace $P(t)$ with 400 and solve for t.

$$400 = \frac{500}{1 + 1.5e^{-0.20273255t}}$$

$$400(1 + 1.5e^{-0.20273255t}) = 500$$

$$1 + 1.5e^{-0.20273255t} = \frac{500}{400}$$

$$1.5e^{-0.20273255t} = 1.25 - 1$$

$$e^{-0.20273255t} = \frac{0.25}{1.5}$$

$$-0.20273255t = \ln\left(\frac{0.25}{1.5}\right) \qquad \text{• Write in logarithmic form.}$$

$$t = \frac{1}{-0.20273255}\ln\left(\frac{0.25}{1.5}\right) \qquad \text{• Solve for } t.$$

$$\approx 8.8$$

According to the logistic model, the coyote population will reach 400 about 8.8 years after the beginning of 2002, which is during the year 2010. The graph of the logistic model is shown in **Figure 7.34.** Note that $P(8.8) \approx 400$ and that as $t \to \infty$, $P(t) \to 500$.

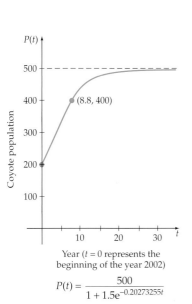

Year ($t = 0$ represents the beginning of the year 2002)

$$P(t) = \frac{500}{1 + 1.5e^{-0.20273255t}}$$

FIGURE 7.34

▶ TRY EXERCISE 48, PAGE 501

TOPICS FOR DISCUSSION

1. Explain the difference between compound interest and simple interest.

2. What is an exponential growth model? Give an example of an application for which the exponential growth model might be appropriate.

3. What is an exponential decay model? Give an example of an application for which the exponential decay model might be appropriate.

4. Consider the exponential model $P(t) = P_0e^{kt}$ and the logistic model $P(t) = \dfrac{c}{1 + ae^{-bt}}$. Explain the similarities and differences between the two models.

EXERCISE SET 7.5

1. **COMPOUND INTEREST** If $8000 is invested at an annual interest rate of 5% and compounded annually, find the balance after

a. 4 years   **b.** 7 years

2. **COMPOUND INTEREST** If $22,000 is invested at an annual interest rate of 4.5% and compounded annually, find the balance after

a. 2 years   **b.** 10 years

3. **COMPOUND INTEREST** If $38,000 is invested at an annual interest rate of 6.5% for 4 years, find the balance if the interest is compounded

a. annually   **b.** daily   **c.** hourly

▶ **4.** **COMPOUND INTEREST** If $12,500 is invested at an annual interest rate of 8% for 10 years, find the balance if the interest is compounded

a. annually   **b.** daily   **c.** hourly

5. **COMPOUND INTEREST** Find the balance if $15,000 is invested at an annual rate of 10% for 5 years, compounded continuously.

▶ **6.** **COMPOUND INTEREST** Find the balance if $32,000 is invested at an annual rate of 8% for 3 years, compounded continuously.

7. **COMPOUND INTEREST** How long will it take $4000 to double if it is invested in a certificate of deposit that pays 7.84% annual interest compounded continuously? Round to the nearest tenth of a year.

8. **COMPOUND INTEREST** How long will it take $25,000 to double if it is invested in a savings account that pays 5.88% annual interest compounded continuously? Round to the nearest tenth of a year.

9. **CONTINUOUS COMPOUNDING INTEREST** Use the Continuous Compounding Interest Formula to derive an expression for the time it will take money to triple when invested at an annual interest rate of r compounded continuously.

▶ **10.** **CONTINUOUS COMPOUNDING INTEREST** How long will it take $1000 to triple if it is invested at an annual interest rate of 5.5% compounded continuously? Round to the nearest year.

11. **CONTINUOUS COMPOUNDING INTEREST** How long will it take $6000 to triple if it is invested in a savings account

that pays 7.6% annual interest compounded continuously? Round to the nearest year.

12. **CONTINUOUS COMPOUNDING INTEREST** How long will it take $10,000 to triple if it is invested in a savings account that pays 5.5% annual interest compounded continuously? Round to the nearest year.

13. **POPULATION GROWTH** The number of bacteria $N(t)$ present in a culture at time t hours is given by

$$N(t) = 2200(2)^t$$

Find the number of bacteria present when

a. $t = 0$ hours   **b.** $t = 3$ hours

14. **POPULATION GROWTH** The population of a town grows exponentially according to the function

$$f(t) = 12,400(1.14)^t$$

for $0 \leq t \leq 5$ years. Find, to the nearest hundred, the population of the town when t is

a. 3 years   **b.** 4.25 years

15. **POPULATION GROWTH** A town had a population of 22,600 in 1990 and a population of 24,200 in 1995.

a. Find the exponential growth function for the town. Use $t = 0$ to represent the year 1990.

b. Use the growth function to predict the population of the town in 2005. Round to the nearest hundred.

16. **POPULATION GROWTH** A town had a population of 53,700 in 1996 and a population of 58,100 in 2000.

a. Find the exponential growth function for the town. Use $t = 0$ to represent the year 1996.

b. Use the growth function to predict the population of the town in 2008. Round to the nearest hundred.

17. **POPULATION GROWTH** The growth of the population of Los Angeles, California, for the years 1992 through 1996 can be approximated by the equation

$$P = 10,130(1.005)^t$$

where $t = 0$ corresponds to January 1, 1992 and P is in thousands.

a. Assuming this growth rate continues, what will be the population of Los Angeles on January 1 in the year 2004?

b. In what year will the population of Los Angeles first exceed 13,000,000?

▶ **18.** ⊙ POPULATION GROWTH The growth of the population of Mexico City, Mexico, for the years 1991 through 1998 can be approximated by the equation

$$P = 20,899(1.027)^t$$

where $t = 0$ corresponds to 1991 and P is in thousands.

a. Assuming this growth rate continues, what will be the population of Mexico City in the year 2003?

b. Assuming this growth rate continues, in what year will the population of Mexico City first exceed 35,000,000?

19. MEDICINE Sodium-24 is a radioactive isotope of sodium that is used to study circulatory dysfunction. Assuming that 4 micrograms of sodium-24 is injected into a person, the amount A in micrograms remaining in that person after t hours is given by the equation $A = 4e^{-0.046t}$.

a. Graph this equation.

b. What amount of sodium-24 remains after 5 hours?

c. What is the half-life of sodium-24?

d. In how many hours will the amount of sodium-24 be 1 microgram?

▶ **20.** ⊙ RADIOACTIVE DECAY Polonium (^{210}Po) has a half-life of 138 days. Find the decay function for the amount of polonium (^{210}Po) that remains in a sample after t days.

21. ⊙ GEOLOGY Geologists have determined that Crater Lake in Oregon was formed by a volcanic eruption. Chemical analysis of a wood chip that is assumed to be from a tree that died during the eruption has shown that it contains approximately 45% of its original carbon-14. Determine how long ago the volcanic eruption occurred. Use 5730 years as the half-life of carbon-14.

22. ⊙ RADIOACTIVE DECAY Use $N(t) = N_0(0.5)^{t/138}$, where t is measured in days, to estimate the percentage of polonium (^{210}Po) that remains in a sample after 2 years. Round to the nearest hundredth of a percent.

23. ⊙ ARCHEOLOGY The Rhind papyrus, named after A. Henry Rhind, contains most of what we know today of ancient Egyptian mathematics. A chemical analysis of a sample from the papyrus has shown that it contains approximately 75% of its original carbon-14. What is the age of the Rhind papyrus? Use 5730 years as the half-life of carbon-14.

▶ **24.** ARCHEOLOGY Determine the age of a bone if it now contains 65% of its original amount of carbon-14. Round to the nearest 100 years.

25. PHYSICS Newton's Law of Cooling states that if an object at temperature T_0 is placed into an environment at constant temperature A, then the temperature of the object, $T(t)$ (in degrees Fahrenheit), after t minutes is given by $T(t) = A + (T_0 - A)e^{-kt}$, where k is a constant that depends on the object.

a. Determine the constant k (to the nearest thousandth) for a canned soda drink that takes 5 minutes to cool from 75°F to 65°F after being placed in a refrigerator that maintains a constant temperature of 34°F.

b. What will be the temperature (to the nearest degree) of the soda drink after 30 minutes?

c. When (to the nearest minute) will the temperature of the soda drink be 36°F?

26. PSYCHOLOGY According to a software company, the users of its typing tutorial can expect to type $N(t)$ words per minute after t hours of practice with the product, according to the function $N(t) = 100(1.04 - 0.99^t)$.

a. How many words per minute can a student expect to type after 2 hours of practice?

b. How many words per minute can a student expect to type after 40 hours of practice?

c. According to the function N, how many hours (to the nearest hour) of practice will be required before a student can expect to type 60 words per minute?

27. PSYCHOLOGY In the city of Whispering Palms, which has a population of 80,000 people, the number of people $P(t)$ exposed to a rumor in t hours is given by the function $P(t) = 80,000(1 - e^{-0.0005t})$.

a. Find the number of hours until 10% of the population have heard the rumor.

b. Find the number of hours until 50% of the population have heard the rumor.

28. LAW A lawyer has determined that the number of people $P(t)$ in a city of 1,200,000 people who have been exposed to a news item after t days is given by the function

$$P(t) = 1,200,000(1 - e^{-0.03t})$$

a. How many days after a major crime has been reported have 40% of the population heard of the crime?

b. A defense lawyer knows it will be very difficult to pick an unbiased jury after 80% of the population have heard of the crime. After how many days will 80% of the population have heard of the crime?

29. Depreciation An automobile depreciates according to the function $V(t) = V_0(1 - r)^t$, where $V(t)$ is the value in dollars after t years, V_0 is the original value, and r is the yearly depreciation rate. A car has a yearly depreciation rate of 20%. Determine, to the nearest 0.1 year, in how many years the car will depreciate to half its original value.

30. Physics The current $I(t)$ (measured in amperes) of a circuit is given by the function $I(t) = 6(1 - e^{-2.5t})$, where t is the number of seconds after the switch is closed.

a. Find the current when $t = 0$.

b. Find the current when $t = 0.5$.

c. Solve the equation for t.

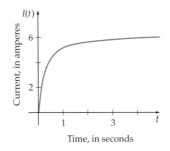

Time, in seconds

31. Air Resistance Assuming that air resistance is proportional to velocity, the velocity v, in feet per second, of a falling object after t seconds is given by $v = 32(1 - e^{-t})$.

a. Graph this equation for $t \geq 0$.

b. Determine algebraically, to the nearest 0.01 second, when the velocity is 20 feet per second.

c. Determine the horizontal asymptote of the graph of v.

d. ✎ Write a sentence that explains the meaning of the horizontal asymptote in the context of this application.

▶ **32. Air Resistance** Assuming that air resistance is proportional to velocity, the velocity v, in feet per second, of a falling object after t seconds is given by $v = 64(1 - e^{-t/2})$.

a. Graph this equation for $t \geq 0$.

b. Determine algebraically, to the nearest 0.1 second, when the velocity is 50 feet per second.

c. Determine the horizontal asymptote of the graph of v.

d. ✎ Write a sentence that explains the meaning of the horizontal asymptote in the context of this application.

33. 🖩 The distance s (in feet) that the object in Exercise 31 will fall in t seconds is given by $s = 32t + 32(e^{-t} - 1)$.

a. Use a graphing utility to graph this equation for $t \geq 0$.

b. Determine, to the nearest 0.1 second, the time it takes the object to fall 50 feet.

c. Calculate the slope of the secant line through $(1, s(1))$ and $(2, s(2))$.

d. ✎ Write a sentence that explains the meaning of the slope of the secant line you calculated in **c.**

34. 🖩 The distance s (in feet) that the object in Exercise 32 will fall in t seconds is given by $s = 64t + 128(e^{-t/2} - 1)$.

a. Use a graphing utility to graph this equation for $t \geq 0$.

b. Determine, to the nearest 0.1 second, the time it takes the object to fall 50 feet.

c. Calculate the slope of the secant line through $(1, s(1))$ and $(2, s(2))$.

d. ✎ Write a sentence that explains the meaning of the slope of the secant line you calculated in **c.**

In Exercises 35 to 40, determine the following constants for the given logistic growth model.
a. The carrying capacity
b. The growth rate constant
c. The initial population P_0

35. $P(t) = \dfrac{1900}{1 + 8.5e^{-0.16t}}$ **36.** $P(t) = \dfrac{32{,}550}{1 + 0.75e^{-0.08t}}$

37. $P(t) = \dfrac{157{,}500}{1 + 2.5e^{-0.04t}}$ **38.** $P(t) = \dfrac{51}{1 + 1.04e^{-0.03t}}$

39. $P(t) = \dfrac{2400}{1 + 7e^{-0.12t}}$ **40.** $P(t) = \dfrac{320}{1 + 15e^{-0.12t}}$

In Exercises 41 to 44, use algebraic procedures to find the logistic growth model for the data.

41. $P_0 = 400$, $P(2) = 780$, and the carrying capacity is 5500.

42. $P_0 = 6200$, $P(8) = 7100$, and the carrying capacity is 9500.

43. $P_0 = 18$, $P(3) = 30$, and the carrying capacity is 100.

44. $P_0 = 3200$, $P(22) \approx 5565$, and the growth rate constant is 0.056.

45. REVENUE The annual revenue R, in dollars, of a new company can be closely modeled by the logistic growth function

$$R(t) = \frac{625,000}{1 + 3.1e^{-0.045t}}$$

where the *natural* number t is the time, in years, since the company was founded.

a. According to the model, what will be the company's annual revenue for its first year and its second year ($t = 1$ and $t = 2$) of operation? Round to the nearest $1000.

b. According to the model, what will the company's annual revenue approach in the long-term future?

46. NEW CAR SALES The number of cars A sold annually by an automobile dealership can be closely modeled by the logistic growth function

$$A(t) = \frac{1650}{1 + 2.4e^{-0.055t}}$$

where the *natural* number t is the time, in years, since the dealership was founded.

a. According to the model, what number of cars will the dealership sell during its first year and its second year ($t = 1$ and $t = 2$) of operation? Round to the nearest unit.

b. According to the model, what will the dealership's annual car sales approach in the long-term future?

47. POPULATION GROWTH The population of wolves in a preserve satisfies a logistic growth model in which $P_0 = 312$ in the year 2002, $c = 1600$, and $P(6) = 416$.

a. Determine the logistic growth model for this population, where t is the number of years after 2002.

b. Use the logistic growth model from part **a.** to predict the size of the wolf population in 2012.

▶ 48. POPULATION GROWTH The population of ground-hogs on a ranch satisfies a logistic growth model in which $P_0 = 240$ in the year 2001, $c = 3400$, and $P(1) = 310$.

a. Determine the logistic growth model for this population, where t is the number of years after 2001.

b. Use the logistic growth model from part **a.** to predict the size of the groundhog population in 2008.

49. POPULATION GROWTH The population of squirrels in a nature preserve satisfies a logistic growth model in which $P_0 = 1500$ in the year 2001. The carrying capacity of the preserve is estimated at 8500 squirrels and $P(2) = 1900$.

a. Determine the logistic growth model for this population, where t is the number of years after 2001.

b. Use the logistic growth model from part **a.** to predict the year in which the squirrel population will first exceed 4000.

50. POPULATION GROWTH The population of walruses on an island satisfies a logistic growth model in which $P_0 = 800$ in the year 2000. The carrying capacity of the island is estimated at 5500 walruses, and $P(1) = 900$.

a. Determine the logistic growth model for this population, where t is the number of years after 2000.

b. Use the logistic growth model from part **a.** to predict the year in which the walrus population will first exceed 2000.

51. LEARNING THEORY The logistic model is also used in learning theory. Suppose that historical records from employee training at a company show that the percent score on a product information test is given by

$$P = \frac{100}{1 + 25e^{-0.095t}}$$

where t is the number of hours of training. What is the number of hours (to the nearest hour) of training needed before a new employee will answer 75% of the questions correctly?

52. LEARNING THEORY A company provides training in the assembly of a computer circuit to new employees. Past experience has shown that the number of correctly assembled circuits per week can be modeled by

$$N = \frac{250}{1 + 249e^{-0.503t}}$$

where t is the number of weeks of training. What is the number of weeks (to the nearest week) of training needed before a new employee will correctly make 140 circuits?

CONNECTING CONCEPTS

53. **MEDICATION LEVEL** A patient is given three dosages of aspirin. Each dosage contains 1 gram of aspirin. The second and third dosages are each taken 3 hours after the previous dosage is administered. The half-life of the aspirin is 2 hours. The amount of aspirin, A, in the patient's body t hours after the first dosage is administered is

$$A(t) = \begin{cases} 0.5^{t/2} & 0 \le t < 3 \\ 0.5^{t/2} + 0.5^{(t-3)/2} & 3 \le t < 6 \\ 0.5^{t/2} + 0.5^{(t-3)/2} + 0.5^{(t-6)/2} & t \ge 6 \end{cases}$$

Find, to the nearest hundredth of a gram, the amount of aspirin in the patient's body when

a. $t = 1$ **b.** $t = 4$ **c.** $t = 9$

54. **MEDICATION LEVEL** Use a graphing calculator and the dosage formula in Exercise 53 to determine when, to the nearest tenth of an hour, the amount of aspirin in the patient's body first reaches 0.25 gram.

Exercises 55 to 57 make use of the factorial function, which is defined as follows. For whole numbers n, the number $n!$ (which is read "n factorial") is given by

$$n! = \begin{cases} n(n-1)(n-2)\cdots 1, & \text{if } n \ge 1 \\ 1, & \text{if } n = 0 \end{cases}$$

Thus, $0! = 1$ and $4! = 4 \cdot 3 \cdot 2 \cdot 1 = 24$.

55. **QUEUEING THEORY** A study shows that the number of people who arrive at a bank teller's window averages 4.1 people every 10 minutes. The probability P that exactly x people will arrive at the teller's window in a given 10-minute period is

$$P(x) = \frac{4.1^x e^{-4.1}}{x!}$$

Find, to the nearest 0.1%, the probability that in a given 10-minute period, exactly

a. 0 people arrive at the window.

b. 2 people arrive at the window.

c. 3 people arrive at the window.

d. 4 people arrive at the window.

e. 9 people arrive at the window.

As $x \to \infty$, what does P approach?

56. **STIRLING'S FORMULA** *Stirling's Formula* (after James Stirling, 1692–1770),

$$n! \approx \left(\frac{n}{e}\right)^n \sqrt{2\pi n}$$

is often used to approximate very large factorials. Use Stirling's Formula to approximate 10!, and then compute the ratio of Stirling's approximation of 10! divided by the actual value of 10!, which is 3,628,800.

57. **RUBIK'S CUBE** The Rubik's cube shown here was invented by Erno Rubik in 1975. The small outer cubes are held together in such a way that they can be rotated around three axes. The total number of positions in which the Rubik's cube can be arranged is

$$\frac{3^8 2^{12} 8! \, 12!}{2 \cdot 3 \cdot 2}$$

If you can arrange a Rubik's cube into a new arrangement every second, how many centuries would it take to place the cube into each of its arrangements? Assume that there are 365 days in a year.

58. **OIL SPILLS** Crude oil leaks from a tank at a rate that depends on the amount of oil that remains in the tank. Because $\frac{1}{8}$ of the oil in the tank leaks out every 2 hours, the volume of oil $V(t)$ in the tank after t hours is given by $V(t) = V_0(0.875)^{t/2}$, where $V_0 = 350{,}000$ gallons is the number of gallons in the tank at the time the tank started to leak ($t = 0$).

a. How many gallons does the tank hold after 3 hours?

b. How many gallons does the tank hold after 5 hours?

c. How long, to the nearest hour, will it take until 90% of the oil has leaked from the tank?

PREPARE FOR SECTION 7.6

59. Determine whether $N(t) = 4 - \ln t$ is an increasing or a decreasing function. [7.2]

60. Determine whether $P(t) = 1 - 2(1.05^t)$ is an increasing or a decreasing function. [7.1]

61. Evaluate $P(t) = \dfrac{108}{1 + 2e^{-0.1t}}$ for $t = 0$. [7.1]

62. Evaluate $N(t) = 840e^{1.05t}$ for $t = 0$. [7.1]

63. Solve $10 = \dfrac{20}{1 + 2.2e^{-0.05t}}$ for t. Round to the nearest tenth. [7.4]

64. Determine the horizontal asymptote of the graph of $P(t) = \dfrac{55}{1 + 3e^{-0.08t}}$ for $t \geq 0$. [7.5]

PROJECTS

A DECLINING LOGISTIC MODEL If $P_0 > c$ (which implies that $-1 < a < 0$), then the logistic function $P(t) = \dfrac{c}{1 + ae^{-bt}}$ decreases as t increases. Biologists often use this type of logistic function to model populations that decrease over time. See the following figure.

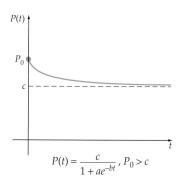

$$P(t) = \frac{c}{1 + ae^{-bt}}, \; P_0 > c$$

1. **A DECLINING FISH POPULATION** A biologist finds that the fish population in a small lake can be closely modeled by the logistic function

$$P(t) = \frac{1000}{1 + (-0.3333)e^{-0.05t}}$$

where t is the time, in years, since the lake was first stocked with fish.

a. What was the fish population when the lake was first stocked with fish?

b. According to the logistic model, what will the fish population approach in the long-term future?

2. **A DECLINING DEER POPULATION** The deer population in a reserve is given by the logistic function

$$P(t) = \frac{1800}{1 + (-0.25)e^{-0.07t}}$$

where t is the time, in years, since July 1, 2001.

a. What was the deer population on July 1, 2001? What was the deer population on July 1, 2003?

b. According to the logistic model, what will the deer population approach in the long-term future?

3. **MODELING WORLD RECORD TIMES IN THE MEN'S MILE RACE** In the early 1950s, many people speculated that no runner would ever run a mile race in under 4 minutes. During the period from 1913 to 1945, the world record in the mile event had been reduced from 4.14.4 (4 minutes, 14.4 seconds) to 4.01.4, but no one seemed capable of running a sub-four-minute mile. Then, in 1954, Roger Bannister broke through the four-minute barrier by running a mile in 3.59.6. In 1999, the current record of 3.43.13 was established. It is fun to think about future record times in the mile race. Will they ever go below 3 minutes, 30 seconds? Below 3 minutes, 20 seconds? What about a sub-three-minute mile?

A declining logistic function that closely models the world record times *WR*, in seconds, in the men's mile run from 1913 ($t = 0$) to 1999 ($t = 86$) is given by

$$WR(t) = \frac{199.13}{1 + (-0.21726)e^{-0.0079889t}}$$

a. Use the above logistic model to predict the world record time for the men's mile run in the year 2020 and the year 2050.

b. According to the logistic function, what time will the world record in the men's mile event approach but never break through?

MODELING DATA WITH EXPONENTIAL AND LOGARITHMIC FUNCTIONS

- ● **ANALYZE SCATTER PLOTS**
- ● **APPLICATIONS**
- ● **USE REGRESSION TO FIND A LOGISTIC GROWTH MODEL**

● ANALYZE SCATTER PLOTS

In Section 1.7 we used linear and quadratic functions to model several data sets. However, in some applications, data can be modeled more closely by using exponential or logarithmic functions. For instance, **Figure 7.35** illustrates some scatter plots that can be effectively modeled by exponential and logarithmic functions.

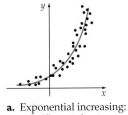

a. Exponential increasing:
$y = ab^x, a > 0, b > 1$

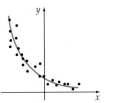

b. Exponential decreasing:
$y = ab^x, a > 0, 0 < b < 1$

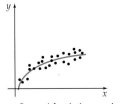

c. Logarithmic increasing:
$y = a + b \ln x, b > 0$

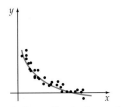

d. Logarithmic decreasing:
$y = a + b \ln x, b < 0$

FIGURE 7.35
Exponential and Logarithmic Models

The terms *concave upward* and *concave downward* are often used to describe a graph. For instance, **Figures 7.36a** and **7.36b** show the graphs of two increasing functions that join the points *P* and *Q*. The graphs of *f* and *g* differ in that they bend in different directions. We can distinguish between these two types of "bending" by examining the positions of *tangent lines* to the graphs. In **Figures 7.36c** and **7.36d**, tangent lines (in red) have been drawn to the graphs of *f* and *g*. The graph of *f* lies above its tangent lines and the graph of *g* lies below its tangent lines. The function *f* is said to be concave upward, and *g* is concave downward.

a.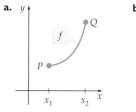

b.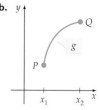

c. f is concave upward.   **d.** g is concave downward.

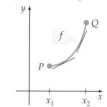   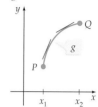

FIGURE 7.36

Definition of Concavity

If the graph of f lies above all of its tangents on an interval $[x_1, x_2]$, then f is **concave upward** on $[x_1, x_2]$.

If the graph of f lies below all of its tangents on an interval $[x_1, x_2]$, then f is **concave downward** on $[x_1, x_2]$.

An examination of the graphs in **Figure 7.35** shows that the graphs of all exponential functions of the form $y = ab^x, a > 0, b > 0, b \neq 1$ are concave upward. The graphs of increasing logarithmic functions are concave downward, and the graphs of decreasing logarithmic functions are concave upward.

In Example 1 we analyze scatter plots by determining whether the shape of the scatter plot can best be approximated by an increasing or a decreasing function, and by a function that is concave upward or concave downward.

❓ **QUESTION** Is the graph of $y = 5 - 2 \ln x$ concave upward or concave downward?

❓ **ANSWER** The equation $y = 5 - 2 \ln x$ has the form $y = a + b \ln x$, with $a > 0$ and $b < 0$. The graph of $y = 5 - 2 \ln x$ is concave upward because the b-value, -2, is less than zero. See **Figure 7.35d.**

> **EXAMPLE 1**   **Analyze Scatter Plots**

For each of the following data sets, determine whether the most suitable model of the data would be an increasing exponential function or an increasing logarithmic function.

$$A = \{(1, 0.6), (2, 0.7), (2.8, 0.8), (4, 1.3), (6, 1.5),$$
$$(6.5, 1.6), (8, 2.1), (11.2, 4.1), (12, 4.6), (15, 8.2)\}$$

$$B = \{(1.5, 2.8), (2, 3.5), (4.1, 5.1), (5, 5.5), (5.5, 5.7), (7, 6.1),$$
$$(7.2, 6.4), (8, 6.6), (9, 6.9), (11.6, 7.4), (12.3, 7.5), (14.7, 7.9)\}$$

See Section 1.7 if you need to review the steps needed to create a scatter plot on a TI-83 calculator.

Solution

For each set construct a scatter plot of the data. See **Figure 7.37.**

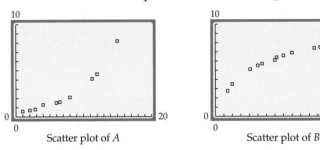

Scatter plot of *A*   Scatter plot of *B*

FIGURE 7.37

The scatter plot of *A* suggests that *A* is an increasing function that is concave upward. Thus *A* can be effectively modeled by an increasing exponential function.

The scatter plot of *B* suggests that *B* is an increasing function that is concave downward. Thus *B* can be effectively modeled by an increasing logarithmic function.

▶ **TRY EXERCISE 4, PAGE 512**

● **APPLICATIONS**

The methods used to model data using exponential or logarithmic functions are similar to the methods used in Section 1.7 to model data using linear or quadratic functions. Here is a summary of the modeling process.

The Modeling Process

Use a graphing utility to:

1. **Construct a *scatter plot* of the data** to determine which type of function will effectively model the data.

2. **Find the *regression equation*** of the modeling function and the correlation coefficient for the regression.

3. **Examine the *correlation coefficient*** and *view a graph* that displays both the modeling function and the scatter plot to determine how well your function fits the data.

In the following example we use the modeling process to find an exponential function that closely models the value of a diamond as a function of its weight.

EXAMPLE 2 | **Model an Application with an Exponential Function**

A diamond merchant has determined the values of several white diamonds that have different weights (measured in carats), but are *similar in quality*. See **Table 7.13**.

TABLE 7.13

0.50 ct	0.75 ct	1.00 ct	1.25 ct	1.50 ct	1.75 ct	2.00 ct	3.00 ct	4.00 ct
$4,600	$5,000	$5,800	$6,200	$6,700	$7,300	$7,900	$10,700	$14,500

Find a function that models the values of the diamonds as a function of their weights and use the function to predict the value of a 3.5-carat diamond of similar quality.

Solution

1. **Construct a scatter plot of the data.**

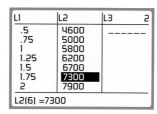
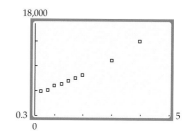

FIGURE 7.38

From the scatter plot in **Figure 7.38** it appears that the data can be closely modeled by an exponential function of the form $y = ab^x$, $a > 0$ and $b > 1$.

2. **Find the regression equation.** The calculator display in **Figure 7.39** shows that the exponential regression equation is $y \approx 4067.6(1.3816)^x$, where x is the carat weight of the diamond and y is the value of the diamond.

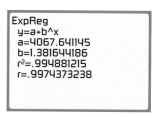

FIGURE 7.39
ExpReg display (DiagnosticOn)

Continued ▶

INTEGRATING TECHNOLOGY

Most graphing utilities have built-in routines that can be used to determine the exponential or logarithmic regression function that best models a set of data. On a TI-83, the ExpReg instruction is used to find the exponential regression function and the LnReg instruction is used to find the logarithmic regression function. The TI-83 does not show the value of the regression coefficient r unless the DiagnosticOn command has been entered. The DiagnosticOn command is in the CATALOG menu.

3. **Examine the correlation coefficient.** The correlation coefficient $r \approx 0.9974$ is close to 1. This indicates that the exponential regression function $y \approx 4067.6(1.3816)^x$ provides a good fit for the data. The graph in **Figure 7.40** also shows that the exponential regression function provides a good model for the data.

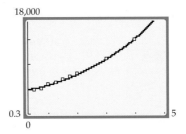

FIGURE 7.40

To estimate the value of a 3.5-carat diamond, replace x in the exponential regression function with 3.5.

$$y \approx 4067.6(1.3816)^{3.5} \approx \$12{,}610$$

According to the exponential regression function, the value of a 3.5-carat diamond of similar quality is about $12,610.

▶ **TRY EXERCISE 22, PAGE 513**

In the next example we consider a data set that can be effectively modeled by more than one type of function.

EXAMPLE 3   **Choosing the Best Model**

 Table 7.14 shows the winning times in the women's Olympic 100-meter freestyle event for the years 1968 to 2000.

TABLE 7.14   Women's Olympic 100-Meter Freestyle, 1968 to 2000

Year	Time (in seconds)	Year	Time (in seconds)
1968	60.0	1988	54.93
1972	58.59	1992	54.64
1976	55.65	1996	54.50
1980	54.79	2000	53.83
1984	55.92		

Source: Time Almanac 2002.

a. Determine whether the data in **Table 7.14** can best be modeled by an exponential function or a logarithmic function.

b. Use the function you chose in part **a.** to predict the winning time in the women's Olympic 100-meter freestyle event for the year 2008.

Solution

a. Construct a scatter plot of the data. In this example we have represented the year 1968 by $x = 68$, the year 2000 by $x = 100$, and the winning time by y.

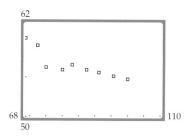

FIGURE 7.41

From the scatter plot in **Figure 7.41**, it appears that the data can be effectively modeled by a decreasing exponential function and also by a decreasing logarithmic function. Use a graphing utility to determine both an exponential regression function and a logarithmic regression function for the data. **Figure 7.42** shows the exponential regression function, and **Figure 7.43** shows the logarithmic regression function.

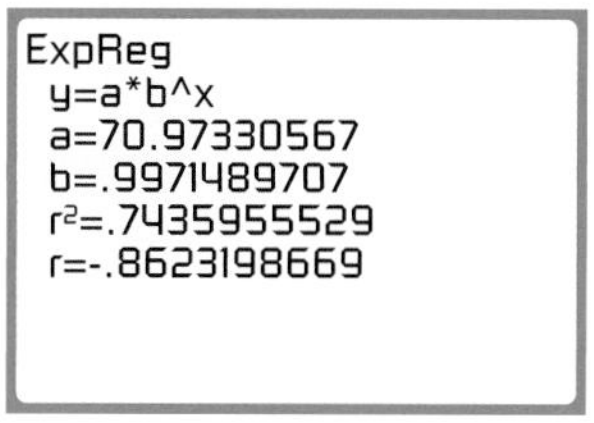

```
ExpReg
 y=a*b^x
 a=70.97330567
 b=.9971489707
 r²=.7435955529
 r=-.8623198669
```

FIGURE 7.42

```
LnReg
 y=a+blnx
 a=116.7153463
 b=-13.75559414
 r²=.7708582677
 r=-.877985346
```

FIGURE 7.43

In this example the regression coefficients are both negative. In such cases, the regression function that has a correlation coefficient closer to -1 provides the better fit for the given data. Thus the logarithmic model provides a slightly better fit for the data in this example. The logarithmic regression function is $y \approx 116.72 - 13.756 \ln x$. The graph of $y \approx 116.72 - 13.756 \ln x$, along with a scatter plot of the data, is shown in **Figure 7.44.**

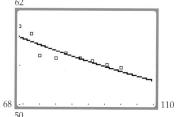

FIGURE 7.44

b. To predict the winning time for the women's Olympic 100-meter freestyle event in the year 2008, replace x in the logarithmic regression function with 108.

$$y \approx 116.72 - 13.756 \ln(108) \approx 52.31$$

According to the logarithmic regression function, the winning time for the women's Olympic 100-meter freestyle event in the year 2008 will be about 52.31 seconds.

▶ **TRY EXERCISE 24, PAGE 513**

● USE REGRESSION TO FIND A LOGISTIC GROWTH MODEL

If a scatter plot of a set of data suggests that the data can be effectively modeled by a logistic growth model, then you can use the logistic regression feature of a graphing utility to find the logistic growth model that provides the best fit for the data. This process is illustrated in Example 4.

EXAMPLE 4 Use Logistic Regression to Find a Logistic Growth Model

Table 7.15 shows the population of deer in an animal preserve for the years 1990 to 2004.

TABLE 7.15 Deer Population at the Wild West Animal Preserve

Year	Population	Year	Population	Year	Population
1990	320	1995	1150	2000	2620
1991	410	1996	1410	2001	2940
1992	560	1997	1760	2002	3100
1993	730	1998	2040	2003	3300
1994	940	1999	2310	2004	3460

Use a graphing utility to find a logistic regression model that approximates the deer population as a function of the year. Use the model to predict the deer population in the year 2010.

Solution

INTEGRATING TECHNOLOGY

On a TI-83 graphing calculator, the logistic growth model is given in the form

$$y = \frac{c}{1 + ae^{-bx}}$$

Think of the variable x as the time t and the variable y as $P(t)$.

1. **Construct a scatter plot of the data.** Enter the data into a graphing utility, and then use the utility to display a scatter plot of the data. In this example we represent the year 1990 by $x = 0$, the year 2004 by $x = 14$, and the deer population by y.

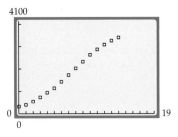

FIGURE 7.45

Figure 7.45 shows that the data can be closely approximated by a logistic growth model.

2. **Find the regression function.** Use a graphing utility to perform a logistic regression on the data. On a TI-83 graphing calculator, select B: Logistic, which is in the STAT CALC menu. The logistic regression function for the data is $y \approx \dfrac{3965.3}{1 + 11.445e^{-0.31152x}}$. See **Figure 7.46.**

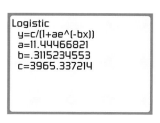

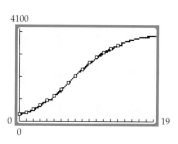

FIGURE 7.46   **FIGURE 7.47**

3. **Examine the fit.** A TI-83 calculator does not give a correlation coefficient for a logistic regression. However, **Figure 7.47** shows that the graph of $y \approx \dfrac{3965.3}{1 + 11.445e^{-0.31152x}}$ provides a good fit for the data. The logistic model predicts that in the year 2010 ($x = 20$), the deer population will be about 3878.

$$y \approx \frac{3965.3}{1 + 11.445e^{-0.31152(20)}} \approx 3878$$

 TRY EXERCISE 26, PAGE 514

TOPICS FOR DISCUSSION

1. A student tries to determine the exponential regression equation for the following data.

x	1	2	3	4	5
y	8	2	0	−1.5	−2

The student's calculator displays an ERROR message. Explain why the calculator was unable to determine the exponential regression equation for the data.

2. Consider the logarithmic model $h(x) = 6 - 2 \ln x$.

   a. Is h an increasing or a decreasing function?

   b. Is h concave up or concave down on the interval $(0, \infty)$?

   c. Find, if possible, $h(0)$ and $h(e)$.

   d. Does h have a horizontal asymptote? Explain.

EXERCISE SET 7.6

 In Exercises 1 to 6, use a scatter plot of the given data to determine which of the following types of functions might provide a suitable model of the data:

- an increasing exponential function
  $y = ab^x$, $a > 0$, $b > 1$ (See **Figure 7.35a.**)
- an increasing logarithmic function
  $y = a + b \ln x$, $b > 0$ (See **Figure 7.35c.**)
- a decreasing exponential function
  $y = ab^x$, $a > 0$, $0 < b < 1$ (See **Figure 7.35b.**)
- a decreasing logarithmic function
  $y = a + b \ln x$, $b < 0$ (See **Figure 7.35d.**)

(*Note:* Some data sets can be closely modeled by more than one type of function.)

1. {(1, 3), (1.5, 4), (2, 6), (3, 13), (3.5, 19), (4, 27)}

2. {(1.0, 1.12), (2.1, 0.87), (3.2, 0.68), (3.5, 0.63), (4.4, 0.52)}

3. {(1, 2.4), (2, 1.1), (3, 0.5), (4, 0.2), (5, 0.1)}

▶ **4.** {(5, 2.3), (7, 3.9), (9, 4.5), (12, 5.0), (16, 5.4), (21, 5.8), (26, 6.1)}

5. {(1, 2.5), (1.5, 1.7), (2, 0.7), (3, −0.5), (3.5, −1.3), (4, −1.5)}

6. {(1, 3), (1.5, 3.8), (2, 4.4), (3, 5.2), (4, 5.8), (6, 6.6)}

 In Exercises 7 to 10, use a graphing utility to find the exponential regression function for the data. State the correlation coefficient *r*. Round *a*, *b*, and *r* to the nearest hundred thousandth.

7. {(10, 6.8), (12, 6.9), (14, 15.0), (16, 16.1), (18, 50.0), (19, 20.0)}

8. {(2.6, 16.2), (3.8, 48.8), (5.1, 160.1), (6.5, 590.2), (7, 911.2)}

9. {(0, 1.83), (1, 0.92), (2, 0.51), (3, 0.25), (4, 0.13), (5, 0.07)}

10. {(4.5, 1.92), (6.0, 1.48), (7.5, 1.14), (10.2, 0.71), (12.3, 0.49)}

 In Exercises 11 to 14, use a graphing utility to find the logarithmic regression function for the data. State the correlation coefficient *r*. Round *a*, *b*, and *r* to the nearest hundred thousandth.

11. {(5, 2.7), (6, 2.5), (7.2, 2.2), (9.3, 1.9), (11.4, 1.6), (14.2, 1.3)}

12. {(11, 15.75), (14, 15.52), (17, 15.34), (20, 15.18), (23, 15.05)}

13. {(3, 16.0), (4, 16.5), (5, 16.9), (7, 17.5), (8, 17.7), (9.8, 18.1)}

14. {(8, 67.1), (10, 67.8), (12, 68.4), (14, 69.0), (16, 69.4)}

 In Exercises 15 to 18, use a graphing utility to find the logistic regression function for the data. Round the constants *a*, *b*, and *c* to the nearest hundred thousandth.

15. {(0, 81), (2, 87), (6, 98), (10, 110), (15, 125)}

16. {(0, 175), (5, 195), (10, 217), (20, 264), (35, 341)}

17. {(0, 955), (10, 1266), (20, 1543), (30, 1752)}

18. {(0, 1588), (5, 2598), (10, 3638), (25, 5172)}

19. INTEREST RATES ON AUTO LOANS The following table shows the annual interest rates for new car loans in 2002 based on the length (term) of the loan.

Term t, in months	12	24	36	48
Annual interest rate, r	5.72%	5.97%	6.23%	6.50%

a. Find an exponential model for the data in the table and use the model to predict the interest rate *r*, to the nearest 0.01%, on an auto loan with a term of 60 months. Round *a*, *b*, and *r* to the nearest hundred thousandth.

b. According to your model in part **a.**, what is the term of a loan with a 7.00% interest rate? Round to the nearest month.

20. GENERATION OF GARBAGE According to the U.S. Environmental Protection Agency, the amount of garbage generated per person has been increasing over the last few decades. The following table shows the per capita garbage, in pounds per day, generated in the United States.

Year, t	1960	1970	1980	1990	2000
Pounds per day, p	2.66	3.27	3.61	4.00	4.30

Represent the year 1960 by $t = 60$.

a. Use a graphing utility to find a linear model and a logarithmic model for the data. Use *t* as the independent variable (domain) and *p* as the dependent variable (range).

b. Examine the correlation coefficients of the two regression models to determine which model provides a better fit for the data.

c. Use the model you selected in part **b.** to predict the amount of garbage that will be generated per capita per day in 2005. Round to the nearest hundredth of a pound.

21. HYPOTHERMIA The following table shows the time *T*, in hours, before a scuba diver wearing a 3-millimeter-thick wet suit reaches hypothermia (95°F) for various water temperatures *F*, in degrees Fahrenheit.

Water temperature, °F	Time *T*, hours
41	1.1
46	1.4
50	1.8
59	3.7

a. Find an exponential regression model for the data. Round the constants *a* and *b* to the nearest hundred thousandth.

b. Use the model from part **a.** to estimate the time it takes for the diver to reach hypothermia in water that has a temperature of 65°F. Round to the nearest tenth of an hour.

▶ **22.** ATMOSPHERIC PRESSURE The following table shows the earth's atmospheric pressure *P* (in newtons per square centimeter) at an altitude of *a* kilometers. Find a suitable function that models the atmospheric pressure as a function of the altitude. Use the function to estimate the atmospheric pressure at an altitude of 24 kilometers. Round to the nearest tenth of a newton per square centimeter.

Altitude *a*, kilometers	Pressure *P*, newtons/cm²
0	10.3
2	8.0
4	6.4
6	5.1
8	4.0
10	3.2
12	2.5
14	2.0
16	1.6
18	1.3

23. HYPOTHERMIA The following table shows the time *T*, in hours, before a scuba diver wearing a 4-millimeter-thick wet suit reaches hypothermia (95°F) for various water temperatures *F*, in degrees Fahrenheit.

Water temperature, °F	Time *T*, hours
41	1.5
46	1.9
50	2.4
59	5.2

a. Find an exponential regression model for the data. Round the constants *a* and *b* to the nearest hundred thousandth.

b. Use the model from part **a.** to estimate the time it takes for the diver to reach hypothermia in water that has a temperature of 65°F. Round to the nearest tenth of an hour. How much greater is this result compared with the answer to Exercise 21**b.**?

▶ **24.** 400-METER RACE The following table lists the progression of world record times in the men's 400-meter race for the years from 1948 to 2002.

World Record Times in the Men's 400-Meter Race, 1948 to 2002

Year	Time, in seconds	Year	Time, in seconds
1948	45.9	1964	44.9
1950	45.8	1967	44.5
1955	45.4	1968	44.1
1956	45.2	1968	43.86
1960	44.9	1988	43.29
1960	44.9	1999	43.18
1963	44.9		

Source: Track and Field Statistics, http://trackfield.brinkster.net/Main.asp.

a. Determine whether the data can best be modeled by a decreasing exponential function or a decreasing logarithmic function. Let *x* = 48 represent the year 1948.

b. Use the function you chose in part **a.** to predict the world record time in the men's 400-meter race for the year 2008. Round to the nearest hundredth of a second.

25. INTERNET VIRUSES The CERT Coordination Center (CERT/CC) is an organization that, among other activities, monitors the number of Internet virus incidents. The data in the table below were compiled from CERT/CC and show the number of virus incidents for various years.

Year	Number of incidents
1996	2573
1997	2134
1998	3734
1999	9859
2000	21,756
2001	52,658
2002	82,094

a. Find an exponential regression model for these data. Use $t = 0$ to correspond to 1996. Round each constant to the nearest thousandth.

b. Use the model to predict the year in which the number of incidents will first exceed one million.

▶ **26.** POPULATION OF HAWAII The following table shows the population of the state of Hawaii for selected years from 1950 to 2002.

Population of the State of Hawaii

Year	Population	Year	Population
1950	499,000	1985	1,039,698
1955	529,000	1990	1,113,491
1960	642,000	1995	1,196,854
1965	704,000	2000	1,212,670
1970	762,920	2001	1,227,024
1975	875,052	2002	1,244,898
1980	967,710		

Source: economagic.com, http://www.economagic.com/em-cgi/data.exe/beapi/a15300.

a. Use the logistic regression feature of a graphing utility to find a logistic growth model that approximates the population of the state of Hawaii as a function of the year. Use $t = 0$ to represent the year 1950.

b. Use the model from part **a.** to predict the population of the state of Hawaii for the year 2010. Round to the nearest ten thousand.

c. What is the carrying capacity of the model? Round to the nearest thousand.

27. OPTOMETRY The *near point p* of a person is the closest distance at which the person can see an object distinctly. As one grows older, one's near point increases. The table below shows data for the average near point of various people with normal eyesight.

Age y, years	Near point p (cm)
15	11
20	13
25	15
30	17
35	20
40	23
50	26

a. Find an exponential regression model for these data. Round each constant to the nearest thousandth.

b. What near point does this equation predict for a person 60 years old? Round to the nearest centimeter.

28. CHEMISTRY The amount of oxygen x, in milliliters per liter, that can be absorbed by water at a certain temperature T, in degrees Fahrenheit, is given in the following table.

Temperature, °F	Oxygen absorbed, ml/L
32	10.5
38	8.4
46	7.6
52	7.1
58	6.8
64	6.5

a. Find a logarithmic regression equation for these data. Round each constant to the nearest thousandth.

b. Using your model, how much oxygen, to the nearest tenth of a milliliter per liter, can be absorbed in water that is 50°F?

29. THE HENDERSON-HASSELBACH FUNCTION The scientists Henderson and Hasselbach determined that the pH of blood is a function of the ratio q of the amounts of bicarbonate and carbonic acid in the blood.

a. Use a graphing utility and the data in the following table to determine a linear model and a logarithmic model for the data. Use q as the independent variable (domain) and pH as the dependent variable (range). State the correlation coefficient for each model. Round a and b to 5 decimal places and r to 6 decimal places. Which model provides the better fit for the data?

q	7.9	12.6	31.6	50.1	79.4
pH	7.0	7.2	7.6	7.8	8.0

b. Use the model you chose in part **a.** to find the q-value associated with a pH of 8.2. Round to the nearest tenth.

30. WORLD POPULATION The following table lists the years in which the world's population first reached 3, 4, 5, and 6 billion.

World Population Milestones

Year	Population
1960	3 billion
1974	4 billion
1987	5 billion
1999	6 billion

Source: Time Almanac 2002, p. 708.

a. Find an exponential model for the data in the table. Let $x = 0$ represent the year 1960.

b. Use the model to predict the year in which the world's population will first reach 7 billion.

31. PANDA POPULATION One estimate gives the world panda population as 3200 in 1980 and 590 in 2000.

a. Find an exponential model for the data and use the model to predict the year in which the panda population p will be reduced to 200. (Let $t = 0$ represent the year 1980.)

b. Because the exponential model in part **a.** fits the data perfectly, does this mean that the model will accurately predict future panda populations? Explain.

32. OLYMPIC RECORDS The following table shows the Olympic gold medal distances for the women's high jump from 1968 to 2000.

Women's Olympic High Jump, 1968 to 2000

Year	Distance	Year	Distance
1968	5 ft 11$\frac{3}{4}$ in.	1988	6 ft 8 in.
1972	6 ft 3$\frac{5}{8}$ in.	1992	6 ft 7$\frac{1}{2}$ in.
1976	6 ft 4 in.	1996	6 ft 8$\frac{3}{4}$ in.
1980	6 ft 5$\frac{1}{2}$ in.	2000	6 ft 7 in.
1984	6 ft 7$\frac{1}{2}$ in.		

Source: Time Almanac 2002.

Represent the year 1968 by 68.

a. Use a graphing utility to determine a linear model and a logarithmic model for the data, with the distance measured in inches. State the correlation coefficient r for each model.

b. Use the correlation coefficient for each of the models in part **a.** to determine which model provides the better fit for the data.

c. Use the model you selected in part **b.** to predict the women's Olympic gold medal high jump distance in 2012. Round to the nearest tenth of an inch.

33. NUMBER OF AUTOMOBILES In 1900, the number of automobiles in the United States was around 8000. By 2000, the number of automobiles in the United States had reached 200 million.

a. Find an exponential model for the data and use the model to predict the number of automobiles, to the nearest 100,000, in the United States in 2010. Use $t = 0$ to represent the year 1900.

b. According to the model, in what year will the number of automobiles in the United States first reach 300 million?

34. TEMPERATURE OF COFFEE A cup of coffee is placed in a room that maintains a constant temperature of 70°F. The following table shows both the coffee temperature T after t minutes and the difference between the coffee temperature and the room temperature after t minutes.

Time t (minutes)	0	5	10	15	20	25
Coffee temp. T (°F)	165°	140°	121°	107°	97°	89°
T − 70°		95°	70°	51°	37°	27°	19°

a. Use a graphing utility to find an exponential model for the difference $T - 70°$ as a function of t.

b. Use the model to predict how long it will take (to the nearest minute) for the coffee to cool to 80°F.

35. OLYMPIC DISTANCES The following table shows the winning Olympic distances for the men's shot put for the years 1948 to 2000.

Men's Olympic Shot Put, 1948 to 2000

Year	Distance	Year	Distance
1948	56 ft 2 in.	1976	69 ft $\frac{3}{4}$ in.
1952	57 ft 1$\frac{1}{2}$ in.	1980	70 ft $\frac{1}{2}$ in.
1956	60 ft 11 in.	1984	69 ft 9 in.
1960	64 ft 6$\frac{3}{4}$ in.	1988	73 ft 8$\frac{3}{4}$ in.
1964	66 ft 8$\frac{1}{4}$ in.	1992	71 ft 2$\frac{1}{2}$ in.
1968	67 ft 4$\frac{3}{4}$ in.	1996	70 ft 11$\frac{1}{4}$ in.
1972	69 ft 6 in.	2000	69 ft 10$\frac{1}{4}$ in.

Source: Time Almanac 2002

Represent the year 1948 by $t = 48$.

a. Use the regression features of a graphing utility to determine a logistic growth model and a logarithmic model for the data.

b. Use graphs of the models in part **a.** to determine which model provides the better fit for the data.

c. Use the model you selected in part **b.** to predict the men's shot put distance for the year 2008. Round to the nearest hundredth of a foot.

36. WORLD POPULATION The following table lists the years in which the world's population first reached 3, 4, 5, and 6 billion.

World Population Milestones

Year	Population
1960	3 billion
1974	4 billion
1987	5 billion
1999	6 billion

Source: Time Almanac 2002, p. 708.

a. Find a logistic growth model, $P(t)$, for the data in the table. Let t represent the number of years after 1960 ($t = 0$ represents the year 1960).

b. According to the logistic growth model, what will the world's population approach as $t \to \infty$? Round to the nearest billion.

37. DESALINATION The following table shows the amount of fresh water w (in cubic yards) produced from saltwater after t hours of a desalination process.

t	1	2.5	3.5	4.0	5.1	6.5
w	18.2	46.6	57.4	61.5	68.7	76.2

a. Use a graphing utility to find a linear model and a logarithmic model for the data.

b. Examine the correlation coefficients of the two regression models to determine which model provides the better fit for the data. State the correlation coefficient r for each model.

c. Use the model you selected in part **b.** to predict the amount of fresh water that will be produced after 10 hours of the desalination process. Round to the nearest tenth of a cubic yard.

38. A CORRELATION COEFFICIENT OF 1 A scientist uses a graphing utility to model the data set $\{(2, 5), (4, 6)\}$ with a logarithmic function. The following display shows the results.

What is the significance of the fact that the correlation coefficient for the regression equation is $r = 1$?

```
LnReg
 y=a+blnx
 a=4
 b=1.442695041
 r²=1
 r=1
```

CONNECTING CONCEPTS

39. **DUPLICATE DATA POINTS** An engineer needs to model the data in set A with an exponential function.

$$A = \{(2, 5), (3, 10), (4, 17), (4, 17), (5, 28)\}$$

Because the ordered pair $(4, 17)$ is listed twice, the engineer decides to eliminate one of these ordered pairs and model the data in set B.

$$B = \{(2, 5), (3, 10), (4, 17), (5, 28)\}$$

Determine whether A and B both have the same exponential regression function.

40. **DOMAIN ERROR** A scientist needs to model the data in set A.

$$A = \{(0, 1.2), (1, 2.3), (2, 2.8), (3, 3.1), (4, 3.3), (5, 3.4)\}$$

The scientist views a scatter plot of the data and decides to model the data with a logarithmic function of the form $y = a + b \ln x$.

a. When the scientist attempts to use a graphing calculator to determine the logarithmic regression equation, the calculator displays the message

"ERR:DOMAIN"

Explain why the calculator was unable to determine the logarithmic regression equation for the data.

b. Explain what the scientist could do so that the data in set A could be modeled by a logarithmic function of the form $y = a + b \ln x$.

41. **POWER FUNCTIONS** A function that can be written in the form $y = ax^b$ is said to be a **power function.** Some data sets can best be modeled by a power function.

On a TI-83, the PwrReg instruction is used to produce a power regression function for a set of data.

a. Use a graphing utility to find an exponential regression function and a power regression function for the following data. State the correlation coefficient r for each model.

x	1	2	3	4	5	6
y	2.1	5.5	9.8	14.6	20.1	25.8

b. Which of the two regression functions provides the better fit for the data?

42. **PERIOD OF A PENDULUM** The following table shows the time t (in seconds) of the period of a pendulum (the time it takes the pendulum to complete a swing to the left and back) of length l (in feet).

a. Use a graphing utility to determine the equation of the best model for the data. Your model must be a power function or an exponential function.

Length l	1	2	3	4	6	8
Time t	1.11	1.57	1.92	2.25	2.72	3.14

b. According to the model you chose in part **a.**, what is the length of a pendulum, to the nearest tenth of a foot, that has a period of 12 seconds?

PROJECTS

1. **A MODELING PROJECT** The purpose of this Project is for you to find data that can be modeled by an exponential or a logarithmic function. Choose data from a *real-life* situation that you find interesting. Search for the data in a magazine, a newspaper, an almanac, or on the Internet. If you wish, you can collect your data by performing an experiment.

a. List the source of your data. Include the date, page number, and any other specifics about the source. If your data were collected by performing an experiment, then provide all the details of the experiment.

b. Explain what you have chosen as your variables. Which variable is the dependent variable and which variable is the independent variable?

c. Use the three-step modeling process to find a regression equation that models the data.

d. Graph the regression equation on the scatter plot of the data. What is the correlation coefficient for the model? Do you think that your regression equation accurately models your data? Explain.

e. Use the regression equation to predict the value of
- the dependent variable for a specific value of the independent variable.
- the independent variable for a specific value of the dependent variable.

f. Write a few comments about what you have learned from this Project.

EXPLORING CONCEPTS WITH TECHNOLOGY

TABLE 7.16

T	V
90	700
100	500
110	350
120	250
130	190
140	150
150	120

Using a Semilog Graph to Model Exponential Decay

Consider the data in **Table 7.16,** which shows the viscosity V of SAE 40 motor oil at various temperatures T. The graph of these data is shown below, along with a curve that passes through the points. The graph in **Figure 7.48** appears to have the shape of an exponential decay model.

One way to determine whether the graph in **Figure 7.48** is the graph of an exponential function is to plot the data on *semilog* graph paper. On this graph paper, the horizontal axis remains the same, but the vertical axis uses a logarithmic scale.

The data in **Table 7.16** are graphed again in **Figure 7.49,** but this time the vertical axis is a natural logarithm axis. This graph is approximately a straight line.

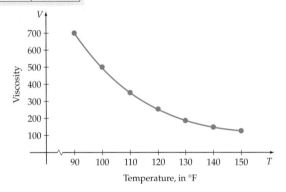

FIGURE 7.48

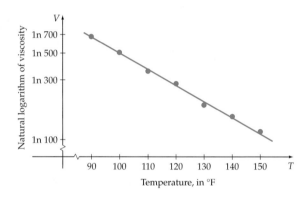

FIGURE 7.49

The slope of the line in **Figure 7.49,** to the nearest ten-thousandth, is

$$m = \frac{\ln 500 - \ln 120}{100 - 150} \approx -0.0285$$

Using this slope and the point-slope formula with V replaced by $\ln V$, we have

$$\ln V - \ln 120 = -0.0285(T - 150)$$
$$\ln V \approx -0.0285T + 9.062 \qquad (1)$$

Equation (1) is the equation of the line on a semilog coordinate grid.

Now solve Equation (1) for V.

$$e^{\ln V} = e^{-0.0285T + 9.062}$$
$$V = e^{-0.0285T} e^{9.062}$$
$$V \approx 8621 e^{-0.0285T} \qquad (2)$$

TABLE 7.17

t	A
1	91.77
4	70.92
8	50.30
15	27.57
20	17.95
30	7.60

Equation (2) is a model of the data in the rectangular coordinate system shown in **Figure 7.48.**

1. A chemist wishes to determine the decay characteristics of iodine-131. A 100-mg sample of iodine-131 is observed over a 30-day period. **Table 7.17** shows the amount A (in milligrams) of iodine-131 remaining after t days.

   a. Graph the ordered pairs (t, A) on semilog paper. (*Note:* Semilog paper comes in different varieties. Our calculations are based on semilog paper that has a natural logarithm scale on the vertical axis.)

b. Use the points $(4, 4.3)$ and $(15, 3.3)$ to approximate the slope of the line that passes through the points.

c. Using the slope calculated in part **b.** and the point $(4, 4.3)$, determine the equation of the line.

d. Solve the equation you derived in part **c.** for A.

e. Graph the equation you derived in part **d.** in a rectangular coordinate system.

f. What is the half-life of iodine-131?

TABLE 7.18

t	B
0	15.5
1	15.7
2	15.9
3	16.2
4	16.7

2. The live birth rates B per thousand births in the United States are given in **Table 7.18** for the years 1986 through 1990 ($t = 0$ corresponds to 1986).

a. Graph the ordered pairs $(t, \ln B)$. (You will need to adjust the scale so that you can discriminate between plotted points. A suggestion is given in **Figure 7.50**.)

b. Use the points $(1, 2.754)$ and $(3, 2.785)$ to approximate the slope of the line that passes through the points.

c. Using the slope calculated in part **b.** and the point $(1, 2.754)$, determine the equation of the line.

d. Solve the equation you derived in part **c.** for B.

e. Graph the equation you derived in part **d.** in a rectangular coordinate system.

f. If the birth rate continues as predicted by your model, in what year will the birth rate be 17.5 per thousand?

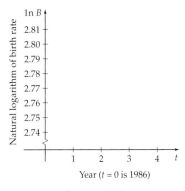

FIGURE 7.50

The difference in graphing strategies between Exercise 1 and Exercise 2 is that in Exercise 1, semilog paper was used. When a point is graphed on this coordinate paper, the y-coordinate is $\ln y$. In Exercise 2, graphing a point $(x, \ln y)$ in a rectangular coordinate system has the same effect as graphing (x, y) in a semilog coordinate system.

CHAPTER 7 SUMMARY

7.1 Exponential Functions and Their Applications

• For all positive real numbers b, $b \neq 1$, the exponential function defined by $f(x) = b^x$ has the following properties:

1. f has the set of real numbers as its domain.

2. f has the set of positive real numbers as its range.

3. f has a graph with a y-intercept of $(0, 1)$.

4. f has a graph asymptotic to the x-axis.

5. f is a one-to-one function.

6. f is an increasing function if $b > 1$.

7. f is a decreasing function if $0 < b < 1$.

• As n increases without bound, $(1 + 1/n)^n$ approaches an irrational number denoted by e. The value of e accurate to eight decimal places is 2.71828183.

• The function defined by $f(x) = e^x$ is called the natural exponential function.

7.2 Logarithmic Functions and Their Applications

- *Definition of a Logarithm* If $x > 0$ and b is a positive constant ($b \neq 1$), then

$$y = \log_b x \quad \text{if and only if} \quad b^y = x$$

- For all positive real numbers b, $b \neq 1$, the function defined by $f(x) = \log_b x$ has the following properties:

  1. f has the set of positive real numbers as its domain.

  2. f has the set of real numbers as its range.

  3. f has a graph with an x-intercept of $(1, 0)$.

  4. f has a graph asymptotic to the y-axis.

  5. f is a one-to-one function.

  6. f is an increasing function if $b > 1$.

  7. f is a decreasing function if $0 < b < 1$.

- The exponential form of $y = \log_b x$ is $b^y = x$.

- The logarithmic form of $b^y = x$ is $y = \log_b x$.

- *Basic Logarithmic Properties*

  1. $\log_b b = 1$ **2.** $\log_b 1 = 0$ **3.** $\log_b (b^p) = p$

- The function $f(x) = \log_{10} x$ is the common logarithmic function. It is customarily written as $f(x) = \log x$.

- The function $f(x) = \log_e x$ is the natural logarithmic function. It is customarily written as $f(x) = \ln x$.

7.3 Logarithms and Logarithmic Scales

- If b, M, and N are positive real numbers ($b \neq 1$), and p is any real number, then

$$\log_b(MN) = \log_b M + \log_b N$$

$$\log_b \frac{M}{N} = \log_b M - \log_b N$$

$$\log_b(M^p) = p \log_b M$$

$$\log_b M = \log_b N \quad \text{implies} \quad M = N$$

$$M = N \quad \text{implies} \quad \log_b M = \log_b N$$

$$b^{\log_b p} = p \quad (\text{for } p > 0)$$

- *Change-of-Base Formula* If x, a, and b are positive real numbers with $a \neq 1$ and $b \neq 1$, then

$$\log_b x = \frac{\log_a x}{\log_a b}$$

- An earthquake with an intensity of I has a Richter scale magnitude of $M = \log\left(\dfrac{I}{I_0}\right)$, where I_0 is the measure of the intensity of a zero-level earthquake.

- The pH of a solution with a hydronium-ion concentration of H^+ mole per liter is given by $pH = -\log[H^+]$.

7.4 Exponential and Logarithmic Equations

- *Equality of Exponents Theorem* If b is a positive real number ($b \neq 1$) such that $b^x = b^y$, then $x = y$.

- Exponential equations of the form $b^x = b^y$ can be solved by using the Equality of Exponents Theorem.

- Exponential equations of the form $b^x = c$ can be solved by taking either the common logarithm or the natural logarithm of each side of the equation.

- Logarithmic equations can often be solved by using the properties of logarithms and the definition of a logarithm.

7.5 Exponential Growth and Decay

- The function defined by $N(t) = N_0 e^{kt}$ is called an exponential growth function if k is a positive constant, and it is called an exponential decay function if k is a negative constant.

- *The Compound Interest Formula* A principal P invested at an annual interest rate r, expressed as a decimal and compounded n times per year for t years, produces the balance

$$A = P\left(1 + \frac{r}{n}\right)^{nt}$$

- *Continuous Compounding Interest Formula* If an account with principal P and annual interest rate r is compounded continuously for t years, then the balance is $A = Pe^{rt}$.

- *The Logistic Model* The magnitude of a population at time t is given by

$$P(t) = \frac{c}{1 + ae^{-bt}}$$

where $P_0 = P(0)$ is the population at time $t = 0$, c is the carrying capacity of the population, and b is a constant called the growth rate constant.

7.6 Modeling Data with Exponential and Logarithmic Functions

- If the graph of f lies above all of its tangents on $[x_1, x_2]$, then f is concave upward on $[x_1, x_2]$.

- If the graph of f lies below all of its tangents on $[x_1, x_2]$, then f is concave downward on $[x_1, x_2]$.

- *The Modeling Process* Use a graphing utility to

  1. construct a scatter plot of the data to determine which type of function will best model the data.

  2. find the regression equation of the modeling function and the correlation coefficient for the regression.

  3. examine the correlation coefficient and view a graph that displays both the function and the scatter plot to determine how well the function fits the data.

CHAPTER 7 TRUE/FALSE EXERCISES

In Exercises 1 to 14, answer true or false. If the statement is false, give an example or state a reason to demonstrate that the statement is false.

1. If $7^x = 40$, then $\log_7 40 = x$.

2. If $\log_4 x = 3.1$, then $4^{3.1} = x$.

3. If $f(x) = \log x$ and $g(x) = 10^x$, then $f[g(x)] = x$ for all real numbers x.

4. If $f(x) = \log x$ and $g(x) = 10^x$, then $g[f(x)] = x$ for all real numbers x.

5. The exponential function $h(x) = b^x$ is an increasing function.

6. The logarithmic function $j(x) = \log_b x$ is an increasing function.

7. The exponential function $h(x) = b^x$ is a one-to-one function.

8. The logarithmic function $j(x) = \log_b x$ is a one-to-one function.

9. The graph of $f(x) = \dfrac{2^x + 2^{-x}}{2}$ is symmetric with respect to the y-axis.

10. The graph of $f(x) = \dfrac{2^x - 2^{-x}}{2}$ is symmetric with respect to the origin.

11. If $x > 0$ and $y > 0$, then $\log(x + y) = \log x + \log y$.

12. If $x > 0$, then $\log x^2 = 2 \log x$.

13. If M and N are positive real numbers, then
$$\ln \frac{M}{N} = \ln M - \ln N$$

14. For all $p > 0$, $e^{\ln p} = p$.

CHAPTER 7 REVIEW EXERCISES

In Exercises 1 to 12, solve each equation. Do not use a calculator.

1. $\log_5 25 = x$

2. $\log_3 81 = x$

3. $\ln e^3 = x$

4. $\ln e^\pi = x$

5. $3^{2x+7} = 27$

6. $5^{x-4} = 625$

7. $2^x = \dfrac{1}{8}$

8. $27(3^x) = 3^{-1}$

9. $\log x^2 = 6$

10. $\dfrac{1}{2} \log |x| = 5$

11. $10^{\log 2x} = 14$

12. $e^{\ln x^2} = 64$

In Exercises 13 to 22, sketch the graph of each function.

13. $f(x) = (2.5)^x$

14. $f(x) = \left(\dfrac{1}{4}\right)^x$

15. $f(x) = 3^{|x|}$

16. $f(x) = 4^{-|x|}$

17. $f(x) = 2^x - 3$

18. $f(x) = 2^{(x-3)}$

19. $f(x) = \dfrac{1}{3} \log x$

20. $f(x) = 3 \log x^{1/3}$

21. $f(x) = -\dfrac{1}{2} \ln x$

22. $f(x) = -\ln |x|$

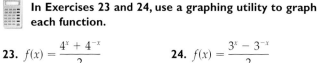 **In Exercises 23 and 24, use a graphing utility to graph each function.**

23. $f(x) = \dfrac{4^x + 4^{-x}}{2}$

24. $f(x) = \dfrac{3^x - 3^{-x}}{2}$

In Exercises 25 to 28, change each logarithmic equation to its exponential form.

25. $\log_4 64 = 3$

26. $\log_{1/2} 8 = -3$

27. $\log_{\sqrt{2}} 4 = 4$

28. $\ln 1 = 0$

In Exercises 29 to 32, change each exponential equation to its logarithmic form.

29. $5^3 = 125$

30. $2^{10} = 1024$

31. $10^0 = 1$

32. $8^{1/2} = 2\sqrt{2}$

In Exercises 33 to 36, write the given logarithm in terms of logarithms of x, y, and z.

33. $\log_b \dfrac{x^2 y^3}{z}$

34. $\log_b \dfrac{\sqrt{x}}{y^2 z}$

35. $\ln xy^3$

36. $\ln \dfrac{\sqrt{xy}}{z^4}$

In Exercises 37 to 40, write each logarithmic expression as a single logarithm with a coefficient of 1.

37. $2 \log x + \dfrac{1}{3} \log (x + 1)$

38. $5 \log x - 2 \log (x + 5)$

39. $\dfrac{1}{2} \ln 2xy - 3 \ln z$

40. $\ln x - (\ln y - \ln z)$

In Exercises 41 to 44, use the change-of-base formula and a calculator to approximate each logarithm accurate to six significant digits.

41. $\log_5 101$

42. $\log_3 40$

43. $\log_4 0.85$

44. $\log_8 0.3$

In Exercises 45 to 60, solve each equation for x. Give exact answers. Do not use a calculator.

45. $4^x = 30$

46. $5^{x+1} = 41$

47. $\ln 3x - \ln(x - 1) = \ln 4$

48. $\ln 3x + \ln 2 = 1$

49. $e^{\ln(x+2)} = 6$

50. $10^{\log(2x+1)} = 31$

51. $\dfrac{4^x + 4^{-x}}{4^x - 4^{-x}} = 2$

52. $\dfrac{5^x + 5^{-x}}{2} = 8$

53. $\log(\log x) = 3$

54. $\ln(\ln x) = 2$

55. $\log \sqrt{x - 5} = 3$

56. $\log x + \log(x - 15) = 1$

57. $\log_4(\log_3 x) = 1$

58. $\log_7(\log_5 x^2) = 0$

59. $\log_5 x^3 = \log_5 16x$

60. $25 = 16^{\log_4 x}$

61. EARTHQUAKE MAGNITUDE Determine, to the nearest 0.1, the Richter scale magnitude of an earthquake with an intensity of $I = 51{,}782{,}000 I_0$.

62. EARTHQUAKE MAGNITUDE A seismogram has an amplitude of 18 millimeters and a time delay of 21 seconds. Find, to the nearest tenth, the Richter scale magnitude of the earthquake that produced the seismogram.

63. COMPARISON OF EARTHQUAKES An earthquake had a Richter scale magnitude of 7.2. Its aftershock had a Richter scale magnitude of 3.7. Compare the intensity of the earthquake to the intensity of the aftershock by finding, to the nearest unit, the ratio of the larger intensity to the smaller intensity.

64. COMPARISON OF EARTHQUAKES An earthquake has an intensity 600 times the intensity of a second earthquake. Find, to the nearest tenth, the difference between the Richter scale magnitudes of the earthquakes.

65. CHEMISTRY Find the pH of tomatoes that have a hydronium-ion concentration of 6.28×10^{-5}. Round to the nearest tenth.

66. CHEMISTRY Find the hydronium-ion concentration of rainwater that has a pH of 5.4.

67. COMPOUND INTEREST Find the balance when $16,000 is invested at an annual rate of 8% for 3 years if the interest is compounded

   a. monthly   **b.** continuously

68. COMPOUND INTEREST Find the balance when $19,000 is invested at an annual rate of 6% for 5 years if the interest is compounded

   a. daily   **b.** continuously

69. DEPRECIATION The scrap value S of a product with an expected life span of n years is given by $S(n) = P(1 - r)^n$, where P is the original purchase price of the product and r is the annual rate of depreciation. A taxicab is purchased for $12,400 and is expected to last 3 years. What is its scrap value if it depreciates at a rate of 29% per year?

70. MEDICINE A skin wound heals according to the function given by $N(t) = N_0 e^{-0.12t}$, where N is the number of square centimeters of unhealed skin t days after the injury, and N_0 is the number of square centimeters covered by the original wound.

   a. What percentage of the wound will be healed after 10 days?

   b. How many days, to the nearest day, will it take for 50% of the wound to heal?

   c. How long, to the nearest day, will it take for 90% of the wound to heal?

In Exercises 71 to 74, find the exponential growth/decay function $N(t) = N_0 e^{kt}$ that satisfies the given conditions.

71. $N(0) = 1, N(2) = 5$ **72.** $N(0) = 2, N(3) = 11$

73. $N(1) = 4, N(5) = 5$ **74.** $N(-1) = 2, N(0) = 1$

75. POPULATION GROWTH

 a. Find the exponential growth function for a city whose population was 25,200 in 2002 and 26,800 in 2003. Use $t = 0$ to represent the year 2002.

 b. Use the growth function to predict, to the nearest hundred, the population of the city in 2009.

76. CARBON DATING Determine, to the nearest ten years, the age of a bone if it now contains 96% of its original amount of carbon-14. The half-life of carbon-14 is 5730 years.

77. ACTIVE MILITARY DUTY PERSONNEL The following table shows the number of U.S. military personnel on active duty for each year from 1991 to 2000. (*Source: Time Almanac 2003* with *Information Please.*)

Active Military Duty Personnel, 1991–2000

1991	1,985,555	1996	1,471,722
1992	1,807,177	1997	1,438,562
1993	1,705,103	1998	1,406,830
1994	1,610,490	1999	1,385,703
1995	1,518,224	2000	1,384,338

 a. Use a graphing utility to find a linear model, an exponential model, and a logarithmic model for the number of active-duty personnel, P, as a function of the year. Represent the year 1991 by $t = 91$.

 b. Examine the correlation coefficients of the three regression models to determine which model provides the best fit.

 c. Use the model you selected in part **b.** to predict, to the nearest 10,000, the number of active duty military personnel for the year 2006.

78. MORTALITY RATE The following table shows the infant mortality rate in the United States for selected years from 1960 to 2000. (*Source: The World Almanac 2003.*)

U.S. Infant Mortality Rate, 1960–2000 (per 1000 live births)

Year	Rate
1960	26.0
1970	20.0
1980	12.6
1990	9.2
1995	7.6
1999	7.1
2000	6.9

 a. Use a graphing utility to find a linear model, an exponential model, and a logarithmic model for the infant mortality rate, R, as a function of the year. Represent the year 1960 by $t = 60$.

 b. Examine the correlation coefficients of the three regression models to determine which model provides the best fit.

 c. Use the model you selected in part **b.** to predict, to the nearest 0.1, the infant mortality rate in 2008.

79. LOGISTIC GROWTH The population of coyotes in a national park satisfies the logistic model with $P_0 = 210$ in 1992, $c = 1400$, and $P(3) = 360$ (the population in 1995).

 a. Determine the logistic model.

 b. Use the model to predict, to the nearest 10, the coyote population in 2005.

80. Consider the logistic function

$$P(t) = \frac{128}{1 + 5e^{-0.27t}}$$

 a. Find P_0.

 b. What does $P(t)$ approach as $t \to \infty$?

CHAPTER 7 TEST

1. a. Write $\log_b(5x - 3) = c$ in exponential form.

   b. Write $3^{x/2} = y$ in logarithmic form.

2. Write $\log_b \dfrac{z^2}{y^3 \sqrt{x}}$ in terms of logarithms of x, y, and z.

3. Write $\log(2x + 3) - 3\log(x - 2)$ as a single logarithm with a coefficient of 1.

4. Use the change-of-base formula and a calculator to approximate $\log_4 12$. Round your result to the nearest ten thousandth.

5. Graph: $f(x) = 3^{-x/2}$

6. Graph: $f(x) = -\ln(x + 1)$

7. Solve: $5^x = 22$. Round your solution to the nearest ten thousandth.

8. Find the *exact* solution of $4^{5-x} = 7^x$.

9. Solve: $\log(x + 99) - \log(3x - 2) = 2$

10. Solve: $\ln(2 - x) + \ln(5 - x) = \ln(37 - x)$

11. Find the balance on $20,000 invested at an annual interest rate of 7.8% for 5 years:

   a. compounded monthly.

   b. compounded continuously.

12. Find the time required for money invested at an annual rate of 4% to double in value if the investment is compounded monthly. Round to the nearest hundredth of a year.

13. a. What, to the nearest tenth, will an earthquake measure on the Richter scale if it has an intensity of $I = 42,304,000I_0$?

   b. Compare the intensity of an earthquake that measures 6.3 on the Richter scale to the intensity of an earthquake that measures 4.5 on the Richter scale by finding the ratio of the larger intensity to the smaller intensity. Round to the nearest whole number.

14. a. Find the exponential growth function for a city whose population was 34,600 in 1996 and 39,800 in 1999. Use $t = 0$ to represent the year 1996.

   b. Use the growth function to predict the population of the city in 2006. Round to the nearest thousand.

15. Determine, to the nearest ten years, the age of a bone if it now contains 92% of its original amount of carbon-14. The half-life of carbon-14 is 5730 years.

16. a. Use a graphing utility to find the exponential regression function for the following data.

$$\{(2.5, 16), (3.7, 48), (5.0, 155), (6.5, 571), (6.9, 896)\}$$

   b. Use the function to predict, to the nearest whole number, the y-value associated with $x = 7.8$.

17. The following table shows the interest rates paid by a bank on certificates of deposit (CDs) of different terms in May of 2003.

CD term, t years	0.5	1.0	2.0	5.0
Rate	1.38%	1.66%	2.01%	3.20%

   a. Find the *logarithmic regression function* for the data and use the function to predict, to the nearest 0.01%, the interest rate on a CD invested for 3.5 years.

   b. According to your function, for how long (to the nearest 0.1 year) would you need to invest to receive a 2.5% interest rate?

18. The population of raccoons in a state park satisfies a logistic growth model with $P_0 = 160$ in 1999 and $P(1) = 190$. A park ranger has estimated the carrying capacity of the park to be 1100 raccoons. Use these data to

   a. find the logistic growth model for the raccoon population.

   b. predict the raccoon population in 2006.

CUMULATIVE REVIEW EXERCISES

1. Given $f(x) = \cos x$ and $g(x) = x^2 + 1$, find $(f \circ g)(x)$.

2. Given $f(x) = 2x + 8$, find $f^{-1}(x)$.

3. For the right triangle shown at the right, find $\sin\theta$, $\cos\theta$, and $\tan\theta$.

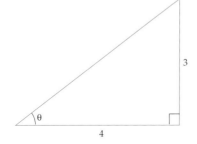

4. For the right triangle shown at the right, find a. Round to the nearest tenth.

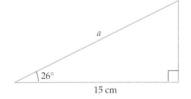

5. What are the amplitude, period, and phase shift of the graph of $y = 4\cos\left(2x - \dfrac{\pi}{2}\right)$?

6. What is the amplitude and period of the graph of $y = \sin x - \cos x$?

7. Is the sine function an even function, an odd function, or neither an even nor an odd function?

8. Verify the identity $\dfrac{1}{\sin x} - \sin x = \cos x \cot x$.

9. Evaluate $\tan\left(\sin^{-1}\left(\dfrac{12}{13}\right)\right)$.

10. Solve $2\cos^2 x + \sin x - 1 = 0$ for $0 \le x < 2\pi$.

11. Find the magnitude and direction angle for the vector $\langle -3, 4 \rangle$. Round the angle to the nearest tenth of a degree.

12. Find the angle between the vectors $\mathbf{v} = \langle 2, -3 \rangle$ and $\mathbf{w} = \langle -3, 4 \rangle$. Round to the nearest tenth of a degree.

13. An airplane is traveling with an airspeed of 400 mph at a heading of 48°. A wind of 55 mph is blowing at a heading of 115°. Find the ground speed and the course of the plane.

14. For triangle ABC, $B = 32°$, $a = 42$ feet, and $b = 50$ feet. Find the measure of angle A. Round to the nearest degree.

15. Use De Moivre's Theorem to find $(1 - i)^8$.

16. Find the two square roots of the imaginary number i.

17. Transform the point $(1, 1)$ in a rectangular coordinate system to a point in a polar coordinate system.

18. Sketch the graph of the polar equation $r = 2 - 2\cos\theta$.

19. Solve $5^x = 10$. Round to the nearest hundredth.

20. The half-life of an isotope of polonium is 138 days. If an ore sample originally contained 3 milligrams of polonium, how many milligrams of polonium remain after 100 days? Round to the nearest tenth of a milligram.

SOLUTIONS TO THE TRY EXERCISES

Exercise Set 1.1, page 12

6. $6(5s - 11) - 12(2s + 5) = 0$

$$30s - 66 - 24s - 60 = 0$$
$$6s - 126 = 0$$
$$s = \frac{126}{6}$$
$$s = 21$$

22. $x^2 + x - 2 = 0$

$$a = 1 \qquad b = 1 \qquad c = -2$$
$$x = \frac{-1 \pm \sqrt{1^2 - 4(1)(-2)}}{2(1)}$$
$$= \frac{-1 \pm \sqrt{1 + 8}}{2} = \frac{-1 \pm 3}{2}$$
$$x = \frac{-1 + 3}{2} = 1 \quad \text{or} \quad x = \frac{-1 - 3}{2} = -2$$

The solutions are 1 and -2.

36. $12w^2 - 41w + 24 = 0$

$$(4w - 3)(3w - 8) = 0$$
$$4w - 3 = 0 \quad \text{or} \quad 3w - 8 = 0$$
$$w = \frac{3}{4} \qquad\qquad w = \frac{8}{3}$$

50. $3(x + 7) \leq 5(2x - 8)$

$$3x + 21 \leq 10x - 40$$
$$-7x \leq -61$$
$$x \geq \frac{61}{7}$$

58. $\qquad 12x^2 + 8x \geq 15$

$$12x^2 + 8x - 15 \geq 0$$
$$(6x - 5)(2x + 3) \geq 0$$
$$x = \frac{5}{6} \quad \text{and} \quad x = -\frac{3}{2} \quad \bullet \text{ Critical values}$$

Use a test value from each of the intervals $\left(-\infty, -\frac{3}{2}\right)$, $\left(-\frac{3}{2}, \frac{5}{6}\right)$, and $\left(\frac{5}{6}, \infty\right)$ to determine where $12x^2 + 8x - 15$ is positive.

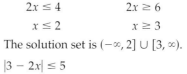

The solution set is $\left(-\infty, -\frac{3}{2}\right] \cup \left[\frac{5}{6}, \infty\right)$.

70. $|2x - 5| \geq 1$

$$2x - 5 \leq -1 \quad \text{or} \quad 2x - 5 \geq 1$$
$$2x \leq 4 \qquad\qquad 2x \geq 6$$
$$x \leq 2 \qquad\qquad x \geq 3$$

The solution set is $(-\infty, 2] \cup [3, \infty)$.

72. $|3 - 2x| \leq 5$

$$-5 \leq 3 - 2x \leq 5$$
$$-8 \leq -2x \leq 2$$
$$4 \geq x \geq -1$$

The solution set is $[-1, 4]$.

Exercise Set 1.2, page 26

6. $d = \sqrt{(x_2 - x_1)^2 + (y_2 - y_1)^2}$

$$d = \sqrt{[-10 - (-5)]^2 + (14 - 8)^2}$$
$$= \sqrt{(-5)^2 + 6^2} = \sqrt{25 + 36}$$
$$= \sqrt{61}$$

26.

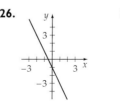

30.

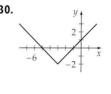

32.

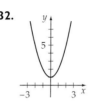

40. y-intercept: $\left(0, -\frac{15}{4}\right)$

x-intercept: $(5, 0)$

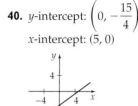

64. $r = \sqrt{(1 - (-2))^2 + (7 - 5)^2}$

$$= \sqrt{9 + 4} = \sqrt{13}$$

Using the standard form

$$(x - h)^2 + (y - k)^2 = r^2$$

with $h = -2$, $k = 5$, and $r = \sqrt{13}$ yields

$$(x + 2)^2 + (y - 5)^2 = \left(\sqrt{13}\right)^2$$

66. $x^2 + y^2 - 6x - 4y + 12 = 0$

$$x^2 - 6x + y^2 - 4y = -12$$
$$x^2 - 6x + 9 + y^2 - 4y + 4 = -12 + 9 + 4$$
$$(x - 3)^2 + (y - 2)^2 = 1^2$$

center $(3, 2)$, radius 1

Exercise Set 1.3, page 42

2. Given $g(x) = 2x^2 + 3$

  a. $g(3) = 2(3)^2 + 3 = 18 + 3 = 21$

  b. $g(-1) = 2(-1)^2 + 3 = 2 + 3 = 5$

  c. $g(0) = 2(0)^2 + 3 = 0 + 3 = 3$

  d. $g\left(\dfrac{1}{2}\right) = 2\left(\dfrac{1}{2}\right)^2 + 3 = \dfrac{1}{2} + 3 = \dfrac{7}{2}$

  e. $g(c) = 2(c)^2 + 3 = 2c^2 + 3$

  f. $g(c + 5) = 2(c + 5)^2 + 3 = 2c^2 + 20c + 50 + 3$

$$= 2c^2 + 20c + 53$$

10. a. Because $0 \le 0 \le 5$, $Q(0) = 4$.

  b. Because $6 < e < 7$, $Q(e) = -e + 9$.

  c. Because $1 < n < 2$, $Q(n) = 4$.

  d. Because $1 < m \le 2$, $8 < m^2 + 7 \le 11$. Thus

$$Q(m^2 + 7) = \sqrt{(m^2 + 7) - 7} = \sqrt{m^2} = m$$

14. $x^2 - 2y = 2$ • Solve for y.

$$-2y = -x^2 + 2$$
$$y = \frac{1}{2}x^2 - 1$$

y is a function of x because each x value will yield one and only one y value.

28. Domain is the set of all real numbers.

40. Domain is the set of all real numbers.

48. a. $[0, \infty]$

  b. Since $31{,}250$ is between $27{,}950$ and $67{,}700$ use
  $T(x) = 0.27(x - 27{,}950) + 3892.50$. Then,
  $T(31{,}250) = 0.27(31{,}250 - 27{,}950) + 3892.50 =$
  $\$4783.50$.

  c. Since $\$72{,}000$ is between $\$67{,}700$ and $\$141{,}250$ use
  $T(x) = 0.30(x - 67{,}700) + 14{,}625$. Then,
  $T(72{,}000) = 0.30(72{,}000 - 67{,}700) + 14{,}625 = \$15{,}915$.

50. a. This is the graph of a function. Every vertical line intersects the graph in at most one point.

  b. This is not the graph of a function. Some vertical lines intersect the graph at two points.

  c. This is not the graph of a function. The vertical line at $x = -2$ intersects the graph at more than one point.

  d. This is the graph of a function. Every vertical line intersects the graph at exactly one point.

66. $v(t) = 44{,}000 - 4200t$, $0 \le t \le 8$

68. a. $V(x) = (30 - 2x)^2 x$

$$= (900 - 120x + 4x^2)x$$
$$= 900x - 120x^2 + 4x^3$$

  b. Domain: $\{x \mid 0 < x < 15\}$

72. $d(A, B) = \sqrt{1 + x^2}$. The time required to swim from A to B at 2 mph is $\dfrac{\sqrt{1 + x^2}}{2}$ hours.

$d(B, C) = 3 - x$. The time required to run from B to C at 8 mph is $\dfrac{3 - x}{8}$ hours.

Thus the total time to reach point C is

$$t = \frac{\sqrt{1 + x^2}}{2} + \frac{3 - x}{8} \text{ hours}$$

Exercise Set 1.4, page 60

14. The graph is symmetric with respect to the x-axis, because replacing y with $-y$ leaves the equation unaltered. The graph is not symmetric with respect to the y-axis, because replacing x with $-x$ alters the equation.

24. The graph is symmetric with respect to the origin because $(-y) = (-x)^3 - (-x)$ simplifies to $-y = -x^3 + x$, which is equivalent to the original equation $y = x^3 - x$.

44. Even, because $h(-x) = (-x)^2 + 1 = x^2 + 1 = h(x)$.

58.

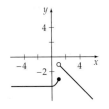

68.

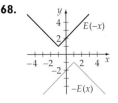

70.

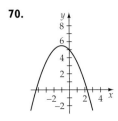

72. a.

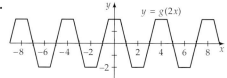

b.

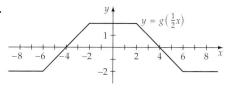

Exercise Set 1.5, page 73

10. $f(x) + g(x) = \sqrt{x-4} - x$ Domain: $\{x \mid x \geq 4\}$
$f(x) - g(x) = \sqrt{x-4} + x$ Domain: $\{x \mid x \geq 4\}$
$f(x)g(x) = -x\sqrt{x-4}$ Domain: $\{x \mid x \geq 4\}$
$\dfrac{f(x)}{g(x)} = -\dfrac{\sqrt{x-4}}{x}$ Domain: $\{x \mid x \geq 4\}$

14. $(f+g)(x) = (x^2 - 3x + 2) + (2x - 4) = x^2 - x - 2$
$(f+g)(-7) = (-7)^2 - (-7) - 2 = 49 + 7 - 2 = 54$

30. $\dfrac{f(x+h) - f(x)}{h} = \dfrac{[4(x+h) - 5] - (4x - 5)}{h}$

$= \dfrac{4x + 4(h) - 5 - 4x + 5}{h}$

$= \dfrac{4(h)}{h} = 4$

38. $(g \circ f)(x) = g[f(x)] = g[2x - 7]$
$\qquad = 3[2x - 7] + 2 = 6x - 19$

$(f \circ g)(x) = f[g(x)] = f[3x + 2]$
$\qquad = 2[3x + 2] - 7 = 6x - 3$

50. $(f \circ g)(4) = f[g(4)]$
$\qquad = f[4^2 - 5(4)]$
$\qquad = f[-4] = 2(-4) + 3 = -5$

66. a. $l = 3 - 0.5t$ for $0 \leq t \leq 6$. $l = -3 + 0.5t$ for $t > 6$. In either case, $l = |3 - 0.5t|$. $w = |2 - 0.2t|$ as in Example 7.

b. $A(t) = |3 - 0.5t||2 - 0.2t|$

c. A is decreasing on $[0, 6]$ and on $[8, 10]$.
A is increasing on $[6, 8]$ and on $[10, 14]$.

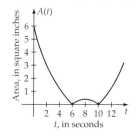

d. The highest point on the graph of A for $0 \leq t \leq 14$ occurs when $t = 0$ seconds.

72. a. On $[2, 3]$,

$a = 2$

$\Delta t = 3 - 2 = 1$

$s(a + \Delta t) = s(3) = 6 \cdot 3^2 = 54$

$s(a) = s(2) = 6 \cdot 2^2 = 24$

$\text{Average velocity} = \dfrac{s(a + \Delta t) - s(a)}{\Delta t}$

$= \dfrac{s(3) - s(2)}{1}$

$= 54 - 24 = 30$ feet per second

This is identical to the slope of the line through $(2, f(2))$ and $(3, f(3))$ because

$m = \dfrac{s(3) - s(2)}{3 - 2} = s(3) - s(2) = 54 - 24 = 30$

b. On $[2, 2.5]$,

$a = 2$

$\Delta t = 2.5 - 2 = 0.5$

$s(a + \Delta t) = s(2.5) = 6(2.5)^2 = 37.5$

$\text{Average velocity} = \dfrac{s(2.5) - s(2)}{0.5}$

$= \dfrac{37.5 - 24}{0.5}$

$= \dfrac{13.5}{0.5} = 27$ feet per second

c. On $[2, 2.1]$,

$a = 2$

$\Delta t = 2.1 - 2 = 0.1$

$s(a + \Delta t) = s(2.1) = 6(2.1)^2 = 26.46$

$\text{Average velocity} = \dfrac{s(2.1) - s(2)}{0.1}$

$= \dfrac{26.46 - 24}{0.1}$

$= \dfrac{2.46}{0.1} = 24.6$ feet per second

d. On $[2, 2.01]$,

$a = 2$

$\Delta t = 2.01 - 2 = 0.01$

$s(a + \Delta t) = s(2.01) = 6(2.01)^2 = 24.2406$

$\text{Average velocity} = \dfrac{s(2.01) - s(2)}{0.01}$

$= \dfrac{24.2406 - 24}{0.01}$

$= \dfrac{0.2406}{0.01} = 24.06$ feet per second

e. On [2, 2.001],

$a = 2$

$\Delta t = 2.001 - 2 = 0.001$

$s(a + \Delta t) = s(2.001) = 6(2.001)^2 = 24.024006$

$$\text{Average velocity} = \frac{s(2.001) - s(2)}{0.001}$$

$$= \frac{24.024006 - 24}{0.001}$$

$$= \frac{0.024006}{0.001} = 24.006 \text{ feet per second}$$

f. On [2, 2 + Δt],

$$\frac{s(2 + \Delta t) - s(2)}{\Delta t} = \frac{6(2 + \Delta t)^2 - 24}{\Delta t}$$

$$= \frac{6(4 + 4(\Delta t) + (\Delta t)^2) - 24}{\Delta t}$$

$$= \frac{24 + 24(\Delta t) + 6(\Delta t)^2 - 24}{\Delta t}$$

$$= \frac{24\Delta t + 6(\Delta t)^2}{\Delta t} = 24 + 6(\Delta t)$$

As Δt approaches zero, the average velocity approaches 24 feet per second.

Exercise Set 1.6, page 87

10. Because the graph of the given function is a line that passes through $(0, 6)$, $(2, 3)$, and $(6, -3)$, the graph of the inverse will be a line that passes through $(6, 0)$, $(3, 2)$, and $(-3, 6)$. See the following figure. Notice that the line shown below is a reflection of the line given in Exercise 10 across the line given by $y = x$.

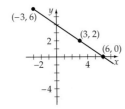

20. Check to see if $f[g(x)] = x$ for all x in the domain of g and $g[f(x)] = x$ for all x in the domain of f. The following shows that $f[g(x)] = x$ for all real numbers x.

$f[g(x)] = f[2x + 3]$

$$= \frac{1}{2}(2x + 3) - \frac{3}{2}$$

$$= x + \frac{3}{2} - \frac{3}{2}$$

$$= x$$

The following shows that $g[f(x)] = x$ for all real numbers x.

$$g[f(x)] = g\left[\frac{1}{2}x - \frac{3}{2}\right]$$

$$= 2\left(\frac{1}{2}x - \frac{3}{2}\right) + 3$$

$$= x - 3 + 3$$

$$= x$$

Thus f and g are inverses.

28. $f(x) = 4x - 8$

$y = 4x - 8$ • Replace $f(x)$ by y.

$x = 4y - 8$ • Interchange x and y.

$x + 8 = 4y$ • Solve for y.

$$\frac{1}{4}(x + 8) = y$$

$$y = \frac{1}{4}x + 2$$

$$f^{-1}(x) = \frac{1}{4}x + 2$$ • Replace y by $f^{-1}(x)$.

34. $f(x) = \dfrac{x}{x - 2}, x \neq 2$

$y = \dfrac{x}{x - 2}$ • Replace $f(x)$ by y.

$x = \dfrac{y}{y - 2}$ • Interchange x and y.

$x(y - 2) = y$

$xy - 2x = y$ • Solve for y.

$xy - y = 2x$

$y(x - 1) = 2x$

$$y = \frac{2x}{x - 1}$$

$$f^{-1}(x) = \frac{2x}{x - 1}, x \neq 1$$ • Replace y by $f^{-1}(x)$ and indicate any restrictions.

40. $f(x) = \sqrt{4 - x}, x \leq 4$

$y = \sqrt{4 - x}$ • Replace $f(x)$ by y.

$x = \sqrt{4 - y}$ • Interchange x and y.

$x^2 = 4 - y$ • Solve for y.

$x^2 - 4 = -y$

$-x^2 + 4 = y$

$f^{-1}(x) = -x^2 + 4, x \geq 0$ • Replace y by $f^{-1}(x)$ and indicate any restrictions.

The range of f is $\{y \mid y \geq 0\}$. Therefore, the domain of f^{-1} is $\{x \mid x \geq 0\}$, as indicated above.

50. $K(x) = 1.3x - 4.7$

$\quad y = 1.3x - 4.7 \qquad$ • Replace $K(x)$ by y.

$\quad x = 1.3y - 4.7 \qquad$ • Interchange x and y.

$\quad x + 4.7 = 1.3y \qquad$ • Solve for y.

$\quad \dfrac{x + 4.7}{1.3} = y$

$\quad K^{-1}(x) = \dfrac{x + 4.7}{1.3} \qquad$ • Replace y by $K^{-1}(x)$.

The function $K^{-1}(x) = \dfrac{x + 4.7}{1.3}$ can be used to convert a United Kingdom men's shoe size to its equivalent U.S. shoe size.

Exercise Set 1.7, page 98

18. Enter the data in the table. Then use your calculator to find the linear regression equation.

a. The linear regression equation is

$\quad y = -72.06131724x + 14926.16191$

b. Evaluate the linear regression equation when $x = 55$.

$\quad y = -72.06131724(55) + 14926.16191$

$\quad \approx 10{,}962.79$

The approximate trade-in value of the car is \$10,963.

32. Enter the data in the table. Then use your calculator to find the quadratic regression model.

a. $y = 0.05208x^2 - 3.56026x + 82.32999$

b. The speed at which the bird has minimum oxygen consumption is the x-coordinate of the vertex of the graph of the regression equation. Recall that the x-coordinate of the vertex is given by $x = -\dfrac{b}{2a}$.

$\quad x = -\dfrac{b}{2a} = -\dfrac{-3.56026}{2(0.05208)}$

$\quad \approx 34$

The speed that minimizes oxygen consumption is approximately 34 kilometers per hour.

Exercise Set 2.1, page 124

2. The measure of the complement of an angle of $87°$ is

$\quad (90° - 87°) = 3°$

The measure of the supplement of an angle of $87°$ is

$\quad (180° - 87°) = 93°$

14. Because $765° = 2 \cdot 360° + 45°$, α is coterminal with an angle that has a measure of $45°$. α is a Quadrant I angle.

32. $-45° = -45°\left(\dfrac{\pi \text{ radians}}{180°}\right) = -\dfrac{\pi}{4}$ radian

40. $\dfrac{\pi}{4}$ radians $= \dfrac{\pi}{4}$ radians $\left(\dfrac{180°}{\pi \text{ radians}}\right) = 45°$

62. $s = r\theta = 5\left(144° \cdot \dfrac{\pi}{180°}\right) = 4\pi \approx 12.57$ meters

66. Let θ_2 be the angle through which the pulley with a diameter of 0.8 meter turns. Let θ_1 be the angle through which the pulley with a diameter of 1.2 meters turns. Let $r_2 = 0.4$ meter be the radius of the smaller pulley, and let $r_1 = 0.6$ meter be the radius of the larger pulley.

$$\theta_1 = 240° = \dfrac{4}{3}\pi \text{ radians}$$

Thus $\quad r_2\theta_2 = r_1\theta_1$

$$0.4\theta_2 = 0.6\left(\dfrac{4}{3}\pi\right)$$

$$\theta_2 = \dfrac{0.6}{0.4}\left(\dfrac{4}{3}\pi\right) = 2\pi \text{ radians or } 360°$$

68. The earth makes 1 revolution ($\theta = 2\pi$) in 1 day.

$\quad t = 24 \cdot 3600 = 86{,}400$ seconds

$\quad \omega = \dfrac{\theta}{t} = \dfrac{2\pi}{86{,}400} \approx 7.27 \times 10^{-5}$ radian/second

74. $C = 2\pi r = 2\pi(18 \text{ inches}) = 36\pi$ inches

Thus one conversion factor is (36π inches/1 rev).

$\quad \dfrac{500 \text{ rev}}{1 \text{ minute}} = \dfrac{500 \text{ rev}}{1 \text{ minute}}\left(\dfrac{36\pi \text{ inches}}{1 \text{ rev}}\right) = \dfrac{18{,}000\pi \text{ inches}}{1 \text{ minute}}$

Now convert inches to miles and minutes to hours.

$\dfrac{18{,}000\pi \text{ inches}}{1 \text{ minute}}$

$= \dfrac{18{,}000\pi \text{ inches}}{1 \text{ minute}}\left(\dfrac{1 \text{ foot}}{12 \text{ inches}}\right)\left(\dfrac{1 \text{ mile}}{5280 \text{ feet}}\right)\left(\dfrac{60 \text{ minutes}}{1 \text{ hour}}\right)$

≈ 54 miles per hour

Exercise Set 2.2, page 136

6.

$x = \sqrt{8^2 - 5^2}$

$x = \sqrt{64 - 25} = \sqrt{39}$

$\sin\theta = \dfrac{y}{r} = \dfrac{5}{8}$

$\csc\theta = \dfrac{r}{y} = \dfrac{8}{5}$

$\cos\theta = \dfrac{x}{r} = \dfrac{\sqrt{39}}{8}$

$\sec\theta = \dfrac{r}{x} = \dfrac{8}{\sqrt{39}} = \dfrac{8\sqrt{39}}{39}$

$\tan\theta = \dfrac{y}{x} = \dfrac{5}{\sqrt{39}} = \dfrac{5\sqrt{39}}{39}$

$\cot\theta = \dfrac{x}{y} = \dfrac{\sqrt{39}}{5}$

20. Because $\tan \theta = \dfrac{y}{x} = \dfrac{4}{3}$, let $y = 4$ and $x = 3$.

$$r = \sqrt{3^2 + 4^2} = 5$$

$$\sec \theta = \frac{r}{x} = \frac{5}{3}$$

38. $\sin \dfrac{\pi}{3} \cos \dfrac{\pi}{4} - \tan \dfrac{\pi}{4} = \dfrac{\sqrt{3}}{2} \cdot \dfrac{\sqrt{2}}{2} - 1$

$$= \frac{\sqrt{6}}{4} - 1 = \frac{\sqrt{6} - 4}{4}$$

64.

$$\tan 68.9° = \frac{h}{116}$$

$$h = 116 \tan 68.9°$$

$$h \approx 301 \text{ meters (3 significant digits)}$$

116 m

68.9°

66.

5°
80 ft
d

$$\sin 5° = \frac{80}{d}$$

$$d = \frac{80}{\sin 5°}$$

$$d \approx 917.9 \text{ feet}$$

Change 9 miles per hour to feet per minute.

$$r = 9 \frac{\text{miles}}{\text{hour}} = \frac{9 \text{ miles}}{1 \text{ hour}} \cdot \frac{5280 \text{ feet}}{1 \text{ mile}} \cdot \frac{1 \text{ hour}}{60 \text{ minutes}}$$

$$= \frac{9(5280)}{60} \frac{\text{feet}}{\text{minute}} = 792 \frac{\text{feet}}{\text{minute}}$$

$$t = \frac{d}{r}$$

$$t \approx \frac{917.9 \text{ feet}}{792 \text{ feet per minute}} \approx 1.2 \text{ minutes (to the nearest}$$
$$\text{tenth of a minute)}$$

74.

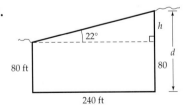

22°
h
80 ft
80
d
240 ft

$$\tan 22° = \frac{h}{240}$$

$$h = 240 \tan 22°$$

$$d = 80 + h$$

$$d = 80 + 240 \tan 22°$$

$$d \approx 180 \text{ feet (2 significant digits)}$$

Exercise Set 2.3, page 149

6. $x = -6, y = -9, r = \sqrt{(-6)^2 + (-9)^2} = \sqrt{117} = 3\sqrt{13}$

$$\sin \theta = \frac{y}{r} = \frac{-9}{3\sqrt{13}} = -\frac{3}{\sqrt{13}} = -\frac{3\sqrt{13}}{13} \qquad \csc \theta = -\frac{\sqrt{13}}{3}$$

$$\cos \theta = \frac{x}{r} = \frac{-6}{3\sqrt{13}} = -\frac{2}{\sqrt{13}} = -\frac{2\sqrt{13}}{13} \qquad \sec \theta = -\frac{\sqrt{13}}{2}$$

$$\tan \theta = \frac{y}{x} = \frac{-9}{-6} = \frac{3}{2} \qquad\qquad\qquad\qquad \cot \theta = \frac{2}{3}$$

18. $\sec \theta = \dfrac{2\sqrt{3}}{3} = \dfrac{r}{x}$ • Let $r = 2\sqrt{3}$ and $x = 3$.

$$y = \pm\sqrt{(2\sqrt{3})^2 - 3^2} = \pm\sqrt{3}$$

$$y = -\sqrt{3} \text{ because } y < 0 \text{ in Quadrant IV.}$$

$$\sin \theta = \frac{-\sqrt{3}}{2\sqrt{3}} = -\frac{1}{2}$$

26. $\theta' = 255° - 180° = 75°$

38. $\cos 300° > 0, \theta' = 360° - 300° = 60°$

Thus $\cos 300° = \cos 60° = \dfrac{1}{2}$.

Exercise Set 2.4, page 160

10. $t = -\dfrac{7\pi}{4}; W(t) = P(x, y)$ where

$$y = \sin t \qquad\qquad x = \cos t$$

$$= \sin\left(-\frac{7\pi}{4}\right) \qquad = \cos\left(-\frac{7\pi}{4}\right)$$

$$= \sin \frac{\pi}{4} \qquad\qquad = \cos \frac{\pi}{4}$$

$$= \frac{\sqrt{2}}{2} \qquad\qquad = \frac{\sqrt{2}}{2}$$

$$W\left(-\frac{7\pi}{4}\right) = \left(\frac{\sqrt{2}}{2}, \frac{\sqrt{2}}{2}\right)$$

16. The reference angle for $-\dfrac{5\pi}{6}$ is $\dfrac{\pi}{6}$.

$$\sec\left(-\frac{5\pi}{6}\right) = -\sec \frac{\pi}{6} \qquad \bullet \sec t < 0 \text{ for } t$$
$$\text{in Quadrant III}$$

$$= -\frac{2\sqrt{3}}{3}$$

44. $F(-x) = \tan(-x) + \sin(-x)$

$\quad\quad = -\tan x - \sin x$ • tan x and sin x are

$\quad\quad = -(\tan x + \sin x)$ odd functions.

$\quad\quad = -F(x)$

Because $F(-x) = -F(x)$, the function defined by $F(x) = \tan x + \sin x$ is an odd function.

50.

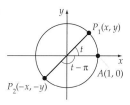

$\tan t = \dfrac{y}{x}$

$\tan(t - \pi) = \dfrac{-y}{-x} = \dfrac{y}{x}$ • From the unit circle

Therefore, $\tan t = \tan(t - \pi)$.

68. $\dfrac{1}{1 - \sin t} + \dfrac{1}{1 + \sin t} = \dfrac{1 + \sin t + 1 - \sin t}{(1 - \sin t)(1 + \sin t)}$

$\quad\quad\quad\quad = \dfrac{2}{1 - \sin^2 t}$

$\quad\quad\quad\quad = \dfrac{2}{\cos^2 t}$

$\quad\quad\quad\quad = 2\sec^2 t$

74. $1 + \tan^2 t = \sec^2 t$

$\quad\quad\quad \tan^2 t = \sec^2 t - 1$

$\quad\quad\quad \tan t = \pm\sqrt{\sec^2 t - 1}$

Because $\dfrac{3\pi}{2} < t < 2\pi$, $\tan t$ is negative. Thus $\tan t = -\sqrt{\sec^2 t - 1}$.

78. March 5 is represented by $t = 2$.

$T(2) = -41\cos\left(\dfrac{\pi}{6} \cdot 2\right) + 36$

$\quad\quad = -41\cos\left(\dfrac{\pi}{3}\right) + 36$

$\quad\quad = -41(0.5) + 36$

$\quad\quad = 15.5°\text{F}$

July 20 is represented by $t = 6.5$.

$T(6.5) = -41\cos\left(\dfrac{\pi}{6} \cdot 6.5\right) + 36$

$\quad\quad\quad \approx -41(-0.9659258263) + 36$

$\quad\quad\quad \approx 75.6°\text{F}$

Exercise Set 2.5, page 170

20. $y = -\dfrac{3}{2}\sin x$

$a = \left|-\dfrac{3}{2}\right| = \dfrac{3}{2}$

period $= 2\pi$

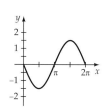

30. $y = \sin\dfrac{3\pi}{4}x$

$a = 1$

period $= \dfrac{2\pi}{b} = \dfrac{2\pi}{\dfrac{3\pi}{4}} = \dfrac{8}{3}$

32. $y = \cos 3\pi x$

$a = 1$

period $= \dfrac{2\pi}{b} = \dfrac{2\pi}{3\pi} = \dfrac{2}{3}$

38. $y = \dfrac{1}{2}\sin\dfrac{\pi x}{3}$

$a = \dfrac{1}{2}$

period $= \dfrac{2\pi}{b} = \dfrac{2\pi}{\dfrac{\pi}{3}} = 6$

46. $y = -\dfrac{3}{4}\cos 5x$

$a = \left|-\dfrac{3}{4}\right| = \dfrac{3}{4}$

period $= \dfrac{2\pi}{b} = \dfrac{2\pi}{5}$

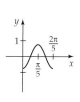

52. $y = -\left|3\sin\dfrac{2}{3}x\right|$

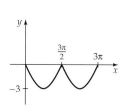

Exercise Set 2.6, page 178

22. $y = \dfrac{1}{3}\tan x$

period $= \dfrac{\pi}{b} = \pi$

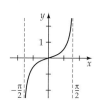

30. $y = -3 \tan 3x$

$\text{period} = \dfrac{\pi}{b} = \dfrac{\pi}{3}$

32. $y = \dfrac{1}{2} \cot 2x$

$\text{period} = \dfrac{\pi}{b} = \dfrac{\pi}{2}$

38. $y = 3 \csc \dfrac{\pi x}{2}$

$\text{period} = \dfrac{2\pi}{b} = \dfrac{2\pi}{\dfrac{\pi}{2}} = 4$

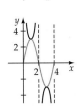

42. $y = \sec \dfrac{x}{2}$

$\text{period} = \dfrac{2\pi}{b} = \dfrac{2\pi}{\dfrac{1}{2}} = 4\pi$

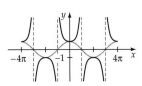

Exercise Set 2.7, page 187

20. $y = \cos\left(2x - \dfrac{\pi}{3}\right)$

$a = 1$

$\text{period} = \pi$

$\text{phase shift} = -\dfrac{c}{b}$

$= -\dfrac{-\dfrac{\pi}{3}}{2} = \dfrac{\pi}{6}$

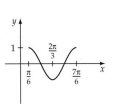

22. $y = \tan(x - \pi)$

$\text{period} = \pi$

$\text{phase shift} = -\dfrac{c}{b}$

$= -\dfrac{-\pi}{1} = \pi$

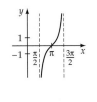

40. $y = 2\sin\left(\dfrac{\pi x}{2} + 1\right) - 2$

$a = 2$

$\text{period} = 4$

$\text{phase shift} = -\dfrac{c}{b}$

$= -\dfrac{1}{\dfrac{\pi}{2}} = -\dfrac{2}{\pi}$

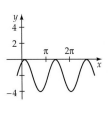

42. $y = -3\cos(2\pi x - 3) + 1$

$a = 3$

$\text{period} = 1$

$\text{phase shift} = -\dfrac{c}{b} = \dfrac{3}{2\pi}$

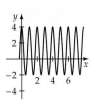

48. $y = \csc \dfrac{x}{3} + 4$

$\text{period} = 6\pi$

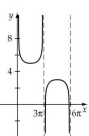

52. a. $\text{Phase shift} = -\dfrac{c}{b} = -\dfrac{\left(-\dfrac{7}{12}\pi\right)}{\left(\dfrac{\pi}{6}\right)} = 3.5 \text{ months,}$

$\text{period:} \dfrac{2\pi}{b} = \dfrac{2\pi}{\left(\dfrac{\pi}{6}\right)} = 12 \text{ months}$

b. First graph $y_1 = 2.7\cos\left(\dfrac{\pi}{6}t\right)$. Because the phase shift is 3.5 months, shift the graph of y_1 3.5 units to the right to produce the graph of y_2. Now shift the graph of y_2 upward 4 units to produce the graph of S.

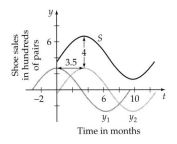

c. 3.5 months after January 1 is the middle of April.

54. $y = \dfrac{x}{2} + \cos x$

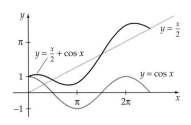

58. $y = -\sin x + \cos x$

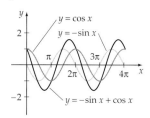

78. $y = x \cos x$

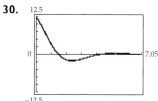

Exercise Set 2.8, page 196

20. Amplitude $= 3$, frequency $= \dfrac{1}{\pi}$, period $= \pi$.

Because $\dfrac{2\pi}{b} = \pi$, we have $b = 2$. Thus $y = 3 \cos 2t$.

28. Amplitude $= |-1.5| = 1.5$

$f = \dfrac{1}{2\pi} \sqrt{\dfrac{k}{m}} = \dfrac{1}{2\pi} \sqrt{\dfrac{3}{27}} = \dfrac{1}{2\pi} \cdot \dfrac{1}{3} = \dfrac{1}{6\pi}$, period $= 6\pi$

$y = a \cos 2\pi f t = -1.5 \cos\left[2\pi\left(\dfrac{1}{6\pi}\right)t\right] = -1.5 \cos\dfrac{1}{3}t$

30.

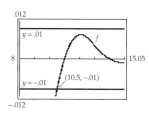

a. f has pseudoperiod $\dfrac{2\pi}{1} = 2\pi$.

$10 \div (2\pi) \approx 1.59$

Thus f completes only one full oscillation on $0 \le t \le 10$.

b. The following graph of f shows that $|f(t)| < 0.01$ for $t > 10.5$.

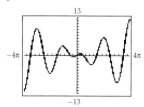

Exercise Set 3.1, page 211

2. The equation $\tan 2x = 2 \tan x$ is not an identity. To verify that it is not an identity, let $x = \dfrac{\pi}{6}$. Then

$$\tan 2x = \tan 2\left(\dfrac{\pi}{6}\right) = \tan\dfrac{\pi}{3} = \sqrt{3}$$

whereas $2 \tan x = 2 \tan\dfrac{\pi}{6} = \dfrac{2\sqrt{3}}{3}$. Because $\sqrt{3} \ne \dfrac{2\sqrt{3}}{3}$, we have shown that the equation is not an identity.

22. $\sin^4 x - \cos^4 x = (\sin^2 x + \cos^2 x)(\sin^2 x - \cos^2 x)$
$= 1(\sin^2 x - \cos^2 x) = \sin^2 x - \cos^2 x$

34. $\dfrac{2 \sin x \cot x + \sin x - 4 \cot x - 2}{2 \cot x + 1}$

$= \dfrac{(\sin x)(2 \cot x + 1) - 2(2 \cot x + 1)}{2 \cot x + 1}$

$= \dfrac{(2 \cot x + 1)(\sin x - 2)}{2 \cot x + 1} = \sin x - 2$

44. $\dfrac{\dfrac{1}{\sin x} + \dfrac{1}{\cos x}}{\dfrac{1}{\sin x} - \dfrac{1}{\cos x}} = \dfrac{\dfrac{1}{\sin x} + \dfrac{1}{\cos x}}{\dfrac{1}{\sin x} - \dfrac{1}{\cos x}} \cdot \dfrac{\sin x \cos x}{\sin x \cos x}$

$= \dfrac{\cos x + \sin x}{\cos x - \sin x}$

$= \dfrac{\cos x + \sin x}{\cos x - \sin x} \cdot \dfrac{\cos x - \sin x}{\cos x - \sin x}$

$= \dfrac{\cos^2 x - \sin^2 x}{\cos^2 x - 2 \sin x \cos x + \sin^2 x}$

$= \dfrac{\cos^2 x - \sin^2 x}{1 - 2 \sin x \cos x}$

54. $\dfrac{\dfrac{1}{\tan x} + \cot x}{\dfrac{1}{\tan x} + \tan x} = \dfrac{\dfrac{1}{\tan x} + \cot x}{\dfrac{1}{\tan x} + \tan x} \cdot \dfrac{\tan x}{\tan x}$

$= \dfrac{1 + 1}{1 + \tan^2 x} = \dfrac{2}{\sec^2 x}$

Exercise Set 3.2, page 222

4. Use the identity $\cos(\alpha - \beta) = \cos \alpha \cos \beta + \sin \alpha \sin \beta$ with $\alpha = 120°$ and $\beta = 45°$.

$\cos(120° - 45°) = \cos 120° \cos 45° + \sin 120° \sin 45°$

$= \left(-\dfrac{1}{2}\right)\left(\dfrac{\sqrt{2}}{2}\right) + \left(\dfrac{\sqrt{3}}{2}\right)\left(\dfrac{\sqrt{2}}{2}\right)$

$= -\dfrac{\sqrt{2}}{4} + \dfrac{\sqrt{6}}{4}$

$= \dfrac{\sqrt{6} - \sqrt{2}}{4}$

20. The value of a given trigonometric function of θ, measured in degrees, is equal to its cofunction of $90° - \theta$. Thus

$$\cos 80° = \sin(90° - 80°)$$
$$= \sin 10°$$

26. $\sin x \cos 3x + \cos x \sin 3x + \sin(x + 3x) = \sin 4x$

38. $\tan \alpha = \dfrac{24}{7}$, with $0° < \alpha < 90°$, $\sin \alpha = \dfrac{24}{25}$, $\cos \alpha = \dfrac{7}{25}$

$\sin \beta = -\dfrac{8}{17}$, with $180° < \beta < 270°$

$\cos \beta = -\dfrac{15}{17}$, $\tan \beta = \dfrac{8}{15}$

a. $\sin(\alpha + \beta) = \sin \alpha \cos \beta + \cos \alpha \sin \beta$

$$= \left(\dfrac{24}{25}\right)\left(-\dfrac{15}{17}\right) + \left(\dfrac{7}{25}\right)\left(-\dfrac{8}{17}\right)$$

$$= -\dfrac{360}{425} - \dfrac{56}{425} = -\dfrac{416}{425}$$

b. $\cos(\alpha + \beta) = \cos \alpha \cos \beta - \sin \alpha \sin \beta$

$$= \left(\dfrac{7}{25}\right)\left(-\dfrac{15}{17}\right) - \left(\dfrac{24}{25}\right)\left(-\dfrac{8}{17}\right)$$

$$= -\dfrac{105}{425} + \dfrac{192}{425} = \dfrac{87}{425}$$

c. $\tan(\alpha - \beta) = \dfrac{\tan \alpha - \tan \beta}{1 + \tan \alpha \tan \beta}$

$$= \dfrac{\dfrac{24}{7} - \dfrac{8}{15}}{1 + \left(\dfrac{24}{7}\right)\left(\dfrac{8}{15}\right)} = \dfrac{\dfrac{24}{7} - \dfrac{8}{15}}{1 + \dfrac{192}{105}} \cdot \dfrac{105}{105}$$

$$= \dfrac{360 - 56}{105 + 192} = \dfrac{304}{297}$$

50. $\cos(\theta + \pi) = \cos \theta \cos \pi - \sin \theta \sin \pi$

$$= (\cos \theta)(-1) - (\sin \theta)(0) = -\cos \theta$$

62. $\cos 5x \cos 3x + \sin 5x \sin 3x = \cos(5x - 3x) = \cos 2x$

$$= \cos(x + x) = \cos x \cos x - \sin x \sin x$$
$$= \cos^2 x - \sin^2 x$$

74. $\sin(\theta + 2\pi) = \sin \theta \cos 2\pi + \cos \theta \sin 2\pi$

$$= (\sin \theta)(1) + (\cos \theta)(0) = \sin \theta$$

Exercise Set 3.3, page 230

2. $2 \sin 3\theta \cos 3\theta = \sin[2(3\theta)] = \sin 6\theta$

26. $\cos \theta = \dfrac{24}{25}$ with $270° < \theta < 360°$

$$\sin \theta = -\sqrt{1 - \left(\dfrac{24}{25}\right)^2} \qquad \tan \theta = \dfrac{-\dfrac{7}{25}}{\dfrac{24}{25}}$$

$$= -\dfrac{7}{25} \qquad\qquad = -\dfrac{7}{24}$$

$\sin 2\theta = 2 \sin \theta \cos \theta \qquad \cos 2\theta = \cos^2\theta - \sin^2\theta$

$$= 2\left(-\dfrac{7}{25}\right)\left(\dfrac{24}{25}\right) \qquad = \left(\dfrac{24}{25}\right)^2 - \left(-\dfrac{7}{25}\right)^2$$

$$= -\dfrac{336}{625} \qquad\qquad = \dfrac{527}{625}$$

$\tan 2\theta = \dfrac{2 \tan \theta}{1 - \tan^2\theta}$

$$= \dfrac{2\left(-\dfrac{7}{24}\right)}{1 - \left(-\dfrac{7}{24}\right)^2}$$

$$= \dfrac{-\dfrac{7}{12}}{1 - \dfrac{49}{576}} \cdot \dfrac{576}{576} = -\dfrac{336}{527}$$

54. $\dfrac{1}{1 - \cos 2x} = \dfrac{1}{1 - 1 + 2 \sin^2x}$

$$= \dfrac{1}{2 \sin^2x} = \dfrac{1}{2} \csc^2x$$

72. $\cos^2 \dfrac{x}{2} = \left[\pm\sqrt{\dfrac{1 + \cos x}{2}}\right]^2 = \dfrac{1 + \cos x}{2}$

$$= \dfrac{1 + \cos x}{2} \cdot \dfrac{\sec x}{\sec x} = \dfrac{\sec x + 1}{2 \sec x}$$

78. $\tan^2 \dfrac{x}{2} = \left(\dfrac{1 - \cos x}{\sin x}\right)^2 = \dfrac{(1 - \cos x)^2}{\sin^2x} = \dfrac{(1 - \cos x)^2}{1 - \cos^2x}$

$$= \dfrac{(1 - \cos x)^2}{(1 - \cos x)(1 + \cos x)} = \dfrac{1 - \cos x}{1 + \cos x}$$

$$= \dfrac{\dfrac{1}{\cos x} - \dfrac{\cos x}{\cos x}}{\dfrac{1}{\cos x} + \dfrac{\cos x}{\cos x}} = \dfrac{\sec x - 1}{\sec x + 1}$$

Exercise Set 3.4, page 239

22. $\cos 3\theta + \cos 5\theta = 2 \cos \dfrac{3\theta + 5\theta}{2} \cos \dfrac{3\theta - 5\theta}{2}$

$\qquad\qquad\quad = 2 \cos 4\theta \cos(-\theta) = 2 \cos 4\theta \cos \theta$

36. $\sin 5x \cos 3x = \dfrac{1}{2}[\sin(5x + 3x) + \sin(5x - 3x)]$

$\qquad\qquad\quad = \dfrac{1}{2}(\sin 8x + \sin 2x)$

$\qquad\qquad\quad = \dfrac{1}{2}(2 \sin 4x \cos 4x + 2 \sin x \cos x)$

$\qquad\qquad\quad = \sin 4x \cos 4x + \sin x \cos x$

44. $\dfrac{\cos 5x - \cos 3x}{\sin 5x + \sin 3x} = \dfrac{-2 \sin \dfrac{5x + 3x}{2} \sin \dfrac{5x - 3x}{2}}{2 \sin \dfrac{5x + 3x}{2} \cos \dfrac{5x - 3x}{2}}$

$\qquad\qquad\qquad = -\dfrac{\sin 4x \sin x}{\sin 4x \cos x} = -\tan x$

62. $a = 1, b = \sqrt{3}, k = \sqrt{(\sqrt{3})^2 + (1)^2} = 2$. Thus α is a first-quadrant angle.

$\sin \alpha = \dfrac{\sqrt{3}}{2}$ and $\cos \alpha = \dfrac{1}{2}$

Thus $\alpha = \dfrac{\pi}{3}$.

$y = k \sin(x + \alpha)$

$y = 2 \sin\left(x + \dfrac{\pi}{3}\right)$

70. From Exercise 62, we know that

$y = \sin x + \sqrt{3} \cos x = 2 \sin\left(x + \dfrac{\pi}{3}\right)$

Thus the graph of $y = \sin x + \sqrt{3} \cos x$ is the graph of $y = 2 \sin x$ shifted $\dfrac{\pi}{3}$ units to the left.

Exercise Set 3.5, page 253

2. $y = \sin^{-1} \dfrac{\sqrt{2}}{2}$ implies

$\sin y = \dfrac{\sqrt{2}}{2}$ for $-\dfrac{\pi}{2} \le y \le \dfrac{\pi}{2}$

Thus $y = \dfrac{\pi}{4}$.

24. Because $\tan(\tan^{-1} x) = x$ for all real numbers x, we have

$\tan\left[\tan^{-1}\left(\dfrac{1}{2}\right)\right] = \dfrac{1}{2}$.

46. Let $x = \cos^{-1} \dfrac{3}{5}$. Thus

$\cos x = \dfrac{3}{5}$ and $\sin x = \sqrt{1 - \left(\dfrac{3}{5}\right)^2} = \dfrac{4}{5}$

$y = \tan\left(\cos^{-1} \dfrac{3}{5}\right) = \tan x = \dfrac{\sin x}{\cos x} = \dfrac{\dfrac{4}{5}}{\dfrac{3}{5}} = \dfrac{4}{3}$

54. $y = \cos\left(\sin^{-1} \dfrac{3}{4} + \cos^{-1} \dfrac{5}{13}\right)$

Let $\alpha = \sin^{-1} \dfrac{3}{4}$, $\sin \alpha = \dfrac{3}{4}$, $\cos \alpha = \sqrt{1 - \left(\dfrac{3}{4}\right)^2} = \dfrac{\sqrt{7}}{4}$.

$\beta = \cos^{-1} \dfrac{5}{13}$, $\cos \beta = \dfrac{5}{13}$, $\sin \beta = \sqrt{1 - \left(\dfrac{5}{13}\right)^2} = \dfrac{12}{13}$.

$y = \cos(\alpha + \beta)$

$\quad = \cos \alpha \cos \beta - \sin \alpha \sin \beta$

$\quad = \dfrac{\sqrt{7}}{4} \cdot \dfrac{5}{13} - \dfrac{3}{4} \cdot \dfrac{12}{13} = \dfrac{5\sqrt{7}}{52} - \dfrac{36}{52} = \dfrac{5\sqrt{7} - 36}{52}$

64. $\sin^{-1} x + \cos^{-1} \dfrac{4}{5} = \dfrac{\pi}{6}$

$\sin^{-1} x = \dfrac{\pi}{6} - \cos^{-1} \dfrac{4}{5}$

$\sin(\sin^{-1} x) = \sin\left(\dfrac{\pi}{6} - \cos^{-1} \dfrac{4}{5}\right)$

$x = \sin \dfrac{\pi}{6} \cos\left(\cos^{-1} \dfrac{4}{5}\right) - \cos \dfrac{\pi}{6} \sin\left(\cos^{-1} \dfrac{4}{5}\right)$

$\quad = \dfrac{1}{2} \cdot \dfrac{4}{5} - \dfrac{\sqrt{3}}{2} \cdot \dfrac{3}{5} = \dfrac{4 - 3\sqrt{3}}{10}$

72. Let $\alpha = \cos^{-1} x$ and $\beta = \cos^{-1}(-x)$. Thus $\cos \alpha = x$ and $\cos \beta = -x$. We know $\sin \alpha = \sqrt{1 - x^2}$ and $\sin \beta = \sqrt{1 - x^2}$ because α is in Quadrant I and β is in Quadrant II.

$\cos^{-1} x + \cos^{-1}(-x)$

$\quad = \alpha + \beta$

$\quad = \cos^{-1}[\cos(\alpha + \beta)]$

$\quad = \cos^{-1}(\cos \alpha \cos \beta - \sin \alpha \sin \beta)$

$\quad = \cos^{-1}[x(-x) - \sqrt{1 - x^2} \cdot \sqrt{1 - x^2}]$

$\quad = \cos^{-1}(-x^2 - 1 + x^2)$

$\quad = \cos^{-1}(-1) = \pi$

76. Recall that the graph of $y = f(x - a)$ is a horizontal shift of the graph of $y = f(x)$. Therefore, the graph of $y = \cos^{-1}(x - 1)$ is the graph of $y = \cos^{-1}x$ shifted 1 unit to the right.

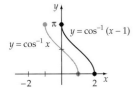

84. a.

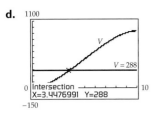

b. Although the water rises 0.1 foot in each case, there is more surface area (and thus more volume of water) at the 4.9- to 5.0-foot level near the diameter of the cylinder than at the 0.1- to 0.2-foot level near the bottom.

c.

$$V(4) = 12\left[25\cos^{-1}\left(\frac{5 - (4)}{5}\right) - [5 - (4)]\sqrt{10(4) - (4)^2}\right]$$

$$= 12\left[25\cos^{-1}\left(\frac{1}{5}\right) - \sqrt{24}\right]$$

$$\approx 352.04 \text{ cubic feet}$$

d.

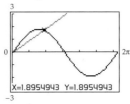

When $V = 288$ cubic feet, $x \approx 3.45$ feet.

Exercise Set 3.6, page 266

14.
$$2\cos^2x + 1 = -3\cos x$$
$$2\cos^2x + 3\cos x + 1 = 0$$
$$(2\cos x + 1)(\cos x + 1) = 0$$
$$2\cos x + 1 = 0 \quad \text{or} \quad \cos x + 1 = 0$$
$$\cos x = -\frac{1}{2} \qquad\qquad \cos x = -1$$
$$x = \frac{2\pi}{3}, \frac{4\pi}{3} \qquad\qquad x = \pi$$

The solutions in the interval $0 \le x < 2\pi$ are $\frac{2\pi}{3}$, π, and $\frac{4\pi}{3}$.

52.
$$\sin x + 2\cos x = 1$$
$$\sin x = 1 - 2\cos x$$
$$(\sin x)^2 = (1 - 2\cos x)^2$$
$$\sin^2x = 1 - 4\cos x + 4\cos^2x$$
$$1 - \cos^2x = 1 - 4\cos x + 4\cos^2x$$
$$0 = \cos x(5\cos x - 4)$$
$$\cos x = 0 \qquad \text{or} \qquad 5\cos x - 4 = 0$$
$$x = 90°, 270° \qquad\qquad \cos x = \frac{4}{5}$$
$$x \approx 36.9°, 323.1°$$

The solutions in the interval $0 \le x < 360°$ are $90°$ and $323.1°$. (*Note*: $x = 270°$ and $x = 36.9°$ are extraneous solutions. Neither of these satisfies the original equation.)

56. $2\cos^2x - 5\cos x - 5 = 0$
$$\cos x = \frac{5 \pm \sqrt{(-5)^2 - 4(2)(-5)}}{2(2)} = \frac{5 \pm \sqrt{65}}{4}$$
$$\cos x \approx 3.27 \quad \text{or} \quad \cos x \approx -0.7656$$
$$\text{no solution} \qquad\qquad x \approx 140.0°, 220.0°$$

The solutions in the interval $0° \le x < 360°$ are $140°$ and $220°$.

66. $\cos 2x = -\frac{\sqrt{3}}{2}$
$$2x = \frac{5\pi}{6} + 2k\pi \quad \text{or} \quad 2x = \frac{7\pi}{6} + 2k\pi, k \text{ an integer}$$
$$x = \frac{5\pi}{12} + k\pi \quad \text{or} \quad x = \frac{7\pi}{12} + k\pi, k \text{ an integer}$$

84. $2\sin x\cos x - 2\sqrt{2}\sin x - \sqrt{3}\cos x + \sqrt{6} = 0$
$$2\sin x(\cos x - \sqrt{2}) - \sqrt{3}(\cos x - \sqrt{2}) = 0$$
$$(\cos x - \sqrt{2})(2\sin x - \sqrt{3}) = 0$$
$$\cos x = \sqrt{2} \quad \text{or} \quad \sin x = \frac{\sqrt{3}}{2}$$
$$\text{no solution} \qquad\qquad x = \frac{\pi}{3}, \frac{2\pi}{3}$$

The solutions in the interval $0 \le x < 2\pi$ are $\frac{\pi}{3}$ and $\frac{2\pi}{3}$.

86. The following graph shows that the solutions in the interval $[0, 2\pi)$ are $x = 0$ and $x = 1.895$.

92. When $\theta = 45°$, d attains its maximum of 4394.5 feet.

94. a. The following work shows the sine regression procedure for a TI-83 calculator. Note that each time (given in hours:minutes) must be converted to hours. Thus 17:03 is $17 + \dfrac{3}{60} = 17.05$ hours.

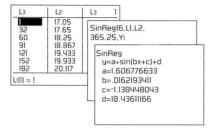

The sine regression function is
$Y_1 \approx 1.6068 \sin(0.01622x - 1.1384) + 18.436$

b. Using the "value" command in the CALC menu gives
$Y_1(141) \approx 19.901719$

$\approx 19\!:\!54$ (to the nearest minute)

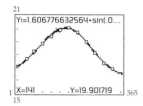

Exercise Set 4.1, page 286

4.

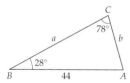

$A = 180° - 78° - 28° = 74°$

$$\frac{b}{\sin B} = \frac{c}{\sin C} \qquad\qquad \frac{a}{\sin A} = \frac{c}{\sin C}$$

$$\frac{b}{\sin 28°} = \frac{44}{\sin 78°} \qquad\qquad \frac{a}{\sin 74°} = \frac{44}{\sin 78°}$$

$$b = \frac{44 \sin 28°}{\sin 78°} \approx 21 \qquad a = \frac{44 \sin 74°}{\sin 78°} \approx 43$$

18. $\dfrac{a}{\sin A} = \dfrac{b}{\sin B}$

$\dfrac{13.8}{\sin A} = \dfrac{5.55}{\sin 22.6}$

$\sin A = 0.9555$

$A \approx 72.9°$ or $107.1°$

If $A = 72.9°$, $C \approx 180° - 72.9° - 22.6° = 84.5°$.

$\dfrac{c}{\sin 84.5°} = \dfrac{5.55}{\sin 22.6°}$

$c = \dfrac{5.55 \sin 84.5°}{\sin 22.6°} \approx 14.4$

If $A = 107.1°$, $C \approx 180° - 107.1° - 22.6° = 50.3°$.

$\dfrac{c}{\sin 50.3°} = \dfrac{5.55}{\sin 22.6°}$

$c = \dfrac{5.55 \sin 50.3°}{\sin 22.6°} \approx 11.1$

Case 1: $A = 72.9°$, $C = 84.5°$, and $c = 14.4$

Case 2: $A = 107.1°$, $C = 50.3°$, and $c = 11.1$

26. The angle with its vertex at the position of the helicopter measures $180° - (59.0° + 77.2°) = 43.8°$. Let the distance from the helicopter to the carrier be x. Using the Law of Sines, we have

$$\frac{x}{\sin 77.2°} = \frac{7620}{\sin 43.8°}$$

$$x = \frac{7620 \sin 77.2°}{\sin 43.8°}$$

$$\approx 10{,}700 \text{ feet} \qquad \text{(three significant digits)}$$

36.

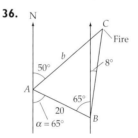

$\alpha = 65°$

$B = 65° + 8° = 73°$

$A = 180° - 50° - 65° = 65°$

$C = 180° - 65° - 73° = 42°$

$$\frac{b}{\sin B} = \frac{c}{\sin C}$$

$$\frac{b}{\sin 73°} = \frac{20}{\sin 42°}$$

$$b = \frac{20 \sin 73°}{\sin 42°}$$

$$b \approx 29 \text{ miles}$$

Exercise Set 4.2, page 296

12. $c^2 = a^2 + b^2 - 2ab \cos C$

$c^2 = 14.2^2 + 9.30^2 - 2(14.2)(9.30) \cos 9.20°$

$c = \sqrt{14.2^2 + 9.30^2 - 2(14.2)(9.30) \cos 9.20°}$

$c \approx 5.24$

18. $\cos A = \dfrac{b^2 + c^2 - a^2}{2bc}$

$\cos A = \dfrac{132^2 + 160^2 - 108^2}{2(132)(160)} \approx 0.7424$

$A \approx \cos^{-1}(0.7424) \approx 42.1°$

26. $K = \dfrac{1}{2} ac \sin B$

$K = \dfrac{1}{2}(32)(25) \sin 127° \approx 320$ square units

28. $A = 180° - 102° - 27° = 51°$

$K = \dfrac{a^2 \sin B \sin C}{2 \sin A}$

$K = \dfrac{8.5^2 \sin 102° \sin 27°}{2 \sin 51°} \approx 21$ square units

36. $s = \dfrac{1}{2}(a + b + c)$

$= \dfrac{1}{2}(10.2 + 13.3 + 15.4) = 19.45$

$K = \sqrt{s(s - a)(s - b)(s - c)}$

$= \sqrt{19.45(19.45 - 10.2)(19.45 - 13.3)(19.45 - 15.4)}$

≈ 66.9 square units

52.

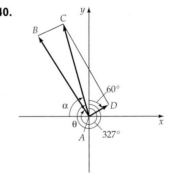

$\alpha = 270° - 254° = 16°$

$A = 16° + 90° + 32° = 138°$

$b = 4 \cdot 16 = 64$ miles

$c = 3 \cdot 22 = 66$ miles

$a^2 = b^2 + c^2 - 2bc \cos A$

$a^2 = 64^2 + 66^2 - 2(64)(66) \cos 138°$

$a = \sqrt{64^2 + 66^2 - 2(64)(66) \cos 138°}$

$a \approx 120$ miles

60. $S = \dfrac{1}{2}(324 + 412 + 516) = 626$

$K = \sqrt{626(626 - 324)(626 - 412)(626 - 516)}$

$= \sqrt{4,450,284,080}$

Cost $= 4.15\left(\sqrt{4,450,284,080}\right) \approx \$276,848$, or \$277,000 to the nearest \$1000

Exercise Set 4.3, page 313

6. $a = 3 - 3 = 0$

$b = 0 - (-2) = 2$

A vector equivalent to P_1P_2 is $\mathbf{v} = \langle 0, 2 \rangle$.

8. $\|\mathbf{v}\| = \sqrt{6^2 + 10^2}$

$= \sqrt{36 + 100} = \sqrt{136} = 2\sqrt{34}$

$\theta = \tan^{-1}\dfrac{10}{6} = \tan^{-1}\dfrac{5}{3} \approx 59.0°$

Thus $\mathbf{v}$ has a direction of about 59° as measured from the positive x-axis.

A unit vector in the direction of $\mathbf{v}$ is

$\mathbf{u} = \left\langle \dfrac{6}{2\sqrt{34}}, \dfrac{10}{2\sqrt{34}} \right\rangle = \left\langle \dfrac{3\sqrt{34}}{34}, \dfrac{5\sqrt{34}}{34} \right\rangle$

20. $\dfrac{3}{4}\mathbf{u} - 2\mathbf{v} = \dfrac{3}{4}\langle -2, 4 \rangle - 2\langle -3, -2 \rangle$

$= \left\langle -\dfrac{3}{2}, 3 \right\rangle - \langle -6, -4 \rangle$

$= \left\langle \dfrac{9}{2}, 7 \right\rangle$

26. $3\mathbf{u} + 2\mathbf{v} = 3(3\mathbf{i} - 2\mathbf{j}) + 2(-2\mathbf{i} + 3\mathbf{j})$

$= (9\mathbf{i} - 6\mathbf{j}) + (-4\mathbf{i} + 6\mathbf{j})$

$= (9 - 4)\mathbf{i} + (-6 + 6)\mathbf{j}$

$= 5\mathbf{i} + 0\mathbf{j}$

$= 5\mathbf{i}$

36. $a_1 = 2 \cos \dfrac{8\pi}{7} \approx -1.8$

$a_2 = 2 \sin \dfrac{8\pi}{7} \approx -0.9$

$\mathbf{v} = a_1\mathbf{i} + a_2\mathbf{j} \approx -1.8\mathbf{i} - 0.9\mathbf{j}$

40.

$\mathbf{AB} = 18 \cos 123°\mathbf{i} + 18 \sin 123°\mathbf{j} \approx -9.8\mathbf{i} + 15.1\mathbf{j}$

$\mathbf{AD} = 4 \cos 30°\mathbf{i} + 4 \sin 30°\mathbf{j} \approx 3.5\mathbf{i} + 2\mathbf{j}$

$\mathbf{AC} = \mathbf{AB} + \mathbf{AD} \approx -9.8\mathbf{i} + 15.1\mathbf{j} + 3.5\mathbf{i} + 2\mathbf{j}$

$= -6.3\mathbf{i} + 17.1\mathbf{j}$

$\|\mathbf{AC}\| = \sqrt{(-6.3)^2 + (17.1)^2} \approx 18$

$\alpha = \tan^{-1}\left|\dfrac{17.1}{-6.3}\right| = \tan^{-1}\dfrac{17.1}{6.3} \approx 70°$

$\theta \approx 270° + 70° = 340°$

The course of the boat is about 18 miles per hour at an approximate heading of 340°.

42. $\alpha = \theta$

$\sin \alpha = \dfrac{120}{800}$

$\alpha \approx 8.6°$

50. $\mathbf{v} \cdot \mathbf{w} = (5\mathbf{i} + 3\mathbf{j}) \cdot (4\mathbf{i} - 2\mathbf{j})$

$= 5(4) + 3(-2) = 20 - 6 = 14$

60. $\cos \theta = \dfrac{\mathbf{v} \cdot \mathbf{w}}{\|\mathbf{v}\|\,\|\mathbf{w}\|}$

$\cos \theta = \dfrac{(3\mathbf{i} - 4\mathbf{j}) \cdot (6\mathbf{i} - 12\mathbf{j})}{\sqrt{3^2 + (-4)^2}\,\sqrt{6^2 + (-12)^2}}$

$\cos \theta = \dfrac{3(6) + (-4)(-12)}{\sqrt{25}\,\sqrt{180}}$

$\cos \theta = \dfrac{66}{5\sqrt{180}} \approx 0.9839$

$\theta \approx 10.3°$

62. $\text{proj}_{\mathbf{w}}\, \mathbf{v} = \dfrac{\mathbf{v} \cdot \mathbf{w}}{\|\mathbf{w}\|}$

$\text{proj}_{\mathbf{w}}\, \mathbf{v} = \dfrac{\langle -7, 5 \rangle \cdot \langle -4, 1 \rangle}{\sqrt{(-4)^2 + 1^2}} = \dfrac{33}{\sqrt{17}} = \dfrac{33\sqrt{17}}{17} \approx 8.0$

70. $W = \|\mathbf{F}\|\,\|\mathbf{s}\| \cos \alpha$

$W = 100 \cdot 25 \cdot \cos 42°$

$W \approx 1858$ foot-pounds

Exercise Set 5.1, page 328

8. $6 - \sqrt{-1} = 6 - i$

18. $(5 - 3i) - (2 + 9i) = 5 - 3i - 2 - 9i = 3 - 12i$

34. $\left(5 + 2\sqrt{-16}\right)\left(1 - \sqrt{-25}\right) = [5 + 2(4i)](1 - 5i)$

$= (5 + 8i)(1 - 5i)$

$= 5 - 25i + 8i - 40i^2$

$= 5 - 25i + 8i - 40(-1)$

$= 5 - 25i + 8i + 40$

$= 45 - 17i$

48. $\dfrac{8 - i}{2 + 3i} = \dfrac{8 - i}{2 + 3i} \cdot \dfrac{2 - 3i}{2 - 3i}$

$= \dfrac{16 - 24i - 2i + 3i^2}{2^2 + 3^2}$

$= \dfrac{16 - 24i - 2i + 3(-1)}{4 + 9}$

$= \dfrac{16 - 26i - 3}{13}$

$= \dfrac{13 - 26i}{13} = \dfrac{13(1 - 2i)}{13}$

$= 1 - 2i$

60. $\dfrac{1}{i^{83}} = \dfrac{1}{i^{80} \cdot i^3} = \dfrac{1}{i^3} = \dfrac{1}{-i}$

$= \dfrac{1}{-i} \cdot \dfrac{i}{i} = \dfrac{i}{-i^2} = \dfrac{i}{-(-1)} = \dfrac{i}{1} = i$

Exercise Set 5.2, page 336

12. $r = \sqrt{1^2 + (\sqrt{3})^2} = 2$

$\alpha = \tan^{-1}\left|\dfrac{\sqrt{3}}{1}\right| = \tan^{-1}\sqrt{3} = 60°$

$\theta = \alpha = 60°, z = 2 \text{ cis } 60°$

26. $z = 4\left(\cos \dfrac{5\pi}{3} + i \sin \dfrac{5\pi}{3}\right) = 4\left(\dfrac{1}{2} - \dfrac{\sqrt{3}}{2}i\right) = 2 - 2i\sqrt{3}$

52. $\sqrt{3} - i = 2 \text{ cis}(-30°); 1 + i\sqrt{3} = 2 \text{ cis } 60°$

$\left(\sqrt{3} - i\right)\left(1 + i\sqrt{3}\right) = 2 \text{ cis}(-30°) \cdot 2 \text{ cis } 60°$

$= 2 \cdot 2 \text{ cis}(-30° + 60°)$

$= 4 \text{ cis } 30°$

$= 4(\cos 30° + i \sin 30°)$

$= 4\left(\dfrac{\sqrt{3}}{2} + i\,\dfrac{1}{2}\right)$

$= 2\sqrt{3} + 2i$

56. $z_1 = 1 + i$

$r_1 = \sqrt{1^2 + 1^2} = \sqrt{2}$

$\alpha_1 = \tan^{-1}\left|\dfrac{1}{1}\right| = 45°; \theta_1 = 45°$

$z_1 = \sqrt{2}(\cos 45° + i \sin 45°) = \sqrt{2} \text{ cis } 45°$

$z_2 = 1 - i$

$r_2 = \sqrt{1^2 + (-1)^2} = \sqrt{2}$

$\alpha_2 = \tan^{-1}\left|\dfrac{-1}{1}\right| = 45°; \theta_2 = 315°$

$z_2 = \sqrt{2}(\cos 315° + i \sin 315°) = \sqrt{2} \text{ cis } 315°$

$\dfrac{z_1}{z_2} = \dfrac{\sqrt{2} \text{ cis } 45°}{\sqrt{2} \text{ cis } 315°}$

$= \text{cis}(-270°)$

$= \cos 270° - i \sin 270° = 0 - (-i) = i$

Exercise Set 5.3, page 342

6. $(2 \text{ cis } 330°)^4 = 2^4 \text{ cis}(4 \cdot 330°)$

$= 16 \text{ cis } 1320°$

$= 16 \text{ cis } 240°$

$= 16(\cos 240° + i \sin 240°)$

$= 16\left[-\dfrac{1}{2} + i\left(-\dfrac{\sqrt{3}}{2}\right)\right]$

$= -8 - 8i\sqrt{3}$

14. $\left(-\dfrac{\sqrt{2}}{2} + i\dfrac{\sqrt{2}}{2}\right)^{12} = [1(\cos 135° + i \sin 135°)]^{12}$

$\qquad = 1^{12}[\cos(12 \cdot 135°) + i \sin (12 \cdot 135°)]$

$\qquad = \cos 1620° + i \sin 1620°$

$\qquad = -1 + 0i, \text{ or } -1$

26. $-2 + 2i\sqrt{3} = 4(\cos 120° + i \sin 120°)$

$w_k = 4^{1/3}\left(\cos \dfrac{120° + 360°k}{3} + i \sin \dfrac{120° + 360°k}{3}\right),$

$\qquad\qquad\qquad\qquad\qquad\qquad k = 0, 1, 2$

$w_0 = 4^{1/3}\left(\cos \dfrac{120°}{3} + i \sin \dfrac{120°}{3}\right) \approx 1.216 + 1.020i$

$w_1 = 4^{1/3}\left(\cos \dfrac{120° + 360°}{3} + i \sin \dfrac{120° + 360°}{3}\right)$

$\qquad \approx -1.492 + 0.543i$

$w_2 = 4^{1/3}\left(\cos \dfrac{120° + 360° \cdot 2}{3} + i \sin \dfrac{120° + 360° \cdot 2}{3}\right)$

$\qquad \approx 0.276 - 1.563i$

28. $-1 + i\sqrt{3} = 2(\cos 120° + i \sin 120°)$

$w_k = 2^{1/2}\left(\cos \dfrac{120° + 360°k}{2} + i \sin \dfrac{120° + 360°k}{2}\right),$

$\qquad\qquad\qquad\qquad\qquad\qquad k = 0, 1$

$w_0 = 2^{1/2}(\cos 60° + i \sin 60°) = \dfrac{\sqrt{2}}{2} + \dfrac{\sqrt{6}}{2}i$

$\qquad\qquad\qquad\qquad\qquad\qquad (k = 0)$

$w_1 = 2^{1/2}\left(\cos \dfrac{120° + 360°}{2} + i \sin \dfrac{120° + 360°}{2}\right)$

$\qquad\qquad\qquad\qquad\qquad\qquad (k = 1)$

$\qquad = 2^{1/2}(\cos 240° + i \sin 240°)$

$\qquad = -\dfrac{\sqrt{2}}{2} - \dfrac{\sqrt{6}}{2}i$

Exercise Set 6.1, page 358

4. Comparing $x^2 = 4py$ with $x^2 = -\dfrac{1}{4}y$, we have $4p = -\dfrac{1}{4}$

or $p = -\dfrac{1}{16}$.

$\qquad$ vertex $(0, 0)$

$\qquad$ focus $\left(0, -\dfrac{1}{16}\right)$

$\qquad$ directrix $y = \dfrac{1}{16}$

20. $x^2 + 5x - 4y - 1 = 0$

$\qquad x^2 + 5x = 4y + 1$

$\qquad x^2 + 5x + \dfrac{25}{4} = 4y + 1 + \dfrac{25}{4}$ • Complete the square.

$\qquad \left(x + \dfrac{5}{2}\right)^2 = 4\left(y + \dfrac{29}{16}\right)$ • $h = -\dfrac{5}{2}, k = -\dfrac{29}{16}$

$\qquad\qquad 4p = 4$ • Compare to

$\qquad\qquad p = 1$ $\quad (x - h)^2 = 4p(y - k)^2.$

vertex $\left(-\dfrac{5}{2}, -\dfrac{29}{16}\right)$

focus $(h, k + p) = \left(-\dfrac{5}{2}, -\dfrac{13}{16}\right)$

directrix $y = k - p = -\dfrac{45}{16}$

28. vertex $(0, 0)$, focus $(5, 0)$; $p = 5$ because focus is $(p, 0)$.

$\qquad y^2 = 4px$

$\qquad y^2 = 4(5)x$

$\qquad y^2 = 20x$

30. vertex $(2, -3)$, focus $(0, -3)$

$\qquad (h, k) = (2, -3)$, so $h = 2$ and $k = -3$.

$\qquad$ Focus is $(h + p, k) = (2 + p, -3) = (0, -3)$.

$\qquad$ Therefore, $2 + p = 0$ and $p = -2$.

$\qquad (y - k)^2 = 4p(x - h)$

$\qquad (y + 3)^2 = 4(-2)(x - 2)$

$\qquad (y + 3)^2 = -8(x - 2)$

38. $\qquad x^2 = 4py$

$\qquad 40.5^2 = 4p(16)$

$\qquad\qquad p = \dfrac{40.5^2}{64}$

$\qquad\qquad p \approx 25.6 \text{ feet}$

Exercise Set 6.2, page 370

20. $25x^2 + 12y^2 = 300$

$\qquad \dfrac{x^2}{12} + \dfrac{y^2}{25} = 1$ $\quad$ • $a^2 = 25, b^2 = 12, c^2 = 25 - 12$

$\qquad\qquad\qquad\qquad\qquad a = 5, b = 2\sqrt{3}, c = \sqrt{13}$

center $(0, 0)$

vertices $(0, 5)$ and $(0, -5)$

foci $\left(0, \sqrt{13}\right)$ and $\left(0, -\sqrt{13}\right)$

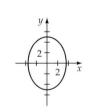

26. $\qquad 9x^2 + 16y^2 + 36x - 16y - 104 = 0$

$\qquad\qquad 9x^2 + 36x + 16y^2 - 16y - 104 = 0$

$\qquad\qquad 9(x^2 + 4x) + 16(y^2 - y) = 104$

$\qquad 9(x^2 + 4x + 4) + 16\left(y^2 - y + \dfrac{1}{4}\right) = 104 + 36 + 4$

$\qquad\qquad 9(x + 2)^2 + 16\left(y - \dfrac{1}{2}\right)^2 = 144$

$\qquad\qquad \dfrac{(x + 2)^2}{16} + \dfrac{\left(y - \dfrac{1}{2}\right)^2}{9} = 1$

center $\left(-2, \dfrac{1}{2}\right)$

$a = 4, b = 3,$

$c = \sqrt{4^2 - 3^2} = \sqrt{7}$

vertices $\left(2, \dfrac{1}{2}\right)$ and $\left(-6, \dfrac{1}{2}\right)$,

foci $\left(-2 + \sqrt{7}, \dfrac{1}{2}\right)$ and

$\left(-2 - \sqrt{7}, \dfrac{1}{2}\right)$

42. Center $(-4, 1) = (h, k)$. Therefore, $h = -4$ and $k = 1$. Length of minor axis is 8, so $2b = 8$ or $b = 4$. The equation of the ellipse is of the form

$\dfrac{(x - h)^2}{a^2} + \dfrac{(y - k)^2}{b^2} = 1$

$\dfrac{(x + 4)^2}{a^2} + \dfrac{(y - 1)^2}{16} = 1$ • $h = -4, k = 1, b = 4$

$\dfrac{(0 + 4)^2}{a^2} + \dfrac{(4 - 1)^2}{16} = 1$ • The point $(0, 4)$ is on the graph. Thus $x = 0$ and $y = 4$ satisfy the equation.

$\dfrac{16}{a^2} + \dfrac{9}{16} = 1$ • Solve for a^2.

$\dfrac{16}{a^2} = \dfrac{7}{16}$

$a^2 = \dfrac{256}{7}$

$\dfrac{(x + 4)^2}{256/7} + \dfrac{(y - 1)^2}{16} = 1$

48. Because the foci are $(0, -3)$ and $(0, 3)$, $c = 3$ and the center is $(0, 0)$, the midpoint of the line segment between $(0, -3)$ and $(0, 3)$.

$e = \dfrac{c}{a}$

$\dfrac{1}{4} = \dfrac{3}{a}$ • $e = \dfrac{1}{4}$

$a = 12$

$3^2 = 12^2 - b^2$ • $c^2 = a^2 - b^2$

$b^2 = 144 - 9 = 135$ • Solve for b^2.

The equation of the ellipse is $\dfrac{x^2}{135} + \dfrac{y^2}{144} = 1$.

56. The mean distance is $a = 67.08$ million miles.

Aphelion $= a + c = 67.58$ million miles

Thus $c = 67.58 - a = 0.50$ million miles.

$b = \sqrt{a^2 - c^2} = \sqrt{67.08^2 - 0.50^2} \approx 67.078$

An equation of the orbit of Venus is

$\dfrac{x^2}{67.08^2} + \dfrac{y^2}{67.078^2} = 1$

58. The length of the semimajor axis is 50 feet. Thus

$c^2 = a^2 - b^2$

$32^2 = 50^2 - b^2$

$b^2 = 50^2 - 32^2$

$b = \sqrt{50^2 - 32^2}$

$b \approx 38.4$ feet

Exercise Set 6.3, page 385

4. $\dfrac{y^2}{25} - \dfrac{x^2}{36} = 1$

$a^2 = 25 \quad b^2 = 36 \quad c^2 = a^2 + b^2 = 25 + 36 = 61$

$a = 5 \quad\quad b = 6 \quad\quad c = \sqrt{61}$

Transverse axis is on y-axis because y^2 term is positive.

center $(0, 0)$

foci $\left(0, \sqrt{61}\right)$ and $\left(0, -\sqrt{61}\right)$

asymptotes $y = \dfrac{5}{6}x$ and

$y = -\dfrac{5}{6}x$

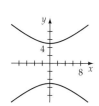

vertices $(0, 5)$ and $(0, -5)$

26. $16x^2 - 9y^2 - 32x - 54y + 79 = 0$

$16(x^2 - 2x + 1) - 9(y^2 + 6y + 9) = -79 + 16 - 81$

$= -144$

$\dfrac{(y + 3)^2}{16} - \dfrac{(x - 1)^2}{9} = 1$

Transverse axis is parallel to y-axis because y^2 term is positive. Center is at $(1, -3)$; $a^2 = 16$ so $a = 4$.

vertices $(h, k + a) = (1, 1)$

$(h, k - a) = (1, -7)$

$c^2 = a^2 + b^2 = 16 + 9 = 25$

$c = \sqrt{25} = 5$

foci $(h, k + c) = (1, 2)$

$(h, k - c) = (1, -8)$

Because $b^2 = 9$ and $b = 3$, the asymptotes are

$y + 3 = \dfrac{4}{3}(x - 1)$ and

$y + 3 = -\dfrac{4}{3}(x - 1)$.

48. Because the vertices are $(2, 3)$ and $(-2, 3)$, $a = 2$ and the center is $(0, 3)$.

$$e = \frac{c}{a} \qquad c^2 = a^2 + b^2$$

$$\qquad\qquad 5^2 = 2^2 + b^2$$

$$\frac{5}{2} = \frac{c}{2} \qquad b^2 = 25 - 4 = 21$$

$$c = 5$$

Substituting into the standard equation yields

$$\frac{x^2}{4} - \frac{(y - 3)^2}{21} = 1.$$

54. a. Because the transmitters are 300 miles apart, $2c = 300$ and $c = 150$.

$$2a = \text{rate} \times \text{time}$$

$$2a = 0.186 \times 800 = 148.8 \text{ miles}$$

Thus $a = 74.4$ miles.

$$b = \sqrt{c^2 - a^2}$$

$$= \sqrt{150^2 - 74.4^2} \approx 130.25 \text{ miles}$$

The ship is located on the hyperbola given by

$$\frac{x^2}{74.4^2} - \frac{y^2}{130.25^2} = 1$$

b. The ship will reach the coastline when $x < 0$ and $y = 0$. Thus

$$\frac{x^2}{74.4^2} - \frac{0^2}{130.25^2} = 1$$

$$\frac{x^2}{74.4^2} = 1$$

$$x^2 = 74.4^2$$

$$x = -74.4$$

The ship reaches the coastline 74.4 miles to the left of the origin at the point $(-74.4, 0)$.

Exercise Set 6.4, page 396

8.
$$xy = -10$$

$$xy + 10 = 0$$

$$A = 0, B = 1, C = 0, F = 10$$

$$\cot 2\alpha = \frac{A - C}{B} = \frac{0 - 0}{1} = 0$$

Thus $2\alpha = 90°$ and $\alpha = 45°$.

$$A' = A \cos^2 \alpha + B \cos \alpha \sin \alpha + C \sin^2 \alpha$$

$$= 0\left(\frac{\sqrt{2}}{2}\right)^2 + 1\left(\frac{\sqrt{2}}{2}\right)\left(\frac{\sqrt{2}}{2}\right) + 0\left(\frac{\sqrt{2}}{2}\right)^2 = \frac{1}{2}$$

$$C' = A \sin^2 \alpha - B \cos \alpha \sin \alpha + C \cos^2 \alpha$$

$$= 0\left(\frac{\sqrt{2}}{2}\right)^2 - 1\left(\frac{\sqrt{2}}{2}\right)\left(\frac{\sqrt{2}}{2}\right) + 0\left(\frac{\sqrt{2}}{2}\right)^2 = -\frac{1}{2}$$

$$F' = F = 10$$

$$\frac{1}{2}x'^2 - \frac{1}{2}y'^2 + 10 = 0 \text{ or } \frac{y'^2}{20} - \frac{x'^2}{20} = 1$$

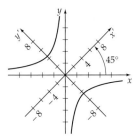

18. $x^2 + 4xy + 4y^2 - 2\sqrt{5}x + \sqrt{5}y = 0$

$$A = 1, B = 4, C = 4, D = -2\sqrt{5}, E = \sqrt{5}, F = 0$$

$$\cot 2\alpha = \frac{A - C}{B} = \frac{1 - 4}{4} = -\frac{3}{4}$$

$$\csc^2 2\alpha = \cot^2 2\alpha + 1$$

$$\csc^2 2\alpha = \left(-\frac{3}{4}\right)^2 + 1 = \frac{25}{16}$$

$$\csc 2\alpha = +\sqrt{\frac{25}{16}} = \frac{5}{4} \quad (2\alpha \text{ is in Quadrant II})$$

$$\sin 2\alpha = \frac{1}{\csc 2\alpha} = \frac{4}{5}$$

$$\sin^2 2\alpha + \cos^2 2\alpha = 1$$

$$\cos^2 2\alpha = 1 - \sin^2 2\alpha$$

$$\cos^2 2\alpha = 1 - \left(\frac{4}{5}\right)^2 = \frac{9}{25}$$

$$\cos 2\alpha = -\sqrt{\frac{9}{25}} = -\frac{3}{5} \quad (2\alpha \text{ is in Quadrant II})$$

$$\sin \alpha = \sqrt{\frac{1 - \left(-\frac{3}{5}\right)}{2}} = \frac{2\sqrt{5}}{5}$$

$$\cos \alpha = \sqrt{\frac{1 + \left(-\frac{3}{5}\right)}{2}} = \frac{\sqrt{5}}{5}$$

$$\alpha \approx 63.4°$$

$$A' = A \cos^2 \alpha + B \cos \alpha \sin \alpha + C \sin^2 \alpha$$

$$= 1\left(\frac{\sqrt{5}}{5}\right)^2 + 4\left(\frac{\sqrt{5}}{5}\right)\left(\frac{2\sqrt{5}}{5}\right) + 4\left(\frac{2\sqrt{5}}{5}\right)^2 = 5$$

$$C' = A \sin^2 \alpha - B \cos \alpha \sin \alpha + C \cos^2 \alpha$$

$$= 1\left(\frac{2\sqrt{5}}{5}\right)^2 - 4\left(\frac{\sqrt{5}}{5}\right)\left(\frac{2\sqrt{5}}{5}\right) + 4\left(\frac{\sqrt{5}}{5}\right)^2 = 0$$

$$D' = D \cos \alpha + E \sin \alpha$$

$$= -2\sqrt{5}\left(\frac{\sqrt{5}}{5}\right) + \sqrt{5}\left(\frac{2\sqrt{5}}{5}\right) = 0$$

$E' = -D \sin \alpha + E \cos \alpha$

$$= 2\sqrt{5}\left(\frac{2\sqrt{5}}{5}\right) + \sqrt{5}\left(\frac{\sqrt{5}}{5}\right) = 5$$

$5x'^2 + 5y' = 0$ or $y' = -x'^2$

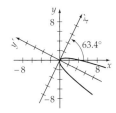

24. Use y_1 and y_2 as in Equations (11) and (12) that precede Example 4. Store the following constants.

$A = 2, B = -8, C = 8, D = 20, E = -24, F = -3$

Graph y_1 and y_2 on the same screen to produce the following parabola.

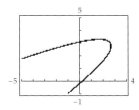

30. Because

$$B^2 - 4AC = \left(-10\sqrt{3}\right)^2 - 4(11)(1)$$
$$= 300 - 44 > 0$$

the graph is a hyperbola.

Exercise Set 6.5, page 409

14.

16. $r = 5 \cos 3\theta$
Because 3 is odd, this is a rose with three petals.

20. Because $|a| = |b| = 2$, the graph of $r = 2 - 2 \csc \theta$, $-\pi \le \theta \le \pi$, is a cardioid.

26. $r = 4 - 4 \sin \theta$

$\theta \min = 0$
$\theta \max = 2\pi$
$\theta \text{ step} = 0.1$

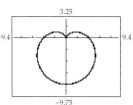

32. $r = -4 \sec \theta$

$\theta \min = 0$
$\theta \max = 2\pi$
$\theta \text{ step} = 0.1$

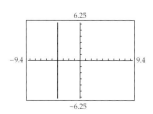

44. $x = r \cos \theta$ $\qquad\qquad$ $y = r \sin \theta$

$$= (2)\left[\cos\left(-\frac{\pi}{3}\right)\right] \qquad = (2)\left[\sin\left(-\frac{\pi}{3}\right)\right]$$

$$= (2)\left(\frac{1}{2}\right) = 1 \qquad\quad = (2)\left(-\frac{\sqrt{3}}{2}\right) = -\sqrt{3}$$

The rectangular coordinates of the point are $\left(1, -\sqrt{3}\right)$.

48. $r = \sqrt{x^2 + y^2}$ $\qquad$ $\alpha = \tan^{-1}\left|\frac{y}{x}\right|$ $\quad$ • α is the reference angle for θ.

$$= \sqrt{(12)^2 + (-5)^2}$$
$$= \sqrt{144 + 25}$$
$$= \sqrt{169} \qquad\qquad = \tan^{-1}\left|\frac{-5}{12}\right| \approx 22.6°$$
$$= 13$$

θ is a Quadrant IV angle. Thus
$\theta \approx 360° - 22.6° = 337.4°$.

The approximate polar coordinates of the point are
$(13, 337.4°)$.

56. $\qquad\qquad r = \cot \theta$

$$r = \frac{\cos\theta}{\sin\theta} \qquad • \cot\theta = \frac{\cos\theta}{\sin\theta}$$

$$r \sin\theta = \cos\theta$$

$$r(r \sin\theta) = r \cos\theta \qquad • \text{Multiply both sides by } r.$$

$$\left(\sqrt{x^2 + y^2}\right)y = x \qquad • y = r \sin\theta;\ x = r\cos\theta$$

$$(x^2 + y^2)y^2 = x^2 \qquad • \text{Square each side.}$$

$$y^4 + x^2y^2 - x^2 = 0$$

64. $\qquad\qquad 2x - 3y = 6$

$$2r\cos\theta - 3r\sin\theta = 6 \qquad • x = r\cos\theta;\ y = r\sin\theta$$

$$r(2\cos\theta - 3\sin\theta) = 6$$

$$r = \frac{6}{2\cos\theta - 3\sin\theta}$$

Exercise Set 6.6, page 416

2. $r = \dfrac{8}{2 - 4\cos\theta}$ • Divide numerator and denominator by 2.

$r = \dfrac{4}{1 - 2\cos\theta}$

$e = 2$, so the graph is a hyperbola. The transverse axis is on the polar axis because the equation involves $\cos\theta$. Let $\theta = 0$.

$r = \dfrac{8}{2 - 4\cos 0} = \dfrac{8}{2 - 4} = -4$

Let $\theta = \pi$.

$r = \dfrac{8}{2 - 4\cos\pi} = \dfrac{8}{2 + 4} = \dfrac{4}{3}$

The vertices are $(-4, 0)$ and $\left(\dfrac{4}{3}, \pi\right)$.

4. $r = \dfrac{6}{3 + 2\cos\theta}$

$r = \dfrac{2}{1 + \left(\dfrac{2}{3}\right)\cos\theta}$ • Divide numerator and denominator by 3.

$e = \dfrac{2}{3}$, so the graph is an ellipse. The major axis is on the polar axis because the equation involves $\cos\theta$. Let $\theta = 0$.

$r = \dfrac{6}{3 + 2\cos 0} = \dfrac{6}{3 + 2} = \dfrac{6}{5}$

Let $\theta = \pi$.

$r = \dfrac{6}{3 + 2\cos\pi} = \dfrac{6}{3 - 2} = 6$

The vertices on the major axis are $\left(\dfrac{6}{5}, 0\right)$ and $(6, \pi)$. Let $\theta = \dfrac{\pi}{2}$.

$r = \dfrac{6}{3 + 2\cos\left(\dfrac{\pi}{2}\right)} = \dfrac{6}{3 + 0} = 2$

Let $\theta = \dfrac{3\pi}{2}$.

$r = \dfrac{6}{3 + 2\cos\left(\dfrac{3\pi}{2}\right)} = \dfrac{6}{3 + 0} = 2$

Two additional points on the ellipse are $\left(2, \dfrac{\pi}{2}\right)$ and $\left(2, \dfrac{3\pi}{2}\right)$.

24. $e = 1$, $r\cos\theta = -2$

Thus the directrix is $x = -2$. The distance from the focus (pole) to the directrix is $d = 2$.

$r = \dfrac{ed}{1 - e\cos\theta}$ • Use Eq (2), since the vertical directrix is to the left of the pole.

$r = \dfrac{(1)(2)}{1 - (1)\cos\theta}$ • $e = 1$, $d = 2$

$r = \dfrac{2}{1 - \cos\theta}$

Exercise Set 6.7, page 424

6. Plotting points for several values of t yields the following graph.

12. $x = 3 + 2\cos t$, $y = -1 - 3\sin t$, $0 \le t < 2\pi$

$\cos t = \dfrac{x - 3}{2}$, $\sin t = -\dfrac{y + 1}{3}$

$\cos^2 t + \sin^2 t = 1$

$\left(\dfrac{x - 3}{2}\right)^2 + \left(-\dfrac{y + 1}{3}\right)^2 = 1$

$\dfrac{(x - 3)^2}{4} + \dfrac{(y + 1)^2}{9} = 1$

14. $x = 1 + t^2$, $y = 2 - t^2$, $t \in R$

$x = 1 + t^2$

$t^2 = x - 1$

$y = 2 - (x - 1)$

$y = -x + 3$

Because $x = 1 + t^2$ and $t^2 \ge 0$ for all real numbers t, $x \ge 1$ for all t. Similarly, $y \le 2$ for all t.

26. The maximum height of the cycloid is $2a = 2(3) = 6$.
The cycloid intersects the x-axis every
$2\pi a = 2\pi(3) = 6\pi$ units.

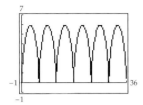

32.

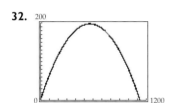

The maximum height is about 195 feet when $t \approx 3.50$ seconds. The range is 1117 feet when $t \approx 6.99$ seconds.

Exercise Set 7.1, page 444

2. $f(3) = 5^3 = 5 \cdot 5 \cdot 5 = 125$

$f(-2) = 5^{-2} = \dfrac{1}{5^2} = \dfrac{1}{5 \cdot 5} = \dfrac{1}{25}$

22. The graph of $f(x) = \left(\dfrac{5}{2}\right)^x$ has a y-intercept of $(0, 1)$

and the graph passes through $\left(1, \dfrac{5}{2}\right)$. Plot a few addi-

tional points, such as $\left(-1, \dfrac{2}{5}\right)$ and $\left(2, \dfrac{25}{4}\right)$. Because the

base $\dfrac{5}{2}$ is greater than 1, we know that the graph must

have all the properties of an increasing exponential function. Draw a smooth increasing curve through the points. The graph should be asymptotic to the negative portion of the x-axis, as shown in the following figure.

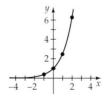

28. Because $F(x) = 6^{x+5} = f(x + 5)$, the graph of $F(x)$ can be produced by shifting the graph of f horizontally to the left 5 units.

30. Because $F(x) = -\left[\left(\dfrac{5}{2}\right)^x\right] = -f(x)$, the graph of $F(x)$ can be produced by reflecting the graph of f across the x-axis.

44. a. $A(45) = 200e^{-0.014(45)}$

≈ 106.52

After 45 minutes the patient will have about 107 milligrams of medication in his or her bloodstream.

b. Use a graphing calculator to graph $y = 200e^{-0.014x}$ and $y = 50$ in the same viewing window as shown below.

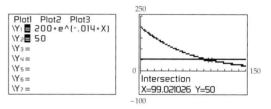

The x-coordinate (which represents time in minutes) of the point of intersection is about 99.02. Thus it will take about 99 minutes before the patient's medication level is reduced to 50 milligrams.

54. a. $P(0) = \dfrac{3600}{1 + 7e^{-0.05(0)}}$

$= \dfrac{3600}{1 + 7}$

$= \dfrac{3600}{8}$

$= 450$

Immediately after the lake was stocked, the lake contained 450 bass.

b. $P(12) = \dfrac{3600}{1 + 7e^{-0.05(12)}}$

≈ 743.54

After 1 year (12 months) there were about 744 bass in the lake.

c. As $t \to \infty$, $7e^{-0.05t} = \dfrac{7}{e^{0.05t}}$ approaches 0. Thus as $t \to \infty$,

$P(t) = \dfrac{3600}{1 + 7e^{-0.05t}}$ will approach $\dfrac{3600}{1 + 0} = 3600$. As

time goes by the bass population will increase, approaching 3600.

Exercise Set 7.2, page 458

4. The exponential form of $\log_b x = y$ is $b^y = x$. Thus the exponential form of $\log_4 64 = 3$ is $4^3 = 64$.

12. The logarithmic form of $b^y = x$ is $y = \log_b x$. Thus the logarithmic form of $5^3 = 125$ is $3 = \log_5 125$.

28. $\log 1{,}000{,}000 = \log_{10} 10^6 = 6$

32. To graph $y = \log_6 x$, use the equivalent exponential
equation $x = 6^y$. Choose some y-values, such as -1, 0, 1,
and calculate the corresponding x-values. This yields the
ordered pairs $\left(\dfrac{1}{6}, -1\right)$, $(1, 0)$, and $(6, 1)$. Plot these
ordered pairs and draw a smooth curve through the
points to produce the following graph.

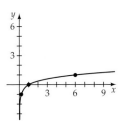

40. $\log_4(5 - x)$ is defined only for $5 - x > 0$, which is
equivalent to $x < 5$. Using interval notation, the domain
of $k(x) = \log_4(5 - x)$ is $(-\infty, 5)$.

50. The graph of $f(x) = \log_6(x + 3)$ can be produced by shifting the graph of $f(x) = \log_6 x$ (from Exercise 32) 3 units
to the left. See the following figure.

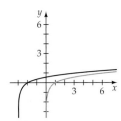

70. a. $S(0) = 5 + 29\ln(0 + 1) = 5 + 0 = 5$. When starting, the
student had an average typing speed of 5 words
per minute. $S(3) = 5 + 29\ln(3 + 1) \approx 45.2$. After
3 months the student's average typing speed was
about 45 words per minute.

b. Use the intersection feature of a graphing utility to
find the x-coordinate of the point of intersection of
the graphs of $y = 5 + 29\ln(x + 1)$ and $y = 65$.

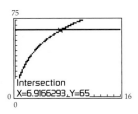

The graphs intersect at about $(6.9, 65)$. The student
will achieve a typing speed of 65 words per minute in
about 6.9 months.

Exercise Set 7.3, page 471

2. $\ln \dfrac{z^3}{\sqrt{xy}} = \ln z^3 - \ln \sqrt{xy}$

$\qquad = \ln z^3 - \ln(xy)^{1/2}$

$\qquad = 3\ln z - \dfrac{1}{2}\ln(xy)$

$\qquad = 3\ln z - \dfrac{1}{2}(\ln x + \ln y)$

$\qquad = 3\ln z - \dfrac{1}{2}\ln x - \dfrac{1}{2}\ln y$

10. $3\log_2 t - \dfrac{1}{3}\log_2 u + 4\log_2 v = \log_2 t^3 - \log_2 u^{1/3} + \log_2 v^4$

$\qquad\qquad = \log_2 \dfrac{t^3}{u^{1/3}} + \log_2 v^4$

$\qquad\qquad = \log_2 \dfrac{t^3 v^4}{u^{1/3}}$

$\qquad\qquad = \log_2 \dfrac{t^3 v^4}{\sqrt[3]{u}}$

16. $\log_5 37 = \dfrac{\log 37}{\log 5} \approx 2.2436$

24. $\log_8(5 - x) = \dfrac{\ln(5 - x)}{\ln 8}$, so enter $\dfrac{\ln(5 - x)}{\ln 8}$ into Y1 on a
graphing calculator.

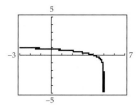

48. $\text{pH} = -\log[H^+] = -\log(1.26 \times 10^{-3}) \approx 2.9$

50. $\text{pH} = -\log[H^+]$

$\quad\; 5.6 = -\log[H^+]$

$\;\, -5.6 = \log[H^+]$

$\;\, 10^{-5.6} = H^+$

The hydronium-ion concentration is $10^{-5.6} \approx 2.51 \times 10^{-6}$
mole per liter.

56. $M = \log\left(\dfrac{I}{I_0}\right) = \log\left(\dfrac{398{,}107{,}000 I_0}{I_0}\right) = \log(398{,}107{,}000)$

$\qquad \approx 8.6$

58. $\log\left(\dfrac{I}{I_0}\right) = 9.5$

$\dfrac{I}{I_0} = 10^{9.5}$

$I = 10^{9.5}I_0$

$I \approx 3{,}162{,}277{,}660 I_0$

60. In Example 7 we noticed that if an earthquake has a Richter scale magnitude of M_1 and a smaller earthquake has a Richter scale magnitude of M_2, then the first earthquake is $10^{M_1 - M_2}$ times as intense as the smaller earthquake. In this exercise, $M_1 = 9.5$ and $M_2 = 8.3$. Thus $10^{M_1 - M_2} = 10^{9.5 - 8.3} = 10^{1.2} \approx 15.8$. The 1960 earthquake in Chile was about 15.8 times as intense as the San Francisco earthquake of 1906.

64. $M = \log A + 3\log 8t - 2.92$

$\quad = \log 26 + 3\log[8 \cdot 17] - 2.92 \qquad$ • Substitute 26 for

$\quad \approx 1.4150 + 6.4006 - 2.92 \qquad\qquad$ A and 17 for t.

$\quad \approx 4.9$

Exercise Set 7.4, page 483

2. $3^x = 243$

$3^x = 3^5$

$x = 5$

10. $\qquad 6^x = 50$

$\log(6^x) = \log 50$

$x \log 6 = \log 50$

$\qquad x = \dfrac{\log 50}{\log 6} \approx 2.18$

18. $\qquad\qquad 3^{x-2} = 4^{2x+1}$

$\qquad\qquad \log 3^{x-2} = \log 4^{2x+1}$

$\qquad\qquad (x-2)\log 3 = (2x+1)\log 4$

$\qquad\qquad x\log 3 - 2\log 3 = 2x\log 4 + \log 4$

$\qquad x\log 3 - 2\log 3 - 2x\log 4 = \log 4$

$\qquad\qquad x\log 3 - 2x\log 4 = \log 4 + 2\log 3$

$\qquad\qquad x(\log 3 - 2\log 4) = \log 4 + 2\log 3$

$\qquad\qquad\qquad x = \dfrac{\log 4 + 2\log 3}{\log 3 - 2\log 4}$

$\qquad\qquad\qquad x \approx -2.141$

22. $\log(x^2 + 19) = 2$

$\qquad x^2 + 19 = 10^2$

$\qquad x^2 + 19 = 100$

$\qquad\qquad x^2 = 81$

$\qquad\qquad x = \pm 9$

A check shows that 9 and -9 are both solutions of the original equation.

26. $\log_3 x + \log_3(x + 6) = 3$

$\qquad \log_3[x(x + 6)] = 3$

$\qquad\qquad 3^3 = x(x + 6)$

$\qquad\qquad 27 = x^2 + 6x$

$\qquad x^2 + 6x - 27 = 0$

$\qquad (x + 9)(x - 3) = 0$

$x = -9 \quad$ or $\quad x = 3$

Because $\log_3 x$ is defined only for $x > 0$, the only solution is $x = 3$.

36. $\ln x = \dfrac{1}{2}\ln\left(2x + \dfrac{5}{2}\right) + \dfrac{1}{2}\ln 2$

$\qquad = \dfrac{1}{2}\left[\ln\left(2x + \dfrac{5}{2}\right) + \ln 2\right]$

$\ln x = \dfrac{1}{2}\ln\left[2\left(2x + \dfrac{5}{2}\right)\right]$

$\ln x = \dfrac{1}{2}\ln(4x + 5)$

$\ln x = \ln(4x + 5)^{1/2}$

$\qquad x = \sqrt{4x + 5}$

$\qquad x^2 = 4x + 5$

$\qquad 0 = x^2 - 4x - 5$

$\qquad 0 = (x - 5)(x + 1)$

$x = 5 \quad$ or $\quad x = -1$

Check: $\quad \ln 5 = \dfrac{1}{2}\ln\left(10 + \dfrac{5}{2}\right) + \dfrac{1}{2}\ln 2$

$\qquad\qquad 1.6094 \approx 1.2629 + 0.3466$

Because $\ln(-1)$ is not defined, -1 is not a solution. Thus the only solution is $x = 5$.

40. $\qquad \dfrac{10^x + 10^{-x}}{2} = 8$

$\qquad 10^x + 10^{-x} = 16$

$\qquad 10^x(10^x + 10^{-x}) = (16)10^x \qquad$ • Multiply each side

$\qquad\qquad 10^{2x} + 1 = 16(10^x) \qquad$ by 10^x.

$\qquad 10^{2x} - 16(10^x) + 1 = 0$

$\qquad\qquad u^2 - 16u + 1 = 0 \qquad$ • Let $u = 10^x$.

$\qquad\qquad u = \dfrac{16 \pm \sqrt{16^2 - 4(1)(1)}}{2} = 8 \pm 3\sqrt{7}$

$\qquad\qquad 10^x = 8 \pm 3\sqrt{7} \qquad$ • Replace u with 10^x.

$\qquad \log 10^x = \log(8 \pm 3\sqrt{7})$

$\qquad\qquad x = \log(8 \pm 3\sqrt{7}) \approx \pm 1.20241$

68. a.
$$t = \frac{9}{24}\ln\frac{24+v}{24-v}$$

$$1.5 = \frac{9}{24}\ln\frac{24+v}{24-v}$$

$$4 = \ln\frac{24+v}{24-v}$$

$$e^4 = \frac{24+v}{24-v} \qquad \bullet\ N = \ln M \text{ means } e^N = M.$$

$$(24-v)e^4 = 24+v$$

$$-v - ve^4 = 24 - 24e^4$$

$$v(-1 - e^4) = 24 - 24e^4$$

$$v = \frac{24 - 24e^4}{-1 - e^4} \approx 23.14$$

The velocity is about 23.14 feet per second.

b. The vertical asymptote is $v = 24$.

c. Due to the air resistance, the object cannot reach or exceed a velocity of 24 feet per second.

Exercise Set 7.5, page 498

4. a. $P = 12{,}500$, $r = 0.08$, $t = 10$, $n = 1$.
$$A = 12{,}500\left(1 + \frac{0.08}{1}\right)^{10} \approx \$26{,}986.56$$

b. $n = 365$
$$A = 12{,}500\left(1 + \frac{0.08}{365}\right)^{3650} \approx \$27{,}816.82$$

c. $n = 8760$
$$A = 12{,}500\left(1 + \frac{0.08}{8760}\right)^{87600} \approx \$27{,}819.16$$

6. $P = 32{,}000$, $r = 0.08$, $t = 3$.
$$A = Pe^{rt} = 32{,}000e^{3(0.08)} \approx \$40{,}679.97$$

10. $t = \dfrac{\ln 3}{r} \qquad r = 0.055$

$$t = \frac{\ln 3}{0.055}$$

$t \approx 20$ years (to the nearest year)

18. a. $P(12) = 20{,}899(1.027)^{12} \approx 28{,}722$ thousands, or $28{,}772{,}000$.

b. P is in thousands, so
$$35{,}000 = 20{,}899(1.027)^t$$

$$\frac{35{,}000}{20{,}899} = 1.027^t$$

$$\ln\left(\frac{35{,}000}{20{,}899}\right) = t\ln 1.027$$

$$\frac{\ln\left(\dfrac{35{,}000}{20{,}899}\right)}{\ln 1.027} = t$$

$$19.35 \approx t$$

According to the growth function, the population will first exceed 35 million in 19.35 years—that is, in the year $1991 + 19 = 2010$.

20.
$$N(t) = N_0 e^{kt}$$

$$N(138) = N_0 e^{138k}$$

$$0.5N_0 = N_0 e^{138k}$$

$$0.5 = e^{138k}$$

$$\ln 0.5 = 138k$$

$$k = \frac{\ln 0.5}{138} \approx -0.005023$$

$$N(t) = N_0(0.5)^{t/138} \approx N_0 e^{-0.005023t}$$

24.
$$N(t) = N_0(0.5)^{t/5730}$$

$$0.65N_0 = N_0(0.5)^{t/5730}$$

$$0.65 = (0.5)^{t/5730}$$

$$\ln 0.65 = \ln(0.5)^{t/5730}$$

$$t = 5730\frac{\ln 0.65}{\ln 0.5} \approx 3600$$

The bone is approximately 3600 years old.

32. a.

b. Here is an algebraic solution. An approximate solution can be obtained from the graph.

$$v = 64(1 - e^{-t/2})$$

$$50 = 64(1 - e^{-t/2})$$

$$\frac{50}{64} = (1 - e^{-t/2})$$

$$1 - \frac{50}{64} = e^{-t/2}$$

$$\ln\left(1 - \frac{50}{64}\right) = -\frac{t}{2}$$

$$t = -2\ln\left(1 - \frac{50}{64}\right) \approx 3.0$$

The velocity is 50 feet per second in approximately 3.0 seconds.

c. As $t \to \infty$, $e^{-t/2} \to 0$. Therefore, $64(1 - e^{-t/2}) \to 64$. The horizontal asymptote is $v = 64$.

d. Because of the air resistance, the velocity of the object will never reach or exceed 64 feet per second.

48. a. Represent the year 2001 by $t = 0$; then the year 2002 will be represented by $t = 1$. Use the following substitutions: $P_0 = 240$, $P(1) = 310$, $c = 3400$, and

$$a = \frac{c - P_0}{P_0} = \frac{3400 - 240}{240} \approx 13.16667.$$

$$P(t) = \frac{c}{1 + ae^{-bt}}$$

$$P(1) = \frac{3400}{1 + 13.16667e^{-b(1)}}$$

$$310 = \frac{3400}{1 + 13.16667e^{-b}}$$

$$310(1 + 13.16667e^{-b}) = 3400$$

$$1 + 13.16667e^{-b} = \frac{3400}{310}$$

$$13.16667e^{-b} = \frac{3400}{310} - 1$$

$$13.16667e^{-b} \approx 9.96774$$

$$e^{-b} \approx \frac{9.96774}{13.16667}$$

$$-b \approx \ln \frac{9.96774}{13.16667}$$

$$b \approx 0.27833$$

Using $a = 13.16667$, $b = 0.27833$, and $c = 3400$ gives the following logistic model.

$$P(t) \approx \frac{3400}{1 + 13.16667e^{-0.27833t}}$$

b. Because 2008 is 7 years past 2001, the year 2008 is represented by $t = 7$.

$$P(7) \approx \frac{3400}{1 + 13.16667e^{-0.27833(7)}} \approx 1182$$

According to the model there will be about 1182 groundhogs in 2008.

Exercise Set 7.6, page 512

4. The following scatter plot suggests that the data can be modeled by an increasing function that is concave down. Thus the most suitable model for the data is an increasing logarithmic function.

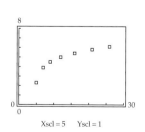

Xscl = 5   Yscl = 1

22. From the scatter plot in the following figure, it appears that the data can be closely modeled by a decreasing exponential function of the form $y = ab^x$, with $a > 0$ and $b < 1$.

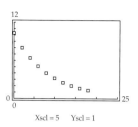

Xscl = 5   Yscl = 1

The calculator display in the following figure shows that the exponential regression equation is $y \approx 10.1468(0.89104)^x$, where x is the altitude in kilometers and y is the pressure in newtons per square centimeter.

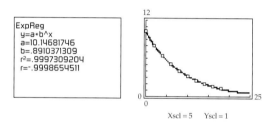

Xscl = 5   Yscl = 1

The correlation coefficient $r \approx -0.99987$ is close to -1. This indicates that the function $y \approx 10.1468(0.89104)^x$ provides a good fit for the data. The graph of y shown above also indicates that the regression function provides a good model for the data. When $x = 24$ kilometers, the atmospheric pressure is about $10.1468(0.89104)^{24} \approx 0.6$ newton per square centimeter.

24. a. Use a graphing utility to perform an exponential regression and a logarithmic regression. For the given data, the logarithmic function $y = 61.735786 - 4.1044761 \ln x$ provides a slightly better fit than does the exponential regression function, as determined by comparing the correlation coefficients. See the calculator displays below.

Xscl = 10   Yscl = 1

b. To predict the world record time in 2008, evaluate $y = 61.735786 - 4.104476 \ln x$ at $x = 108$. The graph on the right above shows that the predicted world record time in the men's 400-meter race for the year 2008 is about 42.52 seconds.

26. a. Use a graphing utility to perform a logistic regression on the data. The following figure shows the results obtained by using a TI-83 graphing calculator.

```
Logistic
 y=c/(1+ae^(-bx))
 a=2.245804979
 b=.0434109644
 c=1541897.467
```

The logistic regression function for the data is

$$P(t) \approx \frac{1,541,897}{1 + 2.24580e^{-0.043411t}}.$$

b. The year 2010 is represented by $t = 60$.

$$P(60) = \frac{1,541,897}{1 + 2.24580e^{-0.043411(60)}} \approx 1,320,000$$

The logistic regression function predicts that Hawaii's population will be about 1,320,000 in 2010.

c. The carrying capacity, to the nearest thousand, of the logistic model is 1,542,000 people.

ANSWERS TO SELECTED EXERCISES

Exercise Set 1.1, page 12

1. 15 **3.** −4 **5.** $\frac{9}{2}$ **7.** $\frac{2}{9}$ **9.** 12 **11.** 16 **13.** 75 **15.** $\frac{1}{2}$ **17.** 1200 **19.** −3, 5 **21.** $\frac{-1 \pm \sqrt{5}}{2}$ **23.** $\frac{-2 \pm \sqrt{2}}{2}$

25. $\frac{5 \pm \sqrt{61}}{6}$ **27.** $\frac{-3 \pm \sqrt{41}}{4}$ **29.** $-\frac{\sqrt{2}}{2}, -\sqrt{2}$ **31.** $\frac{3 \pm \sqrt{29}}{2}$ **33.** −3, 5 **35.** $\frac{3}{8}, -24$ **37.** $0, \frac{7}{3}$ **39.** 2, 8 **41.** $(-\infty, 4)$

43. $(-\infty, -6)$ **45.** $(-\infty, -3]$ **47.** $\left[-\frac{13}{8}, \infty\right)$ **49.** $(-\infty, 2)$ **51.** $(-\infty, -7) \cup (0, \infty)$ **53.** $(-5, -2)$ **55.** $(-\infty, 4] \cup [7, \infty)$

57. $\left[-\frac{1}{2}, \frac{4}{3}\right]$ **59.** $(-4, 4)$ **61.** $(-8, 10)$ **63.** $(-\infty, -33) \cup (27, \infty)$ **65.** $\left(-\infty, -\frac{3}{2}\right) \cup \left(\frac{5}{2}, \infty\right)$ **67.** $(-\infty, -8] \cup [2, \infty)$

69. $\left[-\frac{4}{3}, 8\right]$ **71.** $(-\infty, -4] \cup \left[\frac{28}{5}, \infty\right)$ **73.** $(-\infty, \infty)$ **75.** 4 **77.** 3.5 centimeters by 10 centimeters **79.** 100 feet by 150 feet

81. At least 58 checks **83.** At least 34 sales **85.** $20 \le C \le 40$ **87.** 22 **89.** $(0, 210)$ **91.** 1 second $< t < 3$ seconds

93. a. $|s - 4.25| \le 0.01$ **b.** $4.24 < s < 4.26$

Prepare for Section 1.2, page 14

94. $-\frac{3}{2}$ **95.** $5\sqrt{2}$ **96.** No **97.** 16 **98.** 2 **99.** 5

Exercise Set 1.2, page 26

1. 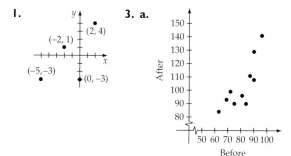 **3. a.** 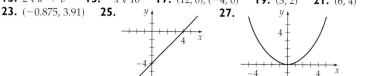 **b.** 23.4 beats per minute **5.** $7\sqrt{5}$ **7.** $\sqrt{1261}$ **9.** $\sqrt{89}$ **11.** $\sqrt{38 - 12\sqrt{6}}$
13. $2\sqrt{a^2 + b^2}$ **15.** $-x\sqrt{10}$ **17.** $(12, 0), (-4, 0)$ **19.** $(3, 2)$ **21.** $(6, 4)$
23. $(-0.875, 3.91)$ **25.** **27.**

29. 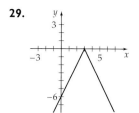 **31.** **33.** 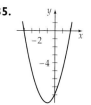 **35.** 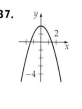 **37.**

39. $(6, 0), \left(0, \frac{12}{5}\right)$ 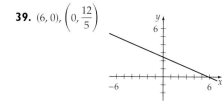 **41.** $(5, 0); \left(0, \sqrt{5}\right), \left(0, -\sqrt{5}\right)$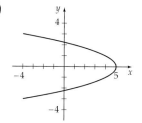

43. $(-4, 0)$; $(0, 4)$, $(0, -4)$ **45.** $(\pm 2, 0)$, $(0, \pm 2)$ **47.** $(\pm 4, 0)$, $(0, \pm 4)$ **49.** center $(0, 0)$, radius 6

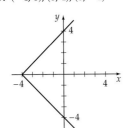   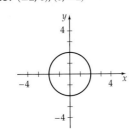   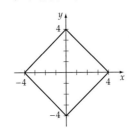

51. center $(1, 3)$, radius 7   **53.** center $(-2, -5)$, radius 5   **55.** center $(8, 0)$, radius $\dfrac{1}{2}$ **57.** $(x - 4)^2 + (y - 1)^2 = 2^2$

59. $\left(x - \dfrac{1}{2}\right)^2 + \left(y - \dfrac{1}{4}\right)^2 = \left(\sqrt{5}\right)^2$ **61.** $(x - 0)^2 + (y - 0)^2 = 5^2$ **63.** $(x - 1)^2 + (y - 3)^2 = 5^2$ **65.** center $(3, 0)$, radius 2

67. center $(7, -4)$, radius 3   **69.** center $\left(-\dfrac{1}{2}, 0\right)$, radius 4   **71.** center $\left(\dfrac{1}{2}, -\dfrac{3}{2}\right)$, radius $\dfrac{5}{2}$ **73.** $(x + 1)^2 + (y - 7)^2 = 25$

75. $(x - 7)^2 + (y - 11)^2 = 121$ **77.**   **79.**   **81.**   **83.**   **85.**

87. $(13, 5)$ **89.** $(7, -6)$ **91.** $x^2 - 6x + y^2 - 8y = 0$ **93.** $9x^2 + 25y^2 = 225$ **95.** $(x + 3)^2 + (y - 3)^2 = 3^2$

Prepare for Section 1.3, page 28

97. -4 **98.** $D = \{-3, -2, -1, 0, 2\}$; $R = \{1, 2, 4, 5\}$ **99.** $\sqrt{58}$ **100.** $x \geq 3$ **101.** $-2, 3$ **102.** 13

Exercise Set 1.3, page 42

1. a. 5   **b.** -4 **c.** -1 **d.** 1   **e.** $3k - 1$ **f.** $3k + 5$ **3. a.** $\sqrt{5}$ **b.** 3   **c.** 3   **d.** $\sqrt{21}$ **e.** $\sqrt{r^2 + 2r + 6}$ **f.** $\sqrt{c^2 + 5}$

5. a. $\dfrac{1}{2}$ **b.** $\dfrac{1}{2}$ **c.** $\dfrac{5}{3}$ **d.** 1   **e.** $\dfrac{1}{c^2 + 4}$ **f.** $\dfrac{1}{|2 + h|}$ **7. a.** 1   **b.** 1   **c.** -1 **d.** -1 **e.** 1   **f.** -1 **9. a.** -11 **b.** 6

c. $3c + 1$ **d.** $-k^2 - 2k + 10$ **11.** Yes   **13.** No   **15.** No   **17.** Yes   **19.** No   **21.** Yes   **23.** Yes   **25.** Yes   **27.** all real numbers

29. all real numbers   **31.** $\{x \mid x \neq -2\}$ **33.** $\{x \mid x \geq -7\}$ **35.** $\{x \mid -2 \leq x \leq 2\}$ **37.** $\{x \mid x > -4\}$

39.   **41.**   **43.**   **45.**

47. a. $1.05   **b.** 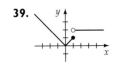   **49.** a, b, and d.   **51.** decreasing on $(-\infty, 0]$; increasing on $[0, \infty)$

53. increasing on $(-\infty, \infty)$ **55.** decreasing on $(-\infty, -3]$; increasing on $[-3, 0]$; decreasing on $[0, 3]$; increasing on $[3, \infty)$ **57.** constant on $(-\infty, 0]$; increasing on $[0, \infty)$

59. decreasing on $(-\infty, 0]$; constant on $[0, 1]$; increasing on $[1, \infty)$ **61.** g and F

63. a. $w = 25 - l$ **b.** $A = 25l - l^2$ **65.** $v(t) = 80{,}000 - 6500t, 0 \leq t \leq 10$

67. a. $C(x) = 2000 + 22.80x$ **b.** $R(x) = 37.00x$ **c.** $P(x) = 14.20x - 2000$

69. $h = 15 - 5r$ **71.** $d = \sqrt{(3t)^2 + 50^2}$ **73.** $d = \sqrt{(45 - 8t)^2 + (6t)^2}$

75. a. $L(x) = \left(\dfrac{1}{4\pi} + \dfrac{1}{16}\right)x^2 - \dfrac{5}{2}x + 25$ **b.** 25, 17.27, 14.09, 15.46, 21.37, 31.83   **c.** $[0, 20]$

77. a. $A(x) = \sqrt{900 + x^2} + \sqrt{400 + (40 - x)^2}$ **b.** 74.72, 67.68, 64.34, 64.79, 70   **c.** $[0, 40]$ **79.** 275, 375, 385, 390, 394

81. $c = -2$ or $c = 3$ **83.** 1 is not in the range of f.

85. 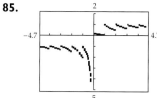 **87.** 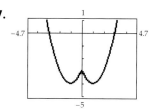 **89.** 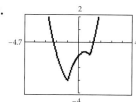 **91.** 4 **93.** 2 **95. a.** 36
b. 13 **c.** 12 **d.** 30
e. $13k - 2$ **f.** $8k - 11$
97. $4\sqrt{21}$ **99.** 1, -3

101.

Prepare for Section 1.4, page 49

105. 3, -1, -3, -3, -1 **106.** $(0, b)$ **107.** $(0, 0)$

Exercise Set 1.4, page 60

1. 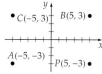 **3.** 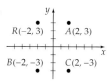 **5.** 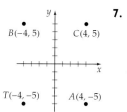 **7.**   **9.**   **11.**

13. a. No **b.** Yes **15. a.** No **b.** No **17. a.** Yes **b.** Yes **19. a.** Yes **b.** Yes **21. a.** Yes **b.** Yes **23.** No **25.** Yes
27. Yes **29.** Yes **31.**   **33.**   **35.**   **37.**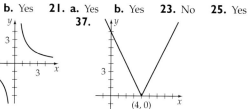

39. 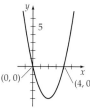 **41.**   **43.** even **45.** odd **47.** even **49.** even **51.** even **53.** even **55.** neither
57. a., b. 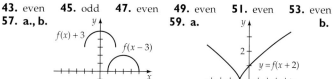 **59. a.** 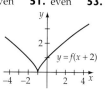 **b.**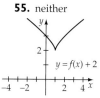

61. a. $(-5, 5), (-3, -2), (-2, 0)$ **b.** $(-2, 6), (0, -1), (1, 1)$ **63. a.** 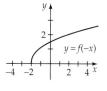 **b.**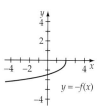

65. a. $(1, 3), (-2, -4)$ **b.** $(-1, -3), (2, 4)$ **67. a., b.**   **69.**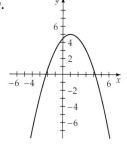

71. a.

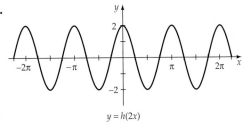

b.

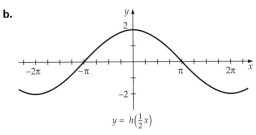

73. a.

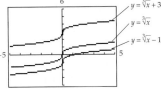

$$y = h(2x)$$

b.

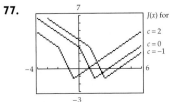

$$y = h\left(\tfrac{1}{2}x\right)$$

75.

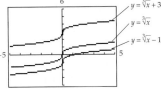

77.

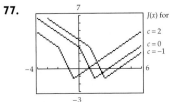

79.

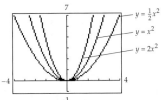

81.

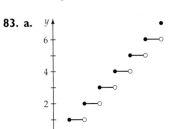

83. a.

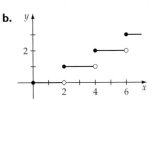

b.

c.

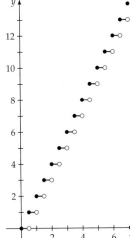

85. a. $f(x) = \dfrac{2}{(x+1)^2 + 1} + 1$ **b.** $f(x) = -\dfrac{2}{(x-2)^2 + 1}$

Prepare for Section 1.5, page 64

87. $x^2 + 1$ **88.** $6x^3 - 11x^2 + 7x - 6$ **89.** $18a^2 - 15a + 2$ **90.** $2h^2 + 3h$ **91.** all real numbers except $x = 1$ **92.** $[4, \infty)$

Exercise Set 1.5, page 73

1. $f(x) + g(x) = x^2 - x - 12$, Domain is the set of all real numbers.
$f(x) - g(x) = x^2 - 3x - 18$, Domain is the set of all real numbers.
$f(x) \cdot g(x) = x^3 + x^2 - 21x - 45$, Domain is the set of all real numbers.
$\dfrac{f(x)}{g(x)} = x - 5$, Domain $\{x \mid x \neq -3\}$

3. $f(x) + g(x) = 3x + 12$, Domain is the set of all real numbers.
$f(x) - g(x) = x + 4$, Domain is the set of all real numbers.
$f(x) \cdot g(x) = 2x^2 + 16x + 32$, Domain is the set of all real numbers.
$\dfrac{f(x)}{g(x)} = 2$, Domain $\{x \mid x \neq -4\}$

5. $f(x) + g(x) = x^3 - 2x^2 + 8x$, Domain is the set of all real numbers.
$f(x) - g(x) = x^3 - 2x^2 + 6x$, Domain is the set of all real numbers.
$f(x) \cdot g(x) = x^4 - 2x^3 + 7x^2$, Domain is the set of all real numbers.
$\dfrac{f(x)}{g(x)} = x^2 - 2x + 7$, Domain $\{x \mid x \neq 0\}$

7. $f(x) + g(x) = 4x^2 + 7x - 12$, Domain is the set of all real numbers.
$f(x) - g(x) = x - 2$, Domain is the set of all real numbers.
$f(x) \cdot g(x) = 4x^4 + 14x^3 - 12x^2 - 41x + 35$, Domain is the set of all real numbers.
$\dfrac{f(x)}{g(x)} = 1 + \dfrac{x - 2}{2x^2 + 3x - 5}$, Domain $\left\{ x \mid x \neq 1, x \neq -\dfrac{5}{2} \right\}$

9. $f(x) + g(x) = \sqrt{x - 3} + x$, Domain $\{x \mid x \geq 3\}$
$f(x) - g(x) = \sqrt{x - 3} - x$, Domain $\{x \mid x \geq 3\}$
$f(x) \cdot g(x) = x\sqrt{x - 3}$, Domain $\{x \mid x \geq 3\}$
$\dfrac{f(x)}{g(x)} = \dfrac{\sqrt{x - 3}}{x}$, Domain $\{x \mid x \geq 3\}$

11. $f(x) + g(x) = \sqrt{4 - x^2} + 2 + x$, Domain $\{x \mid -2 \leq x \leq 2\}$
$f(x) - g(x) = \sqrt{4 - x^2} - 2 - x$, Domain $\{x \mid -2 \leq x \leq 2\}$
$f(x) \cdot g(x) = (\sqrt{4 - x^2})(2 + x)$, Domain $\{x \mid -2 \leq x \leq 2\}$
$\dfrac{f(x)}{g(x)} = \dfrac{\sqrt{4 - x^2}}{2 + x}$ Domain $\{x \mid -2 < x \leq 2\}$

13. 18 **15.** $-\dfrac{9}{4}$ **17.** 30 **19.** 12 **21.** 300 **23.** $-\dfrac{384}{125}$ **25.** $-\dfrac{5}{2}$ **27.** $-\dfrac{1}{4}$ **29.** 2 **31.** $2x + h$ **33.** $4x + 2h + 4$

35. $-8x - 4h$ **37.** $(g \circ f)(x) = 6x + 3$ **39.** $(g \circ f)(x) = x^2 + 4x + 1$ **41.** $(g \circ f)(x) = -5x^3 - 10x$ **43.** $(g \circ f)(x) = \dfrac{1 - 5x}{x + 1}$
$(f \circ g)(x) = 6x - 16$ $(f \circ g)(x) = x^2 + 8x + 11$ $(f \circ g)(x) = -125x^3 - 10x$ $(f \circ g)(x) = \dfrac{2}{3x - 4}$

45. $(g \circ f)(x) = \dfrac{\sqrt{1 - x^2}}{|x|}$ **47.** $(g \circ f)(x) = -\dfrac{2|5 - x|}{3}$ **49.** 66 **51.** 51 **53.** -4 **55.** 41 **57.** $-\dfrac{3848}{625}$ **59.** $6 + 2\sqrt{3}$
$(f \circ g)(x) = \dfrac{1}{x - 1}$ $(f \circ g)(x) = \dfrac{3|x|}{|5x + 2|}$

61. $16c^2 + 4c - 6$ **63.** $9k^4 + 36k^3 + 45k^2 + 18k - 4$ **65. a.** $A(t) = \pi(1.5t)^2$, $A(2) = 9\pi$ square feet ≈ 28.27 square feet **b.** $V(t) = 2.25\pi t^3$, $V(3) = 60.75\pi$ cubic feet ≈ 190.85 cubic feet **67. a.** $d(t) = \sqrt{(48 - t)^2 - 4^2}$ **b.** $s(35) = 13$ feet, $d(35) \approx 12.37$ feet **69.** $(Y \circ F)(x)$ converts x inches to yards. **71. a.** 99.8; this is identical to the slope of the line through $(0, C(0))$ and $(1, C(1))$. **b.** 156.2 **c.** -49.7 **d.** -30.8 **e.** -16.4 **f.** 0

Prepare for Section 1.6, page 76

83. $y = -\dfrac{2}{5}x + 3$ **84.** $y = \dfrac{1}{x - 1}$ **85.** -1 **86.** $>$ **87.** $b \leq 0$ **88.** $\{x \mid x \geq -2\}$

Exercise Set 1.6, page 87

1. 3 **3.** -3 **5.** 3 **7.** range **9.** Yes **11.** Yes **13.** Yes

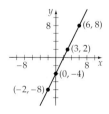

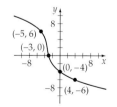

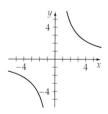

15. No

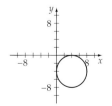

17. Yes **19.** Yes **21.** No **23.** $\{(1, -3), (2, -2), (5, 1), (-7, 4)\}$ **25.** $\{(1, 0), (2, 1), (4, 2), (8, 3), (16, 4)\}$

27. $f^{-1}(x) = \dfrac{1}{2}x - 2$ **29.** $f^{-1}(x) = \dfrac{1}{3}x + \dfrac{7}{3}$ **31.** $f^{-1}(x) = -\dfrac{1}{2}x + \dfrac{5}{2}$ **33.** $f^{-1}(x) = \dfrac{x}{x-2}, x \neq 2$ **35.** $f^{-1}(x) = \dfrac{x+1}{1-x}, x \neq 1$

37. $f^{-1}(x) = \sqrt{x-1}, x \geq 1$ **39.** $f^{-1}(x) = x^2 + 2, x \geq 0$ **41.** $f^{-1}(x) = \sqrt{x+4} - 2, x \geq -4$ **43.** $f^{-1}(x) = -\sqrt{x+5} - 2, x \geq -5$

45. $V^{-1}(x) = \sqrt[3]{x}$. V^{-1} finds the length of a side of a cube given the volume.

47. Yes. Yes. A conversion function is a nonconstant linear function. All nonconstant linear functions have inverses that are also functions.

49. $s^{-1}(x) = \dfrac{1}{2}x - 12$ **51.** $E^{-1}(s) = 20s - 50,000$. From the monthly earnings s the executive can find $E^{-1}(s)$, the value of the software sold.

53. 44205833; $f^{-1}(x) = \dfrac{1}{2}x + \dfrac{1}{2}, f^{-1}(44205833) = 22102917$

55. Because the function is increasing and 4 is between 2 and 5, c must be between 7 and 12.   **57.** between 2 and 5

59. between 3 and 7   **61.** $f^{-1}(x) = \dfrac{x-b}{a}, a \neq 0$ **63.** The reflection of f across the line given by $y = x$ yields f. Thus f is its own inverse.

65. Yes   **67.** No

Prepare for Section 1.7, page 90

69. slope: $-\dfrac{1}{3}$; y-intercept: $(0, 4)$ **70.** slope: $\dfrac{3}{4}$; y-intercept: $(0, -3)$ **71.** $y = -0.45x + 2.3$ **72.** $y = -\dfrac{2}{3}x - 2$ **73.** 19   **74.** 3

Exercise Set 1.7, page 98

1. no linear relationship   **3.** linear   **5.** Figure A   **7.** $y = 2.00862069x + 0.5603448276$ **9.** $y = -0.7231182796x + 9.233870968$
11. $y = 2.222641509x - 7.364150943$ **13.** $y = 1.095779221x^2 - 2.69642857x + 1.136363636$
15. $y = -0.2987274717x^2 - 3.20998141x + 3.416463667$ **17. a.** $y = 23.55706665x - 24.4271215$ **b.** 1247.7 centimeters
19. a. $y = 1.671510024x + 16.32830605$ **b.** 46.4 centimeters   **21. a.** $y = 0.1628623408x - 0.6875682232$ **b.** 25   **23.** No, because the linear correlation coefficient is close to 0.   **25. a.** Yes, there is a strong linear correlation.   **b.** $y = -0.9033088235x + 78.62573529$
c. 56 years   **27.** $r^2 \approx 0.667$. The coefficient of determination means that approximately 66.7% of the variation in EPA mileage estimates can be attributed to the horsepower of the car.   **29.** $y = -0.6328671329x^2 + 33.6160839x - 379.4405594$
31. a. $y = -0.0165034965x^2 + 1.366713287x + 5.685314685$ **b.** 32.8 miles per gallon
33. a.   5 pound: $s = 0.6130952381t^2 - 0.0714285714t + 0.1071428571$
           10 pound: $s = 0.6091269841t^2 - 0.0011904762t - 0.3$
           15 pound: $s = 0.5922619048t^2 + 0.3571428571t - 1.520833333$
b. All the regression equations are approximately the same. Therefore, the equations of motion of the three masses are the same.
35. quadratic; r^2 is closer to 1 for the quadratic model.

Chapter 1 True/False Exercises, page 106

1. False. Let $f(x) = x^2$. Then $f(3) = f(-3) = 9$, but $3 \neq -3$.   **2.** False. Consider $f(x) = x^2$, which does not have an inverse function.
3. False. Let $f(x) = x$ and $g(x) = 2x$. Then $(f \circ g)(0) = 0$ and $(g \circ f)(0) = 0$, but f and g are not inverse functions.   **4.** True   **5.** False. Let
$f(x) = 3x$. $[f(x)]^2 = 9x^2$, whereas $f[f(x)] = f(3x) = 3(3x) = 9x$.   **6.** False. Let $f(x) = x^2$. Then $f(1) = 1, f(2) = 4$. Thus $\dfrac{f(2)}{f(1)} = 4 \neq \dfrac{2}{1}$.   **7.** True
8. False. Let $f(x) = |x|$. Then $f(-1 + 3) = f(2) = 2$. $f(-1) + f(3) = 1 + 3 = 4$.   **9.** True   **10.** True   **11.** True   **12.** True   **13.** True

Chapter 1 Review Exercises, page 107

1. $-\dfrac{9}{4}$ [1.1]   **2.** -4 [1.1]   **3.** -2 [1.1]   **4.** $-\dfrac{2}{3}$ [1.1]   **5.** $-3, 6$ [1.1]   **6.** $\dfrac{1}{2}, 4$ [1.1]   **7.** $\dfrac{-1 \pm \sqrt{13}}{6}$ [1.1]   **8.** $\dfrac{-3 \pm \sqrt{41}}{4}$ [1.1]

9. $c \geq -6$ [1.1]   **10.** $a > 1$ [1.1]   **11.** $(-\infty, -3] \cup [4, \infty)$ [1.1]   **12.** $-\dfrac{1}{2} < x < 1$ [1.1]   **13.** $(-\infty, 1) \cup (4, \infty)$ [1.1]   **14.** $-1 \leq x \leq \dfrac{5}{3}$ [1.1]

15. 10 [1.2]   **16.** $3\sqrt{31}$ [1.2]   **17.** 5 [1.2]   **18.** $7\sqrt{5}$ [1.2]   **19.** $\sqrt{181}$ [1.2]   **20.** $4\sqrt{5}$ [1.2]   **21.** $\left(-\dfrac{1}{2}, 10\right)$ [1.2]   **22.** $(2, -2)$ [1.2]

23.   [1.4]   **24. a.**   **b.**   **c.**   [1.4]

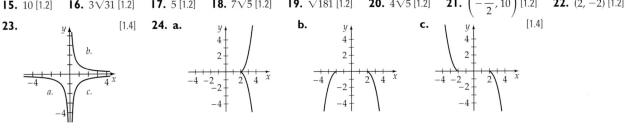

25. y-axis [1.4]   **26.** x-axis [1.4]   **27.** origin [1.4]   **28.** x-axis, y-axis, origin [1.4]   **29.** x-axis, y-axis, origin [1.4]   **30.** origin [1.4]
31. x-axis, y-axis, origin [1.4]   **32.** origin [1.4]   **33.** center $(3, -4)$, radius 9 [1.2]   **34.** $C(-5, -2)$, $r = 3$ [1.2]
35. $(x - 2)^2 + (y + 3)^2 = 5^2$ [1.2]   **36.** $(x + 5)^2 + (y - 1)^2 = 64$ [1.2]   **37. a.** 2   **b.** 10   **c.** $3t^2 + 4t - 5$
d. $3x^2 + 6xh + 3h^2 + 4x + 4h - 5$ **e.** $9t^2 + 12t - 15$ **f.** $27t^2 + 12t - 5$ [1.3]   **38. a.** $\sqrt{55}$ **b.** $\sqrt{39}$ **c.** 0   **d.** $\sqrt{64 - x^2}$
e. $2\sqrt{64 - t^2}$ **f.** $\sqrt{64 - 4t^2}$ [1.3]   **39. a.** 5   **b.** -11 **c.** $x^2 - 12x + 32$ **d.** $x^2 + 4x - 8$ [1.5]   **40. a.** 79   **b.** 56
c. $2x^2 - 4x + 9$ **d.** $2x^2 + 6$ [1.5]   **41.** $8x + 4h - 3$ [1.5]   **42.** $3x^2 + 3xh + h^2 - 1$ [1.5]   **43.** $(-\infty, \infty)$ [1.3]
44. $x \leq 6$ [1.3]   **45.** $[-5, 5]$ [1.3]   **46.** All real numbers except -3 and 5 [1.3]

47. [1.4]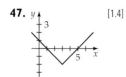
increasing on $[3, \infty)$
decreasing on $(-\infty, 3]$

48. [1.4]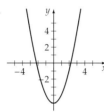
decreasing on $(-\infty, 0]$
increasing on $[0, \infty)$

49. [1.4]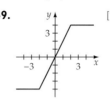
increasing on $[-2, 2]$
constant on $(-\infty, -2] \cup [2, \infty)$

50. [1.4]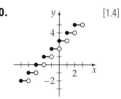
constant on $[n, n + 1)$, where n is an integer

51. [1.4]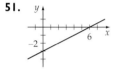
increasing on $(-\infty, \infty)$

52. [1.4]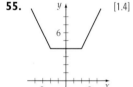
increasing on $(-\infty, \infty)$

53. [1.4]
a. domain $(-\infty, \infty)$
range $\{y \mid y \leq 4\}$
b. even

54. [1.3/1.4]
a. domain $(-\infty, \infty)$
range $(-\infty, \infty)$
b. neither

55. [1.4]
a. domain $(-\infty, \infty)$
range $\{y \mid y \geq 4\}$
b. even

56. [1.3/1.4]
a. domain $[-4, 4]$
range $[0, 4]$
b. even

57. [1.4]
a. domain $(-\infty, \infty)$
range $(-\infty, \infty)$
b. odd

58. [1.3/1.4]
a. domain $(-\infty, \infty)$
range: even integers
b. neither

59. $f(x) + g(x) = x^2 + x - 6$, domain: $(-\infty, \infty)$
$f(x) - g(x) = x^2 - x - 12$, domain: $(-\infty, \infty)$
$f(x) \cdot g(x) = x^3 + 3x^2 - 9x - 27$, domain: $(-\infty, \infty)$
$\dfrac{f(x)}{g(x)} = x - 3$, domain $\{x \mid x \neq -3\}$ [1.5]

60. $(f + g)(x) = x^3 + x^2 - 2x + 12$, domain: $(-\infty, \infty)$
$(f - g)(x) = x^3 - x^2 + 2x + 4$, domain: $(-\infty, \infty)$
$(fg)(x) = x^5 - 2x^4 + 4x^3 + 8x^2 - 16x + 32$, domain: $(-\infty, \infty)$
$\left(\dfrac{f}{g}\right)(x) = x + 2$, domain: $(-\infty, \infty)$ [1.5]

61. Yes [1.6]   **62.** Yes [1.6]

63. Yes [1.6]   **64.** No [1.6]   **65.** $f^{-1}(x) = \dfrac{x + 4}{3}$ [1.6]   **66.** $g^{-1}(x) = -\dfrac{1}{2}x + \dfrac{3}{2}$ [1.6]   **67.** $h^{-1}(x) = -2x - 4$ [1.6]   **68.** $k^{-1}(x) = \dfrac{1}{x}$ [1.6]

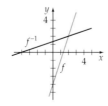 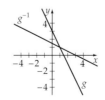 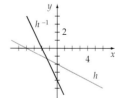

69. 25, 25 [1.3]   **70.** -5, 5 [1.3]   **71.** $t \approx 3.7$ seconds [1.3]   **72. a.** 150 feet   **b.** 525 feet [1.3]
73. a. $y = 0.018024687x + 0.00050045744$ **b.** Yes. $r \approx 0.999$, which is very close to 1.   **c.** 1.8 seconds [1.7]
74. a. $y = 0.0047952048x^2 - 1.756843157x + 180.4065934$ **b.** The graph of the regression equation never crosses the x-axis. Therefore, the model predicts the can will never be empty.   **c.** A regression model only approximates a situation. [1.7]

Chapter I Test, page 109

1. $\frac{1}{2}$ [1.1]   **2.** $x \le 1$ [1.1]   **3.** $-\frac{1}{2}, 2$ [1.1]   **4.** $-\frac{2}{3}, 1$ [1.1]   **5.** $\left(-\infty, -\frac{2}{5}\right) \cup (2, \infty)$ [1.1]   **6.** $\sqrt{85}$ [1.2]   **7.** midpoint $(1, 1)$; length $2\sqrt{13}$ [1.2]

8. $(0, \sqrt{2}), (0, -\sqrt{2}), (-4, 0)$ [1.2]   **9.**   [1.2]   **10.** center $(2, -1)$; radius 3 [1.2]   **11.** -4 [1.3]

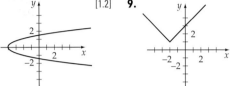

12. domain $\{x \mid x \ge 4 \text{ or } x \le -4\}$ [1.3]

13.   [1.4]   **14.**   [1.4]   **15.** b [1.4]   **16.** $x^2 + x - 3; \dfrac{x^2 - 1}{x - 2}, x \ne 2$ [1.5]   **17.** $2x + h$ [1.5]

$y = -f(x + 2) - 1$

increasing on $(-\infty, 2]$
decreasing on $[2, \infty)$

18. $x - 2\sqrt{x - 2} - 1$ [1.5]   **19.** $f^{-1}(x) = \dfrac{x}{1 - x}$ [1.6]   **20. a.** $y = -7.98245614x + 767.122807$ **b.** 56.7 Calories [1.7]

Exercise Set 2.1, page 124

1. $75°, 165°$ **3.** $19°45', 109°45'$ **5.** $33°26'45'', 123°26'45''$ **7.** $\dfrac{\pi}{2} - 1, \pi - 1$ **9.** $\dfrac{\pi}{4}, \dfrac{3\pi}{4}$ **11.** $\dfrac{\pi}{10}, \dfrac{3\pi}{5}$ **13.** Quadrant III, 250°

15. Quadrant II, 105°   **17.** Quadrant IV, 296°   **19.** $24°33'36''$ **21.** $64°9'28.8''$ **23.** $3°24'7.2''$ **25.** $25.42°$ **27.** $183.56°$ **29.** $211.78°$

31. $\dfrac{\pi}{6}$ **33.** $\dfrac{\pi}{2}$ **35.** $\dfrac{11\pi}{12}$ **37.** $\dfrac{7\pi}{3}$ **39.** $\dfrac{13\pi}{4}$ **41.** $36°$ **43.** $30°$ **45.** $67.5°$ **47.** $660°$ **49.** $85.94°$ **51.** 2.32

53. $472.69°$ **55.** $4, 229.18°$ **57.** $2.38, 136.63°$ **59.** 6.28 inches   **61.** 18.33 centimeters   **63.** 3π **65.** $\dfrac{5\pi}{12}$ radians or 75°

67. $\dfrac{\pi}{30}$ radian per second   **69.** $\dfrac{5\pi}{3}$ radians per second   **71.** $\dfrac{10\pi}{9}$ radians per second ≈ 3.49 radians per second   **73.** 40 mph

75. 1885 feet   **77. a.** 1160 miles per hour   **b.** 0.29 radian per hour   **c.** 10:59 A.M.   **79.** 840,000 miles   **81. a.** 3.9 radians per hour
b. 27,300 kilometers per hour   **83. a.** B **b.** Both points have the same linear velocity.   **85. a.** 1.15 statute miles   **b.** 10%
87. 13 square inches   **89.** 4680 square centimeters   **91.** ≈ 23.1 square inches   **93.** 1780 miles

Prepare for Section 2.2, page 128

95. $\dfrac{\sqrt{3}}{3}$ **96.** $\sqrt{2}$ **97.** 2   **98.** $\dfrac{\sqrt{3}}{3}$ **99.** 3.54   **100.** 10.39

Exercise Set 2.2, page 136

1. $\sin \theta = \dfrac{12}{13}$ $\csc \theta = \dfrac{13}{12}$ **3.** $\sin \theta = \dfrac{4}{7}$ $\csc \theta = \dfrac{7}{4}$ **5.** $\sin \theta = \dfrac{5\sqrt{29}}{29}$ $\csc \theta = \dfrac{\sqrt{29}}{5}$ **7.** $\sin \theta = \dfrac{\sqrt{21}}{7}$ $\csc \theta = \dfrac{\sqrt{21}}{3}$

$\cos \theta = \dfrac{5}{13}$ $\sec \theta = \dfrac{13}{5}$ $\cos \theta = \dfrac{\sqrt{33}}{7}$ $\sec \theta = \dfrac{7\sqrt{33}}{33}$ $\cos \theta = \dfrac{2\sqrt{29}}{29}$ $\sec \theta = \dfrac{\sqrt{29}}{2}$ $\cos \theta = \dfrac{2\sqrt{7}}{7}$ $\sec \theta = \dfrac{\sqrt{7}}{2}$

$\tan \theta = \dfrac{12}{5}$ $\cot \theta = \dfrac{5}{12}$ $\tan \theta = \dfrac{4\sqrt{33}}{33}$ $\cot \theta = \dfrac{\sqrt{33}}{4}$ $\tan \theta = \dfrac{5}{2}$ $\cot \theta = \dfrac{2}{5}$ $\tan \theta = \dfrac{\sqrt{3}}{2}$ $\cot \theta = \dfrac{2\sqrt{3}}{3}$

9. $\sin \theta = \dfrac{2\sqrt{30}}{15}$ $\csc \theta = \dfrac{\sqrt{30}}{4}$ **11.** $\sin \theta = \dfrac{\sqrt{3}}{2}$ $\csc \theta = \dfrac{2\sqrt{3}}{3}$ **13.** $\sin \theta = \dfrac{6\sqrt{61}}{61}$ $\csc \theta = \dfrac{\sqrt{61}}{6}$ **15.** $\dfrac{3}{4}$ **17.** $\dfrac{4}{5}$ **19.** $\dfrac{3}{4}$ **21.** $\dfrac{12}{13}$

$\cos \theta = \dfrac{\sqrt{105}}{15}$ $\sec \theta = \dfrac{\sqrt{105}}{7}$ $\cos \theta = \dfrac{1}{2}$ $\sec \theta = 2$ $\cos \theta = \dfrac{5\sqrt{61}}{61}$ $\sec \theta = \dfrac{\sqrt{61}}{5}$

$\tan \theta = \dfrac{2\sqrt{14}}{7}$ $\cot \theta = \dfrac{\sqrt{14}}{4}$ $\tan \theta = \sqrt{3}$ $\cot \theta = \dfrac{\sqrt{3}}{3}$ $\tan \theta = \dfrac{6}{5}$ $\cot \theta = \dfrac{5}{6}$

23. $\dfrac{13}{5}$ **25.** $\dfrac{3}{2}$ **27.** $\sqrt{2}$ **29.** $-\dfrac{3}{4}$ **31.** $\dfrac{5}{4}$ **33.** $\sqrt{3} - \sqrt{6}$ **35.** $\sqrt{3}$ **37.** $\dfrac{3\sqrt{2} + 2\sqrt{3}}{6}$ **39.** $\dfrac{3 - \sqrt{3}}{3}$ **41.** $2\sqrt{2} - \sqrt{3}$
43. 0.6249   **45.** 0.4488   **47.** 0.8221   **49.** 1.0053   **51.** 0.4816   **53.** 1.0729   **55.** 0.3153   **57.** 9.5 feet

61. 20 kilometers (to the nearest kilometer) **63.** 5.1 feet **65.** 1.7 miles **67.** 1400 feet **69.** 686,000,000 kilometers **73.** 612 feet
75. 560 feet (to the nearest foot) **77. a.** 559 feet **b.** 193 feet **79.** ≈5.2 meters **81.** 49.9 meters ≤ h ≤ 52.1 meters **83.** ≈8.5 feet

Prepare for Section 2.3, page 141

85. $-\dfrac{4}{3}$ **86.** $\dfrac{\sqrt{5}}{2}$ **87.** 60 **88.** $\dfrac{\pi}{5}$ **89.** π **90.** $\sqrt{34}$

Exercise Set 2.3, page 149

1. $\sin\theta = \dfrac{3\sqrt{13}}{13}$ $\csc\theta = \dfrac{\sqrt{13}}{3}$ **3.** $\sin\theta = \dfrac{3\sqrt{13}}{13}$ $\csc\theta = \dfrac{\sqrt{13}}{3}$ **5.** $\sin\theta = -\dfrac{5\sqrt{89}}{89}$ $\csc\theta = -\dfrac{\sqrt{89}}{5}$

$\cos\theta = \dfrac{2\sqrt{13}}{13}$ $\sec\theta = \dfrac{\sqrt{13}}{2}$ $\cos\theta = -\dfrac{2\sqrt{13}}{13}$ $\sec\theta = -\dfrac{\sqrt{13}}{2}$ $\cos\theta = -\dfrac{8\sqrt{89}}{89}$ $\sec\theta = -\dfrac{\sqrt{89}}{8}$

$\tan\theta = \dfrac{3}{2}$ $\cot\theta = \dfrac{2}{3}$ $\tan\theta = -\dfrac{3}{2}$ $\cot\theta = -\dfrac{2}{3}$ $\tan\theta = \dfrac{5}{8}$ $\cot\theta = \dfrac{8}{5}$

7. $\sin\theta = 0$ $\csc\theta$ is undefined. **9.** Quadrant I **11.** Quadrant IV **13.** Quadrant III **15.** $\dfrac{\sqrt{3}}{3}$ **17.** -1 **19.** $-\dfrac{\sqrt{3}}{3}$ **21.** $\dfrac{2\sqrt{3}}{3}$
$\cos\theta = -1$ $\sec\theta = -1$
$\tan\theta = 0$ $\cot\theta$ is undefined.

23. $-\dfrac{\sqrt{3}}{3}$ **25.** $20°$ **27.** $9°$ **29.** $\dfrac{\pi}{5}$ **31.** $\pi - \dfrac{8}{3}$ **33.** $34°$ **35.** $65°$ **37.** $-\dfrac{\sqrt{2}}{2}$ **39.** 1 **41.** $-\dfrac{2\sqrt{3}}{3}$ **43.** $\dfrac{\sqrt{2}}{2}$ **45.** $\sqrt{2}$

47. $\cot 540°$ is undefined. **49.** 0.798636 **51.** -0.438371 **53.** -1.26902 **55.** -0.587785 **57.** -1.70130 **59.** -3.85522 **61.** 0

63. 1 **65.** $-\dfrac{3}{2}$ **67.** 1 **69.** $30°, 150°$ **71.** $150°, 210°$ **73.** $225°, 315°$ **75.** $\dfrac{3\pi}{4}, \dfrac{7\pi}{4}$ **77.** $\dfrac{5\pi}{6}, \dfrac{11\pi}{6}$ **79.** $\dfrac{\pi}{3}, \dfrac{2\pi}{3}$

Prepare for Section 2.4, page 150

91. Yes **92.** Yes **93.** No **94.** 2π **95.** even function **96.** neither

Exercise Set 2.4, page 160

1. $\left(\dfrac{\sqrt{3}}{2}, \dfrac{1}{2}\right)$ **3.** $\left(-\dfrac{\sqrt{3}}{2}, -\dfrac{1}{2}\right)$ **5.** $\left(\dfrac{1}{2}, -\dfrac{\sqrt{3}}{2}\right)$ **7.** $\left(\dfrac{\sqrt{3}}{2}, -\dfrac{1}{2}\right)$ **9.** $(-1, 0)$ **11.** $\left(-\dfrac{1}{2}, -\dfrac{\sqrt{3}}{2}\right)$ **13.** $-\dfrac{\sqrt{3}}{3}$ **15.** $-\dfrac{1}{2}$

17. $-\dfrac{2\sqrt{3}}{3}$ **19.** -1 **21.** $-\dfrac{2\sqrt{3}}{3}$ **23.** 0.9391 **25.** -1.1528 **27.** -0.2679 **29.** 0.8090 **31.** 48.0889 **33. a.** 0.9 **b.** -0.4

35. a. -0.8 **b.** 0.6 **37.** 0.4, 2.7 **39.** 3.4, 6.0 **41.** odd **43.** neither **45.** even **47.** odd **57.** $\sin t$ **59.** $\sec t$ **61.** $-\tan^2 t$

63. $-\cot t$ **65.** $\cos^2 t$ **67.** $2\csc^2 t$ **69.** $\csc^2 t$ **71.** 1 **73.** $\sqrt{1 - \cos^2 t}$ **75.** $\sqrt{1 + \cot^2 t}$ **77.** 750 miles **79.** $-\dfrac{\sin^2 t}{\cos t}$

81. $\csc t \sec t$ **83.** $1 - 2\sin t + \sin^2 t$ **85.** $1 - 2\sin t\cos t$ **87.** $\cos^2 t$ **89.** $2\csc t$ **91.** $(\cos t - \sin t)(\cos t + \sin t)$

93. $(\tan t + 2)(\tan t - 3)$ **95.** $(2\sin t + 1)(\sin t - 1)$ **97.** $(\cos t - \sin t)(\cos t + \sin t)$ **99.** $\dfrac{\sqrt{2}}{2}$ **101.** $-\dfrac{\sqrt{3}}{3}$

Prepare for Section 2.5, page 162

107. 0.7 **108.** -0.7 **109.** Reflect the graph of $y = f(x)$ across the x-axis. **110.** Contract each point on the graph of $y = f(x)$ toward the y-axis by a factor of $\dfrac{1}{2}$. **111.** 6π **112.** 5π

Exercise Set 2.5, page 170

1. $2, 2\pi$ **3.** $1, \pi$ **5.** $\dfrac{1}{2}, 1$ **7.** $2, 4\pi$ **9.** $\dfrac{1}{2}, 2\pi$ **11.** $1, 8\pi$ **13.** $2, 6$ **15.** $3, 3\pi$ **17.** **19.**

21. **23.** **25.** **27.** **29.**

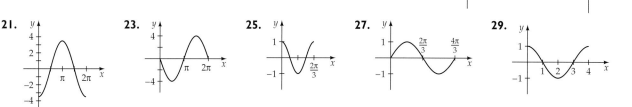

31.   **33.** 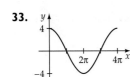 **35.** 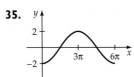 **37.**   **39.**

41.   **43.** 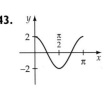 **45.** 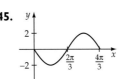 **47.** 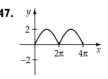 **49.**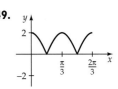

51. 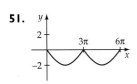 **53.** 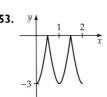 **55.** $y = \cos 2x$ **57.** $y = 2 \sin \dfrac{2}{3}x$ **59.** $y = -2 \cos \pi x$

61. 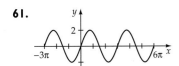 **63.** 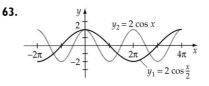 **65.**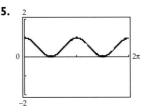

67. 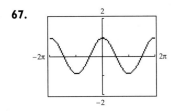 **69.** 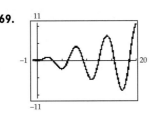 **71.**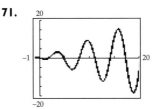

73. 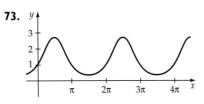 **75.** $f(t) = 60 \cos \dfrac{\pi}{10}t$ **77.** $y = 2 \sin \dfrac{2}{3}x$ **79.** $y = 4 \sin \pi x$ **81.** $y = 3 \cos 4x$

83. $y = 3 \cos \dfrac{4\pi}{5}x$

maximum $= e$, minimum $= \dfrac{1}{e} \approx 0.3679$, period $= 2\pi$

Prepare for Section 2.6, page 172

85. 1.7   **86.** 0.6   **87.** Stretch each point on the graph of $y = f(x)$ away from the x-axis by a factor of 2.

88. Shift the graph of $y = f(x)$ to the right 2 units and up 3 units.   **89.** 2π **90.** $\dfrac{4}{3}\pi$

Exercise Set 2.6, page 178

1. $\dfrac{\pi}{2} + k\pi$, k an integer   **3.** $\dfrac{\pi}{2} + k\pi$, k an integer   **5.** 2π **7.** π **9.** 2π **11.** $\dfrac{2\pi}{3}$ **13.** $\dfrac{\pi}{3}$ **15.** 8π **17.** 1   **19.** 4

21.  **23.**   **25.** 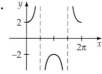 **27.** 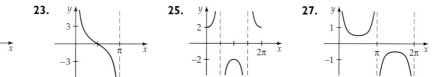 **29.**

31.   **33.** 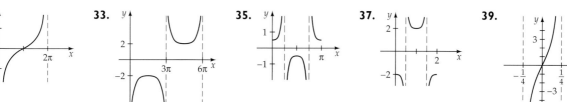 **35.**   **37.**   **39.**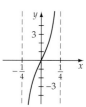

41. 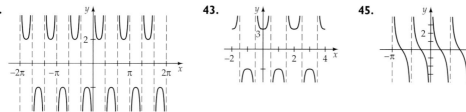 **43.** 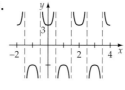 **45.** 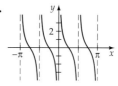 **47.**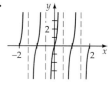

49. $y = \cot \dfrac{3}{2}x$ **51.** $y = \csc \dfrac{2}{3}x$ **53.** $y = \sec \dfrac{3}{4}x$ **55.** 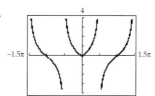 **57.**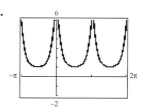

59.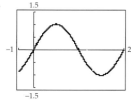

Note: The display screen at the left fails to show that on $[-1, 2\pi]$ the function is undefined at $x = \dfrac{\pi}{2}$ and $x = \dfrac{3\pi}{2}$.

61. 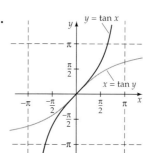 **63.** $y = \tan 3x$

65. $y = \sec \dfrac{8}{3}x$ **67.** $y = \cot \dfrac{\pi}{2}x$ **69.** $y = \csc \dfrac{4\pi}{3}x$

Prepare for Section 2.7, page 180

71. amplitude 2, period π **72.** amplitude $\dfrac{2}{3}$, period 6π **73.** amplitude 4, period 1 **74.** 2 **75.** -3 **76.** y-axis

Exercise Set 2.7, page 187

1. $2, \dfrac{\pi}{2}, 2\pi$ **3.** $1, \dfrac{\pi}{8}, \pi$ **5.** $4, -\dfrac{\pi}{4}, 3\pi$ **7.** $\dfrac{5}{4}, \dfrac{2\pi}{3}, \dfrac{2\pi}{3}$ **9.** $\dfrac{\pi}{8}, \dfrac{\pi}{2}$ **11.** $-3\pi, 6\pi$ **13.** $\dfrac{\pi}{16}, \pi$ **15.** $-12\pi, 4\pi$

17.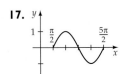

19.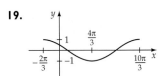

21.

23.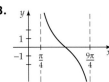

25.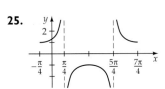

27.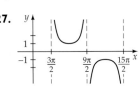

29.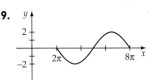

31.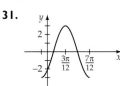

33.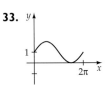

35.

37.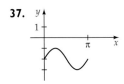

39.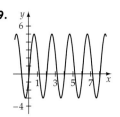

41.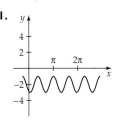

43.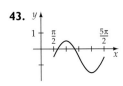

45.

47.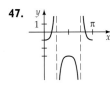

49.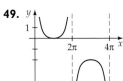

51. a. 7.5 months, 12 months   **b.**

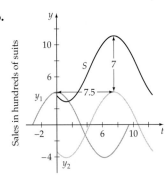

c. August

53.

55.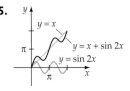

57.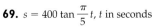

59. $y = \sin\left(2x - \dfrac{\pi}{3}\right)$ **61.** $y = \csc\left(\dfrac{x}{2} - \pi\right)$ **63.** $y = \sec\left(x - \dfrac{\pi}{2}\right)$ **65.** ≈ 25 parts per million

67. $s = 7\cos 10\pi t + 5$ **69.** $s = 400\tan\dfrac{\pi}{5}t,\ t$ in seconds   **71.** $y = 3\cos\dfrac{\pi}{6}t + 9,\ 12$ feet at 6:00 P.M.   **73.**

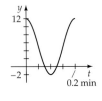

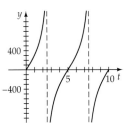

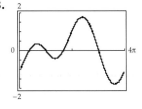

75. 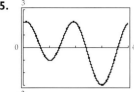   **77.** 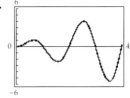   **79.** 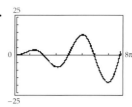   **81.**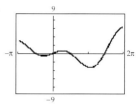

83. 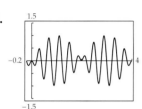   **85.** 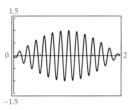   **87.** $y = 2\sin\left(2x - \dfrac{2\pi}{3}\right)$ **89.** $y = \tan\left(\dfrac{x}{2} - \dfrac{\pi}{4}\right)$

91. $y = \sec\left(\dfrac{x}{2} - \dfrac{3\pi}{8}\right)$ **93.** 1   **95.** $\cos^2 x + 2$ **97.**   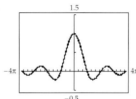   **99.**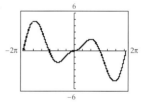

The graph above does not
show that the function is
undefined at $x = 0$.

Prepare for Section 2.8, page 191

101. $\dfrac{3}{2\pi}$ **102.** $\dfrac{5}{2}$ **103.** 4   **104.** 3   **105.** 4   **106.** $y = 4\cos \pi x$

Exercise Set 2.8, page 196

1. $2, \pi, \dfrac{1}{\pi}$ **3.** $3, 3\pi, \dfrac{1}{3\pi}$ **5.** $4, 2, \dfrac{1}{2}$ **7.** $\dfrac{3}{4}, 4, \dfrac{1}{4}$

9. $y = 4\cos 3\pi t$ **11.** $y = \dfrac{3}{2}\cos \dfrac{4\pi}{3}t$ **13.** $y = 2\sin 2t$ **15.** $y = \sin \pi t$ **17.** $y = 2\sin 2\pi t$

    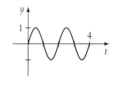   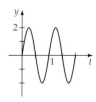

19. $y = \dfrac{1}{2}\cos 4t$ **21.** $y = 2.5\cos \pi t$ **23.** $y = \dfrac{1}{2}\cos \dfrac{2\pi}{3}t$ **25.** $y = 4\cos 4t$ **27.** $4\pi, \dfrac{1}{4\pi}, 2$ feet; $y = -2\cos \dfrac{t}{2}$
29. a. 3   **b.** 59.8 seconds   **31. a.** 10   **b.** 71.0 seconds   **33. a.** 10   **b.** 9.1 seconds   **35. a.** 10   **b.** 6.1 seconds
37. The new period is 3 times the original period.   **39.** Yes   **41.** Yes

Chapter 2 True/False Exercises, page 200

1. False; the initial side must be along the positive x-axis.   **2.** True   **3.** True   **4.** False; in the third quadrant $\cos \theta < 0$ and $\tan \theta > 0$.
5. False; $\sec^2\theta - \tan^2\theta = 1$ is an identity.   **6.** False; the tangent function has no amplitude.   **7.** False; the period is 2π.   **8.** True

9. False; $\sin 45° + \cos(90° - 45°) = \sin 45° + \cos 45° = \dfrac{\sqrt{2}}{2} + \dfrac{\sqrt{2}}{2} = \sqrt{2}$.   **10.** False; $\sin\left(\dfrac{\pi}{2} + \dfrac{\pi}{2}\right) = \sin \pi = 0$, and

$\sin\dfrac{\pi}{2} + \sin\dfrac{\pi}{2} = 1 + 1 = 2$.   **11.** False; $\sin^2\dfrac{\pi}{6} = \left(\dfrac{1}{2}\right)^2 = \dfrac{1}{4}$, and $\sin\left(\dfrac{\pi}{6}\right)^2 = \sin\dfrac{\pi^2}{36} \approx 0.2707$.   **12.** False; the phase shift is $\dfrac{\pi/3}{2} = \dfrac{\pi}{6}$.

13. True **14.** False; 1 radian $\approx 57.3°$. **15.** False; the graph lies on or between the graphs of $y = 2^{-x}$ and $y = -2^{-x}$.
16. False; $|f(t)| \to 0$ as $t \to \infty$.

Chapter 2 Review Exercises, page 200

1. complement measures 25°; supplement measures 115° [2.1] **2.** 80° [2.3] **3.** 114.59° [2.1] **4.** $\dfrac{7\pi}{4}$ [2.1] **5.** 3.93 meters [2.1] **6.** 0.3 [2.1]

7. 55 radians per second [2.1] **8.** $\dfrac{\sqrt{5}}{3}$ [2.2] **9.** $\dfrac{\sqrt{5}}{2}$ [2.2] **10.** $\dfrac{2}{3}$ [2.2] **11.** $\dfrac{3\sqrt{5}}{5}$ [2.2] **12.** $\sin\theta = -\dfrac{3\sqrt{10}}{10}$ $\csc\theta = -\dfrac{\sqrt{10}}{3}$ [2.3]

$\cos\theta = \dfrac{\sqrt{10}}{10}$ $\sec\theta = \sqrt{10}$

$\tan\theta = -3$ $\cot\theta = -\dfrac{1}{3}$

13. a. $-\dfrac{2\sqrt{3}}{3}$ **b.** 1 **c.** -1 **d.** $-\dfrac{1}{2}$ [2.3] **14. a.** -0.5446 **b.** 0.5365 **c.** -3.2361 **d.** 3.0777 [2.3] **15. a.** $-\dfrac{1}{2}$ **b.** $\dfrac{\sqrt{3}}{3}$ [2.3]

16. a. $-\dfrac{2\sqrt{3}}{3}$ **b.** 2 [2.3] **17. a.** $\dfrac{\sqrt{2}}{2}$ **b.** -1 [2.3] **18. a.** $(-1, 0)$ **b.** $\left(\dfrac{1}{2}, -\dfrac{\sqrt{3}}{2}\right)$ **c.** $\left(-\dfrac{\sqrt{2}}{2}, -\dfrac{\sqrt{2}}{2}\right)$ **d.** $(1, 0)$ [2.4]

19. even [2.4] **22.** $\sec^2\phi$ [2.4] **23.** $\tan\phi$ [2.4] **24.** $\sin\phi$ [2.4] **25.** $\tan^2\phi$ [2.4] **26.** $\csc^2\phi$ [2.4] **27.** 0 [2.4] **28.** $3, \pi, \dfrac{\pi}{2}$ [2.5]

29. no amplitude, $\dfrac{\pi}{3}, 0$ [2.6] **30.** $2, \dfrac{2\pi}{3}, -\dfrac{\pi}{9}$ [2.5] **31.** $1, \pi, \dfrac{\pi}{3}$ [2.5] **32.** no amplitude, $\dfrac{\pi}{2}, \dfrac{3\pi}{8}$ [2.6] **33.** no amplitude, $2\pi, \dfrac{\pi}{4}$ [2.6]

34. [2.5] **35.** [2.5] **36.** [2.5] **37.** [2.7]

38. [2.7] **39.** [2.7] **40.** [2.6] **41.** [2.6]

42. [2.7] **43.** [2.7] **44.** [2.7] **45.** [2.7]

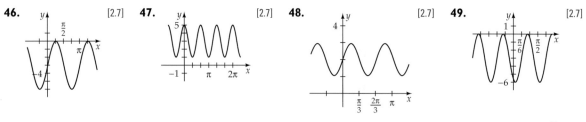

46. [2.7] **47.** [2.7] **48.** [2.7] **49.** [2.7]

50. [2.7] **51.** [2.7] **52.** 0.089 miles [2.2] **53.** 12.3 feet [2.2] **54.** $\dfrac{\sqrt{3}}{3}$ [2.2]

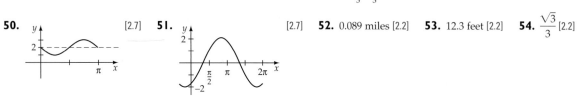

55. 46 feet [2.2]   **56.** $2.5, \dfrac{\pi}{25}, \dfrac{25}{\pi}$ [2.8]   **57.** amplitude $= 0.5, f = \dfrac{1}{\pi}, p = \pi, y = -0.5\cos 2t$ [2.8]   **58.** 7.2 seconds [2.8]

Chapter 2 Test, page 202

1. $\dfrac{5\pi}{6}$ [2.1]   **2.** $\dfrac{\pi}{12}$ [2.1]   **3.** 13.1 centimeters [2.1]   **4.** 12π radians/second [2.1]   **5.** 80 centimeters/second [2.1]   **6.** $\dfrac{\sqrt{58}}{7}$ [2.2]

7. 1.0864 [2.2]   **8.** $\dfrac{\sqrt{3}-6}{6}$ [2.3]   **9.** $\left(\dfrac{\sqrt{3}}{2}, -\dfrac{1}{2}\right)$ [2.4]   **10.** $\sin^2 t$ [2.4]   **11.** $\dfrac{\pi}{3}$ [2.6]   **12.** amplitude 3, period π, phase shift $-\dfrac{\pi}{4}$ [2.7]

13. period 3, phase shift $-\dfrac{1}{2}$ [2.7]   **14.**   [2.5]   **15.**   [2.6]

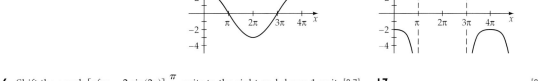

16. Shift the graph [of $y = 2\sin(2x)$] $\dfrac{\pi}{4}$ units to the right and down 1 unit. [2.7]   **17.**   [2.7]

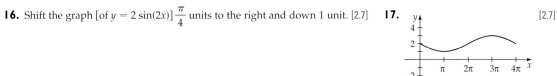

18.   [2.7]   **19.** 25.5 meters [2.2]   **20.** $y = 13\sin\dfrac{2\pi}{5}t$ [2.8]

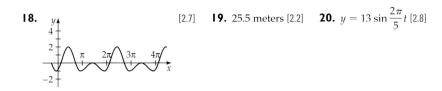

Cumulative Review Exercises, page 203

1. $5\sqrt{2}$ [1.1]   **2.** $\dfrac{\sqrt{3}}{2}$ [1.1]   **3.** y-intercept $(0, -9)$, x-intercepts $(-3, 0)$ and $(3, 0)$ [1.2]   **4.** odd function [1.4]   **5.** $f^{-1}(x) = \dfrac{3x}{2x-1}$ [1.6]
6. $(-\infty, 4) \cup (4, \infty)$ [1.3]   **7.** $2, -3$ [1.1]   **8.** Shift the graph of $y = f(x)$ horizontally 3 units to the right. [1.4]   **9.** Reflect the graph of $y = f(x)$ across the y-axis. [1.4]   **10.** $\dfrac{5\pi}{3}$ [2.1]   **11.** $225°$ [2.1]   **12.** 1 [2.3]   **13.** $\dfrac{\sqrt{3}+1}{2}$ [2.2]   **14.** $\dfrac{5}{4}$ [2.2]   **15.** negative [2.3]   **16.** $30°$ [2.4]
17. $\dfrac{\pi}{3}$ [2.4]   **18.** $(-\infty, \infty)$ [2.4]   **19.** $[-1, 1]$ [2.4]   **20.** $\dfrac{3}{5}$ [2.2]

Exercise Set 3.1, page 211

1. If $x = \dfrac{\pi}{4}$, the left side is 2 and the right side is 1.   **3.** If $x = 0°$, the left side is $\dfrac{\sqrt{3}}{2}$ and the right side is $\dfrac{2+\sqrt{3}}{2}$.   **5.** If $x = 0$, the left side is -1 and the right side is 1.   **7.** If $x = 0$, the left side is -1 and the right side is 1.   **9.** If $x = \dfrac{\pi}{4}$, the left side is 2 and the right side is 1.
67. identity   **69.** identity   **71.** identity   **73.** not an identity

Prepare for Section 3.2, page 214

83. Both functional values equal $\dfrac{1}{2}$.   **84.** Both functional values equal $\dfrac{1}{2}$.   **85.** For each of the given values of θ, the functional values are equal.   **86.** For each of the given values of θ, the functional values are equal.   **87.** Both functional values equal $\dfrac{\sqrt{3}}{3}$.   **88.** 0

Exercise Set 3.2, page 222

1. $\dfrac{\sqrt{6}+\sqrt{2}}{4}$ **3.** $\dfrac{\sqrt{6}+\sqrt{2}}{4}$ **5.** $2 - \sqrt{3}$ **7.** $\dfrac{-\sqrt{6}+\sqrt{2}}{4}$ **9.** $-\dfrac{\sqrt{6}+\sqrt{2}}{4}$ **11.** $2 + \sqrt{3}$ **13.** 0   **15.** $\dfrac{1}{2}$ **17.** $\sqrt{3}$ **19.** $\cos 48°$

21. $\cot 75°$ **23.** $\csc 65°$ **25.** $\sin 5x$ **27.** $\cos x$ **29.** $\sin 4x$ **31.** $\cos 2x$ **33.** $\sin x$ **35.** $\tan 7x$ **37. a.** $-\dfrac{77}{85}$ **b.** $\dfrac{84}{85}$ **c.** $\dfrac{77}{36}$

39. a. $-\dfrac{63}{65}$ **b.** $-\dfrac{56}{65}$ **c.** $-\dfrac{63}{16}$ **41. a.** $\dfrac{63}{65}$ **b.** $\dfrac{56}{65}$ **c.** $\dfrac{33}{56}$ **43. a.** $-\dfrac{77}{85}$ **b.** $-\dfrac{84}{85}$ **c.** $-\dfrac{13}{84}$ **45. a.** $-\dfrac{33}{65}$ **b.** $-\dfrac{16}{65}$ **c.** $\dfrac{63}{16}$

47. a. $-\dfrac{56}{65}$ **b.** $-\dfrac{63}{65}$ **c.** $\dfrac{16}{63}$ **73.** $-\cos\theta$ **75.** $\tan\theta$ **77.** $\sin\theta$ **79.** identity   **81.** identity

Prepare for Section 3.3, page 224

91. $2\sin\alpha\cos\alpha$ **92.** $\cos^2\alpha - \sin^2\alpha$ **93.** $\dfrac{2\tan\alpha}{1-\tan^2\alpha}$ **94.** For each of the given values of α, the functional values are equal.

95. Let $\alpha = 45°$; then the left side of the equation is 1, and the right side of the equation is $\sqrt{2}$.   **96.** Let $\alpha = 60°$; then the left side of the equation is $\dfrac{\sqrt{3}}{2}$, and the right side of the equation is $\dfrac{1}{4}$.

Exercise Set 3.3, page 230

1. $\sin 4\alpha$ **3.** $\cos 10\beta$ **5.** $\cos 6\alpha$ **7.** $\tan 6\alpha$ **9.** $\dfrac{\sqrt{2+\sqrt{3}}}{2}$ **11.** $\sqrt{2}+1$ **13.** $-\dfrac{\sqrt{2+\sqrt{2}}}{2}$ **15.** $\dfrac{\sqrt{2-\sqrt{2}}}{2}$ **17.** $\dfrac{\sqrt{2-\sqrt{2}}}{2}$

19. $\dfrac{\sqrt{2-\sqrt{3}}}{2}$ **21.** $-2-\sqrt{3}$ **23.** $\dfrac{\sqrt{2+\sqrt{3}}}{2}$ **25.** $\sin 2\theta = -\dfrac{24}{25}$, $\cos 2\theta = \dfrac{7}{25}$, $\tan 2\theta = -\dfrac{24}{7}$

27. $\sin 2\theta = -\dfrac{240}{289}$, $\cos 2\theta = \dfrac{161}{289}$, $\tan 2\theta = -\dfrac{240}{161}$ **29.** $\sin 2\theta = -\dfrac{336}{625}$, $\cos 2\theta = -\dfrac{527}{625}$, $\tan 2\theta = \dfrac{336}{527}$

31. $\sin 2\theta = \dfrac{240}{289}$, $\cos 2\theta = -\dfrac{161}{289}$, $\tan 2\theta = -\dfrac{240}{161}$ **33.** $\sin 2\theta = -\dfrac{720}{1681}$, $\cos 2\theta = \dfrac{1519}{1681}$, $\tan 2\theta = -\dfrac{720}{1519}$

35. $\sin 2\theta = \dfrac{240}{289}$, $\cos 2\theta = -\dfrac{161}{289}$, $\tan 2\theta = -\dfrac{240}{161}$ **37.** $\sin\dfrac{\alpha}{2} = \dfrac{5\sqrt{26}}{26}$, $\cos\dfrac{\alpha}{2} = \dfrac{\sqrt{26}}{26}$, $\tan\dfrac{\alpha}{2} = 5$

39. $\sin\dfrac{\alpha}{2} = \dfrac{5\sqrt{34}}{34}$, $\cos\dfrac{\alpha}{2} = -\dfrac{3\sqrt{34}}{34}$, $\tan\dfrac{\alpha}{2} = -\dfrac{5}{3}$ **41.** $\sin\dfrac{\alpha}{2} = \dfrac{\sqrt{5}}{5}$, $\cos\dfrac{\alpha}{2} = \dfrac{2\sqrt{5}}{5}$, $\tan\dfrac{\alpha}{2} = \dfrac{1}{2}$

43. $\sin\dfrac{\alpha}{2} = \dfrac{\sqrt{2}}{10}$, $\cos\dfrac{\alpha}{2} = -\dfrac{7\sqrt{2}}{10}$, $\tan\dfrac{\alpha}{2} = -\dfrac{1}{7}$ **45.** $\sin\dfrac{\alpha}{2} = \dfrac{\sqrt{17}}{17}$, $\cos\dfrac{\alpha}{2} = \dfrac{4\sqrt{17}}{17}$, $\tan\dfrac{\alpha}{2} = \dfrac{1}{4}$

47. $\sin\dfrac{\alpha}{2} = \dfrac{5\sqrt{34}}{34}$, $\cos\dfrac{\alpha}{2} = -\dfrac{3\sqrt{34}}{34}$, $\tan\dfrac{\alpha}{2} = -\dfrac{5}{3}$ **95.** identity   **97.** not an identity

Prepare for Section 3.4, page 233

107. $\sin\alpha\cos\beta$ **108.** $\cos\alpha\cos\beta$ **109.** Both functional values equal $-\dfrac{1}{2}$.   **110.** $\sin x + \cos x$ **111.** Answers will vary.   **112.** 2

Exercise Set 3.4, page 239

1. $\sin 3x - \sin x$ **3.** $\dfrac{1}{2}(\sin 8x - \sin 4x)$ **5.** $\sin 8x + \sin 2x$ **7.** $\dfrac{1}{2}(\cos 4x - \cos 6x)$ **9.** $\dfrac{1}{4}$ **11.** $-\dfrac{\sqrt{2}}{4}$ **13.** $-\dfrac{1}{4}$ **15.** $\dfrac{\sqrt{3}-2}{4}$

17. $2\sin 3\theta\cos\theta$ **19.** $2\cos 2\theta\cos\theta$ **21.** $-2\sin 4\theta\sin 2\theta$ **23.** $2\cos 4\theta\cos 3\theta$ **25.** $2\sin 7\theta\cos 2\theta$ **27.** $-2\sin\dfrac{3}{2}\theta\sin\dfrac{1}{2}\theta$

29. $2\sin\dfrac{3}{4}\theta\sin\dfrac{\theta}{4}$ **31.** $2\cos\dfrac{5}{12}\theta\sin\dfrac{1}{12}\theta$ **49.** $y = \sqrt{2}\sin(x - 135°)$ **51.** $y = \sin(x - 60°)$ **53.** $y = \dfrac{\sqrt{2}}{2}\sin(x - 45°)$

55. $y = 3\sqrt{2}\sin(x + 135°)$ **57.** $y = \pi\sqrt{2}\sin(x - 45°)$ **59.** $y = \sqrt{2}\sin\left(x + \dfrac{3\pi}{4}\right)$ **61.** $y = \sin\left(x + \dfrac{\pi}{6}\right)$ **63.** $y = 20\sin\left(x + \dfrac{2\pi}{3}\right)$

65. $y = 5\sqrt{2}\sin\left(x + \dfrac{3\pi}{4}\right)$

67. 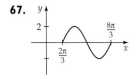   **69.**   **71.**   **73.** 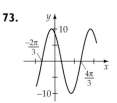   **75.**

77. a. $p(t) = \cos(2\pi \cdot 1336t) + \cos(2\pi \cdot 770t)$ **b.** $p(t) = 2\cos(2106\pi t)\cos(566\pi t)$ **c.** 1053 cycles per second   **79.** identity   **81.** identity
83. identity

Prepare for Section 3.5, page 241

99. A one-to-one function is a function for which each range value (y-value) is paired with one and only one domain value (x-value).
100. If every horizontal line intersects the graph of a function at most once, then the function is a one-to-one function.
101. $f[g(x)] = x$ **102.** $f[f^{-1}(x)] = x$ **103.** The graph of f^{-1} is the reflection of the graph of f across the line given by $y = x$.   **104.** No

Exercise Set 3.5, page 253

1. $\dfrac{\pi}{2}$ **3.** $\dfrac{5\pi}{6}$ **5.** $-\dfrac{\pi}{4}$ **7.** $\dfrac{\pi}{3}$ **9.** $\dfrac{\pi}{3}$ **11.** $-\dfrac{\pi}{4}$ **13.** $-\dfrac{\pi}{3}$ **15.** $\dfrac{2\pi}{3}$ **17.** $\dfrac{\pi}{6}$ **19.** $\dfrac{\pi}{6}$ **21.** $\dfrac{1}{2}$ **23.** 2   **25.** $\dfrac{3}{5}$ **27.** 1

29. $\dfrac{1}{2}$ **31.** $\dfrac{\pi}{6}$ **33.** $\dfrac{\pi}{4}$ **35.** not defined   **37.** 0.4636   **39.** $-\dfrac{\pi}{6}$ **41.** $\dfrac{\sqrt{3}}{3}$ **43.** $\dfrac{4\sqrt{15}}{15}$ **45.** $\dfrac{24}{25}$ **47.** $\dfrac{13}{5}$ **49.** 0   **51.** $\dfrac{24}{25}$

53. $\dfrac{2 + \sqrt{15}}{6}$ **55.** $\dfrac{1}{5}\left(3\sqrt{7} - 4\sqrt{3}\right)$ **57.** $\dfrac{12}{13}$ **59.** 2   **61.** $\dfrac{2 - \sqrt{2}}{2}$ **63.** $\dfrac{7\sqrt{2}}{10}$ **65.** $\cos\dfrac{5\pi}{12} \approx 0.2588$ **67.** $\sqrt{1 - x^2}$ **69.** $\dfrac{\sqrt{x^2 - 1}}{|x|}$

75.   **77.**   **79.**   **81.**   **83. a.** 0.1014   **b.** 0.1552

85. f and g have the same graph in Quadrant I
$f(x) \neq g(x)$ **87.**   **89.**

91.   **93.**   **99.** $y = \dfrac{1}{3}\tan 5x$ **101.** $y = 3 + \cos\left(x - \dfrac{\pi}{3}\right)$

Prepare for Section 3.6, page 255

103. $x = \dfrac{5 \pm \sqrt{73}}{6}$ **104.** $1 - \cos^2 x$ **105.** $\dfrac{5}{2}\pi, \dfrac{9}{2}\pi$, and $\dfrac{13}{2}\pi$ **106.** $(x + 1)\left(x - \dfrac{\sqrt{3}}{2}\right)$ **107.**   **108.** 0, 1

Exercise Set 3.6, page 266

1. $\dfrac{\pi}{4}, \dfrac{7\pi}{4}$ **3.** $\dfrac{\pi}{3}, \dfrac{4\pi}{3}$ **5.** $\dfrac{\pi}{4}, \dfrac{\pi}{2}, \dfrac{3\pi}{4}, \dfrac{3\pi}{2}$ **7.** $\dfrac{\pi}{2}, \dfrac{3\pi}{2}$ **9.** $\dfrac{\pi}{6}, \dfrac{\pi}{4}, \dfrac{3\pi}{4}, \dfrac{11\pi}{6}$ **11.** $\dfrac{\pi}{4}, \dfrac{3\pi}{4}$ **13.** $\dfrac{\pi}{6}, \dfrac{\pi}{4}, \dfrac{5\pi}{6}$ **15.** $\dfrac{\pi}{6}, \dfrac{5\pi}{6}, \dfrac{7\pi}{6}, \dfrac{11\pi}{6}$

17. $0, \dfrac{\pi}{4}, \dfrac{3\pi}{4}, \pi, \dfrac{5\pi}{4}, \dfrac{7\pi}{4}$ **19.** $\dfrac{\pi}{6}, \dfrac{5\pi}{6}, \dfrac{4\pi}{3}, \dfrac{5\pi}{3}$ **21.** $0, \dfrac{\pi}{2}, \pi, \dfrac{3\pi}{2}$ **23.** $41.4°, 318.6°$ **25.** no solution   **27.** $68.0°, 292.0°$ **29.** no solution

31. $12.8°, 167.2°$ **33.** $15.5°, 164.5°$ **35.** $0°, 33.7°, 180°, 213.7°$ **37.** no solution   **39.** no solution   **41.** $0°, 120°, 240°$ **43.** $70.5°, 289.5°$

45. $68.2°, 116.6°, 248.2°, 296.6°$ **47.** $19.5°, 90°, 160.5°, 270°$ **49.** $60°, 90°, 300°$ **51.** $53.1°, 180°$ **53.** $72.4°, 220.2°$ **55.** $50.1°, 129.9°, 205.7°,$

$334.3°$ **57.** no solution   **59.** $22.5°, 157.5°$ **61.** $\dfrac{\pi}{8} + \dfrac{k\pi}{2}$, where k is an integer   **63.** $\dfrac{\pi}{10} + \dfrac{2k\pi}{5}$, where k is an integer

65. $0 + 2k\pi, \dfrac{\pi}{3} + 2k\pi, \pi + 2k\pi, \dfrac{5\pi}{3} + 2k\pi$, where k is an integer   **67.** $\dfrac{\pi}{2} + k\pi, \dfrac{5\pi}{6} + k\pi$, where k is an integer

69. $0 + 2k\pi$, where k is an integer   **71.** $0, \pi$ **73.** $0, \dfrac{\pi}{6}, \dfrac{\pi}{2}, \dfrac{5\pi}{6}, \pi, \dfrac{7\pi}{6}, \dfrac{3\pi}{2}, \dfrac{11\pi}{6}$ **75.** $0, \dfrac{\pi}{2}, \dfrac{3\pi}{2}$ **77.** $0, \dfrac{\pi}{3}, \dfrac{2\pi}{3}, \pi, \dfrac{4\pi}{3}, \dfrac{5\pi}{3}$ **79.** $\dfrac{4\pi}{3}, \dfrac{5\pi}{3}$

81. $0, \dfrac{\pi}{4}, \dfrac{3\pi}{4}, \pi, \dfrac{5\pi}{4}, \dfrac{7\pi}{4}$ **83.** $\dfrac{\pi}{6}, \dfrac{5\pi}{6}, \pi$ **85.** 0.7391 **87.** $-3.2957, 3.2957$ **89.** 1.16 **91.** $14.99°$ and $75.01°$

The sine regression functions in Exercises 93, 95, and 97 were obtained on a TI-83 calculator by using an iteration factor of 16. The use of a different iteration factor may produce a sine regression function that varies from the regression functions listed below.
93. a. $y \approx 1.1213 \sin(0.01595x + 1.8362) + 6.6257$ **b.** $6{:}49$ **95. a.** $y \approx 49.572 \sin(0.21261x - 2.1922) + 49.490$ **b.** 2%
97. a. $y \approx 35.185 \sin(0.30395x - 2.1630) + 2.1515$ **b.** $24.8°$ **99.** day 74 to day 268 = 195 days   **101. b.** $42°$ and $79°$ **c.** $60°$

103. $\dfrac{\pi}{6}, \dfrac{\pi}{2}$ **105.** $\dfrac{5\pi}{3}, 0$ **107.** $0, \dfrac{\pi}{4}, \dfrac{\pi}{2}, \dfrac{3\pi}{4}, \pi, \dfrac{5\pi}{4}, \dfrac{3\pi}{2}, \dfrac{7\pi}{4}$ **109.** $0, \dfrac{\pi}{2}, \pi, \dfrac{3\pi}{2}$ **111.** $\dfrac{\pi}{6}, \dfrac{\pi}{2}, \dfrac{5\pi}{6}, \dfrac{7\pi}{6}, \dfrac{3\pi}{2}, \dfrac{11\pi}{6}$

113. 0.93 foot, 1.39 feet

Chapter 3 True/False Exercises, page 274

1. False; $\dfrac{\tan 45°}{\tan 60°} \neq \dfrac{45°}{60°}$.   **2.** False; if $y = 0$, $\tan \dfrac{x}{y}$ is undefined.   **3.** False. Let $x = 1$. Then $\sin^{-1} 1 = \dfrac{\pi}{2}$ and $\csc^{-1} 1 \approx 1.18$.   **4.** False; if

$\alpha = \dfrac{\pi}{2}$, then $\sin 2\alpha = \sin \pi = 0$ but $2 \sin \dfrac{\pi}{2} = 2$.   **5.** False; $\sin(30° + 60°) \neq \sin 30° + \sin 60°$.   **6.** False; $\sin x = 0$ has an infinite number of

solutions $x = k\pi$, but $\sin x = 0$ is not an identity.   **7.** False; $\tan 45° = \tan 225°$ but $45° \neq 225°$.

8. False; $\cos^{-1}\left(\cos \dfrac{3\pi}{2}\right) = \cos^{-1}(0) = \dfrac{\pi}{2} \neq \dfrac{3\pi}{2}$.   **9.** False; $\cos(\cos^{-1} 2) \neq 2$, because $\cos^{-1} 2$ is undefined.   **10.** False; if $\alpha = 1$, then

$\csc^{-1} \dfrac{1}{\alpha} = \csc^{-1} 1 = \dfrac{\pi}{2} \neq \dfrac{1}{\csc 1} \approx 0.8415$.   **11.** False; $\sin(180° - \theta) = \sin \theta$.   **12.** False because $\sin^2 \theta \geq 0$ for all θ, but $\sin \theta^2$ can be < 0.

Chapter 3 Review Exercises, page 274

1. $\dfrac{\sqrt{6} - \sqrt{2}}{4}$ [3.2]   **2.** $\sqrt{3} - 2$ [3.2]   **3.** $\dfrac{\sqrt{6} - \sqrt{2}}{4}$ [3.2]   **4.** $\sqrt{2} - \sqrt{6}$ [3.2]   **5.** $-\dfrac{\sqrt{6} + \sqrt{2}}{4}$ [3.2]   **6.** $\dfrac{\sqrt{2} + \sqrt{6}}{4}$ [3.2]   **7.** $\dfrac{\sqrt{2} - \sqrt{2}}{2}$ [3.3]

8. $-\dfrac{\sqrt{2} - \sqrt{3}}{2}$ [3.3]   **9.** $\sqrt{2} + 1$ [3.3]   **10.** $\dfrac{\sqrt{2} + \sqrt{2}}{2}$ [3.3]   **11. a.** 0 **b.** $\sqrt{3}$ **c.** $\dfrac{1}{2}$ [3.2/3.3]   **12. a.** 0 **b.** -2 **c.** $\dfrac{1}{2}$ [3.2/3.3]

13. a. $\dfrac{\sqrt{3}}{2}$ **b.** $-\sqrt{3}$ **c.** $-\dfrac{\sqrt{2} - \sqrt{3}}{2}$ [3.2/3.3]   **14. a.** $\dfrac{\sqrt{6} - \sqrt{2}}{4}$ **b.** $-\sqrt{3}$ **c.** 1 [3.2/3.3]   **15.** $\sin 6x$ [3.3]   **16.** $\tan 3x$ [3.2]

17. $\sin 3x$ [3.2]   **18.** $\cos 4\theta$ [3.3]   **19.** $\tan 2\theta$ [3.3]   **20.** $\tan \theta$ [3.3]   **21.** $2 \sin 3\theta \sin \theta$ [3.4]   **22.** $-2 \cos 4\theta \sin \theta$ [3.4]

23. $2 \sin 4\theta \cos 2\theta$ [3.4]   **24.** $2 \cos 3\theta \sin 2\theta$ [3.4]   **43.** $\dfrac{13}{5}$ [3.5]   **44.** $\dfrac{4}{5}$ [3.5]   **45.** $\dfrac{56}{65}$ [3.5]   **46.** $\dfrac{7}{25}$ [3.5]   **47.** $\dfrac{3}{2}$ [3.6]   **48.** $\dfrac{4}{5}$ [3.6]

49. $30°, 150°, 240°, 300°$ [3.6]   **50.** $0°, 45°, 135°$ [3.6]   **51.** $\dfrac{\pi}{2} + 2k\pi, 3.8713 + 2k\pi, 5.553 + 2k\pi$, where k is an integer. [3.6]   **52.** $-\dfrac{\pi}{4} + k\pi,$

$1.2490 + k\pi$, where k is an integer. [3.6]   **53.** $\dfrac{\pi}{12}, \dfrac{5\pi}{12}, \dfrac{13\pi}{12}, \dfrac{17\pi}{12}$ [3.6]   **54.** $\dfrac{7\pi}{12}, \dfrac{19\pi}{12}, \dfrac{3\pi}{4}, \dfrac{7\pi}{4}$ [3.6]

55. $y = 2 \sin\left(x + \dfrac{\pi}{6}\right)$ [3.4]   **56.** $y = 2\sqrt{2} \sin\left(x + \dfrac{5\pi}{4}\right)$ [3.4]   **57.** $y = 2 \sin\left(x + \dfrac{4\pi}{3}\right)$ [3.4]   **58.** $y = \sin\left(x - \dfrac{\pi}{6}\right)$ [3.4]

        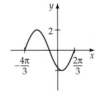

59. [3.5]   **60.** [3.5]   **61.** [3.5]   **62.** [3.5]

63. a. $y \approx 1.5698 \sin(0.01632x + 1.8145) + 5.9360$ **b.** $5{:}22$ [3.6]

Chapter 3 Test, page 277

5. $\dfrac{-\sqrt{6}+\sqrt{2}}{4}$ [3.2]   **6.** $-\dfrac{\sqrt{2}}{10}$ [3.2]   **8.** $\sin 9x$ [3.2]   **9.** $-\dfrac{7}{25}$ [3.3]   **12.** $\dfrac{2-\sqrt{3}}{4}$ [3.4]   **13.** $y=\sin\left(x+\dfrac{5\pi}{6}\right)$ [3.4]   **14.** 0.701 [3.5]

15. $\dfrac{5}{13}$ [3.5]   **16.**

[3.5]   **17.** $41.8°, 138.2°$ [3.6]   **18.** $0, \dfrac{\pi}{6}, \pi, \dfrac{11\pi}{6}$ [3.6]   **19.** $\dfrac{\pi}{2}, \dfrac{2\pi}{3}, \dfrac{4\pi}{3}$ [3.6]

20. a. $y \approx 38.965 \sin(0.40089x + 2.9143) + 33.197$ **b.** $63.9°$ [3.6]

Cumulative Review Exercises, page 278

1. $x > -3$ [1.1]   **2.** Shift the graph of $y = f(x)$ horizontally 1 unit to the left and up 2 units. [1.4]   **3.** Reflect the graph of $y = f(x)$ across the

x-axis. [1.4]   **4.** odd function [1.4/2.5]   **5.** $f^{-1}(x) = \dfrac{x}{x-5}$ [1.6]   **6.** $\dfrac{4\pi}{3}$ [2.1]   **7.** $300°$ [2.1]   **8.** $\dfrac{\sqrt{3}}{2}$ [2.2]   **9.** $\dfrac{2\sqrt{3}}{3}$ [2.2]   **10.** $\dfrac{2\sqrt{5}}{5}$ [2.2]

11. positive [2.3]   **12.** $50°$ [2.4]   **13.** $\dfrac{\pi}{3}$ [2.4]   **14.** $x = \dfrac{1}{2}, y = \dfrac{\sqrt{3}}{2}$ [2.4]   **15.** $0.43, \pi, \dfrac{\pi}{12}$ [2.7]   **16.** $\dfrac{\pi}{6}$ [3.5]   **17.** 2.498 [3.5]

18. $[-1, 1]$ [3.5]   **19.** $\left(-\dfrac{\pi}{2}, \dfrac{\pi}{2}\right)$ [3.5]   **20.** $\dfrac{\pi}{2}, \dfrac{7\pi}{6}, \dfrac{11\pi}{6}$ [3.6]

Exercise Set 4.1, page 286

1. $C = 77°, b \approx 16, c \approx 17$ **3.** $B = 38°, a \approx 18, c \approx 10$ **5.** $C \approx 15°, B \approx 33°, c \approx 7.8$ **7.** $C = 32.6°, c \approx 21.6, a \approx 39.8$
9. $B = 47.7°, a \approx 57.4, b \approx 76.3$ **11.** $A \approx 58.5°, B \approx 7.3°, a \approx 81.5$ **13.** $C = 59°, B = 84°, b \approx 46$ or $C = 121°, B = 22°, b \approx 17$
15. No triangle is formed.   **17.** No triangle is formed.   **19.** $C = 19.8°, B = 145.4°, b \approx 10.7$ or $C = 160.2°, B = 5.0°, b \approx 1.64$
21. No triangle is formed.   **23.** $C = 51.21°, A = 11.47°, c \approx 59.00$ **25.** ≈ 68.8 miles   **27.** 231 yards   **29.** ≈ 110 feet   **31.** ≈ 130 yards

33. ≈ 96 feet   **35.** ≈ 33 feet   **37.** ≈ 8.1 miles   **39.** ≈ 1200 miles   **41.** ≈ 260 meters   **45.**

minimum value of $L \approx 11.19$ meters

Prepare for Section 4.2, page 289

46. 20.7 **47.** 25.5 square inches   **48.** $C = \cos^{-1}\left(\dfrac{a^2+b^2-c^2}{2ab}\right)$ **49.** 12.5 meters   **50.** 6 **51.** $c^2 = a^2 + b^2$

Exercise Set 4.2, page 296

1. ≈ 13 **3.** ≈ 150 **5.** ≈ 29 **7.** ≈ 9.5 **9.** ≈ 10 **11.** ≈ 40.1 **13.** ≈ 90.7 **15.** $\approx 39°$ **17.** $\approx 90°$ **19.** $\approx 47.9°$ **21.** $\approx 116.67°$
23. $\approx 80.3°$ **25.** ≈ 140 square units   **27.** ≈ 53 square units   **29.** ≈ 81 square units   **31.** ≈ 299 square units   **33.** ≈ 36 square units
35. ≈ 7.3 square units   **37.** ≈ 710 miles   **39.** ≈ 74 feet   **41.** $\approx 60.9°$ **43.** ≈ 350 miles   **45.** 40 centimeters
47. ≈ 9.7 inches, 25 inches   **49.** ≈ 55 centimeters   **51.** ≈ 2800 feet   **53.** $\approx 47,500$ square meters   **55.** 162 square inches
57. $\approx \$41,000$ **59.** ≈ 6.23 acres   **61.** Triangle DEF has an incorrect dimension.   **63. a.** 64 square units   **b.** 55 square units
65. $\approx 12.5°$ **67.** ≈ 52.0 centimeters   **73.** ≈ 140 cubic inches

Prepare for Section 4.3, page 300

75. 1 **76.** -6.691 **77.** $30°$ **78.** $157.6°$ **79.** $\dfrac{\sqrt{5}}{5}$ **80.** $\dfrac{14\sqrt{17}}{17}$

Exercise Set 4.3, page 313

1. $a = 7, b = -1; \langle 7, -1\rangle$ **3.** $a = -7, b = -5; \langle -7, -5\rangle$ **5.** $a = 0, b = 8; \langle 0, 8\rangle$ **7.** $5, \approx 126.9°, \left\langle -\dfrac{3}{5}, \dfrac{4}{5}\right\rangle$

9. $\approx 44.7, \approx 296.6°, \left\langle \dfrac{\sqrt{5}}{5}, \dfrac{-2\sqrt{5}}{5}\right\rangle$ **11.** $\approx 4.5, \approx 296.6°, \left\langle \dfrac{\sqrt{5}}{5}, \dfrac{-2\sqrt{5}}{5}\right\rangle$ **13.** $\approx 45.7, \approx 336.8°, \left\langle \dfrac{7\sqrt{58}}{58}, \dfrac{-3\sqrt{58}}{58}\right\rangle$ **15.** $\langle -6, 12\rangle$

17. $\langle -1, 10 \rangle$ **19.** $\left\langle -\frac{11}{6}, \frac{7}{3} \right\rangle$ **21.** $2\sqrt{5}$ **23.** $2\sqrt{109}$ **25.** $-8i + 12j$ **27.** $14i - 6j$ **29.** $\frac{11}{12}i + \frac{1}{2}j$ **31.** $\sqrt{113}$

33. $a_1 \approx 4.5, a_2 \approx 2.3, 4.5i + 2.3j$ **35.** $a_1 \approx 2.8, a_2 \approx 2.8, 2.8i + 2.8j$ **37.** ≈ 380 miles per hour

39. ≈ 250 miles per hour at a heading of 86°   **41.** ≈ 293 pounds   **43.** ≈ 24.7 pounds   **45.** -3 **47.** 0   **49.** 1   **51.** 0   **53.** $\approx 79.7°$

55. 45°   **57.** 90°, orthogonal   **59.** 180°   **61.** $\frac{46}{5}$ **63.** $\frac{14\sqrt{29}}{29} \approx 2.6$ **65.** $\sqrt{5} \approx 2.2$ **67.** $-\frac{11\sqrt{5}}{5} \approx -4.9$ **69.** ≈ 954 foot-pounds

71. ≈ 779 foot-pounds   **73.**

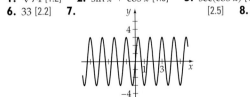

$\langle 6, 9 \rangle$

75. The vector from $P_1(3, -1)$ to $P_2(5, -4)$ is equivalent to $2i - 3j$.
77. $v \cdot w = 0$; the vectors are perpendicular.
79. $\langle 7, 2 \rangle$ is one example.   **81.** 6.4 pounds   **83.** No
87. The same amount of work is done.

Chapter 4 True/False Exercises, page 317

1. False; we cannot solve a triangle using the Law of Cosines if we are only given two sides and the angle opposite one of the given sides.
2. True   **3.** True   **4.** False; $2i \neq 2j$.   **5.** True   **6.** True   **7.** True   **8.** True   **9.** False; $v \cdot v = a^2 + b^2$.
10. False; let $v = i + j$ and $w = i - j$. Then $v \cdot w = 1 - 1 = 0$.

Chapter 4 Review Exercises, page 317

1. $B = 51°, a \approx 11, c \approx 18$ [4.1]   **2.** $A \approx 8.6°, a \approx 1.77, b \approx 11.5$ [4.1]   **3.** $B \approx 48°, C \approx 95°, A \approx 37°$ [4.2]   **4.** $A \approx 47°, B \approx 76°, C \approx 58°$ [4.2]
5. $c \approx 13, A \approx 55°, B \approx 90°$ [4.2]   **6.** $a \approx 169, B \approx 37°, C \approx 61°$ [4.2]   **7.** No triangle is formed. [4.1]   **8.** No triangle is formed. [4.1]
9. $C = 45°, a \approx 29, b \approx 35$ [4.1]   **10.** $A = 115°, a \approx 56, b \approx 26$ [4.1]   **11.** ≈ 360 square units [4.2]   **12.** ≈ 31 square units [4.2]
13. ≈ 920 square units [4.2]   **14.** ≈ 46 square units [4.2]   **15.** ≈ 790 square units [4.2]   **16.** ≈ 210 square units [4.2]
17. ≈ 170 square units [4.2]   **18.** ≈ 190 square units [4.2]   **19.** $a_1 = 5, a_2 = 3, \langle 5, 3 \rangle$ [4.3]   **20.** $a_1 = 1, a_2 = 6, \langle 1, 6 \rangle$ [4.3]

21. $\approx 4.5, 153.4°$ [4.3]   **22.** $\approx 6.7, 333.4°$ [4.3]   **23.** $\approx 3.6, 123.7°$ [4.3]   **24.** $\approx 8.1, 240.3°$ [4.3]   **25.** $\left\langle -\frac{8\sqrt{89}}{89}, \frac{5\sqrt{89}}{89} \right\rangle$ [4.3]

26. $\left\langle \frac{7\sqrt{193}}{193}, -\frac{12\sqrt{193}}{193} \right\rangle$ [4.3]   **27.** $\frac{5\sqrt{26}}{26}i + \frac{\sqrt{26}}{26}j$ [4.3]   **28.** $\frac{3\sqrt{34}}{34}i - \frac{5\sqrt{34}}{34}j$ [4.3]   **29.** $\langle -7, -3 \rangle$ [4.3]   **30.** $\langle 18, 7 \rangle$ [4.3]

31. $-6i - \frac{17}{2}j$ [4.3]   **32.** $-\frac{13}{6}i - \frac{47}{6}j$ [4.3]   **33.** 420 miles per hour [4.3]   **34.** $\approx 7°$ [4.3]   **35.** 18 [4.3]   **36.** -21 [4.3]   **37.** -9 [4.3]

38. 20 [4.3]   **39.** $\approx 86°$ [4.3]   **40.** $\approx 138°$ [4.3]   **41.** $\approx 125°$ [4.3]   **42.** $\approx 157°$ [4.3]   **43.** $\frac{10\sqrt{41}}{41}$ [4.3]   **44.** $\frac{27\sqrt{29}}{29}$ [4.3]

45. ≈ 662 foot-pounds [4.3]

Chapter 4 Test, page 318

1. $B = 94°, a \approx 48, b \approx 51$ [4.1]   **2.** $\approx 11°$ [4.1]   **3.** ≈ 14 [4.2]   **4.** $\approx 48°$ [4.2]   **5.** ≈ 39 square units [4.2]   **6.** ≈ 93 square units [4.2]
7. ≈ 260 square units [4.2]   **8.** $\sqrt{13}$ [4.3]   **9.** $-9.193i - 7.713j$ [4.3]   **10.** $-19i - 29j$ [4.3]   **11.** -1 [4.3]   **12.** 103° [4.3]
13. ≈ 27 miles [4.2]   **14.** ≈ 21 miles [4.1]   **15.** $\approx \$65,800$ [4.2]

Cumulative Review Exercises, page 319

1. $\sqrt{74}$ [1.2]   **2.** $\sin x + \cos x$ [1.5]   **3.** $\sec(\cos x)$ [1.5]   **4.** $f^{-1}(x) = 2x + 6$ [1.6]   **5.** Shifted 2 units to the right and 3 units up [1.4]
6. 33 [2.2]   **7.**   [2.5]   **8.**   [2.6]   **9.**   [2.7]

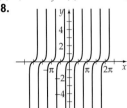

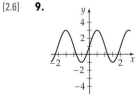

10. amplitude: 3, period: 6π, phase shift: $\dfrac{3\pi}{2}$ [2.7]   **11.** amplitude: $\sqrt{2}$, period: 2π, phase shift: $\dfrac{\pi}{4}$ [2.7]   **12.** 9.5 [4.2]   **13.** See [3.1].   **14.** $\dfrac{\pi}{3}$ [3.5]

15. $\dfrac{5}{12}$ [3.5]   **16.** $0, \dfrac{\pi}{6}, \dfrac{5\pi}{6}, \pi$ [3.6]   **17.** magnitude: 5, direction: 323.1° [4.3]   **18.** 60.3° [4.3]   **19.** 289.5° [4.3]

20. speed: 474 mph, heading: 70° [4.3]

Exercise Set 5.1, page 328

1. $9i$ **3.** $7i\sqrt{2}$ **5.** $4 + 9i$ **7.** $5 + 7i$ **9.** $8 - 3i\sqrt{2}$ **11.** $11 - 5i$ **13.** $-7 + 4i$ **15.** $8 - 5i$ **17.** -10 **19.** $-2 + 16i$ **21.** -40

23. -10 **25.** $19i$ **27.** $20 - 10i$ **29.** $22 - 29i$ **31.** 41 **33.** $12 - 5i$ **35.** $-114 + 42i\sqrt{2}$ **37.** $-6i$ **39.** $3 - 6i$ **41.** $\dfrac{7}{53} - \dfrac{2}{53}i$

43. $1 + i$ **45.** $\dfrac{15}{41} - \dfrac{29}{41}i$ **47.** $\dfrac{5}{13} + \dfrac{12}{13}i$ **49.** $2 + 5i$ **51.** $-16 - 30i$ **53.** $-11 - 2i$ **55.** $-i$ **57.** -1 **59.** $-i$ **61.** -1

63. $\dfrac{1}{2} + \dfrac{\sqrt{3}}{2}i$ **65.** $-\dfrac{3}{2} + \dfrac{\sqrt{3}}{2}i$ **67.** $\dfrac{1}{2} + \dfrac{1}{2}i$ **69.** $(x + 4i)(x - 4i)$ **71.** $(z + 5i)(z - 5i)$ **73.** $(2x + 9i)(2x - 9i)$ **79.** 0

Prepare for Section 5.2, page 330

88. $1 + 3i$ **89.** $\dfrac{1}{2} + \dfrac{1}{2}i$ **90.** $2 - 3i$ **91.** $3 + 5i$ **92.** $-\dfrac{1}{2} \pm \dfrac{\sqrt{3}}{2}i$ **93.** $-3i, 3i$

Exercise Set 5.2, page 336

1–7.

$|-2 - 2i| = 2\sqrt{2}$
$|\sqrt{3} - i| = 2$
$|-2i| = 2$
$|3 - 5i| = \sqrt{34}$

9. $\sqrt{2}$ cis 315°   **11.** 2 cis 330°   **13.** 3 cis 90°   **15.** 5 cis 180°   **17.** $\sqrt{2} + i\sqrt{2}$

19. $\dfrac{\sqrt{2}}{2} - \dfrac{\sqrt{2}}{2}i$ **21.** $-3\sqrt{2} + 3i\sqrt{2}$ **23.** 8 **25.** $-\sqrt{3} + i$ **27.** $-3i$ **29.** $-4\sqrt{2} + 4i\sqrt{2}$ **31.** $\dfrac{9\sqrt{3}}{2} - \dfrac{9}{2}i$

33. $\approx -0.832 + 1.819i$ **35.** 6 cis 255°   **37.** 12 cis 335°   **39.** 10 cis $\dfrac{16\pi}{15}$ **41.** 24 cis 6.5   **43.** $-4 - 4i\sqrt{3}$ **45.** $3i$ **47.** $-\dfrac{3\sqrt{3}}{2} + \dfrac{3i}{2}$

49. $\approx -2.081 + 4.546i$ **51.** $\approx 2.732 - 0.732i$ **53.** $6 + 0i = 6$ **55.** $-\dfrac{1}{2} + \dfrac{\sqrt{3}}{2}i$ **57.** $0 - \sqrt{2}i = -\sqrt{2}i$ **59.** $16 - 16i$

61. $-\dfrac{3}{8} + \dfrac{\sqrt{3}}{8}i$ **63.** $59.0 + 43.0i$ **65.** r^2 or $a^2 + b^2$

Prepare for Section 5.3, page 338

67. i **68.** 2   **69.** 3   **70.** $2\sqrt{2}$ cis $\dfrac{\pi}{4}$ **71.** $-\sqrt{3} + i$ **72.** 1

Exercise Set 5.3, page 342

1. $-128 - 128i\sqrt{3}$ **3.** $-16 + 16i\sqrt{3}$ **5.** $16\sqrt{2} + 16i\sqrt{2}$ **7.** $64 + 0i = 64$ **9.** $0 - 32i = -32i$ **11.** $1024 - 1024i$ **13.** $0 - 1i = -i$

15. $3 + 0i = 3$ **17.** $2 + 0i = 2$ **19.** $0.809 + 0.588i$ **21.** $1 + 0i = 1$ **23.** $1.070 + 0.213i$ **25.** $-0.276 + 1.563i$
$\quad\;\; -3 + 0i = -3$ $\qquad\; 1 + i\sqrt{3}$ $\qquad\;\;\; -0.309 + 0.951i$ $\qquad -\dfrac{1}{2} + \dfrac{\sqrt{3}}{2}i$ $\qquad -0.213 + 1.070i$ $\qquad -1.216 - 1.020i$
$\qquad\qquad\qquad\quad\; -1 + i\sqrt{3}$ $\qquad\;\;\; -1 + 0i = -1$ $\qquad -1.070 - 0.213i$ $\qquad 1.492 - 0.543i$
$\qquad\qquad\qquad\quad\; -2 + 0i = -2$ $\qquad -0.309 - 0.951i$ $\qquad -\dfrac{1}{2} - \dfrac{\sqrt{3}}{2}i$ $\qquad 0.213 - 1.070i$
$\qquad\qquad\qquad\quad\; -1 - i\sqrt{3}$ $\qquad\;\;\; 0.809 - 0.588i$
$\qquad\qquad\qquad\qquad 1 - i\sqrt{3}$

27. $2\sqrt{2} + 2i\sqrt{6}$ **29.** 2 cis 60°   **31.** cis 67.5°   **33.** 3 cis 0°   **35.** 3 cis 45°   **37.** $\sqrt[4]{2}$ cis 75°   **39.** $\sqrt[3]{2}$ cis 80°
$\quad\; -2\sqrt{2} - 2i\sqrt{6}$ $\qquad$ 2 cis 180° $\qquad$ cis 157.5° $\qquad$ 3 cis 120° $\qquad$ 3 cis 135° $\qquad$ $\sqrt[4]{2}$ cis 165° $\qquad$ $\sqrt[3]{2}$ cis 200°
$\qquad\qquad\qquad\qquad$ 2 cis 300° $\qquad$ cis 247.5° $\qquad$ 3 cis 240° $\qquad$ 3 cis 225° $\qquad$ $\sqrt[4]{2}$ cis 255° $\qquad$ $\sqrt[3]{2}$ cis 320°
$\qquad\qquad\qquad\qquad\qquad\qquad\qquad$ cis 337.5° $\qquad\qquad\qquad\qquad$ 3 cis 315° $\qquad$ $\sqrt[4]{2}$ cis 345°

Chapter 5 True/False Exercises, page 346

1. False; $\sqrt{(-1)^2} = \sqrt{1} = 1$ **2.** True   **3.** True   **4.** True   **5.** False; $\dfrac{1 + i}{1 - i} = i$ **6.** True   **7.** True   **8.** False; $z^2 = r^2(\cos 2\theta + i \sin 2\theta)$

9. True   **10.** False; $\cos \pi + i \sin \pi = -1$ **11.** False; $\cos \pi + i \sin \pi = -1$ **12.** True

Chapter 5 Review Exercises, page 346

1. $3 - 8i, 3 + 8i$ [5.1]   **2.** $6 + 2i, 6 - 2i$ [5.1]   **3.** $-2 + i\sqrt{5}, -2 - i\sqrt{5}$ [5.1]   **4.** $-5 - 3i\sqrt{3}, -5 + 3i\sqrt{3}$ [5.1]   **5.** -4 [5.1]
6. 9 [5.1]   **7.** $5 + 2i.$ [5.1]   **8.** $-3 + 4i$ [5.1]   **9.** $-3 + 3i$ [5.1]   **10.** $-5 - 15i$ [5.1]   **11.** $25 - 19i$ [5.1]   **12.** $29 - 2i$ [5.1]

13. $\dfrac{8}{25} - \dfrac{6}{25}i$ [5.1]   **14.** $\dfrac{26}{53} + \dfrac{15}{53}i$ [5.1]   **15.** $-2 - 2i$ [5.1]   **16.** $2 - 11i$ [5.1]   **17.** $6 + 6i$ [5.1]   **18.** $-5 - 6i$ [5.1]   **19.** 7 [5.1]

20. $11 + i\sqrt{5}$ [5.1]   **21.** $-i$ [5.1]   **22.** i [5.1]   **23.** 1 [5.1]   **24.** -1 [5.1]   **25.** 8 [5.1]   **26.** $\sqrt{13}$ [5.1]   **27.** $\sqrt{41}$ [5.1]   **28.** $\sqrt{2}$ [5.1]

29. $2\sqrt{2}$ cis $315°$ [5.2]   **30.** 2 cis$(150°)$ [5.2]   **31.** $\approx \sqrt{13}$ cis $146.3°$ [5.2]   **32.** $\sqrt{17}$ cis$(345.96°)$ [5.2]   **33.** $\dfrac{5\sqrt{2}}{2} - \dfrac{5\sqrt{2}}{2}i \approx 3.536 - 3.536i$ [5.2]

34. $-3 - 3i\sqrt{3}$ [5.2]   **35.** $\approx -0.832 + 1.819i$ [5.2]   **36.** $-1.27 + 2.72i$ [5.2]   **37.** $0 - 30i = -30i$ [5.2]   **38.** $-5\sqrt{2} - 5i\sqrt{2}$ [5.2]

39. $\approx -8.918 + 8.030i$ [5.2]   **40.** $\approx -0.968 + 3.881i$ [5.2]   **41.** $\approx -6.012 - 13.742i$ [5.2]   **42.** $\approx 13.2 + 27.0i$ [5.2]   **43.** 3 cis $(-100°)$ [5.2]

44. 3 cis $110°$ [5.2]   **45.** 5 cis $(-59°)$ [5.2]   **46.** $\sqrt{2}$ cis $(-50°)$ [5.2]   **47.** $\dfrac{5}{3}$ cis 1.9 [5.2]   **48.** $\dfrac{1}{2}$ cis (-4) [5.2]

49. $-\dfrac{243\sqrt{2}}{2} - \dfrac{243\sqrt{2}}{2}i \approx -171.827 - 171.827i$ [5.3]   **50.** -1 [5.3]   **51.** $64 - 64\sqrt{3}\,i \approx 64 - 110.851i$ [5.3]   **52.** $32,768i$ [5.3]
53. $-16\sqrt{2} + 16i\sqrt{2} \approx -22.627 + 22.627i$ [5.3]   **54.** $-237 + 3116i$ [5.3]   **55.** 3 cis $30°$, 3 cis $150°$, 3 cis $270°$ [5.3]
56. $\sqrt[4]{8}$ cis $22.5°$, $\sqrt[4]{8}$ cis $112.5°$, $\sqrt[4]{8}$ cis $202.5°$, $\sqrt[4]{8}$ cis $292.5°$ [5.3]   **57.** 4 cis $0°$, 4 cis $90°$, 4 cis $180°$, 4 cis $270°$ [5.3]
58. 2 cis $45°$, 2 cis $117°$, 2 cis $189°$, 2 cis $261°$, 2 cis $333°$ [5.3]   **59.** 3 cis $0°$, 3 cis $90°$, 3 cis $180°$, 3 cis $270°$ [5.3]
60. 5 cis $60°$, 5 cis $180°$, 5 cis $300°$ [5.3]

Chapter 5 Test, page 347

1. $6 + 3i$ [5.1]   **2.** $3i\sqrt{2}$ [5.1]   **3.** $10 - i$ [5.1]   **4.** $-9 + 9i$ [5.1]   **5.** -6 [5.1]   **6.** $-i$ [5.1]   **7.** $5 + 16i$ [5.1]   **8.** $3 - 54i$ [5.1]

9. $16 + 30i$ [5.1]   **10.** $-5 - 4i$ [5.1]   **11.** $-\dfrac{13}{25} - \dfrac{34}{25}i$ [5.1]   **12.** $2 + 4i$ [5.1]   **13.** $\sqrt{34}$ [5.1]   **14.** $3\sqrt{2}$ cis $315°$ [5.2]   **15.** 6 cis $270°$ [5.2]

16. $-2 + 2i\sqrt{3}$ [5.2]   **17.** $-\dfrac{5\sqrt{2}}{2} - \dfrac{5i\sqrt{2}}{2}$ [5.2]   **18.** $6\sqrt{2} + 6i\sqrt{2}$ [5.2]   **19.** $\approx -11.472 + 16.383i$ [5.2]   **20.** -4 [5.2]   **21.** $-6i$ [5.2]

22. $16,777,216$ [5.3]   **23.** 2 cis $0°$, 2 cis $60°$, 2 cis $120°$, 2 cis $180°$, 2 cis $240°$, 2 cis $300°$ [5.3]   **24.** $\sqrt[3]{2}$ cis $40°$, $\sqrt[3]{2}$ cis $160°$, $\sqrt[3]{2}$ cis $280°$ [5.3]

25. 2 cis $36°$, 2 cis $108°$, 2 cis $180°$, 2 cis $252°$, 2 cis $324°$ [5.3]

Cumulative Review Exercises, page 348

1. $[-2, 3]$ [1.1]   **2.** All real numbers except -2 and 2 [1.3]   **3.** -2 [1.3]   **4.** $(f \circ g)(x) = \sin(x^2 - 1)$ [1.5]

5. $\dfrac{3}{2}$ [1.6]   **6.** $270°$ [2.1]   **7.** 25.4 centimeters [2.2]   **8.** $a = -1, b = 1$ [2.4]   **9.**   [2.5]   **10.**   [2.6]

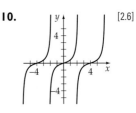

11. See [3.1].   **12.** $-\sin(x)$ [3.2]   **13.** $\dfrac{56}{65}$ [3.2]   **14.** $\dfrac{33}{65}$ [3.5]   **15.** $0, \dfrac{\pi}{6}, \pi, \dfrac{11\pi}{6}$ [3.6]   **16.** 170 centimeters [4.2]   **17.** $64.7°$ [4.3]

18. 1449 foot-pounds [0.0]   **19.** $2\sqrt{2}$ cis$(45°)$ [5.2]   **20.** $-3, \dfrac{3}{2} + \dfrac{3\sqrt{3}}{2}i, \dfrac{3}{2} - \dfrac{3\sqrt{3}}{2}i$ [5.3]

Exercise Set 6.1, page 358

1. vertex: $(0, 0)$

focus: $(0, -1)$

directrix: $y = 1$

3. vertex: $(0, 0)$

focus: $\left(\dfrac{1}{12}, 0\right)$

directrix: $x = -\dfrac{1}{12}$

5. vertex: $(2, -3)$

focus: $(2, -1)$

directrix: $y = -5$

7. vertex: $(2, -4)$

focus: $(1, -4)$

directrix: $x = 3$

9. vertex: $(-4, 1)$

focus: $\left(-\dfrac{7}{2}, 1\right)$

directrix: $x = -\dfrac{9}{2}$

11. vertex: $(2, 2)$

focus: $\left(2, \dfrac{5}{2}\right)$

directrix: $y = \dfrac{3}{2}$

13. vertex: $(-4, -10)$

focus: $\left(-4, -\dfrac{39}{4}\right)$

directrix: $y = -\dfrac{41}{4}$

15. vertex: $\left(-\dfrac{7}{4}, \dfrac{3}{2}\right)$

focus: $\left(-2, \dfrac{3}{2}\right)$

directrix: $x = -\dfrac{3}{2}$

17. vertex: $(-5, -3)$

focus: $\left(-\dfrac{9}{2}, -3\right)$

directrix: $x = -\dfrac{11}{2}$

19. vertex: $\left(-\dfrac{3}{2}, \dfrac{13}{12}\right)$

focus: $\left(-\dfrac{3}{2}, \dfrac{1}{3}\right)$

directrix: $y = \dfrac{11}{6}$

21. vertex: $\left(2, -\dfrac{5}{4}\right)$

focus: $\left(2, -\dfrac{3}{4}\right)$

directrix: $y = -\dfrac{7}{4}$

23. vertex: $\left(\dfrac{9}{2}, -1\right)$

focus: $\left(\dfrac{35}{8}, -1\right)$

directrix: $x = \dfrac{37}{8}$

25. vertex: $\left(1, \dfrac{1}{9}\right)$

focus: $\left(1, \dfrac{31}{36}\right)$

directrix: $y = -\dfrac{23}{26}$

27. $x^2 = -16y$ **29.** $(x + 1)^2 = 4(y - 2)$ **31.** $(x - 3)^2 = 4(y + 4)$

33. $(x + 4)^2 = 4(y - 1)$ **35.** vertex: $(250, 20)$, focus: $\left(\dfrac{3240}{13}, 20\right)$ **37.** on axis 4 feet above vertex **39.** 6.0 inches **41. a.** 5900 square feet

b. 56,800 square feet **43.** $a = 1.5$ inches **45.** $(-0.3660, -0.3660)$ and $(1.3660, 1.3660)$ **47.** $(-1.5616, 3.8769)$ and $(2.5616, 12.1231)$

49. 4 **51.** $4|p|$ **53.** **55.** **57.** $x^2 + y^2 - 8x - 8y - 2xy = 0$

Prepare for Section 6.2, page 360

58. midpoint: $(2, 3)$; **59.** $-8, 2$ **60.** $1 \pm \sqrt{3}$ **61.** $x^2 - 8x + 16 = (x - 4)^2$ **62.** $C(-1, -2), r = 3$ **63.**
length: $2\sqrt{13}$

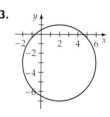

Exercise Set 6.2, page 370

1. vertices: $(0, 5), (0, -5)$
center: $(0, 0)$

foci: $(0, 3), (0, -3)$

3. vertices: $(3, 0), (-3, 0)$
center: $(0, 0)$

foci: $\left(\sqrt{5}, 0\right), \left(-\sqrt{5}, 0\right)$

5. vertices: $(0, 3), (0, -3)$
center: $(0, 0)$

foci: $\left(0, \sqrt{2}\right), \left(0, -\sqrt{2}\right)$

7. vertices: $(0, 4), (0, -4)$
center: $(0, 0)$

foci: $\left(0, \dfrac{\sqrt{55}}{2}\right), \left(0, -\dfrac{\sqrt{55}}{2}\right)$

9. vertices: $(8, -2), (-2, -2)$
center: $(3, -2)$
foci: $(6, -2), (0, -2)$

11. vertices: $(-2, 5), (-2, -5)$
center: $(-2, 0)$
foci: $(-2, 4), (-2, -4)$

13. vertices: $\left(1 + \sqrt{21}, 3\right), \left(1 - \sqrt{21}, 3\right)$
center: $(1, 3)$
foci: $\left(1 + \sqrt{17}, 3\right), \left(1 - \sqrt{17}, 3\right)$

15. vertices: $(1, 2), (1, -4)$
center: $(1, -1)$
foci: $\left(1, -1 + \dfrac{\sqrt{65}}{3}\right), \left(1, -1 - \dfrac{\sqrt{65}}{3}\right)$

17. vertices: $(2, 0), (-2, 0)$
center: $(0, 0)$
foci: $(1, 0), (-1, 0)$

19. vertices: $(0, 5), (0, -5)$
center: $(0, 0)$
foci: $(0, 3), (0, -3)$

21. vertices: $(0, 4), (0, -4)$
center: $(0, 0)$
foci: $\left(0, \dfrac{\sqrt{39}}{2}\right), \left(0, -\dfrac{\sqrt{39}}{2}\right)$

23. vertices: $(3, 6), (3, 2)$
center: $(3, 4)$
foci: $\left(3, 4 + \sqrt{3}\right), \left(3, 4 - \sqrt{3}\right)$

25. vertices: $(-1, -3), (5, -3)$
center: $(2, -3)$
foci: $(0, -3), (4, -3)$

27. vertices: $(2, 4), (2, -4)$
center: $(2, 0)$
foci: $\left(2, \sqrt{7}\right), \left(2, -\sqrt{7}\right)$

29. vertices: $(-1, 6), (-1, -4)$
center: $(-1, 1)$
foci: $(-1, 4), (-1, -2)$

31. vertices: $\left(\dfrac{11}{2}, -1\right), \left(\dfrac{1}{2}, -1\right)$
center: $(3, -1)$
foci: $\left(3 + \dfrac{\sqrt{17}}{2}, -1\right), \left(3 - \dfrac{\sqrt{17}}{2}, -1\right)$

33. $\dfrac{x^2}{25} + \dfrac{y^2}{9} = 1$ **35.** $\dfrac{x^2}{36} + \dfrac{y^2}{16} = 1$ **37.** $\dfrac{x^2}{36} + \dfrac{y^2}{81/8} = 1$ **39.** $\dfrac{(x + 2)^2}{16} + \dfrac{(y - 4)^2}{7} = 1$ **41.** $\dfrac{(x - 2)^2}{25/24} + \dfrac{(y - 4)^2}{25} = 1$

43. $\dfrac{(x - 5)^2}{16} + \dfrac{(y - 1)^2}{25} = 1$ **45.** $\dfrac{x^2}{25} + \dfrac{y^2}{21} = 1$ **47.** $\dfrac{x^2}{20} + \dfrac{y^2}{36} = 1$ **49.** $\dfrac{(x - 1)^2}{25} + \dfrac{(y - 3)^2}{21} = 1$ **51.** $\dfrac{x^2}{80} + \dfrac{y^2}{144} = 1$

53. 41 centimeters from the emitter   **55.** $\dfrac{x^2}{884.74^2} + \dfrac{y^2}{883.35^2} = 1$ **57.** 40 feet   **59.** $\dfrac{\left(x - \dfrac{9\sqrt{15}}{2}\right)^2}{324} + \dfrac{y^2}{81/4} = 1$ **61.** 1512

63. a. $\dfrac{x^2}{307.5^2} + \dfrac{y^2}{255^2} = 1$ **b.** 246,300 square feet

65. $y = \dfrac{-36 \pm \sqrt{1296 - 36(16x^2 - 108)}}{18}$ **67.** $y = \dfrac{54 \pm \sqrt{2916 - 36(16x^2 - 64x + 1)}}{18}$ **69.** $y = \dfrac{-18 \pm \sqrt{324 - 36(4x^2 + 24x + 44)}}{18}$

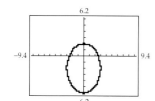 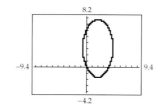 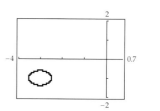

71. $\dfrac{x^2}{36} + \dfrac{y^2}{27} = 1$ **73.** $\dfrac{(x-1)^2}{16} + \dfrac{(y-2)^2}{12} = 1$ **75.** $\dfrac{9}{2}$ **79.** $x = \pm \dfrac{9\sqrt{5}}{5}$

Prepare for Section 6.3, page 374

83. midpoint: $(1, -1)$; length: $2\sqrt{13}$ **84.** $-4, 2$ **85.** $\sqrt{2}$ **86.** $4(x^2 + 6x + 9) = 4(x + 3)^2$ **87.** 8

88.

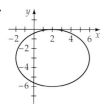

Exercise Set 6.3 page 385

1. center: $(0, 0)$

vertices: $(\pm 4, 0)$

foci: $(\pm\sqrt{41}, 0)$

asymptotes: $y = \pm\dfrac{5}{4}x$

3. center: $(0, 0)$

vertices: $(0, \pm 2)$

foci: $(0, \pm\sqrt{29})$

asymptotes: $y = \pm\dfrac{2}{5}x$

5. center: $(0, 0)$

vertices: $(\pm\sqrt{7}, 0)$

foci: $(\pm 4, 0)$

asymptotes: $y = \pm\dfrac{3\sqrt{7}}{7}x$

7. center: $(0, 0)$

vertices: $\left(\pm\dfrac{3}{2}, 0\right)$

foci: $\left(\pm\dfrac{\sqrt{73}}{2}, 0\right)$

asymptotes: $y = \pm\dfrac{8}{3}x$

9. center: $(3, -4)$
vertices: $(7, -4), (-1, -4)$
foci: $(8, -4), (-2, -4)$

asymptotes: $y + 4 = \pm\dfrac{3}{4}(x - 3)$

11. center: $(1, -2)$
vertices: $(1, 0), (1, -4)$
foci: $\left(1, -2 \pm 2\sqrt{5}\right)$

asymptotes: $y + 2 = \pm\dfrac{1}{2}(x - 1)$

13. center: $(-2, 0)$
vertices: $(1, 0), (-5, 0)$
foci: $\left(-2 \pm \sqrt{34}, 0\right)$

asymptotes: $y = \pm\dfrac{5}{3}(x + 2)$

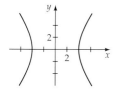

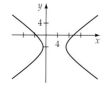

15. center: $(1, -1)$

vertices: $\left(\dfrac{7}{3}, -1\right), \left(-\dfrac{1}{3}, -1\right)$

foci: $\left(1 \pm \dfrac{\sqrt{97}}{3}, -1\right)$

asymptotes: $y + 1 = \pm\dfrac{9}{4}(x - 1)$

17. center: $(0, 0)$

vertices: $(\pm 3, 0)$

foci: $\left(\pm 3\sqrt{2}, 0\right)$

asymptotes: $y = \pm x$

19. center: $(0, 0)$

vertices: $(0, \pm 3)$

foci: $(0, \pm 5)$

asymptotes: $y = \pm\dfrac{3}{4}x$

21. center: $(0, 0)$

vertices: $\left(0, \pm\dfrac{2}{3}\right)$

foci: $\left(0, \pm\dfrac{\sqrt{5}}{3}\right)$

asymptotes: $y = \pm 2x$

23. center: $(3, 4)$

vertices: $(3, 6), (3, 2)$

foci: $\left(3, 4 \pm 2\sqrt{2}\right)$

asymptotes: $y - 4 = \pm(x - 3)$

25. center: $(-2, -1)$

vertices: $(-2, 2), (-2, -4)$

foci: $\left(-2, -1 \pm \sqrt{13}\right)$

asymptotes: $y + 1 = \pm\dfrac{3}{2}(x + 2)$

27. $y = \dfrac{-6 \pm \sqrt{36 + 4(4x^2 + 32x + 39)}}{-2}$

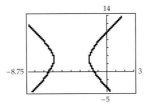

29. $y = \dfrac{64 \pm \sqrt{4096 + 64(9x^2 - 36x + 116)}}{-32}$

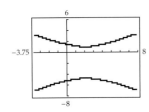

31. $y = \dfrac{18 \pm \sqrt{324 + 36(4x^2 + 8x - 6)}}{-18}$

33. $\dfrac{x^2}{9} - \dfrac{y^2}{7} = 1$ **35.** $\dfrac{y^2}{20} - \dfrac{x^2}{5} = 1$

37. $\dfrac{y^2}{9} - \dfrac{x^2}{36/7} = 1$ **39.** $\dfrac{y^2}{16} - \dfrac{x^2}{64} = 1$ **41.** $\dfrac{(x - 4)^2}{4} - \dfrac{(y - 3)^2}{5} = 1$ **43.** $\dfrac{(x - 4)^2}{144/41} - \dfrac{(y + 2)^2}{225/41} = 1$ **45.** $\dfrac{(y - 2)^2}{3} - \dfrac{(x - 7)^2}{12} = 1$

47. $\dfrac{(y - 7)^2}{1} - \dfrac{(x - 1)^2}{3} = 1$ **49.** $\dfrac{x^2}{4} - \dfrac{y^2}{12} = 1$ **51.** $\dfrac{(x - 4)^2}{36/7} - \dfrac{(y - 1)^2}{4} = 1$ and $\dfrac{(y - 1)^2}{36/7} - \dfrac{(x - 4)^2}{4} = 1$ **53. a.** $\dfrac{x^2}{2162.25} - \dfrac{y^2}{13,462.75} = 1$

b. 221 miles   **55.** $y^2 - x^2 = 10,000^2$, hyperbola   **57.** ellipse   **59.** parabola   **61.** parabola

63. ellipse

65. $\dfrac{x^2}{1} - \dfrac{y^2}{3} = 1$ **67.** $\dfrac{y^2}{9} - \dfrac{x^2}{7} = 1$ **69.** $x = \pm\dfrac{16\sqrt{41}}{41}$ **73.**

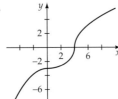

Prepare for Section 6.4, page 388

75. $\cos\alpha\cos\beta - \sin\alpha\sin\beta$ **76.** $\sin\alpha\cos\beta + \cos\alpha\sin\beta$ **77.** $\dfrac{\pi}{6}$ **78.** $150°$ **79.** hyperbola

80.

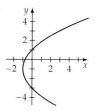

Exercise Set 6.4, page 396

1. $45°$ **3.** $36.9°$ **5.** $73.5°$ **7.** $45°, \dfrac{x'^2}{8} - \dfrac{y'^2}{8} = 1$ **9.** $18.4°, \dfrac{x'^2}{9} + \dfrac{y'^2}{3} = 1$ **11.** $26.6°, \dfrac{x'^2}{1/2} - \dfrac{y'^2}{1/3} = 1$

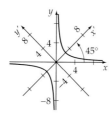 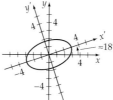 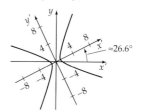

13. $30°, y' = x'^2 + 4$ **15.** $36.9°, y'^2 = 2(x' - 2)$ **17.** $36.9°, 15x'^2 - 10y'^2 + 6x' + 28y' + 11 = 0$

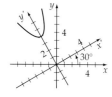 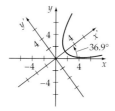 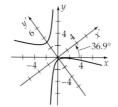

19.

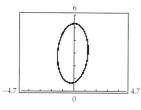

21.

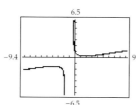

23.

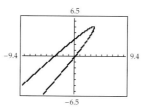

25. $y = \dfrac{\sqrt{3} + 2\sqrt{2}}{2\sqrt{3} - \sqrt{2}}x$ and $y = \dfrac{\sqrt{3} - 2\sqrt{2}}{2\sqrt{3} + \sqrt{2}}x$ **27.** $\left(\dfrac{3\sqrt{15}}{5}, \dfrac{\sqrt{15}}{5}\right)$ and $\left(-\dfrac{3\sqrt{15}}{5}, -\dfrac{\sqrt{15}}{5}\right)$ **29.** hyperbola   **31.** parabola   **33.** parabola

35. hyperbola   **39.** $9x^2 - 4xy + 6y^2 = 100$

Prepare for Section 6.5, page 397

45. odd   **46.** even   **47.** $\dfrac{2\pi}{3}, \dfrac{5\pi}{3}$ **48.** $240°$ **49.** r^2 **50.** $(4.2, 2.6)$

Exercise Set 6.5, page 409

1–7.

9.

$0 \le \theta \le 2\pi$

11.

$0 \le \theta \le \pi$

13.

15.

$0 \le \theta \le 2\pi$

17.

$0 \le \theta \le \pi$

19.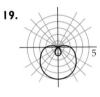

$0 \le \theta \le 2\pi$

21.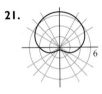

$0 \le \theta \le 2\pi$

23.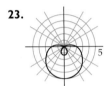

$0 \le \theta \le 2\pi$

25. 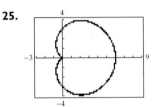 $0 \le \theta \le 2\pi$

27. 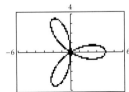 $0 \le \theta \le \pi$ **29.** 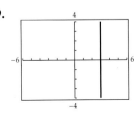 $0 \le \theta \le \pi$ **31.** 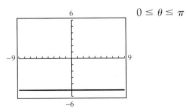 $0 \le \theta \le \pi$

33. 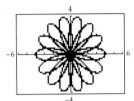 $0 \le \theta \le 4\pi$ **35.** 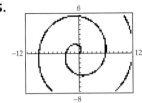 $0 \le \theta \le 6\pi$ **37.** 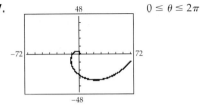 $0 \le \theta \le 2\pi$

39. 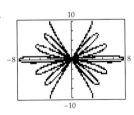 $0 \le \theta \le 2\pi$ **41.** $(2, -60°)$ **43.** $\left(\dfrac{3}{2}, -\dfrac{3\sqrt{3}}{2}\right)$ **45.** $(0, 0)$ **47.** $(5, 53.1°)$ **49.** $x^2 + y^2 - 3x = 0$

51. $x = 3$ **53.** $x^2 + y^2 = 16$ **55.** $x^4 - y^2 + x^2 y^2 = 0$ **57.** $y^2 + 4x - 4 = 0$ **59.** $y = 2x + 6$ **61.** $r = 2 \csc \theta$ **63.** $r = 2$

65. $r \cos^2 \theta = 8 \sin \theta$ **67.** $r^2(\cos 2\theta) = 25$ **69.** 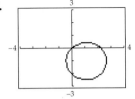   **71.**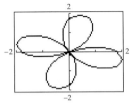

73. 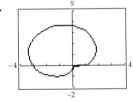   **75.** 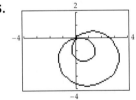   **79.** 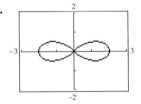   **81.**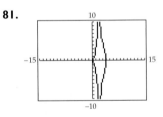

$0 \le \theta \le 4\pi$ $0 \le \theta \le 2\pi$

83.

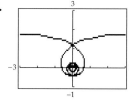

$$-4\pi \le \theta \le 4\pi$$

85.

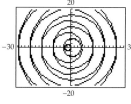

$$-30 \le \theta \le 30$$

87. a.

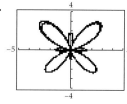

$$0 \le \theta \le 5\pi$$

b.

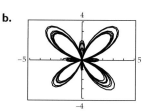

$$0 \le \theta \le 20\pi$$

Prepare for Section 6.6, page 410

88. $\dfrac{3}{5}$ **89.** $x = -1$ **90.** $y = \dfrac{2}{1 - 2x}$ **91.** $\dfrac{3\pi}{2}$ **92.** $e > 1$ **93.** $\dfrac{4}{2 - \cos x}$

Exercise Set 6.6, page 416

1. hyperbola   **3.** ellipse   **5.** parabola   **7.** hyperbola   **9.** ellipse with two holes

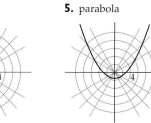
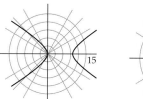
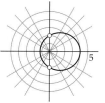

11. ellipse with two holes   **13.** parabola   **15.** $3x^2 - y^2 + 16x + 16 = 0$ **17.** $16x^2 + 7y^2 + 48y - 64 = 0$

19. $x^2 - 6y - 9 = 0$

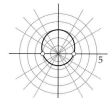

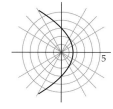

21. $r = \dfrac{2}{1 - 2\cos\theta}$ **23.** $r = \dfrac{2}{1 + \sin\theta}$ **25.** $r = \dfrac{8}{3 - 2\sin\theta}$ **27.** $r = \dfrac{6}{2 + 3\cos\theta}$ **29.** $r = \dfrac{4}{1 - \cos\theta}$ **31.** $r = \dfrac{3}{1 - 2\sin\theta}$

33.

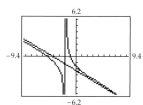

Rotate the graph in Exercise 1 counterclockwise $\dfrac{\pi}{6}$ radian

35.

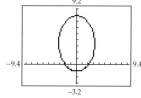

Rotate the graph in Exercise 3 counterclockwise π radians.

37.

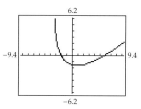

Rotate the graph in Exercise 5 clockwise $\dfrac{\pi}{6}$ radian.

39.

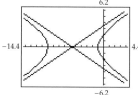

Rotate the graph in Exercise 7 clockwise π radians.

41.

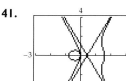

43.

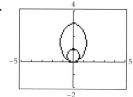

45.

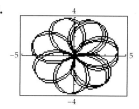

$$0 \le \theta \le 12\pi$$

Prepare for Section 6.7, page 417

49. $y^2 + 3y + \dfrac{9}{4} = \left(y + \dfrac{3}{2}\right)^2$ **50.** $y = 4t^2 + 4t + 1$ **51.** ellipse   **52.** 1   **53.** $t = e^y$ **54.** domain: $(-\infty, \infty)$;
range: $[-3, 3]$

Exercise Set 6.7, page 424

1. 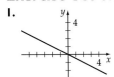   **3.** 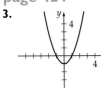   **5.** 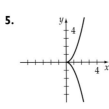   **7.** 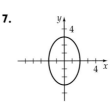   **9.**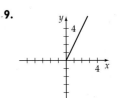

11. $x^2 - y^2 - 1 = 0$
$x \geq 1$
$y \in R$ 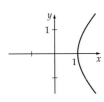   **13.** $y = -2x + 7$
$x \leq 2$
$y \geq 3$ 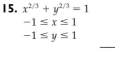   **15.** $x^{2/3} + y^{2/3} = 1$
$-1 \leq x \leq 1$
$-1 \leq y \leq 1$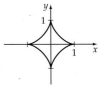

17. $y = x^2 - 1$
$x \geq 0$
$y \geq -1$ **19.** $y = \ln x$
$x > 0$
$y \in R$ 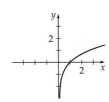   **21.** $C_1: y = -2x + 5, x \geq 2; C_2: y = -2x + 5, x \in R$.
C_2 is a line. C_1 is a ray.

23. 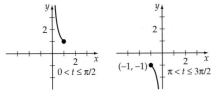   **25.** 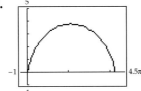   **27.**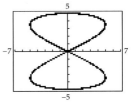

$Xscl = 2\pi$

29. 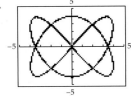   **31.** 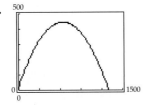   **33.**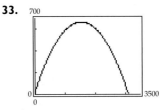

Max height (nearest foot) of 462 feet
is attained when $t \approx 5.38$ seconds.

Range (nearest foot) of 1295 feet is
attained when $t \approx 10.75$ seconds.

Max height (nearest foot) of 694 feet
is attained when $t \approx 6.59$ seconds.

Range (nearest foot) of 3084 feet is
attained when $t \approx 13.17$ seconds.

37. $x = a \cos \theta + a\theta \sin \theta$

$y = a \sin \theta - a\theta \cos \theta$ **39.** $x = (b - a) \cos \theta + a \cos\left(\dfrac{b - a}{a}\theta\right)$

$y = (b - a) \sin \theta - a \sin\left(\dfrac{b - a}{a}\theta\right)$

Chapter 6 True/False Exercises, page 428

1. False; a parabola has no asymptotes.   **2.** True   **3.** False; by keeping the foci fixed and varying asymptotes, we can make the conjugate axis any size needed.   **4.** False; $\dfrac{x^2}{25} + \dfrac{y^2}{9} = 1$ and $\dfrac{x^2}{36} + \dfrac{y^2}{20} = 1$ have the same c's but different a's.   **5.** False; parabolas have no asymptotes.   **6.** True   **7.** False; the graph of a parabola can be a function.   **8.** False; $x = \cos t$, $y = \sin t$ graphs to be a circle.   **9.** True   **10.** True

Chapter 6 Review Exercises, page 428

1. center: $(0, 0)$
vertices: $(\pm 2, 0)$
foci: $(\pm 2\sqrt{2}, 0)$
asymptotes: $y = \pm x$
[6.3]

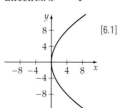

2. vertex: $(0, 0)$
focus: $(4, 0)$
directrix: $x = -1$
[6.1]

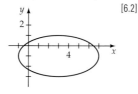

3. center: $(3, -1)$
vertices: $(-1, -1), (7, -1)$
foci: $\left(3 \pm 2\sqrt{3}, -1\right)$
[6.2]

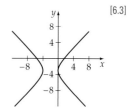

4. center: $(-2, -3)$
vertices: $(0, -3), (-4, -3)$
foci: $\left(-2 + \sqrt{7}, -3\right), \left(-2 - \sqrt{7}, -3\right)$
asymptotes: $y + 3 = \pm\dfrac{\sqrt{3}}{2}(x + 2)$
[6.3]

5. vertex: $(-2, 1)$
focus: $\left(-\dfrac{29}{16}, 1\right)$
directrix: $x = -\dfrac{35}{16}$
[6.1]

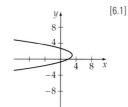

6. vertex: $(3, 1)$
focus: $\left(\dfrac{21}{8}, 1\right)$
directrix: $x = \dfrac{27}{8}$
[6.1]

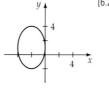

7. center: $(-2, 1)$
vertices: $(-2, -2), (-2, 4)$
foci: $\left(-2, 1 \pm \sqrt{5}\right)$
[6.2]

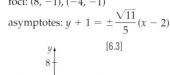

8. center: $(2, -1)$
vertices: $(7, -1), (-3, -1)$
foci: $(8, -1), (-4, -1)$
asymptotes: $y + 1 = \pm\dfrac{\sqrt{11}}{5}(x - 2)$
[6.3]

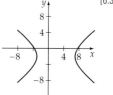

9. center: $\left(1, \dfrac{2}{3}\right)$
vertices: $\left(-5, \dfrac{2}{3}\right), \left(7, \dfrac{2}{3}\right)$
foci: $\left(1 \pm 2\sqrt{13}, \dfrac{2}{3}\right)$
asymptotes: $y - \dfrac{2}{3} = \pm\dfrac{2(x - 1)}{3}$
[6.3]

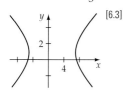

10. center: $\left(-2, \dfrac{1}{2}\right)$
vertices: $\left(2, \dfrac{1}{2}\right), \left(-6, \dfrac{1}{2}\right)$
foci: $\left(-2 + \sqrt{7}, \dfrac{1}{2}\right), \left(-2 - \sqrt{7}, \dfrac{1}{2}\right)$
[6.2]

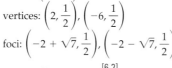
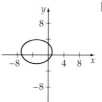

11. vertex: $\left(-\dfrac{7}{2}, -1\right)$
focus: $\left(-\dfrac{7}{2}, -3\right)$
directrix: $y = 1$
[6.1]

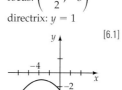

12. vertex: $(3, 3)$
focus: $\left(3, \dfrac{21}{4}\right)$
directrix: $y = -\dfrac{3}{4}$
[6.1]

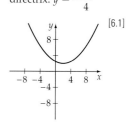

13. $\dfrac{(x - 2)^2}{25} + \dfrac{(y - 3)^2}{16} = 1$ [6.2]   **14.** $\dfrac{(x - 1)^2}{9} - \dfrac{(y - 1)^2}{7} = 1$ [6.3]   **15.** $\dfrac{(x + 2)^2}{4} - \dfrac{(y - 2)^2}{5} = 1$ [6.3]   **16.** $(y + 3)^2 = -8(x - 4)$ [6.1]

17. $x^2 = \dfrac{3(y + 2)}{2}$ or $(y + 2)^2 = 12x$ [6.1]   **18.** $\dfrac{(x + 2)^2}{9} + \dfrac{(y + 1)^2}{5} = 1$ [6.2]   **19.** $\dfrac{x^2}{36} - \dfrac{y^2}{4/9} = 1$ [6.3]   **20.** $y = (x - 1)^2$ [6.1]

21. $x'^2 + 2y'^2 - 4 = 0$, ellipse [6.4]   **22.** $6(x')^2 + \dfrac{\sqrt{2}}{2}x' + \dfrac{9\sqrt{2}}{2}y' - 12 = 0$, parabola [6.4]   **23.** $x'^2 - 4y' + 8 = 0$, parabola [6.4]

24. $\frac{1}{2}(x')^2 - \frac{1}{2}(y')^2 - \sqrt{2}x' - 1 = 0$, hyperbola [6.4]   **25.** [6.5]   **26.** [6.5]   **27.** [6.5]

28. [6.5]   **29.** [6.5]   **30.** [6.5]   **31.** [6.5] 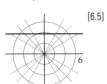   **32.** [6.5]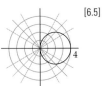

33. [6.5]   **34.** [6.5]   **35.** $r\sin^2\theta = 16\cos\theta$ [6.5]   **36.** $r + 4\cos\theta + 3\sin\theta = 0$ [6.5]   **37.** $3r\cos\theta - 2r\sin\theta = 6$ [6.5]

38. $r^2\sin 2\theta = 8$ [6.5]   **39.** $y^2 = 8x + 16$ [6.5]   **40.** $x^2 - 3x + y^2 + 4y = 0$ [6.5]   **41.** $x^4 + y^4 + 2x^2y^2 - x^2 + y^2 = 0$ [6.5]

42. $y = (\tan 1)x$ [6.5]   **43.** [6.6]   **44.** [6.6]   **45.** [6.6] 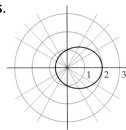   **46.** [6.6]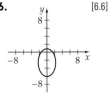

47. $y = \frac{3}{4}x + \frac{5}{2}$ [6.7] 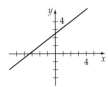   **48.** $y = 2x + 1, x \le 1$ [6.7] 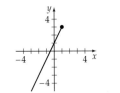   **49.** $\frac{x^2}{16} + \frac{y^2}{9} = 1$ [6.7]   **50.** $\frac{x^2}{1} - \frac{y^2}{16} = 1$ [6.7]

51. $y = -2x, x > 0$ [6.7]    **52.** $(x - 1)^2 + (y - 2)^2 = 1$ [6.7]   **53.** $y = 2^{-x^2}, x \ge 0$ [6.7]   **54.** [6.7]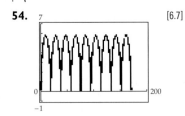

55. [6.4] 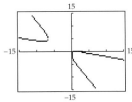   **56.** [6.5] 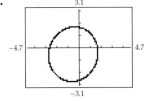   **57.** [6.7]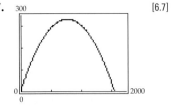

Max. height (nearest foot) of 278 feet is attained when $t \approx 4.17$ seconds.

Chapter 6 Test, page 430

1. focus: $(0, 2)$
vertex: $(0, 0)$
directrix: $y = -2$ [6.1]

2. [6.2]

3. vertices: $(3, 4)$, $(3, -6)$
foci: $(3, 3)$, $(3, -5)$ [6.2]

4. $\dfrac{x^2}{45} + \dfrac{(y + 3)^2}{9} = 1$ [6.2]

5. 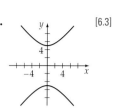 [6.3]

6. vertices: $(6, 0)$, $(-6, 0)$
foci: $(-10, 0)$, $(10, 0)$
asymptotes: $y = \pm\dfrac{4x}{3}$ [6.3]

7. 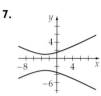 [6.3]

8. $73.15°$ [6.4]

9. ellipse [6.4]

10. $P(2, 300°)$ [6.5]

11. 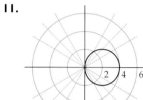 [6.5]

12. 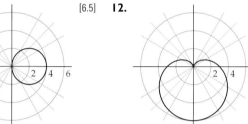 [6.5]

13. 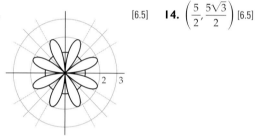 [6.5]

14. $\left(\dfrac{5}{2}, \dfrac{5\sqrt{3}}{2}\right)$ [6.5]

15. $y^2 - 8x - 16 = 0$ [6.5]

16. $x^2 + 8y - 16 = 0$ [6.6]

17. $(x + 3)^2 = \dfrac{1}{2}y$

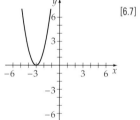 [6.7]

18. $\dfrac{x^2}{16} + \dfrac{(y - 2)^2}{1} = 1$

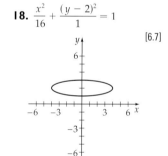 [6.7]

19. 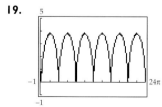 [6.7]
$Xscl = 2\pi$

20. $256\sqrt{3}$ feet ≈ 443 feet [6.7]

Cumulative Review Exercises, page 431

1. $-2 \pm \sqrt{10}$ [1.1]
2. origin [1.4]
3. $(g \circ f)(x) = 3\sin(x) - 2$ [1.5]
4. $\dfrac{4\pi}{3}$ [2.1]
5. 11 mph [2.1]
6. $\tan(t) = -\sqrt{3}$ [2.4]

7. 19 centimeters [2.4]
8. amplitude: $\dfrac{1}{2}$, period: 6 [2.5]
9. 3 [2.6]
10. See [3.1].
11. $-\dfrac{63}{65}$ [3.2]
12. $\dfrac{2\sqrt{6}}{5}$ [3.5]
13. $\dfrac{5}{13}$ [3.6]

14. 298 feet [4.1]   **15.** 38° [4.2]   **16.** $-24.6\mathbf{i} + 17.2\mathbf{j}$ [4.3]   **17.** No. $\mathbf{v} \cdot \mathbf{w} = 1 \neq 0$ [4.3]   **18.** $4 \operatorname{cis}\left(\dfrac{2\pi}{3}\right)$ [5.2]

19. 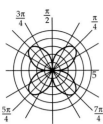   [6.5]   **20.** $y = x^2 + 2x + 2$ [6.7]

Exercise Set 7.1, page 444

1. $f(0) = 1; f(4) = 81$ **3.** $g(-2) = \dfrac{1}{100}; g(3) = 1000$ **5.** $h(2) = \dfrac{9}{4}; h(-3) = \dfrac{8}{27}$ **7.** $j(-2) = 4; j(4) = \dfrac{1}{16}$ **9.** 9.19   **11.** 9.03   **13.** 9.74

15. a. $k(x)$ **b.** $g(x)$ **c.** $h(x)$ **d.** $f(x)$

17.   **19.** 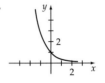   **21.**   **23.**

25. Shift the graph of f vertically upward 2 units.   **27.** Shift the graph of f horizontally to the right 2 units.
29. Reflect the graph of f across the y-axis.   **31.** Stretch the graph of f vertically away from the x-axis by a factor of 2.
33. Reflect the graph of f across the y-axis and then shift this graph vertically upward 2 units.

35. no horizontal asymptote 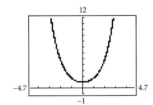   **37.** no horizontal asymptote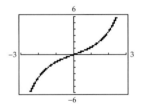

39. horizontal asymptote: $y = 0$ 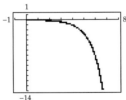   **41.** horizontal asymptote: $y = 10$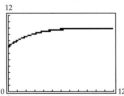

43. a. 122 million connections   **b.** 2006   **45. a.** 233 items per month; 59 items per month   **b.** The demand will approach
25 items per month.   **47. a.** 0.53   **b.** 0.89   **c.** 5.2 minutes   **d.** There is a 98% probability that at least one customer will arrive between
10:00 A.M. and 10:05.2 A.M.   **49. a.** 8.7%   **b.** 2.6%   **51. a.** 6400; 409,600   **b.** 11.6 hours   **53. a.** 515,000 people   **b.** 1997
55. a. 363 beneficiaries; 88,572 beneficiaries   **b.** 13 rounds   **57. a.** 141°F   **b.** after 28.3 minutes   **59. a.** 261.63 vibrations per second
b. No. The function $f(n)$ is not a linear function. Therefore, the graph of $f(n)$ does not increase at a constant rate.
63.   **65.** $(-\infty, \infty)$ **67.** $[0, \infty)$

Prepare for Section 7.2, page 449

68. 4   **69.** 3   **70.** 5   **71.** $f^{-1}(x) = \dfrac{3x}{2 - x}$ **72.** $\{x | x \geq 2\}$ **73.** the set of all positive real numbers

Exercise Set 7.2, page 458

1. $10^1 = 10$ **3.** $8^2 = 64$ **5.** $7^0 = x$ **7.** $e^4 = x$ **9.** $e^0 = 1$ **11.** $\log_3 9 = 2$ **13.** $\log_4 \dfrac{1}{16} = -2$ **15.** $\log_b y = x$ **17.** $\ln y = x$

19. $\log 100 = 2$ **21.** 2   **23.** -5 **25.** 3   **27.** -2 **29.** -4 **31.**

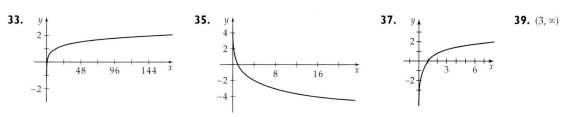

33. **35.** **37.** **39.** $(3, \infty)$

41. $(-\infty, 11)$ **43.** $(-\infty, -2) \cup (2, \infty)$ **45.** $(4, \infty)$ **47.** $(-1, 0) \cup (1, \infty)$ **49.**

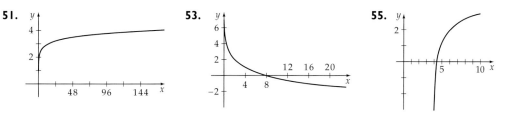

51. **53.** **55.**

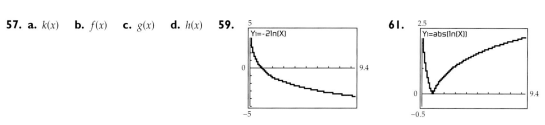

57. a. $k(x)$ **b.** $f(x)$ **c.** $g(x)$ **d.** $h(x)$ **59.** **61.**

63. **65.** **67.**

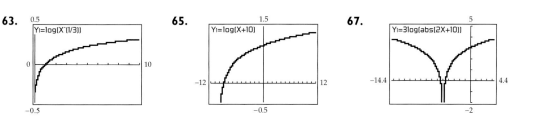

69. a. 2.0%   **b.** 45 months   **71. a.** 3298 units; 3418 units; 3490 units   **b.** 2750 units   **73.** 2.05 square meters
75. a. Answers will vary.   **b.** 96 digits   **c.** 3385 digits   **d.** 6,320,430 digits   **77.** f and g are inverse functions.
79. range of f: $\{y \mid -1 < y < 1\}$; range of g: all real numbers

Prepare for Section 7.3, page 461

81. ≈ 0.77815 for each expression   **82.** ≈ 0.98083 for each expression   **83.** ≈ 1.80618 for each expression
84. ≈ 3.21888 for each expression   **85.** ≈ 1.60944 for each expression   **86.** ≈ 0.90309 for each expression

Exercise Set 7.3, page 471

1. $\log_b x + \log_b y + \log_b z$ **3.** $\ln x - 4\ln z$ **5.** $\dfrac{1}{2}\log_2 x - 3\log_2 y$ **7.** $\dfrac{1}{2}\log_7 x + \dfrac{1}{2}\log_7 z - 2\log_7 y$ **9.** $\log[x^2(x+5)]$ **11.** $\ln(x+y)$
13. $\log[x^3 \cdot \sqrt[3]{y}(x+1)]$ **15.** 1.5395   **17.** 0.8672   **19.** -0.6131 **21.** 0.6447   **23.**

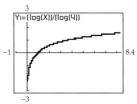

25.   **27.**   **29.**

31. False; $\log 10 + \log 10 = 2$ but $\log(10+10) = \log 20 \neq 2$.   **33.** True   **35.** False; $\log 100 - \log 10 = 1$ but $\log(100 - 10) = \log 90 \neq 1$.
37. False; $\dfrac{\log 100}{\log 10} = \dfrac{2}{1} = 2$ but $\log 100 - \log 10 = 1$.   **39.** False; $(\log 10)^2 = 1$ but $2\log 10 = 2$.   **41.** 2   **43.** 500^{501}
45. $1:870{,}551$; $1:757{,}858$; $1:659{,}754$; $1:574{,}349$; $1:500{,}000$ **47.** 10.4; base   **49.** 3.16×10^{-10} mole per liter
51. a. 82.0 decibels   **b.** 40.3 decibels   **c.** 115.0 decibels   **d.** 152.0 decibels   **53.** 10 times more intense   **55.** 5
57. $10^{6.5}I_0$ or about $3{,}162{,}277.7I_0$ **59.** 100 to 1   **61.** $10^{1.8}$ to 1 or about 63 to 1   **63.** 5.5   **65. a.** $M \approx 6$ **b.** $M \approx 4$ **c.** The results are
close to the magnitudes produced by the amplitude-time-difference formula.

Prepare for Section 7.4, page 474

66. $\log_3 729 = 6$ **67.** $5^4 = 625$ **68.** $\log_a b = x + 2$ **69.** $x = \dfrac{4a}{7b + 2c}$ **70.** $x = \dfrac{3}{44}$ **71.** $x = \dfrac{100(A-1)}{A+1}$

Exercise Set 7.4, page 483

1. 6   **3.** $-\dfrac{3}{2}$ **5.** $-\dfrac{6}{5}$ **7.** 3   **9.** $\dfrac{\log 70}{\log 5}$ **11.** $-\dfrac{\log 120}{\log 3}$ **13.** $\dfrac{\log 315 - 3}{2}$ **15.** $\ln 10$ **17.** $\dfrac{\ln 2 - \ln 3}{\ln 6}$ **19.** $\dfrac{3\log 2 - \log 5}{2\log 2 + \log 5}$
21. 7   **23.** 4   **25.** $2 + 2\sqrt{2}$ **27.** $\dfrac{199}{95}$ **29.** -1 **31.** 3   **33.** 10^{10} **35.** 2   **37.** 5   **39.** $\log(20 + \sqrt{401})$ **41.** $\dfrac{1}{2}\log\left(\dfrac{3}{2}\right)$
43. $\ln(15 \pm 4\sqrt{14})$ **45.** $\ln(1 + \sqrt{65}) - \ln 8$ **47.** 1.61   **49.** 0.96   **51.** 2.20   **53.** -1.93 **55.** -1.34

57. a. 8500, 10,285   **b.** in 6 years   **59. a.** 60°F   **b.** 27 minutes   **61. a.**

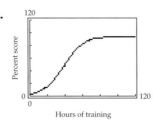

b. 48 hours
c. $P = 100$
d. As the number of hours of
training increases, the test
scores approach 100%.

63. a.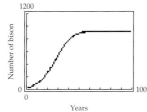
b. in 27 years or the year 2026   **c.** $B = 1000$ **d.** As the number of years increases, the bison population approaches but never reaches or exceeds 1000.

65. a.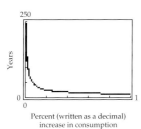
b. 78 years   **c.** 1.9%

67. a. 116 feet per second   **b.** $v = 150$ **c.** The velocity of the package approaches but never reaches or exceeds 150 feet per second.

69. a. 1.72 seconds   **b.** $v = 100$ **c.** The object cannot fall faster than 100 feet per second.

71. a.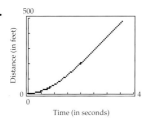
b. 2.6 seconds   **73.** 138

75. The second step; because $\log 0.5 < 0$, the inequality sign must be reversed.   **77.** $x = \dfrac{y}{y - 1}$ **79.** $e^{0.336} \approx 1.4$

Prepare for Section 7.5, page 487
81. 1220.39   **82.** 824.96   **83.** -0.0495 **84.** 1340   **85.** 0.025   **86.** 12.8

Exercise Set 7.5, page 498

1. a. \$9724.05   **b.** \$11,256.80   **3. a.** \$48,885.72   **b.** \$49,282.20   **c.** \$49,283.30   **5.** \$24,730.82   **7.** 8.8 years   **9.** $t = \dfrac{\ln 3}{r}$
11. 14 years   **13. a.** 2200 bacteria   **b.** 17,600 bacteria   **15. a.** $N(t) \approx 22{,}600e^{0.01368t}$ **b.** 27,700   **17. a.** 10,755,000   **b.** 2042

19. a.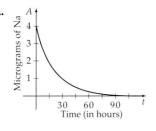
b. 3.18 micrograms   **c.** $\approx$15.07 hours   **d.** $\approx$30.14 hours   **21.** $\approx$6601 years ago

23. $\approx$2378 years old   **25. a.** 0.056   **b.** 42°F   **c.** 54 minutes   **27. a.** 211 hours   **b.** 1386 hours
29. 3.1 years

31. a. 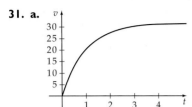   **b.** 0.98 second   **c.** $v = 32$ **d.** As time increases, the velocity approaches but never reaches or exceeds 32 feet per second.

33. a. 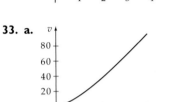   **b.** 2.5 seconds   **c.** ≈24.56 feet per second   **d.** The average speed of the object was approximately 24.56 feet per second during the period from $t = 1$ to $t = 2$ seconds.

35. a. 1900   **b.** 0.16   **c.** 200   **37. a.** 157,500   **b.** 0.04   **c.** 45,000   **39. a.** 2400   **b.** 0.12   **c.** 300

41. $P(t) \approx \dfrac{5500}{1 + 12.75e^{-0.37263t}}$ **43.** $P(t) \approx \dfrac{100}{1 + 4.55556e^{-0.22302t}}$ **45. a.** $158,000, $163,000   **b.** $625,000

47. a. $P(t) \approx \dfrac{1600}{1 + 4.12821e^{-0.06198t}}$ **b.** about 497 wolves   **49. a.** $P(t) \approx \dfrac{8500}{1 + 4.66667e^{-0.14761t}}$ **b.** 2010   **51.** 45 hours

53. a. 0.71 gram   **b.** 0.96 gram   **c.** 0.52 gram   **55. a.** 1.7%   **b.** 13.9%   **c.** 19.0%   **d.** 19.5%   **e.** 1.5%; $P \to 0$
57. 13,715,120,270 centuries

Prepare for Section 7.6, page 503
59. decreasing   **60.** decreasing   **61.** 36   **62.** 840   **63.** 15.8   **64.** $P = 55$

Exercise Set 7.6, page 512
1. increasing exponential function   **3.** decreasing exponential function;   **5.** decreasing logarithmic function
decreasing logarithmic function

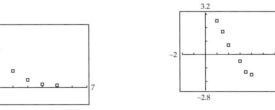

7. $y \approx 0.99628(1.20052)^x$; $r \approx 0.85705$ **9.** $y \approx 1.81505(0.51979)^x$; $r \approx -0.99978$ **11.** $y \approx 4.89060 - 1.35073 \ln x$; $r \approx -0.99921$
13. $y \approx 14.05858 + 1.76393 \ln x$; $r \approx 0.99983$ **15.** $y \approx \dfrac{235.58598}{1 + 1.90188e^{-0.05101x}}$ **17.** $y \approx \dfrac{2098.68307}{1 + 1.19794e^{-0.06004x}}$
19. a. $y \approx 5.48184(1.00356)^x$; 6.78%   **b.** 69 months   **21. a.** $T \approx 0.06273(1.07078)^F$ **b.** 5.3 hours   **23. a.** $T \approx 0.07881(1.07259)^F$
b. 7.5 hours; 2.2 hours   **25. a.** $N(t) \approx 1500.093(1.940)^t$ **b.** 2005   **27. a.** $p \approx 7.862(1.026)^y$ **b.** 36 centimeters
29. a. LinReg: pH $\approx 0.01353q + 7.02852$, $r \approx 0.956627$; LnReg: pH $\approx 6.10251 + 0.43369 \ln q$, $r \approx 0.999998$. The logarithmic model provides a
better fit.   **b.** 126.0   **31. a.** $p \approx 3200(0.91894)^t$; 2012   **b.** No. The model fits the data perfectly because there are only two data points.

33. a. ExpReg: $a \approx 8000(1.10657)^t$; 550,500,000 automobiles   **b.** 2004   **35. a.** logistic: distance in feet $\approx \dfrac{71.84158}{1 + 13.77825e^{-0.07915t}}$;

logarithmic: distance in feet $\approx -22.58293 + 20.91655 \ln t$ **b.** logistic growth model   **c.** 71.65 feet
37. a. LinReg: $w \approx 10.17227t + 16.45111$, $r \approx 0.95601$; LnReg: $w \approx 18.26750 + 31.03499 \ln t$, $r \approx 0.99996$ **b.** The logarithmic model provides
a better fit.   **c.** 89.7 cubic yards   **39.** A and B have different exponential regression functions.   **41. a.** ExpReg: $y \approx 1.81120(1.61740)^x$,
$r \approx 0.96793$; PwrReg: $y \approx 2.09385(x)^{1.40246}$, $r \approx 0.99999$ **b.** The power regression function provides the better fit.

Chapter 7 True/False Exercises, page 521

1. True **2.** True **3.** True **4.** False; f is not defined for negative values of x, and thus $g(f(x))$ is undefined for negative values of x.
5. False; $h(x)$ is not an increasing function for $0 < b < 1$. **6.** False; $j(x)$ is not an increasing function for $0 < b < 1$.
7. True **8.** True **9.** True **10.** True **11.** False; $\log x + \log y = \log(xy)$. **12.** True **13.** True **14.** True

Chapter 7 Review Exercises, page 521

1. 2 [7.2] **2.** 4 [7.2] **3.** 3 [7.2] **4.** π [7.2] **5.** -2 [7.4] **6.** 8 [7.4] **7.** -3 [7.4] **8.** -4 [7.4] **9.** ±1000 [7.4] **10.** $\pm10^{10}$ [7.4]

11. 7 [7.4] **12.** ±8 [7.4] **13.** [7.1] **14.** [7.1] **15.** [7.1]

16. [7.1] **17.** [7.1] **18** [7.1] **19.** [7.2]

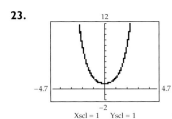

20. [7.2] **21.** [7.2] **22.** [7.2]

23. [7.1] **24.** [7.1] **25.** $4^3 = 64$ [7.2] **26.** $\left(\dfrac{1}{2}\right)^{-3} = 8$ [7.2]

27. $(\sqrt{2})^4 = 4$ [7.2] **28.** $e^0 = 1$ [7.2] **29.** $\log_5 125 = 3$ [7.2] **30.** $\log_2 1024 = 10$ [7.2] **31.** $\log_{10} 1 = 0$ [7.2] **32.** $\log_8 2\sqrt{2} = \dfrac{1}{2}$ [7.2]

33. $2\log_b x + 3\log_b y - \log_b z$ [7.3] **34.** $\dfrac{1}{2}\log_b x - 2\log_b y - \log_b z$ [7.3] **35.** $\ln x + 3\ln y$ [7.3] **36.** $\dfrac{1}{2}\ln x + \dfrac{1}{2}\ln y - 4\ln z$ [7.3]

37. $\log\left(x^2\sqrt[3]{x+1}\right)$ [7.3] **38.** $\log\dfrac{x^5}{(x+5)^2}$ [7.3] **39.** $\ln\dfrac{\sqrt{2xy}}{z^3}$ [7.3] **40.** $\ln\dfrac{xz}{y}$ [7.3] **41.** 2.86754 [7.3] **42.** 3.35776 [7.3]

43. -0.117233 [7.3] **44.** -0.578989 [7.3] **45.** $\dfrac{\ln 30}{\ln 4}$ [7.4] **46.** $\dfrac{\log 41}{\log 5} - 1$ [7.4] **47.** 4 [7.4] **48.** $\dfrac{1}{6}e$ [7.4] **49.** 4 [7.4] **50.** 15 [7.4]

51. $\dfrac{\ln 3}{2\ln 4}$ [7.4] **52.** $\dfrac{\ln(8 \pm 3\sqrt{7})}{\ln 5}$ [7.4] **53.** 10^{1000} [7.4] **54.** $e^{(e^2)}$ [7.4] **55.** 1,000,005 [7.4] **56.** $\dfrac{15 + \sqrt{265}}{2}$ [7.4] **57.** 81 [7.4]
58. $\pm\sqrt{5}$ [7.4] **59.** 4 [7.4] **60.** 5 [7.4] **61.** 7.7 [7.3] **62.** 5.0 [7.3] **63.** 3162 to 1 [7.3] **64.** 2.8 [7.3] **65.** 4.2 [7.3]
66. $\approx 3.98 \times 10^{-6}$ [7.3] **67. a.** \$20,323.79 **b.** \$20,339.99 [7.5] **68. a.** \$25,646.69 **b.** \$25,647.32 [7.5] **69.** \$4,438.10 [7.5]
70. a. 69.9% **b.** 6 days **c.** 19 days [7.5] **71.** $N(t) \approx e^{0.8047t}$ [7.5] **72.** $N(t) \approx 2e^{0.5682t}$ [7.5] **73.** $N(t) \approx 3.783e^{0.0558t}$ [7.5]
74. $N(t) \approx e^{-0.6931t}$ [7.5] **75. a.** $P(t) \approx 25{,}200e^{0.06155789t}$ **b.** 38,800 [7.5] **76.** 340 years [7.5]
77. a. linear: $P \approx -63{,}121t + 7{,}599{,}401$, $r \approx -0.93813$; exponential: $P \approx 64{,}717{,}271(0.96174359)^t$, $r \approx -0.95227$;
logarithmic: $P \approx 29{,}163{,}839 - 6{,}052{,}741 \ln t$, $r \approx -0.94256$ **b.** The exponential equation provides a better fit for the data.
c. 1,040,000 [7.6] **78. a.** linear: $R \approx -0.475297t + 53.1037$, $r \approx -0.98118$; exponential: $R \approx 207.544(0.966206)^t$, $r \approx -0.99660$;
logarithmic: $R \approx 181.202 - 38.0586 \ln t$, $r \approx -0.99073$ **b.** The exponential equation provides a better fit for the data.
c. 5.1 per 1000 live births [7.6] **79. a.** $P(t) \approx \dfrac{1400}{1 + \dfrac{17}{3}e^{-0.22458t}}$ **b.** 1070 [7.5] **80. a.** $21\dfrac{1}{3}$ **b.** $P(t) \to 128$ [7.5]

Chapter 7 Test, page 524

1. a. $b^c = 5x - 3$ [7.2]   **b.** $\log_3 y = \dfrac{x}{2}$ [7.2]   **2.** $2 \log_b z - 3 \log_b y - \dfrac{1}{2} \log_b x$ [7.3]   **3.** $\log \dfrac{2x + 3}{(x - 2)^3}$ [7.3]   **4.** 1.7925 [7.3]

5.   [7.1]   **6.**   [7.2]   **7.** 1.9206 [7.4]   **8.** $\dfrac{5 \ln 4}{\ln 28}$ [7.4]   **9.** 1 [7.4]   **10.** -3 [7.4]

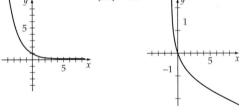

11. a. \$29,502.36   **b.** \$29,539.62 [7.5]   **12.** 17.36 years [7.5]   **13. a.** 7.6   **b.** 63 to 1 [7.3]   **14. a.** $P(t) \approx 34{,}600e^{0.04667108t}$ **b.** 55,000 [7.5]
14. 690 years [7.5]   **16. a.** $y \approx 1.67199(2.47188)^x$ **b.** 1945 [7.6]   **17. a.** $R(t) \approx 1.74830 + 0.78089 \ln x$; 2.73%   **b.** 2.6 years [7.6]
18. a. $P(t) \approx \dfrac{1100}{1 + 5.875e^{-0.20429t}}$ **b.** $\approx$457 raccoons [7.5]

Cumulative Review Exercises, page 525

1. $\cos(x^2 + 1)$ [1.5]   **2.** $f^{-1}(x) = \dfrac{1}{2}x - 4$ [1.6]   **3.** $\sin \theta = \dfrac{3}{5}$, $\cos \theta = \dfrac{4}{5}$, $\tan \theta = \dfrac{3}{4}$ [2.2]   **4.** 16.7 centimeters [2.2]

5. amplitude: 4, period: π, phase shift: $-\dfrac{\pi}{4}$ [2.5]   **6.** amplitude: $\sqrt{2}$, period: 2π [3.4]   **7.** odd [2.4]   **9.** $\dfrac{12}{5}$ [3.5]   **10.** $\dfrac{\pi}{2}, \dfrac{7\pi}{6}, \dfrac{11\pi}{6}$ [3.6]

11. magnitude: 5, direction: 126.9° [4.3]   **12.** 176.8° [4.3]   **13.** ground speed: $\approx$420 mph; course: $\approx$55° [4.1]   **14.** 26° [4.1]   **15.** 16 [5.3]

16. $\dfrac{\sqrt{2}}{2} + i\dfrac{\sqrt{2}}{2}, -\dfrac{\sqrt{2}}{2} - i\dfrac{\sqrt{2}}{2}$ [5.3]   **17.** $\left(\sqrt{2}, 45°\right)$ [6.5]   **18.**   [6.5]   **19.** 1.43 [7.4]   **20.** 1.8 milligrams [7.5]

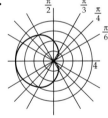

INDEX

Important Formulas

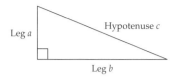

Pythagorean Theorem
$c^2 = a^2 + b^2$

The *distance* between $P_1(x_1, y_1)$ and $P_2(x_2, y_2)$ is

$$d(P_1, P_2) = \sqrt{(x_2 - x_1)^2 + (y_2 - y_1)^2}$$

The *slope m* of a line through $P_1(x_1, y_1)$ and $P_2(x_2, y_2)$ is

$$m = \frac{y_2 - y_1}{x_2 - x_1}, \quad x_1 \neq x_2$$

Quadratic Formula
If $a \neq 0$, the solutions of $ax^2 + bx + c = 0$ are

$$x = \frac{-b \pm \sqrt{b^2 - 4ac}}{2a}$$

Properties of Functions

A *function* is a set of ordered pairs in which no two ordered pairs that have the same first coordinate have different second coordinates.

If a and b are elements of an interval I that is a subset of the domain of a function f, then

- f is an *increasing* function on I if $f(a) < f(b)$ whenever $a < b$.
- f is a *decreasing* function on I if $f(a) > f(b)$ whenever $a < b$.
- f is a *constant* function on I if $f(a) = f(b)$ for all a and b.

A *one-to-one* function satisfies the additional condition that given any y, there is one and only one x that can be paired with that given y.

Graphing Concepts

Odd Functions
A function f is an odd function if $f(-x) = -f(x)$ for all x in the domain of f. The graph of an odd function is symmetric with respect to the origin.

Even Functions
A function is an even function if $f(-x) = f(x)$ for all x in the domain of f. The graph of an even function is symmetric with respect to the y-axis.

Vertical and Horizontal Translations
If f is a function and c is a positive constant, then the graph of

- $y = f(x) + c$ is the graph of $y = f(x)$ shifted up *vertically* c units.
- $y = f(x) - c$ is the graph of $y = f(x)$ shifted down *vertically* c units.
- $y = f(x + c)$ is the graph of $y = f(x)$ shifted left *horizontally* c units.
- $y = f(x - c)$ is the graph of $y = f(x)$ shifted right *horizontally* c units.

Reflections
If f is a function then the graph of

- $y = -f(x)$ is the graph of $y = f(x)$ reflected across the x-axis.
- $y = f(-x)$ is the graph of $y = f(x)$ reflected across the y-axis.

Vertical Shrinking and Stretching
- If $c > 0$ and the graph of $y = f(x)$ contains the point (x, y), then the graph of $y = c \cdot f(x)$ contains the point (x, cy).
- If $c > 1$, the graph of $y = c \cdot f(x)$ is obtained by stretching the graph of $y = f(x)$ away from the x-axis by a factor of c.
- If $0 < c < 1$, the graph of $y = c \cdot f(x)$ is obtained by shrinking the graph of $y = f(x)$ toward the x-axis by a factor of c.

Horizontal Shrinking and Stretching
- If $a > 0$ and the graph of $y = f(x)$ contains the point (x, y), then the graph of $y = f(ax)$ contains the point $\left(\dfrac{1}{a}x, y\right)$.
- If $a > 1$, the graph of $y = f(ax)$ is a *horizontal shrinking* of the graph of $y = f(x)$.
- If $0 < a < 1$, the graph of $y = f(ax)$ is a *horizontal stretching* of the graph of $y = f(x)$.

Definitions of Trigonometric Functions

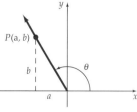

$$\sin \theta = \frac{b}{r} \qquad \csc \theta = \frac{r}{b}$$

$$\cos \theta = \frac{a}{r} \qquad \sec \theta = \frac{r}{a}$$

$$\tan \theta = \frac{b}{a} \qquad \cot \theta = \frac{a}{b}$$

where $r = \sqrt{a^2 + b^2}$

Definitions of Circular Functions

$$\sin t = y \qquad \csc t = \frac{1}{y}$$

$$\cos t = x \qquad \sec t = \frac{1}{x}$$

$$\tan t = \frac{y}{x} \qquad \cot t = \frac{x}{y}$$

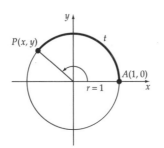

Formulas for Triangles

For any triangle ABC, the following formulas can be used.

Law of Sines

$$\frac{a}{\sin A} = \frac{b}{\sin B} = \frac{c}{\sin C}$$

Law of Cosines

$$c^2 = a^2 + b^2 - 2ab \cos C$$

Area of a Triangle

$$K = \frac{1}{2}ab \sin C \qquad K = \frac{a^2 \sin B \sin C}{2 \sin A}$$

Heron's Formula

$$K = \sqrt{s(s - a)(s - b)(s - c)}, \text{ where } s = \frac{a + b + c}{2}$$

Fundamental Identities

$$\tan \theta = \frac{\sin \theta}{\cos \theta} \qquad \cot \theta = \frac{\cos \theta}{\sin \theta}$$

$$\sin^2 \theta + \cos^2 \theta = 1 \qquad 1 + \tan^2 \theta = \sec^2 \theta$$

$$1 + \cot^2 \theta = \csc^2 \theta$$

Formulas for Negatives

$$\sin(-\theta) = -\sin \theta \qquad \cos(-\theta) = \cos \theta$$

$$\tan(-\theta) = -\tan \theta$$

Reciprocal Identities

$$\csc \theta = \frac{1}{\sin \theta} \qquad \sec \theta = \frac{1}{\cos \theta} \qquad \cot \theta = \frac{1}{\tan \theta}$$

Sum of Two Angle Identities

$$\sin(\alpha + \beta) = \sin \alpha \cos \beta + \cos \alpha \sin \beta$$

$$\cos(\alpha + \beta) = \cos \alpha \cos \beta - \sin \alpha \sin \beta$$

$$\tan(\alpha + \beta) = \frac{\tan \alpha + \tan \beta}{1 - \tan \alpha \tan \beta}$$

Difference of Two Angle Identities

$$\sin(\alpha - \beta) = \sin \alpha \cos \beta - \cos \alpha \sin \beta$$

$$\cos(\alpha - \beta) = \cos \alpha \cos \beta + \sin \alpha \sin \beta$$

$$\tan(\alpha - \beta) = \frac{\tan \alpha - \tan \beta}{1 + \tan \alpha \tan \beta}$$

Double-Angle Identities

$$\sin 2\alpha = 2 \sin \alpha \cos \alpha$$

$$\cos 2\alpha = \cos^2 \alpha - \sin^2 \alpha = 1 - 2 \sin^2 \alpha$$
$$= 2 \cos^2 \alpha - 1$$

$$\tan 2\alpha = \frac{2 \tan \alpha}{1 - \tan^2 \alpha}$$

Half-Angle Identities

$$\sin \frac{\alpha}{2} = \pm \sqrt{\frac{1 - \cos \alpha}{2}} \qquad \cos \frac{\alpha}{2} = \pm \sqrt{\frac{1 + \cos \alpha}{2}}$$

$$\tan \frac{\alpha}{2} = \frac{1 - \cos \alpha}{\sin \alpha} \qquad \sin^2 \alpha = \frac{1 - \cos 2\alpha}{2}$$

$$\cos^2 \alpha = \frac{1 + \cos 2\alpha}{2} \qquad \tan^2 \alpha = \frac{1 - \cos 2\alpha}{1 + \cos 2\alpha}$$

Cofunction Formulas

$$\sin\left(\frac{\pi}{2} - \theta\right) = \cos \theta \qquad \cos\left(\frac{\pi}{2} - \theta\right) = \sin \theta$$

$$\tan\left(\frac{\pi}{2} - \theta\right) = \cot \theta$$